Basic Algebra I

Basic Algebra I

Second Edition

NATHAN JACOBSON
YALE UNIVERSITY

DOVER PUBLICATIONS, INC.
Mineola, New York

Bibliographical Note

This Dover edition, first published in 2009, is an unabridged republication of the second edition of the work, originally published by W. H. Freeman and Company, San Francisco, in 1985.

Library of Congress Cataloging-in-Publication Data

Jacobson, Nathan, 1910–1999.
  Basic algebra / Nathan Jacobson. — Dover ed.
    p. cm.
  Originally published: 2nd ed. San Francisco : W. H. Freeman, 1985.
  ISBN-13: 978-0-486-47189-1
  ISBN-10: 0-486-47189-6
  1. Algebra. I. Title.

QA154.2.J32 2009
512.9—dc22
                                                                2009006506

Manufactured in the United States by Courier Corporation
47189602
www.doverpublications.com

To Florie

Contents

Preface

Since the publication of *Basic Algebra I* in 1974, a number of teachers and students of the text have communicated to the author corrections and suggestions for improvements as well as additional exercises. Many of these have been incorporated in this new edition. Especially noteworthy were the suggestions sent by Mr. Huah Chu of National Taiwan University, Professor Marvin J. Greenberg of the University of California at Santa Cruz, Professor J. D. Reid of Wesleyan University, Tsuneo Tamagawa of Yale University, and Professor F. D. Veldkamp of the University of Utrecht. We are grateful to these people and others who encouraged us to believe that we were on the right track in adopting the point of view taken in *Basic Algebra I*.

Two important changes occur in the chapter on Galois theory, Chapter 4. The first is a completely rewritten section on finite fields (section 4.13). The new version spells out the principal results in the form of formal statements of theorems. In the first edition these results were buried in the account, which was a tour de force of brevity. In addition, we have incorporated in the text the proof of Gauss' formula for the number $N(n, q)$ of monic irreducible polynomials of degree n in a finite field of q elements. In the first edition this formula appeared in an exercise (Exercise 20, p. 145). This has now been altered to ask for $N(2, q)$ and

$N(3, q)$ only. The second important change in Chapter 4 is the addition of section 4.16, "Mod p Reduction," which gives a proof due to John Tate of a theorem of Dedekind's on the existence of certain cycles in the Galois permutation group of the roots of an irreducible monic polynomial $f(x)$ with integer coefficients that can be deduced from the factorization of $f(x)$ modulo a prime p. A number of interesting applications of this theorem are given in the exercises at the end of the section.

In Chapter 5 we have given a new proof of the basic elimination theorem (Theorem 5.6). The new proof is completely elementary, and is independent of the formal methods developed in Chapter 5 for the proof of Tariski's theorem on elimination of quantifiers for real closed fields. Our purpose in giving the new proof is that Theorem 5.6 serves as the main step in the proof of Hilbert's Nullstellensatz given on pp. 424–426 of *Basic Algebra II*. The change has been made for the convenience of readers who do not wish to familiarize themselves with the formal methods developed in Chapter 5.

At the end of the book we have added an appendix entitled "Some Topics for Independent Study," which lists 10 such topics. There is a brief description of each, together with some references to the literature. While some of these might have been treated as integral parts of the text, we feel that students will benefit more by pursuing them on their own.

The items listed account for approximately 10 pages of added text. The remaining 15 or so pages added in this edition can be accounted for by local improvements in the exposition and additional exercises.

The text of the second edition has been completely reset, which presented the chore of proofreading a lengthy manuscript. This arduous task was assumed largely by the following individuals: Huah Chu (mentioned above), Jone-Wen Cohn of Shanghai Normal University, Florence D. Jacobson ("Florie," to whom the book is dedicated), and James D. Reid (also mentioned above). We are deeply indebted to them for their help.

Hamden, Connecticut

Nathan Jacobson
November 1, 1984

Preface to the First Edition

It is more than twenty years since the author began the project of writing the three volumes of *Lectures in Abstract Algebra*. The first and second of these books appeared in 1951 and 1953 respectively, the third in 1964. In the period which has intervened since this work was conceived—around 1950—substantial progress in algebra has occurred even at the level of these texts. This has taken the form first of all of the introduction of some basic new ideas. Notable examples are the development of category theory, which provides a useful framework for a large part of mathematics, homological algebra, and applications of model theory to algebra. Perhaps even more striking than the advent of these ideas has been the acceptance of the axiomatic conceptual method of abstract algebra and its pervading influence throughout mathematics. It is now taken for granted that the methodology of algebra is an essential tool in mathematics. On the other hand, in recent research one can observe a return to the challenge presented by fairly concrete problems, many of which require for their solution tools of considerable technical complexity.

Another striking change that has taken place during the past twenty years—especially since the Soviet Union startled the world by orbiting its "sputniks"—has been the upgrading of training in mathematics in elementary and secondary schools. (Although there has recently been some regression in this process, it is to be hoped that this will turn out to be only a temporary aberration.) The upgrading of school mathematics has had as a corollary a corresponding upgrading of college mathematics. A notable instance of this is the early study of linear algebra, with a view of providing the proper background for the study of multivariable calculus as well as for applications to other fields. Moreover, courses in linear algebra are quite often followed immediately by courses in "abstract" algebra, and so the type of material which twenty years ago was taught at the graduate level is now presented to students with comparatively little experience in mathematics.

The present book, *Basic Algebra* I, and the forthcoming *Basic Algebra* II were originally envisioned as new editions of our *Lectures*. However, as we began to think about the task at hand, particularly that of taking into account the changed curricula in our undergraduate and graduate schools, we decided to organize the material in a manner quite different from that of our earlier books: a separation into two levels of abstraction, the first—treated in this volume—to encompass those parts of algebra which can be most readily appreciated by the beginning student. Much of the material which we present here has a classical flavor. It is hoped that this will foster an appreciation of the great contributions of the past and especially of the mathematics of the nineteenth century. In our treatment we have tried to make use of the most efficient modern tools. This has necessitated the development of a substantial body of foundational material of the sort that has become standard in text books on abstract algebra. However, we have tried throughout to bring to the fore well-defined objectives which we believe will prove appealing even to a student with little background in algebra. On the other hand, the topics considered are probed to a depth that often goes considerably beyond what is customary, and this will at times be quite demanding of talent and concentration on the part of the student. In our second volume we plan to follow a more traditional course in presenting material of a more abstract and sophisticated nature. It is hoped that after the study of the first volume a student will have achieved a level of maturity that will enable him to take in stride the level of abstration of the second volume.

We shall now give a brief indication of the contents and organization of *Basic Algebra* I. The Introduction, on set theory and the number system of the integers, includes material that will be familiar to most readers: the algebra of sets, definition of maps, and mathematical induction. Less familiar, and of paramount importance for subsequent developments, are the concepts of an equivalence relation and quotient sets defined by such relations. We introduce also commutative diagrams and the factorization of a map through an equivalence relation. The fundamental theorem of arithmetic is proved, and a proof of the Recursion Theorem (or definition by induction) is included.

Chapter 1 deals with monoids and groups. Our starting point is the concept of a monoid of transformations and of a group of transformations. In this respect we follow the historical development of the subject. The concept of homomorphism appears fairly late in our discussion, after the reader has had a chance to absorb some of the simpler and more intuitive ideas. However, once the concept of homomorphism has been introduced, its most important ramifications (the fundamental isomorphism theorems and the correspondence between subgroups of a homomorphic image and subgroups containing the kernel) are developed in considerable detail. The concept of a group acting on a set, which now plays such an important role in geometry, is introduced and illustrated with many examples. This leads to a method of enumeration for finite groups, a special case of which is contained in the class equation. These results are applied to derive the Sylow theorems, which constitute the last topic of Chapter 1.

The first part of Chapter 2 repeats in the context of rings many of the ideas that have been developed in the first chapter. Following this, various constructions of new rings from given ones are considered: rings of matrices, fields of fractions of commutative domains, polynomial rings. The last part of the chapter is devoted to the elementary factorization theory of commutative monoids with cancellation property and of commutative domains.

The main objective in Chapter 3 is the structure theory of finitely generated modules over a principal ideal domain and its applications to abelian groups and canonical forms of matrices. Of course, before this can be achieved it is necessary to introduce the standard definitions and concepts on modules. The analogy with the concept of a group acting on a set is stressed, as is the idea that the concept of a module is a natural generalization of the familiar notion of a vector space. The chapter concludes with theorems on the ring of endomorphisms of a finitely generated module over a principal ideal domain, which generalize classical results of Frobenius on the ring of matrices commuting with a given matrix.

Chapter 4 deals almost exclusively with the ramifications of two classical problems: solvability of equations by radicals and constructions with straightedge and compass. The former is by far the more difficult of the two. The tool which was forged by Galois for handling this, the correspondence between subfields of the splitting field of a separable polynomial and subgroups of the group of automorphisms, has attained central importance in algebra and number theory. However, we believe that at this stage it is more effective to concentrate on the problems which gave the original impetus to Galois' theory and to treat these in a thoroughgoing manner. The theory of finite groups which was initiated in Chapter 1 is amplified here by the inclusion of the results needed to establish Galois' criterion for solvability of an equation by radicals. We have included also a proof of the transcendence of π since this is needed to prove the impossibility of "squaring the circle" by straight-edge and compass. (In fact, since it requires very little additional effort, the more general theorem of Lindemann and Weierstrass on algebraic independence of exponentials has been proved.) At the end of the chapter we have undertaken to round out the Galois theory by applying it to derive the main results on finite fields and to prove the theorems on primitive elements and normal bases as well as the fundamental theorems on norms and traces.

Chapter 5 continues the study of polynomial equations. We now operate in a real closed field—an algebraic generalization of the field of real numbers. We prove a generalization of the "fundamental theorem of algebra": the algebraic closure of $R\sqrt{(-1)}$ for R any real closed field. We then derive Sturm's theorem, which gives a constructive method of determining the number of roots in R of a polynomial equation in one unknown with coefficients in R. The last part of the chapter is devoted to the study of systems of polynomial equations and inequations in several unknowns. We first treat the purely algebraic problem of elimination of unknowns in such a system and then establish a far-reaching generalization of Sturm's theorem that is due to Tarski. Throughout this chapter the emphasis is on constructive methods.

The first part of Chapter 6 covers the basic theory of quadratic forms and alternate forms over an arbitrary field. This includes Sylvester's theorem on the inertial index and its generalization that derives from Witt's cancellation theorem. The important theorem of Cartan-Dieudonné on the generation of the orthogonal group by symmetries is proved. The second part of the chapter is concerned with the structure theory of the so-called classical groups: the full linear group, the orthogonal group, and the sympletic group. In this analysis we have employed a uniform method applicable to all three types of groups. This method was originated by Iwasawa for the full linear group and was extended to orthogonal groups by Tamagawa. The results provide some important classes of simple groups whose orders for finite fields are easy to compute.

Chapter 7 gives an introduction to the theory of algebras, both associative and non-associative. An important topic in the associative theory we consider is the exterior algebra of a vector space. This algebra plays an important role in geometry, and is applied here to derive the main theorems on determinants. We define also the regular representation, trace, and norm of an associative algebra, and prove a general theorem on transitivity of these functions. For non-associative algebras we give definitions and examples of the most important classes of non-associative algebras. We follow this with a completely elementary proof of the beautiful theorem on composition of quadratic forms which is due to Hurwitz, and we conclude the chapter with proofs of Frobenius' theorem on division algebras over the field of real numbers and Wedderburn's theorem on finite division algebras.

Chapter 8 provides a brief introduction to lattices and Boolean algebras. The main topics treated are the Jordan-Hölder theorem on semi-modular lattices; the so-called "fundamental theorem of projective geometry"; Stone's theorem on the equivalence of the concepts of Boolean algebras and Boolean rings, that is, rings all of whose elements are idempotent; and finally the Möbius function of a partially ordered set.

Basic Algebra I is intended to serve as a text for a first course in algebra beyond linear algebra. It contains considerably more material than can be covered in a year's course. Based on our own recent experience with earlier versions of the text, we offer the following suggestions on what might be covered in a year's course divided into either two semesters or three quarters. We have found it possible to cover the Introduction (treated lightly) and nearly all the material of Chapters 1–3 in one semester. We found it necessary to omit the proof of the Recursion Theorem in the Introduction, the section on free groups in Chapter 1, the last section (on "rngs") in Chapter 2, and the last section of Chapter 3. Chapter 4, Galois theory, is an excellent starting point for a second semester's course. In view of the richness of this material not much time will remain in a semester's course for other topics. If one makes some omissions in Chapter 4, for example, the proof of the theorem of Lindemann-Weierstrass, one is likely to have several weeks left after the completion of this material. A number of alternatives for completing the semester may be considered. One possibility would be to pass from the study of equations in one unknown to systems of polynomial equations in several unknowns. One aspect of this is presented in Chapter 5. A part of this chapter would certainly fit in well with Chapter 4. On the other hand, there is something to be said for making an abrupt change in theme. One possibility would be to take up the chapter on algebras. Another would be to study a part of the chapter on quadratic forms and the classical groups. Still another would be to study the last chapter, on lattices and Boolean algebras.

A program for a course for three quarters might run as follows: Introduction and Chapters 1 and 2 for a first quarter; Chapter 3 and a substantial part of Chapter 6 for a second quarter. This will require a bit of filling in of the field theory from Chapter 4 which is needed for Chapter 6. One could conclude with a third quarter's course on Chapter 4, the Galois theory.

It is hoped that a student will round out formal courses based on the text by independent reading of the omitted material. Also we feel that quite a few topics lend themselves to programs of supervised independent study.

We are greatly indebted to a number of friends and colleagues for reading portions of the penultimate version of the text and offering valuable suggestions which were taken into account in preparing the final version. Walter Feit and Richard Lyons suggested a number of exercises in group theory; Abraham

Robinson, Tsuneo Tamagawa, and Neil White have read parts of the book on which they are experts (Chapters 5, 6, and 8 respectively) and detected some flaws which we had not noticed. George Seligman has read the entire manuscript and suggested some substantial improvements. S. Robert Gordon, James Hurley, Florence Jacobson, and David Rush have used parts of the earlier text in courses of a term or more, and have called our attention to numerous places where improvements in the exposition could be made.

A number of people have played an important role in the production of the book, among them we mention especially Florence Jacobson and Jerome Katz, who have been of great assistance in the tedious task of proofreading. Finally, we must add a special word for Mary Scheller, who cheerfully typed the entire manuscript as well as the preliminary version of about the same length.

We are deeply indebted to the individuals we have mentioned—and to others—and we take this opportunity to offer our sincere appreciation and thanks.

Hamden, Connecticut   *Nathan Jacobson*

Basic Algebra I

INTRODUCTION

Concepts from Set Theory.
The Integers

The main purpose of this volume is to provide an introduction to the basic structures of algebra: groups, rings, fields, modules, algebras, and lattices— concepts that give a natural setting for a large body of algebra, including classical algebra. It is noteworthy that many of these concepts have arisen either to solve concrete problems in geometry, number theory, or the theory of algebraic equations, or to afford a better insight into existing solutions of such problems. A good example of the interplay between abstract theory and concrete problems can be seen in the Galois theory, which was created by Galois to answer a concrete question: "What polynomial equations in one unknown have solutions expressible in terms of the given coefficients by rational operations and extraction of roots?" To solve this we must first have a precise formulation of the problem, and this requires the concepts of field, extension field, and splitting field of a polynomial. To understand Galois' solution of the problem of algebraic equations we require the notion of a group and properties of solvable groups. In Galois' theory the results were stated in terms of groups of permutations of the roots. Subsequently, a much deeper understanding of what was involved emerged in passing from permutations of the roots to the more abstract notion of the group of automorphisms of an extension field. All of this will be discussed fully in Chapter 4.

Of course, once the machinery has been developed for treating one set of problems, it is likely to be useful in other circumstances, and, moreover, it generates new problems that appear interesting in their own right.

Throughout this presentation we shall seek to emphasize the relevance of the general theory in solving interesting problems, in particular, problems of classical origin. This will necessitate developing the theory beyond the foundational level to get at some of the interesting theorems. Occasionally, we shall find it convenient to develop some of the applications in exercises. For this reason, as well as others, the working of a substantial number of the exercises is essential for a thorough understanding of the material.

The basic ingredients of the structures we shall study are sets and mappings (or, as we shall call them in this book, maps). It is probable that the reader already has an adequate knowledge of the set theoretic background that is required. Nevertheless, for the purpose of fixing the notations and terminology, and to highlight the special aspects of set theory that will be fundamental for us, it seems desirable to indicate briefly some of the elements of set theory.[1] From the point of view of what follows the ideas that need to be stressed concern equivalence relations and the factorization of a map through an equivalence relation. These will reappear in a multitude of forms throughout our study. In the second part of this introduction we shall deal briefly with the number system $\mathbb{Z}$ of the integers and the more primitive system $\mathbb{N}$ of natural numbers or counting numbers: 0, 1, 2, . . . , which serve as the starting point for the constructive development of algebra. In view of the current emphasis on the development of number systems in primary and secondary schools, it seems superfluous to deal with $\mathbb{N}$ and $\mathbb{Z}$ in a detailed fashion. We shall therefore be content to review in outline the main steps in one of the ways of introducing $\mathbb{N}$ and $\mathbb{Z}$ and to give careful proofs of two results that will be needed in the discussion of groups in Chapter 1. These are the existence of greatest common divisors (g.c.d.'s) of integers and "the fundamental theorem of arithmetic," which establishes the unique factorization of any natural number $\neq 0$, 1 as a product of prime factors. Later (in Chapter 2), we shall derive these results again as special cases of the arithmetic of principal ideal domains.

---

[1] For a general reference book on set theory adequate for our purposes we refer the reader to the very attractive little book, *Naive Set Theory*, by Paul R. Halmos, Van Nostrand Reinhold, 1960.

0.1 THE POWER SET OF A SET

We begin our discussion with a brief survey of some set theoretic notions which will play an essential role in this book.

Let S be an arbitrary set (or collection) or elements which we denote as a, b, c, etc. The nature of these elements is immaterial. The fact that an element a belongs to the set S is indicated by writing $a \in S$ (occasionally $S \ni a$) and the negation of $a \in S$ is written as $a \notin S$. If S is a finite set with elements a_i, $1 \leq i \leq n$, then we write $S = \{a_1, a_2, \ldots, a_n\}$. Any set S gives rise to another set $\mathscr{P}(S)$, the set of subsets of S. Among these are included the set S itself and the vacuous subset or null set, which we denote as $\varnothing$. For example, if S is a finite set of n elements, say, $S = \{a_1, a_2, \ldots, a_n\}$, then $\mathscr{P}(S)$ consists of $\varnothing$, the n sets $\{a_i\}$ containing single elements, $n(n-1)/2$ sets $\{a_i, a_j\}$, $i \neq j$, containing two elements, $\binom{n}{i} = n!/i!\,(n-i)! = n(n-1)\cdots(n-i+1)/1 \cdot 2 \cdots \cdot i$ subsets containing i elements, and so on. Hence the cardinality of $\mathscr{P}(S)$, that is, the number of elements in $\mathscr{P}(S)$ is

$$1 + \binom{n}{1} + \binom{n}{2} + \cdots + \binom{n}{n} = (1+1)^n = 2^n.$$

We shall call $\mathscr{P}(S)$, the *power set* of the set S.[2] Often we shall specify a subset of S by a property or set of properties. The standard way of doing this is to write

$$A = \{x \in S \,|\, \cdots\}$$

(or, if S is clear, $A = \{x \,|\, \cdots\}$) where $\cdots$ lists the properties characterizing A. For example, if $\mathbb{Z}$ denotes the set of integers, the $\mathbb{N} = \{x \in \mathbb{Z} \,|\, x \geq 0\}$ defines the subset of non-negative integers, or natural numbers.

If A and $B \in \mathscr{P}(S)$ (that is, A and B are subsets of S) we say that A is *contained in B* or is a *subset of B* (or B *contains A*) and denote this as $A \subset B$ (or $B \supset A$) if every element a in A is also in B. Symbolically, we can write this as $a \in A \Rightarrow a \in B$ where the $\Rightarrow$ is read as "implies." The statement $A = B$ is equivalent to the two statements $A \supset B$ and $B \supset A$ (symbolically, $A = B \Leftrightarrow A \supset B$ and $B \supset A$ where $\Leftrightarrow$ reads "if and only if"). If $A \subset B$ and $A \neq B$ we write $A \subsetneqq B$ and say that A is a *proper subset* of B. Alternatively, we can write $B \supsetneqq A$.

If A and B are subsets of S, the subset of S of elements c such that $c \in A$ and $c \in B$ is called the *intersection* of A and B. We denote this subset as $A \cap B$. If there are no elements of S contained in both A and B, that is, $A \cap B = \varnothing$,

---

[2] This is frequently called the *Boolean of S*, $\mathscr{B}(S)$, after George Boole who initiated its systematic study. The justification of the terminology "power set" is indicated in the footnote on p. 5.

then A and B are said to be *disjoint* (or *non-overlapping*). The *union* (or *logical sum*) $A \cup B$ of A and B is the subset of elements d such that either $d \in A$ or $d \in B$. An important property connecting $\cap$ and $\cup$ is the distributive law:

(1) $$A \cap (B \cup C) = (A \cap B) \cup (A \cap C)$$

This can be indicated pictorially by

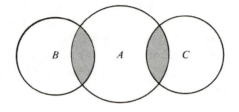

where the shaded region represents (1). To prove (1), let $x \in A \cap (B \cup C)$. Since $x \in (B \cup C)$ either $x \in B$ or $x \in C$, and since $x \in A$ either $x \in (A \cap B)$ or $x \in (A \cap C)$. This shows that $A \cap (B \cup C) \subset (A \cap B) \cup (A \cap C)$. Now let $y \in (A \cap B) \cup (A \cap C)$ so either $y \in A \cap B$ or $y \in A \cap C$. In any case $y \in A$ and $y \in B$ or $y \in C$. Hence $y \in A \cap (B \cup C)$. Thus $(A \cap B) \cup (A \cap C) \subset A \cap (B \cup C)$. Hence we have both $A \cap (B \cup C) \subset (A \cap B) \cup (A \cap C)$ and $(A \cap B) \cup (A \cap C) \subset A \cap (B \cup C)$ and consequently we have (1).

We also have another distributive law which dualizes (1) in the sense that it is obtained from (1) by interchanging $\cup$ and $\cap$:

(2) $$A \cup (B \cap C) = (A \cup B) \cap (A \cup C).$$

It is left to the reader to draw a diagram for this law and carry out the proof. Better still, the reader can show that (2) is a consequence of (1)—and that, by symmetry, (1) is a consequence of (2).

Intersections and unions can be defined for an arbitrary set of subsets of a set S. Let Γ be such a set of subsets ($=$subset of $\mathscr{P}(S)$). Then we define $\bigcap_{A \in \Gamma} A = \{x \,|\, x \in A$ for every A in $\Gamma\}$ and $\bigcup_{A \in \Gamma} A = \{x \,|\, x \in A$ for some A in $\Gamma\}$. If Γ is finite, say, $\Gamma = \{A_1, A_2, \ldots, A_n\}$ then we write also $\bigcap_{i=1}^{n} A_i$ or $A_1 \cap A_2 \cap \cdots \cap A_n$ for the intersection and we use a similar designation for the union. It is easy to see that the distributive laws carry over to arbitrary intersections and unions: $B \cap (\bigcup_{A \in \Gamma} A) = \bigcup_{A \in \Gamma} (B \cap A); \; B \cup (\bigcap_{A \in \Gamma} A) = \bigcap_{A \in \Gamma}(B \cup A)$.

0.2   THE CARTESIAN PRODUCT SET. MAPS

The reader is undoubtedly aware of the central role of the concept of function in mathematics and its applications. The case of interest in beginning calculus real line $\mathbb{R}$; usually, an open or closed interval or the whole of $\mathbb{R}$; and a rule which associates with every element x of this subset a unique real number $f(x)$. Associated with a function as thus "defined" we have the graph in the two-dimensional number space $\mathbb{R}^{(2)}$ consisting of the points $(x, f(x))$. We soon realize that f is determined by its graph and that the characteristic property of the graph is that any line parallel to the y-axis through a point x of the domain of definition (on the x-axis) meets the graph in precisely one point. Equivalently, if (x, y) and (x, y') are on the graph then $y = y'$. It is clear that the notion of a graph satisfying this condition is a precisely defined object whereas the in-tuitive definition of a function by a "rule" is not. We are therefore led to replace the original definition by the definition of a graph.

We shall now proceed along these lines, and we shall also substitute for the word "function" the geometric term "map" which is now more commonly used in the contexts we shall consider. Also, we wish to pass from real-valued func-tions of a real variable to arbitrary maps. First, we need to define the (*Cartesian*) *product set* $S \times T$ of two arbitrary sets S and T. This is the set of pairs (s, t), $s \in S$, $t \in T$. The sets S and T need not be distinct. In the product $S \times T$, the elements (s, t) and (s', t') are regarded as equal if and only if $s = s'$ and $t = t'$. Thus if S consists of m elements $s_1, s_2, \ldots, s_m$ and T consists of n elements $t_1, t_2, \ldots, t_n$, then $S \times T$ consists of the mn elements (s_i, t_j).

We are now ready to define a *map of a set S into a set T*. This consists of the set S, called the *domain* of the map, the set T, called the *co-domain*, and a subset α of $S \times T$ (the *graph*) having the following two properties:

1. For any $s \in S$ there exists a $t \in T$ such that $(s, t) \in \alpha$.
2. If (s, t) and $(s, t') \in \alpha$ then $t = t'$.

The second property is called "single-valuedness." In specifying a definition one often says that "the function is well-defined" when one is assured that condition 2 holds. Together, conditions 1 and 2 state that for every $s \in S$ there is a unique $t \in T$ such that $(s, t) \in \alpha$. The classical notation for this t is $\alpha(s)$. One calls this *the image of s under α*. In many books on algebra (including our previous ones) we find the notations s^α and $s\alpha$ for $\alpha(s)$. This has advantages when we deal with the composite of maps. However, since the consensus clearly favors the classical notation $\alpha(s)$, we have decided to adopt it in this book.

Two maps are regarded as equal if and only if they have the same domain, the same co-domain and the same graphs. The set of maps "from S to T," that is, having domain S and co-domain T will be denoted as T^S.[3]

---

[3] If T consists of two elements $\{0, 1\}$ then we may write $T = 2$ and have the set 2^S of maps of S into $\{0, 1\}$. Such a map is characterized by specifying $A := \{a \in S \mid \alpha(a) = 1\}$. Conversely, given a subset A of S we can define *its characteristic function* χ_A by $\chi_A(a) = 1$ if $a \in A$ and $\chi_A(a) = 0$ if $a \notin A$. In this way one can identify the set 2^S of maps of S into $\{0, 1\}$ with the set of subsets of S, that is, with $\mathcal{P}(S)$. This is the reason for the terminology "power set."

If A is a subset of S, then we write $\alpha(A) = \{\alpha(a) \mid a \in A\}$ and call this the *image of A under α*. In particular, we have $\alpha(S)$, which is called the *image* (or *range*) *of the map*. We shall denote this also as im α. Usually, when the domain and co-domain are clear, we shall speak of the "map α" (or the "function α") even though, strictly speaking, α is just one component of the map.

If S_1 is a subset of S and α is a map of S into T, then we get a map of S_1 to T by restricting the domain to S_1. This is the map of S_1 to T whose graph is the subset of $S_1 \times T$ of elements $(s_1, \alpha(s_1))$, $s_1 \in S_1$. We call this map the *restriction of α to S_1* and denote it as $\alpha \mid S_1$. Turning things around we shall say that a map α of S to T is an *extension of the map β of S_1 to T* if $\beta = \alpha \mid S_1$.

As was mentioned, the terms "map" and "mapping" come from geometry. We shall now give a couple of geometric examples. The first is described by the diagram

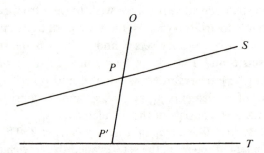

Here the lines S and T are the domain and co-domain respectively, O is a fixed point not on S or T and we "map" the point P on S into the point of intersection P' of the line OP with T. Such mappings, called perspectivities, play an important role in projective geometry. From our point of view, the map consists of the sets S and T and the subset of points (P, P') of $S \times T$. The second example, from Euclidean geometry, is orthogonal projection on a line. Here the domain is the plane, the co-domain is the line, and one maps any point P in the plane on the foot of the perpendicular from P to the given line:

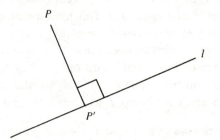

(It is understood that if P is on l then $P' = P$.) As in these examples, it is always a good idea to keep the intuitive picture in mind when dealing with maps, reserving the more precise definition for situations in which a higher degree of rigor appears appropriate. Geometry suggests also denoting a map from S to T by $\alpha: S \to T$, or $S \overset{\alpha}{\to} T$, and indicating the definition of a particular map by $x \to y$ where y is the image of x under the given map: e.g., $P \to P'$ in the foregoing example. In the special case in which the domain and co-domain coincide, one often calls a map from S to S a *transformation* of the set S.

A map $S \overset{\alpha}{\to} T$ is called *surjective* if im $\alpha = T$, that is, if the range coincides with the co-domain. $S \overset{\alpha}{\to} T$ is *injective* if distinct elements of S have distinct images in T, that is, if $s_1 \neq s_2 \Rightarrow \alpha(s_1) \neq \alpha(s_2)$. If α is both injective and surjective, it is called *bijective* (or α is said to be a *one to one correspondence between S and T*). For example, the perspectivity map defined above is bijective.

Let $S \overset{\alpha}{\to} T$ and $T \overset{\beta}{\to} U$. Then we define the map $S \overset{\beta\alpha}{\to} U$ as the map having the domain S, the co-domain U, and the graph the subset of $S \times U$ of elements $(s, \beta(\alpha(s)))$, $s \in S$. Thus, by definition,

$$(\beta\alpha)(s) = \beta(\alpha(s)).$$

We call this the *composite* (or *product*, or sometimes *resultant*) of α and β (β following α).[4] It is often useful to indicate the relation $\gamma = \beta\alpha$ by saying that the triangle

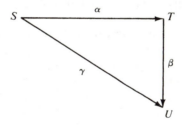

is commutative. Similarly, we express the fact that $\beta\alpha = \delta\gamma$ for $S \overset{\alpha}{\to} T$, $T \overset{\beta}{\to} U$, $S \overset{\gamma}{\to} V$, $V \overset{\delta}{\to} U$ by saying that the rectangle

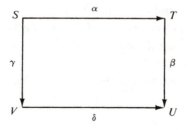

---

[4] Note that the composite is written in the reverse order to that in which the operations are performed: $\beta\alpha$ is α followed by β. To keep the order straight it is good to think of $\beta\alpha$ as β *following* α.

is commutative. In general, commutativity of a diagram of maps, when it makes sense, means that the maps obtained by following the diagram from one initial point to a terminal point along each displayed route are the same. As another example, commutativity of

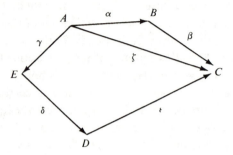

means that $\beta\alpha = \zeta = \varepsilon(\delta\gamma)$.

Composition of maps satisfies the associative law: if $S \xrightarrow{\alpha} T$, $T \xrightarrow{\beta} U$, and $U \xrightarrow{\gamma} V$, then $\gamma(\beta\alpha) = (\gamma\beta)\alpha$. We note first that both of these maps have the same domain S and the same co-domain V. Moreover, for any $s \in S$ we have

$$(\gamma(\beta\alpha))(s) = \gamma((\beta\alpha)(s)) = \gamma(\beta(\alpha(s)))$$

$$((\gamma\beta)\alpha)(s) = (\gamma\beta)(\alpha(s)) = \gamma(\beta(\alpha(s)))$$

so $\gamma(\beta\alpha)$ and $(\gamma\beta)\alpha$ are identical. This can be illustrated by the following diagram:

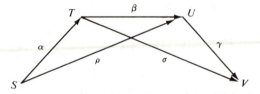

The associative law amounts to the statement that if the triangles STU and TUV are commutative then the whole diagram is commutative.

For any set S one defines the *identity map* 1_S (or 1 if S is clear) as $S \xrightarrow{1_S} S$ where 1_S is the subset of elements (s, s) of $S \times S$. This subset is called the *diagonal* of $S \times S$. If $S \xrightarrow{\alpha} T$ one checks immediately that $1_T\alpha = \alpha = \alpha1_S$. We now state the following important result:

$S \xrightarrow{\alpha} T$ *is bijective if and only if there exists a map* $T \xrightarrow{\beta} S$ *such that* $\beta\alpha = 1_S$ *and* $\alpha\beta = 1_T$.

Proof. Suppose $S \xrightarrow{\alpha} T$ is bijective. Consider the subset β of $T \times S$ of elements $(\alpha(s), s)$. If $t \in T$, surjectivity of α implies there is an s in S such that $\alpha(s) = t$. Hence condition 1 in the definition of a map from T to S holds for the set β of pairs $(\alpha(s), s) \in T \times S$. Condition 2 holds for β by the injectivity of α, since if (t, s_1) and (t, s_2) are in β, then $\alpha(s_1) = t$ and $\alpha(s_2) = t$, so $s_1 = s_2$. Hence we have the map $T \xrightarrow{\beta} S$. If $s \in S$, the facts that $(s, \alpha(s)) \in \alpha$ and $(\alpha(s), s) \in \beta$ imply that $\beta(\alpha(s)) = s$. Thus $\beta\alpha = 1_S$. If $t \in T$, we have $t = \alpha(s)$, $s \in S$, and $(t, s) \in \beta$, so $\beta(t) = s \in S$. Hence $\alpha(\beta(t)) = \alpha(s) = t$, so $\alpha\beta = 1_T$. Conversely, suppose $S \xrightarrow{\alpha} T$, $T \xrightarrow{\beta} S$ satisfy $\beta\alpha = 1_S$, $\alpha\beta = 1_T$. If $t \in T$, let $s = \beta(t)$. Then $\alpha(s) = \alpha(\beta(t)) = t$; hence α is surjective. Next suppose $\alpha(s_1) = \alpha(s_2)$ for $s_i \in S$. Then $s_1 = \beta(\alpha(s_1)) = \beta(\alpha(s_2)) = s_2$, and α is injective. $\square$

The map β satisfying $\beta\alpha = 1_S$ and $\alpha\beta = 1_T$ is unique since if $T \xrightarrow{\beta'} S$ satisfies the same conditions, $\beta'\alpha = 1_S$, $\alpha\beta' = 1_T$, then

$$\beta' = 1_S\beta' = (\beta\alpha)\beta' = \beta(\alpha\beta') = \beta 1_T = \beta.$$

We shall now denote β as α^{-1} and call this the *inverse* of the (bijective) map α. Clearly the foregoing result shows that α^{-1} is bijective and $(\alpha^{-1})^{-1} = \alpha$.

As a first application of the criterion for bijectivity we give a formal proof of a fact which is fairly obvious anyhow: the product of two bijective maps is bijective. For, let $S \xrightarrow{\alpha} T$ and $T \xrightarrow{\beta} U$ be bijective. Then we have the inverses $T \xrightarrow{\alpha^{-1}} S$ and $U \xrightarrow{\beta^{-1}} T$ and the composite map $\alpha^{-1}\beta^{-1}: U \to S$. Moreover,

$$(\beta\alpha)(\alpha^{-1}\beta^{-1}) = ((\beta\alpha)\alpha^{-1})\beta^{-1} = (\beta(\alpha\alpha^{-1}))\beta^{-1} = \beta\beta^{-1} = 1_U.$$

Also,

$$(\alpha^{-1}\beta^{-1})(\beta\alpha) = \alpha^{-1}(\beta^{-1}(\beta\alpha)) = \alpha^{-1}((\beta^{-1}\beta)\alpha) = \alpha^{-1}\alpha = 1_S.$$

Hence $\alpha^{-1}\beta^{-1}$ is an inverse of $\beta\alpha$, that is

$$(3) \qquad\qquad (\beta\alpha)^{-1} = \alpha^{-1}\beta^{-1}.$$

This important formula has been called the "dressing-undressing principle": what goes on in dressing comes off in the reverse order in undressing (e.g., socks and shoes).

It is important to extend the notion of the Cartesian product of two sets to the product of any finite number of sets.[5] If $S_1, S_2, \ldots, S_r$ are any sets, then $\prod S_i$, or $S_1 \times S_2 \times \cdots \times S_r$, is defined to be the set of r-tuples $(s_1, s_2, \ldots, s_r)$ where the ith component $s_i \in S_i$. Equality is defined by $(s_1, s_2, \ldots, s_r) = (s'_1, s'_2, \ldots, s'_r)$ if $s_i = s'_i$ for every i. If all the $S_i = S$ then we write $S^{(r)}$ for $\prod S_i$. The concept of a product set permits us to define the notion of a function of two or more variables. For example, a function of two variables in S with values

---

[5] Also to infinite products. These will not be needed in this volume, so we shall not discuss them here.

in T is a map of $S \times S$ to T. Maps of $S^{(r)}$ to S are called *r-ary compositions* (or *r-ary products*) *on the set S*. The structures we shall consider in the first two chapters of this book (monoids, groups and rings) are defined by certain binary ($=2$-ary) compositions on a set S. At this point we shall be content merely to record the definition and to point out that we have already encountered several instances of binary products. For example, in $\mathscr{P}(S)$, the power set of a set S, we have the binary products $A \cup B$ and $A \cap B$ (that is, $(A, B) \to A \cup B$ and $(A, B) \to A \cap B$).

EXERCISES

1. Consider the maps $f: X \to Y$, $g: Y \to Z$. Prove: (a) f and g injective $\Rightarrow gf$ injective, (b) gf injective $\Rightarrow f$ injective, (c) f and g surjective $\Rightarrow gf$ surjective. (d) gf surjective $\Rightarrow g$ surjective. (e) Give examples of a set X and a map $f: X \to X$ that is injective (surjective) but not surjective (injective). (f) Let gf be bijective. What can be said about f and g respectively (injective, surjective)?

2. Show that $S \xrightarrow{\alpha} T$ is injective if and only if there is a map $T \xrightarrow{\beta} S$ such that $\beta\alpha = 1_S$, surjective if and only if there is a map $T \xrightarrow{\beta} S$ such that $\alpha\beta = 1_T$. In both cases investigate the assertion: if β is unique then α is bijective.

3. Show that $S \xrightarrow{\alpha} T$ is surjective if and only if there exist no maps β_1, β_2 of T into a set U such that $\beta_1 \neq \beta_2$ but $\beta_1\alpha = \beta_2\alpha$. Show that α is injective if and only if there exist no maps γ_1, γ_2 of a set U into S such that $\gamma_1 \neq \gamma_2$ but $\alpha\gamma_1 = \alpha\gamma_2$.

4. Let $S \xrightarrow{\alpha} T$ and let A and B be subsets of S. Show that $\alpha(A \cup B) = \alpha(A) \cup \alpha(B)$. and $\alpha(A \cap B) \subset \alpha(A) \cap \alpha(B)$. Give an example to show that $\alpha(A \cap B)$ need not coincide with $\alpha(A) \cap \alpha(B)$.

5. Let $S \xrightarrow{\alpha} T$, and let A be a subset of S. Let the *complement* of A in S, that is, the set of elements of S not contained in A, be denoted as $\sim A$. Show that, in general, $\alpha(\sim A) \not\subset \sim(\alpha(A))$. What happens if α is injective? Surjective?

0.3 EQUIVALENCE RELATIONS. FACTORING A MAP THROUGH AN EQUIVALENCE RELATION

We say that a (binary) relation is defined on a set S if, given any ordered pair (a, b) of elements of S, we can determine whether or not a is in the given relation to b. For example, we have the relation of order ">" in the set of real numbers. Given two real numbers a and b, presumably we can determine whether or not $a > b$. Another order relation is the lexicographic ordering of words, which determines their position in a dictionary. Still another example of a relation is the first-cousin relation among people (a and b have a common grandparent). To abstract the essential element from these situations and similar ones, we are led to define in a formal way *a (binary) relation R on a set S* to be simply any subset of the product set $S \times S$. If $(a, b) \in R$, then we say that "*a* is in the relation *R* to *b*" and we write *aRb*. Of particular importance for what follows are the equivalence relations, which we now define.

A relation *E* on a set *S* is called an *equivalence relation* if the following conditions hold for any *a, b, c*, in *S*:

1. *aEa* (reflexive property).
2. $aEb \Rightarrow bEa$ (symmetry).
3. *aEb* and $bEc \Rightarrow aEc$ (transitivity).

An example of an equivalence relation is obtained by letting *S* be the set of points in the plane and defining *aEb* if *a* and *b* lie on the same horizontal line. Another example of an equivalence relation *E'* on the same *S* is obtained by stipulating that *aE'b* if *a* and *b* are equidistant from the same point (e.g., the origin *O*).

We shall now show that the concept of an equivalence relation is equivalent to that of a partition of a set. If *S* is a set we define a *partition* $\pi(S)$ of *S* to be a set of non-vacuous subsets of *S* (that is, $\pi(S)$ is a subset of $\mathscr{P}(S)$ not containing $\varnothing$) such that the union of the sets in $\pi(S)$ is the whole of *S* and distinct sets in $\pi(S)$ are disjoint. The subsets making up $\pi(S)$ are called the *blocks* of the partition. We shall now show that with any equivalence relation *E* on *S* we can associate a partition $\pi_E(S)$ and with any partition π we can associate an equivalence relation E_π. Moreover, the relation between *E* and π are reciprocal in the sense that $\pi_{E_\pi} = \pi$ and $E_{\pi_E} = E$. First, suppose *E* is given. If $a \in S$ we let $\bar{a}_E$ (or simply $\bar{a}$) $= \{b \in S \mid bEa\}$. We call $\bar{a}_E$ the *equivalence class (relative to E or E-equivalence class) determined by a*. In the first example considered in the last paragraph, the equivalence class $\bar{a}_E$ is the horizontal line through *a* and in the second, the equivalence class is the circle through *a* having center *O*:

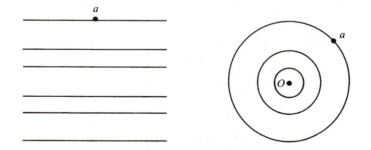

In both examples it is apparent that the set of equivalence classes is a partition of the plane. This is a general phenomenon. Let $\{\bar{a} \mid a \in S\}$ be the set of equivalence classes determined by *E*. Since *aEa*, $a \in \bar{a}$; hence every element of *S* is contained in an equivalence class and so $\bigcup_{a \in S} \bar{a} = S$. We note next that $\bar{a} = \bar{b}$ if and only if aEb. First, let aEb and let $c \in \bar{a}$. Then cEa and so, by condition 3, cEb. Then $c \in \bar{b}$. Then $\bar{a} \subset \bar{b}$. Also, by condition 2, bEa and so $\bar{b} \subset \bar{a}$. Hence $\bar{a} = \bar{b}$. Conversely, suppose $\bar{a} = \bar{b}$. Since $a \in \bar{a} = \bar{b}$ we see that aEb, by the definition of $\bar{b}$. Now suppose $\bar{a}$ and $\bar{b}$ are not disjoint and let $c \in \bar{a} \cap \bar{b}$. Then cEa and cEb. Hence $\bar{a} = \bar{c} = \bar{b}$. We therefore see that distinct sets in the set of equivalence classes are disjoint. Hence $\{\bar{a} | a \in S\}$ is a partition of S. We denote this as π_E.

Conversely, let π be any partition of the set S. Then, if $a \in S$, a is contained in one and only one $A \in \pi$. We define a relation E_π by specifying that $aE_\pi b$ if and only if a and b are contained in the same $A \in \pi$. Clearly this relation is reflexive, symmetric, and transitive. Hence E_π is an equivalence relation. It is clear also that the equivalence class $\bar{a}$ of a relative to E_π is the subset A in the partition π containing a. Hence the partition π_{E_π} associated with E_π is the given π. It is equally clear that if E is a given equivalence relation and $\pi_E = \{\bar{a} | a \in S\}$, then the equivalence relation E_{π_E} in which elements are equivalent if and only if they are contained in the same $\bar{a}$ is the given relation E.

If E is an equivalence relation, the associated partition $\pi = \{\bar{a} | a \in S\}$ is called the *quotient set of S relative to the relation E*. We shall usually denote π as S/E. We emphasize again that S/E is not a subset of S but rather of the power set $\mathscr{P}(S)$ of S. We now call attention to the map v of S into S/E defined by

$$v : a \to \bar{a}.$$

We call this the *natural map* of S to the quotient set S/E. Clearly, v is surjective.

We shall consider next some important connections between maps and equivalence relations. Suppose $S \xrightarrow{\alpha} T$. Then we can define a relation E_α in S by specifying that $aE_\alpha b$ if and only if $\alpha(a) = \alpha(b)$. It is clear that this is an equivalence relation in S. If $c \in T$ we put

(4) $$\alpha^{-1}(c) = \{a \in S | \alpha(a) = c\}$$

and we call this the *inverse image* of the element c. More generally, if C is a subset of T, then we define

(5) $$\alpha^{-1}(C) = \{a \in S | \alpha(a) \in C\}.$$

Clearly, $\alpha^{-1}(C) = \bigcup_{c \in C} \alpha^{-1}(c)$. Also $\alpha^{-1}(c) = \varnothing$ if $c \notin \text{im } \alpha$. On the other hand, if $c = \alpha(a)$ for some $a \in S$, then $\alpha^{-1}(c) = \alpha^{-1}(\alpha(a)) = \{b | \alpha(b) = \alpha(a)\}$ and this is just the equivalence class $\bar{a}_{E_\alpha}$ in S determined by the element a. We shall refer to this subset of S also as the *fiber over the element* $c \in \text{im } \alpha$. The set of these fibers constitutes the partition of S determined by E_α, that is, they are the elements of the quotient set S/E_α.

For example, let α be the orthogonal projection map of the plane onto a line l in the plane, as on page 6. If c is on the line the fiber $\alpha^{-1}(c)$ is the set of points on the line through c perpendicular to l.

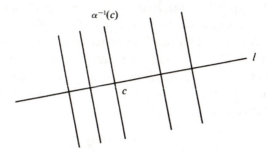

Note that we can define a bijective map of the set of these fibers into l by mapping the fiber $\alpha^{-1}(c)$ into the point c, which is the point of intersection of $\alpha^{-1}(c)$ with the line l.

In the general case α defines a map $\bar{\alpha}$ of S/E_α into T: abbreviating $\bar{a}_{E_\alpha} = \alpha^{-1}(\alpha(a))$ to $\bar{a}$ we simply define $\bar{\alpha}$ by

(6) $$\bar{\alpha}(\bar{a}) = \alpha(a).$$

Since $\bar{a} = \bar{b}$ if and only if $\alpha(a) = \alpha(b)$, it is clear that the right-hand side is independent of the choice of the element a in $\bar{a}$ and so, indeed, we do have a map. We call $\bar{\alpha}$ the map of S/E_α *induced* by α. This is injective since $\bar{\alpha}(\bar{a}) = \bar{\alpha}(\bar{b})$ gives $\alpha(a) = \alpha(b)$ and this implies $\bar{a} = \bar{b}$, by the definition of E_α. Of course, if α is injective to begin with, then $aE_\alpha b$ ($\alpha(a) = \alpha(b)$) implies $a = b$. In this case S/E_α can be identified with S and $\bar{\alpha}$ can be regarded as the same as α.

We now observe that $\bar{\alpha}(v(a)) = \bar{\alpha}(\bar{a}) = \alpha(a)$. Hence we have the factorization

(7) $$\alpha = \bar{\alpha}v$$

of the given map as a product of the natural map v of S to S/E_α and the induced map $\bar{\alpha}$ of S/E_α to T. The map $\bar{\alpha}$ is injective and v is surjective. The relation (7) is equivalent to the commutativity of the diagram

(8)

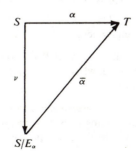

Since v is surjective it is clear that im $\alpha = $ im $\bar{\alpha}$. Hence $\bar{\alpha}$ is bijective if and only if α is surjective. We remark finally that $\bar{\alpha}$ is the only map which can be defined from S/E_α to T to make (8) a commutative diagram. Let $\beta : S/E_\alpha \to T$ satisfy $\beta v = \alpha$. Then $\beta(\bar{a}) = \beta(v(a)) = \alpha(a)$. Hence $\beta = \bar{\alpha}$, by the definition (6).

There is a useful generalization of these simple considerations. Suppose we are given a map $\alpha : S \to T$ and an equivalence relation E on S. We shall say that α is *compatible with E* if aEb for a, b in S implies $\alpha(a) = \alpha(b)$. In this case we can define a map $\bar{\alpha}$ of $\bar{S} = S/E$ to T by $\bar{\alpha} : \bar{a} \equiv \bar{a}_E \to \alpha(a)$. Clearly this is well defined, and if v denotes the natural surjection $a \to \bar{a}$, then $\alpha = \bar{\alpha} v$, that is, we have the commutativity of

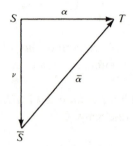

In this case the induced map $\bar{\alpha}$ need not be injective. In fact $\bar{\alpha}$ is injective if and only if $E = E_\alpha$.

The results which we have developed in this section, which at this point may appear to be a tedious collection of trivialities, will play a fundamental role in what follows.

EXERCISES

1. Let $\mathbb{N} = \{0, 1, 2, \ldots\}$. Show that the following are partitions of $\mathbb{N}$:
    (i) $\{0, 2, 4, \ldots, 2k, \ldots\}$, $\{1, 3, 5, \ldots, 2k + 1, \ldots\}$, $(k \in \mathbb{N})$
    (ii) $\{0, 3, 6, \ldots, 3k, \ldots\}$, $\{1, 4, 7, \ldots, 3k + 1, \ldots\}$, $\{2, 5, 8, \ldots, 3k + 2, \ldots\}$.

2. Let $\mathbb{N}$ be as in 1 and let $\mathbb{N}^{(2)} = \mathbb{N} \times \mathbb{N}$. On $\mathbb{N}^{(2)}$ define $(a, b) \sim (c, d)$ if $a + d = b + c$. Verify that $\sim$ is an equivalence relation.

3. Let S be the set of directed line segments PQ (initial point P, terminal point Q) in plane Euclidean geometry. With what equivalence relation on S is the quotient set the usual set of plane vectors?

4. If S and T are sets we define a *correspondence from S to T* to be a subset of $S \times T$. (Note that this encompasses maps as well as relations.) If C is a correspondence from S to T, C^{-1} is defined to be the correspondence from T to S consisting of the

points (t, s) such that $(s, t) \in C$. If C is a correspondence from S to T and D is a correspondence from T to U, the correspondence DC from S to U is defined to be the set of pairs $(s, u) \in S \times U$ for which there exists a $t \in T$ such that $(s, t) \in C$ and $(t, u) \in D$. Verify the associative law for correspondences: $(ED)C = E(DC)$, the identity law $C1_S = C = 1_T C$.

5. Show that the conditions that a relation E on S is an equivalence are: (i) $E \supset 1_S$, (ii) $E = E^{-1}$, (iii) $E \supset EE$.

6. Let C be a binary relation on S. For $r = 1, 2, 3, \ldots$ define $C^r = \{(s, t) | \text{for some } s_1, \ldots, s_{r-1} \in S, \text{ one has } sCs_1, s_1 Cs_2, \ldots, s_{r-1} Ct\}$. Let

$$E = 1_s \cup (C \cup C^{-1}) \cup (C \cup C^{-1})^2 \cup (C \cup C^{-1})^3 \cup \cdots.$$

Show that E is an equivalence relation, and that every equivalence relation on S containing C contains E. E is called the equivalence relation *generated* by C.

7. How many distinct binary relations are there on a set S of 2 elements? of 3 elements? of n elements? How many of these are equivalence relations?

8. Let $S \xrightarrow{\alpha} T \xrightarrow{\beta} U$. Show that if U_1 is a subset of U then $(\beta\alpha)^{-1}(U_1) = \alpha^{-1}(\beta^{-1}(U_1))$.

9. Let $S \xrightarrow{\alpha} T$ and let C and D be subsets of T. Show that $\alpha^{-1}(C \cup D) = \alpha^{-1}(C) \cup \alpha^{-1}(D)$ and $\alpha^{-1}(C \cap D) = \alpha^{-1}(C) \cap \alpha^{-1}(D)$ (cf. exercise 4, p. 10).

10. Let $\mathbb{C}$ be the set of complex numbers, $\mathbb{R}^+$ the set of non-negative real numbers. Let f be the map $z \to |z|$ (the absolute value of z) of $\mathbb{C}$ into $\mathbb{R}^+$. What is the equivalence relation on $\mathbb{C}$ defined by f?

11. Let $\mathbb{C}^*$ denote the set of complex numbers $\neq 0$ and let g be the map $z \to |z|^{-1} z$. What is the equivalence relation on $\mathbb{C}^*$ defined by g?

0.4 THE NATURAL NUMBERS

The system of natural numbers, or counting numbers, $0, 1, 2, 3, \ldots$ is fundamental in algebra in two respects. In the first place, it serves as a starting point for constructing more elaborate systems: the number systems of integers, of rational numbers and ultimately of real numbers, the ring of residue classes modulo an integer, and so on. In the second place, in studying some algebraic structures, certain maps of the set of natural numbers into the given structure play an important role. For example, in a structure S in which an associative binary composition and a unit are defined, any element $a \in S$ defines a map $n \to a^n$ where $a^0 = 1$, $a^1 = a$, and $a^k = a^{k-1}a$. Such maps are useful in studying the structure S.

A convenient and traditional starting point for studying the system $\mathbb{N}$ of natural numbers is an axiomatization of this system due to Peano. From this point of view we begin with a non-vacuous set $\mathbb{N}$, a particular element of $\mathbb{N}$, designated as 0, and a map $a \to a^+$ of $\mathbb{N}$ into itself, called the *successor* map.

Peano's axioms are:

1. $0 \neq a^+$ for any a (that is, 0 is not in the image of $\mathbb{N}$ under $a \to a^+$).
2. $a \to a^+$ is injective.
3. (Axiom of induction.) Any subset of $\mathbb{N}$ which contains 0 and contains the successor of every element in the given subset coincides with $\mathbb{N}$.

Axiom 3 is the basis of proofs by the *first principle of induction*. This can be stated as follows. Suppose that for each natural number n we have associated a statement $E(n)$ (e.g., $0 + 1 + 2 + \cdots + n = n(n + 1)/2$). Suppose $E(0)$ is true and $E(r^+)$ is true *whenever* $E(r)$ *is true*. (The second part is called the inductive step.) Then $E(n)$ is true for all $n \in \mathbb{N}$. This follows directly from axiom 3. Let S be the subset of $\mathbb{N}$ of s for which $E(s)$ is true. Then $0 \in S$ and if $r \in S$, then so does r^+. Hence, by axiom 3, $S = \mathbb{N}$, so $E(n)$ holds for all natural numbers.

Proofs by induction are very common in mathematics and are undoubtedly familiar to the reader. One also encounters quite frequently—without being conscious of it—definitions by induction. An example is the definition mentioned above of a^n by $a^0 = 1$, $a^{r+1} = a^r a$. Definition by induction is not as trivial as it may appear at first glance. This can be made precise by the following

RECURSION THEOREM.   *Let S be a set, φ a map of S into itself, a an element of S. Then there exists one and only one map f from $\mathbb{N}$ to S such that*

$$1.\ f(0) = a, \qquad 2.\ f(n^+) = \varphi(f(n)),\ n \in \mathbb{N}.^6$$

Proof.   Consider the product set $\mathbb{N} \times S$. Let Γ be the set of subsets U of $\mathbb{N} \times S$ having the following two properties: (i) $(0, a) \in U$, (ii) if $(n, b) \in U$ then $(n^+, \varphi(b)) \in U$. Since $\mathbb{N} \times S$ has these properties it is clear that $\Gamma \neq \varnothing$. Let f be the intersection of all the subsets U contained in Γ. We proceed to show that f is the desired function from $\mathbb{N}$ to S. In the first place, it follows by induction that if $n \in \mathbb{N}$, there exists a $b \in S$ such that $(n, b) \in f$. To prove that f is a map of $\mathbb{N}$ to S it remains to show that if (n, b) and $(n, b') \in f$ then $b = b'$. This is equivalent to showing that the subset T of $n \in \mathbb{N}$ such that (n, b) and $(n, b') \in f$ imply $b = b'$ is all of $\mathbb{N}$. We prove this by induction. First, $0 \in T$. Otherwise, we have $(0, a)$ and $(0, a') \in f$ but $a \neq a'$. Then let f' be the subset of f obtained by

---

[6] One is tempted to say that one can *define* f inductively by conditions 1 and 2. However, this does not make sense since in talking about a function on $\mathbb{N}$ we must have an à priori definition of $f(n)$ for every $n \in \mathbb{N}$. A proof of the existence of f must use *all* of Peano's axioms. An example illustrating this is given in exercise 4, p. 19. For a fuller account of these questions we refer the reader to an article, "On mathematical induction," by Leon Henkin in the *American Mathematical Monthly*, vol. 67 (1960), pp. 323–338. Henkin gives a proof of the recursion theorem based on the concept of "partial" functions on $\mathbb{N}$. The proof we shall give is due independently to P. Lorenzen, and to D. Hilbert and P. Bernays (jointly).

deleting the element $(0, a')$ from f. Then it is immediate that f' satisfies the defining conditions (i) and (ii) for the sets $U \in \Gamma$. Hence $f' \supset f$. But $f' \subsetneq f$ since f' was obtained by dropping $(0, a')$ from f. This contradiction proves that $0 \in T$. Now suppose we have a natural number r such that $r \in T$ but $r^+ \notin T$. Let $(r, b) \in f$. Then $(r^+, \varphi(b)) \in f$ and since $r^+ \notin T$, we have a $c \neq \varphi(b)$ such that $(r^+, c) \in f$. Now consider the subset f' of f obtained by deleting (r^+, c). Since $r^+ \neq 0$ and f contains $(0, a)$, f' contains $(0, a)$. The same argument shows that if $n \in \mathbb{N}$ and $n \neq r$ and $(n, d) \in f'$ then $(n^+, \varphi(d)) \in f'$. Now suppose $(r, b') \in f'$ then $b' = b$ and $(r^+, \varphi(b)) \in f'$ since $(r^+, \varphi(b))$ was not deleted in forming f' from f. Thus we see that $f' \in \Gamma$ and this again leads to the contradiction: $f' \supset f$, $f' \subsetneq f$. We have therefore proved that if $r \in T$ then $r^+ \in T$. Hence $T = \mathbb{N}$ by induction, and so we have proved the existence of a function f satisfying the given conditions. To prove uniqueness, let g be any map satisfying the conditions. Then $g \in \Gamma$ so $g \supset f$. But $g \supset f$ for two maps f and g implies $f = g$, by the definition of a map. Hence f is unique. □

Addition and multiplication of natural numbers can be defined by the recursion theorem. Addition of m to n can be defined by taking $a = m$ and φ to be the successor map $n \to n^+$. This amounts to the two formulas:

(a) $$0 + m = m$$

(b) $$n^+ + m = (n + m)^+.$$

For multiplication by m we use $a = 0$ and φ is the map $n \to n + m$. Thus we have

(a) $$0m = 0$$

(b) $$n^+ m = nm + m.$$

It can be proved that we have the associative, commutative, and cancellation laws of addition and multiplication:[7]

A1 $(x + y) + z = x + (y + z)$ (Associative law)

A2 $x + y = y + x$ (Commutative law)

A3 $x + z = y + z \Rightarrow x = y$ (Cancellation law)

M1 $(xy)z = x(yz)$

---

[7]Detailed proofs can be found in E. Landau, *Foundations of Analysis*, 2nd ed., New York, Chelsea Publishing Co., 1960. A sketch of the proofs is given in Paul R. Halmos, *Naive Set Theory*, New York, Van Nostrand Reinhold, 1960.

M2 $\qquad\qquad\qquad xy = yx$

M3 $\qquad\qquad\qquad xz = yz, z \neq 0 \Rightarrow x = y$

We also have the fundamental rule connecting addition and multiplication:

D $\qquad\qquad\qquad z(x + y) = zx + zy \qquad$ (Distributive law)

A fundamental concept for the system $\mathbb{N}$ is the relation of order defined by stating that the natural number a is greater than or equals the natural number b (notation: $a \geq b$ or $b \leq a$) if the equation $a = b + x$ has a solution $x \in \mathbb{N}$. The following are the basic properties of this relation:

O1 $\qquad\qquad\qquad x \geq y \quad$ and $\quad y \geq x \Leftrightarrow x = y.$

O2 $\qquad\qquad\qquad x \geq y \quad$ and $\quad y \geq z \Rightarrow x \geq z.$

O3 $\qquad\qquad$ For any $(x, y) \in \mathbb{N} \times \mathbb{N} \quad$ either $\quad x \geq y \quad$ or $\quad y \geq x.$

We also have the following *well-ordering property* of the set of natural numbers.

O4 $\qquad$ In any non-vacuous subset S of $\mathbb{N}$ there is a least number,
$\qquad\qquad\qquad$ that is, an $l \in S$ such that $l \leq s$ for every $s \in S.$

Proof. Let M be the set of natural numbers m such that $m \leq s$ for every $s \in S$. Then $0 \in M$, and if $s \in S$ then $s^+ \notin M$. Hence $M \neq \mathbb{N}$ and so, by the axiom of induction, there exists a natural number $l \in M$ such that $l^+ \notin M$. Then l is the required number, since $l \leq s$ for every $s \in S$. Moreover, $l \in S$ since otherwise $l < s$ for every $s \in S$ and then $l^+ \leq s$ for every $s \in S$. This contradicts $l^+ \notin M$. $\qquad\qquad\square$

The well-ordering property is the basis of the following *second principle of induction*. Suppose that for every $n \in \mathbb{N}$ we have a statement $E(n)$. Suppose it can be shown that $E(r)$ is true for a particular r if $E(s)$ is true for all $s < r$. (Note that this implies that it can be shown that $E(0)$ is true.) Then $E(n)$ is true for all n. To prove this we must show that the subset F of $\mathbb{N}$ of r such that $E(r)$ is false is vacuous. Now, if F is not vacuous, then, by O4, F contains a least element t. Then $E(t)$ is false but $E(s)$ is true for every $s < t$. This contradicts the hypothesis and proves $F = \varnothing$.

The main relations governing order and addition and order and multiplication are given in the following statements:

OA $\qquad\qquad\qquad a \geq b \Rightarrow a + c \geq b + c.$

OM $\qquad\qquad\qquad a \geq b \Rightarrow ac \geq bc.$

EXERCISES

1. Prove that if $a \geq b$ and $c \geq d$ then $a + c \geq b + d$ and $ac \geq bd$.

2. Prove the following extension of the first principle of induction: Let $s \in \mathbb{N}$ and assume that for every $n \geq s$ we have a statement $E(n)$. Suppose $E(s)$ holds, and if $E(r)$ holds for some $r \geq s$, then $E(r^+)$ holds. Then $E(n)$ is true for all $n \geq s$. State and prove the analogous extension of the second principle of induction.

3. Prove by induction that if c is a real number ≥ -1 and $n \in \mathbb{N}$ then $(1 + c)^n \geq 1 + nc$.

4. (Henkin.) Let $N = \{0, 1\}$ and define $0^+ = 1, 1^+ = 1$. Show that N satisfies Peano's axioms 1 and 3 but not 2. Let φ be the map of N into N such that $\varphi(0) = 1$ and $\varphi(1) = 0$. Show that the recursion theorem breaks down for N and this φ, that is, there exists no map f of N into itself satisfying $f(0) = 0$, $f(n^+) = \phi(f(n))$.

5. Prove A1 and M2.

0.5 THE NUMBER SYSTEM $\mathbb{Z}$ OF INTEGERS

Instead of following the usual procedure of constructing this system by adjoining to $\mathbb{N}$ the negatives of the elements of $\mathbb{N}$ we shall obtain the system of integers in a way that seems more natural and intuitive. Moreover, the method we shall give is analogous to the standard one for constructing the number system $\mathbb{Q}$ of rational numbers from the system $\mathbb{Z}$.

Our starting point is the product set $\mathbb{N} \times \mathbb{N}$. In this set we introduce the relation $(a, b) \sim (c, d)$ if $a + d = b + c$. It is easy to verify that this is an equivalence relation. What we have in mind in making this definition is that the equivalence class $\overline{(a, b)}$ determined by (a, b) is to play the role of the difference of a and b. If we represent the pair (a, b) in the usual way as the point with abscissa a and ordinate b, then $\overline{(a, b)}$ is the set of points with natural number coordinates on the line of slope 1 through (a, b). We call the equivalence classes (a, b) *integers*

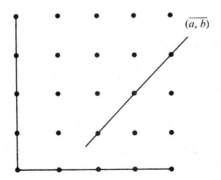

and we denote their totality as $\mathbb{Z}$. As a preliminary to defining addition we note that if $(a, b) \sim (a', b')$ and $(c, d) \sim (c', d')$ then

$$(a + c, b + d) \sim (a' + c', b' + d');$$

for the hypotheses are that $a + b' = a' + b$ and $c + d' = c' + d$. Hence $a + c + b' + d' = a' + c' + b + d$, which means that $(a + c, b + d) \sim (a' + c', b' + d')$. It follows that the integer $\overline{(a + c, b + d)}$ is uniquely determined by $\overline{(a, b)}$ and $\overline{(c, d)}$. We define this integer to be *sum* of the integers $\overline{(a, b)}$ and $\overline{(c, d)}$:

$$\overline{(a, b)} + \overline{(c, d)} = \overline{(a + c, b + d)}.$$

It is easy to verify that the rules A1, A2, and A3 hold. Also we note that $(a, a) \sim (b, b)$ and if we set $0 = \overline{(a, a)}$ (not to be confused with the 0 of $\mathbb{N}$), then

A4 $$0 + x = x \quad \text{for every} \quad x \in \mathbb{Z}.$$

Finally, every integer has a negative: If $x = \overline{(a, b)}$, then we denote $\overline{(b, a)}$ (which is independent of the representative (a, b) in $\overline{(a, b)}$) as $-x$. Then we have

A5 $$x + (-x) = 0.$$

We note next that if $(a, b) \sim (a', b')$ and $(c, d) \sim (c', d')$, then $a + b' = a' + b$, $c + d' = c' + d$. Hence

$$c(a + b') + d(a' + b) + a'(c + d') + b'(c' + d)$$
$$= c(a' + b) + d(a + b') + a'(c' + d) + b'(c + d')$$

so that

$$ac + b'c + a'd + bd + a'c + a'd' + b'c' + b'd$$
$$= a'c + bc + ad + b'd + a'c' + a'd + b'c + b'd'.$$

The cancellation law gives

$$ac + bd + a'd' + b'c' = bc + ad + a'c' + b'd',$$

which shows that $(ac + bd, ad + bc) \sim (a'c' + b'd', a'd' + b'c')$. Hence, if we define

$$\overline{(a, b)}\overline{(c, d)} = \overline{(ac + bd, ad + bc)}$$

we obtain a single-valued product. It can be verified that this is associative and commutative and distributive with respect to addition. The cancellation law holds if the factor z to be cancelled is not 0.

We regard $\overline{(a, b)} \geq \overline{(c, d)}$ if $a + d \geq b + c$. The relation is well defined (that is, it is independent of the choice of the representatives in the equivalence classes). One can verify easily that O1, O2, O3, and OA hold.

The property OM has to be modified to state:

OM'   If $x \geq y$ and $z \geq 0$ then $xz \geq yz$.

We now consider the set $\mathbb{N}'$ of non-negative integers. By definition, this is the subset of $\mathbb{Z}$ of elements $x \geq 0$, hence, of elements x of the form $\overline{(b + u, b)}$. It is immediate that $(b + u, b) \sim (c + u, c)$. Now let u be a natural number (that is, an element of $\mathbb{N}$) and define $u' = \overline{(b + u, b)}$. Our remarks show that $u \rightarrow u'$ defines a map of $\mathbb{N}$ into $\mathbb{Z}$ whose image is $\mathbb{N}'$. Moreover, if $(b + u, b) \sim (c + v, c)$, then $b + u + c = b + c + v$ so $u = v$. Thus $u \rightarrow u'$ is injective. It is left to the reader to verify the following properties:

$$(u + v)' = u' + v'$$

$$(uv)' = u'v'$$

$$u \geq v \Leftrightarrow u' \geq v'.$$

These and the fact that $u \rightarrow u'$ is bijective of $\mathbb{N}$ into $\mathbb{N}'$ imply that these two systems are indistinguishable as far as the basic operations and relation of order are concerned. In view of this situation we can now discard the original system of natural numbers and replace it by the set of non-negative integers, a subset of $\mathbb{Z}$. Also we can appropriate the notations originally used for $\mathbb{N}$ for this subset of $\mathbb{Z}$. Hence from now on we denote the latter as $\mathbb{N}$ and its elements as $0, 1, 2, \ldots$. It is easily seen that the remaining numbers in $\mathbb{Z}$ can be listed as $-1, -2, \ldots$.

EXERCISES

1. Show that $x \geq y \Leftrightarrow -x \leq -y$.

2. Prove that any non-vacuous set S of integers which is bounded below (above), in the sense that there exists an integer b (B) such that $b \leq s$ ($B \geq s$), $s \in S$, has a least (greatest) element.

3. Define $|x| = x$ if $x \geq 0$ and $|x| = -x$ if $x < 0$. Prove that $|xy| = |x||y|$ and $|x + y| \leq |x| + |y|$.

0.6   SOME BASIC ARITHMETIC FACTS ABOUT $\mathbb{Z}$

We shall say that the integer b is a *factor* or *divisor* of the integer a if there exists a $c \in \mathbb{Z}$ such that $a = bc$. Also a is called a *multiple* of b and we denote the relation by $b \,|\, a$. Clearly, this is a transitive relation. If $b \,|\, a$ and $a \,|\, b$, we have $a = bc$ and $b = ad$. Then $a = adc$. If $a \neq 0$ the cancellation law gives $dc = 1$. Then $|d|\,|c| = 1$ and $d = \pm 1, c = \pm 1$. This shows that if $b \,|\, a$ and $a \,|\, b$ and $a \neq 0$, then $b = \pm a$. An integer p is called a *prime* (or *irreducible*) if $p \neq 0, \pm 1$ and the only divisors of p are $\pm p$ and ± 1. If p is a prime so is $-p$.

The starting point for the study of number theory is the fact that every positive integer $\neq 1$ can be written in one and only one way as a product of positive primes: $a = p_1 p_2 \cdots p_s$, p_i primes, $s \geq 1$, and the uniqueness means "uniqueness apart from the order of the factors." This result is called *the fundamental theorem of arithmetic*. We shall now give a proof (due to E. Zermelo) of this result based on mathematical induction.

Let n be an integer > 1. Either n is a prime, or $n = n_1 n_2$ where n_1 and n_2 are > 1 and hence are $< n$. Hence, assuming that every integer > 1 and $< n$ is a product of positive primes, we have that n_1 and n_2 are such products, and consequently $n = n_1 n_2$ is a product of positive primes. Then (by the second principle of induction) every integer > 1 is a product of positive primes. It remains to prove uniqueness of the factorization. Let $n = p_1 p_2 \cdots p_s = q_1 q_2 \cdots q_t$ where the p_i and q_j are positive primes. First suppose $p_1 = q_1$. Cancelling this factor, we obtain $m = p_2 \cdots p_s = q_2 \cdots q_t < n$. If $m = 1$ we are through; otherwise, assuming the property for integers $m \neq 1$, $m < n$, that is, that $p_2, \ldots, p_s$ are the same as $q_2, \ldots, q_t$ except possibly for order, it is clear that this is true also for $p_1, p_2, \ldots, p_s$ and $q_1, q_2, \ldots, q_t$. Thus uniqueness follows for n. Next assume $p_1 \neq q_1$, say $p_1 < q_1$. In this case it is clear that $t > 1$ and $0 < p_1 q_2 \cdots q_t < n = q_1 q_2 \cdots q_t$. Subtracting $p_1 q_2 \cdots q_t$ from n gives

$$m = p_1(p_2 \cdots p_s - q_2 \cdots q_t) = (q_1 - p_1)q_2 \cdots q_t < n.$$

Since $t > 1$, $m > 1$. We obtain two factorizations of m into positive primes by factoring $p_2 \cdots p_s - q_2 \cdots q_t$ and $q_1 - p_1$ into positive primes. In the first p_1 occurs, and in the second the primes occurring are $q_2, \ldots, q_t$ and the primes that divide $q_1 - p_1$. Assuming that the result holds for m, p_1 coincides with one of the primes $q_2, \ldots, q_t$ or it divides $q_1 - p_1$. The latter is excluded since it implies $p_1 \,|\, q_1$, so $p_1 = q_1$. Hence $p_1 = q_j$ for some $j \geq 2$. Writing this q_j as the first factor we obtain a reduction to the previous case.[8]

---

[8] A different proof of this result and generalizations of it will be given in Chapter II.

The fundamental theorem of arithmetic can also be stated in the form:

Any integer $\neq 0$, ± 1 can be written as a product of primes. Apart from order and signs of the factors this factorization is unique.

The result can be stated also in terms of the number system $\mathbb{Q}$ of rational numbers.[9] In this context it states that every rational number $\neq 0$, ± 1 can be written in the form $p_1^{\varepsilon_1} p_2^{\varepsilon_2} \cdots p_s^{\varepsilon_s}$ where the p_i are prime integers and the $\varepsilon_i = \pm 1$. This is unique except for signs and order.

If $n \in \mathbb{Z}$ we can write $n = \pm p_1 p_2 \cdots p_s$ where the p_i are positive primes (assuming always that $n \neq 0$, ± 1). Rearranging the primes, and changing the notation, we have $n = \pm p_1^{k_1} p_2^{k_2} \cdots p_t^{k_t}$ where the p_i are *distinct* positive primes. It follows from the fundamental theorem of arithmetic that if m is a factor of n then m has the form $\pm p_1^{l_1} p_2^{l_2} \cdots p_t^{l_t}$ where the l_i satisfy $0 \leq l_i \leq k_i$. If m and n are two non-zero integers we can write both in terms of the same primes provided we allow the exponents to be 0 (and recall that $a^0 = 1$, if $a \neq 0$); that is, we may assume $m = \pm p_1^{e_1} p_2^{e_2} \cdots p_s^{e_s}$, $n = \pm p_1^{f_1} p_2^{f_2} \cdots p_s^{f_s}$ where the p_i are distinct positive primes and the e_i, $f_i \geq 0$. Now put $g_i = \min (e_i, f_i)$, $h_i = \max (e_i, f_i)$ and consider the two integers

$$(9) \qquad (m, n) \equiv p_1^{g_1} p_2^{g_2} \cdots p_s^{g_s}, \qquad [m, n] \equiv p_1^{h_1} p_2^{h_2} \cdots p_s^{h_s}.$$

It is readily seen that (m, n) is a *greatest common divisor* (g.c.d.) of m and n in the sense that $(m, n) | m$, $(m, n) | n$, and that if d is any integer such that $d | m$ and $d | n$ then $d | (m, n)$. Similarly $[m, n]$ is a *least common multiple* (l.c.m.) of m and n in the sense that $m | [m, n]$, $n | [m, n]$, and if $m | e$ and $n | e$ then $[m, n] | e$. It is clear from (9) that if m and n are positive then

$$(10) \qquad mn = (m, n)[m, n].$$

There is another way of proving the existence of a g.c.d. of two integers which does not require factorizations into primes and which gives the additional information that the g.c.d. can be written in the form $mu + nv$ where $u, v \in \mathbb{Z}$. This is based on

The Division Algorithm in $\mathbb{Z}$. If a and b are integers and $b \neq 0$ then there exist integers q and r, $0 \leq r < |b|$ such that $a = bq + r$.

Proof. Consider the set M of integral multiples $x|b|$ of $|b|$ satisfying $x|b| \leq a$. M is not vacuous since $-|a| |b| \leq -|a| \leq a$. Hence, the set M has a greatest

---

[9]We are assuming the reader is familiar with the construction of $\mathbb{Q}$ from the system $\mathbb{Z}$. A more general situation which covers this will be considered in section 2.9.

number $h|b|$ (exercise 2, p. 21). Then $h|b| \leq a$ so $a = h|b| + r$ where $r \geq 0$. On the other hand, $(h + 1)|b| = h|b| + |b| > h|b|$. Hence $(h + 1)|b| > a$ and $h|b| + |b| > h|b| + r$. Thus, $r < |b|$. We now put $q = h$ if $b > 0$ and $q = -h$ if $b < 0$. Then $h|b| = qb$ and $a = qb + r$ as required. $\square$

Now let $m, n \neq 0 \in \mathbb{Z}$ and let $I = \{mx + ny \,|\, x, y \in \mathbb{Z}\}$. This set includes $|n| > 0$. Hence there is a least positive integer $d = mu + nv \in I$. We claim that d is a g.c.d. of m and n. First, by the division algorithm we can write $m = dq + r$ where $0 \leq r < d$. Then $r = m - dq = m - (mu + nv)q = m(1 - uq) - nvq \in I$. Since d is the least positive integer in I, we must have $r = 0$. Hence $d|m$. Similarly $d|n$. Next suppose $e|m$ and $e|n$. Then $e|mu$ and $e|nv$. Hence $e|mu + nv$. Thus $e|d$.[10]

If d' and d are both g.c.d. of m and n then the second condition defining a g.c.d. gives $d|d'$ and $d'|d$. Hence $d' = \pm d$. If $n \neq 0$ then $d \neq 0$ and we may take $d > 0$. This determination of the greatest common divisor is the one we obtained from the prime factorizations, and we denote this as (m, n).

EXERCISES

1. Show that if p is a prime and $p|ab$ then either $p|a$ or $p|b$.

2. Define g.c.d. and l.c.m. for more than two integers and prove their existence.

3. Show that if k and m are positive integers and $m \neq n^k$ for $n \in \mathbb{Z}$ then $m^{1/k}$ is irrational.

0.7   A WORD ON CARDINAL NUMBERS

We shall have occasion frequently in this book to use the concept of the cardinal number of a set. At this point it will be well to list the main facts on cardinal numbers that will be required. No proofs will be given. These can be found in a number of places, in particular, in Halmos' *Naive Set Theory*.

We begin by saying that two sets *have the same cardinal number* or *cardinality* (or, are *equipotent* or just plain *equivalent*) if there exists a $1-1$ (read "one to

---

[10] There is a third, mechanical way of determining a g.c.d. for two integers, called the *Euclid algorithm*. This is indicated in exercises 11, p. 150.

one") correspondence between them. For example, the sets $\mathbb{N}$, $\mathbb{Z}$ and the set $\mathbb{Q}$ of rational numbers all have the same cardinal number. On the other hand, the set $\mathbb{R}$ of reals has a larger cardinality than $\mathbb{Q}$. As a representative of the class of sets having the same cardinal number we take a particular ordinal number in the class and call this *the cardinal number* of any set in the class. A definition of the ordinal numbers will not be given here, except the finite ones. We define the *ordinal n* for $n \in \mathbb{N}$ to be the subset of $\mathbb{N}$ of natural numbers $< n$. A set is called *finite* if it can be put in 1–1 correspondence with some finite ordinal, that is, with some set of natural numbers less than a given one. Otherwise the set is infinite. In general, we denote the cardinal number of S by $|S|$ and we write $|S| < \infty$ or $|S| = \infty$ according as S is finite or infinite. It is important to know that if m and n are distinct natural numbers then no bijective map between the corresponding ordinals exists. Assuming $m < n$ this is easily proved by induction on n. Another way of saying this is that if S and T are finite sets such that $|S| > |T|$ (in particular, if T is a proper subset of S) then for any surjective map α of S onto T there exist $s_1 \neq s_2$ in S such that $\alpha(s_1) = \alpha(s_2)$. This simple fact, which everyone is aware of, is called the "pigeonhole" principle: if there are more letters than pigeonholes then some pigeonhole must contain more than one letter. This has many important applications in mathematics. The pigeonhole principle is characteristic of finite sets. For any infinite set there always exist bijective maps onto proper subsets. If S and T are finite sets then $|S \times T| = |S||T|$ and $|S^T| = |S|^{|T|}$ where S^T is the set of maps of T into S.

An important result on cardinal numbers of infinite sets is the *Schröder-Bernstein theorem*: If we have injective maps of S into T and of T into S then $|S| = |T|$.

1

Monoids and Groups

The theory of groups is one of the oldest and richest branches of algebra. Groups of transformations play an important role in geometry, and, as we shall see in Chapter 4, finite groups are the basis of Galois' discoveries in the theory of equations. These two fields provided the original impetus for the development of the theory of groups, whose systematic study dates from the early part of the nineteenth century.

A more general concept than that of a group is that of a monoid. This is simply a set which is endowed with an associative binary composition and a unit—whereas groups are monoids all of whose elements have inverses relative to the unit. Although the theory of monoids is by no means as rich as that of groups, it has recently been found to have important "external" applications (notably to automata theory). We shall begin our discussion with the simpler and more general notion of a monoid, though our main target is the theory of groups. It is hoped that the preliminary study of monoids will clarify, by putting into a better perspective, some of the results on groups. Moreover, the results on monoids will be useful in the study of rings, which can be regarded as pairs of monoids having the same underlying set and satisfying some additional conditions (e.g., the distributive laws).

A substantial part of this chapter is foundational in nature. The reader will be confronted with a great many new concepts, and it may take some time to absorb them all. The point of view may appear rather abstract to the uninitiated. We have tried to overcome this difficulty by providing many examples and exercises whose purpose is to add concreteness to the theory. The axiomatic method, which we shall use throughout this book and, in particular, in this chapter, is very likely familiar to the reader: for example, in the axiomatic developments of Euclidean geometry and of the real number system. However, there is a striking difference between these earlier axiomatic theories and the ones we shall encounter. Whereas in the earlier theories the defining sets of axioms are categorical in the sense that there is essentially only one system satisfying them—this is far from true in the situations we shall consider. Our axiomatizations are intended to apply simultaneously to a large number of models, and, in fact, we almost never know the full range of their applicability. Nevertheless, it will generally be helpful to keep some examples in mind.

The principal systems we shall consider in this chapter are: monoids, monoids of transformations, groups, and groups of transformations. The relations among this quartet of concepts can be indicated by the following diagram:

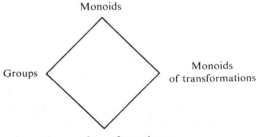

This is intended to indicate that the classes of groups and of monoids of transformations are contained in the class of monoids and the intersection of the first two classes is the class of groups of transformations. In addition to these concepts one has the fundamental concept of homomorphism which singles out the type of mappings that are natural to consider for our systems. We shall introduce first the more intuitive notion of an isomorphism.

At the end of the chapter we shall carry the discussion beyond the foundations in deriving the Sylow theorems for finite groups. Further results on finite groups will be given in Chapter 4 when we have need for them in connection with the theory of equations. Still later, in Chapter 6, we shall study the structure of some classical geometric groups (e.g., rotation groups).

1.1   MONOIDS OF TRANSFORMATIONS
AND ABSTRACT MONOIDS

We have seen in section 0.2 that composition of maps of sets satisfies the asso-
ciative law. If $S \xrightarrow{\alpha} T$, $T \xrightarrow{\beta} U$, and $U \xrightarrow{\gamma} V$, and $\beta\alpha$ is the map from S to U
defined by $(\beta\alpha)(s) = \beta(\alpha(s))$, then we have $\gamma(\beta\alpha) = (\gamma\beta)\alpha$. We recall also that if
1_T is the identity map $t \to t$ on T, then $1_T\alpha = \alpha$ and $\beta 1_T = \beta$ for every $\alpha:S \to T$
and $\beta:T \to U$. Now let us specialize this and consider the set $M(S)$ of transfor-
mations (or maps) of S into itself. For example, let $S = \{1, 2\}$. Here $M(S)$ consists of the four transformations

$$1_S = \begin{pmatrix} 1 & 2 \\ 1 & 2 \end{pmatrix}, \quad \alpha = \begin{pmatrix} 1 & 2 \\ 2 & 1 \end{pmatrix}, \quad \beta = \begin{pmatrix} 1 & 2 \\ 1 & 1 \end{pmatrix}, \quad \gamma = \begin{pmatrix} 1 & 2 \\ 2 & 2 \end{pmatrix}$$

where in each case we have indicated immediately below the element appearing
in the first row its image under the map. It is easy to check that the following
table gives the products in this $M(S)$:

	1	α	β	γ
1	1	α	β	γ
α	α	1	γ	β
β	β	β	β	β
γ	γ	γ	γ	$\gamma.$

(1)

Here, generally, we have put $\rho\sigma$ in the intersection of the row headed by ρ and
the column headed by σ ($\rho, \sigma = 1, \alpha, \beta, \gamma$). More generally, if $S = \{1, 2, \ldots, n\}$
then $M(S)$ consists of n^n transformations, and for a given n, we can write down
a multiplication table like (1) for $M(S)$. Now, for any non-vacuous S, $M(S)$ is an
example of a monoid, which is simply a non-vacuous set of elements, together
with an associative binary composition and a unit, that is, an element 1 whose
product in either order with any element is this element. More formally we give
the following

DEFINITION 1.1.   *A monoid is a triple $(M, p, 1)$ in which M is a non-vacuous
set, p is an associative binary composition (or product) in M, and 1 is an element
of M such that $p(1, a) = a = p(a, 1)$ for all $a \in M$.*

If we drop the hypothesis that p is associative we obtain a system which is
sometimes called a *monad*. On the other hand, if we drop the hypothesis on 1

and so have just a set together with an associative binary composition, then we obtain a *semigroup* (M, p). We shall now abbreviate $p(a, b)$, the product under p of a and b, to the customary ab (or $a \cdot b$). An element 1 of (M, p) such that $a1 = a = 1a$ for all a in M is called a *unit* in (M, p). If $1'$ is another such element then $1'1 = 1$ and $1'1 = 1'$, so $1' = 1$. Hence if a unit exists it is unique, and so we may speak of *the* unit of (M, p). It is clear that a monoid can be defined also as a semi-group containing a unit. However, we prefer to stick to the definition which we gave first. Once we have introduced a monoid $(M, p, 1)$, and it is clear what we have, then we can speak more briefly of "the monoid M," though, strictly speaking, this is the underlying set and is just one of the ingredients of $(M, p, 1)$.

Examples of monoids abound in the mathematics that is already familiar to the reader. We give a few in the following list.

EXAMPLES

1. $(\mathbb{N}, +, 0)$; $\mathbb{N}$, the set of natural numbers, $+$, the usual addition in $\mathbb{N}$, and 0 the first element of $\mathbb{N}$.

2. $(\mathbb{N}, \cdot, 1)$. Here $\cdot$ is the usual product and 1 is the natural number 1.

3. $(\mathbb{P}, \cdot, 1)$; $\mathbb{P}$, the set of positive integers, $\cdot$ and 1 are as in (2).

4. $(\mathbb{Z}, +, 0)$; $\mathbb{Z}$, the set of integers, $+$ and 0 are as usual.

5. $(\mathbb{Z}, \cdot, 1)$; $\cdot$ and 1 are as usual.

6. Let S be any non-vacuous set, $\mathscr{P}(S)$ the set of subsets of S. This gives rise to two monoids $(\mathscr{P}(S), \cup, \varnothing)$ and $(\mathscr{P}(S), \cap, S)$.

7. Let α be a particular transformation of S and define α^k inductively by $\alpha^0 = 1$, $\alpha^r = \alpha^{r-1}\alpha$, $r > 0$. Then $\alpha^k\alpha^l = \alpha^{k+l}$ (which is easy to see and will be proved in section 1.4). Then $\langle \alpha \rangle = \{\alpha^k | k \in \mathbb{N}\}$ together with the usual composition of transformations and $\alpha^0 = 1$ constitute a monoid.

If M is a monoid, a subset N of M is called a *submonoid* of M if N contains 1 and N is closed under the product in M, that is, $n_1 n_2 \in N$ for every $n_i \in N$. For instance, example 2, $(\mathbb{N}, \cdot, 1)$, is a submonoid of $(\mathbb{Z}, \cdot, 1)$; and 3, $(\mathbb{P}, \cdot, 1)$, is a submonoid of $(\mathbb{N}, \cdot, 1)$. On the other hand, the subset $\{0\}$ of $\mathbb{N}$ consisting of 0 only is closed under multiplication, but this is not a submonoid of 2 since it does not contain 1. If N is a submonoid of M, then N together with the product defined in M restricted to N, and the unit, constitute a monoid. It is clear that a submonoid of a submonoid of M is a submonoid of M. A submonoid of the monoid $M(S)$ of all transformations of the set S will be called *a monoid of transformations* (of S). Clearly the definition means that a subset N of $M(S)$ is a monoid of transformations if and only if the identity map is contained in N and the composite of any two maps in N belongs to N.

A monoid is said to be *finite* if it has a finite number of elements. We shall usually call the cardinality of a monoid its *order*, and we shall denote this as $|M|$. In investigating a finite monoid it is useful to have a multiplication table for the products in M. As in the special case which we considered above, if $M = \{a_1 = 1, a_2, \ldots, a_m\}$ the multiplication table has the form

(2)

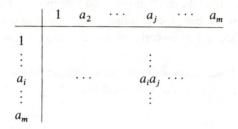

where $a_i a_j$ is tabulated in the intersection of the row headed by a_i and the column headed by a_j.

EXERCISES

1. Let S be a set and define a product in S by $ab = b$. Show that S is a semigroup. Under what condition does S contain a unit?

2. Let $M = \mathbb{Z} \times \mathbb{Z}$ the set of pairs of integers (x_1, x_2). Define $(x_1, x_2)(y_1, y_2) = (x_1 y_1 + 2x_2 y_2, x_1 y_2 + x_2 y_1)$, $1 = (1, 0)$. Show that this defines a monoid. (Observe that the commutative law of multiplication holds.) Show that if $(x_1, x_2) \neq (0, 0)$ then the cancellation law will hold for (x_1, x_2), that is, $(x_1, x_2)(y_1, y_2) = (x_1, x_2)(z_1, z_2) \Rightarrow (y_1, y_2) = (z_1, z_2)$.

3. A machine accepts eight-letter words (defined to be any sequence of eight letters of the alphabet, possibly meaningless), and prints an eight-letter word consisting of the first five letters of the first word followed by the last three letters of the second word. Show that the set of eight-letter words with this composition is a semigroup. What if the machine prints the last four letters of the first word followed by the first four of the second? Is either of these systems a monoid?

4. Let $(M, p, 1)$ be a monoid and let $m \in M$. Define a new product p_m in M by $p_m(a, b) = amb$. Show that this defines a semigroup. Under what condition on m do we have a unit relative to p_m?

5. Let S be a semigroup, u an element not in S. Form $M = S \cup \{u\}$ and extend the product in S to a binary product in M by defining $ua = a = au$ for all $a \in M$. Show that M is a monoid.

1.2 GROUPS OF TRANSFORMATIONS
AND ABSTRACT GROUPS

An element u of a monoid M is said to be *invertible* (or *a unit*[1]) if there exists a v in M such that

(3) $$uv = 1 = vu.$$

If v' also satisfies $uv' = 1 = v'u$ then $v' = (vu)v' = v(uv') = v$. Hence v satisfying (3) is unique. We call this *the inverse* of u and write $v = u^{-1}$. It is clear also that u^{-1} is invertible and $(u^{-1})^{-1} = u$. We now give the following

DEFINITION 1.2. *A group G (or $(G, p, 1)$) is a monoid all of whose elements are invertible.*

We shall call a submonoid of a monoid M (in particular, of a group) a *subgroup* if, regarded as a monoid, it is a group. Since the unit of a submonoid coincides with that of M it is clear that a subset G of M is a subgroup if and only if it has the following closure properties: $1 \in G$, $g_1 g_2 \in G$ for every $g_i \in G$, every $g \in G$ is invertible, and $g^{-1} \in G$.

Let $U(M)$ denote the set of invertible elements of the monoid M and let $u_1, u_2 \in U(M)$. Then

$$(u_1 u_2)(u_2^{-1} u_1^{-1}) = ((u_1 u_2)u_2^{-1})u_1^{-1} = (u_1(u_2 u_2^{-1}))u_1^{-1} = u_1 u_1^{-1} = 1$$

and, similarly, $(u_2^{-1} u_1^{-1})(u_1 u_2) = 1$. Hence $u_1 u_2 \in U(M)$. We saw also that if $u \in U(M)$ then $u^{-1} \in U(M)$, and clearly $1 \cdot 1 = 1$ shows that $1 \in U(M)$. Thus we see that $U(M)$ is a subgroup of M. We shall call this the *group of units* or *invertible elements of M*. For example, if $M = (\mathbb{Z}, \cdot, 1)$ then $U(M) = \{1, -1\}$ and if $M = (\mathbb{N}, \cdot, 1)$ then $U(M) = \{1\}$.

We now consider the monoid $M(S)$ of transformations of a non-vacuous set S. What is the associated group of units $U(M(S))$? We have seen (p. 8) that a transformation is invertible if and only if it is bijective. Hence our group is just the set of bijective transformations of S with the composition as the composite of maps and the unit as the identity map. We shall call $U(M(S))$ the *symmetric group of the set S* and denote it as Sym S. In particular, if $S = \{1, 2, \ldots, n\}$ then we shall write S_n for Sym S and call this the *symmetric group* on n letters. We usually call the elements of S_n *permutations* of $\{1, 2, \ldots, n\}$. We can easily list all of these and determine the order of S_n. Using the notation we introduced in the case $n = 2$, we can denote a transformation of $\{1, 2, \ldots, n\}$ by a symbol

---

[1] This term is quite commonly used in this connection. Unfortunately it conflicts with the meaning of *the* unit 1. It will generally be clear from the context which meaning is intended.

(4)
$$\alpha = \begin{pmatrix} 1 & 2 & \cdots & n \\ 1' & 2' & \cdots & n' \end{pmatrix}$$

where this means the transformation sending $i \to i'$, $1 \leq i \leq n$. In order for α to be injective the second line $1', \ldots, n'$ must contain no duplicates, that is, no i can appear twice. This will also assure bijectivity since we cannot have an injective map of $\{1, 2, \ldots, n\}$ on a proper subset. We can now count the number of elements in S_n by observing that we can take the element $1'$ in the symbol (4) to be any one of the n numbers $1, 2, \ldots, n$. This gives n choices for $1'$. Once this has been chosen, to avoid duplication, we must choose $2'$ among the $n - 1$ numbers different from $1'$. This gives $n - 1$ choices for $2'$. After the partners of 1 and 2 have been chosen, we have $n - 2$ choices for $3'$, and so on. Clearly this means we have $n!$ symbols (4) representing the elements of S_n. We have therefore proved

THEOREM 1.1.   *The order of S_n is $n!$.*

This is to be compared with the order n^n of the monoid of transformations of $S = \{1, 2, \ldots, n\}$.

We have called a submonoid of the monoid of transformations of a set, a monoid of transformations. Similarly, a subgroup of the symmetric group of S will be called a *group of transformations* (or *transformation group*). If S is finite we generally use the term *permutation group* for a group of transformations of S. A set G of transformations of a set S is a group of transformations if and only if it consists of bijective maps and G has the following closure properties: $1 = 1_S \in G$, $\alpha\beta \in G$ if α and $\beta \in G$, $\alpha^{-1} \in G$ if $\alpha \in G$.

EXAMPLES

1. $(\mathbb{Z}, +, 0)$ the group of integers under addition.[2] Here the inverse of a is $-a$.

2. $(\mathbb{Q}, +, 0)$ where $\mathbb{Q}$ denotes the set of rational numbers; the composition is addition; the inverse of a is $-a$.

3. $(\mathbb{R}, +, 0)$, $\mathbb{R}$ the set of real numbers, usual $+$ and 0.

4. $(\mathbb{C}, +, 0)$, $\mathbb{C}$ the set of complex numbers; usual $+$ and 0.

5. $(\mathbb{Q}^*, \cdot, 1)$, $\mathbb{Q}^*$, the set of non-zero rational numbers; the composition is multiplication; 1 is the usual 1 and a^{-1} the usual inverse.

6. $(\mathbb{R}^*, \cdot, 1)$, $\mathbb{R}^*$ the set of non-zero real numbers; usual multiplication, 1, and inverses.

---

[2] Throughout this book we use the following notations (which have become standard): $\mathbb{N}$, for the set of natural numbers $0, 1, 2, \ldots$; $\mathbb{Z}$, for the set of integers; $\mathbb{Q}$, for the set of rational numbers; $\mathbb{R}$, for the set of real numbers; $\mathbb{C}$, for the set of complex numbers.

7. $(\mathbb{C}^*, \cdot, 1)$, $\mathbb{C}^*$ the set of non-zero complex numbers; usual multiplication, 1, and inverses.

8. $(\mathbb{R}^{(3)}, +, 0)$, $\mathbb{R}^{(3)}$ the set of triples of real numbers (x, y, z) with addition as $(x_1, y_1, z_1) + (x_2, y_2, z_2) = (x_1 + x_2, y_1 + y_2, z_1 + z_2)$, $0 = (0, 0, 0)$. The inverse of (x, y, z) is $(-x, -y, -z)$. This example can be described also as the group of vectors in three-dimensional Euclidean space with the usual geometric construction of the sum.

9. The set of rotations about a point 0 in the plane; composition as usual. If 0 is taken to be the origin, the rotation through an angle θ can be represented analytically as the map $(x, y) \to (x', y')$ where

$$x' = x \cos \theta - y \sin \theta, \qquad y' = x \sin \theta + y \cos \theta.$$

For $\theta = 0$ we get the identity map, and the inverse of the rotation through the angle θ is the rotation through $-\theta$.

10. The set of rotations together with the set of reflections in the lines through 0. The latter are given analytically by $(x, y) \to (x', y')$ where

$$x' = x \cos \theta + y \sin \theta, \qquad y' = x \sin \theta - y \cos \theta.$$

The product of two reflections is a rotation and the product in either order of a reflection and a rotation is a reflection.

11. Consider the regular n-gon ($=$ polygon of n sides) inscribed in the unit circle in the plane, so that one of the vertices is $(1, 0)$ e.g., a regular pentagon:

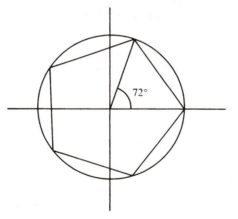

The vertices subtend angles of $0, 2\pi/n, 4\pi/n, \ldots, 2(n-1)\pi/n$ radians with the positive x-axis. The subset of the rotation group which maps our figure into itself consists of the n rotations through angles of $0, 2\pi/n, \ldots, 2(n-1)\pi/n$ radians respectively. These form a subgroup R_n of the rotation group.

12. We now consider the set D_n of rotations and reflections which map the regular n-gon, as in 11, into itself. These form a subgroup of the group defined in 10. We shall call the elements of this group the *symmetries* of the regular n-gon. The reflection in the x-axis is one of our symmetries. Multiplying this on the left by the n rotational symmetries we obtain n distinct reflectional symmetries. This gives them all, for if we let S denote the reflection in the x-axis and T denote any reflectional symmetry then ST is

one of the n-rotational symmetries $R_1, \ldots, R_n$, say R_i. Since $S^2 = 1$, $ST = R_i$ gives $T = SR_i$ which is one of those we counted. Thus D_n consists of n rotations and n reflections and its order is $2n$. The group D_n is called the *dihedral group*. For $n = 3$ and 4 the lines in whose reflections we obtain symmetries of our n-gon are indicated as broken lines in the following figures:

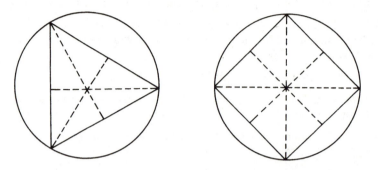

13. Let U_n denote the set of complex numbers which are nth roots of unity in the sense that $z^n = 1$. It is easy to determine these using the polar representation of a complex number: $z = re^{i\theta} = r(\cos\theta + i\sin\theta)$, $r = |z|$, θ, the argument (=angle) of z. If $z_1 = r_1 e^{i\theta_1}$ and $z_2 = r_2 e^{i\theta_2}$ then $z_1 z_2 = r_1 r_2 e^{i(\theta_1 + \theta_2)}$. It follows that if $z^n = 1$ then $|z| = r = 1$ and θ must be one of the angles $\theta = 0$, $2\pi/n$, $4\pi/n$, $\ldots$, $2(n-1)\pi/n$. Since $1^n = 1$, and $z_1^n = 1$ and $z_2^n = 1$ imply $(z_1 z_2)^n = z_1^n z_2^n = 1$ and $(z_1^{-1})^n = (z_1^n)^{-1} = 1$, it is clear that U_n is a subgroup of $\mathbb{C}^*$, the multiplicative group of complex numbers (as in example 7).

14. The rotation group in three-dimensional Euclidean space. This is the set of rotations about the origin 0 in the number space $\mathbb{R}^{(3)}$ of triples (x, y, z), $x, y, z \in \mathbb{R}$. From analytic geometry it is known that these maps are given analytically as $(x, y, z) \to (x', y', z')$ where

$$x' = \lambda_1 x + \mu_1 y + \nu_1 z$$
$$y' = \lambda_2 x + \mu_2 y + \nu_2 z$$
$$z' = \lambda_3 x + \mu_3 y + \nu_3 z$$

and the λ_i, μ_i, ν_i are any real numbers satisfying:

$$\lambda_i^2 + \mu_i^2 + \nu_i^2 = 1, \qquad \lambda_i \lambda_j + \mu_i \mu_j + \nu_i \nu_j = 0 \quad \text{if} \quad i \neq j,$$

$$\begin{vmatrix} \lambda_1 & \mu_1 & \nu_1 \\ \lambda_2 & \mu_2 & \nu_2 \\ \lambda_3 & \mu_3 & \nu_3 \end{vmatrix} = 1.$$

We remark that all the examples 9–14 except 13 are transformation groups. We remark also that in our list of monoids given on p. 29, 1, 2, 3, 5 are not groups and 7 may or may not be a group. The two geometric examples 11 and 12 illustrate a general principle. If G is a transformation group of a set S and A is a subset then the transformations contained in G which map A onto itself ($\sigma(A) = A$) constitute a subgroup G_A of G. The validity of this is immediate.

We shall now consider a general construction of monoids and groups out of given monoids and groups called the direct product. Let $M_1, M_2, \ldots, M_n$ be given monoids and put $M = M_1 \times M_2 \times \cdots \times M_n$. We introduce a product in M by

$$(a_1, a_2, \ldots, a_n)(b_1, b_2, \ldots, b_n) = (a_1 b_1, a_2 b_2, \ldots, a_n b_n)$$

where $a_i, b_i \in M_i$ and put

$$1 = (1_1, 1_2, \ldots, 1_n)$$

1_i, the unit of M_i. Then, writing (a_i) for $(a_1, \ldots, a_n)$ etc., we have $((a_i)(b_i))(c_i) = ((a_i b_i)c_i)$ and $(a_i)((b_i)(c_i)) = (a_i(b_i c_i))$. Hence the associative law holds. Also $1(a_i) = (a_i) = (a_i)1$ so 1 is the unit. Hence we have a monoid. This is called the *direct product* $M_1 \times M_2 \times \cdots \times M_n$ *of the monoids* M_i. If every M_i is a group G_i, then $G_1 \times G_2 \times \cdots \times G_n$ is a group since in this case (a_i) has the inverse (a_i^{-1}). Then $G_1 \times G_2 \times \cdots \times G_n$ is called the *direct product of the groups* G_i. A special case of this construction is given in example 8 above. This can be regarded as a direct product of $(\mathbb{R}, +, 0)$ with itself taken three times. As in this example, it should be noted that we do not require the M_i (or the G_i) to be distinct. In fact, we obtain an interesting case if we take all the $M_i = N$, a fixed monoid. Then we obtain the direct product of N with itself taken n times or the n-fold *direct power* of N. We shall usually denote this as $N^{(n)}$.

EXERCISES

1. Determine $\alpha\beta$, $\beta\alpha$ and α^{-1} in S_5 if

$$\alpha = \begin{pmatrix} 1 & 2 & 3 & 4 & 5 \\ 2 & 3 & 1 & 5 & 4 \end{pmatrix}, \qquad \beta = \begin{pmatrix} 1 & 2 & 3 & 4 & 5 \\ 1 & 3 & 4 & 5 & 2 \end{pmatrix}$$

2. Verify that the permutations

$$1 = \begin{pmatrix} 1 & 2 & 3 \\ 1 & 2 & 3 \end{pmatrix}, \qquad \begin{pmatrix} 1 & 2 & 3 \\ 2 & 3 & 1 \end{pmatrix}, \qquad \begin{pmatrix} 1 & 2 & 3 \\ 3 & 1 & 2 \end{pmatrix}$$

    form a subgroup of S_3.

3. Determine a multiplication table for S_3.

4. Let G be the set of pairs of real numbers (a, b) with $a \neq 0$ and define: $(a, b)(c, d) = (ac, ad + b)$, $1 = (1, 0)$. Verify that this defines a group.

5. Let G be the set of transformations of the real line $\mathbb{R}$ defined by $x \to x' = ax + b$ where a and b are real numbers and $a \neq 0$. Verify that G is a transformation group of $\mathbb{R}$.

6. Verify that the set of translations $x \to x' = x + b$ is a subgroup of the group defined in exercise 5.

7. Show that if an element a of a monoid has a *right inverse* b, that is, $ab = 1$; and a left inverse c, that is, $ca = 1$; then $b = c$, and a is invertible with $a^{-1} = b$. Show that a is invertible with b as inverse if and only if $aba = a$ and $ab^2a = 1$.

8. Let α be a rotation about the origin in the plane and let ρ be the reflection in the x-axis. Show that $\rho\alpha\rho^{-1} = \alpha^{-1}$.

9. Let G be a non-vacuous subset of a monoid M. Show that G is a subgroup if and only if every $g \in G$ is invertible in M and $g_1^{-1}g_2 \in G$ for any $g_1, g_2 \in G$.

10. Let G be a semigroup having the following properties: (a) G contains a *right unit* 1_r, that is, an element satisfying $a1_r = a$, $a \in G$, (b) every element $a \in G$ has a right inverse relative to 1_r $(ab = 1_r)$. Show that G is a group.[3]

11. Show that in a group, the equations $ax = b$ and $ya = b$ are solvable for any a, $b \in G$. Conversely, show that any semigroup having this property contains a unit and is a group.

12. Show that both cancellation laws hold in a group, that is, $ax = ay \Rightarrow x = y$ and $xa = ya \Rightarrow x = y$. Show that any finite semigroup in which both cancellation laws hold is a group (*Hint:* Use the pigeon-hole principle and exercise 11.)

13. Show that any finite group of even order contains an element $a \neq 1$ such that $a^2 = 1$.

14. Show that a group G cannot be a union of two proper subgroups.

15. Let G be a finite set with a binary composition and unit. Show that G is a group if and only if the multiplication table (constructed as for monoids) has the following properties:
    (i) every row and every column contains every element of G,
    (ii) for every pair of elements $x \neq 1$, $y \neq 1$ of G, let R be any rectangle in the body of the table having 1 as one of its vertices, x a vertex in the same row as 1, y a vertex in the same column as 1, then the fourth vertex of the rectangle depends only on the pair (x, y) and not on the position of 1.

1.3 ISOMORPHISM. CAYLEY'S THEOREM

At this point the reader may be a bit overwhelmed by the multitude of examples of monoids and groups. It may therefore be somewhat reassuring to know that

---

[3] The semigroups satisfying (a) and (b'), which is (b) with "right inverse" replaced by "left inverse," need not be groups. Their structure has been determined by A. H. Clifford in *Annals of Mathematics*, vol. 34 (1933), pp. 865–871.

certain groups which look different can be regarded as essentially the same— that is, they are "isomorphic" in a sense which we shall define. Also we shall see that every monoid is isomorphic to a monoid of transformations, and every group is isomorphic to a group of transformations. Thus we obtain essentially all monoids (groups) in the class of monoids (groups) of transformations. This result for groups is due to Cayley. We give first

DEFINITION 1.3.   *Two monoids $(M, p, 1)$ and $(M', p', 1')$ are said to be* isomorphic *if there exists a bijective map η of M to M' such that*

(5) $\eta(1) = 1'$, $\eta(xy) = \eta(x)\eta(y)$, $x, y \in M$.

The fact that M is isomorphic to M' will be indicated by $M \cong M'$. The map η satisfying the conditions (5) is called an *isomorphism* of M onto M'. Actually, the first condition in (5) is superfluous. For, if η satisfies the second condition, then we have $\eta(x)\eta(1) = \eta(x) = \eta(1)\eta(x)$. Since η is surjective, this shows that $\eta(1)$ acts as the unit $1'$ in M', and since we know that the unit is unique, we have $\eta(1) = 1'$. Nevertheless, we prefer to include the first condition in (5) as part of the definition, since this will be needed in a more general context which we shall consider later.

Perhaps the first significant example of isomorphism between groups which was discovered was one between the additive group of real numbers and the multiplicative group of positive reals. We denote these as $(\mathbb{R}, +, 0)$ and $(\mathbb{R}^+, \cdot, 1)$ respectively. An isomorphism of $(\mathbb{R}, +, 0)$ and $(\mathbb{R}^+, \cdot, 1)$ is the exponential map $x \to e^x$. This is bijective with inverse $y \to \log y$ (the natural logarithm) and we have the "functional equation"

$$e^{x+y} = e^x e^y$$

which is just the second condition in (5) since $+$ is the composition in $(\mathbb{R}, +, 0)$.

If M and M' are isomorphic there may exist many isomorphisms between these monoids. For instance, if a is any positive real number $\neq 1$, the map $x \to a^x$ is an isomorphism between the groups we have just considered. It is clear that isomorphism is an equivalence relation: any monoid is isomorphic to itself (with respect to the identity map) and if $\eta : M \to M'$ is an isomorphism, then applying η^{-1} to the second condition in (5) gives $xy = \eta^{-1}(\eta(x)\eta(y))$. Hence if we write $\eta(x) = x'$, $\eta(y) = y'$, then $\eta^{-1}(x')\eta^{-1}(y') = \eta^{-1}(x'y')$, and this holds for all x', $y' \in M'$ since η is surjective. Thus η^{-1} is an isomorphism from M' to M. Finally, if ζ is an isomorphism of M' to M'' then $(\zeta\eta)(xy) = \zeta(\eta(xy)) = \zeta(\eta(x)\eta(y)) = \zeta(\eta(x))\zeta(\eta(y))$. Thus $\zeta\eta : M \to M''$ is an isomorphism.

We shall now prove the result which was mentioned before.

CAYLEY'S THEOREM FOR MONOIDS AND GROUPS. (1) *Any monoid is isomorphic to a monoid of transformations.* (2) *Any group is isomorphic to a transformation group.*

Proof. (1) Let $(M, p, 1)$ be a monoid. Then we shall set up an isomorphism of $(M, p, 1)$ with a monoid of transformations of the set M itself. For any $a \in M$, we define the map $a_L : x \to ax$ of M into M. We call a_L the *left translation* (or *left multiplication*) defined by a. We claim first that the set $M_L = \{a_L | a \in M\}$ is a monoid of transformations, which, we have seen, means that the identity map is in the set M_L and this set is closed under the composite product of maps. Since 1_L is $x \to 1x = x$, $1_L = 1(= 1_M) \in M_L$. Also $a_L b_L$ is the map $x \to a(bx)$. By the associative law, $a(bx) = (ab)x$, and this is $(ab)_L x$. Thus $a_L b_L = (ab)_L \in M_L$. We note next that the map $a \to a_L$ is an isomorphism of $(M, p, 1)$ with the monoid of transformations M_L. The equations $1_L = 1$ and $a_L b_L = (ab)_L$ are the conditions (5) for $a \to a_L$, and, obviously, this map is surjective. Moreover, it is also injective; for, if $a_L = b_L$ then, in particular, $a = a_L 1 = b_L 1 = b$. Hence $a \to a_L$ is an isomorphism.

(2) Now let $(G, p, 1)$ be a group. Then everything will follow from the proof of (1) if we can show that G_L is a group of transformations. This requires two additional facts beyond those we obtained in the preceding argument: the maps a_L are bijective and G_L is closed under inverses. Both follow from $1_L = (a^{-1}a)_L = (a^{-1})_L a_L$ and $1_L = a_L (a^{-1})_L$ which show that a_L has the inverse $(a^{-1})_L$ and this is in G_L. □

It should be noted that if M (or G) is finite then M_L acts in the finite set M. In particular, if $|G| = n$, then G_L is a subgroup of S_n, the symmetric group on a set of n elements. Hence we have the

COROLLARY. *Any finite group of order n is isomorphic to a subgroup of the symmetric group S_n.*

EXAMPLES

1. Let $(\mathbb{R}, +, 0)$ be the additive group of reals. If $a \in \mathbb{R}$, the left translation a_L is $x \to a + x$.

2. Let G be the group of pairs of real numbers (a, b), $a \neq 0$ with product $(a, b)(c, d) = (ac, ad + b)$, $1 = (1, 0)$ (exercise 4, p. 36). Here $(a, b)_L$ is the map

$$(x, y) \to (ax, ay + b).$$

Another transformation group isomorphic to G is the group of transformations of $\mathbb{R}$

consisting of the maps $x \to ax + b$, $a \neq 0$. The map sending (a, b) into the transformation $T_{(a,b)}$ defined as $x \to ax + b$, $a \neq 0$, is an isomorphism.

EXERCISES

1. Use a multiplication table for S_3 (exercise 3, p. 36) and the isomorphism $a \to a_L$ (a_L the left translation defined by a) to obtain a subgroup of S_6 isomorphic to S_3.

2. Show that the two groups given in examples 11 and 13 on pages 33 and 34 are isomorphic. Obtain a subgroup of S_n isomorphic to these groups.

3. Let G be a group. Define the *right translation* a_R for $a \in G$ as the map $x \to xa$ in G. Show that $G_R = \{a_R\}$ is a transformation group of the set G and $a \to a_R^{-1}$ is an isomorphism of G with G_R.

4. Is the additive group of integers isomorphic to the additive group of rationals (examples 1 and 2 on p. 32)?

5. Is the additive group of rationals isomorphic to the multiplicative group of non-zero rationals (examples 2 and 5 on p. 32)?

6. In $\mathbb{Z}$ define $a \circ b = a + b - ab$. Show that $(\mathbb{Z}, \circ, 0)$ is a monoid and that the map $a \to 1 - a$ is an isomorphism of the multiplicative monoid $(\mathbb{Z}, \cdot, 1)$ with $(\mathbb{Z}, \circ, 0)$.

1.4 GENERALIZED ASSOCIATIVITY. COMMUTATIVITY

Let $a_1, a_2, \ldots, a_n$ be a finite sequence of elements of a monoid M. We can determine from this sequence a number of products obtained by iterating the given binary composition of M. For instance, if $n = 4$, we have the following possibilities:

$$((a_1a_2)a_3)a_4, \ (a_1(a_2a_3))a_4, \ (a_1a_2)(a_3a_4), \ a_1((a_2a_3)a_4), \ a_1(a_2(a_3a_4)).$$

In general, we obtain the products of $a_1, a_2, \ldots, a_n$ by partitioning this sequence into two subsequences $a_1, \ldots, a_m$ and $a_{m+1}, \ldots, a_n$, $1 \leq m \leq n - 1$. Assuming we already know how to obtain the products of $a_1, \ldots, a_m$ and $a_{m+1}, \ldots, a_n$, we apply the binary composition to these results to obtain an element of M which is a product associated with the sequence $a_1, a_2, \ldots, a_n$. Varying m in the range $1, \ldots, n - 1$ and taking all the products for the subsequences, we obtain the various products for $a_1, a_2, \ldots, a_n$. Now we claim that the associative law guarantees that all of these products are equal. This is, of course, clear for $n = 1$, if we understand that the "product" in this case is just a_1. To prove the assertion in general we use induction on n and we first prove a little lemma.

LEMMA.   Define $\prod_1^n a_i$ by $\prod_1^1 a_i = a_1$, $\prod_1^{r+1} a_i = (\prod_1^r a_j)a_{r+1}$. Then

$$\prod_1^n a_i \prod_1^m a_{n+j} = \prod_1^{n+m} a_k.$$

Proof.   By definition this holds if $m = 1$. Assume it true for $m = r$ and consider the case $m = r + 1$. Here

$$\prod_1^n a_i \prod_1^{r+1} a_{n+j} = \prod_1^n a_i \left(\left(\prod_1^r a_{n+j} \right) a_{n+r+1} \right)$$

$$= \left(\prod_1^n a_i \prod_1^r a_{n+j} \right) a_{n+r+1}$$

$$= \left(\prod_1^{n+r} a_k \right) a_{n+r+1}$$

$$= \prod_1^{n+r+1} a_k. \quad \square$$

Now consider any product associated with the sequence $a_1, a_2, \ldots, a_n$. This has the form uv where u is a product associated with $a_1, \ldots, a_m$ and v is a product associated with $a_{m+1}, \ldots, a_n$. By induction on n we may assume that $u = \prod_1^m a_i$ and $v = \prod_1^{n-m} a_{m+j}$. Then, by the lemma, $uv = \prod_1^n a_k$. Thus all products determined by the squence $a_1, \ldots, a_n$ are equal ($= \prod_1^n a_k$). From now on we shall denote this uniquely determined product as $a_1 a_2 \cdots a_n$, omitting all parentheses.

If all the $a_i = a$, we denote $a_1 a_2 \cdots a_n$ as a^n and call this the nth *power* of a. It is clear by counting that

(6) $a^m a^n = a^{m+n}, \qquad (a^m)^n = a^{mn}.$

Also, if we define $a^0 = 1$, then it is immediate that (6) is valid for all $m, n \in \mathbb{N}$.

If a is an invertible element of M, then we define a^{-n} for $n \in \mathbb{N}$ by $a^{-n} = (a^{-1})^n = a^{-1}a^{-1} \cdots a^{-1}$ (n times). It is clear that $a^{-n} = (a^n)^{-1}$ and one can prove easily that (6) holds for all $m, n \in \mathbb{Z}$. This is left to the reader to check.

If a and b are elements of a monoid M, it may very well happen that $ab \neq ba$. For example, in the monoid $M(S)$, $S = \{1, 2\}$, whose multiplication table is (1) we have $\alpha\beta = \gamma$ whereas $\beta\alpha = \beta$. If $ab = ba$ in M then a and b are said to *com-*

mute and if this happens for all a and b in M then M is called a *commutative monoid*. Commutative groups are generally called *abelian groups* after Niels Hendrik Abel, a great Norwegian mathematician of the early nineteenth century.[4] We shall adopt this terminology in what follows.

If $a \in M$ we define the *centralizer* $C(a)$—or $C_M(a)$ if we need to indicate M— as the subset of M of elements b which commute with a. This is a submonoid of M. For, $1 \in C(a)$ since $1a = a = a1$ and if $b_1, b_2 \in C(a)$ then

$$(b_1 b_2)a = b_1(b_2 a) = b_1(ab_2) = (b_1 a)b_2 = (ab_1)b_2 = a(b_1 b_2).$$

Also, if $b \in C(a)$ and b is invertible then $b^{-1} \in C(a)$, since multiplication of $ab = ba$ on the left and on the right by b^{-1} gives $b^{-1}a = ab^{-1}$. This shows also that if $M = G$ is a group then $C(a)$ is a subgroup.

It is immediate that if $\{M_\alpha\}$ is a set of submonoids of a monoid then $\bigcap M_\alpha$ is a submonoid. Similarly, the intersection of any set of subgroups of a group is a subgroup.

*If A is a subset of M we define the *centralizer* of A as $C(A) = \bigcap_{a \in A} C(a)$.* Clearly this is a submonoid and it is a subgroup if M is a group. The submonoid $C(M)$ is called the *center* of M.

Suppose we have elements $a_1, a_2, \ldots, a_n \in M$ such that $a_i a_j = a_j a_i$ for all i, j and consider any product $a_{1'} a_{2'} \cdots a_{n'}$ where $1', 2', \ldots, n'$ is a permutation of $1, 2, \ldots, n$. Suppose a_n occurs in the hth place in $a_{1'} a_{2'} \cdots a_{n'}$, that is, $a_{h'} = a_n$. Then, since the $a_i \in C(a_n)$, $a_{(h+1)'} \cdots a_{n'} \in C(a_n)$ and so

$$a_{1'} a_{2'} \cdots a_{h'} \cdots a_{n'} = a_{1'} \cdots a_{(h-1)'} a_{(h+1)'} \cdots a_{n'} a_n.$$

The sequence of numbers $1', \ldots, (h-1)', (h+1)', \ldots, n'$ is a permutation of $1, 2, \ldots, n-1$. Hence, using induction, we may assume that

$$a_{1'} \cdots a_{(h-1)'} a_{(h+1)'} \cdots a_{n'} = a_1 a_2 \cdots a_{n-1}.$$

This implies that $a_{1'} a_{2'} \cdots a_{n'} = a_1 a_2 \cdots a_n$. Thus the product $a_1 a_2 \cdots a_n$ is invariant under all permutations of the arguments. In particular, if $ab = ba$, then

(7) $$\qquad\qquad (ab)^n = a^n b^n, \qquad n = 0, 1, 2, \ldots.$$

Since $a^{-n} = (a^{-1})^n$ it is clear that (7) holds also for negative integers if a and b are invertible.

If M is commutative, one frequently denotes the composition in M as $+$ and writes $a + b$ for ab. Also one writes 0 for 1. Then $+$ is called addition and 0

---

[4] An attractive biography of Abel's life has been written by Oystein Ore, *Niels Hendrik Abel*, Minneapolis, University of Minnesota Press, 1957.

the zero element. Also in this additive notation one writes $-a$ for a^{-1} and calls this the negative of a. The nth power a^n becomes na, the nth *multiple* of a. The rules for powers become the following rules for multiples:

$$(8) \qquad\qquad ma + na = (m + n)a, \qquad m(na) = (mn)a$$

$$(9) \qquad\qquad\qquad n(a + b) = na + nb.$$

These are valid for all integral m and n if M is an abelian group.

EXERCISES

1. Let A be a monoid, $M(A)$ the monoid of transformations of A into itself, A_L the set of left translations a_L, and A_R the set of right translations a_R. Show that A_L (respectively A_R) is the centralizer of A_R (respectively A_L) in $M(A)$ and that $A_L \cap A_R = \{c_R = c_L | c \in C\}$, C the center of A.

2. Show that if $n \geq 3$, then the center of S_n is of order 1.

3. Show that any group in which every a satisfies $a^2 = 1$ is abelian. What if $a^3 = 1$ for every a?

4. For a given binary composition define a *simple product* of the sequence of elements $a_1, a_2, \ldots, a_n$ inductively as either $a_1 u$ where u is a simple product of $a_2, \ldots, a_n$ or as va_n where v is a simple product of $a_1, \ldots, a_{n-1}$. Show that any product of $\geq 2^r$ elements can be written as a simple product of r elements (which are themselves products).

1.5 SUBMONOIDS AND SUBGROUPS GENERATED BY A SUBSET. CYCLIC GROUPS

Given a subset S of a monoid M or of a group G, one often needs to consider the "smallest" submonoid of M or subgroup of G containing S. What we want to have is a submonoid (or subgroup) containing the given set and contained in every submonoid (subgroup) containing this set. If such an object exists it is unique; for the stated properties imply that if $H(S)$ and $H'(S)$ both satisfy the conditions, then we have $H(S) \supset H'(S)$ and $H'(S) \supset H(S)$. Hence $H(S) = H'(S)$. Existence can also be established immediately in the following way. Let S be a given subset of a monoid M (or of a group G) and let $\{M_\alpha\}$ ($\{G_\alpha\}$) be the set of all submonoids of M (subgroups of G) which contain the set S. Form the intersection $\langle S \rangle$ of all these M_α (G_α). This is a submonoid (subgroup) since the intersection of submonoids (subgroups) is a submonoid (subgroup). Of course, $\langle S \rangle \supset S$. Moreover, if N is any submonoid of M (or subgroup of G) containing S, then N is one of the M_α (G_α) and so N contains $\langle S \rangle$ which is the intersection of all the M_α (G_α). We shall call $\langle S \rangle$ the *submonoid (subgroup) generated by S*. If S is a finite set, say, $S = \{s_1, s_2, \ldots, s_r\}$, then we write $\langle s_1, s_2, \ldots, s_r \rangle$ in place of the more cumbersome $\langle \{s_1, s_2, \ldots, s_r\} \rangle$. An important situation occurs when $\langle S \rangle = M$ (or G). In this case we say that the monoid M (group G) is *generated by the subset S*, or S is a set of *generators* for M (or G). This simply means that no proper submonoid of M (subgroup of G) contains the set S.

The reader may feel somewhat uncomfortable with the non-constructive nature of our definition of $\langle S \rangle$. Modern mathematics is full of such definitions, and so one has to learn to cope with them, and to use them with ease. Nevertheless, it is nice and often useful to have constructive definitions when these are available. This is the case with $\langle S \rangle$, as we shall now show. We consider first the case of monoids. What do the elements of $\langle S \rangle$ look like? Since $\langle S \rangle$ is a submonoid containing S, clearly $\langle S \rangle$ contains 1 and every product of the form $s_1 s_2 \cdots s_r$ where the s_i are elements of S (which need not be distinct). Thus

$$(10) \qquad \langle S \rangle \supset \langle S \rangle' \equiv \{1, s_1 s_2 \cdots s_r | s_i \in S\}.$$

Here the notation indicates that $\langle S \rangle'$ is the subset of the given monoid M consisting of 1 and every product of a finite number of elements of S. Now we claim that, in fact, $\langle S \rangle = \langle S \rangle'$. To see this we observe that $\langle S \rangle'$ contains S, since we are allowing $r = 1$ in (10). Also $\langle S \rangle'$ contains the unit, and the product of any two elements of the form $s_1 \cdots s_r$, $s_i \in S$, is again an element of this form. Hence $\langle S \rangle'$ is a submonoid of M and since $\langle S \rangle' \supset S$ we have $\langle S \rangle' \supset \langle S \rangle$. Since previously we had $\langle S \rangle \supset \langle S \rangle'$, $\langle S \rangle = \langle S \rangle'$. Thus a constructive definition of $\langle S \rangle$ is that this is just the subset of M consisting of 1 and all finite products of elements of the set S.

In the group case we let $\langle S \rangle'$ be the subset of the given group G consisting of 1 and all finite products of elements of S or the inverses of elements of S. In other words,

$$(11) \qquad \langle S \rangle' = \{1, s_1 s_2 \cdots s_r | s_i \text{ or } s_i^{-1} \in S\}.$$

It is immediate that $\langle S \rangle \supset \langle S \rangle'$, that $\langle S \rangle' \supset S$ and $\langle S \rangle'$ is a subgroup. Hence $\langle S \rangle' = \langle S \rangle$.

We now restrict our attention to groups, and we consider the simplest possible groups—those with a single generator. We have $G = \langle a \rangle$, and we call G *cyclic* with generator a. The preceding discussion (or the power rules) show that $\langle a \rangle = \{a^k | k \in \mathbb{Z}\}$ and this is an abelian group. One example of a cyclic group is the additive group of integers $(\mathbb{Z}, +, 0)$ which is generated by 1 (or by -1).

We now consider the map

$$n \to a^n$$

of $\mathbb{Z}$ into $\langle a \rangle$. Since $\langle a \rangle = \{a^k\}$ this map is surjective. Also we have $m + n \to a^{m+n} = a^m a^n$, $0 \to 1$. Hence if our map is injective it will be an isomorphism. Now suppose $n \to a^n$ is not an isomorphism. Then $a^m = a^n$ for some $m \neq n$. We may assume $n > m$. Then $a^{n-m} = a^n a^{-m} = a^m a^{-m} = 1$; so there exist positive integers p such that $a^p = 1$. Let r be the least such positive integer. Then we claim that

(12) $\langle a \rangle = \{1, a, a^2, \ldots, a^{r-1}\}$

and the elements listed in (12) are distinct, so $|\langle a \rangle| = r$. Let a^m be any element of $\langle a \rangle$. By the division algorithm for integers, we can write $m = rq + p$ where $0 \leq p < r$. Then we have $a^m = a^{rq+p} = (a^r)^q a^p = 1^q a^p = a^p$. Hence $a^m = a^p$ is one of the elements displayed in (12). Next we note that if $k \neq l$ are in the range $0, 1, \ldots, r - 1$ then $a^k \neq a^l$. Otherwise, taking $l > k$ we obtain $a^{l-k} = 1$ and $0 < l - k < r$ contrary to the choice of r. We now see that if $n \to a^n$ is not an isomorphism, then $\langle a \rangle$ is a finite group. Accordingly, any infinite cyclic group is isomorphic to $(\mathbb{Z}, +, 0)$ and so any two infinite cyclic groups are isomorphic.

We shall show next that any two finite cyclic groups of the same order are isomorphic. Suppose $\langle b \rangle$ has order r. Then, as in the case of $\langle a \rangle$, we have $\langle b \rangle = \{1, b, \ldots, b^{r-1}\}$, where r is the smallest positive integer such that $b^r = 1$. We now observe that if h is any integer such that $a^h = 1$, then $r \mid h$ (r is a divisor of h). We have $h = qr + s$, $0 \leq s < r$, so $1 = a^h = (a^r)^q a^s = 1^q a^s = a^s$. Since r was the least positive integer satisfying $a^r = 1$ we must have $s = 0$ and so $h = qr$. We now claim that if m and n are any two integers such that $a^m = a^n$ then also $b^m = b^n$. For, $a^m = a^n$ gives $a^{m-n} = 1$; hence $m - n = qr$. Then $b^{m-n} = (b^r)^q = 1^q = 1$ and $b^m = b^n$. By symmetry $b^m = b^n$ implies $a^m = a^n$. It is now clear that we have a 1–1 correspondence between $\langle a \rangle$ and $\langle b \rangle$ pairing a^n and b^n. Since $a^m a^n = a^{m+n}$ is paired with $b^{m+n} = b^m b^n$, $a^n \to b^n$ is an isomorphism of $\langle a \rangle$ and $\langle b \rangle$.

Our analysis has proved the following

THEOREM 1.2.   *Any two cyclic groups of the same order (finite or infinite) are isomorphic.*

We have seen that $(\mathbb{Z}, +, 0)$ can serve as *the* model of a cyclic group of infinite order. If r is any positive integer, the multiplicative group U_r of the complex rth roots of unity (example 13, p. 34) can serve as a model for cyclic groups of order r. The elements of this group are the complex numbers

$e^{2k\pi i/r} = \cos 2k\pi/r + i \sin 2k\pi/r$, $k = 0, 1, \ldots, r - 1$. Since $e^{i\theta}e^{i\theta'} = e^{i(\theta + \theta')}$ it is clear that $a = e^{2\pi i/r}$ generates U_r.

We can use the notion of a cyclic group to obtain a classification of the elements of any group G. If $a \in G$ we say that a is of *infinite order* or of *finite order* r according as the subgroup $\langle a \rangle$ is infinite or finite of order r. In the first case $a^m \neq 1$ for $m \neq 0$. In the second case we have $a^r = 1$ and r is the least positive integer having this property. Also, if $a^m = 1$ then m is a multiple of r. We shall denote the order of a by $o(a)$ (finite or infinite). It is clear that if $o(a) = r = st$ where s and t are positive integers then $o(a^s)$ is t. More generally, one sees easily that if $o(a) = r < \infty$ then $o(a^k)$ for any integer $k \neq 0$ is $[r, k]/k = r/(r, k)$ where as usual $[,]$ denotes the l.c.m. and $(,)$ denotes the g.c.d. (exercise 4, p. 47).

Cyclic groups are the simplest kind of groups. It is therefore not surprising that most questions on groups are easy to answer for this class. For example, one can determine all the subgroups of a cyclic group. This is generally an arduous task for most groups. We shall now prove

THEOREM 1.3. *Any subgroup of a cyclic group $\langle a \rangle$ is cyclic. If $\langle a \rangle$ is infinite, the subgroups $\neq 1$ are infinite and $s \to \langle a^s \rangle$ is a bijective map of $\mathbb{N}$ with the set of subgroups of $\langle a \rangle$. If $\langle a \rangle$ is finite of order r, then the order of every subgroup is a divisor of r, and for every positive divisor q of r there is one and only one subgroup of order q.*

Proof. Let H be a subgroup of $\langle a \rangle$. If $H = 1$ $(=\{1\})$ then $H = \langle 1 \rangle$. Now let $H \neq 1$. Then there exists an $n \neq 0$ in $\mathbb{Z}$ such that $a^n \in H$. Since also $a^{-n} = (a^n)^{-1} \in H$ we may assume $n > 0$. Now let s be the smallest positive integer such that $a^s \in H$. Then we claim $H = \langle a^s \rangle$. Let $a^m \in H$ and write $m = qs + t$ where $0 \leq t < s$. Then $a^t = a^m(a^s)^{-q} \in H$, and, since s was the least positive integer such that $a^s \in H$, we must have $t = 0$. Then $a^m = (a^s)^q \in \langle a^s \rangle$. Since a^m was any element of H we have $H = \langle a^s \rangle$, which proves the first statement of the theorem.

If $\langle a \rangle$ is infinite we saw that for distinct integers m and n, $a^m \neq a^n$. Hence for any positive s, the elements a^{ms}, $m = 0, \pm 1, \pm 2, \ldots$ are distinct, so $\langle a^s \rangle$ is an infinite group. Moreover, s is the smallest positive integer such that $a^s \in \langle a^s \rangle$. Thus every subgroup $\neq 1$ is infinite and we have the 1–1 correspondence $s \to \langle a^s \rangle$ between the set of positive integers and the set of subgroups $\neq 1$ of $\langle a \rangle$.

Now suppose $\langle a \rangle$ is of finite order r, so $\langle a \rangle = \{1, a, \ldots, a^{r-1}\}$. We have seen that if H is a subgroup $\neq 1$ of $\langle a \rangle$, then $H = \langle a^s \rangle$ where s is the smallest positive integer such that $a^s \in H$. We claim that $s \mid r$. For, writing $r = qs + t$ with $0 \leq t < s$, we have $1 = a^r = (a^s)^q a^t$, so $a^t = (a^s)^{-q} \in H$. The minimality of s then forces $t = 0$ and so $r = qs$. We can now list the elements of H as

(13) $$\{1, a^s, \ldots, a^{(q-1)s}\}$$

and $a^{sq} = a^r = 1$. This applies to $H = 1$ if we take $s = r$. In this way we obtain a bijective map $s \to \langle a^s \rangle$ of the set of positive divisors s of r onto the set of subgroups of $\langle a \rangle$. The order of the subgroup $\langle a^s \rangle$ corresponding to s is $q = r/s$ and as s runs through the positive divisors of r, so does q. Hence the order of every subgroup is a divisor of r and for every positive $q|r$ we have one and only one subgroup of this order. This completes the proof. $\square$

We note again that the subgroup of order q of the finite cyclic group $\langle a \rangle$ of order r can be displayed as in (13). There is another characterization of this subgroup which is often useful, namely:

COROLLARY. *If $\langle a \rangle$ has order $r < \infty$, then the subgroup H of order $q|r$ is the set of elements $b \in \langle a \rangle$ such that $b^q = 1$.*

Proof. Any element of H has the form a^{ks} where $s = r/q$. Then $(a^{ks})^q = a^{kr} = 1$. Conversely, let $b = a^m$ satisfy $b^q = 1$. Then $a^{mq} = 1$ and hence $mq = kr$. Then $m = ks$ so $b = (a^s)^k \in H$. $\square$

After cyclic groups the next simplest type of groups are the finitely generated abelian ones, (that is, abelian groups with a finite number of generators). These include the finite abelian groups. We shall determine the structure of this class of groups in Chapter 3, obtaining a complete classification by means of numerical invariants. Independently of the structure theory, we shall now derive a criterion for a finite abelian group to be cyclic. This result will be needed to prove an important theorem on fields (Theorem 2.18, p. 128) To state our criterion we require the concept of the *exponent*, exp G, of a finite group G, which we define to be the smallest positive integer e such that $x^e = 1$ for all $x \in G$. For example, exp $S_3 = 6 = |S_3|$. The result we wish to prove is

THEOREM 1.4. *Let G be a finite abelian group. Then G is cyclic if and only if* exp $G = |G|$.

The proof will be based on two lemmas that are of independent interest.

LEMMA 1. *Let g and h be elements of an abelian group G having finite relatively prime orders m and n respectively (that is, $(m, n) = 1$). Then $o(gh) = mn$.*

Proof. Suppose $(gh)^r = 1$. Then $k = g^r = h^{-r} \in \langle g \rangle \cap \langle h \rangle$. Then $o(k)|m$ and $o(k)|n$ and hence $o(k) = 1$. Thus $(gh)^r = 1 \Rightarrow g^r = 1 = h^r$. Then $m|r$ and $n|r$ and hence $mn = [m, n]|r$. On the other hand, $(gh)^{mn} = g^{mn}h^{mn} = 1$. Hence $o(gh) = mn$. $\square$

LEMMA 2. *Let G be a finite abelian group, g an element of G of maximal order. Then* $\exp G = o(g)$.

Proof. We have to show that $h^{o(g)} = 1$ for every $h \in G$. Write $o(g) = p_1^{e_1} \cdots p_s^{e_s}$, $o(h) = p_1^{f_1} \cdots p_s^{f_s}$, where the p_i are distinct primes and $e_i \geq 0$, $f_i \geq 0$. If $h^{o(g)} \neq 1$, then some $f_i > e_i$ and we may assume $f_1 > e_1$. Put $g' = g^{p_1^{e_1}}$, $h' = h^{p_2^{f_2} \cdots p_s^{f_s}}$. Then $o(g') = p_2^{e_2} \cdots p_s^{e_s}$ and $o(h') = p_1^{f_1}$. Hence, by Lemma 1, $o(g'h') = p_1^{f_1}p_2^{e_2} \cdots p_s^{e_s} > o(g)$. This contradicts the maximality of $o(g)$. $\square$

We can now give the

Proof of Theorem 1.4. First suppose $G = \langle g \rangle$. Then $|G| = o(g)$ and hence $\exp G = |G|$. Conversely, let G be any finite abelian group such that $\exp G = |G|$. By Lemma 2 we have an element g such that $\exp G = o(g)$. Then $|G| = o(g) = |\langle g \rangle|$. Hence $G = \langle g \rangle$. $\square$

EXERCISES

1. As in section 1.4, let $C(A)$ denote the centralizer of the subset A of a monoid M (or a group G). Note that $C(C(A)) \supset A$ and if $A \subset B$ then $C(A) \supset C(B)$. Show that these imply that $C(C(C(A))) = C(A)$. Without using the explicit form of the elements of $\langle A \rangle$ show that $C(A) = C(\langle A \rangle)$. (*Hint*: Note that if $c \in C(A)$ then $A \subset C(c)$ and hence $\langle A \rangle \subset C(c)$.) Use the last result to show that if a monoid (or a group) is generated by a set of elements A which pair-wise commute, then the monoid (group) is commutative.

2. Let M be a monoid generated by a set S and suppose every element of S is invertible. Show that M is a group.

3. Let G be an abelian group with a finite set of generators which is *periodic* in the sense that all of its elements have finite order. Show that G is finite.

4. Show that if g is an element of a group and $o(g) = n$ then g^k, $k \neq 0$, has order $[n, k]/k = n/(n, k)$. Show that the number of generators of $\langle g \rangle$ is the number of positive integers $< n$ which are relatively prime to n. This number is denoted as $\varphi(n)$ and φ is called the *Euler φ-function*.

5. Show that any finitely generated subgroup of the additive group of rationals $(\mathbb{Q}, +, 0)$ is cyclic. Use this to prove that this group is not isomorphic to the direct product of two copies of it.

6. Let a, b be as in Lemma 1. Show that $\langle a \rangle \cap \langle b \rangle = 1$ and $\langle a, b \rangle = \langle ab \rangle$.

7. Show that if $o(a) = n = rs$, where $(r, s) = 1$, then $\langle a \rangle \cong \langle b \rangle \times \langle c \rangle$, where $o(b) = r$ and $o(c) = s$. Hence prove that any finite cyclic group is isomorphic to a direct product of cyclic groups of prime power orders.

1.6   CYCLE DECOMPOSITION OF PERMUTATIONS

A permutation γ of $\{1, 2, \ldots, n\}$ which permutes a sequence of elements i_i, $i_2, \ldots, i_r$, $r > 1$, cyclically in the sense that

$$(14) \qquad \gamma(i_1) = i_2, \qquad \gamma(i_2) = i_3, \ldots, \gamma(i_{r-1}) = i_r, \qquad \gamma(i_r) = i_1$$

and fixes (that is, leaves unchanged) the other numbers in $\{1, 2, \ldots, n\}$ is called a *cycle* or an *r-cycle*. We denote this as

$$(15) \qquad \gamma = (i_1 i_2 \cdots i_r).$$

It is clear that we can equally well write

$$\gamma = (i_2 i_3 \cdots i_r i_1) = (i_3 i_4 \cdots i_r i_1 i_2), \text{ etc.}$$

The permutation γ^2 maps i_1 into i_3, i_2 into $i_4, \ldots, i_r$ into i_2 etc., and, in general, for $1 \leq k \leq r$,

$$(16) \qquad \begin{aligned} \gamma^k(i_j) &= i_{j+k} && \text{if} \quad j + k \leq r \\ \gamma^k(i_j) &= i_{j+k-r} && \text{if} \quad j + k > r. \end{aligned}$$

Clearly this shows that $\gamma^r = 1$ but $\gamma^k \neq 1$ if $1 \leq k < r$. Hence γ is of order r.

Two cycles γ and γ' are said to be *disjoint* if their symbols contain no common letters. In this case it is clear that any number moved by one of these transformations is fixed by the other. Hence if i is any number such that $\gamma(i) \neq i$ then $\gamma\gamma'(i) = \gamma(i)$, and since also $\gamma^2(i) \neq \gamma(i)$, $\gamma'\gamma(i) = \gamma(i)$. Similarly, if $\gamma'(i) \neq i$ then $\gamma'\gamma(i) = \gamma'(i) = \gamma\gamma'(i)$. Also if $\gamma(i) = i = \gamma'(i)$ then $\gamma\gamma'(i) = \gamma'\gamma(i)$. Thus $\gamma\gamma' = \gamma'\gamma$, that is, any two disjoint cycles commute. Let α be a product of disjoint cycles, that is,

$$(17) \qquad \alpha = (i_1 i_2 \cdots i_r)(j_1 j_2 \cdots j_s) \cdots (l_1 l_2 \cdots l_u).$$

Let m be the least common multiple of $r, s, \ldots, u$. Then we claim that m is the order of α. Putting $\gamma_1 = (i_1 \cdots i_r)$, $\gamma_2 = (j_1 \cdots j_s), \ldots, \gamma_k = (l_1 \cdots l_u)$ we have $\alpha^m = \gamma_1^m \gamma_2^m \cdots \gamma_k^m = 1$. On the other hand, α permutes $i_1, \ldots, i_r$ and so do its powers and the restriction of α to $\{i_1, \ldots, i_r\}$ is γ_1. Hence if $\alpha^n = 1$ then $\gamma_1^n =$

1 and so n is divisible by r. Similarly, n is divisible by $s, \ldots, u$ and so n is divisible by the least common multiple of $r, s, \ldots, u$. Hence the least common multiple of these numbers is the order of α.

It is convenient to extend the definition of cycles and the cycle notation to 1-cycles where we adopt the convention that for any i, (i) is the identity mapping. With this convention we can see that every permutation is a product of disjoint cycles. For example, if

$$\alpha = \begin{pmatrix} 1 & 2 & 3 & 4 & 5 & 6 & 7 & 8 \\ 3 & 6 & 5 & 4 & 8 & 2 & 7 & 1 \end{pmatrix}$$

then

$$\alpha(1) = 3, \alpha(3) = 5, \alpha(5) = 8, \alpha(8) = 1; \alpha(2) = 6, \alpha(6) = 2: \alpha(4) = 4, \alpha(7) = 7$$

from which one deduces that

$$\alpha = (7)(4)(26)(1358).$$

In general, for any α we can begin with any number in $1, 2, \ldots, n$, say i_1, and form $\alpha(i_1) = i_2, \alpha(i_2) = i_3, \ldots$, until we reach a number that occurs previously in this list. The first such repetition occurs when $i_{r+1} = \alpha(i_r) = i_1$; for, we have $i_k = \alpha^{k-1}(i_1)$ and if $i_k = i_l$ for $l > k$ then $\alpha^{l-k}(i_1) = i_1$. Thus the sequence $i_1, i_2, \ldots, i_r$ is permuted cyclically by α. If $r < n$ we choose a j_1 not in $\{i_1, i_2, \ldots, i_r\}$. If $\alpha^m(j_1) = \alpha^q(i_1)$ then $j_1 = \alpha^{q-m}(i_1) \in \{i_1, i_2, \ldots, i_r\}$ contrary to our choice of j_1. Hence we obtain a new sequence of numbers $j_1, j_2, \ldots, j_s$ permuted cyclically by α and having no elements in common with the first. Continuing in this way we ultimately exhaust the set $\{1, 2, \ldots, n\}$. It is clear, on comparing the images of any i under the two maps α and $(l_1 \cdots l_u) \cdots (i_1 \cdots i_r)$ that

$$\alpha = (l_1 \cdots l_u) \cdots (i_1 \cdots i_r),$$

a product of disjoint cycles. The different cycles occurring in such a factorization commute and we may add or drop trivial one-cycles. Apart from order of the factors and inclusion or omission of 1-cycles this factorization is unique. For, if we have one which is essentially different from the one displayed above (or 17)), then for some $i, j, i \neq j$, which occur in the order i followed by j in one of the cycles in (17), we have that this is not the case in the other one. The first factorization then shows that $\alpha(i) = j$ and the second that $\alpha(i) \neq j$. This contradiction proves our assertion.

A cycle of the form (ab) is called a *transposition*. It is easy to verify that

(18) $$(i_1 i_2 \cdots i_r) = (i_1 i_r) \cdots (i_1 i_3)(i_1 i_2),$$

a product of $r - 1$ transpositions. It follows that any $\alpha \in S_n$ is a product of transpositions. In fact, if α factors as a product of disjoint cycles as in (17), then α is a product of $(r - 1) + (s - 1) + \cdots + (u - 1)$ transpositions. We denote this number, which is uniquely determined by α, as $N(\alpha)$. It is clear that $N(1) = 0$. There is no uniqueness of factorization of a permutation as a product of transpositions. For example, we have $(123) = (13)(12) = (12)(23) = (23)(13)$. However, as we shall now show, there is one common feature of all the factorizations of a given α as a product of transpositions. The number of factors occurring all have the same parity: that is, their number is either always even or always odd. Our proof of this fact will be based on a simple formula, which is anyhow worth noting:

$$(19) \qquad (ab)(ac_1 \cdots c_h bd_1 \cdots d_k) = (bd_1 \cdots d_k)(ac_1 \cdots c_h).$$

Here we are allowing h or k to be 0, meaning thereby that no c's or no d's occur. Comparing images of any i in $\{1, 2, .., n\}$ shows that (19) holds. Since $(ab)^{-1} = (ab)$ multiplying both sides of (19) on the left by (ab) gives:

$$(20) \qquad (ab)(bd_1 \cdots d_k)(ac_1 \cdots c_h) = (ac_1 \cdots c_h bd_1 \cdots d_k).$$

If N is defined as above, we have $N((ac_1 \cdots c_h bd_1 \cdots d_k)) = h + k + 1$ and $N((bd_1 \cdots d_k)(ac_1 \cdots c_h)) = h + k$. It follows that $N((ab)(\alpha)) = N(\alpha) - 1$ if a and b occur in the same cycle in the decomposition of α into disjoint cycles and $N((ab)\alpha) = N(\alpha) + 1$ if a and b occur in different cycles. Hence if α is a product of m transpositions then, since $N(1) = 0$, $N(\alpha) = \sum_{i=1}^{m} \varepsilon_i$ where $\varepsilon_i = \pm 1$. Changing an $\varepsilon_i = -1$ to 1 amounts to adding 2 to the sum and so does not change the parity. If we make this change for every $\varepsilon_i = -1$ the final sum we obtain is m. Hence m and $N(\alpha)$ have the same parity. Hence the number of factors in any two factorizations of α as a product of transpositions have the same parity, namely, the parity of $N(\alpha)$.

We call α *even* or *odd* according as α factors as a product of an even or an odd number of transpositions (equivalently: $N(\alpha)$ is even or odd.) We define the *sign* of α, $sg\,\alpha$, by

$$(21) \qquad sg\,\alpha = 1 \text{ if } \alpha \text{ is even}, \qquad sg\,\alpha = -1 \text{ if } \alpha \text{ is odd}$$

Then $sg\,1 = 1$ and if $\alpha = (ab) \cdots (kl)$, $\beta = (pq) \cdots (uv)$, $\alpha\beta = (ab) \cdots (kl)(pq) \cdots (uv)$. Hence $\alpha\beta$ is even if and only if both α and β are even or both are odd while $\alpha\beta$ is odd if one of the factors is even and the other is odd. It follows that

$$(22) \qquad sg\,\alpha\beta = (sg\,\alpha)(sg\,\beta).$$

It is clear also that the subset A_n of even permutations is a subgroup of S_n.

This is called the *alternating group* (of degree n). Suppose we list its elements as

$$\alpha_1, \alpha_2, \ldots, \alpha_m.$$

Then if $n \geq 2$ we have m different odd permutations

$$\alpha_1(ab), \alpha_2(ab), \ldots, \alpha_m(ab)$$

and this catches them all, since if β is odd $\beta(ab)$ is even so $\beta(ab) = \alpha_i$ for some i and $\beta = \alpha_i(ab)$. Hence $|S_n| = 2m = 2|A_n|$ and so $|A_n| = n!/2$ if $n \geq 2$.

EXERCISES

1. Write $(456)(567)(671)(123)(234)(345)$ as a product of disjoint cycles.

2. Show that if $n \geq 3$ then A_n is generated by the 3-cycles (abc).

3. Determine the sign of the permutation
$$\begin{pmatrix} 1 & 2 & \cdots & n-1 & n \\ n & n-1 & \cdots & 2 & 1 \end{pmatrix}.$$

4. Show that if α is any permutation then
$$\alpha(i_1 i_2 \cdots i_r)\alpha^{-1} = (\alpha(i_1)\alpha(i_2) \cdots \alpha(i_r)).$$

5. Show that S_n is generated by the $n-1$ transpositions $(12), (13), \ldots, (1n)$ and also by the $n-1$ transpositions $(12), (23), \ldots, (n-1\,n)$.

1.7 ORBITS. COSETS OF A SUBGROUP

Let G be a group of transformations of a set S. Then G defines an equivalence relation on S by the rule that $x \sim_G y$ (read: x is *G-equivalent* to y) if $y = \alpha(x)$ for some $\alpha \in G$. That this relation is reflexive, symmetric, and transitive is immediate from the definition of a transformation group: $x = 1_S(x)$, also if $y = \alpha(x)$ then $x = \alpha^{-1}(y)$, and if $y = \alpha(x)$ and $z = \beta(y)$ then $z = (\beta\alpha)(x)$. Moreover, $1_S \in G$ and α^{-1} and $\beta\alpha \in G$, if α and $\beta \in G$. The G-equivalence class determined by an element x is the set $Gx = \{\alpha(x) | \alpha \in G\}$ and this is called the *G-orbit* of $x \in S$. For example, if G is the group of rotations about the origin in a plane, then the orbit of a point P is the circle through P with center at the origin. As with any equivalence relation, the set of orbits constitute a partition of the set S. It may happen that there is just one orbit, that is, $S = Gx$ for some x (and hence for every x). In this case we say that G is a *transitive* group of transformations of the set S. It is clear that S_n is transitive on $\{1, 2, \ldots, n\}$. The reader will have no difficulty showing that this is true also of the alternating group A_n if $n \geq 3$. On the other hand, if $\alpha \in S_n$ and $\alpha = (i_1 \cdots i_r)(j_1 \cdots j_s) \cdots (l_1 \cdots l_u)$ is the factorization of α into disjoint cycles, where we have included the 1-cycles, and every letter in $\{1, 2, \ldots, n\}$ appears once and only once among $i_1, \ldots, i_r, j_1, \ldots, j_s, \ldots, l_1, \ldots, l_u$, then the sets

$$\{i_1, \ldots, i_r\}, \{j_1, \ldots, j_s\}, \ldots, \{l_1, \ldots, l_u\}$$

are the orbits in $\{1, 2, \ldots, n\}$ determined by the cyclic subgroup $\langle \alpha \rangle$ of S_n. Observe that this gives another interpretation of the number $N(\alpha)$ which we used in section 1.6, namely, $N(\alpha) = \sum(k - 1)$ where k runs over the cardinal numbers of the orbits determined by $\langle \alpha \rangle$.

Now let G be any group and let H be a subgroup of G. We recall that we have the transformation groups G_L of left translations g_L $(x \to gx)$ and G_R of right translations g_R both acting in G. Since $y = gx$ and $y = xg$ are solvable for g for any given y and x it is clear that G_L and G_R are transitive groups. Now let $H_L(G)$ denote the subset of G_L of maps h_L (in G) for $h \in H$. Since H is a subgroup of G and $g \to g_L$ is an isomorphism, $H_L(G)$ is a subgroup of G_L and hence $H_L(G)$ is a transformation group of the set G. What are the orbits in the set G determined by $H_L(G)$? If $x \in G$ then it is clear that its $H_L(G)$-orbit is

$$(23) \qquad\qquad Hx = \{hx \mid h \in H\}.$$

In the group theory literature this is sometimes called the left coset of x relative to the subgroup H and sometimes the right coset of x relative to H. The majority opinion seems to favor the second terminology. Accordingly, we shall adopt it here and call Hx the *right coset* of x relative to H. We have the partition $G = \bigcup_{x \in G} Hx$. Moreover, any two right cosets Hx and Hy have the same cardinality since the map $(x^{-1}y)_R : z \to z(x^{-1}y)$ is bijective from Hx to Hy. Since $H = H1$ is one of the right cosets we have $|Hx| = |H|$.

In particular, suppose G is a finite group and $|G| = n$ and $|H| = m$. We have the partition

$$(24) \qquad\qquad G = Hx_1 \cup Hx_2 \cup \cdots \cup Hx_r,$$

where we have displayed the distinct cosets, so $Hx_i \cap Hx_j = \emptyset$ if $i \neq j$. We call the number r of these cosets the *index* of H in G and denote this as $[G:H]$. Since $|Hx_i| = m$, we have by (24) that $n = mr$. This proves a fundamental theorem which is due to Lagrange:

THEOREM 1.5. *The order of a subgroup H of a finite group G is a factor of* *the order of G. More precisely, we have*

$$|G| = |H|[G:H].$$

We also have the following

COROLLARY.   *If G is a finite group of order n, then $x^n = 1$ for every $x \in G$.*

Proof.   Let m be the order of $\langle x \rangle$. Then $x^m = 1$ and $n = mr$, so $x^n = 1$.   □

The results on right cosets have their counterparts for left cosets. These are the orbits in G determined by the transformation group $H_R(G)$. The orbit of x in this case is $xH = \{xh \,|\, h \in H\}$ and this is called the *left coset* of x relative to H. If Hx is a right coset the set of inverses $(hx)^{-1} = x^{-1}h^{-1}$ of the elements of Hx is the left coset $x^{-1}H$. It is immediate that the map $Hx \to x^{-1}H$ is a bijective map of the set of right cosets onto the set of left cosets. It follows that these two sets (of left and right cosets) have the same cardinal number. As in the case of finite groups, we call this the *index of H in G* and denote it as $[G:H]$.

EXERCISES

1. Determine the cosets of $\langle \alpha \rangle$ in S_4 where $\alpha = (1234)$.

2. Show that if G is finite and H and K are subgroups such that $H \supset K$ then $[G:K] = [G:H][H:K]$.

3. Let H_1 and H_2 be subgroups of G. Show that any right coset relative to $H_1 \cap H_2$ is the intersection of a right coset of H_1 with a right coset of H_2. Use this to prove *Poincaré's Theorem* that if H_1 and H_2 have finite index in G then so has $H_1 \cap H_2$.

4. Let G be a finitely generated group, H a subgroup of finite index. Show that H is finitely generated.

5. Let H and K be two subgroups of a group G. Show that the set of maps $x \to hxk$, $h \in H$, $k \in K$ is a group of transformations of the set G. Show that the orbit of x relative to this group is the set $HxK = \{hxk \,|\, h \in H, k \in K\}$. This is called the *double coset of x relative to the pair* (H, K). Show that if G is finite then $|HxK| = |H|[K:x^{-1}Hx \cap K] = |K|[H:xKx^{-1} \cap H]$.

6. Let H be a subgroup of the finite group G. Show that there exists a subset $\{z_1, \ldots, z_r\}$ of G which is simultaneously a set of representatives of the left and of the right cosets of H in G, that is, G is a disjoint union of the z_iH and also of the Hz_i, $1 \leq i \leq r$. (Hint: For any $g \in G$, write $HgH = \bigcup_1^s x_jgH$, where the $x_j \in H$

and $x_j gH \cap x_k gH = \varnothing$ if $j \neq k$. Note that the number of right cosets of H contained in HgH is s and write $HgH = \bigcup_1^s Hgy_j$, where $y_j \in H$. Put $z_j = x_j gy_j$ and show that $HgH = \bigcup z_j H = \bigcup Hz_j$.)

1.8 CONGRUENCES. QUOTIENT MONOIDS AND GROUPS

In elementary number theory two integers a and b are defined to be *congruent modulo the integer m* and this is denoted as $a \equiv b \pmod{m}$ if $a - b$ is a multiple of $m: a - b = km$, $k \in \mathbb{Z}$.[5] The relation between a and b thus defined for fixed m is an equivalence relation; for, we have $a \equiv a \pmod m$ since $a - a = 0 = 0m$, $a \equiv b \pmod m$ implies $b \equiv a \pmod m$ since $a - b = km$ implies $b - a = (-k)m$ and $a \equiv b \pmod m$ and $b \equiv c \pmod m$ imply $a \equiv c \pmod m$ since $a - b = km$ and $b - c = lm$ imply $a - c = (k + l)m$. In the additive group $(\mathbb{Z}, +, 0)$ congruences mod m can be added, that is, if $a \equiv a' \pmod m$ and $b \equiv b' \pmod m$ then $a + b \equiv a' + b' \pmod m$. This follows since $a - a' = km, b - b' = lm$ imply $a + b - (a' + b') = (k + l)m$. Also in the monoid $(\mathbb{Z}, \cdot, 1)$ congruences mod m can be multiplied: $a \equiv a' \pmod m$, $b \equiv b' \pmod m$ imply $ab \equiv a'b' \pmod m$, since $a = a' + km$, $b = b' + lm$ imply $ab = a'b' + (a'l + b'k + klm)m$. Congruences mod m in $(\mathbb{Z}, +, 0)$ and in $(\mathbb{Z}, \cdot, 1)$ are examples of a general notion which we shall now define.

DEFINITION 1.4. *Let $(M, \cdot, 1)$ be a monoid. A* congruence *(or* congruence relation*) $\equiv$ in M is an equivalence relation in M such that for any a, a', b, b' such that $a \equiv a'$ and $b \equiv b'$ one has $ab \equiv a'b'$. (In other words, congruences are equivalence relations which can be multiplied.)*

Let $\equiv$ be a congruence in the monoid M and consider the quotient set $\bar{M} = M/\equiv$ of M relative to $\equiv$. We recall that $\bar{M}$ is the subset of the power set $\mathscr{P}(M)$ consisting of the equivalence classes $\bar{a} = \{b \in M \mid b \equiv a\}$. For example, in $(\mathbb{Z}, +, 0)$ if we define $\equiv \pmod m$ as above, then $\bar{a} = \{a + km \mid k \in \mathbb{Z}\}$. Since congruences can be multiplied it is clear in the general case that, if $\bar{a} = \bar{a'}$ and $\bar{b} = \bar{b'}$, then $\overline{ab} = \overline{a'b'}$. Hence

$$(\bar{a}, \bar{b}) \to \overline{ab}$$

is a well-defined map of $\bar{M} \times \bar{M}$ into $\bar{M}$; that is, this is a binary composition on $\bar{M}$. We denote this again as $\cdot$, and we shall now show that $(\bar{M}, \cdot, \bar{1})$ is a monoid. We note first that $(\bar{a}\bar{b})\bar{c} = \bar{a}(\bar{b}\bar{c})$, since the left-hand side is $\overline{ab}\bar{c} = \overline{(ab)c}$ and the

---

[5] It is interesting to read the discussion of congruences for integers at the beginning of the great classic on number theory, *Disquisitiones Arithmeticae*, by Carl Friedrich Gauss. This work, published in 1801, was written when Gauss was nineteen. English translation by A.A. Clarke, Yale University Press, New Haven, 1966.

right-hand side is $\overline{a}\overline{bc} = \overline{a(bc)}$. Hence $(\overline{a}\overline{b})\overline{c} = \overline{a}(\overline{b}\overline{c})$ follows from the associative law in M. Also $\overline{a}\overline{1} = \overline{a1} = \overline{a}$ and $\overline{1}\overline{a} = \overline{1a} = \overline{a}$ so $\overline{1}$ is a unit. The monoid $(\overline{M}, \cdot, \overline{1})$ is called the *quotient monoid of M relative to the congruence* $\equiv$.

In the special case $M = (\mathbb{Z}, +, 0)$ in which $\equiv$ is $\equiv (\mathrm{mod}\ m)$ where $m > 0$, any $a \in \mathbb{Z}$ can be written as $a = qm + r$ where $0 \leq r < m$, which means that $a \equiv r\ (\mathrm{mod}\ m)$. If r_1 and r_2 both satisfy $0 \leq r_i < m$ then $r_1 \equiv r_2\ (\mathrm{mod}\ m)$ implies that $r_1 = r_2$. Hence in this case the quotient monoid, which we shall denote as $\mathbb{Z}/\mathbb{Z}m$ (a special case of a general notation that will be introduced below), consists of m elements:

$$\overline{0} = \{0, \pm m, \pm 2m, \pm 3m, \ldots\}$$
$$\overline{1} = \{1, 1 \pm m, 1 \pm 2m, 1 \pm 3m, \ldots\}$$
$$\vdots$$
$$\overline{m-1} = \{m-1, m-1 \pm m, m-1 \pm 2m, \ldots\}.$$

In the multiplicative case of $M = (\mathbb{Z}, \cdot, 1)$ we also have this same set of elements as the underlying set for the monoid $(\mathbb{Z}/\mathbb{Z}m, \cdot, \overline{1})$.

We can say a good deal more if $M = G$ is a group and $\equiv$ is a congruence on G. In the first place, in this case the quotient monoid $(\overline{G}, \cdot, \overline{1})$ is a group since $\overline{a}\overline{a^{-1}} = \overline{1} = \overline{a^{-1}}\overline{a}$. Hence every $\overline{a}$ is invertible and its inverse is $\overline{a^{-1}}$. Next we can determine all congruences on a group—or, more precisely, we can reduce the problem of determining the congruences to that of determining certain kinds of subgroups of the given group which we specify in the following

DEFINITION 1.5. *A subgroup K of a group G is said to be* normal *(sometimes called* invariant, *and in the older literature,* self-conjugate) *if*

$$g^{-1}kg \in K$$

for every $g \in G$ and $k \in K$.

We have the following fundamental connection between congruences on a group G and normal subgroups of G.

THEOREM 1.6. *Let G be a group and $\equiv$ a congruence on G. Then the congruence class $K = \overline{1}$ of the unit is a normal subgroup of G and for any $g \in G$, $\overline{g} = Kg = gK$, the right or the left coset of g relative to K. Conversely let K be any normal subgroup of G, then $\equiv$ defined by:*

$$a \equiv b\ (\mathrm{mod}\ K) \quad if \quad a^{-1}b \in K$$

is a congruence relation in G whose associated congruence classes are the left (or right) cosets gK.

Proof. Suppose first that we have a congruence $\equiv$ on G and let $K = \bar{1}$. If $k_1, k_2 \in K$, then $k_1 k_2 \in K$ since $\overline{k_1 k_2} = \bar{k}_1 \bar{k}_2 = \bar{1}\bar{1} = \bar{1}$. Also $1 \in K$ and $k_1^{-1} \in K$ since, as we showed above, $\overline{k_1^{-1}} = \bar{k}_1^{-1} = \bar{1}^{-1} = \bar{1}$. Hence K is a subgroup of G. Next let g be any element of G and consider the congruence class $\bar{g}$. If $a \in \bar{g}$ then $g^{-1}a$ and $ag^{-1} \in K$ since $\overline{g^{-1}a} = \overline{g^{-1}}\bar{a} = \bar{g}^{-1}\bar{g} = \bar{1} = K$ and, similarly. $ag^{-1} \in K$. It follows that $a \in Kg$ and $a \in gK$. Conversely, let $a \in Kg$. Then $a = kg, k \in K$, and $\bar{a} = \overline{kg} = \bar{1}\bar{g} = \bar{g}$ so $a \equiv g$. The same thing holds if $a \in gK$. Thus

(25) $$\bar{g} = gK = Kg, \qquad g \in G.$$

It follows that K is normal in the sense of the foregoing definition. This can be seen directly, or better still, it can be seen by observing that $gK = Kg$ for all g and a subgroup K is equivalent to normality. If this holds, then for any $g \in G$ and any $k \in K$, $kg \in gK$, so kg has the form gk', $k' \in K$. Then $g^{-1}kg \in K$, so K is normal. On the other hand, if K is normal, a reversal of the steps shows that $kg \in gK$ for $k \in K$, $g \in G$. Hence $Kg \subset gK$. Replacing g by g^{-1} in the definition of normality, we obtain $Kg^{-1} \subset g^{-1}K$, which implies that $gK \subset Kg$. Hence $Kg = gK$ for every g in G.

Conversely, let K be a normal subgroup of G and define $a \equiv b \pmod{K}$ to mean $a^{-1}b \in K$. This is equivalent to saying that $b \in aK$, or that b is in the orbit of a relative to the transformation group $K_R(G)$. We showed in the last section that the relation we are considering is an equivalence relation in G for any subgroup K of G. We now proceed to show that normality of K insures that equivalences can be multiplied and hence that $a \equiv b \pmod{K}$ is a congruence. Thus let $a \equiv g \pmod{K}$ and $b \equiv h \pmod{K}$. Then $a = gk_1$, $b = hk_2$, $k_i \in K$, and since $Kh = hK$, $k_1h = hk_3$, $k_3 \in K$. Then $ab = gk_1hk_2 = ghk_3k_2$ so $ab \equiv gh \pmod{K}$. Thus $\equiv \pmod{K}$ is a congruence relation in G. For this congruence we have $\bar{1} = \{k \,|\, 1^{-1}k \in K\} = K$ and for any g, $\bar{g} = \{a \,|\, g^{-1}a \in K\} = gK$. This completes our verification. $\square$

We shall now write G/K for $\bar{G} = G/\equiv \pmod{K}$ and call this the *factor group* (or *quotient group*) of G relative to the normal subgroup K. By definition, the product in G/K is

(26) $$(gK)(hK) = ghK,$$

$K = 1K$ is the unit, and the inverse of gK is $g^{-1}K$.

Every group $\neq 1$ has two normal subgroups: G and 1. G is called *simple* if these are its only normal subgroups. Equivalently, G is simple if the only congruences on G are the two trivial ones: $=$, and the one in which any two elements are equivalent. It is clear from the definition that any subgroup of an abelian group is normal. It follows easily that the only simple abelian groups are the cyclic groups of prime order. It is left to the reader to prove this. We remark also that if C is the center of G then every subgroup of C is normal in G.

There is another way of looking at factor groups in terms of multiplication of subsets of a group. If A and B are subsets of a group G (similarly of a monoid) one defines

$$AB = \{ab \mid a \in A, b \in B\}.$$

With this definition of product and $1 = \{1\}$, the set of non-vacuous subsets of G is a monoid, since $(AB)C$ is the set of elements $(ab)c$ and $A(BC)$ is the set of elements $a(bc)$, $a \in A$, $b \in B$, $c \in C$. Hence, associativity follows from the associative law in G. Also $1A = A = A1$. It is clear that a subset H of G is a subgroup if and only if: (1) $H^2 \subset H$, (2) $1 \in H$, (3) $H^{-1} \equiv \{h^{-1} \mid h \in H\} \subset H$, and (1) and (2) together imply that $H^2 = H$. It is clear also that the coset Hg (respectively gH) is the product of H and $\{g\}$ (of $\{g\}$ and H). A subgroup K is normal if and only if any of the following equivalent conditions hold: $g^{-1}Kg \subset K$, $Kg = gK$, $g^{-1}Kg = K$ for all $g \in G$. In this case, the product for sets as just defined gives $(gK)(hK) = g(Kh)K = g(hK)K = ghK^2 = ghK$. Thus the product in G/K as defined by (26) coincides with the set product of gK and hK.

EXERCISES

1. Determine addition tables for $(\mathbb{Z}/\mathbb{Z}3, +)$ and $(\mathbb{Z}/\mathbb{Z}6, +)$. Determine all the subgroups of $(\mathbb{Z}/\mathbb{Z}6, +)$.

2. Determine a multiplication table for $(\mathbb{Z}/\mathbb{Z}6, \cdot)$.

3. Let G be the group of pairs of real numbers (a, b) $a \neq 0$, with the product $(a, b)(c, d) = (ac, ad + b)$ (exercise 4, p. 36). Verify that $K = \{(1, b) \mid b \in \mathbb{R}\}$ is a normal subgroup of G. Show that $G/K \cong (\mathbb{R}^*, \cdot, 1)$ the multiplicative group of non-zero reals.

4. Show that any subgroup of index two is normal. Hence prove that A_n is normal in S_n.

5. Verify that the intersection of any set of normal subgroups of a group is a normal subgroup. Show that if H and K are normal subgroups, then HK is a normal subgroup.

6. Let G_1 and G_2 be simple groups. Show that every normal subgroup of $G = G_1 \times G_2$, $\neq G$, $\neq 1$ is isomorphic to either G_1 or G_2.

7. Let $\equiv$ be an equivalence relation on a monoid M. Show that $\equiv$ is a congruence if and only if the subset of $M \times M$ defining $\equiv$ (p. 10) is a submonoid of $M \times M$.

8. Let $\{\equiv_i\}$ be a set of congruences on M. Define the *intersection* as the intersection of the corresponding subsets of $M \times M$. Verify that this is a congruence on M.

9. Let G_1 and G_2 be subgroups of a group G and let α be the map of $G_1 \times G_2$ into G defined by $\alpha(g_1, g_2) = g_1 g_2$. Show that the fiber over $g_1 g_2$—that is, $\alpha^{-1}(g_1 g_2)$—is the set of pairs $(g_1 k, k^{-1} g_2)$ where $k \in K = G_1 \cap G_2$. Hence show that all fibers have the same cardinality, namely, that of K. Use this to show that if G_1 and G_2 are finite than

$$|G_1 G_2| = \frac{|G_1||G_2|}{|G_1 \cap G_2|}.$$

10. Let G be a finite group, A and B non-vacuous subsets of G. Show that $G = AB$ if $|A| + |B| > |G|$.

11. Let G be a group of order $2k$ where k is odd. Show that G contains a subgroup of index 2. (*Hint*: Consider the permutation group G_L of left translations and use exercise 13, p. 36.)

1.9 HOMOMORPHISMS

In dealing with mathematical structures such as monoids, groups, vector spaces, topological spaces, etc., it is important to specify the types of maps which in some sense are natural in the particular context. For vector spaces these are the linear maps, and for topological spaces they are the continuous ones. Nearly all the interesting results in linear algebra concern linear transformations, or equivalently, matrices. In fact, there is not much one can say about vector spaces that does not involve explicitly the notion of a linear transformation or matrix.[6] The natural maps for monoids (and for groups) are called homomorphisms. These are obtained simply by dropping the requirement of bijectivity in the definition of an isomorphism. The concept of homomorphism was a rather late bloomer in the theory of groups, and it became an important tool for the study of groups only comparatively recently—during the past forty or fifty years. The concept is applicable to all types of algebraic structures. In the case of monoids we can state the definition formally as follows:

DEFINITION 1.6.   *If M and M' are monoids, then a map η of M into M' is*

---

[6] Perhaps the deepest result of linear algebra not using linear transformations is the theorem on the invariance of dimensionality (any two bases have the same cardinality).

called a homomorphism *if*

$$\eta(ab) = \eta(a)\eta(b), \qquad \eta(1) = 1', \qquad a, b \in M.$$

If M' is a group the second condition is superfluous. For, if the first holds, we have $\eta(1) = \eta(1^2) = \eta(1)^2$ and multiplying by $\eta(1)^{-1}$ we obtain $1' = \eta(1)$. We have already encountered several instances of homomorphisms which may not be isomorphisms. One of these is the map

$$\eta_a : n \to a^n$$

of the additive group of integers into any group G, determined by a fixed element $a \in G$. Since $\eta_a(n + m) = a^{n+m} = a^n a^m = \eta_a(n)\eta_a(m)$, this is a homomorphism of $(\mathbb{Z}, +, 0)$ into G. Another example we had is the map

$$\alpha \to sg\ \alpha$$

of the symmetric group S_n into the multiplicative group $\{1, -1\}$. That this is a homomorphism is clear from (22). Some additional examples of homomorphisms (and of one fake) are given in the following list.

EXAMPLES

1. Let M and M' be monoids and map every $a \in M$ into the unit $1'$ of M'. This is a homomorphism of M into M'.

2. Let M be the multiplicative monoid of integers: $M = (\mathbb{Z}, \cdot, 1)$. Map every $a \in M$ into 0. This satisfies $\eta(ab) = \eta(a)\eta(b)$ but it is not a homomorphism since $1 \to 0$ ($\neq 1$).

3. Let $G = (\mathbb{R}, +, 0)$, $G' = (\mathbb{C}^*, \cdot, 1)$ the multiplicative group of non-zero complex numbers. Let $\eta : \theta \to e^{i\theta}$. This is a homomorphism of G into G'.

4. Let G be the group of pairs (a, b), $a \neq 0$, given in exercise 4, p. 36, and map G into $G' = (\mathbb{R}^*, \cdot, 1)$ by $(a, b) \to a$. This is a homomorphism.

5. Let G be a transformation group of a set S and let T be a subset of S which is stabilized by G in the sense that $\alpha(T) \subset T$ for every $\alpha \in G$. Let $\alpha | T$ be the restriction of α to T. Then $\alpha \to \alpha | T$ is a homomorphism of G into Sym T. This is called the *restriction homomorphism*.

We emphasize that—as in the foregoing examples—a homomorphism η need not be surjective or injective. If, by chance, η is surjective then we call it an *epimorphism*, and if it is injective then we call it a *monomorphism*. Of course, if it is bijective, then η is an isomorphism.

If η is a homomorphism of the monoid M into the monoid M', then induction shows that for any $a \in M$ and $k \in \mathbb{N}$, $\eta(a^k) = \eta(a)^k$. If a is invertible, application of η to $aa^{-1} = 1 = a^{-1}a$ gives $\eta(a)\eta(a^{-1}) = 1' = \eta(a^{-1})\eta(a)$. Hence $a' = \eta(a)$

is invertible in M' and $\eta(a^{-1}) = \eta(a)^{-1}$. It then follows that $\eta(a^k) = \eta(a)^k$ for all $k \in \mathbb{Z}$. Another useful result which we have to refer to frequently enough to warrant stating as a theorem is

THEOREM 1.7. *Let η and ζ be homomorphisms of a monoid M (or group G) into a monoid M' and let S be a set of generators for M (for the group G). Suppose $\eta(s) = \zeta(s)$ for all $s \in S$. Then $\eta = \zeta$.*

Proof. We consider first the case of monoids and let

$$M_1 = \{a \in M \,|\, \eta(a) = \zeta(a)\}.$$

Then $1 \in M_1$ since $\eta(1) = 1' = \zeta(1)$ and $M_1 \supset S$. Also if $a, b \in M_1$, then $ab \in M_1$ since $\eta(ab) = \eta(a)\eta(b) = \zeta(a)\zeta(b) = \zeta(ab)$. Thus M_1 is a submonoid, and since it contains a set of generators, $M_1 = M$. Hence $\eta(a) = \zeta(a)$ for all a, and so $\eta = \zeta$. The proof is similar in the case of a group G. In this case the argument shows that the subset $G_1 = \{a \in G \,|\, \eta(a) = \zeta(a)\}$ is a submonoid. But if $a \in G_1$, then $\eta(a^{-1}) = \eta(a)^{-1} = \zeta(a)^{-1} = \zeta(a^{-1})$. Hence $a^{-1} \in G_1$ and G_1 is a subgroup. Then $G_1 = G$ since G_1 contains a set of generators of G (as a group). $\square$

A homomorphism of M into itself is called an *endomorphism* and an iso-morphism of M to M is called an *automorphism* of M. The identity map is an automorphism. Theorem 1.7 applied to any endomorphism η and to $\zeta = 1$ shows that if η is an endomorphism of a monoid or a group and η is the iden-tity map on a set of generators then $\eta = 1$. We remark also that if η is an en-domorphism, then the set of fixed elements under η ($\eta(a) = a$) is a submonoid if M is a monoid and a subgroup if $M = G$ is a group. This is clear from the proof of Theorem 1.7.

Let $\eta : M \to M'$ and $\zeta : M' \to M''$ be homomorphisms of monoids. Then for $a, b \in M$, $\zeta\eta(ab) = \zeta(\eta(ab)) = \zeta(\eta(a)\eta(b)) = (\zeta\eta(a))(\zeta\eta(b))$. Also $\zeta\eta(1) = \zeta(1') = 1''$, the unit of M''. Hence $\zeta\eta : M \to M''$ is a homomorphism. If η is bijective then, as we saw before, η^{-1} is an isomorphism of M' into M. It is clear that the identity map is an automorphism. Hence the set, Aut M, of automorphisms of a monoid is a group of transformations of the monoid. We call this the *group of automorphisms* of M. We remark also that the larger set, End M, of endomor-phisms is a monoid of transformations, the *endomorphism monoid* of M.

Let M be a monoid, $\equiv$ a congruence on M and $\bar{M}$ the quotient monoid determined by $\equiv$. Then the natural map $v : a \to \bar{a}$ (the congruence class of a) is a homomorphism, since, $v(1) = \bar{1}$ is the unit of $\bar{M}$ and $v(ab) = \overline{ab} = \bar{a}\bar{b} =$

$v(a)v(b)$, by definition of the product in $\bar{M}$. We shall now derive the main result on homomorphisms of monoids and groups which we state as the

FUNDAMENTAL THEOREM OF HOMOMORPHISMS OF MONOIDS AND GROUPS.

Let η be a homomorphism of a monoid M into a monoid M'. Then the image $\eta(M)$ is a submonoid of M' and if M is a group, $\eta(M)$ is a subgroup of M'. The equivalence relation E_η determined by the map η ($aE_\eta b$ means $\eta(a) = \eta(b)$) is a congruence in M and we have a unique homomorphism $\bar{\eta}$ of the quotient monoid $\bar{M} = M/E_\eta$ into M' making

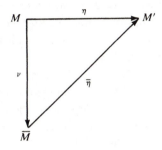

commutative. v is an epimorphism and $\bar{\eta}$ is a monomorphism. In the case of groups, $\bar{1} = K = \eta^{-1}(1')$ is a normal subgroup of M, $\bar{M} = M/K$, v is $a \to aK$, and $\bar{\eta}$ is $aK \to \eta(a)$.

Proof. As happens frequently at the foundational level, the proof is not much longer than the statement of the theorem and it amounts merely to a direct verification of the various assertions. Let $\eta : M \to M'$ be a homomorphism of monoids. Then $1' = \eta(1) \in \eta(M)$, and $\eta(a)\eta(b) = \eta(ab)$ shows that $\eta(M)$ is closed under the product in M'. Hence $\eta(M)$ is a submonoid. If M is a group, $\eta(a)$ is invertible with inverse $\eta(a^{-1})$, and so $\eta(M)$ is a subgroup of M'. Now consider the equivalence relation E_η in M. Suppose $a_1 E_\eta a_2$ and $b_1 E_\eta b_2$, which means that $\eta(a_1) = \eta(a_2)$ and $\eta(b_1) = \eta(b_2)$. Then $\eta(a_1 b_1) = \eta(a_1)\eta(b_1) = \eta(a_2)\eta(b_2) = \eta(a_2 b_2)$ so $a_1 b_1 E_\eta a_2 b_2$. Thus E_η is a congruence. Our results on maps of sets (section 0.3) show that we have a unique induced map $\bar{\eta}$ of $\bar{M} = M/E_\eta$ into M' such that $\bar{\eta}v = \eta$. We have seen that v is a homomorphism. All that remains (for the case of monoids) is to show that $\bar{\eta}$ is a homomorphism. We have $\bar{\eta}(\bar{a}) = \eta(a)$. Then $\bar{\eta}(\bar{a}\bar{b}) = \bar{\eta}(\overline{ab}) = \eta(ab) = \eta(a)\eta(b) = \bar{\eta}(\bar{a})\bar{\eta}(\bar{b})$ and $\bar{\eta}(\bar{1}) = \eta(1) = 1'$, which is what we needed. We saw in section 0.3 that v is surjective and $\bar{\eta}$ is injective. Hence these are respectively an epimorphism and monomorphism of M and $\bar{M}$. Now suppose M and M' are groups. Since E_η is a congruence in the group M, we know that the congruence class K of 1 is a normal subgroup of M and the congruence class of any a is $Ka = aK$ (section 1.8). By definition, the congruence class of 1 is

$$K = \{a \in M \mid \eta(a) = \eta(1) = 1'\},$$

that is, $K = \eta^{-1}(1')$. The rest is clear by Theorem 1.6. $\square$

In the foregoing discussion we have derived the results on groups as consequences of results on monoids. For the latter the concepts of congruence and quotient monoid defined by a congruence are essential. On the other hand, the basic results on group homomorphisms can also be derived directly without recourse to congruences. We proceed to do this. This will help clarify the situation in the most important case of group homomorphisms.

We start from scratch and consider a homomorphism η of a group G into a group G'. Then it is immediate that the image im G is a subgroup of G'. Next we consider $K = \eta^{-1}(1')$, which is analogous to the null space of a linear map of one vector space into a second one. Direct verification shows that K is a normal subgroup of G. We call this the *kernel* of η and denote it also as ker η. We observe first that η is injective if and only if ker $\eta = 1$; for, if ker $\eta \neq 1$ then we have $b \neq 1$ in G such that $\eta(b) = 1' = \eta(1)$. On the other hand, if η is not injective then we have $a \neq b$ in G with $\eta(a) = \eta(b)$. Then $a^{-1}b \neq 1$ and $\eta(a^{-1}b) = \eta(a)^{-1}\eta(b) = 1'$, so ker $\eta \neq 1$.

Now let L be a normal subgroup of G contained in K. Then we can form the factor group $\bar{G} = G/L$ consisting of the cosets $aL = La$, $a \in G$, with multiplication $(aL)(bL) = abL$ and unit $\bar{1} = L$ (see the last paragraph on p. 56). This definition shows that the map $v:a \rightarrow aL$ is a homomorphism of G onto $\bar{G} = G/L$. Now suppose $aL = bL$. Then $b = al$, $l \in L$, and $\eta(b) = \eta(a)\eta(l) = \eta(a)1'$ (since $L \subset$ ker η) $= \eta(a)$. Hence we have a well-defined map $\bar{\eta}:aL \rightarrow \eta(a)$ of G/L into G'. Since $\bar{\eta}((aL)(bL)) = \bar{\eta}(abL) = \eta(ab) = \eta(a)\eta(b) = \bar{\eta}(aL)\bar{\eta}(bL)$, $\bar{\eta}$ is a homomorphism. We call $\bar{\eta}$ the homomorphism of $\bar{G} = G/L$ *induced* by η. If $a \in G$ then $\bar{\eta}v(a) = \bar{\eta}(aL) = \eta(a)$. Thus $\eta = \bar{\eta}v$, which means that we have a commutative diagram as on the preceding page.

Evidently im $\bar{\eta} =$ im η. What is the kernel of $\bar{\eta}$? By definition, this is the set of cosets aL such that $\bar{\eta}(aL) = 1'$. Since $\bar{\eta}(aL) = \eta(a)$, the condition is $\eta(a) = 1'$. Hence ker $\bar{\eta} = \{aL \mid a \in$ ker $\eta\} =$ ker η/L (Clearly L is a normal subgroup of K.) Since a homomorphism is injective if and only if its kernel is 1, $\bar{\eta}$ is injective if and only if $L =$ ker η.

The facts we have listed go beyond those stated in the "Fundamental Theorem" in the replacement of $K =$ ker η by any normal subgroup L of G contained in K. Now suppose $K = L$ and η is surjective. Then the homomorphism $\bar{\eta}$ of $G = G/K$ into G' is surjective and injective, hence an isomorphism. We therefore have the

COROLLARY. *If G is a group and η is an epimorphism of G onto the group G' with kernel K, then the induced map $\bar{\eta}: aK \to \eta(a)$ is an isomorphism. Thus any homomorphic image of a group G is isomorphic to a factor group G/K by a normal subgroup K.*

EXERCISES

1. Let $G = (\mathbb{Q}, +, 0)$, $K = \mathbb{Z}$. Show that $G/K \cong$ the group of complex numbers of the form $e^{2\pi i \theta}$, $\theta \in \mathbb{Q}$, under multiplication.

2. Let G be the set of triples of integers (k, l, m) and define $(k_1, l_1, m_1)(k_2, l_2, m_2) = (k_1 + k_2 + l_1 m_2, l_1 + l_2, m_1 + m_2)$. Verify that this defines a group with unit $(0, 0, 0)$. Show that $C = \{(k, 0, 0) | k \in \mathbb{Z}\}$ is a normal subgroup and that $G/C \cong$ the group $\mathbb{Z}^{(2)} = \{(l, m) | l, m \in \mathbb{Z}\}$ with the usual addition as composition.

3. Show that $a \to a^{-1}$ is an automorphism of a group G if and only if G is abelian, and if G is abelian, then $a \to a^k$ is an endomorphism for every $k \in \mathbb{Z}$.

4. Determine Aut G for (i) G an infinite cyclic group, (ii) a cyclic group of order six, (iii) for any finite cyclic group.

5. Determine Aut S_3.

6. Let $a \in G$, a group, and define the *inner automorphism* (or *conjugation*) I_a to be the map $x \to axa^{-1}$ in G. Verify that I_a is an automorphism. Show that $a \to I_a$ is a homomorphism of G into Aut G with kernel the center C of G. Hence conclude that Inn $G \equiv \{I_a | a \in G\}$ is a subgroup of Aut G with Inn $G \cong G/C$. Verify that Inn G is a normal subgroup of Aut G. Aut G/Inn G is called the *group of outer automorphisms*.

7. Let G be a group, G_L the set of left translations a_L, $a \in G$. Show that G_L Aut G is a group of transformations of the set G and that this contains G_R. G_L Aut G is called the *holomorph of G* and is denoted as Hol G. Show that if G is finite, then $|\text{Hol } G| = |G| |\text{Aut } G|$.

8. Let G be a group such that Aut $G = 1$. Show that G is abelian and that every element of G satisfies the equation $x^2 = 1$. Show that if G is finite then $|G| = 1$ or 2. (*Hint*: Use the procedure of finding a base for a vector space to show that G contains elements $a_1, a_2, \ldots, a_r$ such that every element of G can be written in one and only one way in the form $a_1^{k_1} a_2^{k_2} \cdots a_r^{k_r}$, $k_i = 0, 1$. Then show that there exists an automorphism interchanging a_1 and a_2.)

9. Let α be an automorphism of a group G which fixes only the unit of G ($\alpha(a) = a \Rightarrow a = 1$). Show that $a \rightarrow \alpha(a)a^{-1}$ is injective. Hence show that if G is finite, then every element of G has the form $\alpha(a)a^{-1}$.

10. Let G and α be as in 8, G finite, and assume $\alpha^2 = 1$. Show that G is abelian of odd order.

11. Let G be a finite group, α an automorphism of G, and set

$$I = \{g \in G | \, \alpha(g) = g^{-1}\}.$$

Suppose $|I| > \frac{3}{4}|G|$. Show that G is abelian. If $|I| = \frac{3}{4}|G|$, show that G has an abelian subgroup of index 2.

1.10 SUBGROUPS OF A HOMOMORPHIC IMAGE. TWO BASIC ISOMORPHISM THEOREMS

We shall establish a 1–1 correspondence between the set of subgroups of a homomorphic image $\bar{G}$ of a group G and the set of subgroups of G containing the kernel of a given homomorphism. Since any homomorphic image is isomorphic to a factor group we may assume $\bar{G} = G/K$, K a normal subgroup of G. Then we have

THEOREM 1.8. *Let K be a normal subgroup of G, H a subgroup of G containing K. Then $\bar{H} = H/K$ is a subgroup of $\bar{G} = G/K$ and the map $H \rightarrow \bar{H}$ is a bijective map of the set of subgroups of G containing K with the set of subgroups of $\bar{G}$. $H(\supset K)$ is normal in G if and only if $\bar{H}$ is normal in $\bar{G}$. In this case,*

$$\frac{G}{H} \cong \frac{\bar{G}}{\bar{H}} = \frac{G/K}{H/K}.$$

Proof. The fact that H/K is a subgroup of G/K is clear from the definition of G/K. Now let H_1 and H_2 be two subgroups of G containing K and suppose $H_1/K = H_2/K$. Then for any $h_1 \in H_1$, $h_1 K \in H_2/K$, so $h_1 K = h_2 K$ for some $h_2 \in H_2$. Then $h_2^{-1}h_1 \in K$, so $h_1 = h_2 k$, $k \in K$. Since $K \subset H_2$ this shows that $h_1 \in H_2$. Thus $H_1 \subset H_2$ and, similarly, $H_2 \subset H_1$. Hence $H_1 = H_2$, and we have shown that $H \rightarrow H/K$ is injective. To see that it is surjective let $\bar{H}$ be a subgroup of $\bar{G}$, so that $\bar{H}$ is a collection of cosets. Let H be the union in G of these cosets. If h_1, $h_2 \in H$, $h_1 K$, $h_2 K \in \bar{H}$ and $h_1 h_2 K = (h_1 K)(h_2 K) \in \bar{H}$. Hence $h_1 h_2 \in H$. Similarly $h_1^{-1}K = (h_1 K)^{-1} \in \bar{H}$, so $h_1^{-1} \in H$. Hence H is a subgroup of G. Clearly $\bar{H} = H/K$. It is evident that if H is normal in G, then $\bar{H}$ is normal in $\bar{G}$. Conversely, if $\bar{H}$ is normal in $\bar{G}$, then for any $h \in H$, $g \in G$, $(g^{-1}hg)K = (gK)^{-1}(hK)(gK) = h'K$ for some $h' \in H$. It follows that $g^{-1}hg \in H$ and H is normal in G. If this condition is satisfied we can form the factor group $\bar{G}/\bar{H}$ and we have the natural homomorphism $\bar{v}:\bar{g} \to \bar{g}\bar{H}$ of $\bar{G}$ with $\bar{G}/\bar{H}$. We also have the natural homomorphism $g \to \bar{g}$ of G with $\bar{G}$. Hence we have the homomorphism $g \to \bar{g}\bar{H}$ of G with $\bar{G}/\bar{H}$. The kernel is the set of $g \in G$ such that $\bar{g} \in \bar{H}$, that is, the set of g such that $gK = hK$ for some $h \in H$. This is just the subgroup H. Hence, by the fundamental theorem of homomorphisms, $gH \to \bar{g}\bar{H}$ is an isomorphism of G/H with $\bar{G}/\bar{H}$. $\square$

It is sometimes useful to state Theorem 1.8 in what appears to be a slightly more general form, as follows:

THEOREM 1.8'. *Let η be an epimorphism of G onto G' and let Λ be the set of subgroups H of G containing $K = \ker \eta$. Then the map $H \to \eta(H)$ of Λ gives a $1-1$ correspondence between the set Λ and the complete set of subgroups of G'. H is normal in G if and only if $\eta(H)$ is normal in G'. In this case*

(27)
$$gH \to \eta(g)\eta(H)$$

is an isomorphism of G/H with $G'/\eta(H)$.

This can either be proved directly in a manner similar to the proof of Theorem 1.8, or, it can be deduced from Theorem 1.8 via the isomorphism $gK \to \eta(g)$ of G/K with G'. We leave the details to the reader.

The isomorphism (27) is often called the *first isomorphism theorem* for groups. There is also a basic *second isomorphism theorem*. This is

THEOREM 1.9. *Let H and K be subgroups of G, K normal in G. Then $HK = \{hk \,|\, h \in H, k \in K\}$ is a subgroup of G containing K, $H \cap K$ is normal in H and the map*

(28)
$$hK \to h(K \cap H), \qquad h \in H$$

is an isomorphism of HK/K with $H/(K \cap H)$.

Proof. Since K is normal we have $hK = Kh$, $h \in H$. Since $HK = \bigcup_{h \in H} hK$ and $KH = \bigcup_{h \in H} Kh$, clearly $HK = KH$. Then $(HK)^2 = HKHK = H^2 K^2 = HK$. Also $1 \in HK$ and if $hk \in HK$ ($h \in H$, $k \in K$) then $(hk)^{-1} = k^{-1}h^{-1} \in KH = HK$. Hence HK is a subgroup of G. Clearly, $HK \supset 1K = K$ and K is normal in HK. We now consider the restriction $v' = v|H$ where $v:g \to gK$. The image of v' is the set of cosets hK, $h \in H$. Since any coset of the form hkK, $h \in H$, $k \in K$, coincides with hK, it is clear that im v' is HK/K. The kernel of this homomorphism is the set of $h \in H$ such that $hK = K$, the unit of HK/K. Since $hK = K$ if and only if $h \in K$, we see that $\ker v' = H \cap K$ and so this is a normal subgroup of H, and by the fundamental theorem of homomorphisms, $h(H \cap K) \to hK$ is an isomorphism of $H/(H \cap K)$ with HK/K. The inverse is $hK \to h(H \cap K)$ as given in (28). $\square$

The proofs of the theorems in this section illustrate the power of the fundamental theorem. As another illustration of this and also of the use of the subgroup correspondence of Theorem 1.8, we shall now give a quick re-derivation of the results on cyclic groups. Everything will follow from the determination of the subgroups of $(\mathbb{Z}, +, 0)$ and their inclusion relations. Let K be a subgroup $\neq 0$ of $\mathbb{Z}$. Then if $n \in K$ so does $-n$; hence K contains positive integers and consequently K contains a least positive integer k. Now let n be any element of K. Then the division algorithm in $\mathbb{Z}$ permits us to write $n = qk + r$ where $0 \leq r < k$. Clearly $qk \in K$ and since $n \in K$, $r = n - qk \in K$. This forces $r = 0$, since k is the least positive integer in K. Thus we see that every element of K is a multiple of k and, of course, every multiple of k is in K. Hence $K = \mathbb{Z}k = \{mk \mid m \in \mathbb{Z}\}$. Conversely, it is clear that for any $k \geq 0$, $\mathbb{Z}k$ is a subgroup. This includes the subgroup 0 as $\mathbb{Z}0$. Thus the set of subgroups of $\mathbb{Z}$ are the various sets $\mathbb{Z}k$, $k \in \mathbb{N}$. Suppose $k, l \in \mathbb{N}$ and $\mathbb{Z}l \supset \mathbb{Z}k$. Then $k \in \mathbb{Z}l$ so $k = lm$ and $l \mid k$. The converse is clear. Hence

(29) $$\mathbb{Z}l \supset \mathbb{Z}k \Leftrightarrow l \mid k.$$

Next we note that if $k = 0$ then $\mathbb{Z}/\mathbb{Z}k \cong \mathbb{Z}$ and if $k > 0$ then $\mathbb{Z}/\mathbb{Z}k$ is just the set of congruence classes modulo the integer k, and these are

$$\bar{0} = \mathbb{Z}k, \qquad \bar{1} = \{1 + mk \mid m \in \mathbb{Z}\}, \qquad \bar{2} = \{2 + mk \mid m \in \mathbb{Z}\},$$

$$\ldots, \overline{k-1} = \{(k-1) + mk \mid m \in \mathbb{Z}\}.$$

Thus the order of $\mathbb{Z}/\mathbb{Z}k$ is k. Clearly $\mathbb{Z}/\mathbb{Z}k$ is cyclic with $\bar{1}$ as generator.

Now let $G = \langle a \rangle$, so that G is a cyclic group with generator a. Since $a^m a^n = a^{m+n}$ we have the epimorphism of $(\mathbb{Z}, +, 0)$ into G sending $n \to a^n$. Hence $G \cong \mathbb{Z}/\mathbb{Z}k$ for some $k \in \mathbb{N}$. If $k = 0$, $G \cong \mathbb{Z}$ and if $k > 0$, G is finite of order k. Hence it is clear that any two cyclic groups of the same order are isomorphic.

We can also determine the subgroups of $\mathbb{Z}/\mathbb{Z}k$. If $k = 0$ we are dealing with $\mathbb{Z}$ and we have the determination which we made: the subgroups are $\mathbb{Z}l$, $l \geq 0$, and $\mathbb{Z}l$ is cyclic with generator l. If $k > 0$ it follows from Theorem 1.8 that the subgroups of $\mathbb{Z}/\mathbb{Z}k$ have the form $\mathbb{Z}l/\mathbb{Z}k$ where $l \geq 0$ and $\mathbb{Z}l \supset \mathbb{Z}k$. Then $l \mid k$, say, $k = lm$. Now $(\mathbb{Z}/\mathbb{Z}k)/(\mathbb{Z}l/\mathbb{Z}k) \cong \mathbb{Z}/\mathbb{Z}l$ so $|\mathbb{Z}l/\mathbb{Z}k| = |\mathbb{Z}/\mathbb{Z}k|/|\mathbb{Z}/\mathbb{Z}l| = k/l = m$. It follows that the cyclic group $\mathbb{Z}/\mathbb{Z}k$ of order k has one and only one subgroup of order m for each divisor m of k. Moreover, this subgroup, $\mathbb{Z}l/\mathbb{Z}k$, is cyclic with $l + \mathbb{Z}k$ as generator.

EXERCISES

1. Show that $\mathbb{Z}l \cap \mathbb{Z}k = \mathbb{Z}[l, k]$ and $\mathbb{Z}l + \mathbb{Z}k = \{a + b \,|\, a \in \mathbb{Z}l,\, b \in \mathbb{Z}k\} = \mathbb{Z}(l, k)$.

2. Let $\{H_\alpha\}$ be a collection of subgroups containing the normal subgroup K. Show that $\bigcap (H_\alpha/K) = (\bigcap H_\alpha)/K$.

1.11   FREE OBJECTS. GENERATORS AND RELATIONS

The method used in the last section of studying cyclic groups by considering these as a homomorphic images of the "universal" cyclic group $(\mathbb{Z}, +, 0)$ can be generalized to obtain the structure of finitely generated abelian groups. We shall carry out this program in Chapter 3. At this point we shall define these universal finitely generated abelian groups, called free abelian groups, and consider also their analogues for commutative monoids, for arbitrary monoids, and for arbitrary groups.

We construct first for any positive integer r and abelian group $\mathbb{Z}^{(r)}$ with r generators $x_1, x_2, \ldots, x_r$ such that if G is any abelian group and $a_1, a_2, \ldots, a_r$ are elements of G then there exists a unique homomorphism of $\mathbb{Z}^{(r)}$ into G sending

$$x_i \to a_i, \quad 1 \le i \le r.$$

Let $\mathbb{Z}^{(r)}$ be the r-fold direct power of $\mathbb{Z}$: $\mathbb{Z}^{(r)}$ is the set of r-tuples $(n_1, n_2, \ldots, n_r)$ of integers n_i with addition by components, $(m_i) + (n_i) = (m_i + n_i)$ and $0 = (0, 0, \ldots, 0)$. This is an abelian group. Put

$$(30) \qquad\qquad x_i = (0, \ldots, 0, \overset{i}{1}, 0, \ldots, 0), \qquad 1 \le i \le r.$$

Then $(n_1, n_2, \ldots, n_r) = \sum_1^r n_i x_i$, so the x_i generate $\mathbb{Z}^{(r)}$. Now let $a_1, a_2, \ldots, a_r$ be a sequence of r elements of any abelian group G and consider the map

$$(31) \qquad\qquad \eta : (n_1, n_2, \ldots, n_r) \to a_1^{\,n_1} a_2^{\,n_2} \cdots a_r^{\,n_r}.$$

Since the a_i commute, we have

$$(a_1^{\,m_1} a_2^{\,m_2} \cdots a_r^{\,m_r})(a_1^{\,n_1} a_2^{\,n_2} \cdots a_r^{\,n_r}) = a_1^{\,m_1 + n_1} a_2^{\,m_2 + n_2} \cdots a_r^{\,m_r + n_r}$$

which implies that η is a homomorphism of $\mathbb{Z}^{(r)}$ into G. Moreover,

$$\eta(x_i) = \eta(0, \ldots, 0, \overset{i}{1}, 0, \ldots, 0) = a_1^{\,0} \cdots a_{i-1}^{\,0} a_i^{\,1} a_{i+1}^{\,0} \cdots a_r^{\,0} = a_i$$

and, since the x_i generate $\mathbb{Z}^{(r)}$, there is only one homomorphism of $\mathbb{Z}^{(r)}$ sending $x_i \to a_i$, $1 \le i \le r$ (see Theorem 1.7). We shall call $\mathbb{Z}^{(r)}$ the *free abelian group with r (free) generators x_i*.

Identical considerations apply to commutative monoids. Let $\mathbb{N}^{(r)}$ be the r-fold direct power of the monoid $(\mathbb{N}, +, 0)$. This is a commutative monoid generated by the r elements x_i, as in (30). Moreover, as in the group case, if $a_1, a_2, \ldots, a_r$ are elements of a commutative monoid M, there exists a unique homomorphism of $\mathbb{N}^{(r)}$ into M such that $x_i \to a_i$, $1 \le i \le r$. We call $\mathbb{N}^{(r)}$ the *free commutative monoid with r (free) generators* x_i.

We shall now drop the requirement of commutativity in these considerations. We seek to construct first a monoid, then a group, generated by r elements x_i such that if a_i are any r elements of a monoid M (group G), then there exists a unique homomorphism of the constructed monoid (group) sending $x_i \to a_i$, $1 \le i \le r$.

We consider first the monoid case. Put $X^1 = X = \{x_1, x_2, \ldots, x_r\}$. $X^j = X \times X \times \cdots \times X$, j times, where $j = 2, 3, \ldots$. Let $FS^{(r)}$ denote the disjoint union of the sets $X^1, X^2, \ldots$. The elements of $FS^{(r)}$ are "words in the alphabet X," that is, they are sequences $(x_{i_1}, x_{i_2}, \ldots, x_{i_m})$, $x_{i_j} \in X$, $m = 1, 2, 3, \ldots$. We introduce a multiplication in $FS^{(r)}$ by juxtaposition, that is,

$$(32) \qquad (x_{i_1}, x_{i_2}, \ldots, x_{i_m})(x_{j_1}, x_{j_2}, \ldots, x_{j_n}) = (x_{i_1}, \ldots, x_{i_m}, x_{j_1}, \ldots, x_{j_n}).$$

This is clearly an associative product, but we have no unit. However, we can adjoin one and call it 1 (see exercise 5, p. 30) to obtain a monoid $FM^{(r)}$. It is clear from (32) that $(x_{i_1}, \ldots, x_{i_m}) = x_{i_1} \cdots x_{i_m}$; hence $FM^{(r)}$ is generated by the x_i. Now let $a_1, a_2, \ldots, a_r$ be any r elements of any monoid M. Then since we have a unique way of writing an element $\ne 1$ of $FM^{(r)}$ as $(x_{i_1}, \ldots, x_{i_m})$,

$$\eta : 1 \to 1, \qquad (x_{i_1}, \ldots, x_{i_m}) \to a_{i_1} \cdots a_{i_m}$$

is a well defined map of $FM^{(r)}$. It is clear from (32) that this is a homomorphism of $FM^{(r)}$ sending $x_i \to a_i$, $1 \le i \le r$. Since the x_i generate $FM^{(r)}$ this is the only homomorphism having this property. We call $FM^{(r)}$ the *free monoid (freely) generated by the r elements* x_i *(or the monoid of words in the* x_i*)*.

To obtain a construction of a free group we observe first that the subgroup of a group generated by a subset X coincides with the submonoid generated by the union of X and the set of inverses of the elements of X. This suggests forming the set $X \cup X'$ where X is the given set $\{x_1, x_2, \ldots, x_r\}$ and X' is another set $\{x_1', x_2', \ldots, x_r'\}$ disjoint to X and in 1–1 correspondence $x_i \leftrightarrow x_i'$ with X. Form the free monoid $FM^{(2r)}$ generated by $X \cup X'$. Now suppose G is a group, and $a_1, a_2, \ldots, a_r$ is a sequence of elements of G. Then we have a unique homomorphism η of $FM^{(2r)}$ into G sending $x_i \to a_i$, $x_i' \to a_i^{-1}$, $1 \le i \le r$. By the fundamental theorem of homomorphisms, we obtain a congruence E_η on $FM^{(2r)}$ by specifying that $aE_\eta b$ means that $\eta(a) = \eta(b)$. Then $x_i x_i' E_\eta 1$ and $x_i' x_i E_\eta 1$. This suggests that we consider the set Γ of all the congruences $\equiv_\alpha$ on $FM^{(2r)}$

in which $x_i x_i' \equiv_\alpha 1$ and $x_i' x_i \equiv_\alpha 1$ for $1 \le i \le r$, and form their intersection $\equiv$. By definition, $a \equiv b$ means $a \equiv_\alpha b$ for every $\equiv_\alpha$. This is again a congruence (exercises 8, p. 57) and so we can form the quotient monoid $FM^{(2r)}/\equiv$, which we shall denote as $FG^{(r)}$. We observe first that $FG^{(r)}$ is a group generated by the congruence classes $\bar{x}_i$, $1 \le i \le r$. This is clear since the congruence class $\bar{x}_i$ has the inverse $\bar{x}_i'$ in $FG^{(r)}$ and $FG^{(r)}$ is generated as monoid by the elements $\bar{x}_i$ and $\bar{x}_i'$. Again, let G be a group, $a_1, a_2, \ldots, a_r$ a sequence of elements of G. We have the unique homomorphism η of $FM^{(2r)}$ into G sending $x_i \to a_i$, $x_i' \to a_i^{-1}$, $1 \le i \le r$ which gives a congruence E_η on $FM^{(2r)}$ such that $x_i x_i' E_\eta 1$ and $x_i' x_i E_\eta 1$. Then $a \equiv b$ on $FM^{(2r)}$ implies $a E_\eta b$ and hence we obtain a well defined map of $FG^{(r)}$ sending the element $\bar{a}$ into $\eta(a)$. This is a homomorphism of $FG^{(r)}$ mapping $\bar{x}_i \to a_i$, $1 \le i \le \dot{r}$. Since the $\bar{x}_i$ generate $FG^{(r)}$ this is the only homomorphism which does this.

To summarize: given the set $X = \{x_1, \ldots, x_r\}$ we have obtained a map $x_i \to \bar{x}_i$ of X into a group $FG^{(r)}$ such that if G is any group and $x_i \to a_i$, $1 \le i \le r$ is any map of X into G then we have a unique homomorphism of $FG^{(r)}$ into G, making the following diagram commutative:

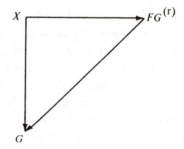

We shall now show that the map $x_i \to \bar{x}_i$ is injective. We do this by taking G in the foregoing diagram to be the free abelian group $\mathbb{Z}^{(r)}$ generated by the elements $(0, \ldots, 0, 1, 0, \ldots, 0)$ and choose the vertical arrow to be the map sending $x_i \to (0, \ldots, 1, 0, \ldots, 0)$. Since this is injective, and injectivity of the composite $\beta\alpha$ of two maps implies injectivity of α, it follows that $x_i \to \bar{x}_i$ is injective. Our last step is to identify x_i with its image $\bar{x}_i$. We can then say that $FG^{(r)}$ is generated by the x_i. Moreover, if $a_i \in G$ then we have a unique homomorphism of $FG^{(r)}$ into G such that $x_i \to a_i$, $1 \le i \le r$. We call $FG^{(r)}$ the *free group (freely) generated by the r elements x_i.*[7]

A group G is said to be *finitely generated* if it contains a finite set of generators $\{a_i | 1 \le i \le r\}$. Then we have the homomorphism η of $FG^{(r)}$ sending $x_i \to a_i$. Since the a_i generate G, this is an epimorphism and $G \cong FG^{(r)}/K$ where K is

---

[7]Another construction of free groups is given on p. 89 of *Basic Algebra* II.

the kernel of η. The normal subgroup K is called the *set of relations connecting the generators* a_i. If S is a subset of a group, we can define the *normal subgroup generated by* S to be the intersection of all normal subgroups of the group containing S. This is a normal subgroup containing S and contained in every normal subgroup containing S. If S is a subset of $FG^{(r)}$ we say that G is *defined by the relations* S if $G \cong FG^{(r)}/K$ where K is the normal subgroup generated by S. If S is finite, then we say that G is a *finitely presented group*.

As an example, we shall now show that the dihedral group D_n consisting of the n rotations and the n reflections mapping a regular n-gon into itself (example 12, p. 34) is defined by the relations

$$(33) \qquad\qquad x^n,\ y^2,\ xyxy$$

in the free group generated by x and y. It is clear that D_n is generated by the rotation R through an angle of $2\pi/n$ and the reflection S in the x-axis. We have the relations

$$(34) \qquad\qquad R^n = 1, \qquad S^2 = 1, \qquad SRS = R^{-1}.$$

Hence D_n is a homomorphic image of $FG^{(2)}/K$ where K is the normal subgroup generated by the elements (33). We shall now show that $|FG^{(2)}/K| \leq 2n$ which will imply that $D_n \cong FG^{(2)}/K$. Let $\bar{x} = xK$, $\bar{y} = yK$ in $FG^{(2)}/K$. Then, since x^n, y^2, and $xyxy \in K$ we have $\bar{x}^n = 1$, $\bar{y}^2 = 1$, $\bar{x}\bar{y}\bar{x}\bar{y} = 1$. Then $\bar{y}\bar{x} = \bar{x}^{-1}\bar{y}$ which implies that $\bar{y}\bar{x}^k = \bar{x}^{-k}\bar{y}$. From this we see that the product of any two of the elements $\bar{x}^k$, $\bar{x}^k\bar{y}$, $k = 0, 1, \ldots, n - 1$, is one of these elements. Also, 1 is included in the displayed set of elements and the set is closed under inverses. Hence it is a subgroup of $FG^{(2)}/K$. Since it contains the generators $\bar{x}$ and $\bar{y}$, $FG^{(2)}/K = \{\bar{x}^k, \bar{x}^k\bar{y} | 0 \leq k \leq n - 1\}$. Thus $|FG^{(2)}/K| \leq 2n$ and $D_n \cong FG^{(2)}/K$.

EXERCISES

1. Let S be a subset of a group G such that $g^{-1}Sg \subset S$ for any $g \in G$. Show that the subgroup $\langle S \rangle$ generated by S is normal. Let T be any subset of G and let $S = \bigcup_{g \in G} g^{-1}Tg$. Show that $\langle S \rangle$ is the normal subgroup generated by T.

2. Let G be the group defined by the following relations in $FG^{(3)}$: $x_2x_1 = x_3x_1x_2$, $x_3x_1 = x_1x_3$, $x_3x_2 = x_2x_3$. Show that G is isomorphic to the group defined in exercise 2, p. 62.

The following three exercises are taken from Burnside's *The Theory of Groups of Finite Order*, 2nd ed., 1911. (Dover reprint, pp. 464–465.)

3. Using the generators (12), $(13), \ldots, (1n)$ (see exercise 5, p. 51) for S_n, show that S_n is defined by the following relations on $x_1, x_2, \ldots, x_{n-1}$ in $FG^{(n-1)}$:

$$x_i^2, (x_i x_j)^3, (x_i x_j x_i x_k)^2, i, j, k \neq.$$

4. Using the generators (12), $(23), \ldots, (n-1n)$ for S_n show that this group is defined by $x_1, \ldots, x_{n-1}$ subjected to the relations:

$$x_i^2, (x_i x_{i+1})^3, (x_i x_j)^2, j > i + 1.$$

5. Show that A_n can be defined by the following relations on $x_1, x_2, \ldots, x_{n-2}$:

$$x_1^3; x_i^2, i > 1; (x_i x_{i+1})^3; (x_i x_j)^2, j > i + 1.$$

1.12   GROUPS ACTING ON SETS

Historically, the theory of groups dealt at first only with transformation groups. The concept of an abstract group was introduced later in order to focus attention on those properties of transformation groups that concern the resultant composition only and do not refer to the set on which the transformations act. However, in geometry one is interested primarily in transformation groups, and even in the abstract theory it often pays to switch back from the abstract point of view to the concrete one of transformation groups. For one thing, the use of transformation groups provides a counting technique that plays an important role in the theory of finite groups. We have already seen one instance of this in the proof of Lagrange's theorem. We shall see other striking examples of results obtained by counting arguments in this section and the next.

It is useful to have a vehicle for passing from the abstract point of view to the concrete one of transformations. This is provided by the concept of a group acting on a set which we proceed to define. The idea is a simple one. We begin with an abstract group G and we are interested in the various "realizations" of G by groups of transformations. At first one is tempted to consider only those realizations which are "faithful" in the sense that they are isomorphisms of G with groups of transformations. Experience soon shows that it is preferable to broaden the outlook to encompass also homomorphisms of G into transformation groups.

We now consider a group G and a homomorphism T of G into Sym S, the group of bijective transformations of a set S. Writing the transformation corresponding to $g \in G$ as $T(g)$, the conditions on T are:

1. $T(1) = 1$ $(= 1_S$,   the identity map of S).
2. $T(g_1 g_2) = T(g_1) T(g_2)$, $g_i \in G$.

The first of these can be omitted if we assume, as we are doing, that every $T(g)$ is bijective. On the other hand, if we retain condition 1, then the hypothesis that $T(g)$ is bijective is redundant. For, if T is a map of the group G into the monoid

$M(S)$ of transformations of S satisfying both conditions, then T is a homo-morphism of G into $M(S)$. Hence the image of G is a subgroup of $M(S)$ and so this is contained in Sym S. It is useful to regard the image $T(g)x$ of x under the transformation $T(g)$ corresponding to g as simply a product gx of the element $g \in G$ with the element $x \in S$. Thus we obtain a map

$$(g, x) \to gx \qquad (\equiv T(g)x)$$

of $G \times S$ into S. What are its properties? Clearly, conditions 1 and 2 imply respectively:

(i) $$\qquad\qquad\qquad\qquad\qquad 1x = x, \qquad x \in S$$

(ii) $$\qquad\qquad\qquad\qquad\qquad (g_1 g_2)x = g_1(g_2 x).$$

We shall now reverse the order and put the following

DEFINITION 1.7.   *A group G is said to* act *(or* operate*) on the set S if there exists a map $(g, x) \to gx$ of $G \times S$ into S satisfying* (i) *and* (ii).

We have seen that a homomorphism T of G into $M(S)$ defines an action of G on S simply by putting

$$gx = T(g)x.$$

Conversely, suppose G acts on S. Then we define $T(g)$ to be the map $x \to gx$, $x \in S$. Then (i) and (ii) imply 1 and 2 so $T: g \to T(g)$ is a homomorphism of G into Sym S.

We shall refer to T as *the homomorphism* associated with the action and to $T(G)$ as the *associated transformation group*. If T is a monomorphism then we shall say that *G acts effectively on the set S*. Also the kernel of T will be called the *kernel of the action*. Thus G acts effectively if and only if the kernel of the action is 1.

EXAMPLES

1. Let $S = G$, the underlying set of the group G. Define gx for $g \in G$ and $x \in S$ to be the product in G of g and x. Then (i) and (ii) are clear. This action is called *the action of G on itself by left translations* (or *left multiplications*). This is the action which was used to prove Cayley's theorem. The point of the proof of that theorem was that this action is effective.

2. Next we define an action of G on itself by right translations. Again we take the set S to be the set G. In order to avoid confusion with the group product gx we now

denote the action of $g \in G$ on $x \in S$ by $g \circ x$ and we define this to be xg^{-1}. Then we have $1 \circ x = x1 = x$ and $(g_1g_2) \circ x = x(g_1g_2)^{-1} = xg_2^{-1}g_1^{-1} = g_1 \circ (g_2 \circ x)$. Hence we do indeed have an action of G on itself. We call this action G the *action by right translations*. This is effective.

3. Another action of G on itself is the *action by conjugations*. This time we denote the action of $g \in G$ on $x \in S$ $(= G)$ by ^{g}x which we define to be gxg^{-1}. Then $^{1}x = x$ and $^{g_1g_2}x = (g_1g_2)x(g_1g_2)^{-1} = g_1(g_2xg_2^{-1})g_1^{-1} = {}^{g_1}({}^{g_2}x)$. The kernel of this action is the set of c such that $^{c}x = x$ for all x. This means $cxc^{-1} = x$ or $cx = xc$. Hence the kernel is the center C and the action is effective if and only if the center is trivial ($C = 1$).

4. If we have an action of G on a set S we have an action of any subgroup H of G on S by restriction. In particular, we have the actions of H on G by left and by right translations.

5. Let H be a subgroup and let G/H denote the set of left cosets xH, $x \in G$. We used this notation previously only when H was normal in G and G/H denoted the factor group. We shall call G/H the *(left) coset space of G relative to H*. If $g \in G$ we take $g(xH)$ to be the set product of $\{g\}$ with xH, so $g(xH) = gxH$. It is clear that this defines an action of G on G/H. The kernel of this action is the set of g such that $gxH = xH$ for all $x \in G$, which is equivalent to $x^{-1}gx \in H$ for all x. This is equivalent to $g \in xHx^{-1}$ for all x or $g \in \bigcap_{x \in G} xHx^{-1}$. We see easily that the right-hand side is the largest normal subgroup of G contained in H. Hence the action of G on G/H is effective if and only if H contains no subgroup $\neq 1$ which is normal in G.

6. As in 5 we obtain an action of G on the set $G \backslash H$ of right cosets Hx by $g \circ (Hx) = (Hx)g^{-1} = Hxg^{-1}$.

7. Suppose we have an action of G on a set S and T is a subset stabilized by the action in the sense that $gT \subset T$ for every $g \in G$. Then restricting the action to T gives an action of G on T. For example, consider the action of G on itself by conjugation. If K is a normal subgroup of G then $^{g}K = K$, $g \in G$, so we have an action of G on K by restricting the conjugation action to K.

8. If G acts on a set S, then we have an induced action on the power set $\mathscr{P}(S)$. Here, if A is a non-vacuous subset we define $gA = \{gx \mid x \in A\}$ and if $A = \varnothing$ we put $g\varnothing = \varnothing$. Then $1A = A$ and $(g_1g_2)A = g_1(g_2A)$, so we have defined an action of G on $\mathscr{P}(S)$. It is clear that $|gA| = |A|$. Hence we have induced actions also on the subsets of S of a fixed cardinality.

There is a natural definition of equivalence of actions of a fixed group G: we say that two actions of G on S and S' respectively are *equivalent* if there exists a bijective map $x \to x'$ of S onto S' such that

$$(35) \qquad (gx)' = gx', \qquad g \in G, x \in S.$$

If we denote $x \to x'$ by α and the transformations $x \to gx$ and $x' \to gx'$ by $T(g)$ and $T'(g)$ respectively, then (35) means the same thing as

$$(36) \qquad \alpha T(g) = T'(g)\alpha, \qquad g \in G.$$

In other words, for every $g \in G$ we have the commutativity of the diagram

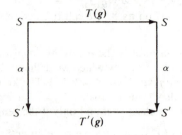

Since α is bijective (36) can be written also as

$$(36') \qquad\qquad T'(g) = \alpha T(g)\alpha^{-1}, \qquad g \in G.$$

As an example of equivalence we consider the two actions of G on itself by left and by right translations. Here the map $x \to x^{-1}$ is an equivalence since $(gx)^{-1} = x^{-1}g^{-1} = g \circ x^{-1}$.

The equivalence relation on a set S defined by a transformation group of S carries over to actions. If G acts on S we define $x \sim_G y$ for x, $y \in S$ to mean that $y = gx$ for some $g \in G$. Evidently this means the same thing as equivalence relative to the transformation group $T(G)$, as we defined it before. As before we obtain a partition of S into orbits, where the *G-orbit* of x is $Gx = \{gx \,|\, g \in G\}$. We denote the quotient set consisting of these orbits by S/G.

If H is a subgroup of G then the H-orbits of the action of H on G by left (right) translations are the right (left) cosets of H. Now let G act on itself by conjugations. In this case the orbit of $x \in G$ is $^Gx = \{gxg^{-1} \,|\, g \in G\}$. This is called the *conjugacy class* of the element x. Of course, we have a partition of G into the distinct conjugacy classes. It is worth noting that ^Gx consists of a single element, $^Gx = \{x\}$, if and only if x is in the center. Thus the center is the union of the set of conjugacy classes which consist of single elements of G.

As an example of a decomposition into conjugacy classes we consider the problem of determining this decomposition for S_n. We have noted before (exercise 4, p. 51) that if $\beta \in S_n$ then $\beta(i_1 i_2 \cdots i_r)\beta^{-1} = (\beta(i_1), \beta(i_2), \ldots, \beta(i_r))$. It follows that if α is a product of cycles $\gamma_1, \gamma_2, \ldots$ as in (17) then $\beta\alpha\beta^{-1} = (\beta\gamma_1\beta^{-1})(\beta\gamma_2\beta^{-1})\cdots$. Hence if $\alpha = (i_1 \cdots i_r)\cdots(l_1 \cdots l_u)$ then

$$(37) \qquad\qquad \beta\alpha\beta^{-1} = (\beta(i_1), \ldots, \beta(i_r))\cdots(\beta(l_1), \ldots, \beta(l_u)).$$

It is convenient to assume that $r \geq s \geq \cdots \geq u$ and that the decomposition into disjoint cycles displays every number in $\{1, 2, \ldots, n\}$ once and only once. In this way we can associate with α a set of positive integers $(r, s, \ldots, u)$ satisfying

$$(38) \qquad\qquad r \geq s \geq \cdots \geq u, \qquad r + s + \cdots + u = n.$$

We call such a sequence $(r, s, \ldots, u)$ a *partition of n*. It is clear from (37) that two permutations are conjugate if and only if they determine the same partition. It follows that the conjugacy classes are in $1-1$ correspondence with the different partitions of n. Hence if $p(n)$ denotes the number of distinct partitions of n, then there are $p(n)$ conjugacy classes in S_n. The function of positive integers $p(n)$ is an interesting arithmetic function. Its first few values are

$$p(2) = 2, \ p(3) = 3, \ p(4) = 5, \ p(5) = 7, \ p(6) = 11.$$

If there is just one orbit in the action of a group G on a set S, that is, if $S = Gx$ for some $x \in S$ (and hence for every $x \in S$), then we say that G *acts transitively* on S. It is clear that the actions of G on itself by translations are transitive. More generally, if H is a subgroup the action of G on the coset space G/H (set of left cosets) is transitive, since for any xH and yH we have $gxH = yH$ for $g = yx^{-1}$. We are now going to show that in essence these are the only transitive actions of a group G. To see this we need to introduce the *stabilizer*, Stab x, of an element $x \in S$, which we define to be the set of elements $g \in G$ such that $gx = x$. It is clear that this is a subgroup of G. For example, in the action of G on G by conjugation, Stab $x = C(x)$, the centralizer of x in G. If $y = ax$ then $gy = y$ is equivalent to $gax = ax$ and to $(a^{-1}ga)x = x$. Hence Stab $x = a^{-1}$ (Stab $y)a$. It follows that if G acts transitively on S then all stabilizers of elements of S are conjugate: Stab $y = a(\text{Stab } x)a^{-1}$.

We shall now prove the following result, which gives an internal characterization of transitive actions.

THEOREM 1.10.   *Let G act transitively on S and let $H = \text{Stab } x$ for $x \in S$. Then the action of G on S is equivalent to the action of G on the coset space G/H.*

Proof.   Consider the map $\alpha: g \to gx$ of G into S. This is surjective since G is transitive on S. Hence we have an induced bijective map $\bar{\alpha}$ of the quotient set $\bar{G}$ of G defined by α. We recall that $\bar{G}$ is the set of equivalence classes in G defined by $\bar{g} = \{a | \alpha(a) = \alpha(g)\} = \{a | ax = gx\}$. Now $ax = gx$ is equivalent to $g^{-1}ax = x$, that is, to $g^{-1}a \in \text{Stab } x$. Hence $\bar{g}$ is the coset $g(\text{Stab } x)$ of Stab x and so we have the bijective map $\bar{\alpha}: g(\text{Stab } x) \to gx$. It remains to see that this is an equivalence of actions. This requires verifying that if $g' \in G$ then $g'(g \text{ Stab } x) \to g'(gx)$ by $\bar{\alpha}$. This is clear since these are respectively $(g'g)\text{Stab } x$ and $(g'g)x$. $\square$

From the point of view of finite groups one of the most important conclusions that can be drawn from the preceding theorem is that if G is a finite group acting transitively on a set S then $|S| = [G:\text{Stab } x]$ for any $x \in S$. This shows that $|S|$ is finite and this number is a divisor of $|G|$. More generally, we can apply this to any action of a finite group G on a finite set S. We have the partition

$$(39) \qquad\qquad S = O_1 \cup O_2 \cup \cdots \cup O_r$$

where the O_i are the different orbits of elements of S under the action of G. Then G acts transitively in O_i, so if $x_i \in O_i$ then $|O_i| = [G:\text{Stab } x_i]$. Hence we have the following enumeration of the elements of S,

$$(40) \qquad\qquad |S| = \sum [G:\text{Stab } x_i],$$

where the summation is taken over a set $\{x_1, x_2, \ldots, x_r\}$ of representatives of the orbits. It is important to take note that all the terms $[G:\text{Stab } x_i]$ on the right-hand side are divisors of $|G|$. Another useful remark that is applicable to any group is

$$(41) \qquad\qquad \text{Stab } axa^{-1} = a(\text{Stab } x)a^{-1}$$

The proof is clear.

An important special case of (40) is obtained by letting G act on itself by conjugations. Then (40) specializes to

$$(42) \qquad\qquad |G| = \sum [G:C(x_i)]$$

where $C(x_i)$ is the centralizer of x_i and $\{x_i\}$ is a set of representatives of the conjugacy classes of G. This formula is called the *class equation of the finite group* G. We can modify the formula slightly by collecting the classes consisting of the x_i such that $C(x_i) = G$. These are just the elements of the center C of G, and their classes contain a single element. Hence we have

$$(42') \qquad\qquad |G| = |C| + \sum [G:C(y_i)]$$

where y_j runs through a set of representatives of the conjugacy classes which contain more than one element.

The type of counting of elements of a finite group given in (40) and (42) is an important tool in the study of finite groups. Some instances of this will be encountered in the next section when we consider the Sylow theorems. At this point we illustrate the method by using the class equation to prove

THEOREM 1.11. *Any finite group G of prime power order has a center $C \neq 1$.*

Proof. The left hand side of (42') is divisible by the prime p and every term on the right-hand side is a power of p. Moreover, since $C(y_j) \neq G$, $[G:C(y_j)] > 1$, so $[G:C(y_j)]$ is divisible by p. Then (41') shows that $|C|$ is divisible by p and so $C \neq 1$. $\square$

There is a useful distinction we can make for transitive actions called primitivity and imprimitivity. This has to do with the induced action on the power set $\mathscr{P}(S)$. We shall say that a partition $\pi(S)$ of S is *stabilized* by the action of G on S if $gA \in \pi(S)$ for every $g \in G$ and $A \in \pi(S)$. There are two partitions which trivially have this property: $\pi_1(S) = \{S\}$ and $\pi_0(S)$ consisting of the set of subsets $\{x\}$, $x \in S$. Now we shall call the action *primitive* if π_1 and π_0 are the only partitions of S stabilized by G. We have the partition of S into the orbits relative to G and this partition is stabilized by G since $gA = A$ for every orbit A and every $g \in G$. If the orbits consist of single points, then G acts trivially in the sense that $gx = x$, $g \in G$, $x \in S$; if there is just one orbit then G is transitive. Hence if we have a non-trivial and intransitive action of G on S then this action is imprimitive. The interesting situation is that in which G acts transitively on a set with more than one element. In this case we have the following criterion.

THEOREM 1.12. *If G acts transitively on a set S with $|S| > 1$, then G acts primitively if and only if the stabilizer, Stab x, of any $x \in S$ is a maximal subgroup of G, that is, there exists no subgroup H such that Stab $x \subsetneq H \subsetneq G$.*

Proof. We observe first that G acts imprimitively on a set S if and only if there exists a proper subset A of S with $|A| \geq 2$ such that for any $g \in G$ either $gA = A$ or $gA \cap A = \emptyset$. If this condition holds, then for any $g_1, g_2 \in G$ we have either $g_1 A = g_2 A$ or $g_1 A \cap g_2 A = \emptyset$. Let B be the complement in S of $\bigcup_{g \in G} gA$. Then $g_1 B \cap g_2 A = \emptyset$ for every $g_1, g_2 \in G$, which implies that $gB = B$ for every $g \in G$. It follows that the set of (distinct) subsets gA, $g \in G$, together with B constitute a non-trivial partition of S which is stabilized by G. Conversely, suppose G acts imprimitively on S so that we have a partition $\pi(S)$ that contains a proper subset A with $|A| \geq 2$ such that $\pi(S)$ is stabilized by G. Then if $g \in G$ either $gA = A$ or $gA \cap A = \emptyset$.

Now suppose Stab x for some $x \in S$ is not maximal, and let H be a subgroup such that Stab $x \subsetneq H \subsetneq G$. Since we are assuming that G acts transitively on S, this action is equivalent to the usual one on the coset space $G/\text{Stab } x$. Since equivalent actions are either both primitive or both imprimitive, it suffices to show that the action of G on $G/\text{Stab } x$ is imprimitive. Now consider the set A of cosets of the form h Stab x, $h \in H$. Since Stab $x \subsetneq H \subsetneq G$ we have $|A| \geq 2$ and A is a proper subset of $G/\text{Stab } x$. If $h' \in H$ then $h' A$ is the set of cosets $h'h$ Stab x, $h \in H$, and so $h'A = A$. On the other hand, if $g \notin H$, then gh_1 Stab $x \neq h_2$ Stab x for every $h_1, h_2 \in H$. Otherwise, we have $gh_1 k_1 = h_2 k_2$, where $h_1, h_2 \in H$, k_1,

$k_2 \in$ Stab x. This implies that $g = h_2 k_2 k_1{}^{-1} h_1{}^{-1} \in H$, contrary to our hypothesis. We now see that gA, which is the set of cosets of the form gh Stab x, $h \in H$, has vacuous intersection with A if $g \notin H$. Thus $gA \cap A = \varnothing$ in this case. It follows as above that G acts imprimitively on $G/$Stab x, hence on S.

Next assume that G is transitive but not primitive on S. Then we have a subset A of S, $A \neq S$, $|A| \geq 2$, such that for any $g \in G$, either $gA = A$ or $gA \cap A = \varnothing$. Let $x \in A$ and let $H = \{h \in G \,|\, hA = A\}$. Then H is a subgroup of G and $H \supset$ Stab x since $gx = x \Rightarrow gA \cap A \neq \varnothing \Rightarrow gA = A$. Since $A \neq S$ and G is transitive on S, there exists a $g \in G$ such that $gx \notin A$. Then $gA \neq A$ and $g \notin H$. Hence $G \neq H$. Now let $y \in A$, $y \neq x$ (existence clear since $|A| \geq 2$). Then we have a $g \in G$ such that $gx = y$. Then $(gA \cap A) \ni y$ and, consequently, $gA = A$ but $gx \neq x$. Thus $g \in H$, $\notin$ Stab x, and so $H \neq$ Stab x. Hence Stab x is not a maximal subgroup of G. This completes the proof. $\square$

EXERCISES

1. Let $\gamma = (12 \cdots n)$ in S_n. Show that the conjugacy class of γ in S_n has cardinality $(n-1)!$. Show that the centralizer $C(\gamma) = \langle \gamma \rangle$.

2. Determine representatives of the conjugacy classes in S_5 and the number of elements in each class. Use this information to prove that the only normal subgroups of S_5 are 1, A_5, S_5.

3. Let the partition associated with a conjugacy class be $(n_1, n_2, \ldots, n_q)$ where

    $$n_1 = \cdots = n_{q_1} > n_{q_1 + 1} = \cdots = n_{q_1 + q_2} > n_{q_1 + q_2 + 1} = \cdots .$$

    Show that the number of elements in this conjugacy class is

    $$\frac{n!}{\prod q_i! \prod n_j}.$$

4. Show that if a finite group G has a subgroup H of index n then H contains a normal subgroup of G of index a divisor of $n!$. (*Hint:* Consider the action of G on G/H by left translations.)

5. Let p be the smallest prime dividing the order of a finite group. Show that any subgroup H of G of index p is normal.

6. Show that every group of order p^2, p a prime, is abelian. Show that up to isomorphism there are only two such groups.

7. Let H be a proper subgroup of a finite group G. Show that $G \neq \bigcup_{g \in G} gHg^{-1}$.

8. Let G act on S, H act on T, and assume $S \cap T = \varnothing$. Let $U = S \cup T$ and define
    for $g \in G$, $h \in H$, $s \in S$, $t \in T$; $(g, h)s = gs$, $(g, h)t = ht$. Show that this defines an
    action of $G \times H$ on U.

9. A group H is said to *act on a group* K *by automorphisms* if we have an action
    of H on K and for every $h \in H$ the map $k \to hk$ of K is an automorphism. Suppose
    this is the case and let G be the product set $K \times H$. Define a binary composition
    in $K \times H$ by

    $$(k_1, h_1)(k_2, h_2) = (k_1(h_1 k_2), h_1 h_2)$$

    and define $1 = (1, 1)$—the units of K and H respectively. Verify that this defines
    a group such that $h \to (1, h)$ is a monomorphism of H into $K \times H$ and $k \to (k, 1)$
    is a monomorphism of K into $K \times H$ whose image is a normal subgroup. G is
    called a *semi-direct product of* K *and* H. Note that if H and K are finite then
    $|K \times H| = |K||H|$.

10. Let G be a group, H a transformation group acting on a set S and let G^S denote
    the set of maps of S into G. Then G^S is a group (the S-direct power of G) if we
    define $(f_1 f_2)(s) = f_1(s)f_2(s), f_i \in G^S, s \in S$. If $h \in H$ and $f \in G^S$ define hf by $(hf)(s) = f(h^{-1}s)$. Verify that this defines an action of H on G^S by automorphism. The
    semi-direct product of H and G^S is called the *(unrestricted) wreath product* $G \wr H$
    of G with H.

11. Let G, H, S be as in exercise 10 and suppose G acts on a set T. Let $(f, h) \in G \wr H$
    where f is a map of S into G. If (f_1, h_1), (f_2, h_2) are two such elements, the product
    in $G \wr H$ is $(f_1(h_1 f_2), h_1 h_2)$. If $(t, s) \in T \times S$ define $(f, h)(t, s) = (f(s)t, hs)$. Verify that
    this defines an action of $G \wr H$ on $T \times S$. Note that if everything is finite then
    $|G \wr H| = |G|^{|S|}|H|$ and the *degree of the action*, defined to be the cardinality of the
    set on which the action takes place, is the product of the degrees of the actions
    of H and of G.

12. Let G act on S. Then the action is called *k-fold transitive* for $k = 1, 2, 3, \ldots$, if
    given any two elements $(x_1, \ldots, x_k), (y_1, \ldots, y_k)$ in $S^{(k)}$, where the x_i and the y_i are
    distinct, there exists a $g \in G$ such that $gx_i = y_i$, $1 \le i \le k$. Show that if the action
    of G is doubly transitive then it is primitive.

13. Show that if the action of G on S is primitive and effective then the induced action
    on S by any normal subgroup $H \neq 1$ of G is transitive.

1.13 SYLOW'S THEOREMS

We have seen that the order of a subgroup of a finite group G is a factor of
$|G|$ and if G is cyclic, there is one and only one subgroup of order any given
divisor of $|G|$. A natural question is: If k divides $|G|$ is there always a subgroup
of G of order k? A little experimenting shows that this is not so. For example,
the alternating group A_4, whose order is 12, contains no subgroup of order 6.
Moreover, we shall show later (in Chapter 4) that A_n for $n \ge 5$ is simple, that

is, contains no normal subgroup $\neq 1$, A_n. Since any subgroup of index two is normal, it follows that A_n, $n \geq 5$, contains no subgroup of order $n!/4$. The main positive result of the type we are discussing was discovered by Sylow. This states that if a prime power p^k divides the order of a finite group G, then G contains a subgroup of order p^k. Sylow also proved a number of other important results on the subgroups of order p^m where p^m is the highest power of p dividing $|G|$. We shall now consider these results.

We prove first

SYLOW I. *If p is a prime and p^k, $k \geq 0$, divides $|G|$ (assumed finite), then G contains a subgroup of order p^k.*

Proof. We shall prove the result by induction on $|G|$. It is clear if $|G| = 1$, and we may assume it holds for every group of order $< |G|$. We first prove a special case of the theorem (which goes back to Cauchy): if G is finite abelian and p is a prime divisor of $|G|$ then G contains an element of order p. To prove this we take an element $a \neq 1$ in G. If the order r of a is divisible by p, say $r = pr'$, then $b = a^{r'}$ has order p. On the other hand, if the order r of a is prime to p, then the order $|G|/r$ of $G/\langle a \rangle$ is divisible by p and is less than $|G|$. Hence this factor group contains an element $b\langle a \rangle$ of order p. We claim that the order s of b is divisible by p, for we have $(b\langle a \rangle)^s = b^s\langle a \rangle = 1(=\langle a \rangle)$. Hence the order p of $b\langle a \rangle$ is a divisor of s. Now, since b has order divisible by p, we obtain an element of order p as before. After this preliminary result we can quickly give the proof. We consider the class equation (41'): $|G| = |C| + \sum [G:C(y_j)]$. If $p \nmid |C|$ then $p \nmid [G:C(y_j)]$ for some j. Then $p^k \mid |C(y_j)|$ and the subgroup $C(y_j)$ has order $< |G|$ since y_j is not in the center. Then, by the induction hypothesis, $C(y_j)$ contains a subgroup of order p^k. Next suppose $p \mid |C|$. Then, by Cauchy's result, C contains an element c of order p. Now $\langle c \rangle$ is a normal subgroup of G of order p, and the order $|G|/p$ of $G/\langle c \rangle$ is divisible by p^{k-1}. Hence, by induction, $G/\langle c \rangle$ contains a subgroup of order p^{k-1}. This subgroup has the form $H/\langle c \rangle$ where H is a subgroup of G containing $\langle c \rangle$. Then

$$|H| = [H:\langle c \rangle]|\langle c \rangle| = p^{k-1}p = p^k. \quad \square$$

Let p^m be the largest power of p dividing $|G|$. Then Sylow I proves the existence of subgroups of order p^m of G. Such subgroups are called *Sylow p-subgroups* of G. The next Sylow theorem concerns these.

SYLOW II. (1) *Any two Sylow p-subgroups of G are conjugate in G; that is, if P_1 and P_2 are Sylow p-subgroups, then there exists an $a \in G$ such that $P_2 = aP_1a^{-1}$.* (2) *The number of Sylow p-subgroups is a divisor of the index* *of any Sylow p-subgroup and is* $\equiv 1$ (mod p). (3) *Any subgroup of order* p^k *is contained in a Sylow subgroup.*

We shall obtain the proof by considering the action of G on the set Π of Sylow p-subgroups by conjugation. More generally, we note that if H is a subgroup of a group G and $g \in G$ then gHg^{-1} is a subgroup. It follows that we have an action of G on the set Γ of subgroups of G by conjugation: $^gH = gHg^{-1}$. The stabilizer of H under this action is the subgroup $N(H)$ (or $N_G(H)$) $= \{g \in G \,|\, gHg^{-1} = H\}$. This is called the *normalizer of* H *in* G. Evidently $H \subset N(H)$ and hence H is a normal subgroup of $N(H)$. The orbit of H under the conjugation action of G is $\{gHg^{-1} \,|\, g \in G\}$. The counting formula on p. 74 shows that $|\{gHg^{-1} \,|\, g \in G\}| = [G:N(H)]$. If G is finite then $[G:N(H)] \,|\, [G:H]$ since $G \supset N(H) \supset H$ and hence $[G:H] = [G:N(H)][N(H):H]$.

Now let G be finite and let Π denote the set of Sylow p-subgroups of G. If $P \in \Pi$ then $gPg^{-1} \in \Pi$, so we have an action of G on Π induced by the conjugation action on Γ. We shall require the following

LEMMA. *Let P be a Sylow p-subgroup of G, H a subgroup of order p^j contained in $N(P)$. Then $H \subset P$.*

Proof. Since H is a subgroup of $N(P)$ and P is a normal subgroup of $N(P)$, HP is a subgroup and $HP/P \cong H/(H \cap P)$ (by the first isomorphism theorem, p. 64). Thus HP/P is isomorphic to a factor group of H and so it has order p^k. Then $|HP| = p^k|P|$. Since P is a Sylow p-subgroup, $k = 0$, $HP = P$ and so $H \subset P$. $\square$

Evidently P is a Sylow p-subgroup of $N(P)$. Moreover, it is clear from the foregoing lemma that P is the only Sylow p-subgroup of $N(P)$.

We are now ready to give the

Proof of Sylow II. Let Π be the set of Sylow p-subgroups and let G act on Π by conjugation. Let Σ be one of the orbits under this action. Now let $P \in \Sigma$ and restrict the action of G on Σ to an action of P on Σ. Then we have a decomposition of Σ into P-orbits, one of which is $\{P\}$. Moreover, $\{P\}$ is the only P-orbit in Σ of cardinality one. For, if $\{P'\}$ is such a P-orbit then $P \subset N(P')$, so $P = P'$ since P' is the only Sylow p-subgroup of $N(P')$. Now every P-orbit has cardinality a power of p since this cardinality is a divisor of $|P|$. Hence $|\Sigma| \equiv 1$ (mod p). We show next that $\Sigma = \Pi$. Otherwise, we have a $P \in \Pi$, $\notin \Sigma$. Applying the foregoing argument to this P we see that there are no P-orbits in Σ of cardinality one. This gives $|\Sigma| \equiv 0 \pmod{p}$ contrary to $|\Sigma| \equiv 1 \pmod{p}$. Hence $\Sigma = \Pi$, which means G acts transitively on Π. Hence (1) is proved. We also have $|\Pi| \equiv 1 \pmod{p}$, which is the second assertion in (2). The first is clear also, since $|\Pi| = [G{:}N(P)]$. Now let H be a subgroup of G of order p^k and restrict the action of G on Π to H. Since the H-orbits have cardinality a power of p and since $|\Pi| \equiv 1 \pmod{p}$, there exists an orbit $\{P\}$ containing one element. Then $H \subset N(P)$ and so $H \subset P$, by the lemma. This proves (3). $\qquad\square$

EXERCISES

1. Show that if P is a Sylow subgroup then $N(N(P)) = N(P)$.

2. Show that there are no simple groups of order 148 or of order 56.

3. Show that there is no simple group of order pq, p, and q primes (cf. exercise 5, p. 77).

4. Show that every non-abelian group of order 6 is isomorphic to S_3.

5. Determine the number of non-isomorphic groups of order 15.

An element of order 2 in a group is called an *involution*. An important insight into the structure of a finite group is obtained by studying its involutions and their centralizers. The next five exercises give a program for characterizing S_5 in this way. These were communicated to me by Walter Feit who attributes the first four to Richard Brauer—though he notes that John Thompson first recognized the importance of the result in 9. In all of these exercises, as well as in the rest of this set, G is a finite group.

6. Let u and v be distinct involutions in G. Show that $\langle u, v \rangle$ is (isomorphic to) a dihedral group.

7. Let u and v be involutions in G. Show that if uv is of odd order then u and v are conjugate in G $(v = gug^{-1})$.

8. Let u and v be involutions in G such that uv has even order $2n$, so $w = (uv)^n$ is an involution. Show that $u, v \in C(w)$.

9. Suppose G contains exactly two conjugacy classes of involutions. Let u_1 and u_2 be non-conjugate involutions in G. Let $c_i = |C(u_i)|$, $i = 1, 2$. Let S_i, $i = 1, 2$, be the set of ordered pairs (x, y) with x conjugate to u_1, y conjugate to u_2, and $(xy)^n = u_i$ for some n. Let $s_i = |S_i|$. Prove that $|G| = c_1 s_2 + c_2 s_1$. (*Hint:* Count the number of ordered pairs (x, y) with x conjugate to u_1 and y conjugate to u_2 in two ways. First, this number is $(|G|/c_1)(|G|/c_2)$. Since x is not conjugate to y, exercises 7 and 8 imply that for $n = o(xy)/2$, $(xy)^n$ is conjugate to either u_1 or u_2. This implies that $(|G|/c_1)(|G|/c_2) = (|G|/c_1)s_1 + (|G|/c_2)s_2$.)

10. (An abstract characterization of S_5.) Let G contain exactly two conjugacy classes of involutions and let u_1 and u_2 be representatives of these classes. Suppose $C_1 = C(u_1) \cong \langle u_1 \rangle \times S_3$ and $C_2 = C(u_2)$ is a dihedral group of order 8. Then $G \cong S_5$.

Sketch of proof.

(i) Since some involution is in the center of a Sylow subgroup, C_2 is a Sylow 2-subgroup.

(ii) Replacing u_1 by a conjugate, one may assume $u_1 \in C_2$; and then $u_2 \in C_1$.

(iii) C_2 contains three classes of involutions. If x is an involution in C_2, $x \neq u_2$ then x is conjugate to xu_2. Since G contains two classes of involutions, deduce that either $s_2 = 0$ or $s_2 = 4$ and C_2 contains a non-cyclic group V of order 4 such that all involutions in V are conjugate to u_2 in G.

(iv) C_1 contains three conjugacy classes of involutions. If x is an involution in C_1, $x \neq u_1$, then x is not conjugate to xu_1 in C_1. Since G contains two classes of involutions (iii) implies that for any involution x in C_1, $x \neq u_1$, exactly one of x and xu_1 is conjugate to u_1. Hence deduce that $s_1 = 9$ (in the notation of exercise 9).

(v) Use exercise 10 to show that either $s_2 = 4$, $|G| = 120$ or $s_2 = 0$, $|G| = 72$.

(vi) Show that $|G| \neq 72$ as follows. Let P be a Sylow 3-group of C_1. Assume $|G| = 72$. Let Q be a Sylow subgroup of G containing P. Then $|Q| = 9$ and $\langle C_1, Q \rangle \subset N(P)$. Then $36 \,|\, |N(P)|$. Hence there exists H with $C(P) \subset H$ and $|H| = 36$. This implies that $u_1 \in H$ and since u_2 is a square, $u_2 \in H$. Since $[G:H] = 2$, $H \lhd G$ and so H contains all involutions in G. Then $C_2 \cap H$ contains all involutions in C_2. This is impossible as $|C_2 \cap H| = 4$ and C_2 contains five involutions.

(vii) By (iii), C_2 contains a non-cyclic group V of order 4 such that $u_2 \in V$ and all the involutions in V are conjugate in G. Let x be an element of G such that $x^{-1}u_2x \neq u_2$, $x^{-1}u_2x \in V$. Then $x^{-1}C_2x \neq C_2$ and $u_2 \in C(x^{-1}u_2x) = x^{-1}C_2x$.

(viii) $C(V) = V$. $N(V)$ contains at least two Sylow 2-subgroups of G, by (vii).

(ix) $N(V)/V \cong \text{Aut } V \cong S_3$. Hence $|N(V)| = 24$.

(x) $[G:N(V)] = 5$. Show that G acts effectively on the coset space $G/N(V)$ and hence that $G \cong S_5$.

The next four exercises are designed to prove the following extension of Sylow's first theorem. If p is a prime and $p^k \,|\, |G|$, then the number of subgroups of order p^k is congruent 1 (mod p). The theorem is due to Frobenius. The proof we shall indicate is a very slick one due to P. X. Gallagher (*Archiv der Mathematik*, vol. XXIII (1967), p. 469). It is based on the action of G on the set S of subsets of cardinality p^k. This type of proof of Sylow's theorem has had a curious history. It seems to have been discovered by G. A. Miller more than fifty years ago (*Annals of Math.*, vol. 16 (1915), pp. 169–171). However, it seems to have been totally forgotten until it was rediscovered by H. Wielandt in 1959.

11. Let $|G| = p^k m$ where p is a prime, and let n denote number of subgroups of G of order p^k. Let S be the set of subsets of G of cardinality p^k and let G act on S by left translation. If $A \in S$, let $H_A = \text{Stab } A$. Then H_A acts on A by left translations. Note that the orbits in A under the action of H_A are collections of right cosets. Hence prove that $|H_A| \,|\, p^k$.

12. Let S_0 be the subset of $A \in S$ such that $|H_A| = p^k$, and $\bar{S}_0$ the subset of $B \in S$ such that $|H_B| = p^l$, $l < k$. Note that the orbit of any B under the action of G on S has cardinality divisible by pm and hence prove that

$$|S| \equiv |S_0| \pmod{pm}.$$

13. Let $A \in S_0$ and let $x \in A$. Then $H_A x \subset A$ and since $|H_A| = p^k = |A|$, $H_A x = A$. Thus A is a right coset of H_A, a subgroup of order p^k. Conversely, let H be any subgroup of order p^k, Hx one of its right cosets. Then $H(Hx) = Hx$ so Stab Hx contains H. Then, by exercise 11, Stab $Hx = H$ and so $Hx \in S_0$. Conclude from this that

$$|S_0| = nm$$

where n is the number of subgroups of order p^k.

14. Note that $|S|$ depends only on $|G|$ and p^k, and that by exercises 12 and 13, $n = |S_0|/m \equiv |S|/m \pmod{p}$. Hence the congruence class of $n \pmod{p}$ depends only on $|G|$ and p^k, and not on G. Now look at a cyclic group of order $|G|$. In this case there is exactly one subgroup of order p^k. Hence $n \equiv 1 \pmod{p}$.

The next two exercises are designed to construct a group isomorphic to any Sylow p-subgroup of S_n, p a prime not exceeding n.

15. Show that the order of the Sylow p-subgroup of S_n is $p^{v_p(n!)}$ where

$$v_p(n!) = \left[\frac{n}{p}\right] + \left[\frac{n}{p^2}\right] + \left[\frac{n}{p^3}\right] + \cdots$$

where $[k/l]$ denotes the largest integer $\leq k/l$. Show also that if we write

$$n = a_0 + a_1 p + a_2 p^2 + \cdots + a_k p^k$$

where $0 \leq a_i < p$ (note that this is the representation of n using the base p), then

$$v_p(n!) = \sum_{i=1}^{k} a_i(1 + p + \cdots + p^{i-1}).$$

16. Let Z_p denote the subgroup of S_p generated by the cycle $(12 \cdots p)$. Note that the wreath product $Z_p \wr Z_p$ has order p^{p+1} and is isomorphic to a subgroup of S_{p^2} (exercises 10 and 11, p. 79). Define Z_p^{lr}, $r \geq 1$, inductively by $Z_p^{l1} = Z_p$, $Z_p^{lk+1} = Z_p^{lk} \wr Z_p$. Show that Z_p^{lr} has order $p^{(p^{r-1} + p^{r-2} + \cdots + 1)}$ and is isomorphic to a subgroup of S_{p^r}. Hence show that if $n = a_0 + a_1 p + \cdots + a_k p^k$, $0 \leq a_i < p$, then any Sylow p-subgroup of S_n is isomorphic to

$$Z_p^{l1} \times \cdots \times Z_p^{l1} \times Z_p^{l2} \times \cdots \times Z_p^{l2} \times \cdots \times Z_p^{lk} \times \cdots \times Z_p^{lk}.$$
$$\underbrace{\qquad\qquad}_{a_1} \qquad \underbrace{\qquad\qquad}_{a_2} \qquad\qquad \underbrace{\qquad\qquad}_{a_k}$$

2

Rings

In this chapter we begin the study of a second type of algebraic structure, called a ring. The prototype for these structures is the ring $\mathbb{Z}$ of integers, which in the last chapter we regarded from the monoid point of view as providing the two monoids $(\mathbb{Z}, +, 0)$ and $(\mathbb{Z}, \cdot, 1)$. The ring theoretic way of viewing $\mathbb{Z}$ treats these two structures simultaneously and relates the two by means of the distributive law. Unlike the theory of groups, which had essentially one source— namely, the study of bijective transformations relative to the resultant composition—the theory of rings has been fused out of a number of special theories. For this reason it will appear less orderly and unified than the theory of groups. However, the multitude of examples, including many familiar to the reader, should be convincing evidence of the richness of this branch of algebra. In the next chapter we shall see that rings also arise in a manner analogous to that of transformation groups, namely, as rings of endomorphisms of abelian groups. Moreover, we have the concept of a module, which for rings is the exact analogue of the concept of a group acting on a set.

We begin our discussion with definitions and examples of the various types of rings: domains, division rings, commutative rings, and fields. After this we study the basic notions of ideals, quotient rings, and homomorphisms, which are analogous, respectively, to normal subgroups, factor groups, and homomorphisms for groups. In the second half of the chapter we restrict our attention mainly to commutative rings, first considering constructions and characterizations of certain extensions of these: fields of fractions of commutative domains, polynomial rings in an indeterminate x. After this we consider the elementary factorization theory of commutative domains. Applications, especially to number theory, will be indicated from time to time. The last section, which may be regarded as optional, will be devoted to "rings without unit" and the imbedding of these in "rings," which we consider always as having a unit.

A good deal of this material will seem familiar. However, the student should note that our point of view has some differences from those which he may have encountered before. For example, polynomials are treated formally rather than functionally, and matrices are allowed to have entries in any ring, rather than just in the ring $\mathbb{R}$ of real numbers. Also we emphasize the basic homomorphism properties associated with certain constructions of extensions of a given ring. In important instances these properties give a characterization of the extension and play an important role in what follows.

2.1 DEFINITION AND ELEMENTARY PROPERTIES

DEFINITION 2.1.   *A* ring *is a structure consisting of a non-vacuous set R together with two binary compositions* $+$, $\cdot$ *in R and two distinguished elements* $0, 1 \in R$ *such that*

1. $(R, +, 0)$ *is an abelian group.*
2. $(R, \cdot, 1)$ *is a monoid.*
3. *The distributive laws*

   D
   $$a(b + c) = ab + ac$$
   $$(b + c)a = ba + ca$$

hold for all $a, b, c \in R$.[1]

---

[1] The term "ring" appears to have been used first by A. Fraenkel, who gave a set of axioms for this concept in an article in *Journal für die reine und angewandte Mathematik*, vol. 145 (1914). However, his definition was marred by the inclusion of some ad hoc assumptions that are not appropriate for a general theory. The concept as defined here is due to Emmy Noether, who formulated it in a paper in *Mathematische Annalen*, vol. 83 (1921). Before this the term "Zahlring" had occurred in algebraic number theory.

Thus the assumptions included under 1 and 2 are that $a + b$ and $ab \in R$, and the following conditions hold:

A1 $$(a + b) + c = a + (b + c).$$

A2 $$a + b = b + a.$$

A3 $$a + 0 = a = 0 + a.$$

A4 For each a there is an inverse $-a$ such that $a + (-a) = 0 = -a + a$.

M1 $$(ab)c = a(bc).$$

M2 $$a1 = a = 1a.$$

The structure $(R, +, 0)$ is called the *additive group* of R and $(R, \cdot, 1)$ is called the *multiplicative monoid* of R. A subset S of a ring R is a *subring* if S is a subgroup of the additive group and also a submonoid of the multiplicative monoid of R. Clearly the intersection of any set of subrings of R is a subring. Hence if A is a subset of R one can define the *subring generated by A* to be the intersection of all subrings of R which contain A. This is characterized by the properties: it is a subring, it contains A, and it is contained in every subring containing A.

EXAMPLES

1. $\mathbb{Z}, +, \cdot, 0, 1$ as usual. We noted in the Introduction that this is a ring.

2. $\mathbb{Q}$ the rational numbers with usual $+, \cdot, 0, 1$.

3. $\mathbb{R}$ the ring of real numbers.

4. $\mathbb{C}$ the ring of complex numbers. $\mathbb{R}, \mathbb{Q}$, and $\mathbb{Z}$ are subrings of $\mathbb{C}$.

5. The set $\mathbb{Z}[\sqrt{2}]$ of real numbers of the form $m + n\sqrt{2}$, $m, n \in \mathbb{Z}$. Clearly the difference of two numbers in $\mathbb{Z}[\sqrt{2}]$ is in $\mathbb{Z}[\sqrt{2}]$. Also $1 \in \mathbb{Z}[\sqrt{2}]$ and if $m, n, m', n' \in \mathbb{Z}$ then $(m + n\sqrt{2})(m' + n'\sqrt{2}) = (mm' + 2nn') + (mn' + nm')\sqrt{2} \in \mathbb{Z}[\sqrt{2}]$. Hence $\mathbb{Z}[\sqrt{2}]$ is a subring of $\mathbb{R}$.

6. Same as (5) with $\mathbb{Z}$ replaced by $\mathbb{Q}$. The same calculations show that this is a subring of $\mathbb{R}$.

7. Similarly, we check that $\mathbb{Z}[\sqrt{-1}]$ and $\mathbb{Q}[\sqrt{-1}]$—the sets of complex numbers $m + n\sqrt{-1}$, where, in the first case $m, n \in \mathbb{Z}$, and in the second $m, n \in \mathbb{Q}$—are subrings of $\mathbb{C}$. These are the subrings generated by $\mathbb{Z}$ and $\sqrt{-1}$, and by $\mathbb{Q}$ and $\sqrt{-1}$, respectively. The first of these is called the *ring of Gaussian integers*.

8. The set Γ of real-valued continuous functions on the interval $[0, 1]$ where we define $f + g$ and fg as usual by $(f + g)(x) = f(x) + g(x)$, $(fg)(x) = f(x)g(x)$. Let 0 and 1 be the constant functions 0 and 1, respectively. Then $(\Gamma, +, \cdot, 0, 1)$ is a ring.

9. The set $\{0, 1, 2\}$ with the indicated 0 and 1, and with addition and multiplication defined by the tables:

+	0	1	2		·	0	1	2
0	0	1	2		0	0	0	0
1	1	2	0		1	0	1	2
2	2	0	1		2	0	2	1

is a ring. This can be verified directly. It will be clear without such direct verification soon (perhaps it is already).

A number of elementary properties of rings are consequences of the fact that a ring is an abelian group relative to addition and a monoid relative to multiplication. For example, we have $-(a + b) = -a - b \equiv -a + (-b)$ and if na is defined for $n \in \mathbb{Z}$ as before, then the rules for multiples (or powers) in an abelian group,

$$n(a + b) = na + nb$$

$$(n + m)a = na + ma$$

$$(nm)a = n(ma)$$

hold. We also have the generalized associative laws for addition and multiplication and the generalized commutative law for addition (see pp. 40 and 41). There are also a number of simple consequences of the distributive laws which we now note. In the first place, induction on m and n gives the generalization

$$(a_1 + a_2 + \cdots + a_m)(b_1 + b_2 + \cdots + b_n)$$
$$= a_1 b_1 + a_1 b_2 + \cdots + a_1 b_n + a_2 b_1 + a_2 b_2 + \cdots + a_2 b_n + \cdots$$
$$+ a_m b_1 + a_m b_2 + \cdots + a_m b_n,$$

or

$$\left(\sum_1^m a_i \right) \left(\sum_1^n b_j \right) = \sum_{i=1, j=1}^{m, n} a_i b_j.$$

We note next that

$$a0 = 0 = 0a$$

for all a; for we have $a0 = a(0 + 0) = a0 + a0$. Addition of $-a0$ gives $a0 = 0$. Similarly, $0a = 0$. We have the equation

$$0 = 0b = (a + (-a))b = ab + (-a)b,$$

which shows that

$$(-a)b = -ab.$$

Similarly, $a(-b) = -ab$; consequently

$$(-a)(-b) = -a(-b) = -(-ab) = ab.$$

If a and b commute, that is, $ab = ba$, then $a^m b^n = b^n a^m$. Also, by induction we can prove the *binomial theorem*

(1) $$(a + b)^n = a^n + \binom{n}{1}a^{n-1}b + \binom{n}{2}a^{n-2}b^2 + \cdots + b^n$$

where the binomial coefficient

(2) $$\binom{n}{i} = \frac{n!}{i!(n-i)!}.$$

The inductive step of the proof comes from the formula

$$\binom{r}{k} + \binom{r}{k-1} = \frac{r!}{k!(r-k)!} + \frac{r!}{(k-1)!(r-k+1)!}$$

$$= \frac{(r+1)!}{k!(r-k+1)!} = \binom{r+1}{k}.$$

The reader should carry out the proof and note just how the commutative law of multiplication intervenes.

EXERCISES

1. Let C be the set of real-valued continuous functions on the real line $\mathbb{R}$. Show that C with the usual addition of functions and 0 is an abelian group, and that C with product $(f \cdot g)(x) = f(g(x))$ and 1 the identity map is a monoid. Is C with these compositions and 0 and 1 a ring?

2. Show that in a ring R, $a(b - c) = ab - ac$ where $b - c \equiv b + (-c)$ and $n(ab) = (na)b = a(nb)$ if $n \in \mathbb{Z}$.

3. Show that if all the axioms for a ring except commutativity of addition are assumed, then commutativity follows, and hence we have a ring.

4. Let I be the set of complex numbers of the form $m + n\sqrt{-3}$ where either $m, n \in \mathbb{Z}$ or both m and n are halves of odd integers. Show that I is a subring of $\mathbb{C}$.

5. If a and b are elements of a ring, define $a^{(0)} = a, a' = [a, b] \equiv ab - ba$ and inductively $a^{(k)} = [a^{(k-1)}, b]$ (note that for the sake of simplicity we do not indicate the dependence of $a^{(k)}$ on b). Prove the following formula:

$$\sum_{i=0}^{k} b^i a b^{k-i} = \sum_{j=0}^{k} \binom{k+1}{j+1} b^{k-j} a^{(j)}.$$

2.2 TYPES OF RINGS

We obtain various types of rings by imposing special conditions on the multiplicative monoid. For example, a ring R is called *commutative* if $(R, \cdot, 1)$ is commutative. All the examples listed in the preceding section have this property. Examples of non-commutative rings will be given in the next two sections. A ring is called a *domain* (also *integral domain*) if the set R^* of non-zero elements of R is a submonoid of $(R, \cdot, 1)$. It is implicit in the definition of a domain R that $R \neq 0$. Besides this, the condition that R is a domain is that $a \neq 0$ and $b \neq 0$ in R imply $ab \neq 0$. Clearly any subring of a domain is a domain. All the examples in section 1 except 8 are domains. On the other hand, in 8 we can take the two elements f and g such that

$$f(x) = \begin{cases} 0 & \text{for} \quad 0 \le x \le \frac{1}{2} \\ x - \frac{1}{2} & \text{for} \quad \frac{1}{2} < x \le 1 \end{cases}$$

$$g(x) = \begin{cases} -x + \frac{1}{2} & \text{for} \quad 0 \le x \le \frac{1}{2} \\ 0 & \text{for} \quad \frac{1}{2} < x \le 1. \end{cases}$$

Then $f \neq 0$ (the constant function 0) and $g \neq 0$ but $fg = 0$. Hence the ring of real-valued continuous functions on $[0, 1]$ is not a domain .

If a is an element of a ring R for which there exists a $b \neq 0$ such that $ab = 0$ $(ba = 0)$, then a is called a *left (right) zero divisor*. Clearly 0 is a left and a right zero divisor if R has more than one element. If $a \neq 0$ is a left zero divisor and $ab = 0$ for $b \neq 0$, then b is a non-zero right zero divisor. If is clear from this and the definition of a domain that $R \neq 0$ is a domain if and only if it possesses no zero divisors $\neq 0$ (right or left).

We note also that a ring is a domain if and only if $R \neq 0$ and the restricted cancellation laws hold, that is, $ab = ac$, $a \neq 0$, imply $b = c$, and $ba = ca$, $a \neq 0$, imply $b = c$. For, if R is a domain and $ab = ac$, then $a(b - c) = 0$, so if $a \neq 0$, then $b - c = 0$ and $b = c$. Similarly, $ba = ca$, $a \neq 0$ give $b = c$. Conversely, let R be a ring $\neq 0$ in which these cancellation laws hold. Let $ab = 0$, $a \neq 0$. Then $ab = a0$, so that cancelling gives $b = 0$. Hence R is a domain.

A ring R is called a *division ring* (also *skew field*, *sfield*, or *field*) if the set R^* of non-zero elements is a subgroup of $(R, \cdot, 1)$. This is equivalent to: $1 \neq 0$, and for any $a \neq 0$ there exists a b such that $ab = 1 = ba$. Examples 2, 3, 4, 6, and 9 as well as the second example in 7 are division rings in which multiplication is commutative. Division rings that have this property are called *fields*. We shall give an example of a non-commutative division ring in section 2.4.

It is clear that any division ring is a domain, and since subrings of domains are domains, any subring of a division ring is a domain. The converse does not hold, since $\mathbb{Z}$ is a domain which is not a division ring, and $\mathbb{Z}$ is a subring of the field $\mathbb{Q}$. A subring of a ring which is itself a division ring will be called a *division subring*. If $a \neq 0$ in a division ring R then the equation $ax = b$ has the solution $x = a^{-1}b$. By the restricted cancellation law this is the only solution of the equation. Similarly, $ya = b$ has the unique solution $y = ba^{-1}$.

We have seen that the set of invertible elements of any monoid is a subgroup. In particular, the set U of invertible elements of $(R, \cdot, 1)$ is a subgroup. We shall call the elements of U *units*—even though this conflicts slightly with the designation *the unit* for 1—and U is called the *group of units* (or *invertible elements*) of the ring. For example, the group of units of $\mathbb{Z}$ is $\{1, -1\}$.

EXERCISES

1. Show that any finite domain is a division ring.

2. Show that a domain contains no idempotents ($e^2 = e$) except $e = 0$ and $e = 1$. An element z is called *nilpotent* if $z^n = 0$ for some $n \in \mathbb{Z}^+$. Show that 0 is the only nilpotent in a domain.

3. Let z be an element of a ring for which there exists a $w \neq 0$ such that $zwz = 0$. Show that z is either a left or a right zero divisor.

4. Show that if $1 - ab$ is invertible in a ring then so is $1 - ba$.

5. Show that a function f in the example (8) of section 2.1 is a zero divisor if and only if the set of points x where $f(x) = 0$ contains an open interval. What are the idempotents of this ring? The nilpotents? The units?

6. Let u be an element of a ring that has a right inverse. Prove that the following conditions on u are equivalent: (1) u has more than one right inverse, (2) u is not a unit, (3) u is a left 0 divisor.

7. (Kaplansky.) Prove that if an element of a ring has more than one right inverse then it has infinitely many. Construct a counterexample to show that this does not hold for monoids.

8. Show that an element u of a ring is a unit with $v = u^{-1}$ if and only if either of the following conditions holds: (1) $uvu = u$, $vu^2v = 1$, (2) $uvu = u$ and v is the only element satisfying this condition.

9. (Hua.) Let a and b be elements of a ring such that a, b, and $ab - 1$ are units. Show that $a - b^{-1}$ and $(a - b^{-1})^{-1} - a^{-1}$ are units and the following identity holds:

$$((a - b^{-1})^{-1} - a^{-1})^{-1} = aba - a.$$

10. (Cohn.) Let G be a group, e an element of G and θ a map of the subset $G_1 = \{x \in G \,|\, x \neq 1\}$ into itself satisfying
    (i) $\theta(yxy^{-1}) = y(\theta x)y^{-1}$, $x \in G_1, y \in G$.
    (ii) $\theta^2(x) = x$.
    (iii) $\theta(x^{-1}) = e(\theta x)x^{-1}$.
    (iv) $\theta(xy^{-1}) = (\theta(\theta(x)\theta(y^{-1})))\theta(y^{-1})$, $x, y \in G_1, x \neq y$.
    Show that there exists a unique division ring D such that $D^* = G$ and in G, $\theta x = 1 - x$, $x \in G_1$, $e = -1$.

2.3 MATRIX RINGS

The reader is probably already familiar with matrices and determinants from his study of linear algebra or multivariable calculus. We shall now generalize these notions to the extent which will be needed in our subsequent work: matrices with entries in any ring and determinants of matrices with entries in a commutative ring. For a reader already familiar with matrices and determinants the content of this section can be summarized by saying that the familiar results carry over in this generality.

Let R be a ring, n a positive integer. We shall now define the *ring $M_n(R)$ of $n \times n$ matrices over the ring R*. The underlying set of this ring are the $n \times n$ arrays or *matrices*

$$(3) \qquad A = \begin{bmatrix} a_{11} & a_{12} & \cdots & a_{1n} \\ a_{21} & a_{22} & \cdots & a_{2n} \\ \multicolumn{4}{c}{\dotfill} \\ a_{n1} & a_{n2} & \cdots & a_{nn} \end{bmatrix}$$

of n rows and columns with *entries* (also *elements, coefficients*, or *coordinates*) $a_{ij} \in R$. The element a_{ij} of R in the intersection of the ith row and jth column of A will be referred to as the (i, j)-*entry* of A. Two matrices A and $B = (b_{ij})$ are regarded as equal if and only if $a_{ij} = b_{ij}$ for every i, j, and the set $M_n(R)$ is the complete set of $n \times n$ matrices with entries in R. In short, $M_n(R)$ is the product set of n^2 copies of R.

We define addition of matrices by the formula

$$
\begin{bmatrix} a_{11} & a_{12} & \cdots & a_{1n} \\ a_{21} & a_{22} & \cdots & a_{2n} \\ \hdotsfor{4} \\ a_{n1} & a_{n2} & \cdots & a_{nn} \end{bmatrix} + \begin{bmatrix} b_{11} & b_{12} & \cdots & b_{1n} \\ b_{21} & b_{22} & \cdots & b_{2n} \\ \hdotsfor{4} \\ b_{n1} & b_{n2} & \cdots & b_{nn} \end{bmatrix}
$$

$$
= \begin{bmatrix} a_{11}+b_{11} & a_{12}+b_{12} & \cdots & a_{1n}+b_{1n} \\ a_{21}+b_{21} & a_{22}+b_{22} & \cdots & a_{2n}+b_{2n} \\ \hdotsfor{4} \\ a_{n1}+b_{n1} & a_{n2}+b_{n2} & \cdots & a_{nn}+b_{nn} \end{bmatrix}.
$$

Thus, to obtain the sum we add the entries a_{ij} and b_{ij} in the same position. We define the matrix 0 to be the matrix whose entries are all 0. Then it is easy to verify that with the given addition and 0, $M_n(R)$ is an abelian group. Multiplication of matrices is defined by

$$
\begin{bmatrix} a_{11} & a_{12} & \cdots & a_{1n} \\ a_{21} & a_{22} & \cdots & a_{2n} \\ \hdotsfor{4} \\ a_{n1} & a_{n2} & \cdots & a_{nn} \end{bmatrix} \begin{bmatrix} b_{11} & b_{12} & \cdots & b_{1n} \\ b_{21} & b_{22} & \cdots & b_{2n} \\ \hdotsfor{4} \\ b_{n1} & b_{n2} & \cdots & b_{nn} \end{bmatrix}
$$

$$
= \begin{bmatrix} \sum a_{1k}b_{k1} & \sum a_{1k}b_{k2} & \cdots & \sum a_{1k}b_{kn} \\ \sum a_{2k}b_{k1} & \sum a_{2k}b_{k2} & \cdots & \sum a_{2k}b_{kn} \\ \hdotsfor{4} \\ \sum a_{nk}b_{k1} & \sum a_{nk}b_{k2} & \cdots & \sum a_{nk}b_{kn} \end{bmatrix}.
$$

Thus the product $P = AB$ has as its (i, j)-entry the element

$$p_{ij} = a_{i1}b_{1j} + a_{i2}b_{2j} + \cdots + a_{in}b_{nj}.$$

For example, in the ring $M_3(\mathbb{Z})$ of 3×3 matrices over $\mathbb{Z}$ we have

$$
\begin{pmatrix} 1 & -2 & 3 \\ 0 & 1 & -1 \\ 2 & 5 & -2 \end{pmatrix} \begin{pmatrix} 0 & 3 & 4 \\ 2 & 5 & 1 \\ -1 & -6 & 2 \end{pmatrix} = \begin{pmatrix} -7 & -25 & 8 \\ 3 & 11 & -1 \\ 12 & 43 & 9 \end{pmatrix}.
$$

We define the unit matrix 1 by

$$
1 = \begin{bmatrix} 1 & 0 & \cdots & 0 \\ 0 & 1 & \cdots & 0 \\ \hdotsfor{4} \\ 0 & 0 & \cdots & 1 \end{bmatrix}
$$

that is, we have the unit 1 of R on the "main" diagonal running from the upper left-hand corner to the lower right-hand corner, and all other entries are 0. Then it is immediate that $A1 = A = 1A$ for $A \in M_n(R)$. Also multiplication is associative: the (i, l)-entry of $A(BC)$, $A = (a_{ij})$, $B = (b_{ij})$, $C = (c_{ij})$ is $\sum_{j,k} a_{ij}(b_{jk}c_{kl})$ and the (i, l)-entry of $(AB)C$ is $\sum_{j,k} (a_{ij}b_{jk})c_{kl}$. These are equal by the associativity of multiplication in R. The distributive laws hold, for the (i, j)-entries of $A(B + C)$ and of $AB + AC$ are respectively $\sum_k a_{ik}(b_{kj} + c_{kj})$ and

$$\sum_k (a_{ik}b_{kj} + a_{ik}c_{kj})$$

and these are equal by one of the distributive laws in R. Similarly, we have the other distributive law in $M_n(R)$. Hence we have shown that $(M_n(R), +, \cdot, 0, 1)$ is a ring.

We now define e_{ij} to be the matrix having a lone 1 as its (i, j)-entry and all other entries 0. The n^2 matrices e_{ij}, $1 \leq i, j \leq n$ are customarily called *matrix units*, though they are *not* (except for $n = 1$) units ($=$ invertible elements) of $M_n(R)$. It is easy to verify the following multiplication table:

(4) $e_{ij}e_{kl} = \delta_{jk}e_{il}$

where δ_{jk} is the *Kronecker delta* defined by

(5) $\delta_{jj} = 1, \qquad \delta_{jk} = 0 \quad \text{if} \quad j \neq k.$

Also we have

(6) $1 = e_{11} + e_{22} + \cdots + e_{nn}.$

The e_{ii} are idempotent: $e_{ii}^2 = e_{ii}$, and if $n > 1$, we have $e_{11}e_{12} = e_{12}, e_{12}e_{11} = 0$, which shows that $M_n(R)$ is never commutative if $n > 1$ and $R \neq 0$.

We shall denote the matrix

$$\begin{pmatrix} a_1 & & & \\ & a_2 & & 0 \\ 0 & & \ddots & \\ & & & a_n \end{pmatrix}$$

having the entries $a_1, a_2, \ldots, a_n$ in this order on the main diagonal and 0's elsewhere as $\text{diag}\{a_1, a_2, \ldots, a_n\}$. It is clear that the set of these diagonal matrices is a subring of $M_n(R)$. We now put $a' = \text{diag}\{a, a, \ldots, a\}$. Then $a \to a'$ is injective and we have $(a + b)' = a' + b', (ab)' = a'b', 0' = 0, 1' = 1$. Thus the map $a \to a'$ is both a monomorphism of $(R, +, 0)$ into $(M_n(R), +, 0)$ and of $(R, \cdot, 1)$ into $(M_n(R), \cdot, 1)$. It follows that $R' = \{a' | a \in R\}$ is a subring of $M_n(R)$ and $a \to a'$

regarded as a map of R into R' is an *isomorphism* of rings, where we define this to be a map which is both an isomorphism for the additive groups and an isomorphism for the multiplicative monoids.

We shall now identify R with the isomorphic subring R' of $M_n(R)$, identifying an $a \in R$ with the corresponding diagonal matrix $a' = \text{diag}\{a, a, \ldots, a\}$. This identification is similar to the one which is made in identifying the integers with the rational numbers with denominators 1, and has the effect of embedding R in $M_n(R)$. We now observe that multiplication of a matrix A on the left (right) by $a \in R$ amounts to multiplication of all the entries on the left (right) by a. Hence $ae_{ij} = e_{ij}a$ and this matrix has the element a in the (i, j)-position and 0's elsewhere. Then it is clear that for the matrix A of (3) we have

$$(7) \qquad A = \sum_{i,j} a_{ij}e_{ij}.$$

Thus every matrix is a linear combination of the e_{ij} with "coefficients" $a_{ij} \in R$.

The group of invertible elements of $M_n(R)$ is called the *linear group* $GL_n(R)$. We shall now derive, for the case R commutative, a determinant criterion for a matrix A to be invertible, that is, to belong to $GL_n(R)$. It is assumed that the reader is familiar with the definition of determinants and the elementary facts about them.[2] It is easy to convince ourselves that the main formulas on determinants, which can be found in any text on linear algebra, are valid for determinants of matrices over any commutative ring. Thus if R is commutative we can define for $A = (a_{ij})$ the *determinant*

$$(8) \qquad \det A = \sum_{\pi} (sg\ \pi)a_{1i_1}a_{2i_2} \cdots a_{ni_n}$$

where the summation is taken over all permutations π of $1, 2, \ldots, n$, and $sg\ \pi = 1$ or -1 according as π is even or odd. The *cofactor* of the element a_{ij} in A, as in (3), is $(-1)^{i+j}$ times the determinant of the $n - 1 \times n - 1$ matrix obtained by striking out the ith row and the jth column of A. We recall that we can "expand" a determinant by any row and any column in the sense that we obtain $\det A$ by multiplying the entries of any row (or column) by their cofactors and adding the results. Thus if A_{ij} denotes the cofactor of a_{ij} then we have

$$(9) \qquad \begin{aligned} a_{i1}A_{i1} + a_{i2}A_{i2} + \cdots + a_{in}A_{in} &= \det A \\ a_{1i}A_{1i} + a_{2i}A_{2i} + \cdots + a_{ni}A_{ni} &= \det A. \end{aligned}$$

We recall also that the sum of the products of the elements of any row (column)

---

[2] The principal theorems on determinants will be derived later in this book, using exterior algebras (section 7.2, pp. 416–419).

and the corresponding cofactors of the elements of another row (column) is 0:

(10)
$$a_{i1}A_{j1} + a_{i2}A_{j2} + \cdots + a_{in}A_{jn} = 0, \qquad i \neq j,$$
$$a_{1i}A_{1j} + a_{2i}A_{2j} + \cdots + a_{ni}A_{nj} = 0, \qquad i \neq j.$$

These relations lead us to define the *adjoint* of the matrix $A = (a_{ij})$ to be the matrix whose (i, j)-entry is $\alpha_{ij} = A_{ji}$. Using this definition it is immediate that formulas (9) and (10) are equivalent to the matrix equations

(11)
$$A(\text{adj } A) = \det A = (\text{adj } A)A$$

where det A in the middle is the corresponding element diag $\{\det A, \ldots, \det A\}$ in $M_n(R)$. We recall also the rule for multiplying determinants, which in matrix form is

(12)
$$\det AB = (\det A)(\det B).$$

The multiplication rule (12) and the fact that det $1 = 1$ imply that $A \to \det A$ is a homomorphism of the multiplicative monoid of $M_n(R)$, R commutative, into the multiplicative monoid of R. It is clear that such a homomorphism maps the group $GL_n(R)$ into $U(R)$, the group of units of R: that is, if $A \in GL_n(R)$, then det A is a unit in R. Conversely, suppose $\Delta = \det A$ is a unit. Since R is commutative $aB = Ba$ for every $a \in R$, $B \in M_n(R)$. In particular, $(\text{adj } A)\Delta^{-1} = \Delta^{-1}(\text{adj } A)$ so

$$A(\text{adj } A)\Delta^{-1} = \Delta\Delta^{-1} = (\Delta^{-1} \text{adj } A)A.$$

Thus we see that

(13)
$$(\text{adj } A)\Delta^{-1} = A^{-1}.$$

This result shows that if det A is a unit then A is invertible, moreover, we have the formula (13) for its inverse. The main part of the result we have proved is stated in the following

THEOREM 2.1. *If R is a commutative ring, a matrix $A \in M_n(R)$ is invertible if and only if its determinant is invertible in R.*

A noteworthy special case of the theorem is the

COROLLARY. *If F is a field, $A \in M_n(F)$ is invertible if and only if det $A \neq 0$.*

EXERCISES

1. Show that the matrix

$$\begin{pmatrix} 1 & 4 & 1 \\ 0 & 1 & -1 \\ -3 & -6 & -8 \end{pmatrix}$$

is invertible in $M_3(\mathbb{Z})$ and find its inverse.

2. Prove that if R is a commutative ring then $AB = 1$ in $M_n(R)$ implies $BA = 1$. (This is not always true for non-commutative R.)

3. Verify that for any $p \in R$ and $i \neq j$, $1 + pe_{ij}$ is invertible in $M_n(R)$ with inverse $1 - pe_{ij}$. More generally, show that if z is a nilpotent element of a ring (that is, $z^n = 0$ for some positive integer n), then $1 - z$ is invertible. Also determine its inverse.

4. Show that diag $\{a_1, a_2, \ldots, a_n\}$ is invertible in $M_n(R)$ if and only if every a_i is invertible in R. What is the inverse?

5. Verify that for a, $b \in \mathbb{R}$, $a + b\sqrt{-1} \to \begin{pmatrix} a & b \\ -b & a \end{pmatrix}$ is an isomorphism of $\mathbb{C}$ with a subring of $M_2(\mathbb{R})$.

6. Show that in any ring the set $C(S)$ of elements which commute with every element of a given subset S constitute a subring. If S is taken to be the whole ring, then $C = C(S)$ is called the *center* of the ring. Note that this subring is commutative. Determine $C(S)$ in $M_n(R)$ for $S = \{e_{ij} | i, j = 1, \ldots, n\}$. Also determine the center of $M_n(R)$.

7. Determine $C(S)$ where S is the single matrix $N = e_{12} + e_{23} + \cdots + e_{n-1,n}$.

8. Show that if R is commutative and D is the set of diagonal matrices in $M_n(R)$, then $C(D) = D$.

9. Let S be any ring which contains a set of matrix units, that is, a set of elements $\{e_{ij} | i, j = 1, \ldots, n\}$ such that $e_{ij}e_{kl} = \delta_{jk}e_{il}$ and $\sum_1^n e_{ii} = 1$. For any i, j, $1 \leq i, j \leq n$ and any $a \in S$ define $a_{ij} = \sum_{k=1}^n e_{ki}ae_{jk}$. Show that $a_{ij} \in R \equiv C(\{e_{kl} | k, l = 1, \ldots, n\})$ and that $a = \sum_{i,j} a_{ij}e_{ij}$. Show that if r_{ij} are any elements of R, then $\sum r_{ij}e_{ij} = 0$ only if every $r_{ij} = 0$. Hence show that $S \cong M_n(R)$ ($\cong$ denotes isomorphism).

10. Let R be a ring, R' a set, η a bijective map of R' into R. Show that R' becomes a ring if one defines:

$$a' + b' = \eta^{-1}(\eta(a') + \eta(b')), \qquad 0' = \eta^{-1}(0)$$
$$a'b' = \eta^{-1}(\eta(a')\eta(b')), \qquad 1' = \eta^{-1}(1)$$

and that η is an isomorphism of R' with R. Use this to prove that if u is an invertible element of a ring then $(R, +, \cdot_u, 0, u^{-1})$ where $a\cdot_u b = aub$ is a ring isomorphic to R. Show also that $(R, \oplus, \circ, 1, 0)$ where $a \oplus b = a + b - 1$, $a \circ b = a + b - ab$ is a ring isomorphic to R.

11. Show that the rings $M_{nm}(R)$ and $M_n(M_m(R))$ are isomorphic (*Hint*: Use "block" addition and multiplication of matrices.)

12. Show that if R is a field, $A \in M_n(R)$ is a zero divisor in this ring if and only if A is not invertible. Does this hold for arbitrary commutative R? Explain.

2.4 QUATERNIONS

In 1843, W. R. Hamilton constructed the first example of a division ring in which the commutative law of multiplication does not hold. This was an extension of the field of complex numbers, whose elements were quadruples of real numbers $(\alpha, \beta, \gamma, \delta)$ for which the usual addition and a multiplication were defined so that $1 = (1, 0, 0, 0)$ is the unit and $i = (0, 1, 0, 0)$, $j = (0, 0, 1, 0)$, and $k = (0, 0, 0, 1)$ satisfy $i^2 = j^2 = k^2 = -1 = ijk$.[3] Hamilton called his quadruples quaternions. Previously he had defined complex numbers as pairs of real numbers (α, β) with the product $(\alpha, \beta)(\gamma, \delta) = (\alpha\gamma - \beta\delta, \alpha\delta + \beta\gamma)$. Hamilton's discovery of quaternions led to a good deal of experimentation with other such "hypercomplex" number systems and eventually to a structure theory whose goal was to classify such systems. A good deal of important algebra thus evolved from the discovery of quaternions.

We shall not follow Hamilton's way of introducing quaternions. Instead we shall define this system as a certain subring of the ring $M_2(\mathbb{C})$ of 2×2 matrices with complex number entries. This will have the advantage of reducing the calculations to a single simple verification.

We consider the subset $\mathbb{H}$ of the ring $M_2(\mathbb{C})$ of complex 2×2 matrices that have the form

(14)
$$x = \begin{pmatrix} a & b \\ -\bar{b} & \bar{a} \end{pmatrix} = \begin{pmatrix} \alpha_0 + \alpha_1\sqrt{-1} & \alpha_2 + \alpha_3\sqrt{-1} \\ -\alpha_2 + \alpha_3\sqrt{-1} & \alpha_0 - \alpha_1\sqrt{-1} \end{pmatrix}, \quad \alpha_i \text{ real.}$$

We claim that $\mathbb{H}$ is a subring of $M_2(\mathbb{C})$. Since $\overline{a_1 - a_2} = \bar{a}_1 - \bar{a}_2$ for complex numbers it is clear that $\mathbb{H}$ is closed under subtraction; hence $\mathbb{H}$ is a subgroup of the additive group of $M_2(\mathbb{C})$. We obtain the unit matrix by taking $a = 1$, $b = 0$ in (14). Hence $1 \in \mathbb{H}$. Since

$$\begin{pmatrix} a & b \\ -\bar{b} & \bar{a} \end{pmatrix}\begin{pmatrix} c & d \\ -\bar{d} & \bar{c} \end{pmatrix} = \begin{pmatrix} ac - b\bar{d} & ad + b\bar{c} \\ -\bar{b}c - \bar{a}\bar{d} & -\bar{b}d + \bar{a}\bar{c} \end{pmatrix}$$

---

[3] It seems to have taken Hamilton ten years to arrive at this multiplication table. In fact, he had spent a good deal of effort trying to construct a field of triples of real numbers (which is not possible) before he realized that it was necessary to go to quadruples and to drop the commutativity of multiplication. Perhaps this bit of history may serve as an encouragement to the student who sometimes finds himself on the wrong track in attacking a problem. (See Carl A. Boyer, *A History of Mathematics*, New York, Wiley, 1968, p. 625.)

and $\overline{a_1 a_2} = \bar{a}_1 \bar{a}_2$, the right-hand side has the form

$$\begin{pmatrix} u & v \\ -\bar{v} & \bar{u} \end{pmatrix}$$

where $u = ac - b\bar{d}$, $v = ad + b\bar{c}$. Hence $\mathbb{H}$ is closed under multiplication and so $\mathbb{H}$ is a subring of $M_2(\mathbb{C})$.

We shall now show that $\mathbb{H}$ is a division ring. We note first that

$$\Delta \equiv \det \begin{pmatrix} \alpha_0 + \alpha_1 \sqrt{-1} & \alpha_2 + \alpha_3 \sqrt{-1} \\ -\alpha_2 + \alpha_3 \sqrt{-1} & \alpha_0 - \alpha_1 \sqrt{-1} \end{pmatrix} = \alpha_0{}^2 + \alpha_1{}^2 + \alpha_2{}^2 + \alpha_3{}^2.$$

Since the α_i are real numbers this is real, and is 0 only if every $\alpha_i = 0$, that is, if the matrix is 0. Hence every non-zero element of $\mathbb{H}$ has an inverse in $M_2(\mathbb{C})$. Moreover, we have, by the definition of the adjoint given in section 2.3, that

$$\text{adj} \begin{pmatrix} a & b \\ -\bar{b} & \bar{a} \end{pmatrix} = \begin{pmatrix} \bar{a} & -b \\ \bar{b} & a \end{pmatrix}.$$

Since $\bar{\bar{a}} = a$ this is obtained from the x in (14) by replacing a by $\bar{a}$ and b by $-b$, and so it is contained in $\mathbb{H}$. Thus if the matrix x is $\neq 0$ then its inverse is

$$\begin{pmatrix} \bar{a}\Delta^{-1} & -b\Delta^{-1} \\ \bar{b}\Delta^{-1} & a\Delta^{-1} \end{pmatrix}$$

and this is contained in $\mathbb{H}$. Hence $\mathbb{H}$ is a division ring.

The ring $\mathbb{H}$ contains in its center the field $\mathbb{R}$ of real numbers identified with the set of diagonal matrices $\text{diag}\{\alpha, \alpha\}$, $\alpha \in \mathbb{R}$. $\mathbb{H}$ also contains the matrices

$$i = \begin{pmatrix} \sqrt{-1} & 0 \\ 0 & -\sqrt{-1} \end{pmatrix}, \quad j = \begin{pmatrix} 0 & 1 \\ -1 & 0 \end{pmatrix}, \quad k = \begin{pmatrix} 0 & \sqrt{-1} \\ \sqrt{-1} & 0 \end{pmatrix}.$$

We verify that

(15) $$x = \alpha_0 + \alpha_1 i + \alpha_2 j + \alpha_3 k$$

and if $\alpha_0 + \alpha_1 i + \alpha_2 j + \alpha_3 k = \beta_0 + \beta_1 i + \beta_2 j + \beta_3 k$, $\beta_i \in \mathbb{R}$, then

$$\begin{pmatrix} \alpha_0 + \alpha_1 \sqrt{-1} & \alpha_2 + \alpha_3 \sqrt{-1} \\ -\alpha_2 + \alpha_3 \sqrt{-1} & \alpha_0 - \alpha_1 \sqrt{-1} \end{pmatrix} = \begin{pmatrix} \beta_0 + \beta_1 \sqrt{-1} & \beta_2 + \beta_3 \sqrt{-1} \\ -\beta_2 + \beta_3 \sqrt{-1} & \beta_0 - \beta_1 \sqrt{-1} \end{pmatrix}$$

so $\alpha_i = \beta_i$, $0 \le i \le 3$. Thus any x in $\mathbb{H}$ can be written in one and only one way in the form (15). The product of two elements in $\mathbb{H}$

$$(\alpha_0 + \alpha_1 i + \alpha_2 j + \alpha_3 k)(\beta_0 + \beta_1 i + \beta_2 j + \beta_3 k)$$

is determined by the product and sum in $\mathbb{R}$, the distributive laws and the multiplication table

(16)
$$i^2 = j^2 = k^2 = -1$$
$$ij = -ji = k, jk = -kj = i, ki = -ik = j.$$

Incidentally, because these show that $\mathbb{H}$ is not commutative we have constructed a division ring that is not a field. The ring $\mathbb{H}$ is called the *division ring of real quaternions*.

EXERCISES

1. Define $\bar{x} = \alpha_0 - \alpha_1 i - \alpha_2 j - \alpha_3 k$ for $x = \alpha_0 + \alpha_1 i + \alpha_2 j + \alpha_3 k$. Show that $\overline{x + y} = \bar{x} + \bar{y}$, $\overline{xy} = \bar{y}\bar{x}$, and that $\bar{x} = x$ if $x \in \mathbb{R}$.

2. Show that $x\bar{x} = N(x)$ where $N(x) = \alpha_0^2 + \alpha_1^2 + \alpha_2^2 + \alpha_3^2$. Define $T(x) = 2\alpha_0$. Show that x satisfies the quadratic equation $x^2 - T(x)x + N(x) = 0$.

3. Prove that $N(xy) = N(x)N(y)$.

4. Show that the set $\mathbb{H}_0$ of quaternions $x = \alpha_0 + \alpha_1 i + \alpha_2 j + \alpha_3 k$, whose "coordinates" α_i are rational, form a division subring of $\mathbb{H}$.

5. Verify that the set I of quaternions x in which all the coordinates α_i are either integers or all are halves of odd integers is a subring of $\mathbb{H}$. Is this a division subring? Show that $T(x)$ and $N(x) \in \mathbb{Z}$ for any $x \in I$. Determine the group of units of I.

6. Show that the subring of $M_2(\mathbb{C})$ generated by $\mathbb{C}$ and $\mathbb{H}$ is $M_2(\mathbb{C})$.

7. Let m and n be non-zero integers and let R be the subset of $M_2(\mathbb{C})$ consisting of the matrices of the form
    $$\begin{pmatrix} a + b\sqrt{m} & c + d\sqrt{m} \\ n(c - d\sqrt{m}) & a - b\sqrt{m} \end{pmatrix}$$
    where $a, b, c, d \in \mathbb{Q}$. Show that R is a subring of $M_2(\mathbb{C})$ and that R is a division ring if and only if the only rational numbers x, y, z, t satifying the equation $x^2 - my^2 - nz^2 + mnt^2 = 0$ are $x = y = z = t = 0$. Give a choice of m, n that R is a division ring and a choice of m, n that R is not a division ring.

8. Determine the center of $\mathbb{H}$. Determine the subring $C(i)$ commuting with i.

9. Let S be a division subring of $\mathbb{H}$ which is stabilized by every map $x \to dxd^{-1}, d \neq 0$ in $\mathbb{H}$. Show that either $S = \mathbb{H}$ or S is contained in the center.

10. (Cartan-Brauer-Hua.) Let D be a division ring, C its center and let S be a division subring of D which is stabilized by every map $x \to dxd^{-1}$, $d \neq 0$ in D. Show that either $S = D$ or $S \subset C$.

2.5   IDEALS, QUOTIENT RINGS

We define a congruence $\equiv$ in a ring to be a relation in R which is a congruence for the additive group $(R, +, 0)$ and the multiplicative monoid $(R, \cdot, 1)$. Hence $\equiv$ is an equivalence relation such that $a \equiv a'$ and $b \equiv b'$ imply $a + b \equiv a' + b'$ and $ab \equiv a'b'$. Let $\bar{a}$ denote the congruence class of $a \in R$ and let $\bar{R}$ be the quotient set. As we have seen in section 1.5, we have binary compositions $+$ and $\cdot$ in $\bar{R}$ defined by $\bar{a} + \bar{b} = \overline{a + b}$, $\bar{a}\bar{b} = \overline{ab}$. These define the group $(\bar{R}, +, \bar{0})$ and the monoid $(\bar{R}, \cdot, \bar{1})$. We also have

$$\bar{a}(\bar{b} + \bar{c}) = \bar{a}(\overline{b + c}) = \overline{a(b + c)} = \overline{ab + ac} = \overline{ab} + \overline{ac} = \bar{a}\bar{b} + \bar{a}\bar{c}.$$

Similarly, $(\bar{b} + \bar{c})\bar{a} = \bar{b}\bar{a} + \bar{c}\bar{a}$. Hence $(\bar{R}, +, \cdot, \bar{0}, \bar{1})$ is a ring which we shall call a *quotient* (or *difference*) ring of R.

We recall also that the congruences in $(R, +, 0)$ are obtained from the subgroups I (necessarily normal since $(R, +)$ is commutative) by defining $a \equiv b$ if $a - b \in I$. Then the congruence class $\bar{a}$ is the coset $a + I$. If this is also a congruence for the multiplicative monoid, then for any $a \in R$ and any $b \in I$ we have $a \equiv a$ and $b \equiv 0$, and so $ab \equiv a0 = 0$ and $ba \equiv 0$. In other words, if $a \in R$ and $b \in I$ then ab and $ba \in I$. Conversely, suppose I is a subgroup of the additive group satisfying this condition. Then if $a \equiv a'$ and $b \equiv b'$ (mod I), $a - a' \in I$ so $ab - a'b = (a - a')b \in I$. Also $a'b - a'b' = a'(b - b') \in I$. Hence $ab - a'b' = (ab - a'b) + (a'b - a'b') \in I$. Hence $ab \equiv a'b'$ (mod I). We now give the following

DEFINITION 2.2   *If R is a ring, an ideal I of R is a subgroup of the additive group such that for any $a \in R$ and any $b \in I$, ab and $ba \in I$.*

Our results show that congruences in a ring R are obtained from ideals I of R by defining $a \equiv a'$ if $a - a' \in I$. The corresponding quotient ring $\bar{R}$ will be denoted as R/I and will be called the *quotient ring of R with respect to the ideal* I. The elements of R/I are the cosets $a + I$ and the addition and multiplication in R/I are defined by

(17)
$$(a + I) + (b + I) = (a + b) + I$$
$$(a + I)(b + I) = ab + I.$$

Also I is the 0 and $1 + I$ the unit of R/I.

It is interesting to look at the "algebra" of ideals of a ring R. We note first that the intersection of any set of ideals in R is an ideal. This is immediate from the definition. If S is a subset of R then the intersection (S) of all ideals of R containing S (non-vacuous, since R is such an ideal) is an ideal containing S

and is contained in every ideal containing S. We call (S) *the ideal generated by* S. If S is a finite set, $\{a_1, a_2, \ldots, a_n\}$, then we write $(a_1, a_2, \ldots, a_n)$ for (S). It is not easy to write down all the elements of this ideal. It is clear first that it contains all finite sums of products of the form xa_iy where $x, y \in R$ and there is no way of combining $xa_iy + x'a_iy'$ into a single term. Thus we see that to indicate explicitly all the elements of the ideal $(a_1, a_2, \ldots, a_n)$ we must consider all elements of the form

$$(18) \qquad \sum_{i_1} x_{1i_1} a_1 y_{1i_1} + \sum_{i_2} x_{2i_2} a_2 y_{2i_2} + \cdots + \sum_{i_n} x_{ni_n} a_n y_{ni_n}.$$

Now it is clear that the set I of elements of the form (18) is an ideal. It is clear also that I contains every $a_i = 1a_i1$. Hence

$$I = (a_1, a_2, \ldots, a_n).$$

If I and J are ideals we denote the ideal generated by $I \cup J$ as $I + J$. We claim that this is the set K of elements of the form $a + b$, $a \in I$, $b \in J$. This is clear since K is an ideal containing I and J and is contained in every ideal containing I and J. Another important ideal associated with I and J is the product IJ, defined to be the ideal generated by all the products ab, $a \in I$, $b \in J$. It is easily seen that IJ coincides with the set of elements of the form $a_1b_1 + a_2b_2 + \cdots + a_mb_m$ where $a_i \in I$, $b_i \in J$.

Sometimes we need to consider a sequence of ideals $I_1, I_2, \ldots$ such that $I_1 \subset I_2 \subset \cdots$. We call this an *ascending chain of ideals*. It is useful to observe that for such a chain, $\bigcup I_j$ is an ideal. It suffices to show that $\bigcup I_j$ is closed under subtraction and under left and right multiplication by arbitrary elements of R. To see the first, let $a, b \in \bigcup I_j$. Then $a \in I_j$ for some j and $b \in I_k$ for some k. If l is the greater of j and k then both a and b are in I_l. Hence $a - b \in I_l$ since I_l is an ideal. Also xa and $ax \in I_j$ for any $x \in R$. Thus $a - b \in \bigcup I_j$ and $xa, ax \in \bigcup I_j$ for any a and b in $\bigcup I_j$ and any $x \in R$. Then $\bigcup I_j$ is an ideal.

If R is commutative, our description of the elements of $(a_1, a_2, \ldots, a_n)$ simplifies considerably: namely, this ideal is the set of elements of the form $\sum_1^n x_i a_i$ $(= \sum_1^n a_i x_i)$, $x_i \in R$. This is clear from (18). In particular, the ideal (a) generated by a is the set of elements xa, $x \in R$. This is called the *principal ideal generated by* a.

We can give a neat characterization of fields in terms of ideals: namely, we have

THEOREM 2.2. *Let R be a commutative ring $\neq 0$. Then R is a field if and only if the only ideals in R are R $(= (1))$ and 0 $(= (0))$.*

Proof. Suppose R is a division ring and I is a non-zero ideal in R. If $a \neq 0$

is in I then so is $1 = aa^{-1}$. It is clear that the only ideal of a ring containing 1 is R (since I will then contain every $x = x1$). Hence $I = R$. This proves that the only ideals in a division ring are 0 and R. In particular this holds for fields. Conversely, suppose that R is a commutative ring $\neq 0$ whose only ideals are 0 and R. If $a \neq 0$ is in R then $(a) \neq 0$, so $(a) = R$. It follows that $1 \in (a)$ and hence there is an $x \in R$ such that $ax = 1$. Thus every non-zero element of R is invertible and R is a field. $\square$

EXERCISES

1. Let Γ be the ring of real-valued continuous functions on $[0, 1]$ (example 8, p. 87). Let S be a subset of $[0, 1]$ and let $Z_S = \{f \,|\, f(x) = 0, \, x \in S\}$. Verify that Z_S is an ideal. Let $S_1 = [0, \frac{1}{2}]$, $S_2 = [\frac{1}{2}, 1]$, $I_1 = Z_{S_1}$, $I_2 = Z_{S_2}$. Show that $I_1 I_2 = I_1 \cap I_2 = 0$.

2. Show that the associative law holds for products of ideals: $(IJ)K = I(JK)$ if I, J, and K are ideals.

3. Does the distributive law, $I(J + K) = IJ + IK$ hold?

4. If R is a ring we define a *right* (*left*) *ideal* in R to be a subgroup of the additive group of R such that $ba \in I$ $(ab \in I)$ for every $a \in R$, $b \in I$. Verify that the subset of matrices of the form $\begin{pmatrix} 0 & 0 \\ a & b \end{pmatrix}$ is a right ideal and the subset of the form $\begin{pmatrix} a & 0 \\ b & 0 \end{pmatrix}$ is a left ideal in $M_2(R)$ for any R. Are either of these sets ideals?

5. Prove the following extension of Theorem 2.2. A ring $R \neq 0$ is a division ring if and only if 0 and R are the only left (right) ideals in R.

6. Let R be a commutative ring and let N denote the set of nilpotent elements of R. Show that N is an ideal and R/N contains no non-zero nilpotent elements.

7. Let I be an ideal in R, U the group of units of R. Let U_1 be the subset of elements $a \in U$ such that $a \equiv 1 \pmod{I}$. Show that U_1 is a normal subgroup of U.

8. Let I be an ideal in R and let $M_n(I)$ denote the set of $n \times n$ matrices with entries in I. Show that $M_n(I)$ is an ideal in $M_n(R)$. Prove that every ideal in $M_n(R)$ has the form $M_n(I)$ for some ideal I of R, and that $I \to M_n(I)$ is a bijective map of the set of ideals of R onto the set of ideals of $M_n(R)$.

2.6 IDEALS AND QUOTIENT RINGS FOR $\mathbb{Z}$

After the generalities of the last section we now consider the ideals of $\mathbb{Z}$ and their corresponding quotient rings $\mathbb{Z}/I$. This will lead us to some interesting number theoretic results.

As we have seen in section 1.5 and again in section 1.10, the subgroups of the additive group $(\mathbb{Z}, +, 0)$ are the cyclic groups $\langle k \rangle$ where k is a non-negative integer. Since $\langle k \rangle = \{xk \mid x \in \mathbb{Z}\}$ it is clear that $\langle k \rangle$ is the same thing as the principal ideal (k) of multiples of k. Since any ideal is a subgroup it follows that every ideal in $\mathbb{Z}$ is a principal ideal. Now it is clear that $(l) \supset (k)$ if and only if $k \in (l)$, hence, if and only if $k = lm$, $m \in \mathbb{Z}$. Thus the inclusion relation $(l) \supset (k)$ for the principal ideals (l), (k) is equivalent to the divisibility condition $l \mid k$. A consequence of this is that if $m, n \in \mathbb{Z}$ and (m, n) denotes the ideal generated by m and n, then $(m, n) = (d)$ where d is a g.c.d. of m and n. Since $(m, n) \supset (m)$ and (n), we have $d \mid m$ and $d \mid n$. On the other hand, if $e \mid m$ and $e \mid n$ then $(e) \supset (m)$ and $(e) \supset (n)$. Then $(e) \supset (m, n) = (d)$ so $e \mid d$. Similarly, we see that $(m) \cap (n) = ([m, n])$ where $[m, n]$ is a least common multiple of m and n.

We look next at the quotient ring $\mathbb{Z}/(k)$, which is called the *ring of residues modulo k*. Since $(k) = (-k)$ we may assume $k \geq 0$. If $k = 0$, then $\mathbb{Z}/(k)$ can be identified with $\mathbb{Z}$, and if $k > 0$, the elements of $\mathbb{Z}/(k)$ are the k cosets

$$\bar{0} = (k), \bar{1} = 1 + (k), \bar{2} = 2 + (k), \ldots, \overline{k-1} = k - 1 + (k).$$

Suppose first that k is composite: $k = lm$, $l > 1$, $m > 1$. Then $\bar{l} \neq \bar{0}$ and $\bar{m} \neq \bar{0}$ in $\mathbb{Z}/(k)$ but $\bar{l}\bar{m} = \bar{k} = \bar{0}$. Thus $\mathbb{Z}/(k)$ has non-zero zero divisors if k is composite. Next let $k = p$ be a prime. In this case every $\bar{a} \neq \bar{0}$ in $\mathbb{Z}/(p)$ is invertible. Since $\mathbb{Z}/(k)$ is commutative $(\bar{a}\bar{b} = \overline{ab} = \overline{ba} = \bar{b}\bar{a})$, it follows that $\mathbb{Z}/(p)$ is a field. Given $\bar{a} \neq \bar{0}$, then $p \nmid a$ and 1 is a g.c.d. of p and a. Hence we have integers x and y such that $ax + py = 1$. Then $\bar{1} = \overline{ax + py} = \overline{ax} + \overline{py} = \bar{a}\bar{x}$. Hence $\bar{a}$ is invertible with $\bar{x}$ as inverse.

These simple results are important enough to state as a theorem.

THEOREM 2.3. *The ring $\mathbb{Z}/(k)$ for k composite is not a domain. On the other hand, $\mathbb{Z}/(p)$ for p prime is a field.*

We shall now determine the group $U(\mathbb{Z}/(k))$ of units of $\mathbb{Z}/(k)$. If $k = 0$ then these are 1 and -1. If $k > 0$ we have

THEOREM 2.4. *The group $U(\mathbb{Z}/(k))$, $k > 0$, consists of the classes $\bar{a} = a + (k)$ such that a and k are relatively prime (that is, have 1 as g.c.d.).*

Proof. If $(a, k) = 1$ (equivalently: the ideal $(a, k) = (1)$), then we have integers x and y such that $ax + ky = 1$. Then $\bar{a}\bar{x} = \bar{1}$, so $\bar{a}$ is invertible. Conversely, if $\bar{a}\bar{b} = \bar{1}$, then $\overline{ab} = \bar{1}$, so $ab = 1 + mk$, $m \in \mathbb{Z}$. Clearly this equation shows that any common divisor of a and k divides 1. Hence a and k are relatively prime. $\square$

The foregoing result shows that $|U(\mathbb{Z}/(k))|$ is the number $\varphi(k)$ of positive integers less than k and relatively prime to k. The function φ of positive integers thus defined is called the *Euler φ-function* (see exercises 4, p. 47). For example, if $k = 12$, the units of $\mathbb{Z}/(k)$ are $\bar{1}, \bar{5}, \bar{7}, \overline{11}$, and thus $\varphi(12) = 4$. In the next section we shall indicate in an exercise a formula for computing $\varphi(k)$ from the factorization of k into primes. At this point we note that if p is a prime, then it is clear from the definition that $\varphi(p) = p - 1$. Also it is easy to see that $\varphi(p^e) = p^e - p^{e-1} = p^e(1 - 1/p)$.

We recall that is G is a finite group, then $a^{|G|} = 1$ for every $a \in G$. A consequence of this result and Theorem 2.4 is that if $(a, k) = 1$, then $\bar{a}^{\varphi(k)} = \bar{1}$. The usual way of stating this result is

THEOREM 2.5. (Euler.) *If a is an integer prime to the positive integer k, then* $a^{\varphi(k)} \equiv 1 \pmod{k}$.

For $k = p$ a prime this reduces to an earlier result due to Fermat.

COROLLARY. *If p is a prime and a is an integer not divisible by p then* $a^{p-1} \equiv 1 \pmod{p}$.

This result can also be stated in a slightly different form, namely, that $a^p \equiv a \pmod{p}$. This holds for all a since it is trivial if a is divisible by p. On the other hand, if $a^p \equiv a \pmod{p}$ and $a \not\equiv 0 \pmod{p}$, then $a^{p-1} \equiv 1 \pmod{p}$ by cancellation. Hence the two statements are equivalent.

EXERCISES

1. Write down addition and multiplication tables for $\mathbb{Z}/(5)$ and for $\mathbb{Z}/(6)$.

2. Show that $\mathbb{Z}/(k)$ contains non-zero nilpotent elements ($z^n = 0$, $z \neq 0$) if and only if k is divisible by the square of a prime. Determine the nilpotent elements of $\mathbb{Z}/(180)$.

3. Prove that if D is a finite division ring then $a^{|D|} = a$ for every $a \in D$.

4. Let $A \in GL_2(\mathbb{Z}/(p))$ (that is, A is an invertible 2×2 matrix with entries in $\mathbb{Z}/(p)$). Show that $A^q = 1$ if $q = (p^2 - 1)(p^2 - p)$. Show also that $A^{q+2} = A^2$ for every $A \in M_2(\mathbb{Z}/(p))$.

5. Let T denote the set of triangular matrices $\begin{pmatrix} a & 0 \\ b & c \end{pmatrix}$ where $a, b, c \in \mathbb{Z}$. Verify that T is a subring of $M_2(\mathbb{Z})$. Determine the ideals of T.

2.7 HOMOMORPHISMS OF RINGS. BASIC THEOREMS

In this section we define homomorphism for rings and derive their basic properties. Everything will follow from our earlier results on homomorphisms of monoids and of groups (in sections 1.9 and 1.10) since our starting point is

DEFINITION 2.3.   *A homomorphism of a ring R into a ring R' is a map of R into R' which is a homomorphism of both the additive group and the multiplicative monoid of R into the corresponding objects of R'.*

Recalling that η is a homomorphism of a group G into a group G' if $\eta(ab) = \eta(a)\eta(b)$, we see that the conditions that a map η of a ring R into a ring R' is a homomorphism are

$$\eta(a + b) = \eta(a) + \eta(b), \qquad \eta(ab) = \eta(a)\eta(b), \qquad \eta(1) = 1'$$

where $1'$ is the unit of R'. If I is an ideal in R we have the corresponding congruence in R and the quotient ring $\bar{R} = R/I$. Also we have the natural map $v : a \to \bar{a}$. This is an epimorphism for the additive groups and the multiplicative monoids, hence it is an epimorphism (= surjective homomorphism) of the ring R onto the ring $\bar{R}$. As in the case of groups, we call $K = \eta^{-1}(0')$ the *kernel* of the homomorphism η of R ($0'$ the zero element of R'). Since $a \equiv b \pmod{K}$—that is, $a - b \in K$—is a congruence, the result of section 2.5 shows that K is an ideal in R (a fact, which can be verified directly also). The homomorphism η is a monomorphism (= injective homomorphism) if and only if the kernel is 0. It is clear also that the image under a homomorphism of R into R' is a subring of R', since it is a subgroup of the additive group of R' as well as a submonoid of the multiplicative monoid.

Now suppose η is a homomorphism of the ring R into the ring R' and I is an ideal contained in the kernel of η. Then we know that

$$\bar{\eta} : \bar{a} = a + I \to \eta(a)$$

is a group and a monoid homomorphism, hence it is a ring homomorphism. We call $\bar{\eta}$ the *induced (ring) homomorphism* of R/I into R'. It is clear that we

have the commutativity of

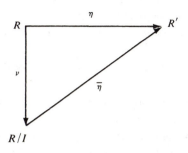

and $\bar{\eta}$ is the only homomorphism from R/I to R' making this diagram com-mutative. Also $\bar{\eta}$ is a monomorphism if and only if I coincides with the kernel of η. In this case we have the

FUNDAMENTAL THEOREM OF HOMOMORPHISMS OF RINGS.
Let η be a homomorphism of a ring R into a ring R', $K = \eta^{-1}(0')$ the kernel. Then K is an ideal in R and we have a unique homomorphism $\bar{\eta}$ of R/K into R' such that $\eta = \bar{\eta}v$ where v is the natural homomorphism of R into R/K. Moreover, v is an epimorphism and $\bar{\eta}$ is a monomorphism.

This, of course, has the immediate

COROLLARY.   *Any homomorphic image of a ring R is isomorphic to a quotient ring R/K of R by an ideal K.*

The subgroup correspondence of a group and a homomorphic image given in Theorem 1.8' is applicable to rings via their additive groups. The result for rings is

THEOREM 2.6.   *Let η be an epimorphism of a ring R onto a ring R', K the kernel. Then in the 1–1 correspondence of the set of subgroups H of $(R, +, 0)$ containing K with the set of subgroups of R' pairing H with $\eta(H)$, H is a subring (ideal) if and only if $\eta(H)$ is a subring (ideal) of R'. Moreover, if I is an ideal of R containing K then*

(19) $a + I \rightarrow \eta(a) + I',\qquad I' = \eta(I)$

is an isomorphism of R/I with R'/I'.

Proof. Since the image under a homomorphism is a subring it is clear that if H is a subring of R then $\eta(H)$ is a subring of R'. If H is an ideal in R, then $\eta(H)$ is a subgroup of the additive group of R'. If $h \in H$ and $x' \in R'$ then there exists an x such that $\eta(x) = x'$. Hence $\eta(h)x' = \eta(h)\eta(x) = \eta(hx) \in \eta(H)$ and similarly $x'\eta(h) \in \eta(H)$. Hence $\eta(H)$ is an ideal. If H' is a subring (ideal) in R' then $\eta^{-1}(H')$ is a subgroup of the additive group of R and it is immediate that this is a subring (ideal) of R. It follows that the $1-1$ correspondence between the set of subgroups of the additive group of R containing K with the set of subgroups of R' induces $1-1$ correspondences between the sets of subrings and also between the sets of ideals contained in the two sets of subgroups. Also we know from the group result that (19) is an isomorphism of the additive groups of R/I and R'/I' if I is an ideal in R containing K and $I' = \eta(I)$. Since

$$(a + I)(b + I) = ab + I \to \eta(ab) + I' = \eta(a)\eta(b) + I'$$
$$= (\eta(a) + I')(\eta(b) + I')$$

(19) is a ring isomorphism. $\square$

The isomorphism of R/I and R'/I' given in the foregoing theorem is sometimes called the *first isomorphism theorem for rings*. We also have, as we have for groups, the

SECOND ISOMORPHISM THEOREM FOR RINGS. *Let R be a ring, S a subring, I an ideal in R. Then $S + I = \{s + i \mid s \in S, \ i \in I\}$ is a subring of R containing I as an ideal, $S \cap I$ is an ideal in S, and we have the isomorphism*

(20) $$s + I \to s + (S \cap I), \qquad s \in S$$

of $(S + I)/I$ with $S/(S \cap I)$.

Proof. Direct verification shows that $S + I$ is a subring. Obviously I is an ideal in $S + I$. We have the homomorphism $s \to s + I$ of S into R/I which is the restriction to S of the natural homomorphism of R into R/I. The image is clearly $(S + I)/I$ and the kernel is the set of s such that $s + I = I$. This is the set $S \cap I$. Hence we have the isomorphism $s + (S \cap I) \to s + I$ of $S/(S \cap I)$ into $(S + I)/I$. The isomorphism (20) is the inverse of this map. $\square$

We shall now apply the fundamental homomorphism theorem of rings to identify the smallest subring of a given ring R, that is, the subring generated by 1. We shall call this the *prime ring* of R (though it may have nothing to do with primes). For our purpose we need to use the ring of integers $\mathbb{Z}$ with unit 1 and for the moment it will be clearer if we use a different symbol, say e, for the unit of R. Consider the map $n \to ne$, $n \in \mathbb{Z}$, of $\mathbb{Z}$ into R. Since

$$(n + m)e = ne + me$$

$$(nm)e = (nm)e^2 = (ne)(me)$$

hold in R (see section 2.1) and $1 \to e$, our map is a homomorphism of $\mathbb{Z}$ into R. The image $\mathbb{Z}e = \{ne \mid n \in \mathbb{Z}\}$ is therefore a subring of R. Moreover, if S is any subring of R then $e \in S$ and so $\mathbb{Z}e \subset S$. Hence it is clear that $\mathbb{Z}e$ is the prime ring. Our homomorphism can also be regarded as one into $\mathbb{Z}e$, in which case it is an epimorphism. Consequently $\mathbb{Z}e \cong \mathbb{Z}/K$ for some ideal K in $\mathbb{Z}$ and we know that $K = (k)$, $k \geq 0$. If $k = 0$ we have $\mathbb{Z}e \cong \mathbb{Z}$ and if $k > 0$ then $\mathbb{Z}e$ is isomorphic to the ring of residues modulo k. We can now safely shift back to the notation 1 for the unit of R and we can identify the prime ring with the ring $\mathbb{Z}$ or $\mathbb{Z}/(k)$ to which it is isomorphic. With this understanding we have the following

THEOREM 2.7. *The prime ring of a ring R is either $\mathbb{Z}$ or the ring $\mathbb{Z}/(k)$ of residues modulo some $k > 0$.*

We recall that if k is composite then $\mathbb{Z}/(k)$ has non-zero zero divisors. Hence if R is a domain then the prime ring is either $\mathbb{Z}$ or $\mathbb{Z}/(p)$ for some prime p. We shall say that R is of *characteristic* k if its prime ring is $\mathbb{Z}/(k)$, $k \geq 0$ (so that $\mathbb{Z}/(0) = \mathbb{Z}$). Hence for a domain the characteristic is either 0 or a prime p. We remark also that if the characteristic of a ring is $k > 0$ then $ka = (k1)a = 0$ for all a in the ring. Clearly, k is the smallest positive integer having this property.

EXERCISES

1. Prove that if η is a homomorphism of the ring R into the ring R' and ζ is a homomorphism of R' into R'' then $\zeta\eta$ is a homomorphism of R into R''.

2. Show that if u is a unit in R and η is a homomorphism of R into R' then $\eta(u)$ is a unit in R'. Suppose η is an epimorphism. Does this imply that η is an epimorphism of the group of units of R onto the group of units of R'?

3. Let I be an ideal in R, n a positive integer. Apply the fundamental theorem on homomorphisms to prove that $M_n(R)/M_n(I) \cong M_n(R/I)$.

4. Show that if R is a commutative ring of prime characteristic p then $a \to a^p$ is an endomorphism of R ($=$ homomorphism of R into R). Is this an automorphism?

5. Let F be a finite field of characteristic p (a prime). Show that $p - 1 \big| |F| - 1$. Hence conclude that if $|F|$ is even then the characteristic is two. (We shall see later that $|F|$ is a power of p.)

6. A ring R is *simple* if $R \neq 0$ and R and 0 are the only ideals in R. Show that the characteristic of a simple ring is either 0 or a prime p.

7. If S is a subset of a ring (field) R then *the subring (subfield) generated by S* is defined to be the intersection of all the subrings (subfields) containing S. If this is R itself then S is called a *set of generators* of the ring R (field R). Show that if η_1 and η_2 are homomorphisms of the ring R (field R) into a second ring (field) and $\eta_1(s) = \eta_2(s)$ for every s in a set of generators of the ring R (field R) then $\eta_1 = \eta_2$.

8. Show that every homomorphism of a division ring into a ring $R \neq 0$ is a monomorphism.

9. If $R_1, R_2, \ldots, R_n$ are rings we define the *direct sum* $R_1 \oplus R_2 \oplus \cdots \oplus R_n$ as for monoids and groups. The underlying set is $R = R_1 \times R_2 \times \cdots \times R_n$. Addition, multiplication, 0, and 1 are defined by

$$(a_1, a_2, \ldots, a_n) + (b_1, b_2, \ldots, b_n) = (a_1 + b_1, a_2 + b_2, \ldots, a_n + b_n)$$

$$(a_1, a_2, \ldots, a_n)(b_1, b_2, \ldots, b_n) = (a_1 b_1, a_2 b_2, \ldots, a_n b_n)$$

$$0 = (0_1, 0_2, \ldots, 0_n)$$

$$1 = (1_1, 1_2, \ldots, 1_n),$$

0_i, 1_i the zero and unit of R_i. Verify that R is a ring. Show that the units of R are the elements $(u_1, u_2, \ldots, u_n)$, u_i a unit of R_i. Hence show that if $U = U(R)$ and $U_i = U(R_i)$ then $U = U_1 \times U_2 \times \cdots \times U_n$, the direct product of the U_i, and that $|U| = \Pi |U_i|$ if the U_i are finite.

10. (Chinese remainder theorem). Let I_1 and I_2 be ideals of a ring R which are *relatively prime* in the sense that $I_1 + I_2 = R$. Show that if a_1 and a_2 are elements of R then there exists an $a \in R$ such that $a \equiv a_i \pmod{I_i}$. More generally, show that if $I_1, \ldots, I_m$ are ideals such that $I_j + \bigcap_{k \neq j} I_k = R$ for $1 \leq j \leq m$, then for any $(a_1, a_2, \ldots, a_m)$, $a_i \in R$, there exists an $a \in R$ such that $a \equiv a_k \pmod{I_k}$ for all k.

11. Use the Chinese remainder theorem and the fundamental theorem of homomorphisms to show that if I_1 and I_2 are relatively prime ideals and $I = I_1 \cap I_2$ then $R/I \cong R/I_1 \oplus R/I_2$.

12. Use exercise 11 to prove that if m and n are relatively prime integers then $\varphi(mn) = \varphi(m)\varphi(n)$, φ the Euler φ-function (p. 105). Show also that if p is a prime then $\varphi(p^e) = p^e - p^{e-1}$. Hence prove that if $n = p_1^{e_1} p_2^{e_2} \cdots p_r^{e_r}$, p_i distinct primes, then

$$\varphi(n) = \prod_{i=1}^{r} (p_i^{e_i} - p_i^{e_i - 1}) = n \prod_{1}^{r} \left(1 - \frac{1}{p_i}\right).$$

13. Show that the only ring homomorphism of $\mathbb{R}$ into $\mathbb{R}$ is the identity.

14. Let R be the ring of real-valued continuous functions on $[0, 1]$ (example 8, p. 87). Note that if $0 \leq t \leq 1$ then the evaluation map $\eta_t : f \rightarrow f(t)$ is a homomorphism of R into $\mathbb{R}$. Show that any homomorphism η of R into $\mathbb{R}$ is of this form. (*Hint:* If $\eta \neq \eta_t$ there is an $f_t \in R$ such that $\eta(f_t) \neq \eta_t(f_t) = f_t(t)$. Then $g_t = f_t - \eta(f_t)1 \in R$ and $g_t(t) \neq 0$ but $\eta(g_t) = 0$. Show that there exist a finite number of t_i such that $g(x) = \sum g_{t_i}^2(x) \neq 0$ for all x. Then $g^{-1} \in R$ but $\eta(g) = 0$.)

15. Define a *maximal ideal* of a ring R to be a proper ideal I such that there exists no proper ideal I' such that $I' \supsetneq I$. Show that an ideal I of a commutative ring R is maximal if and only if R/I is a field.

16. Define a *prime ideal* I of a commutative ring R by the conditions: $I \neq R$ and if $ab \in I$ then either $a \in I$ or $b \in I$. Show that if I is maximal then I is prime.

17. Determine the ideals and the maximal ideals and prime ideals of $\mathbb{Z}/(60)$.

2.8 ANTI-ISOMORPHISMS

Let R be a commutative ring, $M_n(R)$ the ring of $n \times n$ matrices with entries in R. If $A = (a_{ij}) \in M_n(R)$ we define the *transpose* of A (or *transposed matrix*) ^tA to be the matrix having a_{ji} as its (i, j)-entry. This means that ^tA is obtained by reflecting the elements of A in its main diagonal. For example, if

$$A = \begin{pmatrix} 1 & 2 & 3 \\ 2 & -1 & 4 \\ 5 & -1 & 6 \end{pmatrix}$$

then

$$^tA = \begin{pmatrix} 1 & 2 & 5 \\ 2 & -1 & -1 \\ 3 & 4 & 6 \end{pmatrix}.$$

It is clear that $^t(^tA) = A$, so $A \rightarrow {}^tA$ is bijective. Also, if $A = (a_{ij})$ and $B = (b_{ij})$ then $A + B = (a_{ij} + b_{ij})$, so $^t(A + B)$ has $a_{ji} + b_{ji}$ as its (i, j)-entry. Hence $^t(A + B) = {}^tA + {}^tB$. Thus the transpose map $t : A \rightarrow {}^tA$ is an automorphism of the additive group of $M_n(R)$. Clearly $^t1 = 1$. Now consider $P = AB$ whose (i, j)-entry is $p_{ij} = \sum_{k=1}^{n} a_{ik}b_{kj}$. Hence the (i, j)-entry of ^tP is $\sum_{k=1}^{n} a_{jk}b_{ki}$. On the other hand, the (i, j)-entry of $^tB \, ^tA$ is $\sum_{k=1}^{n} b_{ki}a_{jk} = \sum_{k=1}^{n} a_{jk}b_{ki}$, since R is commutative. We have shown that

(21) $$^t(AB) = (^tB)(^tA).$$

A map $x \rightarrow x^*$ of a ring R into itself which is an automorphism of the additive group, sends 1 into 1 and reverses the order of multiplication: $(xy)^* = y^*x^*$ is called an *anti-automorphism* of R. If, in addition, $x** = x$, $x \in R$, then the map is called an *involution*. Our calculations show that this is the case with the transpose map in $M_n(R)$, R commutative.

Another important instance of an involution is the map

(22)
$$x = \alpha_0 + \alpha_1 i + \alpha_2 j + \alpha_3 k \to \bar{x}$$
$$= \alpha_0 - \alpha_1 i - \alpha_2 j - \alpha_3 k, \qquad \alpha_i \in \mathbb{R},$$

in Hamilton's quaternion algebra $\mathbb{H}$. This can be verified directly or it can be deduced from the anti-automorphic character of the transpose map, as we proceed to show. We observe first that if u is an invertible element of a ring then the map $x \to uxu^{-1}$ is an automorphism. As in the case of groups, such automorphisms are called *inner* automorphisms. We note next that if we compose an automorphism with an anti-automorphism in either order the result is an anti-automorphism. As a consequence of these two remarks we see that the map

$$\begin{pmatrix} a & b \\ c & d \end{pmatrix} \to \begin{pmatrix} 0 & -1 \\ 1 & 0 \end{pmatrix}\begin{pmatrix} a & c \\ b & d \end{pmatrix}\begin{pmatrix} 0 & 1 \\ -1 & 0 \end{pmatrix} = \begin{pmatrix} d & -b \\ -c & a \end{pmatrix} \equiv \mathrm{adj}\begin{pmatrix} a & b \\ c & d \end{pmatrix}$$

is an anti-automorphism in $M_2(R)$. Moreover, the formula for $\mathrm{adj}\begin{pmatrix} a & b \\ c & d \end{pmatrix}$ shows that $\mathrm{adj}\left(\mathrm{adj}\begin{pmatrix} a & b \\ c & d \end{pmatrix}\right) = \begin{pmatrix} a & b \\ c & d \end{pmatrix}$. Hence the "adj" map is an involution. We now specialize $R = \mathbb{C}$ and we refer back to the definition of $\mathbb{H}$ as the subring of $M_2(\mathbb{C})$ of matrices of the form $\begin{pmatrix} a & b \\ -\bar{b} & \bar{a} \end{pmatrix}$. We recall also the definitions of i, j, k as

$$i = \begin{pmatrix} \sqrt{-1} & 0 \\ 0 & -\sqrt{-1} \end{pmatrix}, \qquad j = \begin{pmatrix} 0 & 1 \\ -1 & 0 \end{pmatrix}, \qquad k = \begin{pmatrix} 0 & \sqrt{-1} \\ \sqrt{-1} & 0 \end{pmatrix}.$$

Then $\mathrm{adj}\, i = -i$, $\mathrm{adj}\, j = -j$, and $\mathrm{adj}\, k = -k$. Thus the involution $x \to \mathrm{adj}\, x$ in $\mathbb{C}_2$ stabilizes $\mathbb{H}$ and induces the involution $x \to \bar{x}$, as in (22), in $\mathbb{H}$.

A map $x \to x'$ of a ring R into a ring R' is called an *anti-isomorphism* if it is an isomorphism for the additive groups and satisfies

(23) $(xy)' = y'x', \qquad 1 \to 1'$, the unit of R'.

If such a map exists, then R and R' are said to be *anti-isomorphic*. It is sometimes useful to have a ring which is anti-isomorphic to a given ring R. Such a ring can be constructed easily. To do this we take the same underlying set R, the same +, 1 and 0, but we define a new product by simply reversing the factors and then multiplying as in R. Denoting this product as $a \times b$ we have the definition:

(24) $$a \times b = ba.$$

Then

$$(a \times b) \times c = ba \times c = c(ba)$$
$$a \times (b \times c) = a \times cb = (cb)a.$$

and

$$a \times (b + c) = (b + c)a = ba + ca = a \times b + a \times c$$
$$(b + c) \times a = a(b + c) = ab + ac = b \times a + c \times a.$$

Also $a \times 1 = 1a = a = a1 = 1 \times a$. Hence $(R, +, \times, 0, 1)$ is a ring. To distinguish this from $R = (R, +, \cdot, 0, 1)$ we shall denote it as R^0 (read "R opposite") and call it the *opposite (ring) of R*. It is clear that the identity map is an anti-isomorphism of R and R^0. Also any anti-isomorphism of R is the same thing as an isomorphism of R^0.

EXERCISES

1. Show that the identity map in R is an anti-automorphism if and only if R is commutative.

2. Show that $x = \alpha_0 + \alpha_1 i + \alpha_2 j + \alpha_3 k \to x^* = \alpha_0 - \alpha_1 i + \alpha_2 j + \alpha_3 k$ is an involution in $\mathbb{H}$.

3. Let $x \to x'$ be an anti-isomorphism of R onto R'. If $A = (a_{ij})$ let $A^* = {}^t(a'_{ij})$. Verify that $A \to A^*$ is an anti-isomorphism of $M_n(R)$ onto $M_n(R')$.

4. Let $a \to a^*$ be an anti-automorphism of a ring R. Let $H = \{h \mid h^* = h\}$ (called *symmetric* or $*$ - *symmetric elements*) and $K = \{k \mid k^* = -k\}$ (called *skew* or $*$ - *skew elements*). Verify that H and K are subgroups of the additive group of R. Define $\{ab\} = ab + ba$ and $[ab] = ab - ba$. Show that if $a, b, c \in H$ then so do

$$aba, a^n \text{ for } n \in \mathbb{N}, \{ab\}, abc + cba, [[ab]c],$$

and that $[ab] \in K$. Show that if $a, b \in K$ then $[ab] \in K$ and if $a \in H$ and $b \in K$ then $[ab] \in H$.

5.   Let

$$u = \begin{pmatrix} 0 & 1 & 0 \\ 0 & 0 & 1 \\ 0 & 0 & 0 \end{pmatrix}$$

in $M_3(\mathbb{Q})$ and let

$$x = \begin{pmatrix} u & 0 \\ 0 & u^2 \end{pmatrix}, \qquad y = \begin{pmatrix} 0 & 1 \\ 0 & 0 \end{pmatrix}$$

where u is as indicated and 0 and 1 are the 0 and unit matrices in $M_3(\mathbb{Q})$. Hence $x, y \in M_6(\mathbb{Q})$. Verify the following relations

$$x^3 = 0 = y^2, \qquad yx = x^2 y.$$

Let R be the subring of $M_6(\mathbb{Q})$ generated by $\mathbb{Q}$, x and y. Show that every element of R has the form $f(x) + g(x)y$ where $f(x) = a + bx + cx^2$, $g(x) = a' + b'x + c'x^2$, and $a, b, c,\ a', b', c' \in \mathbb{Q}$, and that $(1, x, x^2, y, yx, yx^2)$ is a base for R as vector space over $\mathbb{Q}$. Show that if x' is a nilpotent element of R and y' is an element of R such that $y'^2 = 0$, then $y'x'^2 = 0$. Hence conclude that R has no anti-automorphisms.

6.   Define *anti-homomorphism* of a ring R into a ring R' to be a map η which is a homomorphism of the additive group of R into R' sending 1 into 1 (for 1') and satisfying $\eta(ab) = \eta(b)\eta(a)$. Verify that the composite of a homomorphism (anti-homomorphism) and an anti-homomorphism (homomorphism) is an anti-homo-morphism and the composite of two anti-homomorphisms is a homomorphism.

7.   Define a *Jordan homomorphism* η of a ring R into a ring R' by the conditions: η is an additive group homomorphism, $\eta(1) = 1$, and $\eta(aba) = \eta(a)\eta(b)\eta(a)$. Show that any homomorphism or anti-homomorphism is a Jordan homomorphism. Show that Jordan homomorphisms satisfy:

$$\eta(a^k) = \eta(a)^k, k \in \mathbb{N}$$
$$\eta(abc + cba) = \eta(a)\eta(b)\eta(c) + \eta(c)\eta(b)\eta(a)$$
$$\eta(ab + ba) = \eta(a)\eta(b) + \eta(b)\eta(a).$$

8.   (Jacobson and Rickart.) Show that if η is a Jordan homomorphism of a ring R into a domain D then for any $a, b \in R$ either $\eta(ab) = \eta(a)\eta(b)$ or $\eta(ab) = \eta(b)\eta(a)$.

9.   (Hua.) Let η be a mapping of a ring R into a ring R' such that $\eta(a + b) = \eta(a) + \eta(b)$, $\eta(1) = 1$, and for any a, b in R either $\eta(ab) = \eta(a)\eta(b)$ or $\eta(ab) = \eta(b)\eta(a)$. Prove that η is either a homomorphism or anti-homomorphism.

10. (Jacobson and Rickart.) Prove that any Jordan homomorphism of a ring into a domain is either a homomorphism or an anti-homomorphism.

11. (Hua.) Let η be a map of a division ring D into a division ring D' satisfying the following conditions: (i) η is a homomorphism of the additive groups, (ii) $\eta(1) = 1'$, (iii) if $a \neq 0$ then $\eta(a) \neq 0$ and $\eta(a)^{-1} = \eta(a^{-1})$. Show that η is either a homomor-phism or an anti-homomorphism. (*Hint:* Use Hua's identity, exercise 9, p. 92).

2.9 FIELD OF FRACTIONS OF A COMMUTATIVE DOMAIN

We have seen that any subring of a division ring is a domain. It is natural to ask if the converse holds: namely, can every domain be imbedded in a division ring? By this we mean: given domain D, does there exist a monomorphism of D into some division ring F? If this were the case then D would be isomorphic to a subring D' of F, so that by identifying D with D' we could regard D as a subring of the division ring F. The question we have raised was an open one for some time until it was answered in the negative by A. Malcev, who gave the first example of a domain which cannot be imbedded in a division ring. We shall indicate Malcev's example in some exercises below. Our main concern in this section will be in the most important positive result in this direction, namely, that every *commutative* domain can be imbedded in a field. The method for doing this is exactly the familiar one that is used to construct the field of rational numbers from the ring of integers. To understand why it works it will be well to look first at the relation between a subring D of a field and the subfield F generated by D.

Accordingly, we suppose we have a subring D of a field. Let F be the subfield generated by D. What are the elements of F? First it is clear that if $a, b \in D$ and $b \neq 0$ then $ab^{-1} \in F$. We now make the important observation that F is just the set of elements of this form. First, the following equations show that

$$\{ab^{-1} \,|\, a, b \in D, b \neq 0\}$$

is a subfield of the given field:

$$ab^{-1} + cd^{-1} = adb^{-1}d^{-1} + cbb^{-1}d^{-1} = (ad + bc)(bd)^{-1}$$

$$0 = 0b^{-1}$$

$$-ab^{-1} = (-a)b^{-1}$$

$$(ab^{-1})(cd^{-1}) = acb^{-1}d^{-1} = ac(bd)^{-1}$$

$$1 = aa^{-1}$$

$$(ab^{-1})^{-1} = ba^{-1} \quad \text{if} \quad a \neq 0.$$

(It should be noted that commutativity of multiplication is used in several places in these calculations.) Since F is generated by D, no subfield of F different from F contains D, and since the set of $\{ab^{-1}\}$ contains D as the subset of elements $a1^{-1} = a$, it is clear that

(25) $$F = \{ab^{-1} \,|\, a, b \in D, b \neq 0\}.$$

One more question needs to be raised. When do we have equality, $ab^{-1} = cd^{-1}$, for the elements of the set we have determined? It is clear that this is the case if and only if $ad = bc$, since this relation follows from $ab^{-1} = cd^{-1}$ if we multiply both sides by bd, and $ab^{-1} = cd^{-1}$ results if we multiply both sides of $ad = bc$ by $(bd)^{-1}$.

Suppose now that we are given a commutative domain D. We wish to imbed D in a field. The foregoing remarks indicate that if this can be done, then the elements of a minimal field extension of D are to be obtained from the pairs (a, b), $a, b \in D$, $b \neq 0$. We have in mind that (a, b) is to play the role of ab^{-1}. Hence we adopt the following procedure, which is suggested by the foregoing considerations.

Let D^* denote the set of non-zero elements of D. Then $D^* \neq \varnothing$ since $D \neq 0$. We consider the product set $D \times D^*$ of pairs (a, b), $a \in D$, $b \in D^*$ and we introduce a relation $\sim$ in $D \times D^*$ by defining $(a, b) \sim (c, d)$ if and only if $ad = bc$. Then $(a, b) \sim (a, b)$ since $ab = ba$, and if $(a, b) \sim (c, d)$, then $ad = bc$; hence $cb = da$, and so $(c, d) \sim (a, b)$. Finally, if $(a, b) \sim (c, d)$ and $(c, d) \sim (e, f)$ then $ad = bc$ and $cf = de$. Hence $adf = bcf = bde$. Since $d \neq 0$ and D is commutative, d may be cancelled to give $af = be$, which is the condition that $(a, b) \sim (e, f)$. We have therefore proved that $\sim$ is an equivalence relation. We shall call the equivalence class determined by (a, b) the *fraction* (or *quotient*) a/b. Thus we have $a/b = c/d$ if and only if $ad = bc$. Let $F = \{a/b\}$ the quotient set determined by our equivalence relation in $D \times D^*$.

We shall now introduce an addition, multiplication, 0, and 1 in F to make F a field. We note first that if a/b and c/d are two fractions, then $bd \neq 0$ since $b \neq 0$ and $d \neq 0$. Hence we can form the fraction $(ad + bc)/bd$. Moreover, if $a/b = a'/b'$ and $c/d = c'/d'$, then

$$(26) \qquad (ad + bc)/bd = (a'd' + b'c')/b'd',$$

for, by assumption, $ab' = ba'$ and $cd' = dc'$. Hence

$$ab'dd' = ba'dd' \quad \text{and} \quad cd'bb' = dc'bb'$$

so that

$$ab'dd' + cd'bb' = ba'dd' + dc'bb'$$

or

$$(ad + bc)b'd' = (a'd' + b'c')bd,$$

which implies (26). It is now clear that

$$(27) \qquad a/b + c/d = (ad + bc)/bd$$

defines a (single-valued) composition $+$ in F. Similarly we see that if a/b and c/d are fractions then so is ac/bd. Moreover, if $a/b = a'/b'$ and $c/d = c'/d'$, then $ab' = ba'$ and $cd' = c'd$, so $ab'cd' = ba'c'd$. Hence $ac/bd = a'c'/b'd'$ and so

$$(28) \qquad\qquad (a/b)(c/d) = ac/bd$$

defines a (single-valued) multiplication in F. If we put $0 = 0/1$ and $1 = 1/1$ we obtain $a/b + 0 = a/b + 0/1 = (a1 + b0)/b1 = a/b$ and similarly $0 + a/b = a/b$. Also $(a/b)1 = a/b = 1(a/b)$. A straightforward verification, which is left to the reader, will show that $(F, +, \cdot, 0, 1)$ is a commutative ring. Now suppose $a/b \neq 0$. Then $a \neq 0$, since $0/b = 0/1$ by $01 = 0 = 0b$. Hence b/a is a fraction and $(a/b)(b/a) = ab/ab = 1/1 = 1$. Thus a/b has the inverse $(a/b)^{-1} = b/a$ and hence F is a field.

We now consider the map

$$(29) \qquad\qquad a \to a/1$$

of D into F. Clearly this maps 0 into 0, 1 into 1; and $a + b \to (a + b)/1 = a/1 + b/1$ and $ab \to ab/1 = (a/1)(b/1)$. Hence (29) is a homomorphism. If $a/1 = 0 = 0/1$ then $a1 = 10 = 0$, so $a = 0$. Hence the kernel is 0 and (29) is a monomorphism.

We have therefore proved the following

THEOREM 2.8. *Any commutative domain can be imbedded in a field.*

We shall now identify a with $a/1$ (just as we identify the integer a with the rational number $a/1$). Then D is identified with a subring of F. Moreover, for any element a/b of F we have $a/b = (a/1)(1/b) = (a/1)(b/1)^{-1} = ab^{-1}$ (because of our identification). Thus it is clear that D generates the field F. We shall call F *the field of fractions of D.* The basic homomorphism property of this field is given in

THEOREM 2.9. *Let D be a commutative domain, F its field of fractions. Then any monomorphism η_D of D into a field F' has a unique extension to a monomorphism of η_F of F into F'.*

Proof. We indicate η_D as $a \to a'$. We shall prove first that if η_D can be extended to a homomorphism η_F of F into F' then this can be done in only one way. In other words, we settle the uniqueness question first. Now this part is clear, since if $b \neq 0$ then $b^{-1} \to (b')^{-1}$ under η_F. Hence $ab^{-1} \to a'(b')^{-1}$ under η_F. Since every element of F can be written as ab^{-1} it follows that η_F is determined to be the map $ab^{-1} \to a'(b')^{-1}$. It is now clear that our task is to show that $ab^{-1} \to a'(b')^{-1}$ is a well-defined map and is a monomorphism of F into F' which extends η_D. To prove that this defines a map we assume that $ab^{-1} = cd^{-1}$. Then we have $ad = bc$ and consequently $a'd' = b'c'$ in F'. Hence $a'(b')^{-1} = c'(d')^{-1}$. This shows that $ab^{-1} \to a'(b')^{-1}$ is single-valued. Next we check the homomorphism property. This follows from the following calculations in which $a, b, c, d \in D$ and $b \neq 0, d \neq 0$.

$$ab^{-1} + cd^{-1} = (ad + bc)(bd)^{-1} \to (ad + bc)'((bd)')^{-1}$$
$$= (a'd' + b'c')(b')^{-1}(d')^{-1}$$
$$= a'(b')^{-1} + c'(d')^{-1}$$
$$(ab^{-1})(cd^{-1}) = ac(bd)^{-1} \to a'c'(b')^{-1}(d')^{-1}$$
$$= (a'(b')^{-1})(c'(d')^{-1}).$$

It is clear also that $1 \to 1'$, the unit of F', since D and F have the same unit and η_D is a homomorphism. We note next that $ab^{-1} \to a'(b')^{-1}$ is an extension of η_D since it maps $a = a1^{-1} \to a'(1')^{-1} = a'$. We have seen that any homomorphism of a field is a monomorphism (exercise 8, p. 110). Hence we have proved that η_D can be extended to a monomorphism η_F of F, and we saw at the outset that this is unique. $\square$

EXERCISES

1. What is the field of fractions of a field?

2. Show that if D is a domain and F_1 and F_2 are fields such that D is a subring of each and each is generated by D, then there is a unique isomorphism of F_1 onto F_2 that is the identity map on D.

3. Show that any commutative monoid satisfying the cancellation law ($ab = ac \Rightarrow b = c$) can be imbedded in an abelian group.

4. Show that if $a^m = b^m$ and $a^n = b^n$, for m and n relatively prime positive integers, and a and b in a commutative domain, then $a = b$.

5. Let R be a commutative ring, and S a submonoid of the multiplicative monoid of R. In $R \times S$ define $(a, s) \sim (b, t)$ if there exists a $u \in S$ such that $u(at - bs) = 0$. Show that this is an equivalence relation in $R \times S$. Denote the equivalence class of (a, s) as a/s and the quotient set consisting of these classes as RS^{-1}. Show that

RS^{-1} becomes a ring relative to

$$a/s + b/t = (at + bs)/st$$
$$(a/s)(b/t) = ab/st$$
$$0 = 0/1$$
$$1 = 1/1.$$

Show that $a \to a/1$ is a homomorphism of R into RS^{-1} and that this is a monomorphism if and only if no element of S is a zero divisor in R. Show that the elements $s/1$, $s \in S$, are units in RS^{-1}.

6. (Ore.) Let D be a domain (not necessarily commutative) having the *right common multiple property* that any two non-zero elements $a, b \in D$ have a non-zero right common multiple $m = ab_1 = ba_1$. Consider $D \times D^*$, D^* the set of non-zero elements of D, and define $(a, b) \sim (c, d)$ if for $b_1 \neq 0$ and $d_1 \neq 0$ such that $bd_1 = db_1$ we have $ad_1 = cb_1$. Show that this is independent of the choice of b_1, d_1 and that $\sim$ is an equivalence relation in $D \times D^*$. Let F denote the set of equivalence classes a/b. Show that F becomes a division ring relative to $a/b + c/d = (ad_1 + cb_1)/m$ where $m = bd_1 = db_1 \neq 0$, $0 = 0/1$, $1 = 1/1$, $(a/b)(c/d) = ac_1/db_1$ where $b_1 \neq 0$ and $cb_1 = bc_1$. Show that $a \to a/1$ is a monomorphism of D into F and F is the set of elements $(a/1)(b/1)^{-1}$, $a, b \in D$, $b \neq 0$.

7. (Malcev.) Show that if a_i, b_i, $1 \leq i \leq 4$, are elements of a group satisfying the relations $a_1 a_2 = a_3 b_4$, $a_1 b_2 = a_3 b_4$, $b_1 a_2 = b_3 a_4$, then $b_1 b_2 = b_3 b_4$. Let W be the free monoid generated by elements a_i, b_i, $1 \leq i \leq 4$ (see p. 68), and let $\equiv$ be the smallest congruence relation ($=$ intersection of all congruence relations) in W containing the elements $(a_1 a_2, a_3 a_4)$, $(a_1 b_2, a_3 b_4)$, $(b_1 a_2, b_3 a_4)$. Let $S = W/\equiv$. Show that S satisfies the cancellation laws but that S cannot be imbedded in a group.

8. (Malcev.) Let $\mathbb{Z}[S]$ be the set of integral linear combinations of the elements of the monoid S of exercise 7 with the obvious definitions of equality, addition, multiplication, 0, and 1 (see exercise 8, p. 127). Show that $\mathbb{Z}[S]$ is a domain that cannot be imbedded in a division ring.

2.10   POLYNOMIAL RINGS

For the remainder of this chapter—except in section 2.17 and in an occasional exercise—*all rings will be commutative* and the word "ring" will be synonymous with "commutative ring."

One is often interested in studying a ring R' relative to a given subring R. In this connection we wish to consider subrings of R' generated by R and subsets U of R'. Such a subring will be denoted as $R[U]$ and will be called the subring obtained by "adjoining" the subset U to the subring R. If V is a second subset then $R[U][V]$ is the subring obtained by adjoining V to the subring $R[U]$. We claim that this coincides with $R[U \cup V]$, the subring of R' resulting from the adjunction of $U \cup V$ to R. First, it is clear that $R[U \cup V]$ contains $R[U]$ and

V and, since the subring generated by $R[U]$ and V is contained in every subring containing these sets, we have $R[U \cup V] \supset R[U][V]$. Next, it is clear that $R[U][V]$ contains R and the subset $U \cup V$; hence $R[U][V] \supset R[U \cup V]$. Thus $R[U][V] = R[U \cup V]$.

We are interested primarily in subrings obtained by adjoining finite subsets to the "base" ring R. If $U = \{u_1, u_2, \ldots, u_n\}$ we write $R[u_1, u_2, \ldots, u_n]$ for $R[U]$. Inducting on the foregoing remark we see that

$$(30) \qquad R[u_1, u_2, \ldots, u_n] = R[u_1][u_2] \cdots [u_n],$$

that is, $R[u_1, u_2, \ldots, u_n]$ is obtained from R by a succession of adjunctions of single elements to previously constructed subrings. It is therefore natural to study first subrings of the form $R[u]$. We can immediately write down all the elements of $R[u]$; these are just the *polynomials in u with coefficients in R*, that is, the set of elements of the form

$$(31) \qquad a_0 + a_1 u + a_2 u^2 + \cdots + a_n u^n, \qquad a_i \in R.$$

It is clear that $R[u]$ contains all of these elements. Moreover, if $\sum_0^n a_i u^i$ and $\sum_0^m b_j u^j$ are polynomials in u with coefficients in R and $n \geq m$, then

$$(32) \qquad \begin{aligned} &(a_0 + a_1 u + \cdots + a_n u^n) + (b_0 + b_1 u + \cdots + b_m u^m) \\ &= (a_0 + b_0) + (a_1 + b_1)u + \cdots + (a_m + b_m)u^m \\ &\quad + a_{m+1} u^{m+1} + \cdots + a_n u^n \end{aligned}$$

and, since $(a_i u^i)(b_j u^j) = a_i b_j u^{i+j}$, we have, by the distributive laws,

$$(33) \qquad \begin{aligned} &(a_0 + a_1 u + \cdots + a_n u^n)(b_0 + b_1 u + \cdots + b_m u^m) \\ &= p_0 + p_1 u + \cdots + p_{n+m} u^{n+m} \end{aligned}$$

where

$$(34) \qquad p_i = \sum_{j=0}^{i} a_j b_{i-j} = \sum_{j+k=i} a_j b_k.$$

Moreover, 0 and 1 are polynomials in u and $-\sum_0^n a_i u^i = \sum_0^n (-a_i)u^i$. Thus the set of polynomials in u with coefficients in R form a subring of R'. Hence this set coincides with $R[u]$.

The formulas (32)–(34) show us how to calculate the sum and the product of given polynomials. All of this is simple enough. However, there is one difficulty—that of deciding when two polynomial expressions in u represent the same element. It may happen that we have different-looking expressions for the same element. For example, if $u \in R$ (which is not excluded) then the element $u \in R[u]$ can be represented both as a_0 with $a_0 = u$ and as $a_1 u$ with $a_1 = 1$. Less trivially, taking $R' = \mathbb{C}$ and $R = \mathbb{R}$, $u = \sqrt{-1}$, we have $u^2 = -1$.

We shall now construct a ring $R[x]$ in which the only relations of the form $a_0 + a_1x + \cdots = b_0 + b_1x + \cdots$ are the trivial ones in which $a_i = b_i$ for all i. Heuristically, the ring we seek is the set of expressions $a_0 + a_1x + \cdots + a_nx^n$, $a_i \in R$, where equality is *defined* by equality of the coefficients: $\sum a_ix^i = \sum b_ix^i$ only if $a_i = b_i$ for all i. Addition and multiplication will be given by (32)–(34) with x replacing u. The statement on equality means that we want a polynomial in x to determine the sequence of its coefficients and, of course, these are all 0 from a certain point on. We are therefore led to *identify* a polynomial in x with a sequence $(a_0, a_1, \ldots, a_n, 0, 0, \ldots)$, $a_i \in R$, and to introduce an addition and multiplication for such sequences corresponding to the formulas (32)–(34).

We shall now carry out this program precisely and in detail. Let R be a given ring and let $R[x]$ denote the set of infinite sequences

$$(a_0, a_1, a_2, \ldots)$$

that have only a finite number of non-zero terms a_i. Sequences $(a_0, a_1, a_2, \ldots)$ and $(b_0, b_1, b_2, \ldots)$ are regarded as equal if and only if $a_i = b_i$ for all i. In other words, $R[x]$ is the set of maps $i \to a_i$ of the set $\mathbb{N}$ of non-negative integers into the given ring R such that $a_i = 0$ for sufficiently large i. For the present, x in our notation $R[x]$ is meaningless, but a genuine x will soon make its appearance to justify the notation. We introduce a binary composition in $R[x]$ by

$$(a_0, a_1, a_2, \ldots) + (b_0, b_1, b_2, \ldots) = (a_0 + b_0, a_1 + b_1, a_2 + b_2, \ldots)$$

which evidently is in $R[x]$ and zero element by

$$0 = (0, 0, 0, \ldots).$$

Then it is immediate that $(R[x], +, 0)$ is an abelian group. Next we introduce another binary composition $\cdot$ in $R[x]$ by

(35) $$(a_0, a_1, a_2, \ldots)(b_0, b_1, b_2, \ldots) = (p_0, p_1, p_2, \ldots)$$

where p_i is given by (34). If $a_i = 0$ for $i > n$ and $b_j = 0$ for $j > m$ then $p_k = 0$ for $k > m + n$. Hence the element on the right-hand side of (35) is in $R[x]$. We also put

$$1 = (1, 0, 0, \ldots).$$

Then $(a_0, a_1, \ldots)1 = (a_0, a_1, \ldots) = 1(a_0, a_1, \ldots)$. If $A = (a_0, a_1, \ldots)$, $B = (b_0, b_1, \ldots)$, and $C = (c_0, c_1, \ldots) \in R[x]$, then the $(i+1)$-st term in $(AB)C$ is

$$\sum_{\substack{j+k=m \\ m+l=i}} (a_jb_k)c_l = \sum_{j+k+l=i} (a_jb_k)c_l.$$

Similarly, the corresponding term in $A(BC)$ is

$$\sum_{m+j=i} a_j \left(\sum_{k+l=m} b_k c_l \right) = \sum_{j+k+l=i} a_j(b_k c_l).$$

Hence $(AB)C = A(BC)$ follows from the associative law in R. Similarly, we can verify the distributive laws. Also commutativity of multiplication is clear from the definition of the p_i in (34) and the commutative law in R. Hence $(R[x], +, \cdot, 0, 1)$ is a commutative ring.

We now consider the map

$$a \to a' = (a, 0, 0, \ldots)$$

of R into $R[x]$. It is clear that this is a monomorphism of the ring R into $R[x]$. We shall now identify R with its image in $R[x]$, identifying a with a'. In this way we can regard R as a subring of $R[x]$. Now let x denote the element $(0, 1, 0, 0, \ldots)$ of $R[x]$. The formula for the product and induction on k show that if $k \geq 0$, then

$$x^k = (0, 0, \ldots, \overset{k+1}{0}, 1, 0, \ldots).$$

We have for $a \in R$ (identified with $a' = (a, 0, \ldots)$),

$$ax^k = (0, 0, \ldots, \overset{k+1}{0}, a, 0, \ldots).$$

Hence

$$(a_0, a_1, \ldots, a_n, 0, 0, \ldots) = a_0 + a_1 x + \cdots + a_n x^n$$

and $R[x]$ is the ring obtained by adjoining x to R. We shall call $R[x]$ the *ring of polynomials over R in the indeterminate x*. The foregoing formula and the definition of equality show that if $\sum a_i x^i = \sum b_i x^i$, then $a_i = b_i$ for all i. In particular, $\sum a_i x^i = 0$ implies every $a_i = 0$.

Once we have constructed the ring $R[x]$ we can use it to study any ring $R[u]$, for we shall see that any $R[u]$ is a homomorphic image of $R[x]$. Thus we shall have $R[u] \cong R[x]/I$, I an ideal in $R[x]$. This will imply that the problem of relations in $R[u]$ can be solved by noting that $a_0 + a_1 u + \cdots = b_0 + b_1 u + \cdots$ if and only if $\sum a_i x^i \equiv \sum b_i x^i \pmod{I}$. Hence we shall know the relations if we know the ideal I. The fundamental homomorphism property of $R[x]$ is given in

THEOREM 2.10. *Let R and S be (commutative) rings, η a homomorphism of R into S, u an element of S. Let $R[x]$ be the ring of polynomials over R in the indeterminate x. Then η has one and only one extension to a homomorphism η_u of $R[x]$ into S mapping x into u.*

Proof. If $A = a_0 + a_1 x + \cdots + a_n x^n$ then we simply put

$$\eta_u(A) = a_0' + a_1' u + \cdots + a_n' u^n$$

where, in general, $a' = \eta(a)$. If $B = b_0 + b_1 x + \cdots + b_m x^m$, then $AB = p_0 + p_1 x + \cdots + p_{n+m} x^{n+m}$ where $p_i = \sum_{j+k=i} a_j b_k$. Then

$$\eta_u(AB) = p_0' + p_1' u + \cdots + p_{n+m}' u^{n+m}$$

and

$$p_i' = \sum_{j+k=i} a_j' b_k'$$

since η is a ring homomorphism. On the other hand,

$$\eta_u(A)\eta_u(B) = (a_0' + a_1' u + \cdots + a_n' u^n)(b_0' + b_1' u + \cdots + b_m' u^m)$$
$$= p_0' + p_1' u + \cdots + p_{n+m}' u^{n+m} = \eta_u(AB).$$

Still easier is the verification of $\eta_u(A + B) = \eta_u(A) + \eta_u(B)$, which is left to the reader. Now we have for $a \in R$ that $\eta_u(a) = a' = \eta(a)$, so η_u is an extension of η. Also $\eta_u(1) = \eta(1) = 1$ (the unit of S) and $\eta_u(x) = u$. Hence η_u is a homomorphism of $R[x]$ which extends η and maps x into u. Since $R[x]$ is generated by R and x it is the only homomorphism having this property (exercise 7, p. 110). This completes the proof. $\square$

Now let S be any overring of R—that is, let S be a ring containing R as a subring—and let $u \in S$. Then the theorem shows that we have a unique homomorphism, which is the identity map on R and sends $x \to u$. We shall now write $A(x)$ for $A = a_0 + a_1 x + \cdots + a_n x^n$ and we shall denote the image of $A(x)$ under this homomorphism as $A(u)$. In this way we shall be using the customary functional notations in the present situation, though we are not really dealing with functions. It will be convenient also to speak of "substituting u for x in $A(x)$" when in reality what we are doing is applying the homomorphism of $R[x]$ into S which extends the identity map on R and sends x into u. If I is the kernel of our homomorphism, then $R[u] \cong R[x]/I$. Since the homomorphism is the identity on R, we have $R \cap I = 0$. This result tells us precisely what the rings $R[u]$ obtained by adjoining a single element u to R look like: namely, we have the

COROLLARY. $R[u] \cong R[x]/I$ where x is an indeterminate and I is an ideal in $R[x]$ such that $I \cap R = 0$

Conversely, if I is an ideal in $R[x]$ such that $I \cap R = 0$, then the restriction to R of the natural homomorphism v of $R[x]$ into $R[x]/I$ is a monomorphism. We may identify R with its image (the element $a \in R$ with the coset $a + I$). In this way $R[x]/I \supset R$ as a subring. Since $R[x]$ is generated by R and x, its homomorphic image is generated by R and $u = x + I$. Hence $R[x]/I \cong R[u]$. $\square$

The homomorphism $A(x) \to A(u)$ is a monomorphism if and only if $A(u) = 0$ implies $A(x) = 0$, that is, $a_0 + a_1 u + \cdots + a_n u^n = 0$ implies every $a_i = 0$. In this case u is called *transcendental* over R, otherwise u is *algebraic* over R. The classical case of this is the one in which $S = \mathbb{R}$ (or $\mathbb{C}$) and $R = \mathbb{Q}$. Then a real (or complex) number is called *algebraic* or *transcendental* according as this element of $\mathbb{R}$ (or $\mathbb{C}$) is algebraic or transcendental over $\mathbb{Q}$.

We shall now consider the extension of all of this from one element to a finite number. Reversing somewhat the foregoing order of presentation, we shall launch directly into the generalization of Theorem 2.10, which we state in the following form.

THEOREM 2.11.   *For any ring R and any positive integer r there exists a ring $R[x_1, x_2, \ldots, x_r]$ with the following "universal" property. If S is any ring and η is a homomorphism of R into S and $i \to u_i$ is a map of $\{1, 2, \ldots, r\}$ into S, then there exists a unique extension of η to a homomorphism $\eta_{u_1, \ldots, u_r}$ of $R[x_1, \ldots, x_r]$ into S sending $x_i \to u_i$, $1 \le i \le r$.*

Proof.   We define $R[x_1, \ldots, x_r]$ inductively: $R[x_1]$ is the polynomial ring in an indeterminate x_1 (for x) over R and, generally, $R[x_1, \ldots, x_i]$ is the polynomial ring in an indeterminate x_i over $R[x_1, \ldots, x_{i-1}]$, $1 \le i \le r$. By Theorem 2.10, we have a homomorphism η_{u_1} of $R[x_1]$ into S extending η and sending $x_1 \to u_1$. Using induction, we may assume we have a homomorphism of $R[x_1, \ldots, x_{r-1}]$ extending η and sending $x_i \to u_i$, $1 \le i \le r - 1$. Then Theorem 2.10 provides an extension of this to a homomorphism $\eta_{u_1, \ldots, u_r}$ of

$$R[x_1, \ldots, x_r] = R[x_1, \ldots, x_{r-1}][x_r]$$

into S sending $x_r \to u_r$. Then $\eta_{u_1, \ldots, u_r}$ is a homomorphism extension of η to $R[x_1, \ldots, x_r]$ such that $x_i \to u_i$, $1 \le i \le r$. The uniqueness of $\eta_{u_1, \ldots, u_r}$ is clear since $R[x_1, \ldots, x_r]$ is generated by R and the x's. $\square$

There is essentially only one ring having the property stated in Theorem 2.11. To show this, suppose that $R[y_1, \ldots, y_r]$ is another one. Then we have a homomorphism ζ of $R[x_1, \ldots, x_r]$ into $R[y_1, \ldots, y_r]$ which is the identity on R and sends $x_i \to y_i$, $1 \le i \le r$. We also have a homomorphism ζ' of $R[y_1, \ldots, y_r]$ into

$R[x_1, \ldots, x_r]$ which is the identity on R and sends $y_i \to x_i, 1 \leq i \leq r$. Then $\zeta'\zeta$ is an endomorphism of $R[x_1, \ldots, x_r]$, which is the identity on R and the x's. Hence $\zeta'\zeta$ is the identity automorphism of $R[x_1, \ldots, x_r]$. Similarly, $\zeta\zeta'$ is the identity on $R[y_1, \ldots, y_r]$. Then ζ and ζ' are isomorphisms.

We shall now call $R[x_1, \ldots, x_r]$ the *ring of polynomials over R in r indeterminates* $x_1, \ldots, x_r$. The result just proved shows that how one constructs this ring is only a matter of esthetics, since it is essentially unique. (Another construction will be indicated in exercise 9, at the end of this section.) Though our construction (by *successive* adjunctions of single indeterminates) does not treat the x's symmetrically, the end product is symmetric. In fact, we have the following

THEOREM 2.12. *Let $R[x_1, \ldots, x_r]$ be the polynomial ring in r indeterminates over R and let π be a permutation of $1, 2, \ldots, r$. Then there exists a unique automorphism $\zeta(\pi)$ of $R[x_1, \ldots, x_r]$ which is the identity on R and sends $x_i \to x_{\pi(i)}$, $1 \leq i \leq r$.*

Proof. Theorem 2.11 gives a unique endomorphism $\zeta(\pi)$ satisfying the stated conditions. We have to show that this is an automorphism. Now, if we compare effects on the set of generators $R \cup \{x_1, \ldots, x_r\}$, we see that if π_1 and π_2 are two permutations of $1, \ldots, r$, then $\zeta(\pi_1\pi_2) = \zeta(\pi_1)\zeta(\pi_2)$. Also $\zeta(1) = 1$. Hence $\zeta(\pi)\zeta(\pi^{-1}) = 1 = \zeta(\pi^{-1})\zeta(\pi)$. Thus $\zeta(\pi)$ is an automorphism. $\square$

If $(i_1, \ldots, i_r) \in \mathbb{N}^{(r)}$, that is, we have an r-tuple of non-negative integers, then we can associate with this the *monomial* $x_1^{i_1} \cdots x_r^{i_r}$ in the x's. We have $(x_1^{i_1} \cdots x_r^{i_r})(x_1^{j_1} \cdots x_r^{j_r}) = x_1^{i_1+j_1} \cdots x_r^{i_r+j_r}$. It follows readily from this as in the special case $r = 1$) that $R[x_1, \ldots, x_r]$ is the set of *polynomials* $\sum a_{i_1 \ldots i_r} x_1^{i_1} \cdots x_r^{i_r}$ (finite sum) where the *coefficients* $a_{i_1 \ldots i_r} \in R$. For example, $R[x, y]$ is the set of polynomials

$$a_{00} + a_{10}x + a_{01}y + a_{20}x^2 + a_{11}xy + a_{02}y^2 + \cdots, \qquad a_{ij} \in R.$$

We shall now show that if $(i_1, \ldots, i_r) \neq j_1, \ldots, j_r)$ then the associated monomials $x_1^{i_1} \cdots x_r^{i_r}, x_1^{j_1} \cdots x_r^{j_r}$ are distinct and the only relations $\sum a_{i_1 \ldots i_r} x_1^{i_1} \cdots x_r^{i_r} = 0$ connecting distinct monomials are the trivial ones with every $a_{i_1 \ldots i_r} = 0$. This will follow by showing that if

$$\sum_{(i)} a_{i_1 \ldots i_r} x_1^{i_1} \cdots x_r^{i_r} = 0, \qquad (i) = (i_1, \ldots, i_r)$$

where the summation is taken over a finite number of distinct elements $(i) \in \mathbb{N}^{(r)}$, then every coefficient is 0. Note that this will imply that for $(i) \neq (j)$, $x_1^{i_1} \cdots x_r^{i_r} \neq x_1^{j_1} \cdots x_r^{j_r}$ since, otherwise, we have the non-trivial relation

$1x_1^{i_1} \cdots x_r^{i_r} - 1x_1^{j_1} \cdots x_r^{j_r} = 0$. To prove our assertion we observe that the case $r = 1$ has already been established and we assume the result for $r - 1$ if $r > 1$. We can write

$$\sum_{(i)} a_{i_1 \cdots i_r} x_1^{i_1} \cdots x_r^{i_r} = \sum_{i_r} A_{i_r} x_r^{i_r}$$

where i_r ranges over a finite subset of $\mathbb{N}$ and

$$A_{i_r} = \sum_{(i')} a_{i_1 \cdots i_{r-1} i_r} x_1^{i_1} \cdots x_{r-1}^{i_{r-1}}$$

where $(i') = (i_1, \ldots, i_{r-1})$, and the summation is taken over a finite set of distinct (i'). If $\sum_{(i)} a_{i_1 \cdots i_r} x_1^{i_1} \cdots x_r^{i_r} = 0, \sum A_{i_r} x_r^{i_r} = 0, i_r = 0, 1, 2, \ldots$. Then every $A_{i_r} = 0$ and so, by induction, we conclude that $a_{i_1 \cdots i_{r-1} i_r} = 0$ for any fixed i_r and every (i'). Then $a_{i_1 \cdots i_r} = 0$ for every (i).

As in the case $r = 1$ treated before, we see that for any $R[u_1, \ldots, u_r]$ the homomorphism of $R[x_1, \ldots, x_r]$ into $R[u_1, \ldots, u_r]$ sending $a \to a$, $a \in R$, and $x_i \to u_i, 1 \leq i \leq r$, is an isomorphism if and only if the following independence property holds for the u's: $\sum_{(i)} a_{i_1 \cdots r} u_1^{i_1} \cdots u_r^{i_r} = 0$ only if every $a_{i_1 \cdots i_r} = 0$. If this is the case the r elements $u_1, \ldots, u_r$ are said to be *algebraically independent over R*. It is clear that this property of the x's gives another characterization of the ring $R[x_1, \ldots, x_r]$ as an extension of R.

EXERCISES

1. Show that the complex number $\omega = -\frac{1}{2} + \frac{1}{2}\sqrt{3}i$ ($i = \sqrt{-1}$) is algebraic (over $\mathbb{Q}$). Show that $\mathbb{Q}[\omega] \cong \mathbb{Q}[x]/I$ where I is the principal ideal $(x^2 + x + 1)$.

2. Show that $\sqrt{3} \notin \mathbb{Q}[\sqrt{2}]$ and that the real numbers $1, \sqrt{2}, \sqrt{3}, \sqrt{6}$ are linearly independent over $\mathbb{Q}$. Show that $u = \sqrt{2} + \sqrt{3}$ is algebraic and determine an ideal I such that $\mathbb{Q}[x]/I \cong \mathbb{Q}[u]$.

3. Let I be an ideal in R and let $I[x_1, \ldots, x_r]$ denote the subset of $R[x_1, \ldots, x_r]$ of polynomials whose coefficients are contained in I. Show that $I[x_1, \ldots, x_r]$ is an ideal in the ring $R[x_1, \ldots, x_r]$, and that $R[x_1, \ldots, x_r]/I[x_1, \ldots, x_r] \cong (R/I)[y_1, \ldots, y_r]$ where the y_i are indeterminates over R/I.

4. Let $\Delta = \prod_{i > j}(x_i - x_j)$ in $\mathbb{Z}[x_1, \ldots, x_r]$ and let $\zeta(\pi)$ be the automorphism of $\mathbb{Z}[x_1, \ldots, x_r]$ which maps $x_i \to x_{\pi(i)}, 1 \leq i \leq r$. (Every automorphism of the ring $\mathbb{Z}[x_1, \ldots, x_r]$ is the identity on $\mathbb{Z}$. Why?) Verify that if τ is a transposition then $\Delta \to -\Delta$ under $\zeta(\tau)$. Use this to prove the result given in section 1.6 that if π is a product of an even number of transpositions, then every factorization of π as a product of transpositions contains an even number of transpositions. Show that $\Delta^2 \to \Delta^2$ under every $\zeta(\pi)$.

5. Verify that the constructions in the text of $R[x]$ and $R[x_1, \ldots, x_r]$ are valid also for an R which is not necessarily commutative. Show that in this case the x_i are in the center of $R[x_1, \ldots, x_r]$. State and prove the analogues of Theorems 2.10 and 2.11 for $R[x]$ and $R[x_1, \ldots, x_r]$.

6. Show that the matrix ring $M_n(R[x_1, \ldots, x_r]) \cong M_n(R)[x_1, \ldots, x_r]$, x_i indeterminates in both cases.

7. Let $R[[x]]$ denote the set of unrestricted sequences $(a_0, a_1, a_2, \ldots)$, $a_i \in R$. Show that one gets a ring from $R[[x]]$ if one defines $+, \cdot, 0, 1$ as in the polynomial ring. This is called the ring of *formal power series in one indeterminate*.

8. Let M be a monoid, R a commutative ring, and $R[M]$ the set of maps $m \to f(m)$ of M into R such that $f(m) = 0$ for all but a finite number of m. Define addition, multiplication, 0, and 1 in $R[M]$ by

$$(f + g)(m) = f(m) + g(m)$$
$$(fg)(m) = \sum_{pq = m} f(p)g(q)$$
$$0(m) = 0$$
$$1(1) = 1, \qquad 1(m) = 0 \quad \text{if} \quad m \neq 1.$$

Show that $R[M]$ is a ring. Show that the set of maps a' such that $a'(1) = a$ and $a'(m) = 0$ if $m \neq 1$ is a subring isomorphic to R, and the set of maps m' such that $m'(m) = 1$ and $m'(n) = 0$ if $n \neq m$ is a submonoid of the multiplicative monoid of $R[M]$ isomorphic to M. Identify the subrings and monoids just indicated. Show that R is in the center of $R[M]$ and that every element of $R[M]$ can be written as a linear combination of elements of M with coefficients in R: that is, in the form $\sum r_i m_i, r_i \in R, m_i \in M$. Show that $\sum r_i m_i = 0$ if and only if every $r_i = 0$. Show that if σ is a homomorphism of R into a ring S such that $\sigma(R)$ is contained in the center of S, and if τ is a homomorphism of M into the multiplicative monoid of S, then there exists a unique homomorphism of $R[M]$ into S coinciding with σ on R and with τ on M. If M is a group, $R[M]$ is called the *group algebra of M over R*.

9. Let R be any commutative ring and let $\mathbb{N}^{(r)}$ be the free commutative monoid with r generators x_i as on page 68. Show that $R[\mathbb{N}^{(r)}]$ defined as in exercise 8 is the same thing, as $R[x_1, \ldots, x_r]$, x_i indeterminates.

10. Let $M = FM^{(r)}$ be the free monoid with r generators $x_1, \ldots, x_r$ (p. 68), and construct $R[M]$ as in exercise 8. This is called the *free algebra over R generated by* the x_i. State the basic homomorphism property of this ring.

2.11 SOME PROPERTIES OF POLYNOMIAL RINGS AND APPLICATIONS

Let $R[x]$ be the ring of polynomials in an indeterminate x over the (commutative) ring R. If $f(x) \neq 0$ is in $R[x]$ we can write

(36) $$f(x) = a_0 + a_1 x + \cdots + a_n x^n$$

with $a_n \neq 0$. Then a_n is called the *leading coefficient* of $f(x)$ and n is the *degree*, $\deg f$, of $f(x)$. It will be convenient also to say that the degree of 0 is the symbol $-\infty$ and to adopt the usual conventions that $-\infty < n$ for every $n \in \mathbb{N}$, $-\infty + (-\infty) = -\infty$, $-\infty + n = -\infty$. We remark that $f(x) \in R$ if and only if $\deg f = 0$ or $-\infty$ and $f(x) \in R^*$, the set of non-zero elements of R, if and only if $\deg f = 0$. Also it is clear that

$$(37) \qquad \deg[f(x) + g(x)] \leq \max(\deg f(x), \deg g(x))$$

and equality holds in (37) unless $\deg f = \deg g$. If $g(x) = b_0 + b_1 x + \cdots + b_m x^m$ with $b_m \neq 0$ and $f(x)$ is as in (36) then

$$(38) \qquad f(x)g(x) = a_0 b_0 + (a_0 b_1 + a_1 b_0)x + \cdots + a_n b_m x^{n+m}$$

Hence if either a_n or b_m is not a zero divisor then $a_n b_m \neq 0$ and

$$(39) \qquad \deg f(x)g(x) = \deg f(x) + \deg g(x).$$

If we take into account our convention on $-\infty$, we see that (39) holds for all $f(x)$ and $g(x)$ if $R = D$ is a domain. In the case of a domain the properties of the degree function imply the following

THEOREM 2.13.   *If D is a domain then so is the polynomial $D[x_1, \ldots, x_r]$ in r indeterminates over D. Moreover, the units of $D[x_1, \ldots, x_r]$ are the units of D.*

Proof.   We consider first $D[x]$. If $f(x)g(x) = 0$ then its degree is $-\infty$. By (39), this can happen only if either $\deg f(x) = -\infty$ or $\deg g(x) = -\infty$: that is, if either $f(x) = 0$ or $g(x) = 0$. If $f(x)g(x) = 1$ then the degree relation (39) implies that $\deg f = 0 = \deg g$. Hence if $f(x)$ is a unit in $D[x]$ it is contained in D and its inverse is in D. Thus the units of $D[x]$ are the units of D. The extension of the two statements to $D[x_1, \ldots, x_r]$ is immediate by induction on r. $\square$

We look next at the extension of the familiar division algorithm for polynomials. Generally we are interested in this only when the coefficient ring is a field. However, occasionally we must consider the following more general situation.

THEOREM 2.14.   *Let $f(x)$ and $g(x) \neq 0$ be polynomials in $R[x]$, R a ring, and let m be the degree and b_m the leading coefficient of $g(x)$. Then there exists a $k \in \mathbb{N}$ and polynomials $q(x)$ and $r(x) \in R[x]$ with $\deg r(x) < \deg g(x)$ such that*

$$(40) \qquad b_m^k f(x) = q(x)g(x) + r(x).$$

Proof. If $\deg f < \deg g$ the result is clear on writing $f(x) = 0 \cdot g(x) + f(x)$. Hence suppose $\deg f \geq m = \deg g$. Then put

(41) $$b_m f(x) - a_n x^{n-m} g(x) = f_1(x).$$

Since the coefficients of x^n in $b_m f(x)$ and in $a_n x^{n-m} g(x)$ are both $a_n b_m$ it is clear that $\deg f_1 < \deg f$. Hence we can use induction on the degree of $f(x)$ to obtain a $k_1 \in \mathbb{N}$, $q_1(x)$, $r(x) \in R[x]$ with $\deg r(x) < \deg g(x)$ such that

(42) $$b_m^{k_1} f_1(x) = g(x) q_1(x) + r(x).$$

Then, by (41) and (42),

$$b_m^{k_1+1} f(x) = b_m^{k_1} a_n x^{n-m} g(x) + g(x) q_1(x) + r(x) = g(x) q(x) + r(x)$$

where $q(x) = b_m^{k_1} a_n x^{n-m} + q_1(x)$. $\square$

There are several remarks that are worth making about Theorem 2.14. In the first place, it is easy to see that the proof leads to an algorithm for finding k, $q(x)$, and $r(x)$ in a finite number of steps. This is the usual "long" division for polynomials. We leave it to the reader to convince himself of this by looking at some examples. It is easy to see that we can always take the integer k to be the larger of the two integers 0 and $\deg f - \deg g + 1$. We note also that if b_m is a unit then we can divide out by b_m^k and obtain a relation of the form

(40') $$f(x) = q(x) g(x) + r(x)$$

(not the same q and r as in (40)), where $\deg r(x) < \deg g(x)$. This is always the case if $R = F$ is a field. Moreover, in this case the "quotient" $q(x)$ and "remainder" $r(x)$ are unique. For, if

$$f(x) = q(x) g(x) + r(x) = q_1(x) g(x) + r_1(x)$$

and $\deg r(x)$ and $\deg r_1(x) < \deg g(x)$ then we have

$$[q(x) - q_1(x)] g(x) = r_1(x) - r(x).$$

Hence, if $q(x) \neq q_1(x)$ then the degree of the left-hand side is at least m, and the degree of the right-hand side is less than m. This contradiction shows that $q(x) = q_1(x)$ and hence $r(x) = r_1(x)$. It is clear from this that $g(x)$ is a *divisor* or *factor* of $f(x)$—that is, there exists a $q(x)$ such that $f(x) = g(x) q(x)$ if and only if $r(x) = 0$—and this fact can be ascertained in a finite number of steps by carrying out the division algorithm. Finally, we note that if we pass to the field of fractions, then (40') is equivalent to $f(x)/g(x) = q(x) + r(x)/g(x)$, which may be a form more familiar to the reader.

An important special case of Theorem 2.14 is

COROLLARY 1. (The "remainder theorem.") *If $f(x) \in R[x]$ and $a \in R$ then there exists a unique $q(x) \in R[x]$ such that*

(43) $f(x) = (x - a)q(x) + f(a).$

Proof.   The argument above shows that we have a unique $q(x) \in R[x]$ and an $r \in R$ such that $f(x) = (x - a)q(x) + r$. Substitution of $x = a$ (that is, applying the homomorphism of $R[x]$ into R, which is the identity on R and sends $x \to a$) gives $f(a) = (a - a)q(a) + r = r$. Hence we have (43), and $q(x)$ is unique. $\square$

An immediate corollary of Corollary 1 is

COROLLARY 2.   (The "factor theorem.") $(x - a)|f(x)$ *$((x - a)$ is a factor of $f(x))$ if and only if $f(a) = 0$.*

We shall now apply these results to obtain some important properties of $F[x]$, F a field, and more generally of $F[u]$, a ring generated by F and a single element u. We shall call a domain D a *principal ideal domain* (abbreviated as p.i.d.) if every ideal in D is principal. We recall that this is the case for $D = \mathbb{Z}$ (section 2.6) and we now prove

THEOREM 2.15.   *If F is a field then the ring $F[x]$ of polynomials in one indeterminate x over F is a principal ideal domain.*

Proof.   Let I be an ideal in $F[x]$. If $I = 0$ (the ideal with the single element 0) then we can write $I = (0)$. Now assume $I \neq 0$ and consider the non-zero elements of I. Since these have degrees which are non-negative integers, there exists a $g(x) \neq 0$ in I of minimal degree among the non-zero elements of I. Let $f(x)$ be any element of I. Applying the division algorithm we obtain $f(x) = q(x)g(x) + r(x)$ where $\deg r(x) < \deg g(x)$. Since I is an ideal and $f(x)$ and $g(x)$ are in I then $r(x) = f(x) - q(x)g(x) \in I$. If $r(x) \neq 0$ we have a contradiction to the choice of $g(x)$ as an element $\neq 0$ of least degree in I. Hence $r(x) = 0$ and $f(x) = q(x)g(x)$. This shows that every element of I is a multiple of $g(x) \in I$ and, of course, every such multiple is in I. Hence $I = (g(x))$. Since this holds for every ideal I and since $F[x]$ has no non-zero zero divisors, $F[x]$ is a p.i.d. $\square$

This result does not extend beyond the case of one indeterminate: $F[x_1, x_2, \ldots, x_r]$ is not a p.i.d. if $r > 1$. For example, let I be the set of polynomials in $F[x_1, \ldots, x_r]$ having 0 as constant term: that is, having the form $\sum a_{i_1 \cdots i_r} x_1^{i_1} \cdots x_r^{i_r}$ with $a_{0 \cdots 0} = 0$. It is clear that I is an ideal with the generators $x_1, x_2, \ldots, x_r$. If $I = (a)$ then $a | x_i$ for $1 \leq i \leq r$. Since x_i is an irreducible polynomial, either a is a unit or a is an associate of x_i. Since $r > 1$ and $I \neq (1)$, both of these possibilities are excluded. Thus I is not principal.

In $F[x]$ we have $(f(x)) \supset (g(x))$ if and only if $g(x) = f(x)h(x)$, that is, if and only if $f(x) | g(x)$. If $f(x) | g(x)$ and $g(x) | f(x)$ we have $g(x) = f(x)h(x)$ and $f(x) = g(x)k(x)$ so $g(x) = g(x)k(x)h(x)$. Hence if $g(x) \neq 0$ then $k(x)h(x) = 1$, and k and h are non-zero elements of F. It follows that the generator $g(x)$ of $(g(x)) \neq 0$ is determined up to a unit multiplier. We may therefore normalize the generator so that its leading coefficient is 1, and it is then uniquely determined by this property. Polynomials having leading coefficient 1 will be called *monic*.

We now consider any ring of the form $F[u]$, F a field. We have the epimorphism $f(x) \to f(u)$ of $F[x]$ onto $F[u]$, whose kernel is an ideal I such that $I \cap F = 0$ (section 2.10). Now $I = (g(x))$ and $g(x)$ is not a unit since $I \cap F = 0$. Hence either $g(x) = 0$ or $\deg g(x) > 0$. In the first case $I = 0$, so the epimorphism $f(x) \to f(u)$ is an isomorphism and u is transcendental over F. If $\deg g(x) > 0$ we may assume it to be the monic generator of I. Then we shall call $g(x)$ the *minimum polynomial over F* of the (algebraic) element u. This is the monic polynomial of least degree having u for a *root* in the sense that $g(u) = 0$. Moreover, it is clear that if $f(x)$ is any polynomial such that $f(u) = 0$ then $f(x) \in I = (g(x))$, and $f(x)$ is thus a multiple of $g(x)$. The structure of $F[u]$ depends on the way $g(x)$ factors in $F[x]$. For example, we have

THEOREM 2.16. *Let u be algebraic over F with minimum polynomial $g(x)$. Then $F[u]$ is a field if $g(x)$ is irreducible in $F[x]$ in the sense that we cannot write $g(x) = f(x)h(x)$ where $\deg f(x) > 0$ and $\deg h(x) > 0$. On the other hand, if $g(x)$ is reducible then $F[u]$ is not a domain.*

Proof. We know that any ideal of $F[x]/I$ has the form J/I where J is an ideal of $F[x]$ containing $I = (g(x))$ (Theorem 2.6, p. 107). Then $J = (f(x))$ and $g(x) = f(x)h(x)$. If $g(x)$ is irreducible either $f(x)$ or $h(x)$ is a unit. In the first case, $J = F[x]$; in the second case, $J = I$. Hence $F[u] \cong F[x]/I$ has just two ideals: 0 and the whole ring. This implies that $F[u]$ is a field, by Theorem 2.2, p. 102. Now assume $g(x) = f(x)h(x)$ where $\deg f(x) > 0$ and $\deg h(x) > 0$. Then $\deg f(x)$ and $\deg h(x) < \deg g(x)$. Hence $f(u) \neq 0$ and $h(u) \neq 0$. However, $f(u)h(u) = g(u) = 0$. Thus $F[u]$ has zero divisors $\neq 0$. $\square$

We shall apply next the "factor theorem" to establish the following important result on roots of a polynomial.

THEOREM 2.17   Let $f(x)$ be a polynomial of degree $n > 0$ in $F[x]$, F a field. Then $f(x)$ has at most n distinct roots in F.

Proof.   Let $a_1, a_2, \ldots, a_r$ be distinct roots of $f(x)$. We shall prove by induction on r that $f(x)$ is divisible by $\prod_1^r (x - a_i)$. This has just been proved for $r = 1$. Assume it for $r - 1$. Then $f(x) = \prod_1^{r-1} (x - a_j)h(x)$ in $F[x]$; hence $0 = f(a_r) = \prod_1^{r-1} (a_r - a_j)h(a_r)$. Since every $a_r - a_j \neq 0$ we get $h(a_r) = 0$. Hence $h(x) = (x - a_r)k(x)$, by the case $r = 1$. Then $f(x) = \prod_1^r (x - a_i)k(x)$. Comparison of degrees shows that $r \leq n$. $\square$

As an application of this result and a criterion for a finite abelian group to be cyclic, which we gave in Theorem 1.4 (p. 46), we shall now prove the following beautiful theorem on fields.

THEOREM 2.18.   *Any finite subgroup of the multiplicative group of a field is cyclic.*

Proof.   Let G be a finite subgroup of the multiplicative group F^* of non-zero elements of the field F. Of course, G is abelian since F is a field. The criterion we had was that G is cyclic if and only if $|G| = \exp G$, the smallest integer m such that $a^m = 1$ for every $a \in G$. Since $a^{|G|} = 1$ for every a in a finite group we always have $\exp G \leq |G|$. On the other hand, by Theorem 2.17, $f(x) = x^{\exp G} - 1$ has at most $\exp G$ solutions in F and hence in G. Hence $|G| \leq \exp G$. Thus $\exp G = |G|$ and G is cyclic. $\square$

We remark that the foregoing result is not valid for division rings that are not commutative. For example, let $\mathbb{H}$ be the division ring of quaternions over $\mathbb{R}$. The quaternions ± 1, $\pm i$, $\pm j$, $\pm k$ form a finite non-cyclic subgroup of the multiplicative group of $\mathbb{H}$.

As a special case of Theorem 2.18 we see that if F is a finite field then F^* is cyclic. In particular, the non-zero elements of $\mathbb{Z}/(p)$, p a prime, constitute a cyclic group of order $p - 1$ under multiplication. Some number theoretic consequences of the results we have obtained will be indicated in the following exercises.

EXERCISES

1. Let $f(x) = x^n + a_1 x^{n-1} + \cdots + a_n$, $a_i \in F$, a field, $n > 0$, and let $u = x + (f(x))$ in $F[x]/(f(x))$. Show that every element of $F[u]$ can be written in one and only one way in the form $b_0 + b_1 u + \cdots + b_{n-1} u^{n-1}$, $b_j \in F$.

2. Take $F = \mathbb{Q}$, $f(x) = x^3 + 3x - 2$ in exercise 1. Show that $F[u]$ is a field and express the elements

$$(2u^2 + u - 3)(3u^2 - 4u + 1), \qquad (u^2 - u + 4)^{-1}$$

as polynomials of degree ≤ 2 in u.

3. (a) Show that $\mathbb{Q}[\sqrt{2}]$ and $\mathbb{Q}[\sqrt{3}]$ are not isomorphic.
   (b) Let $\mathbb{F}_p = \mathbb{Z}/(p)$, p a prime, and let $R_1 = \mathbb{F}_p[x]/(x^2 - 2)$, $R_2 = \mathbb{F}_p[x]/(x^2 - 3)$. Determine whether $R_1 \cong R_2$ in each of the cases in which $p = 2$, 5, or 11.

4. Show that $x^3 + x^2 + 1$ is irreducible in $(\mathbb{Z}/(2))[x]$ and that $(\mathbb{Z}/(2))[x]/(x^3 + x^2 + 1)$ is a field with eight elements.

5. Construct fields with 25 and 125 elements.

6. Show that $x^3 - x$ has 6 roots in $\mathbb{Z}/(6)$.

7. Use the Chinese remainder theorem (exercises 10 and 11, p. 110) to show that if F is a field and $f(x) \in F[x]$ is monic and factors as $f(x) = g(x)h(x)$, $(g(x), h(x)) = 1$, then $F[x]/(f(x)) \cong F[x]/(g(x)) \oplus F(x)/(h(x))$. Show also that if $f(x) = \prod_1^n (x - a_i)$ in $F[x]$ where the a_i are distinct then $F[x]/(f(x)) = F \oplus \cdots \oplus F$ (n F's).

8. Show that the quaternion division ring $\mathbb{H}$ contains an infinite number of elements u satisfying $u^2 = -1$.

9. Show that the ideal $(3, x^3 - x^2 + 2x - 1)$ in $\mathbb{Z}[x]$ is not principal.

10. Let I denote the ideal given in exercise 9. Is $\mathbb{Z}[x]/I$ a domain? (*Hint*: Show that $\mathbb{Z}[x]/I \cong \bar{\mathbb{Z}}[x]/\bar{I}$ where $\bar{\mathbb{Z}} = \mathbb{Z}/(3)$ and $\bar{I} = (\bar{x}^3 - \bar{x}^2 + \bar{2}\bar{x} - \bar{1})$, $\bar{x} = x + (3)$.)

11. Let R be a ring without nilpotent elements $\neq 0$ ($z^n = 0$ in $R \Rightarrow z = 0$). Prove that if $f(x) \in R[x]$ is a zero divisor then there exists an element $a \neq 0$ in R such that $af(x) = 0$ (*Note*: This holds without restriction on R.)

12. Let F be a field of q elements, $F^* = \{a_1, \ldots, a_{q-1}\}$ the set of non-zero elements of F. Show that $a_1 a_2 \cdots a_{q-1} = -1$. (*Hint*: Use the proof of Theorem 2.18 and also exercise 5, p. 110, if q is even.)

13. Prove Wilson's theorem: If p is a prime in $\mathbb{Z}$, then $(p - 1)! \equiv -1 \pmod{p}$.

14. Find generators for the cyclic groups $\mathbb{Z}_p^*$ of non-zero elements of $\mathbb{Z}/(p)$ for $p = 3$, 5, 7, and 11.

15. An integer a is called a *quadratic residue modulo the prime p* or *quadratic non-residue* mod p according as the congruence $x^2 \equiv a \pmod{p}$ has or has not a solution. We define the *Legendre symbol* $\left(\dfrac{a}{p}\right)$ by $\left(\dfrac{a}{p}\right) = 0$ if $a \equiv 0 \pmod{p}$, $\left(\dfrac{a}{p}\right) = 1$ if

$a \not\equiv 0 \pmod{p}$ and a is a quadratic residue $\pmod{p}$, $\left(\dfrac{a}{p}\right) = -1$ if a is not a quadratic residue modulo p. Note that $\left(\dfrac{a}{p}\right) = 1$ if and only if $a + (p)$ is a square in the multiplicative group of $\mathbb{Z}/(p)$. Hence show that for $p \neq 2$, $\left(\dfrac{a}{p}\right) = 1$ if and only if $a^{(p-1)/2} \equiv 1 \pmod{p}$. Show that for any integers a and b, $\left(\dfrac{ab}{p}\right) = \left(\dfrac{a}{p}\right)\left(\dfrac{b}{p}\right)$.

16. Let $f(x)$, $g(x) \neq 0$ be elements of $F[x]$ with $\deg g = m$. Show that $f(x)$ can be written in one and only one way in the form $a_0(x) + a_1(x)g(x) + a_2(x)g(x)^2 + \cdots + a_r(x)g(x)^r$ where $\deg a_i(x) < m$.

The following exercise gives an alternative proof of the remainder theorem that has several advantages over the proof in the text; notably, it gives an explicit formula for the quotient and it is valid for non-commutative rings.

17. Let $f(x) = a_0 + a_1 x + \cdots + a_n x^n$. We have the formulas $x^i - a^i = (x^{i-1} + ax^{i-2} + \cdots + a^{i-1})(x - a)$, $i \geq 1$. Left multiplication by a_i and summation on i gives $\sum_0^n a_i x^i - \sum_0^n a_i a^i = \sum_1^n a_i(x^{i-1} + ax^{i-2} + \cdots + a^{i-1})(x - a)$. Hence $f(x) = q(x)(x - a) + f(a)$ where $f(a) = \sum_0^n a_i a^i$ and $q(x) = \sum_1^n q_j x^{j-1}$, $q_j = a_j + a_{j+1}a + \cdots + a_n a^{n-j}$.

2.12 POLYNOMIAL FUNCTIONS

The reader is undoubtedly familiar with the notion of a polynomial function of a real variable which occurs in the calculus. We shall now consider the generalization of such functions to any field F and determine the relation between the ring of polynomial functions and the ring of polynomials in indeterminates over F.

Let S be a non-vacuous set and F a field, and let F^S denote the set of maps $s \to f(s)$ of S into F. As usual, $f = g$ means $f(s) = g(s)$ for all s and addition and multiplication of functions are defined by

$$(f + g)(s) = f(s) + g(s)$$

$$(fg)(s) = f(s)g(s).$$

If $a \in F$ then a defines the *constant function* a such that $a(s) = a$ for all s. In particular we have the constant functions 0 and 1. It is straightforward to verify that $(F^S, +, \cdot, 0, 1)$ is a (commutative) ring. For example, we have

$$((f + g)h)(s) = (f(s) + g(s))h(s) = f(s)h(s) + g(s)h(s) = (fh + gh)(s).$$

Hence $(f + g)h = fh + gh$. If we define $(-f)(s) = -f(s)$ we have $f + (-f) = 0$.

It is immediate also that the map of F into F^S which sends any $a \in F$ into the corresponding constant function is a monomorphism. From now on we identify F with its image, so F^S becomes an extension of the field F.

We now take $S = F$, and so are considering the ring of maps of F into itself. In addition to the constant functions a particularly important map is the identity $s \to s$, which we have usually denoted as 1 (or 1_F). In the present context we shall use the customary calculus notation s for this function as well as for the variable s—with the hope that we will create no more than the usual confusion that results from the double meaning assigned to this symbol. We now consider the subring $F[s]$ generated by F (that is, the field of constant functions) and s (the identity function). The elements of this ring will be called *polynomial functions in one variable over* F. Since the ring $F[s]$ is generated by F and s we have the epimorphism of $F[x]$, x an indeterminate, onto $F[s]$, which is the identity map on F and sends $x \to s$. Here $f(x) \to f(s)$ and $f(s)$ is the function $s \to a_0 + a_1 s + \cdots + a_n s^n$ if $f(x) = a_0 + a_1 x + \cdots + a_n x^n$.

The homomorphism $f(x) \to f(s)$ is an isomorphism if and only if F is infinite. To see this we observe that $f(s) = 0$ in the ring of polynomial functions means that $f(s) = 0$ for all values of the variable s: that is, $f(a) = 0$ for all $a \in F$. We have already seen that if $f(x) \neq 0$ and $\deg f = n$ then $f(x)$ has no more than n distinct roots in F. Thus if F is infinite, then $f(a) = 0$ for all a forces $f = 0$. Hence the kernel of the epimorphism is 0 and $f(x) \to f(s)$ is an isomorphism of $F[x]$ with the ring of polynomial functions. On the other hand, if F is finite—say, if $F = \{a_1, a_2, \ldots, a_q\}$—then the polynomial

$$h(x) = (x - a_1)(x - a_2) \cdots (x - a_q) \neq 0$$

whereas the function

$$h(s) = (s - a_1)(s - a_2) \cdots (s - a_q) = 0.$$

This is clear since $h(a_i) = 0$, $1 \leq i \leq q$. Hence the homomorphism $f(x) \to f(s)$ is not an isomorphism if F is finite. This is clear also by counting: the set of all maps of F into F is finite. Hence $F[s]$ is finite. On the other hand, $F[x]$ is infinite. Hence no isomorphism can exist between $F[x]$ and $F[s]$.

The definition of polynomial functions in several variables is an immediate generalization of the foregoing. Here we take $S = F^{(r)}$, the product set $F \times F \times \cdots \times F$ of r copies of F. Its elements are the finite sequences $(s_1, s_2, \ldots, s_r)$. As before, we have the ring of functions $F^S = F^{F^{(r)}}$, which is an extension of the field F. We now pick out r particular functions, "the projections on the r axes." These are the maps

$$(s_1, s_2, \ldots, s_r) \to s_i, \qquad 1 \leq i \leq r.$$

Again, following tradition, we denote the ith projection, just displayed, as s_i, and we consider the ring $F[s_1, s_2, \ldots, s_r]$ obtained by adjoining these to the field F (of constant functions). The elements of $F[s_1, \ldots, s_r]$ are called *polynomial functions in r variables over F*. If $F[x_1, x_2, \ldots, x_r]$ is the polynomial ring in r indeterminates we have the epimorphism of $F[x_1, \ldots, x_r]$ into $F[s_1, \ldots, s_r]$, sending $a \to a$, $a \in F$, $x_i \to s_i$ the ith projection function. We denote the image of $f(x_1, x_2, \ldots, x_r)$ as $f(s_1, s_2, \ldots, s_r)$. If F is a finite field of q elements, then we see, as in the special case $r = 1$, that $f(x_1, \ldots, x_r) \to f(s_1, \ldots, s_r)$ is not an isomorphism; but if F is infinite it is an isomorphism, as we shall now prove. This assertion is equivalent to the following basic theorem.

THEOREM 2.19.   *If F is an infinite field and $f(x_1, x_2, \ldots, x_r)$ is a polynomial $\neq 0$ in $F[x_1, x_2, \ldots, x_r]$ (x_i indeterminates) then there exist elements $a_1, a_2, \ldots, a_r$ in F such that $f(a_1, a_2, \ldots, a_r) \neq 0$.*

Proof.   The case $r = 1$ has been proved. Hence we assume $r > 1$ and we assume the result for $r - 1$ indeterminates. We write

$$f(x_1, x_2, \ldots, x_r) = B_0 + B_1 x_r + B_2 x_r^2 + \cdots + B_n x_r^n$$

where $B_i \in F[x_1, x_2, \ldots, x_{r-1}]$ and we may assume $B_n \equiv B_n(x_1, \ldots, x_{r-1}) \neq 0$. Then, by the induction hypothesis, we know that there exist $a_i \in F$ such that $B_n(a_1, \ldots, a_{r-1}) \neq 0$. Then

$$f(a_1, \ldots, a_{r-1}, x_r) = B_0(a_1, \ldots, a_{r-1}) + B_1(a_1, \ldots, a_{r-1})x_r$$
$$+ \cdots + B_n(a_1, \ldots, a_{r-1})x_r^n \neq 0$$

in $F[x_r]$. Hence we can choose $x_r = a_r$ so that $f(a_1, \ldots, a_r) \neq 0$. $\square$

We can also easily determine the kernel K of the foregoing epimorphism of $F[x_1, \ldots, x_r]$ into the ring of polynomial functions in the case of a finite F. We sketch the argument for this and leave it to the reader to fill in the details. First, we note that if $|F| = q$ then the foregoing argument will show that if $f(x_1, \ldots, x_r) \in F[x_1, \ldots, x_r]$, and the degree of f in every $x_i < q$, then the corresponding polynomial function $f(s_1, \ldots, s_r) \neq 0$. Next we observe that $x_i^q - x_i \in K$ since $a^q = a$, $a \in F$ (exercise 3, p. 105). The next step is to prove that every polynomial $f(x_1, \ldots, x_r)$ can be written in the form

(44) $$f(x_1, \ldots, x_r) = \sum_1^r f_i(x_1, \ldots, x_r)(x_i^q - x_i) + f_0(x_1, \ldots, x_r)$$

where the degree of f_0 in every x_i is $< q$. This can be seen by expressing every power $x_i^k = (x_i^q - x_i)q_k(x_i) + r_k(x_i)$ where $q_k, r_k \in F[x_i]$ and $\deg r_k < q$. Making these substitutions in every monomial $x_1^{k_1}x_2^{k_2}\cdots x_r^{k_r}$ occurring in $f(x_1,\ldots,x_r)$ we obtain (44). We now see that $f(x_1,\ldots,x_r)\in K$ if and only if $f_0(x_1,\ldots,x_r)=0$. This shows that K is the ideal $(x_1^q-x_1,x_2^q-x_2,\ldots,x_r^q-x_r)$ generated by the $x_i^q-x_i$. Hence the ring of polynomial functions in r variables over a field of q elements is isomorphic to

$$F[x_1,\ldots,x_r]/(x_1^q-x_1,x_2^q-x_2,\ldots,x_r^q-x_r).$$

EXERCISES

1. Prove the following extension of Theorem 2.19. If $f(x_1,\ldots,x_r)\in F[x_1,\ldots,x_r]$, F infinite, and $f(a_1,\ldots,a_r)=0$ for all $(a_1,a_2,\ldots,a_r)$ for which a second polynomial $g(x_1,\ldots,x_r)\neq 0$ has values $g(a_1,a_2,\ldots,a_r)\neq 0$, then

$$f(x_1,\ldots,x_r)=0.$$

In the remainder of the exercises F is a finite field with $|F|=q$.

2. Prove that every function in r variables over F (every element of $F^{F^{(r)}}$) is a polynomial function. (*Hint*: Count both sets.)

3. Define the *degree* of the monomial $x_1^{i_1}\cdots x_r^{i_r}$ to be $\sum_1^r i_j$ and the (*total*) *degree* of the polynomial f as the maximum of the degrees of the monomials occurring in f (that is, monomials having non-zero coefficients $a_{i_1\cdots i_r}$ in $f=\sum a_{i_1\cdots i_r}x_1^{i_1}\cdots x_r^{i_r}$). Show that the method of proving (44) by replacing every $x_i^k=(x_i^q-x_i)q_k(x_i)+r_k(x_i)$ where $\deg r_k<q$ yields a polynomial $f_0(x_1,\ldots,x_0)$ of $\deg\le\deg f$ (as well as of $\deg<q$ in every x_i).

4. Show that if f_0 and g_0 are two polynomials of $\deg<q$ in every x_i, and f_0 and g_0 define the same function, then $f_0=g_0$.

5. Let $f(x_1,\ldots,x_r)$ satisfy $f(0,\ldots,0)=0$ and $f(a_1,\ldots,a_r)\neq 0$ for every $(a_1,\ldots,a_r)\neq(0,\ldots,0)$. Prove that if $g(x_1,\ldots,x_r)=1-f(x_1,\ldots,x_r)^{q-1}$ then

$$g(a_1,\ldots,a_r)=\begin{cases}1 & \text{if } (a_1,\ldots,a_r)=(0,\ldots,0)\\ 0 & \text{otherwise.}\end{cases}$$

6. Show that the g of exercise 5 determines the same polynomial function as

$$f_0(x_1,\ldots,x_r)=(1-x_1^{q-1})(1-x_2^{q-1})\cdots(1-x_r^{q-1}).$$

Hence prove that $\deg g\ge r(q-1)$.

7. (Artin-Chevalley.) Let $f(x_1,\ldots,x_r)$ be a polynomial of degree $n<r$, the number of indeterminates. Assume $f(0,\ldots,0)=0$. Prove that there exist $(a_1,\ldots,a_r)\neq(0,\ldots,0)$ such that $f(a_1,\ldots,a_r)=0$.

2.13   SYMMETRIC POLYNOMIALS

Let R be a ring, $R[x_1, \ldots, x_r]$ the ring of polynomials over R in r indeterminates. We have seen that if π is a permutation $i \to i'$ of $\{1, 2, \ldots, r\}$ then π determines an automorphism $\zeta(\pi)$ of $R[x_1, \ldots, x_r]$ such that $a \to a$, $a \in R$, $x_i \to x_{i'}$, $1 \le i \le r$ (Theorem 2.12, p. 125). A polynomial $f(x_1, \ldots, x_r)$ is said to be *symmetric* (in the x's) if $f(x_1, \ldots, x_r)$ is fixed under $\zeta(\pi)$ for every permutation π. The set of symmetric polynomials is a subring Σ of $R[x_1, \ldots, x_r]$ containing R. The coefficients of the powers of x of the polynomial

$$(45) \qquad g(x) = (x - x_1)(x - x_2) \cdots (x - x_r)$$

are symmetric, for we can extend the automorphism $\zeta(\pi)$ to an automorphism $\zeta'(\pi)$ of $R[x_1, \ldots, x_r; x]$ sending $x \to x$. Then $\zeta'(\pi)(g(x)) = (x - x_{1'})(x - x_{2'}) \cdots (x - x_{r'}) = g(x)$. Hence if we write

$$(46) \qquad g(x) = x^r - p_1 x^{r-1} + p_2 x^{r-2} - \cdots + (-1)^r p_r.$$

where $p_i \in R[x_1, \ldots, x_r]$, then $\zeta(\pi)(p_i) = p_i$ for all π. Thus $p_i \in \Sigma$. Comparing (45) and (46) we obtain

$$(47) \quad p_1 = \sum_1^r x_i, \qquad p_2 = \sum_{i<j} x_i x_j, \qquad p_3 = \sum_{i<j<k} x_i x_j x_k, \ldots, \qquad p_r = x_1 x_2 \cdots x_r.$$

The polynomials p_i are called the *elementary symmetric polynomials* in $x_1, \ldots, x_r$. We shall now prove that $\Sigma = R[p_1, p_2, \ldots, p_r]$ and that the p_i are algebraically independent over R.

The equation $\Sigma = R[p_1, \ldots, p_r]$ means, of course, that every symmetric polynomial can be expressed as a polynomial in the elementary symmetric polynomials p_i with coefficients in R. It suffices to prove this for homogeneous polynomials. By a *homogeneous polynomial* we mean one in which all of the terms $a x_1^{k_1} \cdots x_r^{k_r}$ which occur have the same (*total*) *degree* $k_1 + k_2 + \cdots + k_r$. Any polynomial can be written in one and only one way as a sum of homogeneous polynomials of different degrees. Since the automorphism $\zeta(\pi)$ maps homogeneous polynomials of degree k into homogeneous polynomials of degree k it is clear that if $f(x_1, \ldots, x_r)$ is symmetric then so are its homogeneous parts.

We now suppose that $f(x_1, \ldots, x_r)$ is a homogeneous symmetric polynomial of degree, say m. We introduce the lexicographic ordering in the set of monomials of degree m: that is, we say that $x_1^{k_1} \cdots x_r^{k_r}$ is *higher* than $x_1^{l_1} \cdots x_r^{l_r}$ if $k_1 = l_1, \ldots, k_s = l_s$ but $k_{s+1} > l_{s+1}$ ($s \ge 0$). For example, $x_1^2 x_2 x_3 > x_1 x_2^3 > x_1 x_2^2 x_3$. Let $x_1^{k_1} x_2^{k_2} \cdots x_r^{k_r}$ be the highest monomial occurring in f (with non-zero coefficient). Since f is symmetric it contains all the monomials obtained from $x_1^{k_1} x_2^{k_2} \cdots x_r^{k_r}$ by permuting the x's. Hence $k_1 \geq k_2 \geq k_3 \geq \cdots \geq k_r$.

We now consider the highest monomial in the homogeneous symmetric polynomial $p_1^{d_1} p_2^{d_2} \cdots p_r^{d_r}$, $d_i \geq 0$. We observe that if M_1 and M_2 are monomials of degree m and N is a monomial of degree r then $M_1 > M_2$ implies $NM_1 > NM_2$. Hence if $N_1 > N_2$ then $M_1N_1 > M_2N_2$. Now it is clear that the highest monomial in p_i is $x_1 x_2 \cdots x_i$. It follows that the highest monomial in $p_1^{d_1} p_2^{d_2} \cdots p_r^{d_r}$ is

$$x_1^{d_1 + d_2 + \cdots + d_r} x_2^{d_2 + \cdots + d_r} \cdots x_r^{d_r}.$$

Hence the highest monomial in $p_1^{k_1 - k_2} p_2^{k_2 - k_3} \cdots p_r^{k_r}$ is the same as that in f, so if the coefficient in f of this monomial is a, then the highest monomial in $f_1 = f - a p_1^{k_1 - k_2} p_2^{k_2 - k_3} \cdots p_r^{k_r}$ is less than that of f. We can repeat the process with f_1. Since there are only a finite number of monomials of degree m, a finite number of applications of the process yields a representation of f as a polynomial in $p_1, p_2, \ldots, p_r$.

We show next that the p_i are algebraically independent. Suppose

$$\sum_{(d)} a_{d_1 \cdots d_r} p_1^{d_1} \cdots p_r^{d_r} = 0$$

where this is summed over a finite set of distinct $(d) = (d_1, \ldots, d_r)$, $d_i \in \mathbb{Z}^+$. If the relation is non-trivial we have $a_{d_1 \cdots d_r} \neq 0$ for some (d). For any (d) define $(k) = (k_1, k_2, \ldots, k_r)$ by $k_i = d_i + d_{i+1} + \cdots + d_r$. Then the degree of $p_1^{d_1} \cdots p_r^{d_r}$ in the x's is $m = \sum_1^r k_i = \sum_1^r i d_i$ and the highest monomial of this degree occurring in $p_1^{d_1} \cdots p_r^{d_r}$ is $x_1^{k_1} \cdots x_r^{k_r}$. If $(d') = (d'_1, \ldots, d'_r)$ and $k'_i = d'_i + \cdots + d'_r = k_i$ for $1 \leq i \leq r$ then $d'_i = d_i$, $1 \leq i \leq r$. Thus distinct monomials $p_1^{d_1} \cdots p_r^{d_r}$ in the p's have distinct highest monomials in the x's occurring in them. We now choose among the (d) such that $a_{d_1 \cdots d_r} \neq 0$ the one such that m is maximal and the highest monomial $x_1^{k_1} \cdots x_r^{k_r}$ is maximal. Then expressing our relation in the p's in terms of the x's we get the terms $x_1^{k_1} \cdots x_r^{k_r}$ only once and with non-zero coefficient $a_{d_1 \cdots d_r}$. This contradicts the algebraic independence of the x's.

We have now proved the first two statements in

THEOREM 2.20. *Every symmetric polynomial is expressible as a polynomial in the elementary symmetric polynomials p_i. The elementary symmetric polynomials are algebraically independent over R. Every x_i is algebraic over $R[p_1, p_2, \ldots, p_r]$.*

The last statement is clear since (45) and (46) give

$$g(x_i) = x_i^r - p_1 x_i^{r-1} + p_2 x_i^{r-2} - \cdots + (-1)^r p_r = 0.$$

EXERCISES

1. Express $\sum_{i,j,k \neq} x_i^2 x_j^2 x_k$, $r \geq 5$, in terms of the p's.

2. Let $\Delta = \prod_{i<j}(x_i - x_j)$. Show that Δ^2 is symmetric and express Δ^2 for $r = 3$ in terms of the elementary symmetric polynomials.

3. (*Newton's identities.*) Let $s_k = \sum_{i=1}^{n} x_i^k$. Establish the following relations connecting the symmetric polynomials s_k and the elementary symmetric polynomials p_i: $s_k - p_1 s_{k-1} + p_2 s_{k-2} - \cdots + (-1)^{k-1} p_{k-1} s_1 + (-1)^k k p_k = 0$, $1 \leq k \leq n$, $s_{n+j} - p_1 s_{n+j-1} + \cdots + (-1)^k p_k s_{n+j-k} + \cdots + (-1)^n p_n s_j = 0$, $j > 0$. (Note that these are recursive formulas for expressing the power sums s_k as polynomials in the p_i. On the other hand, they show that $k! p_k$ is a polynomial in $s_1, \ldots, s_k$ with integer coefficients.) (*Sketch of proof.* Write $f(x) = x^n - p_1 x^{n-1} + \cdots + (-1)^n p_n = \prod_1^n (x - x_i) = (x - x_i) q_i(x)$. By exercise 17, p. 134, $q_i(x) = x^{n-1} - (p_1 - x_i) x^{n-2} + \cdots + (-1)^k (p_k - p_{k-1} x_i + p_{k-2} x_i^2 - \cdots + (-1)^k x_i^k) x^{n-k+1} + \cdots$. Formal differentiation (see pp. 230–231) gives $nx^{n-1} - (n-1) p_1 x^{n-2} + \cdots = \sum_1^n q_i(x) = nx^{n-1} - (np_1 - s_1) x^{n-2} + \cdots + (-1)^k (np_k - p_{k-1} s_1 + \cdots + (-1)^k s_k) x^{n-k-1} + \cdots$. Comparison of the coefficients of x^{n-k-1} yields the first set of Newton's identities for $k \leq n - 1$. The remaining identities can be obtained by summing on i the relations $x_i^{n+j} - p_1 x_i^{n+j-1} + \cdots + (-1)^n p_n x_i^j = 0$ for $j \geq 0$.)

2.14   FACTORIAL MONOIDS AND RINGS

In the remainder of this chapter we consider the elementary theory of divisibility in (commutative) domains. In a number of important domains every $a \neq 0$ and not a unit can be written as $a = p_1 p_2 \cdots p_s$, where the p_i are irreducible, and such factorizations are unique up to unit factors and the order of the factors. When this is the case we can determine all the factors (up to unit multipliers) of a and hence we can give a simple condition for $a|b$, that is, for $ax = b$ to be solvable. Since the factorization theory that we shall consider is a purely multiplicative one, mainly concerned with the multiplicative monoid of a domain, it will be clearer to consider first the divisibility theory of monoids.

Let M be a commutative monoid satisfying the cancellation law: $ab = ac$ implies $b = c$. Let U be the subgroup of units of M. If $a, b \in M$, we say that b is a *factor* or *divisor* of a if there exists an element c in M such that $a = bc$. We indicate this by writing $b|a$, and in this case we say that a is a *multiple* of b. The relation of divisibility is transitive and reflexive—if $b|a$ and $c|b$ then $c|a$, and $a|a$—but it is not symmetric. An element u is a unit if and only if $u|1$. The units are trivial factors since they are factors of every element ($a = u(u^{-1}a)$). If $a|b$ and $b|a$ then we shall say that a and b are *associates* and write $a \sim b$. The conditions for this are $b = au$, $a = bv$. Hence $b = bvu$, and thus, by the cancellation law, $vu = 1$ and v and u are units. The converse is immediate, so the condition that $a \sim b$ is that a and b differ by a unit factor. Since the set of units is a subgroup of M, it is clear that the relation of associatesness is an equivalence relation.

If $b|a$ but $a\nmid b$ (a is not a factor of b) then we say that b is a *proper factor* of a. If u is a unit and $u = vw$, then it is immediate that v and w are units. Thus the units of M do not have proper factors. An element $a \in M$ is said to be *irreducible*[4] if a is not a unit and a has no proper factors other than units. If a is not a unit and is not irreducible then $a = bc$ where b and c are proper factors of a. Any associate of an irreducible element is also irreducible.

If an element $a \in M$ has a factorization $a = p_1 p_2 \cdots p_s$, where the p_i are irreducible, then a also has the factorization $a = p'_1 p'_2 \cdots p'_s$ where $p'_i = u_i p_i$ and the u_i are units such that $u_1 u_2 \cdots u_s = 1$. Hence if M has units $\neq 1$ and $s > 1$ we can always alter a factorization in the way indicated to obtain other factorizations into irreducible elements, and since the commutative law holds we can also change the order of the factors. We shall say that a factorization into irreducible elements is *essentially unique* if these are the only changes that can be made in factoring an element into irreducible ones. More precisely, $a = p_1 p_2 \cdots p_s$ is an essentially unique factorization of a into irreducible elements p_i if for any other factorization $a = p'_1 p'_2 \cdots p'_t$, p'_i irreducible, we have $t = s$ and $p'_{i'} \sim p_i$ for a suitable permutation $i \to i'$ of $\{1, 2, \ldots, s\}$. We use this definition to formulate the following

DEFINITION 2.4. *Let M be a commutative monoid satisfying the cancellation law. Then M is called* factorial (*sometimes* Gaussian *or a* unique factorization monoid) *if every non-unit of M has an essentially unique factorization into irreducible elements. A domain D is* factorial *if its monoid D^* of non-zero elements is factorial.*

Our main objective in the remainder of this chapter is to show that a number of important types of domains are factorial. That this is not always the case can be seen in considering the following

EXAMPLE

Let $D = \mathbb{Z}[\sqrt{-5}]$, the set of complex numbers of the form $a + b\sqrt{-5}$, where $a, b \in \mathbb{Z}$. It is easy to check that D is a subring of $\mathbb{C}$. Hence D is a domain. To investigate the arithmetic in D we introduce the norm of an element of this domain: if $r = a + b\sqrt{-5}$, then we define the *norm* $N(r) = r\bar{r} = a^2 + 5b^2$. Since the absolute value of complex numbers is a multiplicative function, N is multiplicative on D: that is, $N(rs) = N(r)N(s)$. Also

---

[4] We use this term rather than "prime," which we have used hitherto in discussing the arithmetic of $\mathbb{Z}$. In the general case prime elements will be defined differently below (p. 142).

$N(r)$ is a positive integer if $r \neq 0$. We use the norm first to determine the units of D. If $rs = 1$ then $N(r)N(s) = 1$, so $N(r) = a^2 + 5b^2 = 1$. Since a and b are integers this holds only if $a = \pm 1$ and $b = 0$. Hence $U = \{1, -1\}$. It follows that the only associates of an element r are r and $-r$. We shall now show that 9 has two factorizations into irreducibles in D which do not differ merely by unit factors. These are:

$$9 = 3 \cdot 3 = (2 + \sqrt{-5})(2 - \sqrt{-5}).$$

All of the factors $3, 2 \pm \sqrt{-5}$ are irreducible, for if $3 = rs, r, s \in D$, then $9 = N(3) = N(r)N(s)$. Hence if r and s are non-units then $N(r) = 3$ and $N(s) = 3$. However, it is clear that $N(r) = a^2 + 5b^2 = 3$ has no integral solution. Thus 3 is irreducible and, similarly, $2 + \sqrt{-5}$ and $2 - \sqrt{-5}$ are irreducible. Also, it is clear that $3, 2 + \sqrt{-5}$ and $3, 2 - \sqrt{5}$ are not associates. Hence 9 does not have an essentially unique factorization into irreducible elements (though it does have factorizations into irreducibles), and $\mathbb{Z}[\sqrt{-5}]$ is therefore not factorial.

In any factorial monoid M one can determine up to unit factors all the factors of a given non-unit a, provided that a factorization of a into irreducible elements is known; for, if $a = p_1 p_2 \cdots p_s$ where the p_i are irreducible, and if $a = bc$ where $b = p'_1 \cdots p'_1, c = p''_1 \cdots p''_u$, and the p'_j and p''_k are irreducible, then

$$a = p_1 p_2 \cdots p_s = p'_1 p'_2 \cdots p'_t p''_1 p''_2 \cdots p''_u.$$

Hence, by the uniqueness property, $p'_j \sim p_{i_j}$ where $i_j \neq i_k$ if $j \neq k$. Hence $b \sim p_{i_1} p_{i_2} \cdots p_{i_t}$. Thus any factor of a is an associate of one of the products of the form $p_{i_1} p_{i_2} \cdots p_{i_t}$ obtained from the factorization $a = p_1 p_2 \cdots p_s$. If we call the number s of irreducible factors in the decomposition $a = p_1 \cdots p_s$ the *length* of a then it is clear that any proper factor of a has smaller length than a. Hence it is clear that any factorial monoid satisfies the following

Divisor chain condition. M contains no infinite sequences of elements $a_1, a_2, \ldots$ such that each a_{i+1} is a proper factor of a_i.

Equivalently, the condition is that if $a_1, a_2, \ldots$ is a sequence of elements of M such that $a_{i+1} | a_i$ then there exists an integer N such that $a_N \sim a_{N+1} \sim a_{N+2} \sim \cdots$.

We obtain next a second necessary condition for factoriality. An element p of M is called a *prime* if p is not a unit and if $p | ab$ implies either $p | a$ or $p | b$. In other words, p is not a unit and $p \nmid a$ and $p \nmid b$ implies $p \nmid ab$. Now let p be an irreducible element in a factorial monoid M and suppose $p | ab$. Then p is not a unit and if a is a unit then $ab \sim b$ so $p | b$. Similarly, if b is a unit then $p | a$. If a and b are non-units we have $a = p_1 \cdots p_s$, $b = p'_1 \cdots p'_t$, p_i, p'_j irreducible. Then $ab = p_1 \cdots p_s p'_1 \cdots p'_t$ and since $p | ab$, either $p \sim p_i$ for some i or $p \sim p'_j$ for some j. Thus either $p | a$ or $p | b$, and we have proved that any factorial monoid satisfies the

Primeness condition.   Every irreducible element of M is prime.

We shall now show that the foregoing two conditions are sufficient for factoriality. We note first that the divisor chain condition insures the existence of a factorization into irreducible elements for any non-unit of M. Let a be a non-unit. We shall show first that a has an irreducible factor. If a is irreducible, there is nothing to prove. Otherwise, let $a = a_1 b_1$ where a_1 is a proper factor of a. Either a_1 is irreducible or $a_1 = a_2 b_2$ where a_2 is a proper factor of a_1. We continue this process and obtain a sequence $a, a_1, a_2, \ldots$ in which each element is a proper factor of the preceding one. By the divisor chain condition this process terminates in a finite number of steps with an irreducible factor a_n of a.

Now put $a_n = p_1$ and write $a = p_1 a'$. If a' is a unit, a is irreducible and we are through. Otherwise, $a' = p_2 a''$ where p_2 is irreducible. Continuing this process, we obtain the sequence $a, a', a'', \ldots$ where each element is a proper factor of the preceding and each $a^{(i-1)} = p_i a^{(i)}$, p_i irreducible. This breaks off with an irreducible element $a^{(s-1)} = p_s$. Then

$$a = p_1 a' = p_1 p_2 a'' = \cdots = p_1 p_2 \cdots p_s$$

and we have the required factorization of a into irreducible factors.

We shall show next that the primeness condition insures the essential uniqueness of factorization into irreducible elements. Let

(48)
$$a = p_1 p_2 \cdots p_s = p_1' p_2' \cdots p_t'$$

be two factorizations of a into irreducible elements. If $s = 1$, $a = p_1$ is irreducible; hence $t = 1$ and $p_1' = p_1$. We shall now use induction and assume that any element which has a factorization as a product of $s - 1$ irreducible elements has essentially only one such factorization. Since p_1 in (48) is irreducible, it is prime by the primeness condition, and it is clear by induction that if p is a prime and $p|a_1 a_2 \cdots a_r$, then $p|a_i$ for some i. Hence $p_1|p_j'$ for some j. By rearranging the p', if necessary, we may assume $p_1|p_1'$. Since p_1' is irreducible this means that $p_1' \sim p_1$ and so $p_1' = p_1 u_1$, u_1 a unit. We substitute this in the second factorization in (48) and cancel p_1 to obtain

$$b \equiv p_2 \cdots p_s = u_1 p_2' \cdots p_t' = p_2'' \cdots p_t''$$

where $p_2'' = u_1 p_1'$ and $p_i'' = p_i'$, $i > 2$, are irreducible. By the induction assumption we have $s - 1 = t - 1$ and for a suitable ordering of the p_j'' we have $p_j \sim p_j''$, $j = 2, \ldots, s$. Then $s = t$ and $p_i \sim p_i'$, $1 \leq i \leq s$.

We have now established the following criterion:

THEOREM 2.21.   *Let M be a commutative monoid satisfying the cancellation law. Then M is factorial if and only if the divisor chain condition and the primeness condition hold in M.*

We shall show next that we can replace the second condition in the foregoing theorem by the condition that every pair of elements of M have a greatest common divisor. An element d is called a *greatest common divisor* (g.c.d.) of a and b if $d|a$ and $d|b$; and if c is any element such that $c|a$ and $c|b$, then $c|d$. If d and d' are two g.c.d.'s of a and b, then the definition shows that $d|d'$ and $d'|d$. Hence $d \sim d'$. Thus, the g.c.d., if it exists, is determined up to a unit multiplier. We shall find it convenient to denote any determination of a g.c.d. of a and b as (a, b). The dual notion of a g.c.d. is a least common multiple. We call m a *least common multiple* (l.c.m.) of a and b if $a|m$ and $b|m$; and if n is any element such that $a|n$ and $b|n$, then $m|n$. We denote any l.c.m. of a and b by $[a, b]$.

We shall now show that in a factorial monoid any two elements a and b have a g.c.d. and an l.c.m. If a is a unit then it is clear that a is a g.c.d. and b is an l.c.m. of a and b. Hence we may assume that a is not a unit. Then we look at a factorization of a as a product of irreducible elements. By replacing associated irreducible factors in such a factorization of a by a single representative one multiplied by unit factors, we obtain a factorization

$$(49) \qquad\qquad a = up_1^{e_1}p_2^{e_2} \cdots p_r^{e_r}$$

where u is a unit, the p_i are irreducible and not associates, and the e_i are positive integers. It is clear now that the factors of a have the form $u'p_1^{e_1'}p_2^{e_2'} \cdots p_r^{e_r'}$ where u' is a unit and the e_i' are integers such that $0 \le e_i' \le e_i$. It is easy to see also that if a and b are two non-units, then we can write these in terms of the same non-associate irreducible elements, that is, we can obtain

$$(50) \qquad\quad a = up_1^{e_1}p_2^{e_2} \cdots p_t^{e_t}, \qquad b = vp_1^{f_1}p_2^{f_2} \cdots p_t^{f_t}$$

where u and v are units, if we allow the e_i and f_i to be non-negative integers. Now consider the element

$$(51) \qquad\qquad d = p_1^{g_1}p_2^{g_2} \cdots p_t^{g_t}, \qquad g_i = \min(e_i, f_i).$$

Clearly $d|a$ and $d|b$. Moreover, if $c|a$ and $c|b$, then $c = wp_1^{k_1}p_2^{k_2} \cdots p_t^{k_t}$ where w is a unit and $0 \le k_i \le e_i, f_i$. Then $k_i \le g_i$ and $c|d$. Thus the element d is a g.c.d. of a and b. In a similar manner one sees that if $h_i = \max(e_i, f_i)$, then

(52) $$m = p_1^{h_1} p_2^{h_2} \cdots p_t^{h_t}$$

is an l.c.m. of a and b.

If a and b have a unit as g.c.d. then we have $(a, b) = 1$ and we say that a and b are *relatively prime*. This is the case if and only if either a or b is a unit or no irreducible factor of either one is a factor of both.

Now let M be a commutative monoid with cancellation law and assume that M satisfies the

G.c.d. condition. Any two elements of M have a g.c.d.

We shall show that this implies that irreducible elements of M are prime. We break the argument up into a number of simple lemmas.

LEMMA 1. *Any finite number of elements $a_1, \ldots, a_r$ of M have a g.c.d., that is, there exists a d in M such that $d \mid a_i$, $1 \leq i \leq r$, and if $e \in M$ satisfies $e \mid a_i$ for $1 \leq i \leq r$, then $e \mid d$.*

Proof. Let $d_1 = (a_1, a_2)$, $d_2 = (d_1, a_3)$, $\ldots$, $d = d_r = (d_{r-1}, a_r)$. Then the definitions show that d is a g.c.d. of $a_1, \ldots, a_r$. $\square$

We denote any g.c.d. of $a_1, \ldots, a_r$ as $(a_1, \ldots, a_r)$.

LEMMA 2. $((a, b), c) \sim (a, (b, c))$.

Proof. Both are g.c.d.'s of a, b, and c. $\square$

LEMMA 3. $c(a, b) \sim (ca, cb)$.

Proof. Let $(a, b) = d$, $(ca, cb) = e$. Then $cd \mid ca$ and $cd \mid cb$, and so $cd \mid (ca, cb)$. Hence $e = cdu$. Now $ca = ex = cdux$. Hence $a = dux$, that is, $du \mid a$. Similarly, $du \mid b$ and so $du \mid d$. Hence u is a unit and $(ca, cb) \sim cd \sim c(a, b)$. $\square$

LEMMA 4. *If $(a, b) \sim 1$ and $(a, c) \sim 1$ then $(a, bc) \sim 1$.*

Proof. If $(a, b) \sim 1$, then Lemma 3 shows that $(ac, bc) \sim c$. It is clear that $(a, ac) \sim a$. Hence

$$1 \sim (a, c) \sim (a, (ac, bc)) \sim ((a, ac), bc) \sim (a, bc). \quad \square$$

We can now prove

LEMMA 5. *The g.c.d. condition implies the primeness condition.*

Proof. Let p be irreducible and suppose $p \nmid a$ and $p \nmid b$. Since p is irreducible these imply that $(p, a) \sim 1$ and $(p, b) \sim 1$. Then Lemma 4 shows that $(p, ab) \sim 1$ and so $p \nmid ab$. Thus if $p \mid ab$ then either $p \mid a$ or $p \mid b$. □

These results yield our second criterion for factoriality:

THEOREM 2.22   *Let M be a commutative monoid satisfying the cancellation law. Then M is factorial if and only if the divisor chain condition and the g.c.d. condition hold in M.*

Proof. Lemma 5 shows that if the indicated conditions hold then the divisor chain condition and primeness condition hold. Hence M is factorial by Theorem 2.21. Conversely, if M is factorial then M satisfies the divisor chain condition and, as we have seen, every pair of elements of M have a g.c.d. □

EXERCISES

1. Show that if M is factorial then $ab \sim [a, b](a, b)$ in M.

2. Let M be a commutative monoid with cancellation law. Show that the relation of associateness $\sim$ is a congruence relation. Let $\bar{M}$ be the corresponding quotient monoid. Show that $\bar{M}$ satisfies the cancellation law and that $\bar{1}$ is the only unit in $\bar{M}$. Show that M is factorial if and only if $\bar{M}$ is factorial.

3. Show that $\mathbb{Z}[\sqrt{-5}]$ satisfies the divisor chain condition.

4. Show that $\mathbb{Z}[x]$ satisfies the divisor chain condition.

5. Let D be the set of expressions $a_1 x^{\alpha_1} + a_2 x^{\alpha_2} + \cdots + a_n x^{\alpha_n}$ where the $a_i \in$ some field F and the α_i are non-negative rational numbers. Define equality and addition in the obvious way and multiplication using the distributive law and $(a_i x^{\alpha_i})(a_j x^{\alpha_j}) = a_i a_j x^{\alpha_i + \alpha_j}$. (This can be done rigorously using the procedure of exercise 8, p. 127.) Show that D is a domain. Show that the divisor chain condition fails in D.

6. Show that any prime is irreducible.

7. Let $\mathbb{Z}[\sqrt{10}]$ be the set of real numbers of the form $a + b\sqrt{10}$ where $a, b \in \mathbb{Z}$. Show that $\mathbb{Z}[\sqrt{10}]$ is not factorial.

8. Let p be a prime of the form $4n + 1$ and let q be a prime such that $\left(\dfrac{q}{p}\right) = -1$ (see

p. 133 for the definition of $\left(\dfrac{q}{p}\right)$). Show that $\mathbb{Z}\left[\sqrt{pq}\right]$ is not factorial.

2.15 PRINCIPAL IDEAL DOMAINS AND EUCLIDEAN DOMAINS

We are now going to apply our results on factorization in monoids to domains. The results are applicable to any commutative domain D, since the set D^* of non-zero elements of D is a submonoid of the multiplicative monoid of D and the cancellation law holds. The concepts and results carry over. We now make the important observation (which we have already made for $\mathbb{Z}$) that the divisibility $b\,|\,a$ is equivalent to the set inclusion $(b) \supset (a)$ for the principal ideals (b) and (a). For, $(b) \supset (a)$ is equivalent to $a \in (b)$ and this is equivalent to $a = bc$, by the definition of (b). Since a and b are associates in D^* if and only if $a\,|\,b$ and $b\,|\,a$, we see that $a \sim b$ if and only if $(a) \supset (b)$ and $(b) \supset (a)$; hence, if and only if $(a) = (b)$. Thus a is a proper factor of b if and only if we have the proper inclusion $(a) \supsetneq (b)$. The divisor chain condition for $M = D^*$ is therefore equivalent to:

The ascending chain condition for principal ideals. D contains no infinite properly ascending chain of principal ideals $(a_1) \subsetneq (a_2) \subsetneq (a_3) \subsetneq \cdots$.

We have defined a principal ideal domain (p.i.d.) to be a domain in which every ideal is principal. We have seen that $\mathbb{Z}$ and $F[x]$ for any field F are p.i.d., and we shall give other examples of p.i.d. below. We shall now show that any p.i.d. D is factorial. We establish first the divisor chain property by proving the ascending chain condition for principal (hence all) ideals. We recall that in any ring the union of an ascending chain of ideals is an ideal (section 2.5, p. 102). Hence if $(a_1) \subset (a_2) \subset (a_3) \subset \cdots$ then $I = \bigcup (a_i)$ is an ideal in D. Consequently, $I = (d)$ for some $d \in I$. Then $d \in (a_n)$ for some n and $I = (d) \subset (a_n)$. Then if $m \geq n$, $(a_m) \supset (a_n) \supset I \supset (a_m)$ so $(a_n) = (a_{n+1}) = \cdots$. This proves that D contains no infinite properly ascending chain of ideals.

To complete the proof of factoriality it is enough to show that D^* satisfies either the primeness condition or the g.c.d. condition. We shall prove both, thereby giving two alternative proofs of factoriality.

Let $a, b \in D$ and consider the ideal (a, b) generated by a and b.[5] Exactly as in the case of $\mathbb{Z}$ (p. 104) we see that if $(a, b) = (d)$ then d is a g.c.d. Since every ideal is principal this shows that every pair of elements of D have a g.c.d.

---

[5] There is no harm in allowing either $a = 0$ or $b = 0$ in these considerations.

We shall give next a direct proof that irreducible elements of a p.i.d. are prime. This will give a proof of factoriality that is independent of the considerations on greatest common divisors that led to Theorem 2.22.

Let p be irreducible in D^* and suppose $p|ab$ but $p\nmid a$, $a, b \in D^*$. The condition p irreducible means that there exists no ideal I such that $D \supsetneq I \supsetneq (p)$. Since $p\nmid a$, $a \notin (p)$ so $(p, a) \supsetneq (p)$ and hence $(p, a) = (1)$. Thus we have $u, v \in D$ such that $up + va = 1$. Then $upb + vab = b$. Since $p|ab$, this implies that $p|b$. Hence p is a prime.

We have now doubly proved:

THEOREM 2.23. *Any principal ideal domain is factorial.*

In particular, this implies that if F is a field, then $F[x]$ is factorial. We remark that it also gives another proof of the fact that $\mathbb{Z}$ is factorial (p. 22).

The notion of a principal ideal domain is a nice abstract concept. However, we need a practical criterion for proving that certain rings are p.i.d. This is provided by the notion of a Euclidean domain, which we now define.

DEFINITION 2.5. *A domain D is called* Euclidean *if there exists a map $\delta : a \to \delta(a)$ of D into the set $\mathbb{N}$ of non-negative integers such that if $a, b \neq 0 \in D$, then there exist $q, r \in D$ such that $a = bq + r$ where $\delta(r) < \delta(b)$.*

The ring $\mathbb{Z}$ becomes Euclidean if one defines $\delta(a) = |a|$. Also the division algorithm for polynomials shows that $F[x]$ is Euclidean for any field F if we define $\delta(f(x)) = 2^{\deg f(x)}$ (where it is understood that $2^{-\infty} = 0$). Another important example of a Euclidean domain is the

Ring of Gaussian integers $\mathbb{Z}[\sqrt{-1}]$. This is the subset of $\mathbb{C}$ of complex numbers of the form $m + ni$ where $m, n \in \mathbb{Z}$ and $i = \sqrt{-1}$. Thus $\mathbb{Z}[\sqrt{-1}]$ can be identified with the set of "lattice" points, that is, points with integral coordinates in the complex plane. It is readily verified that $\mathbb{Z}[i]$ is a subring of $\mathbb{C}$, hence an integral domain. If $a = m + ni$ we put $\delta(a) = a\bar{a} = |a|^2 = m^2 + n^2$. Then $\delta(a) \in \mathbb{N}$ and $\delta(ab) = \delta(a)\delta(b)$. To prove that δ satisfies the condition of the definition of a Euclidean domain, we note that if $b \neq 0$ then $ab^{-1} = \mu + vi$, where μ and v are rational numbers. Now we can find integers u and v such that $|u - \mu| \leq \frac{1}{2}$, $|v - v| \leq \frac{1}{2}$. Set $\varepsilon = \mu - u$, $\eta = v - v$, so that $|\varepsilon| \leq \frac{1}{2}$ and $|\eta| \leq \frac{1}{2}$. Then

$$a = b[(u + \varepsilon) + (v + \eta)i]$$
$$= bq + r$$

where $q = u + vi$ is in $\mathbb{Z}[i]$ and $r = b(\varepsilon + \eta i)$. Since $r = a - bq$, $r \in \mathbb{Z}[i]$. Moreover

$$\delta(r) = |r|^2 = |b|^2(\varepsilon^2 + \eta^2) \le |b|^2(\tfrac{1}{4} + \tfrac{1}{4}) = \tfrac{1}{2}\delta(b).$$

Thus $\delta(r) < \delta(b)$. Hence $\mathbb{Z}[\sqrt{-1}]$ is Euclidean.

The main result on Euclidean domains is the following

THEOREM 2.24. *Euclidean domains are principal.*

Proof. The proof is identical with the one given in the special case $D = F[x]$. Let I be an ideal in a Euclidean domain D. If $I = 0$ we have $I = (0)$. Otherwise, let $b \ne 0$ be an element of I for which $\delta(b)$ is minimal for the non-zero elements of I. Let a be any element of I. Then $a = bq + r$ for some $q, r \in D$ with $\delta(r) < \delta(b)$. Since $r = a - bq \in I$ and $\delta(r) < \delta(b)$ we must have $r = 0$ by the choice of b in I. Hence $a = bq$ so $I = (b)$. $\square$

Since every p.i.d. is factorial we have the

COROLLARY. *Euclidean domains are factorial.*

EXERCISES

1. Let F be a field. Is F a p.i.d.?

2. Show that the set $\mathbb{Z}[\sqrt{2}]$ of real numbers of the form $m + n\sqrt{2}$, $m, n \in \mathbb{Z}$, is a Euclidean domain with respect to the function $\delta(m + n\sqrt{2}) = |m^2 - 2n^2|$.

3. Let D be the set of complex numbers of the form $m + n\sqrt{-3}$ where m and n are either both in $\mathbb{Z}$ or are both halves of odd integers (exercise 4, p. 89). Show that D is a Euclidean domain relative to $\delta(m + n\sqrt{-3}) = m^2 + 3n^2$.

4. Let D be a p.i.d., E a domain containing D as a subring. Show that if d is a g.c.d. of a and b in D, then d is also a g.c.d. of a and b in E.

5. Show that if $a \ne 0$ in a p.i.d. D, then $D/(a)$ is a field if a is a prime and $D/(a)$ is not a domain if a is not prime.

6. Let D be a Euclidean domain whose function δ satisfies: (i) $\delta(ab) = \delta(a)\delta(b)$ and (ii) $\delta(a + b) \le \max(\delta(a), \delta(b))$. Show that either D is a field or $D = F[x]$, F a field, x an indeterminate.

7. Let p be a prime of the form $4n + 1$, $n \in \mathbb{Z}$. Use the criterion of exercise 15, p. 133 to show that $\left(\dfrac{-1}{p}\right) = 1$. Hence prove that p is not a prime in $\mathbb{Z}[i]$, the ring of Gaussian integers.

8. Use exercise 7 to prove that any prime p of the form $4n + 1$ is a sum $a^2 + b^2$, $a, b \in \mathbb{Z}$.

9. Determine the primes ($=$irreducible elements) of $\mathbb{Z}[i]$.

10. Show that a positive integer m is a sum of two squares of integers if and only if the primes of the form $4n + 3$ occurring in the prime decomposition of m occur with even multiplicities.

11. (Euclid's algorithm for finding the g.c.d.) Let a_1, a_2 be non-zero elements of a Euclidean domain. Define a_i and q_i recursively by $a_1 = q_1 a_2 + a_3$, $a_i = q_i a_{i+1} + a_{i+2}$ where $\delta(a_{i+2}) < \delta(a_{i+1})$. Show that there exists an n such that $a_n \neq 0$ but $a_{n+1} = 0$, and that $d = a_n = (a_1, a_2)$. Also use the equations to obtain an expression for d in the form $xa_1 + ya_2$.

12. Apply the foregoing to the polynomials $x^3 + x^2 + x - 3$ and $x^4 - x^3 + 3x^2 + x - 4$ in $\mathbb{Q}[x]$.

The next three exercises are designed to explain one of the mysteries of the integral calculus: the partial fraction decomposition of rational functions.

13. Let F be a field and suppose $f(x)$ is a non-zero polynomial in $F[x]$ which has a factorization $f(x) = f_1(x)f_2(x)$ where $\deg f_i > 0$ and $(f_1, f_2) = 1$. Show that if $\deg g(x) < \deg f(x)$, then there exist $u_i(x) \in F[x]$ such that $g(x) = u_2(x)f_1(x) + u_1(x)f_2(x)$ and $\deg u_i < \deg f_i$. (*Hint:* Existence of $v_1(x)$ and $v_2(x)$ such that $v_2(x)f_1(x) + v_1(x)f_2(x) = g(x)$ is clear. Now divide $v_i(x)$ by $f_i(x)$ obtaining the remainder $u_i(x)$ of degree $< \deg f_i$. Apply degree considerations.) Note that in the field of fractions $F(x)$ of $F[x]$ one has $g(x)/f(x) = u_1(x)/f_1(x) + u_2(x)/f_2(x)$. Use induction to prove that if $f(x) = p_1(x)^{e_1} \cdots p_r(x)^{e_r}$, $p_i(x)$ distinct primes, then $g(x)/f(x) = \sum_1^r g_i(x)/p_i(x)^{e_i}$ where $\deg g_i < \deg p_i^{e_i}$.

14. Show that if $g(x)$, $p(x) \neq 0$ in $F[x]$ then there exist $a_i(x) \in F[x]$ with $\deg a_i < \deg p$ such that

$$g(x) = a_0(x) + a_1(x)p(x) + \cdots + a_{e-1}(x)p(x)^{e-1}.$$

$$\frac{g(x)}{p(x)^e} = \frac{a_0(x)}{p(x)^e} + \frac{a_1(x)}{p(x)^{e-1}} + \cdots + \frac{a_{e-1}(x)}{p(x)}.$$

15. Assuming the result (which will be proved in Chapter 5) that the irreducible polynomials in $\mathbb{R}[x]$ are either linear or quadratic, show that if $f(x)$, $g(x) \in \mathbb{R}[x]$ and $\deg g < \deg f$, then one can decompose the fraction $g(x)/f(x)$ in $\mathbb{R}(x)$ as a sum of of partial fractions of one of the forms $a/(x - r)^e$ or $(bx + c)/(x^2 + sx + t)^e$ where $x^2 + sx + t$ is irreducible. More precisely, suppose $f(x) = \prod_1^m (x - r_i)^{e_i} \prod_1^n (x^2 + s_j x + t_j)^{f_j}$ where the quadratics are irreducible then $g(x)/f(x)$ can be written in the

form

$$\sum_{i=1,k_i=1}^{m,e_i} \frac{a_{ik_i}}{(x-r_i)^{k_i}} + \sum_{j=1,e_j=1}^{n,f_j} \frac{b_{je_j}x+c_{je_j}}{(x^2+s_jx+t_j)^{e_j}}.$$

16. Investigate the uniqueness questions posed by exercises 13–15.

17. Define the Möbius function $\mu(n)$ of positive integers by the following rules: (a) $\mu(1)=1$, (b) $\mu(n)=0$ if n has a square factor, (c) $\mu(n)=(-1)^s$ if $n=p_1p_2\cdots p_s$, p_i distinct primes. Prove that μ is *multiplicative* in the sense that $\mu(n_1n_2)=\mu(n_1)\mu(n_2)$ if $(n_1, n_2)=1$. Also prove that

$$\sum_{d|n}\mu(d)=\begin{cases}1 & \text{if } n=1 \\ 0 & \text{if } n\neq 1.\end{cases}$$

18. Prove the Möbius inversion formula: If $f(n)$ is a function of positive integers with values in a ring and

$$g(n)=\sum_{d|n} f(d)$$

then

$$f(n)=\sum_{d|n}\mu\left(\frac{n}{d}\right)g(d).$$

19. Prove that if $\varphi(n)$ is the Euler φ-function then

$$\varphi(n)=\sum_{d|n}\mu\left(\frac{n}{d}\right)d.$$

20. Let F be a field with q ($<\infty$) elements. Prove that the number of irreducible monic quadratic polynomials with coefficients in F is $q(q-1)/2$ and the number of irreducible cubics with coefficients in F is $q(q^2-1)/3$. (See Corollary 2 to Theorem 4.26, p. 289.)

2.16 POLYNOMIAL EXTENSIONS OF FACTORIAL DOMAINS

In this section we prove the important theorem that states that if D is factorial then so is the domain $D[x]$ of polynomials in an indeterminate x over D.

Let D be factorial. Then any finite set of non-zero elements of D have a g.c.d. We shall find it convenient to define the g.c.d. $(a_1, a_2, \ldots, a_k)$ where $a_i \in D$ to be 0 if all the $a_i=0$, and otherwise to be the g.c.d. of the non-zero a_i. If $f(x)=a_0+a_1x+\cdots+a_nx^n\neq 0$ we define the *content* $c(f)$ of $f(x)$ as $(a_0, a_1, \ldots, a_n)$ ($\neq 0$). If $d=c(f)$ we can write $a_i=da_i'$, $0\leq i\leq n$, and $f(x)=df_1(x)$ where

$$f_1(x)=a_0'+a_1'x+\cdots+a_n'x^n.$$

We have seen in our discussion of g.c.d.'s in monoids (section 2.14) that $(da, db)=d(a, b)$. It follows by induction that $d(b_1, b_2, \ldots, b_r)=(db_1, \ldots, db_r)$.

This evidently implies that the content $c(f_1)$ is 1. A polynomial having this property is called *primitive*. Hence we have the factorization $f(x) = c(f)f_1(x)$ as a product of the content of f and a primitive polynomial. Now let $f(x) = ef_2(x)$ be any factorization of $f(x)$ as a product of a constant e and a primitive polynomial $f_2(x) = a_0'' + a_1''x + \cdots + a_n''x^n$. Then $a_i = a_i''e$ and 1 is a g.c.d. of the a_i''. Hence e is a g.c.d. of the a_i, and so $e \sim c(f)$.

It is useful to extend the factorization of a polynomial as product of an element of D and a primitive polynomial to polynomials with coefficients in the field of fractions. The result we require is

LEMMA 1. *Let D be a factorial domain, F the field of fractions of D, and $f(x) \neq 0 \in F[x]$. Then $f(x) = \gamma f_1(x)$ where $\gamma \in F$ and $f_1(x)$ is a primitive polynomial in $D[x]$. Moreover, this factorization is unique up to unit multipliers in D.*

Proof. Let $f(x) = \alpha_0 + \alpha_1 x + \cdots + \alpha_n x^n$ where the $\alpha_i \in F$ and $\alpha_n \neq 0$. We can write $\alpha_i = a_i b_i^{-1}$, a_i, $b_i \in D$. Then if $b = \prod b_i$, $bf(x) \in D[x]$ so $bf(x) = cf_1(x)$ where $f_1(x) \in D[x]$ and is primitive. Then $f(x) = \gamma\, f_1(x)$ where $\gamma = cb^{-1} \in F$. Now let $f(x) = \delta f_2(x)$ where $\delta \in F$ and $f_2(x) \in D[x]$ and is primitive. Then $\delta = de^{-1}$, d, $e \in D$. Hence we have $cb^{-1}f_1(x) = de^{-1}f_2(x)$ and $cef_1(x) = bdf_2(x)$. The result proved before for polynomials with coefficients in D shows that $f_1(x) \sim f_2(x)$ and $ce \sim bd$. Then we have $bd = uce$, u is a unit in D, and $de^{-1} = ucb^{-1}$. Hence $\delta = u\gamma$ as required. $\square$

As in the case of $D[x]$, we call the element γ, which is determined up to a unit multiplier by $f(x)$, the *content* of $f(x) \in F[x]$. An immediate consequence of Lemma 1 is the

COROLLARY. *Let $f(x)$ and $g(x)$ be primitive in $D[x]$ and assume these are associates in $F[x]$. Then they are associates in $D[x]$.*

Proof. We are given that $f(x) = \alpha g(x)$, $\alpha \neq 0$ in F. Then the uniqueness part of Lemma 1 shows that α is a unit in D. $\square$

The key lemma for proving the factoriality of $D[x]$ is

LEMMA 2 (Gauss' lemma.) *The product of primitive polynomials is primitive.*

Proof. Suppose $f(x)$ and $g(x)$ are primitive but $h(x) = f(x)g(x)$ is not. Then there exists an irreducible element (hence a prime) $p \in D$ such that $p \nmid f(x)$, $p \nmid g(x)$ but $p \mid h(x)$. We now observe that saying that p is a prime is equivalent to saying that $\bar{D} \equiv D/(p)$ is a domain. This is immediate from the definitions. Hence $\bar{D}[x]$

is a domain. We now apply the homomorphism of $D[x]$ onto $\bar{D}[x]$ sending $a \in D$ into its coset $\bar{a} = a + (p)$ and $x \to x$. This gives $\bar{f}(x)\bar{g}(x) = \bar{h}(x) = \bar{0}$ but $\bar{f}(x) \neq 0$, $\bar{g}(x) \neq \bar{0}$. This contradicts the fact that $\bar{D}[x]$ is a domain and hence proves the lemma. $\square$

LEMMA 3. *If $f(x) \in D[x]$ has positive degree and is irreducible in $D[x]$, then $f(x)$ is irreducible in $F[x]$.*

Proof. If $f(x) \in D[x]$ has positive degree and is irreducible in $D[x]$ then $f(x)$ is primitive. Suppose that $f(x)$ is reducible in $F[x]$: $f(x) = \varphi_1(x)\varphi_2(x)$ where $\varphi_i(x) \in F[x]$ and deg $\varphi_i(x) > 0$. We have $\varphi_i(x) = \alpha_i f_i(x)$ where $\alpha_i \in F$ and $f_i(x)$ is primitive in $D[x]$. Then $f(x) = \alpha_1 \alpha_2 f_1(x) f_2(x)$ and $f_1(x) f_2(x)$ is primitive by Gauss' lemma. It follows that $f(x)$ and $f_1(x) f_2(x)$ differ by a unit multiplier in D. Since deg $f_i(x) > 0$ this contradicts the irreducibility of $f(x)$ in $D[x]$. $\square$

We are now ready to prove

THEOREM 2.25. *If D is factorial then so is $D[x]$.*

Proof. Let $f(x) \in D[x]$ be non-zero and not a unit. Then $f(x) = df_1(x)$ where $d \in D$ and $f_1(x)$ is primitive. If deg $f_1(x) > 0$ then $f_1(x)$ is not a unit and if this is not irreducible we have $f_1(x) = f_{11}(x) f_{12}(x)$ where deg $f_{1i}(x) > 0$ so deg $f_{1i}(x) <$ deg $f_1(x)$. Clearly $f_{1i}(x)$ is primitive. Hence using induction on the degree we see that $f_1(x) = q_1(x) q_2(x) \cdots q_t(x)$ where the $q_i(x)$ are irreducible in $D[x]$. If d is not a unit we have $d = p_1 p_2 \cdots p_s$ where the p_i are irreducible in D. Clearly these are then irreducible in $D[x]$. Using the factorizations of d and $f_1(x)$ (when these are not units) we obtain a factorization of $f(x)$ into irreducible factors in $D[x]$. It remains to prove uniqueness up to unit multipliers of any two such factorizations. Suppose first that $f(x)$ is primitive. Then the irreducible factors of $f(x)$ all have positive degree. Thus we have $f(x) = q_1(x) \cdots q_h(x) = q_1'(x) \cdots q_k'(x)$ where the $q_i(x)$ and $q_j'(x)$ are irreducible of positive degree. Then these are irreducible in $F[x]$ by Lemma 3. Since $F[x]$ is factorial we have $h = k$, and by suitably ordering the $q_j'(x)$ we may assume that $q_i(x)$ and $q_i'(x)$ for $1 \leq i \leq h$ are associates in $F[x]$. Then the corollary to Lemma 1 shows that $q_i(x) \sim q_i'(x)$ in $D[x]$. Next suppose that $f(x)$ is not primitive. Since the irreducible factors of positive degree are primitive, their product is primitive. Hence any factorization of $f(x)$ into irreducible elements in $D[x]$ contains factors belonging to D, and their product is the content of $f(x)$. By modifying by a unit multiplier we may assume that this is the same for the two factorizations. Since D is factorial we can pair off the irreducible factors of $f(x)$ belonging to D into associate pairs. The product of the remaining factors, if any, is a primitive polynomial. Since we have taken care of these the proof is complete. □

An immediate consequence of the theorem is that if D is factorial so is the ring $D[x_1, \ldots, x_r]$ of polynomials in r indeterminates over D: for example, $\mathbb{Z}[x_1, \ldots, x_r]$ is factorial and so is $F[x_1, \ldots, x_r]$ for any field F. It is clear from this that the class of factorial domains is more extensive than that of p.i.d. (see p. 131 and also exercise 5 below).

An important consequence of the factoriality of $D[x]$ and of Lemma 3 is the following

COROLLARY. *If D is factorial and $f(x) \in D[x]$ is monic, then any monic factor of $f(x)$ in $F[x]$ is contained in $D[x]$.*

Proof. We can write $f(x) = p_1(x)^{e_1} \cdots p_r(x)^{e_r}$ where the $p_i(x)$ are monic and irreducible in $D[x]$, $p_i(x) \neq p_j(x)$ if $i \neq j$ and $e_i > 0$. Then the monic factors of $f(x)$ in $D[x]$ have the form $p_1(x)^{f_1} \cdots p_r(x)^{f_r}$ with $0 \leq f_i \leq e_i$. If we now pass from $D[x]$ to $F[x]$ then, by Lemma 3, the $p_i(x)$ are irreducible in $F[x]$. Hence $f(x)$ has the same monic factors in $D[x]$ and in $F[x]$. □

EXERCISES

1. Prove that if $f(x)$ is a monic polynomial with integer coefficients then any rational root of $f(x)$ is an integer.

2. Prove the following irreducibility criterion due to Eisenstein. If $f(x) = a_0 + a_1 x + \cdots + a_n x^n \in \mathbb{Z}[x]$ and there exists a prime p such that $p \mid a_i$, $0 \leq i \leq n - 1$, $p \nmid a_n$ and $p^2 \nmid a_0$, then $f(x)$ is irreducible in $\mathbb{Q}[x]$.

3. Show that if p is a prime (in $\mathbb{Z}$) then the polynomial obtained by replacing x by $x + 1$ in $x^{p-1} + x^{p-2} + \cdots + 1 = (x^p - 1)/(x - 1)$ is irreducible in $\mathbb{Q}[x]$. Hence prove that the "cyclotomic" polynomial $x^{p-1} + x^{p-2} + \cdots + 1$ is irreducible in $\mathbb{Q}[x]$.

4. Obtain factorizations into irreducible factors in $\mathbb{Z}[x]$ of the following polynomials: $x^3 - 1$, $x^4 - 1$, $x^5 - 1$, $x^6 - 1$, $x^7 - 1$, $x^8 - 1$, $x^9 - 1$, $x^{10} - 1$.

5. Prove that if D is a domain which is not a field then $D[x]$ is not a p.i.d.

6. Let F be a field and $f(x)$ an irreducible polynomial in $F[x]$. Show that $f(x)$ is irreducible in $F(t)[x]$, t an indeterminate.

2.17 "RNGS" (RINGS WITHOUT UNIT)

In most algebra books a ring is defined to be non-vacuous set R equipped with two binary compositions $+$ and $\cdot$ and an element 0 such that $(R, +, 0)$ is an abelian group, $(R, \cdot)$ is a semigroup (p. 29), and the distributive laws hold. In other words, the existence of a unit for multiplication is not assumed. We shall consider these systems briefly, and so as not to conflict with our old terminology we adopt a different term: *rngs*[6] for the structures which are not assumed to have units. We remark first that the elementary properties of rings which we noted in section 2.1 (generalized associativity, generalized distributivity, rules for multiples, etc.) carry over to rngs. The verification of this is left to the reader. We shall now show that any rng can be imbedded in a ring. This fact permits the reduction of most questions on rngs to the case of rings.

Suppose we are given a rng R. Our procedure for constructing a ring containing R is to take $S = \mathbb{Z} \times R$ the product set of $\mathbb{Z}$ and R. If $m, n \in \mathbb{Z}$ and $a, b \in R$ we define addition in S by

$$(53) \qquad (m, a) + (n, b) = (m + n, a + b)$$

We define $0 = (0, 0)$. Then it is clear that $(S, +, 0)$ is an abelian group: in fact, it is the direct product (also called direct sum) of $(\mathbb{Z}, +, 0)$ and $(R, +, 0)$. We define multiplication in S by

$$(54) \qquad (m, a)(n, b) = (mn, mb + na + ab)$$

where on the right-hand side mb and na denote respectively the mth multiple of b and the nth multiple of a as defined in the additive group $(R, +, 0)$. We have

$$
\begin{aligned}
((m, a)(n, b))(q, c) &= (mn, mb + na + ab)(q, c) \\
&= ((mn)q, (mn)c + q(mb + na + ab) + (mb + na + ab)c) \\
&= ((mn)q, (mn)c + q(mb) + q(na) + q(ab) + (mb)c \\
&\quad + (na)c + (ab)c)
\end{aligned}
$$

$$
\begin{aligned}
(m, a)((n, b)(q, c)) &= (m, a)(nq, nc + qb + bc) \\
&= (m(nq), m(nc) + m(qb) + m(bc) + (nq)a \\
&\quad + a(nc + qb + bc)).
\end{aligned}
$$

It now follows from the associative laws in $\mathbb{Z}$ and in R, the distributive laws in R, and the properties of multiples in R that the associative law of multiplication is valid in S. If we put $1 = (1, 0)$ then we have $1(m, a) = (1, 0)(m, a) = (m, a) = (m, a)(1, 0) = (m, a)1$. Hence $(S, \cdot, 1)$ is a monoid.

---

[6] Suggested pronunciation: rŭngs. This term was suggested to me by Louis Rowen.

Also we have

$$(m, a)[(n, b) + (q, c)] = (m, a)(n + q, b + c)$$
$$= (m(n + q), m(b + c) + (n + q)a + a(b + c))$$
$$(m, a)(n, b) + (m, a)(q, c) = (mn, mb + na + ab) + (mq, mc + qa + ac)$$
$$= (mn + mq, mb + na + ab + mc + qa + ac).$$

Hence $(m, a)[(n, b) + (q, c)] = (m, a)(n, b) + (m, a)(q, c)$. Similarly, the other distributive law holds. Hence $(S, +, \cdot, 0, 1)$ is a ring.

We now consider the subset of elements $(0, a)$ in S. We have $(0, a) + (0, b) = (0, a + b)$, $(0, a)(0, b) = (0, ab)$ and $0 = (0, 0)$ is in this subset. Thus the subset is a subrng isomorphic to R (with the obvious definitions of these terms). We have therefore proved

THEOREM 2.26. *Any rng can be imbedded in a ring.*

We note also that R identified with the corresponding subset of S is an ideal in S since $(m, b)(0, a) = (0, ma + ba)$ and $(0, a)(m, b) = (0, ma + ab)$.

EXERCISES

1. An element a of a rng R is called *right (left) quasi-invertible* (or *right* or *left quasi-regular*) if there exists a b such that $a + b - ab = 0$ $(a + b - ba = 0)$. Show that this is equivalent to saying that $1 - a$ has the right inverse (left inverse) $1 - b$ in $S = \mathbb{Z} \times R$, with the ring structure defined above.

2. (Kaplansky.) Let R be a rng in which every element but one is right quasi-invertible. Show that R has a unit and R is a division ring.

3. Let R be a rng for which there exists a positive integer k such that $ka = 0$ for all $a \in R$. Let $S_k \equiv \mathbb{Z}/(k) \times R$. Write $\bar{m} = m + (k)$ in $\mathbb{Z}/(k)$ and define $(\bar{m}, a) + (\bar{n}, b) = (\bar{m} + \bar{n}, a + b)$, $(\bar{m}, a)(\bar{n}, b) = (\bar{m}\bar{n}, mb + na + ab)$, $0 = (\bar{0}, 0)$, $1 = (\bar{1}, 0)$. Verify that $(S_k, +, \cdot, 0, 1)$ is a ring of characteristic k and that R is imbedded in S_k.

4. Let R be a rng without zero divisors $\neq 0$ (that is, $ab = 0$ in R implies either $a = 0$ or $b = 0$). Assume R contains elements a and $b \neq 0$ such that $ab + kb = 0$ for some positive integer k. Show that $ca + kc = 0 = ac + kc$ for all $c \in R$.

5. Let R be a rng without zero divisors $\neq 0$ and let S be the ring $\mathbb{Z} \times R$ as in the text. Let $Z = \{z \in S \mid za = 0 \text{ for all } a \in R\}$. Show that Z is an ideal in S and S/Z is a domain. Show that $a \to a + Z$ is a monomorphism of R into S/Z.

3

Modules over a
Principal Ideal Domain

The central concept of the axiomatic development of linear algebra is that of a vector space over a field. The axiomatization of linear algebra, which was effected in the 1920's, was motivated to a large extent by the desire to introduce geometric notions in the study of certain classes of functions in analysis. At first one dealt exclusively with vector spaces over the reals or the complexes. It soon became apparent that this restriction was rather artificial, since a large body of the results depended only on the solution of linear equations and thus were valid for arbitrary fields. This led to the study of vector spaces over arbitrary fields and this is what presently constitutes linear algebra.

The concept of a module is an immediate generalization of that of a vector space. One obtains the generalization by simply replacing the underlying field by any ring. Why make this generalization? In the first place, one learns from experience that the internal logical structure of mathematics strongly urges the pursuit of such "natural" generalizations. These often result in an improved insight into the theory which led to them in the first place. A good illustration of this is afforded by the study of a linear transformation in a finite dimensional vector space over a field—a central problem of linear algebra. As we shall see in sections 3.2 and 3.10, given a linear transformation T in a vector space V over F, we can use this to convert V into a module over the polynomial ring $F[\lambda]$, λ an indeterminate.[1] The study of this module will lead to the theory of canonical forms for matrices of a linear transformation and to the solution of the problem of similarity of matrices.

It is an easy step to pass from modules over $F[\lambda]$ to modules over any principal ideal domain. This will give us other applications. In particular, specializing the p.i.d. to be $\mathbb{Z}$, we shall obtain the structure theory of finitely generated abelian groups, hence, of finite abelian groups.

It would be wrong to conclude from these remarks that the historical development of the theory of modules followed the logical path of extension of linear algebra which we have indicated. The concept of a module seems to have made its first appearance in algebra in algebraic number theory—in studying subsets of rings of algebraic numbers closed under addition and multiplication by elements of a specified subring. Modules first became an important tool in algebra in the late 1920's largely due to the insight of Emmy Noether, who was the first to realize the potential of the module concept. In particular she observed that this concept could be used to bridge the gap between two important developments in algebra that had been going on side by side and independently: the theory of representations (= homomorphisms) of finite groups by matrices due to Frobenius, Burnside, and Schur, and the structure theory of algebras due to Molien, Cartan, and Wedderburn. We consider these matters in Vol. II of this work. More recently one has had the development of homological algebra, in which modules also play a central role. This, too, is considered in Vol. II.

The principal topic of this chapter is the study of finitely generated modules over a p.i.d. D and the two special cases, in which D is either $\mathbb{Z}$ or a polynomial ring $F[\lambda]$, F a field. As we have noted, these give, respectively, the structure theory of finitely generated abelian groups and canonical forms for linear transformations. Of course, we shall need to begin with some general theory. However, we shall not develop this much beyond what is actually needed to achieve our immediate objectives. Most of the general theory of modules and other applications are discussed in our second volume.

3.1   RING OF ENDOMORPHISMS OF AN ABELIAN GROUP

Let M be an abelian group. We use the additive notation in M: $+$ for the given binary composition, 0 for the unit, $-a$ for the inverse of a, and ma, $m \in \mathbb{Z}$, for the mth power. Let End M denote the set of endomorphisms of M. By defini-

---

[1] We use λ to denote an indeterminate in the present chapter. We do this in order to reserve x to represent vectors or, more generally, elements of a module.

tion, these are the maps η of M into M such that

(1) $$\eta(x + y) = \eta(x) + \eta(y), \qquad \eta(0) = 0$$

and we have seen that the second condition is a consequence of the first. Hence a map η of M into M is an endomorphism if and only if

(1') $$\eta(x + y) = \eta(x) + \eta(y).$$

We recall that this implies also that $\eta(mx) = m\eta(x)$ for any $m \in \mathbb{Z}$. We recall further that if X is a set of generators for M, then η is determined by its effect on X: that is, if $\eta(x) = \zeta(x)$ for two endomorphisms η and ζ and all x in a set of generators, then $\eta = \zeta$.

Let us look at some

EXAMPLES

1. Let M be an infinite cyclic group $(\mathbb{Z}, +, 0)$. Then 1 is a generator and if $\eta(1) = m$, then $\eta(x) = \eta(x1) = x\eta(1) = xm$. Hence η is the map $x \to mx$, $x \in M$, where $m = \eta(1)$. Moreover, if m is any element of $\mathbb{Z}$, then the map $x \to mx$ is an endomorphism since we have the power rule $m(x + y) = mx + my$. It is clear that $x \to mx$ maps 1 into m. Since endomorphisms are determined by their effects on the generator 1 it is clear we have a 1–1 correspondence between the set End M, $M = (\mathbb{Z}, +, 0)$ and $\mathbb{Z}$, which pairs $\eta \in$ End M with $\eta(1) = m \in \mathbb{Z}$.

2. Let $M = (\mathbb{Z}^{(2)}, +, 0)$, the direct product (or sum) of two copies of $(\mathbb{Z}, +, 0)$. The elements here are the pairs of integers (x, y) and we have $(x, y) = x(1, 0) + y(0, 1)$, so $e = (1, 0)$ and $f = (0, 1)$ generate $\mathbb{Z}^{(2)}$. Hence if $\eta \in$ End $\mathbb{Z}^{(2)}$, then η is determined by the pair of elements $\eta(e)$, $\eta(f)$. Moreover, any pair of elements $(u, v) \in \mathbb{Z}^{(2)} \times \mathbb{Z}^{(2)}$ can be obtained in this way, this is, if (u, v) is given, then there exists an endomorphism η such that $\eta(e) = u$ and $\eta(f) = v$. To see this we let η be the map which sends $(x, y) = xe + yf$ into $xu + yv$. Then $(x', y') \to x'u + y'u$ and $(x + x', y + y') \to (x + x')u + (y + y')v = (xu + yv) + (x'u + y'v)$. Hence η is a homomorphism and $\eta(e) = u$ and $\eta(f) = v$, as required. Thus we have a 1–1 correspondence between End $\mathbb{Z}^{(2)}$ and $\mathbb{Z}^{(2)} \times \mathbb{Z}^{(2)}$, which pairs an endomorphism η with the element $(\eta(e), \eta(f)) \in \mathbb{Z}^{(2)} \times \mathbb{Z}^{(2)}$.

These considerations generalize immediately to $M = \mathbb{Z}^{(n)}$ for any positive integer n and lead to a 1–1 correspondence between End $\mathbb{Z}^{(n)}$ and $\overbrace{\mathbb{Z}^{(n)} \times \cdots \times \mathbb{Z}^{(n)}}^{n}$.

3. Let M be a finite cyclic group. In this case we may take $M = (\mathbb{Z}/(n), +, \bar{0})$ where n is a positive integer, and, in general, $\bar{x}$ is the coset $x + (n)$. Then $\bar{1}$ is a generator and we have a 1–1 correspondence between End $\mathbb{Z}/(n)$ and $\mathbb{Z}/(n)$ sending $\eta \in$ End $\mathbb{Z}/(n)$ into $\eta(\bar{1})$.

We shall now organize End M for any abelian group M into a ring. We know that if $\eta, \zeta \in$ End M, then the composite $\eta\zeta \in$ End M, and we have the associative law $(\eta\zeta)\rho = \eta(\zeta\rho)$. Also, the identity map $1: x \to x$ is an endomorphism. Hence (End M, $\cdot$, 1) is a monoid. All of this holds even if M is not abelian. However, a good deal more can be said in the abelian case: namely, as we shall now show, End M with composite multiplication and an addition and 0, which we shall now define, constitute a ring. If η, $\zeta \in$ End M we define $\eta + \zeta$ by

(2) $$(\eta + \zeta)(x) = \eta(x) + \zeta(x).$$

This map of M into M is an endomorphism since

$$\begin{aligned}
(\eta + \zeta)(x + y) &= \eta(x + y) + \zeta(x + y) \\
&= \eta(x) + \eta(y) + \zeta(x) + \zeta(y) \\
&= \eta(x) + \zeta(x) + \eta(y) + \zeta(y) \\
&= (\eta + \zeta)(x) + (\eta + \zeta)(y).
\end{aligned}$$

We remark that the commutativity of $+$ is used in the passage from the second to the third of these equations. Next we define the map 0 as $x \to 0$, $x \in M$. Evidently this is an endomorphism and $\eta + 0 = \eta = 0 + \eta$ for any $\eta \in$ End M. Let $-\eta$ be the map $x \to -\eta(x)$ so $-\eta$ is the composite of η and the map $x \to -x$, which is an automorphism, since M is abelian. Hence $-\eta \in$ End M, and clearly $\eta + (-\eta) = 0 = -\eta + \eta$. Since $((\eta + \zeta) + \rho)(x) = (\eta + \zeta)(x) + \rho(x) = \eta(x) + \zeta(x) + \rho(x)$ and $(\eta + (\zeta + \rho))(x) = \eta(x) + (\zeta + \rho)(x) = \eta(x) + \zeta(x) + \rho(x)$, associativity holds for the addition composition $+$. Commutativity also holds since $(\eta + \zeta)(x) = \eta(x) + \zeta(x) = \zeta(x) + \eta(x) = (\zeta + \eta)(x)$. Thus we have verified that (End M, $+$, 0) is an abelian group.

Previously, we had that (End M, $\cdot$, 1) is a monoid. Now, we have for η, ζ, $\rho \in$ End M,

$$(\rho(\eta + \zeta))(x) = \rho(\eta(x) + \zeta(x)) = (\rho\eta)(x) + (\rho\zeta)(x) = (\rho\eta + \rho\zeta)(x)$$

Similarly, $((\eta + \zeta)\rho)(x) = \eta(\rho(x)) + \zeta(\rho(x))$. Hence both distributive laws hold in End M, and so we have verified the following basic

THEOREM 3.1 *Let M be an abelian group (written additively) and let End M denote the set of endomorphisms of M. Define $\eta\zeta$ and $\eta + \zeta$ for η, $\zeta \in$ End M by $(\eta\zeta)(x) = \eta(\zeta(x))$ and $(\eta + \zeta)(x) = \eta(x) + \zeta(x)$, 1 and 0 by $1x = x$, $0x = 0$. Then (End M, $+$, $\cdot$, 0, 1) is a ring.*

We shall call (End M, $+$, $\cdot$, 0, 1) or, more briefly, End M, *the ring of endomorphisms of the abelian group M*. We consider again the examples we gave above and we seek to identify the rings End M in these cases.

EXAMPLES

1. $M = (\mathbb{Z}, +, 0)$. We saw that the map $\eta \to \eta(1)$ is a bijective map of End M onto $\mathbb{Z}$. In this map $\eta + \zeta \to (\eta + \zeta)(1) = \eta(1) + \zeta(1)$, $\eta\zeta \to (\eta\zeta)(1) = \eta(\zeta(1)) = \eta(\zeta(1)1) = \zeta(1)\eta(1) = \eta(1)\zeta(1)$ and $1 \to 1(1) = 1$. Hence $\eta \to \eta(1)$ is an isomorphism of End M with the ring of integers $\mathbb{Z}$. Hence we can say that the ring of endomorphisms of an infinite cyclic group is the ring $\mathbb{Z}$.

2. $M = (\mathbb{Z}^{(2)}, +, 0)$. In this case we obtain the bijective map $\eta \to (\eta(e), \eta(f))$ of End M onto $\mathbb{Z}^{(2)} \times \mathbb{Z}^{(2)}$, the set of pairs of elements of $\mathbb{Z}^{(2)}$. Here $e = (1, 0)$ and $f = (0, 1)$. Suppose $\eta(e) = (a, b)$ and $\eta(f) = (c, d)$. Then we evidently have a bijective map

$$(3) \qquad\qquad \eta \to \begin{pmatrix} a & c \\ b & d \end{pmatrix}$$

of End M onto the ring $M_2(\mathbb{Z})$ of 2×2 integral matrices. We claim that this is an isomorphism. Suppose ζ is a second endomorphism and $\zeta(e) = (a', b')$, $\zeta(f) = (c', d')$. Then

$$(4) \qquad\qquad \zeta \to \begin{pmatrix} a' & c' \\ b' & d' \end{pmatrix}.$$

Now $(\eta + \zeta)(e) = \eta(e) + \zeta(e) = (a, b) + (a', b') = (a + a', b + b')$ and similarly $(\eta + \zeta)(f) = (c + c', d + d')$. Hence

$$\eta + \zeta \to \begin{pmatrix} a + a' & c + c' \\ b + b' & d + d' \end{pmatrix}$$

and this is the sum of the matrices in (3) and (4). Next we determine $(\eta\zeta)(e) = \eta(\zeta(e)) = \eta(a', b') = \eta(a'e + b'f) = \eta(a'e) + \eta(b'f) = a'\eta(e) + b'\eta(f) = a'(a, b) + b'(c, d) = (a'a, a'b) + (b'c, b'd) = (aa' + cb', ba' + db')$. Similarly, $(\eta\zeta)(f) = (ac' + cd', bc' + dd')$. Thus

$$\eta\zeta \to \begin{pmatrix} aa' + cb' & ac' + cd' \\ ba' + db' & bc' + dd' \end{pmatrix} = \begin{pmatrix} a & c \\ b & d \end{pmatrix}\begin{pmatrix} a' & c' \\ b' & d' \end{pmatrix}$$

the product of the matrix in (3) followed by the one in (4). Also $1(e) = (1, 0)$ and $1(f) = (0, 1)$ so $1 \to \begin{pmatrix} 1 & 0 \\ 0 & 1 \end{pmatrix}$. Hence we have verified that the map (3) is an isomorphism of End M with the matrix ring $M_2(\mathbb{Z})$.

3. M a cyclic group of order n. One sees, as in 1, that End M is isomorphic to the ring $\mathbb{Z}/(n)$.

The fact that End M is a ring with respect to the compositions and the 0 and 1 that we defined is analogous to the fact that the set of bijective maps of a set with the usual composition and 1 is a group. We now define a *ring of endomorphisms* to be any subring of a ring End M, M an abelian group. We shall now prove the analogue for rings of Cayley's theorem for groups (p. 38).

THEOREM 3.2. *Any ring is isomorphic to a ring of endomorphisms of an abelian group.*[2]

Proof. The idea of the proof is identical with that of Cayley's theorem. Given the ring R we take $M = (R, +, 0)$, the additive group of R, and for any a we call the map $a_L : x \to ax$ the *left multiplication determined by* a.[3] Since $a_L(x + y) = a(x + y) = ax + ay = a_L x + a_L y$, $a_L \in$ End M. Also $(a + b)_L x = (a + b)x = ax + bx = a_L x + b_L x = (a_L + b_L)x$ (by definition of the sum of endomorphisms) and $(ab)_L x = (ab)x = a(bx) = a_L(b_L x) = (a_L b_L)(x)$, $1x = x$. Hence $a \to a_L$ is a homomorphism of the ring R into End M. Since $a_L = b_L$ implies $a = a_L 1 = b_L 1 = b$, $a \to a_L$ is a monomorphism. The image is a subring R_L of End M and we have $R \cong R_L$. $\square$

It is interesting to consider also the right multiplications of a ring. We define $a_R : x \to xa$ and note that this is an endomorphism of $M = (R, +, 0)$ since $(x + y)a = xa + ya$. Also it is immediate that $a \to a_R$ is an anti-homomorphism of R into End M. The image $R_R = \{a_R\}$ is a subring of End M and R and R_R are anti-isomorphic. We note also that the subrings R_L and R_R are the centralizers of each other in End M, that is, we have

THEOREM 3.3 $R_L = C(R_R)$ *and* $R_R = C(R_L)$ *in* End M.

Proof. It is clear from $(ax)b = a(xb)$ that $a_L b_R = b_R a_L$ for any $a, b \in R$. Now let η be an endomorphism of M such that $a_L \eta = \eta a_L$, $a \in R$. Then $\eta(x) = \eta(x1) = \eta(x_L 1) = x_L(\eta(1)) = x\eta(1)$. Hence $\eta = \eta(1)_R \in R_R$. Thus $C(R_L) = R_R$ and, by symmetry $C(R_R) = R_L$. $\square$

EXERCISES

1. Let G be a group (written multiplicatively), and let $F = G^G$ be the set of maps of G into G. If $\eta, \zeta \in F$ define $\eta\zeta$ in the usual way as the composite η following ζ. Define $\eta + \zeta$ by $(\eta + \zeta)(x) = \eta(x)\zeta(x)$. Define $1 : x \to x$, $0 : x \to 1$. Investigate the properties of the structure $(F, +, \cdot, 0, 1)$.

---

[2] This result in a somewhat more special sitution—that of algebras—seems to have been noted first by Poincaré.

[3] We recall that in the group case our preferred terminology was "translation" for such a map.

2. Let M be an abelian group. Observe that Aut M is the group of units (invertible elements) of End M. Use this to show that Aut M for the cyclic group of order n is isomorphic to the group of cosets $\bar{m} = m + (n)$ in $\mathbb{Z}/(n)$ such that $(m, n) = 1$.

3. Determine Aut M for $M = (\mathbb{Z}^{(2)}, +, 0)$.

4. Determine End $(\mathbb{Q}, +, 0)$.

5. In several cases we have considered, we have End $(R, +, 0) \cong R$ for a ring R. Does this hold in general? Does it hold if R is a field?

3.2 LEFT AND RIGHT MODULES

The concept of a left module is the ring analogue of a group acting on a set. As in the group case, this arises in considering a homomorphism of a given ring R into the ring of endomorphisms, End M, of an abelian group M. If η is such a homomorphism, $\eta(a) \in$ End M, so we have

$$\eta(a)(x + y) = \eta(a)(x) + \eta(a)(y), \qquad x, y \in M,$$

and since η is a homomorphism we have

$$\eta(a + b)(x) = (\eta(a) + \eta(b))(x) = \eta(a)(x) + \eta(b)(x)$$

$$\eta(ab)(x) = (\eta(a)\,\eta(b))(x) = \eta(a)(\eta(b)(x))$$

$$\eta(1)(x) = x,$$

$x \in M$, $a, b \in R$. We now consider the map $(a, x) \to \eta(a)(x)$ of $R \times M$ into M and we abbreviate the image $\eta(a)(x)$ as ax. Then the foregoing equations read:

1. $$a(x + y) = ax + ay$$

2. $$(a + b)x = ax + bx$$

3. $$(ab)x = a(bx)$$

4. $$1x = x$$

for $x, y \in M$, $a, b, 1 \in R$. We formalize this in the following

DEFINITION 3.1. *If R is a ring, a* left R-module *is an abelian group M together with a map $(a, x) \to ax$ of $R \times M$ into M satisfying properties 1–4.*

We have seen that a homomorphism η of R into End M gives rise to a left module structure on M by defining $ax = \eta(a)(x)$ for $a \in R$, $x \in M$. Conversely, suppose we are given a left R-module M. For any $a \in R$ we let a_L be the map $x \to ax$ of M into itself. Then the module property 1 states that $a_L \in \text{End } M$. Moreover, it is clear from properties 2–4 that $a \to a_L$ is a homomorphism of R into End M. The module obtained from this homomorphism by the procedure we gave is the given left module. On the other hand, if we begin with a homomorphism η of R into End M and we construct the corresponding left R-module M, then the associated homomorphism $a \to a_L$ coincides with η, since $a_L x = ax = \eta(a)(x)$. Thus it is clear that the concept of a left R-module is equivalent to that of a homomorphism of R into the ring of endomorphisms of some abelian group.

The notion of right R-module is dual to that of left R-module. We give this in

DEFINITION 3.1'.   *A right module for a ring R is an abelian group M together with a map $(x, a) \to xa$ of $M \times R$ into M satisfying for $a, b, 1 \in R$ and $x, y \in M$:*

1'. $(x + y)a = xa + ya$

2'. $x(a + b) = xa + xb$

3'. $x(ab) = (xa)b$

4'. $x1 = x.$

Let a_R denote the map $x \to xa$ in M. Then $a_R \in \text{End } M$ and $a \to a_R$ satisfies $(a + b)_R = a_R + b_R$, $(ab)_R = b_R a_R$, $1_R = 1$, so this is an anti-homomorphism of R into End M (section 2.8, p. 114). Conversely, if η is an anti-homomorphism of R into the endomorphism ring, End M, of an abelian group, M becomes an R-module if we define the action xa, $x \in M$, $a \in R$, to be $\eta(a)(x)$.

Any anti-homomorphism η of a ring R can be regarded as a homomorphism of the opposite ring R^0 of R (p. 113). This is clear since the identity map is an anti-isomorphism of R^0 onto R and the composite of this and η is a homomorphism. It follows from this that if M is a right (left) module for R, and we put $ax = xa$ $(xa = ax)$, we make M into a left (right) R^0-module. If R is commutative, $R^0 = R$ as rings and so any left (right) R-module is also a right (left) R-module in which $ax = xa$. Thus for commutative rings there is no distinction between left and right modules.

We now consider some important instances of modules. We observe first that any abelian group M (written additively) is a $\mathbb{Z}$-module. Here one defines ax in the usual way for $a \in \mathbb{Z}$, $x \in M$. The module conditions 1–4 are clear from the properties of multiples in an abelian group. The observation that abelian groups are $\mathbb{Z}$-modules permits us to subsume the theory of abelian groups in that of modules. The usefulness of this reduction will be apparent in what follows.

A type of module which is very probably familiar to the reader is a vector space V over a field F. We recall that a vector space is defined axiomatically as an abelian group V together with a product $ax \in V$ for $a \in F$, $x \in V$ such that conditions 1–4 hold. Thus V is a left F-module. Now suppose T is a linear transformation in V. We abbreviate $T(x)$ as Tx. Then the defining conditions are that T maps V into V and

(5) $$T(x + y) = Tx + Ty, \qquad T(ax) = a(Tx),$$

$a \in F$, x, $y \in V$. The first of these conditions is that $T \in \text{End } V$ and the second is that $a_L T = T a_L$ for every endomorphism $a_L : x \to ax$, $a \in F$. It follows that the subring $F_L[T]$, generated by $F_L = \{a_L \mid a \in F\}$ and T, is a commutative subring of End V. Since $a \to a_L$ is a homomorphism of F, the basic homomorphism property of $F[\lambda]$, λ an indeterminate, (Theorem 2.10, p. 122) shows that the map

$$a_0 + a_1 \lambda + \cdots + a_m \lambda^m \to a_{0L} + a_{1L} T + \cdots + a_{mL} T^m$$

$(a_i \in F)$ is a homomorphism of $F[\lambda]$ into $F_L[T]$, hence, into End V. Then it is clear that V becomes a left $F[\lambda]$-module if we define

$$(a_0 + a_1 \lambda + \cdots + a_m \lambda^m)x = a_0 x + a_1(Tx) + \cdots + a_m(T^m x)$$

for every $f(\lambda) = a_0 + a_1 \lambda + \cdots + a_m \lambda^m \in F[\lambda]$. We shall see that the theory of a single linear transformation of a finite dimensional vector space can be derived by viewing the vector space as an $F[\lambda]$-module in this way.

As our last example of a module we consider any ring R, and take M to be the additive group $(R, +, 0)$ of R. Let R act on M by left multiplication: ax for $a \in R$ and $x \in M$ is the product as defined in R. Then 1–4 are clear, and so M is a left R-module. Similarly M is a right R-module if we define xa, $x \in M$, $a \in R$, to be the ring product.

EXERCISES

1. Let M be a left R-module and let η be a homomorphism of a ring S into R. Show that M becomes a left S-module if we define $ax = \eta(a)(x)$ for $a \in S$, $x \in M$.

2. Let M be a left R-module and let $B = \{b \in R \mid bx = 0 \text{ for all } x \in M\}$. Verify that B is an ideal in R. Show also that if C is any ideal contained in B then M becomes a left R/C-module by defining $(a + C)x = ax$.

3. Let M be a left R-module, S a subring of R. Show that M is a left S-module if we define bx, $b \in S$, $x \in M$, as given in M as left R-module. (Note that this is a special case of exercise 1). In particular, the ring R can be regarded as a left S-module in this way.

4. Let $V = \mathbb{R}^{(n)}$ the vector space of n-tuples of real numbers with the usual addition and multiplication by elements of $\mathbb{R}$. Let T be the linear transformation of V defined by

$$x = (x_1, x_2, \ldots, x_n) \to Tx = (x_n, x_1, x_2, \ldots, x_{n-1}).$$

   Consider V as left $\mathbb{R}[\lambda]$-module as in the text, and determine: (a) λx, (b) $(\lambda^2 + 2)x$, (c) $(\lambda^{n-1} + \lambda^{n-2} + \cdots + 1)x$. What elements satisfy $(\lambda^2 - 1)x = 0$?

5. Consider the example of exercise 4 and let B be the ideal in $\mathbb{R}[\lambda]$ defined as in exercise 2. Give an explicit description of B.

6. Let M be an abelian group written additively. Show that there is only one way of making M into a left $\mathbb{Z}$-module.

7. Let M be a left $\mathbb{Q}$-module. Show that the given action of $\mathbb{Q}$ is the only one which can be used to make M a left $\mathbb{Q}$-module.

8. Let M be a finite abelian group $\neq 0$. Can M be made into a left $\mathbb{Q}$-module?

3.3 FUNDAMENTAL CONCEPTS AND RESULTS

From now on we shall deal almost exclusively with left modules and we shall refer to these simply as "modules," "R-modules," or "modules over R" (R the given ring). Of course, what we shall say about these will be applicable also to right modules. The modifier "right" will be used when we wish to state results explicitly for these.

Let M be an R-module. The fact that $x \to ax$ is an endomorphism of $(M, +, 0)$ implies that $a0 = 0$ and $a(-x) = -ax$, $x \in M$, $a \in R$. The fact that $a \to a_L$ is a homomorphism of R into $\operatorname{End} M$ gives $0x = 0$, $(-a)x = -ax$. Also, by induction, we have $a(\sum x_i) = \sum ax_i$ and $(\sum a_i)x = \sum a_i x$.

We define a *submodule* N of M to be a subgroup of the additive group $(M, +, 0)$ which is closed under the action of the elements of R: that is, if $a \in R$ and $y \in N$, then $ay \in N$. Explicitly, the conditions for a non-vacuous subset N of M to be a submodule are: (a) if $y_1, y_2 \in N$ then $y_1 + y_2 \in N$, (b) if $y \in N$ and $a \in R$ then $ay \in N$. These are certainly satisfied by submodules. On the other hand, if N satisfies these conditions, then N contains $0 = 0y$, $y \in N$, and N contains $-y = (-1)y$. Thus N is a subgroup of the additive group and hence a submodule of M.

What are the submodules of the types of modules we considered in section 3.2? First, let M be a $\mathbb{Z}$-module. If N is a subgroup of $(M, +, 0)$, and n is a positive integer and $y \in N$, then $ny = y + \cdots + y$ (n terms) $\in N$. Also $0y$ and

$(-n)y \in N$. Hence N is a $\mathbb{Z}$-submodule. The converse is clear. Hence the $\mathbb{Z}$-submodules of M are the subgroups of $(M, +, 0)$. Next let V be a vector space over a field F. Then it is clear from the definitions that the submodules are the subspaces of V. Now let T be a linear transformation in V and regard V as an $F[\lambda]$-module in the manner of section 3.2. In this case the submodules are simply the subspaces W stabilized by T—that is, satisfying $TW(\equiv T(W)) \subset W$—since this condition on a subspace amounts to $\lambda w \in W$ if $w \in W$, and clearly this implies that $(a_0 + a_1\lambda + \cdots + a_n\lambda^n)w \in W$. Finally, we consider the case of R regarded as left R-module ($M = (R, +, 0)$ and the module action is left multiplication). Here the submodules are the subsets of R that are closed under addition and under left multiplication by arbitrary elements of R. Such a subset is called a *left ideal* of R (cf. exercise 4 on p. 103). Similarly, the submodules of R regarded as right R-modules in the usual way are the *right ideals*: subsets closed under addition and under right multiplication by arbitrary elements of R.

If $\{N_\alpha\}$ is a set of submodules of M, then $\bigcap N_\alpha$ is a submodule. Hence if S is a non-vacuous subset, then the intersection $\langle S \rangle$ of all the submodules of M containing S is a submodule of M. We call this the *submodule generated by S*, since it is a submodule containing S and contained in every submodule containing S. It is immediate that $\langle S \rangle$ is the subset of elements of the form $a_1 y_1 + a_2 y_2 + \cdots + a_r y_r$ where the $a_i \in R$ and the $y_i \in S$. If $\{N_\alpha\}$ is a set of submodules, then the submodule generated by $\bigcup N_\alpha$ is the set of sums $y_{\alpha_1} + y_{\alpha_2} + \cdots + y_{\alpha_r}$ where $y_{\alpha_k} \in N_{\alpha_k}$. We call this the *submodule generated by the N_α* and denote it as $\sum N_\alpha$. If $\{N_\alpha\}$ is finite, say, $\{N_1, N_2, \ldots, N_m\}$, then we write either $\sum N_i$ or $N_1 + N_2 + \cdots + N_m$ for the submodule generated by the N_i.

We now consider the factor group $\bar{M} = M/N$ of M relative to a submodule N. Its elements are the cosets $\bar{x} = x + N$ with the addition $(x_1 + N) + (x_2 + N) = x_1 + x_2 + N$, the 0-element N, and $-(x + N) = -x + N$. If $a \in R$ and $x_1 \equiv x_2 \pmod{N}$, that is, $x_2 - x_1 \in N$ then $ax_2 - ax_1 = a(x_2 - x_1) \in N$ so $ax_1 \equiv ax_2 \pmod{N}$. It follows that if we put

$$(6) \qquad a\bar{x} = a(x + N) = ax + N = \overline{ax}$$

then this coset is independent of the choice of the element x in its coset. Hence $(a, \bar{x}) \to \overline{ax}$ is a map of $R \times \bar{M}$ into $\bar{M}$. We also have

$$a(\bar{x}_1 + \bar{x}_2) = a\overline{(x_1 + x_2)} = \overline{ax_1 + ax_2}$$
$$= \overline{ax_1} + \overline{ax_2} = a\bar{x}_1 + a\bar{x}_2$$

and, similarly, $(a + b)\bar{x} = a\bar{x} + b\bar{x}$, $(ab)\bar{x} = a(b\bar{x})$ and $1\bar{x} = \bar{x}$. Thus $\bar{M} = M/N$ with the action (6) is an R-module. We call this the *quotient module M/N* of M with respect to the submodule N.

We define homomorphisms for modules only if the rings over which these are defined are identical. In this case we define a *homomorphism* (*module homomorphism, R-homomorphism, homomorphism over R*) of M into M' to be a map η of M into M' which is a homomorphism of the additive groups and which satisfies $\eta(ax) = a\eta(x)$, $a \in R$, $x \in M$. It is clear from (6) that if N is a submodule of M then the natural map $v: x \to \bar{x} = x + N$ is a module homomorphism of M into $\bar{M}$.

The kernel of a homomorphism of M into M' is defined to be the kernel $\eta^{-1}(0)$ of the group homomorphism. This is a subgroup of M, and since $\eta(y) = 0$ implies $\eta(ay) = a\eta(y) = 0$, ker η is a submodule of M. The image $\eta(M)$ (or im $\eta = \{\eta(x) \,|\, x \in M\}$) is a submodule of M'; for it is a subgroup of M', and if $y \in \eta(M)$, $y = \eta(x)$, $x \in M$, and $ay = a\eta(x) = \eta(ax) \in \eta(M)$. As in the case of groups, it is immediate that if N is a submodule contained in ker η, then the map

(7) $$\bar{\eta}: \bar{x} = x + N \to \eta(x)$$

is a module homomorphism of M/N into M' such that $\eta = \bar{\eta}v$ where v is the homomorphism $x \to \bar{x} = x + N$. Moreover, $\bar{\eta}$ is a monomorphism if and only if $N = $ ker η. In this case we have the *fundamental theorem of homomorphisms for modules* that any homomorphism η can be factored as $\bar{\eta}v$ where v is the natural homomorphism of M onto $\bar{M} = M/$ker η and $\bar{\eta}$ is the induced monomorphism of $\bar{M}$ into M' ($\bar{\eta}: M \to M'$). If η is surjective so is $\bar{\eta}$, and $\bar{\eta}$ is then an isomorphism. Thus any homomorphic image of M is isomorphic to a quotient module.

The results in sections 1.9 and 1.10 on group homomorphisms carry over to modules. It is left to the reader to check this; we shall feel free to use the corresponding module results when we have need for them.

The analogue for modules of cyclic groups are *cyclic modules*. Such a module is generated by a single element and thus has the form $M = Rx = \{ax \,|\, a \in R\}$ where $x \in M$. The role played by the infinite cyclic group $(\mathbb{Z}, +, 0)$ is now taken by R as R-module. This is generated by 1, since $R = R1$. If $M = Rx$ then we have the homomorphism μ_x of R into Rx which sends $a \to ax$. Clearly this is a group homomorphism and $\mu_x(ba) = (ba)x$, and $b\mu_x(a) = b(ax)$. Hence $\mu_x(ba) = b\mu_x(a)$ and μ_x is indeed a module homomorphism of R. Evidently this is surjective and hence $M = Rx \cong R/$ker μ_x. Now ker $\mu_x = \{d \in R \,|\, dx = 0\}$ and, being a submodule of R, it is a left ideal of R. We shall call this the *annihilator of x* (*in R*) and denote it as ann x. In this notation we have the following formula for a cyclic module:

(8) $$Rx \cong R/\text{ann } x.$$

If ann $x = 0$ we have $Rx \cong R$. In the special case $R = \mathbb{Z}$ we have either $\mathbb{Z}x \cong R$, or ann $x = (n)$ where $n > 0$ and n is the smallest positive integer such that $nx = 0$. Clearly this is the order of the element x and of the cyclic group $\langle x \rangle$. Thus ann x for an element x of a module can be regarded as a generalization of the order of an element of a group. For this reason ann x is sometimes called the *order ideal* of the element x.

Now let M and N be modules and let $\mathrm{Hom}(M, N)$ (or $\mathrm{Hom}_R(M, N)$) denote the set of homomorphisms of M into N. This set can be made into an abelian group by defining $\eta + \zeta$ for $\eta, \zeta \in \mathrm{Hom}(M, N)$ by $(\eta + \zeta)(x) = \eta(x) + \zeta(x)$ and 0 by $0(x) = 0$ (the zero element of N). The verification that $\eta + \zeta$, $0 \in \mathrm{Hom}(M, N)$ and that $(\mathrm{Hom}(M, N), +, 0)$ is an abelian group requires only one step more than the corresponding verification that the endomorphisms of an abelian group form an abelian group (p. 160). This is that $(\eta + \zeta)(ax) = a((\eta + \zeta)(x))$, which is clear, since $(\eta + \zeta)(ax) = \eta(ax) + \zeta(ax) = a\eta(x) + a\zeta(x)$ and $a((\eta + \zeta)(x)) = a(\eta(x) + \zeta(x)) = a\eta(x) + a\zeta(x)$. Now consider a third module P, and let $\eta \in \mathrm{Hom}(M, N)$, $\zeta \in \mathrm{Hom}(N, P)$. Then $\zeta\eta$ is a homomorphism of the additive group $(M, +, 0)$ into $(P, +, 0)$, and since $(\zeta\eta)(ax) = \zeta(\eta(ax)) = \zeta(a\eta(x)) = a\zeta(\eta(x)) = a((\zeta\eta)(x))$, $\zeta\eta \in \mathrm{Hom}(M, P)$. As in the special case of End M, we have the distributive laws $(\zeta_1 + \zeta_2)\eta = \zeta_1\eta + \zeta_2\eta$, $\zeta(\eta_1 + \eta_2) = \zeta\eta_1 + \zeta\eta_2$ if $\eta, \eta_1, \eta_2 \in \mathrm{Hom}(M, N)$ and $\zeta, \zeta_1, \zeta_2 \in \mathrm{Hom}(N, P)$. It is clear also that $1_N\eta = \eta = \eta 1_M$, and if Q is a fourth module, then $(\omega\zeta)\eta = \omega(\zeta\eta)$ for $\eta \in \mathrm{Hom}(M, N)$, $\zeta \in \mathrm{Hom}(N, P)$, $\omega \in \mathrm{Hom}(P, Q)$. These results specialize to the conclusion that $(\mathrm{Hom}(M, M), +, \cdot, 0, 1)$ is a ring. We shall denote this ring as $\mathrm{End}_R M$ and call it the *ring of endomorphisms of the module M*.

EXERCISES

1. Determine $\mathrm{Hom}(\mathbb{Z}, \mathbb{Z}/(n))$ and $\mathrm{Hom}(\mathbb{Z}/(n), \mathbb{Z})$, $n > 0$ (as $\mathbb{Z}$-modules).

2. Determine $\mathrm{Hom}(\mathbb{Z}/(m), \mathbb{Z}/(n))$, $m, n > 0$ (as $\mathbb{Z}$-modules).

3. Show that $\mathrm{Hom}(\mathbb{Z}^{(2)}, \mathbb{Z}) \cong (\mathbb{Z}^{(2)}, +, 0)$.

4. Prove that for any R and R-module M, $\mathrm{Hom}(R, M) \cong (M, +, 0)$.

5. Show that $\mathrm{End}_R M$ is the centralizer in End M of the set of group endomorphisms a_L, $a \in R$.

6. Does $a_L \in \mathrm{End}_R M$?

7. A module M is called *irreducible* if $M \neq 0$ and 0 and M are the only submodules of M. Show that M is irreducible if and only if $M \neq 0$ and M is cyclic with every non-zero element as generator.

8. A left (right) ideal I of R is called *maximal* if $R \neq I$ and there exist no left (right) ideals I' such that $R \supsetneq I' \supsetneq I$. Show that a module M is irreducible if and only if $M \cong R/I$ where I is a maximal left ideal of R.

9. (Schur's lemma.) Show that if M_1 and M_2 are irreducible modules, then any non-zero homomorphism of M_1 into M_2 is an isomorphism. Hence show that if M is irreducible then $\operatorname{End}_R M$ is a division ring.

3.4   FREE MODULES AND MATRICES

Let R be a ring and let $R^{(n)}$ be the set of n-tuples $(x_1, x_2, \ldots, x_n)$, $x_i \in R$. As a generalization of the familiar construction of the n dimensional vector space $\mathbb{R}^{(n)}$ we introduce an addition, 0 element in $R^{(n)}$, and a multiplication by elements of R in the following manner:

(9) $(x_1, x_2, \ldots, x_n) + (y_1, y_2, \ldots, y_n) = (x_1 + y_1, x_2 + y_2, \ldots, x_n + y_n)$

(10) $0 = (0, 0, \ldots, 0)$

(11) $a(x_1, x_2, \ldots, x_n) = (ax_1, ax_2, \ldots, ax_n).$

It is clear that $(R^{(n)}, +, 0)$ is an abelian group; this is just a special case of the direct product construction that we gave on p. 35. It is immediate also from (11) that the module conditions 1–4 hold for $R^{(n)}$. Hence $R^{(n)}$ is a module over the ring R. In the special case $n = 1$, $R^{(1)}$ is the same thing as R regarded as left R-module in the usual manner.

Put

$$i$$

(12) $e_i = (0, \ldots, 0, 1, 0, \ldots, 0).$

Then $x_i e_i = (0, \ldots, 0, x_i, 0, \ldots, 0)$ and

(13) $x \equiv (x_1, x_2, \ldots, x_n) = \sum_1^n x_i e_i.$

Hence the n elements e_i generate $R^{(n)}$ as R-module. Moreover, by (13), $\sum x_i e_i = 0$ implies $(x_1, x_2, \ldots, x_n) = 0$, which implies every $x_i = 0$. Equivalently, $\sum x_i e_i = \sum y_i e_i$ implies $x_i = y_i$, $1 \le i \le n$. A set of generators having these properties is called a *base*. The existence of a base of n elements characterizes $R^{(n)}$ in the sense of isomorphism. We shall show this by first establishing another basic property of $R^{(n)}$, namely, if M is any module over R and $(u_1, u_2, \ldots, u_n)$ is an ordered set of n elements of M, then there exists a unique homomorphism μ of $R^{(n)}$ into M sending $e_i \to u_i$, $1 \le i \le n$. To see this we simply define μ by

(14) $$\mu:(x_1, \ldots, x_n) = \sum x_i e_i \to \sum x_i u_i.$$

It is clear that this is single valued, and direct verification shows that it is a module homomorphism. Moreover, we have $\mu(e_i) = u_i$ for all i and since a homomorphism is determined by its action on a set of generators (module analogue of Theorem 1.7, p. 60), it is clear that μ is the only homomorphism of $R^{(n)}$ into M sending e_i into u_i, $1 \le i \le n$.

Now suppose the u_i constitute a base for M in the sense defined above. Then im μ, which is a submodule of M, contains the generators $u_1, \ldots, u_n$. Hence im $\mu = M$. Also, if $x = (x_1, \ldots, x_n) \in \ker \mu$ then $\sum x_i u_i = 0$, so, by the definition of a base, every $x_i = 0$ and $x = 0$. Thus ker $\mu = 0$ and so μ is an isomorphism. We have therefore shown that the existence of a base of n elements for a module M implies that $M \cong R^{(n)}$. In this case we shall say that M is a *free R-module of rank n*.

It may happen that there exist distinct integers m and n such that $R^{(m)} \cong R^{(n)}$. Examples of R for which this occurs are somewhat difficult to construct. In fact, for many important classes of rings one has the familiar result of linear algebra of invariance of base number. In particular, as we shall now show, this holds for all commutative rings.

THEOREM 3.4. *If R is commutative, $R^{(m)} \cong R^{(n)}$ implies $m = n$.*

Proof. In view of the result on free modules, the statement to be proved is equivalent to: if M is a module over a commutative ring R and M has bases of m and of n elements, then $m = n$. Thus let $\{e_i | 1 \le i \le n\}$, $\{f_j | 1 \le j \le m\}$ be bases for M. Then we have

$$f_j = \sum_1^n a_{ji} e_i, \qquad e_i = \sum_1^m b_{ij} f_j$$

where the $a_{ji}, b_{ij} \in R$. Substitution now gives

$$f_j = \sum_{i=1, j'=1}^{n,m} a_{ji} b_{ij'} f_{j'}$$

$$e_i = \sum_{j=1, i'=1}^{m,n} b_{ij} a_{ji'} e_{i'}.$$

Since the f's and the e's form bases we have

(15) $$\sum_{i=1}^n a_{ji} b_{ij'} = \begin{cases} 1 & \text{if } j = j' \\ 0 & \text{if } j \ne j' \end{cases}$$

(16)
$$\sum_{j=1}^{m} b_{ij} a_{ji'} = \begin{cases} 1 & \text{if } i = i' \\ 0 & \text{if } i \neq i' \end{cases}$$

where $j, j' = 1, 2, \ldots, m$; $i, i' = 1, 2, \ldots, n$. Now suppose $m < n$ and consider the two $n \times n$ matrices

$$A = \begin{pmatrix} a_{11} & a_{12} & \cdots & a_{1n} \\ \cdots & \cdots & \cdots & \cdots \\ a_{m1} & a_{m2} & \cdots & a_{mn} \\ 0 & 0 & \cdots & 0 \\ \cdots & \cdots & \cdots & \cdots \\ 0 & 0 & \cdots & 0 \end{pmatrix}$$

$$B = \begin{pmatrix} b_{11} & \cdots & b_{1m} & 0 & \cdots & 0 \\ b_{21} & \cdots & b_{2m} & 0 & \cdots & 0 \\ \cdots & \cdots & \cdots & \cdots & \cdots & \cdots \\ b_{n1} & \cdots & b_{nm} & 0 & \cdots & 0 \end{pmatrix}.$$

Then (16) is equivalent to the matrix condition $BA = 1$. Since R is commutative this implies $AB = 1$ (Theorem 2.1, p. 96 and exercise 2, p. 97). However, it is clear from the form of the matrices A and B that the last $n - m$ rows of AB are 0, so $AB \neq 1$. This contradiction shows that $m \geq n$. By symmetry, $n \geq m$, so $m = n$. $\square$

The foregoing argument shows that if $(e_1, \ldots, e_n)$ and $(f_1, \ldots, f_n)$ are bases and $f_j = \sum_{i=1}^{n} a_{ji} e_i$, $e_i = \sum_{j=1}^{n} b_{ij} f_j$, then $AB = 1 = BA$ for $A = (a_{ij})$, $B = (b_{ij})$. Hence A and B are invertible, that is, $A, B \in GL_n(R)$, the group of $n \times n$ invertible matrices with entries in R. Conversely, suppose $(e_1, \ldots, e_n)$ is a base and $A \in GL_n(R)$. Define $f_j = \sum_{i=1}^{n} a_{ji} e_i$, $1 \leq j \leq n$. Then $(f_1, \ldots, f_n)$ is also a base. First, we have $\sum b_{kj} f_j = \sum_{i,j=1}^{n} b_{kj} a_{ji} e_i = e_k$ since $BA = 1$. Since the e_i generate M, this shows that the f_i also generate M. Next suppose we have a relation $\sum d_j f_j = 0$. Then $\sum_{i,j} d_j a_{ji} e_i = 0$ and $\sum_{j=1}^{n} d_j a_{ji} = 0$, $1 \leq i \leq n$. Hence $\sum_{i,j=1}^{n} d_j a_{ji} b_{ih} = 0$ for all h. Since $AB = 1$ this gives $d_h = 0$ for all h. Hence $(f_1, \ldots, f_n)$ is a base. This result shows that if we are given one ordered base $(e_1, \ldots, e_n)$ for a free module over a commutative ring R, then we obtain all ordered bases $(f_1, \ldots, f_n)$ by applying the matrices $A \in GL_n(R)$ to (e_i) in the sense that we take $f_j = \sum a_{ji} e_i$, $A = (a_{ij})$.

We now drop the restriction that R is commutative, and we consider the additive group $\text{Hom}(R^{(m)}, R^{(n)})$ of (module) homomorphisms of $R^{(m)}$ into $R^{(n)}$ for any m, n. To study this we choose bases $(e_1, \ldots, e_m)$, $(f_1, \ldots, f_n)$ for $R^{(m)}$ and $R^{(n)}$ respectively. If $\eta \in \text{Hom}(R^{(m)}, R^{(n)})$ we tabulate

$$\eta(e_1) = a_{11}f_1 + a_{12}f_2 + \cdots + a_{1n}f_n$$

$$\eta(e_2) = a_{21}f_1 + a_{22}f_2 + \cdots + a_{2n}f_n$$

(17) . . . .

$$\eta(e_m) = a_{m1}f_1 + a_{m2}f_2 + \cdots + a_{mn}f_n$$

and call the $m \times n$ matrix $A = (a_{ij})$ (m rows and n columns) the *matrix of η relative to the (ordered) bases* $(e_1, \ldots, e_m), (f_1, \ldots, f_n)$. The homomorphism η is determined by its matrix relative to the bases $(e_i), (f_j)$. For, if we have (17), and if $x = (x_1, \ldots, x_m) = \sum x_i e_i$, then

$$\eta(x) = \eta\left(\sum x_i e_i\right) = \sum_i x_i \eta(e_i) = \sum_{i,j} x_i a_{ij} f_j.$$

Thus η is the map

(18) $$(x_1, \ldots, x_m) \to (y_1, \ldots, y_n)$$

where

(19) $$y_j = \sum_{i=1}^{m} x_i a_{ij}, \qquad j = 1, \ldots, n.$$

We can express this also in matrix form. In general, if $A = (a_{ij})$ and $B = (b_{ij})$ are $m \times n$ matrices, we define the sum $A + B = (a_{ij} + b_{ij})$: that is, $A + B$ is the matrix whose (i, j)-entry is $a_{ij} + b_{ij}$. If $A = (a_{ij})$ is an $m \times n$ matrix and $B = (b_{jk})$ is an $n \times q$ matrix, then we define the product $P = AB$ as the $m \times q$ matrix whose (i, k)-entry, $1 \leq i \leq m$, $1 \leq k \leq q$, is given by the formula

(20) $$p_{ik} = \sum_{j=1}^{n} a_{ij} b_{jk}.$$

For example, we have

$$\begin{pmatrix} 0 & -1 & 2 \\ 3 & 5 & -4 \end{pmatrix} \begin{pmatrix} 2 & 0 \\ 6 & 3 \\ -1 & 5 \end{pmatrix} = \begin{pmatrix} -8 & 7 \\ 40 & -5 \end{pmatrix}.$$

If we use the definition of the matrix product given by (20) then we can rewrite (18) and (19) as

(21) $$\eta:(x_1, \ldots, x_m) \to (y_1, \ldots, y_n) = (x_1, \ldots, x_m)A.$$

The set $M_{m,n}(R)$ of $m \times n$ matrices with entries taken from R is a group under the addition composition $(a_{ij}) + (b_{ij}) = (a_{ij} + b_{ij})$ and 0 as the $m \times n$ matrix all of whose entries are 0. We shall now show that this group is isomorphic to

$\text{Hom}(R^{(m)}, R^{(n)})$ under the map $\eta \to A$ where A is the matrix of η relative to the bases $(e_i), (f_j)$ for $R^{(m)}$ and $R^{(n)}$ respectively. It is clear that $\eta \to A$ is injective since A determines η by (21), and also our map is surjective, since if A is a given matrix in $M_{m,n}(R)$ we can define $v_i = \sum_{j=1}^{n} a_{ij}f_j$. Then, as we have seen, there exists an $\eta \in \text{Hom}(R^{(m)}, R^{(n)})$ such that $\eta(e_i) = v_i$, $1 \le i \le m$. Clearly, this η has as its associated matrix the given matrix A. Hence $\eta \to A$ is bijective. Now let $\zeta \in \text{Hom}(R^{(m)}, R^{(n)})$ and let $\zeta \to B = (b_{ij})$ so $\zeta(e_i) = \sum b_{ij}f_j$. Then $(\eta + \zeta)(e_i) = \eta(e_i) + \zeta(e_i) = \sum_j a_{ij}f_j + \sum_j b_{ij}f_i = \sum_j (a_{ij} + b_{ij})f_j$. Thus $\eta + \zeta \to A + B$, and $\eta \to A$ is a group isomorphism.

Next let $\rho \in \text{Hom}(R^{(n)}, R^{(q)})$ and let $(g_1, g_2, \ldots, g_q)$ be a base for $R^{(q)}$. Let C be the matrix of ρ relative to the bases $(f_1, \ldots, f_n), (g_1, \ldots, g_q)$ so $\rho(f_j) = \sum_{k=1}^{q} c_{jk}g_k$, $C = (c_{jk})$. As before, let $\eta \in \text{Hom}(R^{(m)}, R^{(n)})$ have the matrix $A = (a_{ij})$ relative to (e_i) and (f_j). Then $\rho\eta \in \text{Hom}(R^{(m)}, R^{(q)})$, and

$$(\rho\eta)(e_i) = \rho(\eta(e_i)) = \rho\left(\sum_j a_{ij}f_j\right) = \sum_j a_{ij}\rho(f_j) = \sum_{j,k} a_{ij}c_{jk}g_k.$$

Thus the matrix of $\rho\eta$ relative to $(e_i), (g_k)$ is AC. We can use this fact to prove that multiplication of rectangular matrices is associative, a fact, which, of course, can be established also directly, as in the special case of square matrices (p. 94). We introduce a fourth free module $R^{(s)}$ with base (h_l) and let $\tau \in \text{Hom}(R^{(q)}, R^{(s)})$. Then $\tau(\rho\eta) = (\tau\rho)\eta \in \text{Hom}(R^{(m)}, R^{(s)})$. We shall now denote the matrix of any homomorphism we are considering relative to the bases we have chosen by putting a superscript $*$ after the symbol for the map, e.g., $\eta^* = A$, $\rho^* = C$. Then we have $(\rho\eta)^* = \eta^*\rho^*$ and hence $\eta^*(\rho^*\tau^*) = \eta^*(\tau\rho)^* = ((\tau\rho)\eta)^* = (\tau(\rho\eta))^* = (\rho\eta)^*\tau^* = (\eta^*\rho^*)\tau^*$. Since η^*, ρ^* and τ^* can be taken to be any $m \times n$, $n \times q$, $q \times s$ matrices this proves associativity for arbitrary matrix multiplications. In the same way one can establish the distributive laws: if A, A_1, $A_2 \in M_{m,n}(R)$ and C, C_1, $C_2 \in M_{n,q}(R)$ then $(A_1 + A_2)C = A_1C + A_2C$ and $A(C_1 + C_2) = AC_1 + AC_2$.

In the special case of $\text{End}_R R^{(n)} = \text{Hom}(R^{(n)}, R^{(n)})$ our result gives an anti-isomorphism $\eta \to \eta^* = A$ of $\text{End}_R R^{(n)}$ with the ring of matrices $M_n(R)$ $(= M_{n,n}(R))$. Here A is the matrix of η relative to the base (e_i): that is, if $\eta(e_i) = \sum a_{ij}e_j$ then $A = (a_{ij})$. If R is commutative we have the anti-automorphism $A \to {}^t A$ (the transpose of A) in the matrix ring $M_n(R)$ (see p. 111). Combining this with the anti-isomorphism $\eta \to \eta^* = A$ we obtain an isomorphism $\eta \to {}^t A$ of $\text{End}_R R^{(n)}$ with $M_n(R)$. This is what we used in the example of $\mathbb{Z}^{(2)}$ which we considered on p. 161.

All of these considerations relating homomorphisms between free modules and matrices should be familar to the reader in the special case of matrices associated with linear maps of vector spaces. The foregoing discussion illustrates the general principle that in many situations the passage from vector spaces to free modules is fairly routine.

EXERCISES

1. Let R be arbitrary and let $(e_1, \ldots, e_n)$ be a base for $R^{(n)}$. Show that $(f_1, \ldots, f_m)$, $f_j = \sum_{j'=1}^n a_{jj'} e_{j'}$ is a base for $R^{(m)}$ if and only if these exists an $n \times m$ matrix B such that $AB = 1_m$, $BA = 1_n$ where $A = (a_{ij})$, 1_m is the usual $m \times m$ unit matrix, and 1_n is the $n \times n$ unit matrix. Hence show that $R^{(m)} \cong R^{(n)}$ if and only if there exists $A \in M_{m,n}(R)$, $B \in M_{n,m}(R)$ such that $AB = 1_m$, $BA = 1_n$.

2. Let $\eta \in \operatorname{End}_R (R^{(n)})$ and let A be the matrix of η relative to the base $(e_1, \ldots, e_n)$. Let $f_i = \sum p_{ij} e_j$ where $P = (p_{ij}) \in GL_n(R)$. Verify that the matrix of η relative to the base $(f_1, \ldots, f_n)$ is PAP^{-1}.

3. Let R_n denote a free right R-module with base $(e_1, \ldots, e_n)$. Let $\eta \in \operatorname{End}_R R_n$ and write $\eta(e_i) = \sum_{j=1}^n e_j a_{ji}$. Show that $\eta \to A = (a_{ij})$ is an isomorphism of $\operatorname{End}_R R_n$ with $M_n(R)$.

4. Let R be commutative. Show that if η is a surjective endomorphism of $R^{(n)}$ then η is bijective. Does the same conclusion hold if η is injective?

5. Let R be commutative and let M and N be R-modules. If $a \in R$ and $\eta \in \operatorname{Hom}(M, N)$ define $a\eta$ by $(a\eta)(x) = a(\eta(x)) = \eta(ax)$. Show that $a\eta \in \operatorname{Hom}(M, N)$ and that this action of R on $\operatorname{Hom}(M, N)$ converts the latter into an R-module. Show that $\operatorname{Hom}(R^{(m)}, R^{(n)})$ is free of rank mn.

6. Let R be commutative and let $(e_1, \ldots, e_n)$ be a base for $R^{(n)}$. Put $f_i = \sum a_{ij} e_j$ where $A = (a_{ij}) \in M_n(R)$. Show that the f_i form a base for a free submodule K of $R^{(n)}$ if and only if $\det A$ is not a zero-divisor. Show that for any $\bar{x} = x + K$ in $R^{(n)}/K$ one has $(\det A)\bar{x} = 0$. (*Hint*: It suffices to show that $(\det A)\bar{e}_i = 0$ for $1 \leq i \leq n$.)

3.5 DIRECT SUMS OF MODULES

We shall now define the module analogue of the direct product of monoids or of groups (p. 35). Let $M_1, M_2, \ldots, M_n$ be modules over the same ring R and let M be the product set $M_1 \times M_2 \times \cdots \times M_n$ of n-tuples $(x_1, x_2, \ldots, x_n)$ where $x_i \in M_i$. As in the special case of the free module $R^{(n)}$, we introduce an addition, a 0 element, and a multiplication by elements in R by

$$(x_1, \ldots, x_n) + (y_1, \ldots, y_n) = (x_1 + y_1, \ldots, x_n + y_n)$$

$$0 = (0, \ldots, 0)$$

$$a(x_1, \ldots, x_n) = (ax_1, \ldots, ax_n), \qquad a \in R.$$

These define a module structure on M. Then M with this structure is called the *direct sum* of the modules M_i and is denoted either as $M_1 \oplus M_2 \oplus \cdots \oplus M_n$ or as $\oplus_1^n M_i$.

A basic homomorphism property of $\oplus_1^n M_i$ is the following result. Suppose we are given homomorphisms η_i, $1 \leq i \leq n$, of M_i into a module N. Then we have the map η of $\oplus M_i$ into N defined by

$$\eta:(x_1, \ldots, x_n) \to \sum_1^n \eta_i(x_i).$$

Since

$$\eta(x_1 + y_1, \ldots, x_n + y_n) = \sum_1^n \eta_i(x_i + y_i) = \sum_1^n \eta_i(x_i) + \sum_1^n \eta_i(y_i)$$

$$= \eta(x_1, \ldots, x_n) + \eta(y_1, \ldots, y_n)$$

and

$$\eta(ax_1, \ldots, ax_n) = \sum_1^n \eta_i(ax_i) = \sum_1^n a\eta_i(x_i) = a \sum_1^n \eta_i(x_i)$$

η is a homomorphism of $\oplus M_i$ into N. We shall use this homomorphism in the proof of the first part of the following theorem, which characterizes by internal properties the direct sum of modules.

THEOREM 3.5.   *Let M be a module and suppose M contains submodules $M_1, \ldots, M_n$ having the following properties:*

(i) $M = M_1 + M_2 + \cdots + M_n$ *(that is, M is generated by the M_i)*
(ii) *for every i, $1 \le i \le n$, we have*

$$M_i \cap (M_1 + \cdots + M_{i-1} + M_{i+1} + \cdots + M_n) = 0.$$

Then the map

$$\iota:(x_1, \ldots, x_n) \to \sum_1^n x_i$$

is an isomorphism of $\oplus M_i$ with M. Conversely, in $\oplus M_i$ let

$$M_i' = \{(0, \ldots, 0, \overset{i}{x_i}, 0, \ldots, 0) \,|\, x_i \in M_i\}.$$

Then M_i' is a submodule of $\oplus M_i$ isomorphic to M_i, and the conditions (i), (ii) hold for these submodules of $\oplus M_i$,

Proof.   Suppose the submodules M_i of M satisfy (i) and (ii), and consider the map $\iota:(x_1, \ldots, x_n) \to \sum_1^n x_i$. Since this is just the map η defined by the isomorphisms $x_i \to x_i$ of M_i onto M_i as above, ι is a homomorphism of $\oplus M_i$ into M. Now ι is surjective; for, if x is any element of M we can write $x = \sum x_i$, $x_i \in M_i$, since $M = \sum M_i$ by condition (i) and $\sum M_i$ is the set of elements of the form $\sum x_i$, $x_i \in M_i$. Then $\iota(x_1, \ldots, x_n) = \sum x_i = x$. To see that ι is injective it suffices to show that its kernel is 0, that is, to prove that if $\iota(x_1, \ldots, x_n) = \sum_1^n x_i = 0$ then every $x_i = 0$. This is clear from (ii) since $\sum_1^n x_i = 0$ gives $-x_i = \sum_{j \neq i} x_j$, hence $x_i \in M_i \cap (\sum_{j \neq i} M_j) = 0$. Thus every $x_i = 0$. We have now proved that ι is an isomorphism. Conversely, consider $\underset{i}{\oplus} M_i$. It is immediate that the map

$\iota_i : x_i \to (0, \ldots, 0, x_i, 0, \ldots, 0)$ is a monomorphism of M_i into M. The image is M_i', so M_i' is a submodule of M isomorphic to M_i. Since

$$(x_1, 0, \ldots, 0) + (0, x_2, 0, \ldots, 0) + \cdots + (0, \ldots, 0, x_n) = (x_1, x_2, \ldots, x_n),$$

(i) holds for the submodules M_i' of M. Since $\sum_{j \neq i} M_j'$ is the set of elements of the form $(x_1, \ldots, x_{i-1}, 0, x_{i+1}, \ldots, x_n)$ it is clear also that (ii) holds. This completes the proof. $\square$

This theorem permits us to identify a module M with $\oplus M_i$ if the M_i are submodules of M satisfying the conditions (i) and (ii). In this case we shall say that M is the (*internal*) *direct sum of its submodules* M_i, and we shall also write $M = \oplus M_i$ or $M = M_1 \oplus M_2 \oplus \cdots \oplus M_n$ whenever conditions (i) and (ii) hold for the submodules M_i.

If a set of submodules M_i, $1 \le i \le n$, satisfy condition (ii) then we shall say that these submodules of M are *independent*. It is immediate that this is the case if and only if every relation of the form $\sum_1^n x_i = 0$, $x_i \in M_i$, implies every $x_i = 0$. Also the M_i are independent if and only if every relation $\sum_1^n x_i = \sum_1^n y_i$, $x_i, y_i \in M_i$, forces $x_i = y_i$, $1 \le i \le n$. It should be noted that the independence conditions are stronger than the condition $M_i \cap M_j = 0$, $i \neq j$, and are even stronger than the set of conditions $M_i \cap (\bigcup_{j \neq i} M_j) = 0$. For example, in the two dimensional vector space $\mathbb{R}^{(2)}$ over $\mathbb{R}$, let

$$X = \{(x, 0) | x \in \mathbb{R}\}, \qquad Y = \{(0, y) | y \in \mathbb{R}\}, \qquad \text{and} \qquad Z = \{(z, z) | z \in \mathbb{R}\}.$$

Pictorially, we have

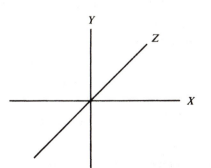

that is, X is the set of vectors having end points on the x-axis, Y is the set having end points on the y-axis and Z is the set having end points on the $45°$ line. Clearly, the intersection of any one of these lines with the union of the other two is the origin. On the other hand, $X + Y = \mathbb{R}^{(2)}$, so $(X + Y) \cap Z = Z$. Hence X, Y, and Z are not independent.

The criteria in terms of elements for independence of submodules have the following consequences:

I. Let $M_1, \ldots, M_n$ be independent submodules of M. Put $N_1 = M_1 + \cdots + M_{r_1}$, $N_2 = M_{r_1+1} + \cdots + M_{r_1+r_2}$, $N_3 = M_{r_1+r_2+1} + \cdots + M_{r_1+r_2+r_3}$, etc. Then $N_1, N_2, \cdots$ are independent.

II. Let $M_1, \ldots, M_n$ be independent and suppose $M_i = M_{i1} \oplus M_{i2} \oplus \cdots \oplus M_{ir_i}$, $1 \le i \le n$, where the M_{ij} are submodules of M_i. Then the submodules $M_{11}, \ldots, M_{1r_1}, M_{21}, \ldots, M_{2r_2}, \ldots, M_{n1}, \ldots, M_{nr_n}$ are independent.

The proof is left to the reader. An immediate consequence of these results is

THEOREM 3.6. *Let* $M = \oplus M_i$, M_i *a submodule (that is, M is the direct sum of the submodules M_i). Put* $N_1 = M_1 + \cdots + M_{r_1}$, $N_2 = M_{r_1+1} + \cdots + M_{r_1+r_2}$, *etc. Then* $M = \oplus N_j$. *Also, if* $M_i = \oplus M_{ij}$, $1 \le i \le n$, $1 \le j \le r_i$, *then* $M = \oplus M_{ij}$.

We omit the proof of this also.

EXERCISES

1. Let V be a vector space over a field F. Show that the non-zero vectors x_i, $1 \le i \le n$, of V are linearly independent if and only if the subspaces Fx_i are independent. Show also that the x_i form a base if and only if $V = \oplus Fx_i$.

2. Let M be a module, and M_i, $1 \le i \le n$, be submodules such that $M = \sum M_i$ and the "triangular" set of conditions

$$M_1 \cap M_2 = 0$$
$$(M_1 + M_2) \cap M_3 = 0$$
$$\vdots$$
$$(M_1 + \cdots + M_{n-1}) \cap M_n = 0$$

hold. Show that $M = \oplus M_i$.

3. Show that $\mathbb{Z}/(p^e)$, p a prime, $e > 0$, regarded as a $\mathbb{Z}$-module is not a direct sum

of any two non-zero submodules. Does this hold for $\mathbb{Z}$? Does it hold for $\mathbb{Z}/(n)$ for other positive integers n?

4. Show that if $M = M_1 \oplus M_2$ then $M_1 \cong M/M_2$ and $M_2 \cong M/M_1$.

5. Let M and N be R-modules, $f : M \to N$, $g : N \to M$ R-module homomorphisms such that $fg(y) = y$ for all $y \in N$. Show that $M = \ker f \oplus \operatorname{Im} g$.

3.6   FINITELY GENERATED MODULES OVER A P.I.D.
PRELIMINARY RESULTS

We are now ready to turn our attention to the main objective of this chapter: the study of finitely generated modules over a principal ideal domain and the applications of this theory to finite abelian groups and to linear transformations. Let M be a module over a p.i.d. D which is generated by a finite set of elements $x_1, x_2, \ldots, x_n$, so $M = \sum_1^n D x_i$. To study M it is natural to introduce the free module $D^{(n)}$ with base $(e_1, e_2, \ldots, e_n)$ and the epimorphism $\eta : \sum_1^n a_i e_i \to \sum_1^n a_i x_i, a_i \in D$, of $D^{(n)}$ onto M. Then $M \cong D^{(n)}/K$ where $K = \ker \eta$. A first result we shall need is that K is finitely generated. This will follow from the following stronger result.

THEOREM 3.7.   *Let D be a p.i.d. and let $D^{(n)}$ be the free module of rank n over D. Then any submodule K of $D^{(n)}$ is free with base of $m \le n$ elements.*

Proof.   Since we are not excluding $K = 0$ we must adopt the convention that the module consisting of 0 alone is "free of rank 0" (with vacuous base). Of course, the result is trivial if $n = 0$. Now suppose $n > 0$ and assume the result holds for any submodule of a free module with a base of $n - 1$ elements over D. Let $D^{(n-1)}$ be the submodule generated by $e_2, \ldots, e_n$. This is free with $(e_2, \ldots, e_n)$ as base; hence, if $K \subset D^{(n-1)}$ then the result holds. Thus we may assume $K \not\subset D^{(n-1)}$. Consider the subset I of D of elements b for which there exists an element of the form $be_1 + y \in K$ where $y \in D^{(n-1)}$. This is an ideal in D and since $K \not\subset D^{(n-1)}$, $I \ne 0$. Hence $I = (d)$ with $d \ne 0$ and we have an element $f_1 = de_1 + y_1 \in K$ where $y_1 \in D^{(n-1)}$. Now consider $L = K \cap D^{(n-1)}$. This is a submodule of $D^{(n-1)}$, so, by induction, it has a base $(f_2, \ldots, f_m)$ of $m - 1 \le n - 1$ elements (where we may have $m - 1 = 0$). We shall now show that $(f_1, f_2, \ldots, f_m)$ is a base for K and this will prove the theorem. First, let $x \in K$. Then $x = be_1 + y$ where $b \in I = (d)$ and $y \in D^{(n-1)}$. Then $b = k_1 d$ and so $x - k_1 f_1 = k_1 de_1 + y - k_1(de_1 + y_1) = y - k_1 y_1 \in L = K \cap D^{(n-1)}$. Hence $x - k_1 f_1 = \sum_2^m k_j f_j$ where the $k_j \in D$ and $x = \sum_1^m k_i f_i$. Thus the f_i generate K. Next suppose $\sum_1^m k_i f_i = 0$. Then $k_1 de_1 + k_1 y_1 + \sum_2^m k_j f_j = 0$. Since y_1 and the f_j,

$j \geq 2$, are in $D^{(n-1)}$, this gives a relation $k_1 d e_1 + \sum_2^n l_k e_k = 0$ with $l_k \in D$. Hence $k_1 d = 0$ and since $d \neq 0$, $k_1 = 0$. Then $\sum_2^m k_j f_j = 0$ and since $(f_2, \ldots, f_m)$ is a base for L, every $k_j = 0$. Thus $(f_1, f_2, \ldots, f_m)$ is a base for K. $\square$

Since any field F is a p.i.d. (whose only ideals are (0) and (1)), the foregoing theorem can be specialized to the case in which $D = F$ is a field. Then it reduces to the following well known result of linear algebra. If V is an n dimensional vector space over F (that is, V is a free F-module of rank n) then any subspace of V is finite dimensional with dimensionality $m \leq n$.

We return now to $M \cong D^{(n)}/K$ and we apply Theorem 3.7 to conclude that K has a base of $m \leq n$ elements. The method we are going to apply will work just as well if we have a finite set of generators, and as a practical matter it is sometimes useful not to have to resort to a base. Hence we assume we have a set of generators $f_1, f_2, \ldots, f_m$ for the submodule K where m may exceed n. We now express these generators in terms of the base $(e_1, e_2, \ldots, e_n)$ in the form

$$
\begin{aligned}
f_1 &= a_{11} e_1 + a_{12} e_2 + \cdots + a_{1n} e_n \\
f_2 &= a_{21} e_1 + a_{22} e_2 + \cdots + a_{2n} e_n \\
&\cdots\cdots\cdots\cdots\cdots\cdots\cdots\cdots\cdots \\
f_m &= a_{m1} e_1 + a_{m2} e_2 + \cdots + a_{mn} e_n.
\end{aligned}
$$

(22)

The $m \times n$ matrix $A = (a_{ki})$ of these relations is called the *relations matrix* of the ordered set of generators $(f_1, \ldots, f_m)$ in terms of the ordered base $(e_1, \ldots, e_n)$. Of course, there is nothing special about our choices of the base (e_i) for $D^{(n)}$ and the generators (f_k) for K. This observation suggests that we see what happens when we change these. Now we know that any other base for $D^{(n)}$ will have the form $(e'_1, \ldots, e'_n)$ where $e'_i = \sum_{j=1}^n p_{ij} e_j$ where $P = (p_{ij})$ is an invertible matrix in the matrix ring $M_n(D)$. We can't make such a sweeping statement about sets of generators for the submodule K. However, it is clear that if $Q = (q_{kl})$ is an invertible matrix in $M_m(D)$ with inverse $Q^{-1} = (q_{kl}^*)$ then $(f'_1, \ldots, f'_m)$, where $f'_k = \sum_{l=1}^m q_{kl} f_l$ is another set of generators for K. For, it is clear that the $f'_k \in K$ and $\sum_k q_{rk}^* f'_k = \sum_{k,l} q_{rk}^* q_{kl} f_l = f_r$ so the f's are in the submodule generated by the f''s. Hence the f''s generate K. What is the relations matrix of the f''s relative to the e''s? We have

$$
f'_k = \sum q_{kl} f_l = \sum_{l,j} q_{kl} a_{lj} e_j = \sum_{l,j,i} q_{kl} a_{lj} p_{ji}^* e'_i
$$

where $(p_{ij}^*) = P^{-1}$. Hence the new relations matrix is

$$
A' = Q A P^{-1}.
$$

We are now led to the problem of making the "right" choices for Q and P to achieve a simple "normal" form for the relations which will yield important information on $M \cong D^{(n)}/K$. Since the matrix problem thus posed is of interest in its own right we shall treat it separately in the next section before returning to our analysis of $D^{(n)}/K$.

EXERCISES

1. Find a base for the submodule of $\mathbb{Z}^{(3)}$ generated by $f_1 = (1, 0, -1)$, $f_2 = (2, -3, 1)$, $f_3 = (0, 3, 1)$, $f_4 = (3, 1, 5)$.

2. Find a base for the submodule of $\mathbb{Q}[\lambda]^{(3)}$ generated by $f_1 = (2\lambda - 1, \lambda, \lambda^2 + 3)$, $f_3 = (\lambda, \lambda, \lambda^2)$, $f_3 = (\lambda + 1, 2\lambda, 2\lambda^2 - 3)$.

3. Find a base for the $\mathbb{Z}$-submodule of $\mathbb{Z}^{(3)}$ consisting of all (x_1, x_2, x_3) satisfying the conditions $x_1 + 2x_2 + 3x_3 = 0$, $x_1 + 4x_2 + 9x_3 = 0$.

3.7   EQUIVALENCE OF MATRICES WITH ENTRIES IN A P.I.D.

Two $m \times n$ matrices with entries in a p.i.d. D are said to be *equivalent* if there exists an invertible matrix P in $M_m(D)$ and an invertible matrix Q in $M_n(D)$ such that $B = PAQ$. It is clear that this defines an equivalence relation in the set $M_{m,n}(D)$ of $m \times n$ matrices with entries in D. We now consider the problem of selecting among the matrices equivalent to a given matrix A one that has a particularly simple "normal" form. The result we shall prove is the following

THEOREM 3.8.   *If $A \in M_{m,n}(D)$, D a p.i.d., then A is equivalent to a matrix which has the "diagonal" form*

$$\mathrm{diag}\{d_1, d_2, \ldots, d_r, 0, \ldots, 0\}$$

(23)
$$\equiv \begin{pmatrix} d_1 & & & & \\ & d_2 & & & 0 \\ & & \ddots & & \\ & & & d_r & \\ 0 & & & & 0 \\ & & & & & \ddots \end{pmatrix}$$

where the $d_i \neq 0$ and $d_i \mid d_j$ if $i \leq j$.

We shall obtain the matrices P and Q which transform A into a matrix of the form (23) as products of matrices of some special forms which we shall now define. Without specifying the size ($m \times m$ or $n \times n$) we introduce first certain invertible (square) matrices with entries in D, which we shall call elementary, and consider the effects of left or right multiplications by these matrices.

First, let $b \in D$ and let $i \neq j$. Put $T_{ij}(b) = 1 + be_{ij}$ where e_{ij} is the matrix with a lone 1 in the (i, j) place, 0's elsewhere. $T_{ij}(b)$ is invertible since

$$T_{ij}(b)T_{ij}(-b) = (1 + be_{ij})(1 - be_{ij}) = 1.$$

Next, let u be an invertible element of D and put $D_i(u) = 1 + (u - 1)e_{ii}$, so $D_i(u)$ is diagonal with ith diagonal entry u and remaining diagonal entries 1. Then $D_i(u)$ is invertible with $D_i(u)^{-1} = D_i(u^{-1})$. Finally, let $P_{ij} = 1 - e_{ii} - e_{jj} + e_{ij} + e_{ji}$. Also this matrix is invertible since $P_{ij}^2 = 1$.

It is easy to verify that

I. Left multiplication of A by the $m \times m$ matrix $T_{ij}(b)$ yields a matrix whose ith row is obtained by multiplying the jth row of A by b and adding it to the ith row of A, and whose remaining rows are the same as in A.

Right multiplication of A by the $n \times n$ matrix $T_{ij}(b)$ gives a matrix whose jth column is b times the ith column of A plus the jth column of A, and whose remaining columns are identical with those of A.

II. Left multiplication of A by the $m \times m$ matrix $D_i(u)$ amounts to the operation of multiplying the ith row of A by u, and leaving the other rows as in A.

Right multiplication of A by the $n \times n$ matrix $D_i(u)$ amounts to multiplying the ith column of A by u, and leaving the remaining columns unaltered.

III. Left multiplication of A by the $m \times m$ matrix P_{ij} amounts to interchanging the ith and jth rows of A, and leaving the other rows as in A.

Right multiplication of A by the $n \times n$ matrix P_{ij} amounts to interchanging the ith and jth columns of A, and leaving the other columns unchanged.

We call the matrices $T_{ij}(b)$, $D_i(u)$, P_{ij} *elementary matrices* of *types* I, II, and III respectively. Left (right) multiplication of A by one of these will be called an *elementary transformation on the rows (columns)* of the corresponding type. Such elementary transformations yield matrices equivalent to A.

We now proceed to the

Proof of Theorem 3.8.   We shall first give a proof in the special case in which D is Euclidean with map δ of D into $\mathbb{N}$ (p. 148). If $A = 0$ there is nothing to prove. Otherwise, let a_{ij} be a non-zero element of A with minimal $\delta(a_{ij})$. Elementary row and column transformations will bring this element to the $(1, 1)$ position. Assume now that it is there. Let $k > 1$ and $a_{1k} = a_{11}b_k + b_{1k}$, where $\delta(b_{1k}) <$

$\delta(a_{11})$. Now subtract the first column times b_k from the kth. This elementary transformation replaces a_{1k} by b_{1k}. If $b_{1k} \neq 0$ we obtain a matrix equivalent to A for which the minimum δ for the non-zero entries is less than that appearing in A. We repeat the original procedure with this new matrix. Similarly, if $a_{k1} = a_{11}b_k + b_{k1}$, where $b_{k1} \neq 0$ and $\delta(b_{k1}) < \delta(a_{11})$, then an elementary transformation of type I on the rows gives an equivalent matrix for which the minimum δ for the non-zero entries has been reduced. Since the "degree" δ is a non-negative integer a finite number of applications of this process yields an equivalent matrix $B = (b_{ij})$ in which $b_{11}|b_{1k}$ and $b_{11}|b_{k1}$ for all k. Then elementary transformation on the rows and columns of type I gives an equivalent matrix of form

(24)
$$\begin{pmatrix} b_{11} & 0 & \cdots & 0 \\ 0 & c_{22} & \cdots & c_{2n} \\ \vdots & & & \vdots \\ 0 & c_{m2} & \cdots & c_{mn} \end{pmatrix}$$

We can also arrange to have $b_{11} \mid c_{kl}$ for every k, l. For if $b_{11} \nmid c_{kl}$ then we add the kth row to the first obtaining the new first row $(b_{11}, c_{k2}, \ldots, c_{kl}, \ldots, c_{kn})$. Repetition of the first process replaces c_{kl} by a non-zero element with a δ less than that of b_{11}. A finite number of steps of the sort indicated will then give a matrix (24) equivalent to A in which $b_{11} \neq 0$ and $b_{11}|c_{kl}$ for every k, l. We now repeat the process on the submatrix (c_{kl}). This gives an equivalent matrix of the form

(25)
$$\begin{pmatrix} b_{11} & 0 & \cdot & \cdots & 0 \\ 0 & c_{22} & 0 & \cdots & 0 \\ 0 & 0 & d_{33} & \cdots & d_{3n} \\ \vdots & & & & \vdots \\ 0 & 0 & d_{m3} & \cdots & d_{mn} \end{pmatrix}$$

in which $c_{22} \nmid d_{pq}$ for all p, q. Moreover, the elementary transformations on the rows and columns of (c_{kl}) which yield (25) do not affect the divisibility condition by b_{11}. Hence $b_{11}|c_{22}$ and $b_{11}|d_{pq}$. Continuing in this way we obtain the equivalent diagonal matrix $\text{diag}\{d_1, d_2, \ldots, d_r, 0, \ldots, 0\}$ with $d_i|d_j$ for $i \leq j$ ($d_1 = b_{11}, d_2 = c_{22}$, etc).

The argument in the general case is quite similar to the foregoing. Here we use induction on the length of a non-zero element of D in place of $\delta(a)$. We define the *length* $l(a)$ of $a \neq 0$ to be the number of prime factors occurring in a factorization $a = p_1 p_2 \cdots p_r$, p_i primes. We also use the convention that $l(u) = 0$ if u is a unit. In addition to the elementary transformations that sufficed in the Euclidean case we shall need to use also multiplications by matrices of the form

(26)

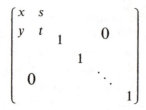

where $\begin{pmatrix} x & s \\ y & t \end{pmatrix}$ is invertible. As in the previous case we may assume that $a_{11} \neq 0$ and $l(a_{11}) \leq l(a_{ij})$ for every $a_{ij} \neq 0$. Assume $a_{11} \nmid a_{1k}$. Interchanging the second and kth column we may assume $a_{11} \nmid a_{12}$. Write $a = a_{11}$, $b = a_{12}$, and let $d = (a, b)$ so $l(d) < l(a)$. There exist elements $x, y \in D$ such that $ax + by = d$. Put $s = bd^{-1}$, $t = -ad^{-1}$. Then we have the matrix equation

$$\begin{pmatrix} -t & s \\ y & -x \end{pmatrix} \begin{pmatrix} x & s \\ y & t \end{pmatrix} = \begin{pmatrix} 1 & 0 \\ 0 & 1 \end{pmatrix}$$

which implies that both matrices are invertible (since D is commutative). Then (26) is invertible. Multiplying A on the right by this gives the matrix whose first row is $(d, 0, a_{13}, \ldots, a_{1n})$ and $l(d) < l(a_{11})$. Similarly, if $a_{11} \nmid a_{k1}$ for some k, elementary transformations together with left multiplication by a suitable matrix (26) yields an equivalent matrix in which the length of some non-zero element is less that $l(a_{11})$. In this way we can arrange to have $a_{11} | a_{1k}$ and $a_{11} | a_{k1}$ for all k. Elementary transformations then give a matrix of the form (24). The rest of the argument is essentially the same as in the Euclidean case. The only difference is that we continue to reduce the length rather than the degree δ. ☐

A matrix equivalent to A having the diagonal form given in Theorem 3.8 is called a *normal form* for A. The diagonal elements of a normal form are called *invariant factors* of A. Clearly any of these can be replaced by an associate (product by a unit). We shall now show that this is the only alteration which can be made in the invariant factors, that is, these are determined up to unit multipliers. We shall obtain this result by deriving formulas for the invariant factors in terms of the elements of A. We recall that the matrix A is said to be of (*determinantal*) *rank* r if there exists a non-zero r-rowed minor in A but every $(r + 1)$-rowed minor of A is 0. Since the i-rowed minors are sums of products of $(i - 1)$-rowed minors by elements of D it is clear that if the rank is r, then for every i, $1 \leq i \leq r$, A has non-zero i-rowed minors. We now have the following result, which gives formulas for the invariant factors.

THEOREM 3.9.   *Let A be an* m × n *matrix with entries in a* p.i.d. *D and suppose the rank of A to be* r. *For each* $i \leq r$ *let* Δ_i *be a g.c.d. of the* i-*rowed minors of A.*

Then any set of invariant factors for A differ by unit multipliers from the elements

(27) $$d_1 = \Delta_1, d_2 = \Delta_2\Delta_1^{-1}, \ldots, d_r = \Delta_r\Delta_{r-1}^{-1}.$$

(Note: It is clear that $\Delta_i = 0$ and $\Delta_{i-1}|\Delta_i$.)

Proof. Let $Q = (q_{kl})$ be an $m \times m$ matrix with entries in D. Then the (k, i)-entry of QA is $\sum_j q_{kj}a_{ji}$. This shows that the rows of QA are linear combinations with coefficients in D of the rows of A. Hence the i-rowed minors of QA are linear combinations of the i-rowed minors of A and so the g.c.d. of the i-rowed minors of A is a divisor of the g.c.d. of the i-rowed minors of QA. Similarly, since the columns of AP, $P \in M_n(D)$, are linear combinations of the columns of A, the g.c.d. of the i-rowed minors of A is a divisor of the g.c.d. of the i-rowed minors of AP. Combining these two facts and using symmetry of the relation of equivalence, we see that if A and B are equivalent the g.c.d. of the i-rowed minors of A and B are the same. Now let $B = \text{diag}\{d_1, d_2, \ldots, d_r, 0, \ldots, 0\}$ be a normal form for A. Then the divisibility conditions $d_i|d_j$ if $i < j$ imply that a g.c.d. of the i-rowed minors of B is $\Delta_i = d_1d_2 \cdots d_i$. Evidently the assertion of the theorem follows from this. $\square$

An immediate consequence of Theorem 3.9 is that the invariant factors are determined up to unit multipliers and two $m \times n$ matrices are equivalent if and only if they have the same invariant factors.

EXERCISES

1. Obtain a normal form for the integral matrix

$$\begin{pmatrix} 6 & 2 & 3 & 0 \\ 2 & 3 & -4 & 1 \\ -3 & 3 & 1 & 2 \\ -1 & 2 & -3 & 5 \end{pmatrix}$$

2. Obtain a normal form for the matrix

$$A = \begin{pmatrix} \lambda - 17 & 8 & 12 & -14 \\ -46 & \lambda + 22 & 35 & -41 \\ 2 & -1 & \lambda - 4 & 4 \\ -4 & 2 & 2 & \lambda - 3 \end{pmatrix}$$

in $M_4(\mathbb{Q}[\lambda])$, λ an indeterminate. Also find invertible matrices P and Q such that PAQ is in normal form.

3. Determine the invariant factors of

$$\begin{pmatrix} \lambda + 1 & 2 & -6 \\ 1 & \lambda & -3 \\ 1 & 1 & \lambda - 4 \end{pmatrix}$$

by using the formulas (27).

4. Prove that if D is Euclidean then any invertible matrix in $M_n(D)$ is a product of elementary matrices. Show also that any elementary matrix of type III is a product of elementary matrices of types I and II. (Consider the case of 2×2 matrices first.) Hence prove that if D is Euclidean any invertible matrix in $M_n(D)$ is a product of elementary matrices of types I and II.[4]

5. Prove that if F is a field any matrix in $M_n(F)$ of determinant 1 is a product of elementary matrices of type I.

6. Let D be a p.i.d. and $a_i \in D$, $1 \le i \le n$. Let d be a g.c.d. of the elements a_i. Show that there exists an invertible matrix Q in $M_n(D)$ such that

$$(a_1, a_2, \ldots, a_n) Q = (d, 0, \ldots, 0).$$

7. Show that if the elements $a_{11}, a_{12}, \ldots, a_{1n}$ are relatively prime then there exist $a_{kj} \in D, 2 \le k \le n, 1 \le j \le n$ such that the square matrix (a_{ij}) is invertible in $M_n(D)$ (D a p.i.d.).

8. Let $A \in M_n(D)$ where D is Euclidean and assume $\det A \ne 0$. Show that there exists an invertible $P \in M_n(D)$ such that PA has the triangular form

$$\begin{bmatrix} d_1 & b_{12} & \cdots & b_{1n} \\ & d_2 & b_{23} & \cdots & b_{2n} \\ & & & \ddots & \\ & 0 & & & d_n \end{bmatrix}$$

where the $d_i \ne 0$ and for any i, $\delta(b_{ji}) < \delta(d_i)$.

9. Show that if $A \in M_{m,n}(D)$, D a p.i.d., then A and ^tA have the same invariant factors.

10. Let R be a ring and define the elementary matrix $T_{ij}(a)$, $i \ne j$, $a \in R$, as above. Verify the following relations:

(i) $(T_{ij}(a))^{-1} = T_{ij}(-a)$.
(ii) $T_{ij}(a) T_{ij}(b) = T_{ij}(a + b)$.
(iii) $(T_{ij}(a), T_{jk}(b)) = T_{jk}(ab)$ if $k \ne i$ where, in general, $(x, y) = x^{-1}y^{-1}xy$.
(vi) $(T_{ij}(a), T_{kl}(b)) = 1$ if $j \ne k$, $i \ne l$.

These are called the *Steinberg relations*.

---

[4] There exist p.i.d. in which not every invertible matrix is a product of elementary ones. An example of this type is given in a paper by P.M. Cohn, *On the structure of GL_2 of a ring*, Institut des Hautes Études Scientifiques, Publication #30 (1966), pp. 5–54.

3.8 STRUCTURE THEOREM FOR FINITELY GENERATED MODULES OVER A P.I.D.

We are now ready to prove the

FUNDAMENTAL STRUCTURE THEOREM FOR FINITELY GENER-ATED MODULES OVER A P.I.D. *If M ($\neq 0$) is a finitely generated module over a* p.i.d. D, M *is a direct sum of cyclic modules:* $M = Dz_1 \oplus Dz_2 \oplus \cdots \oplus Dz_s$ *such that the order ideals* ann z_i *satisfy*

$$(28) \qquad \text{ann } z_1 \supset \text{ann } z_2 \supset \cdots \supset \text{ann } z_s, \qquad \text{ann } z_k \neq D.$$

Remark. If $b \in$ ann z, $b(az) = a(bz) = 0$ for any $a \in D$. Hence ann $az \supset$ ann z. This implies that any two generators of a cyclic D-module have the same annihilator. Thus ann z is independent of the choice of the generator z of Dz.

Proof. We have seen that if $x_1, x_2, \ldots, x_n$ is a set of generators for M we have the epimorphism η of the free module $D^{(n)}$ with base (e_i), $1 \leq i \leq n$, onto M sending $e_i \to x_i$. Then $M \cong D^{(n)}/K$ and K is generated by a finite set of elements $f_1, \ldots, f_m$ such that $f_j = \sum a_{ji} e_i$. Thus we have the relations matrix $A = (a_{ji}) \in M_{m,n}(D)$. We now replace the base (e_i) by (e_i') where $e_i' = \sum_{j=1}^n p_{ij} e_j$, $P = (p_{ij})$ invertible in $M_n(D)$, and we replace the set of generators f_k, $1 \leq k \leq m$, by $f_1', \ldots, f_m'$ where $f_k' = \sum_{l=1}^m q_{kl} f_l$ and $Q = (q_{kl})$ is invertible in $M_m(D)$. Then, as we saw in section 3.6, the new relations matrix is QAP^{-1}. By Theorem 3.8, we can choose P and Q so that $QAP^{-1} = \text{diag}\{d_1, \ldots, d_r, 0, \ldots, 0\}$ where the d's are $\neq 0$ and $d_i | d_j$ if $i \leq j$. This means that the relations connecting the generators f_k' of K to the base (e_i') are

$$(29) \qquad f_1' = d_1 e_1', \ldots, f_r' = d_r e_r', f_{r+1}' = \cdots = f_m' = 0.$$

Now put $y_i = \sum p_{ij} x_j$, $1 \leq i \leq n$. Then $y_1, y_2, \ldots, y_n$ is another set of generators of M which are the images of the base (e_i') under the epimorphism η of $D^{(n)}$ into M. Since $d_i e_i' = f_i' \in K$ for $1 \leq i \leq r$, we have $d_i y_i = 0$ for the corresponding y_i. Now suppose we have a relation $\sum_1^n b_i y_i = 0$ where the $b_i \in D$. Then $\sum b_i e_i' \in K$ and hence we have $\sum b_i e_i' = \sum c_i f_i' = \sum c_i d_i e_i'$. Since $(e_1', e_2', \ldots, e_n')$ is a base for $D^{(n)}$ this implies that $b_i = c_i d_i$, $1 \leq i \leq n$. But then $b_i y_i = c_i d_i y_i = 0$. Thus we have shown that if $\sum b_i y_i = 0$ then every $b_i y_i = 0$. Hence we have

$$M = \sum Dy_i = Dy_1 \oplus Dy_2 \oplus \cdots \oplus Dy_n.$$

Moreover, we have the additional fact that if $b_i y_i = 0$ then $b_i \in (d_i)$. Since $d_i y_i = 0$ we have ann $y_i = (d_i)$. The divisibility conditions on the d_i evidently give the relations

$$(d_1) \supset (d_2) \supset \cdots \supset (d_n).$$

Now it is clear that if d_i is a unit then $d_i y_i = 0$ implies $y_i = 0$. Hence this element can be dropped from the set of generators $\{y_1, y_2, \ldots, y_n\}$. Suppose $d_1, \ldots, d_t$ are units and that $d_{t+1}, d_{t+2}, \ldots$ are not units, and put $z_1 = y_{t+1}, z_2 = y_{t+2}, \ldots,$ $z_s = y_n$ where $s = n - t$. Then we have $M = Dz_1 \oplus Dz_2 \oplus \cdots \oplus Dz_s$ where every $Dz_j \neq 0$ and the conditions (28) hold. $\square$

EXERCISES

1. Determine the structure of $\mathbb{Z}^{(3)}/K$ where K is generated by $f_1 = (2, 1, -3)$, $f_2 = (1, -1, 2)$.

2. Let D be the ring of Gaussian integers $\mathbb{Z}[\sqrt{-1}]$. Determine the structure of $D^{(3)}/K$ where K is generated by $f_1 = (1, 3, 6)$, $f_2 = (2 + 3i, -3i, 12 - 18i)$, $f_3 = (2 - 3i, 6 + 9i, -18i)$, $i = \sqrt{-1}$. Show that $M = D^{(3)}/K$ is finite (of order 352512).

3. Let M be the ideal in $\mathbb{Z}[x]$ generated by 2 and x. Show that M is not a direct sum of cyclic $\mathbb{Z}[x]$-modules.

The remaining exercises are designed to develop a proof of the fundamental structure theorem which does not depend on the normal form of matrices (Theorem 3.8). In these M is a finitely generated module over a p.i.d. D. We use the notion of length of an element of D as defined in section 3.7, extending this to 0 by putting $l(0) = \infty$, which we regard as greater than any integer. Also, if $x \in M$, we define $l(x) = l(d)$ where ann $x = (d)$.

4. Let N be a submodule of M, $x \in M$. Show that: (i) ann $(x + N) \supset$ ann x and ann $(x + N) \supsetneq$ ann x if and only if $Dx \cap N \neq 0$, (ii) $l(x + N) \leq l(x)$ and $l(x + N) < l(x)$ if and only if $Dx \cap N \neq 0$.

5. Let $x_1, x_2, \ldots, x_n$ be a set of n (≥ 1) generators for M and let $y = \sum_1^n a_i x_i$ where the greatest common divisor $(a_1, a_2, \ldots, a_n) = 1$. Show that there exists a set of n generators $y_1, y_2, \ldots, y_n$ with $y_1 = y$ (cf. exercise 7, p. 186). (*Sketch of proof.* Clear for $n = 1$. For $n = 2$, let $b_1, b_2 \in D$ satisfy $a_1 b_1 + a_2 b_2 = 1$. Then $y_1 = y$, $y_2 = -b_2 x_1 + b_1 x_2$ generate M. For $n > 2$, put $d = (a_2, \ldots, a_n)$. The case $d = 0$ (hence $a_2 = \cdots = a_n = 0$) is trivial, so assume $d \neq 0$ and write $a_j = da_j'$. Then $(a_2', \ldots, a_n') = 1$, so, by induction, one has a set of generators $y_2 = \sum_2^n a_j' x_j$, $y_3, \ldots, y_n$ for $N = \sum_2^n Dx_j$. Also $(a_1, d) = 1$ and $y = a_1 x_1 + dy_2$, so the case $n = 2$ shows that there is a z_2 in $P = Dx_1 + Dy_2$ such that $Dy + Dz_2 = P$. Then $y, z_2, y_3, \ldots, y_n$ generate M.)

6. Let $x_1, x_2, \ldots, x_n$ be a set of generators for M such that (i) n is minimal, (ii) $l(x_1)$ is minimal for all sets of n generators for M. Show that $M = Dx_1 \oplus N$ where $N = \sum_2^n Dx_j$ and that ann $x_1 \supset$ ann y for any $y \in N$. This will prove the structure theorem by induction on n. *(Sketch of proof.* If $Dx_1 \cap N \neq 0$, $l(x_1 + N) < l(x_1)$ by exercise 4. Then ann $(x_1 + N) = (a_1) \neq 0$ and $a_1x_1 + a_2x_2 + \cdots + a_nx_n = 0$ for $a_i \in D$. Put $d = (a_1, \ldots, a_n)$, $a_i = da_i'$. Then $(a_1', \ldots, a_n') = 1$, so by exercise 5 we have a set of generators $y_1 = \sum a_i'x_i, y_2, \ldots, y_n$. We have $dy_1 = 0$ and $l(y_1) \leq l(d) \leq l(a_1) \leq l(x_1)$ contrary to the choice of $x_1, \ldots, x_n$. To show ann $x_1 \supset$ ann y for $y \in N$ it suffices to prove ann $x_1 \supset$ ann $x_j, j > 1$ and, by symmetry, it is enough to show ann $x_1 \supset$ ann x_2. Suppose not and let ann $x_1 = (d_1)$, ann $x_2 = (d_2)$. Then $d_2 \neq 0$, so $(d_1, d_2) = d \neq 0$ and $l(d) < l(d_1) = l(x_1)$. Also $d_i = dd_i'$ and $(d_1', d_2') = 1$ so we have a set of generators $y_1, y_2, \ldots, y_n$ with $y_1 = d_1'x_1 + d_2'x_2$. Then $l(y_1) < l(x_1)$, a contradiction.)

3.9 TORSION MODULES AND PRIMARY COMPONENTS. INVARIANCE THEOREM

The decomposition of a finitely generated module over a p.i.d. given by the fundamental structure theorem is generally not unique. For example, if M is free, then any base $(e_1, \ldots, e_n)$ determines such a decomposition, $M = De_1 \oplus \cdots \oplus De_n$ with ann $e_i = 0$, and changing the base to another one $(f_1, \ldots, f_n)$ where the f's are not merely multiples of the e's in some order gives a second direct decomposition, $M = Df_1 \oplus \cdots \oplus Df_n$, different from the first. However, there is something which is invariant about the various decompositions of M into cyclic submodules whose order ideals satisfy the inclusion relations stated in the structure theorem: namely, the sequence of order ideals ann z_1, ann $z_2, \ldots$ is the same for any two such decompositions. Our next main objective is to prove this. However, before launching into the proof it will be useful to introduce the concept of the torsion submodule of a module over a p.i.d. and to develop some of its properties. This will facilitate the proof of the invariance theorem and afford a better insight into the structure of modules over a p.i.d.

Let M be a finitely generated module over a p.i.d. D, and let tor M be the subset of elements $y \in M$ such that $ay = 0$ for some $a \neq 0$ in D. Then $y \in$ tor M if and only if ann $y \neq 0$. If $a_iy_i = 0$, $i = 1, 2$, and $a_i \neq 0$, then $a = a_1a_2 \neq 0$ and $a(y_1 + y_2) = a_2a_1y_1 + a_1a_2y_2 = 0$. This, and the fact that $ay = 0$ implies $a(by) = b(ay) = 0$ shows that tor M is a submodule of M. We call this the *torsion submodule* of M and say that M is a *torsion module* if $M =$ tor M. Now suppose we have the decomposition $M = Dz_1 \oplus Dz_2 \oplus \cdots \oplus Dz_s$, where ann $z_1 \supset$ ann $z_2 \supset \cdots \supset$ ann z_s. Suppose also that ann $z_i \neq 0$ if $i \leq r$ and ann $z_i = 0$ if $r < i \leq s$. Then the z_i, $i \leq r$, are in tor M so $Dz_1 + \cdots + Dz_r \subset$ tor M. On the other hand, suppose $y = b_1z_1 + \cdots + b_sz_s \in$ tor M. Then there exists an $a \neq 0$ such that $0 = ay = ab_1z_1 + \cdots + ab_sz_s = 0$. Then every $ab_iz_i = 0$, which implies that $ab_i = 0$ if $i > r$. Since $a \neq 0$ this gives $b_i = 0$, $i > r$, and $y = \sum_1^r b_j z_j \in Dz_1 + \cdots + Dz_r$. Thus we have

(30) $\text{tor } M = Dz_1 + \cdots + Dz_r.$

It is clear also that $Dz_{r+1} + \cdots + Dz_s = Dz_{r+1} \oplus \cdots \oplus Dz_s$ is a free submodule of M, and $M = \text{tor } M \oplus (Dz_{r+1} + \cdots + Dz_s)$. We therefore have the following

THEOREM 3.10   *Any finitely generated module over a p.i.d. is a direct sum of its torsion submodule and a free submodule.*

If p is a prime we define the *p-component* M_p of M to be the subset of M of elements y such that $p^k y = 0$ for some $k \in \mathbb{N}$. This is contained in tor M and it is a submodule. If $p_1, p_2, \ldots, p_h$ are distinct primes then the corresponding p_i-components are independent. To see this it is enough to show that $M_{p_1} \cap (M_{p_2} + \cdots + M_{p_h}) = 0$. Hence let y be in this intersection. Then $y = y_2 + \cdots + y_h$, $y_i \in M_{p_i}$, and $p_i^{k_i} y_i = 0$ for some $k_i \in \mathbb{N}$. Hence $p_2^{k_2} \cdots p_h^{k_h} y = 0$. On the other hand, we have $p_1^{k_1} y = 0$ since $y \in M_{p_1}$. Then $p_1^{k_1} \in \text{ann } y$ and $p_2^{k_2} \cdots p_h^{k_h} \in \text{ann } y$. Hence $1 = (p_1^{k_1}, p_2^{k_2} \cdots p_h^{k_h}) \in \text{ann } y$ and so $y = 0$. We shall show next that "almost all", meaning all but a finite number, of the p-components are 0 and tor M is a direct sum of these p-components. This will follow from the first part of the following.

LEMMA.   (1) *If* $M = Dx$ *where* ann $x = (d)$ *and* $d = gh$ *with* $(g, h) = 1$, *then* $M = Dy \oplus Dz$ *where* ann $y = (g)$ *and* ann $z = (h)$. (2) *If* $M = Dy + Dz$ *where* ann $y = (g)$, ann $z = (h)$, *and* $(g, h) = 1$, *then* $M = Dx$ *where* ann $x = (gh)$.

Proof.   (1) Put $y = hx$, $z = gx$. Then $Dy + Dz$ contains x since there exist $a, b \in D$ such that $ah + bg = 1$; hence $x = (ah + bg)x = a(hx) + b(gx) = ay + bz$. Since $M = Dx$ it follows that $M = Dy + Dz$. If $u \in Dy \cap Dz$, $gu = 0$ and $hu = 0$ since $gy = ghx = 0$ and $hz = hgx = 0$. Then $u = 1u = ahu + bgu = 0$. Hence $Dy \cap Dz = 0$ and $M = Dy \oplus Dz$. It is clear also that ann $y = (g)$ and ann $z = (h)$. (2) As in (1), we have $M = Dy \oplus Dz$, and if we put $x = y + z$, then $cx = 0$ implies $cy = 0 = cz$. Then c is a multiple of g and of h, hence, of their least common multiple gh. Since $(gh)(y + z) = 0$ we have ann $x = (gh)$. Also, if $ah + bg = 1$ then $y = ahy = ah(y + z) = ahx$. Hence $y \in Dx$. Then $z = x - y \in Dx$ and so $Dx = M$.   □

It follows by induction from the first part of this lemma that if $d = p_1^{e_1} p_2^{e_2} \cdots p_t^{e_t}$ where the p_i are distinct primes and ann $x = (d)$, then $M = Dx_1 \oplus \cdots \oplus Dx_t$

where ann $x_i = (p_i^{e_i})$. This shows that any cyclic torsion module is a direct sum of cyclic modules which are *primary* in the sense that their order ideals have the form (p^e), p a prime. We can use this to prove

THEOREM 3.11. *Let M be a finitely generated torsion module over a p.i.d. Then the primary component $M_p = 0$ for all but a finite number of primes: say, $p_1, p_2, \cdots, p_h$, and $M = M_{p_1} \oplus M_{p_2} \oplus \cdots \oplus M_{p_h}$.*

Proof. Let $x_1, \ldots, x_n$ be a set of generators, so $M = Dx_1 + \cdots + Dx_n$, and let $p_1, \ldots, p_h$ be the distinct prime factors of all the d_i such that ann $x_i = (d_i)$. Then $Dx_i \subset M_{p_1} + \cdots + M_{p_h}$ and so $M = M_{p_1} + \cdots + M_{p_h}$. Since the M_{p_i} are independent we have $M = M_{p_1} \oplus \cdots \oplus M_{p_h}$. Now let p be a prime different from all the p_i, $1 \leq i \leq h$. Then $M_p = M_p \cap (M_{p_1} + \cdots + M_{p_h}) = 0$. Hence every M_p, $p \neq p_i$, is 0. $\square$

We now combine the fundamental structure theorem with the decomposition of a cyclic torsion module into primary ones. This gives the following result:

THEOREM 3.12. *Any finitely generated torsion module is a direct sum of primary cyclic modules.*

Evidently we obtain this result by writing M as a direct sum of cyclic modules as in the main theorem. Then, as we saw above, each of these is a direct sum of primary cyclic submodules. Consequently M is such a direct sum. More precisely, if $M = Dz_1 \oplus \cdots \oplus Dz_r$ and ann $z_i = (d_i)$ satisfies ann $z_1 \supset$ ann $z_2 \supset \cdots \supset$ ann z_r then $d_1|d_2| \cdots |d_r$. Then we may assume that $d_i = p_1^{e_{1i}} \cdots p_h^{e_{hi}}$ where the displayed primes are distinct and $e_{j1} \leq e_{j2} \leq \cdots \leq e_{jr}$, $1 \leq j \leq h$. Then M is a direct sum of cyclic modules with annihilators $(p_j^{e_{ji}})$. We remark also that if the prime powers $p_j^{e_{ji}}$ are given then we can reconstruct the d_i: the last one, d_s, is the least common multiple of all the prime powers that occur. Striking out the prime power factors of d_s, then d_{s-1} is the l.c.m. of the remaining ones, and so on. For example, if $D = \mathbb{Z}$ and the prime power factors of the d_i are 3^2, 3^3, 3^4, 5^2, 5^4, 7, 7^3 then $d_s = 3^4 5^4 7^3$, $d_{s-1} = 3^3 5^2 7$, and $d_{s-2} = d_1 = 3^2$. We note also that if we are given a decomposition of M as a direct sum of primary cyclic submodules, then by forming sums of suitable primary cyclic submodules as in the second part of the foregoing lemma we obtain a direct decomposition into cyclic submodules. In our example let $x_1, x_2, \ldots, x_7$ be generators of the sequence of primary direct summands of M. Then $Dx_3 + Dx_5 + Dx_7 = Dz_3$, $Dx_2 + Dx_4 + Dx_6 = Dz_2$, $Dx_1 = Dz_1$ and ann $z_i = (d_i)$ satisfy ann $z_1 \supset$ ann $z_2 \supset$ ann z_3.

We are now ready to prove the

INVARIANCE THEOREM. *Let* $M = Dz_1 \oplus Dz_2 \oplus \cdots \oplus Dz_s = Dw_1 \oplus$ $Dw_2 \oplus \cdots \oplus Dw_t$, *where* ann $z_1 \supset$ ann $z_2 \supset \cdots \supset$ ann z_s *and* ann $w_1 \supset$ ann w_2 $\supset \cdots \supset$ ann w_t *and none of the components are* 0. *Then* $s = t$ *and* ann $z_i =$ ann w_i, $1 \leq i \leq s$.

Proof. I. *Reduction to torsion modules.* Suppose that ann $z_i \neq 0$ for $i \leq r$ and $= 0$ for $i > r$, and that ann $w_j \neq 0$ for $j \leq u$ and $= 0$ for $j > u$. Then

$$\text{tor } M = Dz_1 \oplus \cdots \oplus Dz_r = Dw_1 \oplus \cdots \oplus Dw_u,$$

by (30). Also $M/\text{tor } M \cong Dz_{r+1} \oplus \cdots \oplus Dz_s \cong Dw_{u+1} \oplus \cdots \oplus Dw_t$ and these are free modules of ranks $s - r$ and $t - u$ respectively. The theorem on invariance of rank for free modules over a commutative ring (Theorem 3.4) shows that $s - r = t - u$. Thus the number of ann $z_i = 0$ is the same as the number of ann $w_j = 0$. It remains to prove the theorem for tor M, for which we have the displayed direct decompositions into cyclic submodules.

II. *Reduction to primary torsion modules.* We now assume M is a torsion module and we decompose the cyclic summands Dz_i and Dw_j as direct sums of primary cyclic submodules. The foregoing considerations on decomposition into primary cyclic submodules imply that the theorem will follow for torsion modules if we can show that any two decompositions of M as direct sums of primary cyclic submodules have the same set of order ideals. This amounts to showing that for any prime power p^e the number of cyclic direct summands with order ideal (p^e) is the same for the two decompositions. Now if we fix p and form the sum of the cyclic summands in each decomposition having order ideals of the form (p^e), $e = 1, 2, \ldots$, then both of these sums coincide with the p-component M_p. Hence it suffices to prove the result for each M_p, that is, we may assume $M = M_p$ is primary.

III. *Proof in the primary case.* We now assume $M = M_p$. Then ann $z_i = (p^{e_i})$, ann $w_i = (p^{f_i})$ and, since ann $z_1 \supset$ ann $z_2 \supset \cdots \supset$ ann z_s and ann $w_1 \supset$ ann $w_2 \supset \cdots \supset$ ann w_t, we have $e_1 \leq e_2 \leq \cdots \leq e_s$ and $f_1 \leq f_2 \leq \cdots \leq f_t$. We now observe that for any $k \in \mathbb{N}$, $p^k M = \{p^k x \,|\, x \in M\}$ is a submodule and $M \supset pM \supset$ $p^2 M \supset \cdots$. Let $M^{(k)} = p^k M / p^{k+1} M$. Any coset of this D-module has the form $p^k x + p^{k+1} M$ and satisfies $p(p^k x + p^{k+1} M) = p^{k+1} M = 0$ in $M^{(k)}$. Thus the ideal (p) annihilates $M^{(k)}$ so $M^{(k)}$ can be regarded in a natural way as $\bar{D} = D/(p)$ module (exercise 2, p. 165). Since p is a prime, $\bar{D}$ is a field, and so $M^{(k)}$ is a vector space over $\bar{D}$. We can relate its dimensionality to the e_i and f_j in the following way. We have $p^k M = 0$ if $k \geq e_s$ and $p^k M = Dp^k z_{q+1} + Dp^k z_{q+2} + \cdots +$

$Dp^k z_s$ if e_{q+1} is the first $e_i > k$. Then the cosets $p^k z_{q+1} + p^{k+1}M, \ldots, p^k z_s + p^{k+1}M$ form a base for $M^{(k)}$ as vector space over $\bar{D}$. Hence we see that the dimensionality of this space is the same as the number of $e_i > k$. Similarly, the dimensionality is the number of $f_j > k$. We therefore conclude that for any $k \in \mathbb{N}$ the number of $e_i > k$ is the same as the number of $f_j > k$. This forces $s = t$ and $e_i = f_i$, $1 \le i \le s$, which completes the proof of the theorem. $\square$

We shall now call the sequence of order ideals, ann z_1, ann $z_2, \ldots$, whose uniqueness has just been proved, the *invariant factor ideals* of the module M. Our proof shows also that if M is a torsion module the order ideals of the primary cyclic submodules in any two decompositions of M as direct sum of such submodules are invariant. We call these the *elementary divisor ideals* of M. It is clear that any two finitely generated modules over a p.i.d. are isomorphic if and only if they have the same invariant factor ideals. Similarly, for torsion modules, isomorphism holds if and only if the two modules have the same elementary divisor ideals.

In the special case $D = \mathbb{Z}$ any ideal has a unique non-negative generator, and if $D = F[\lambda]$, F a field, then any ideal is either generated by 0 or by a monic polynomial. It is natural in these cases to replace the invariant factor ideals and elementary divisor ideals by these normalized generators. One calls these the *invariant factors* and *elementary divisors* of the module.

EXERCISES

1. Let $D = \mathbb{R}[\lambda]$ and suppose M is a direct sum of cyclic modules whose order ideals are the ideals generated by the polynomials $(\lambda - 1)^3$, $(\lambda^2 + 1)^2$, $(\lambda - 1)(\lambda^2 + 1)^4$, $(\lambda + 2)(\lambda^2 + 1)^2$. Determine the elementary divisors and invariant factors of M.

2. Show that a torsion module M over a p.i.d. D is irreducible (definition in exercise 7, p. 169) if and only if $M = Dz$ and ann $z = (p)$, p a prime. Show that if M is finitely generated then M is *indecomposable* in the sense that M is not a direct sum of two non-zero submodules if and only if $M = Dz$ where ann $z = 0$ or ann $z = (p^e)$, p a prime.

3. Define the *rank* of a finitely generated module M over a p.i.d. D to be the rank of the free module $M/\text{tor } M$. (This is free since it is isomorphic to F if $M = \text{tor } M \oplus F$, F free as in Theorem 3.10.) Show that if $M \cong D^{(n)}/K$ then rank $M = n - \text{rank } K$. Show also that if N is a submodule of M then N and M/N are finitely generated and rank $M = \text{rank } N + \text{rank } M/N$.

4. Let M be a torsion module for the p.i.d. D with invariant factor ideals $(d_1) \supset$ $(d_2) \supset \ldots \supset (d_s)$. Show that any homomorphic image $\bar{M}$ of M is a torsion module with invariant factor ideals $(\bar{d}_1) \supset (\bar{d}_2) \supset \cdots \supset (\bar{d}_t)$ satisfying the conditions: $t \leq s$, $\bar{d}_t | d_s, \bar{d}_{t-1} | d_{s-1}, \ldots, \bar{d}_1 | d_{s-t+1}$. (*Hint*: Suppose first that M is primary.)

5. Let $A, B \in M_n(D)$ satisfy det $AB \neq 0$ (D a p.i.d.). Let diag$\{a_1, a_2, \ldots, a_n\}$, diag$\{b_1, b_2, \ldots, b_n\}$, diag$\{c_1, c_2, \ldots, c_n\}$ be normal forms for A, B and AB respectively (so $a_i | a_{i+1}$, etc.). Prove that $a_i | c_i$ and $b_i | c_i$ for $1 \leq i \leq n$.

6. Show that the assertion made in exercise 4 on a homomorphic image $\bar{M}$ of M holds for any submodule N of M.

7. Call a submodule N of M *pure* if for any $y \in N$ and $a \in D$, $ax = y$ is solvable in M if and only if it is solvable in N. Show that if N is a direct summand then N is pure. Show that if N is a pure submodule of M and ann $(x + N) = (d)$ then x can be chosen in its coset $x + N$ so that ann $x = (d)$.

8. Show that if N is a pure submodule of a finitely generated torsion module M over a p.i.d., then N is a direct summand of M.

9. Let M be a finitely generated torsion module over a p.i.d. Show that any cyclic submodule Dz such that ann $z \subset$ ann x for every $x \in M$, is a pure submodule. Hence by exercise 8, Dz is a direct summand.

3.10 APPLICATIONS TO ABELIAN GROUPS AND TO LINEAR TRANSFORMATIONS

We first specialize the structure theory of finitely generated modules M over a p.i.d. D to the case $D = \mathbb{Z}$. Then M is any abelian group with a finite set of generators. In particular, M can be any finite group. The main structure theorem now states that any finitely generated abelian group M is a direct sum of cyclic groups: $M = \langle z_1 \rangle \oplus \langle z_2 \rangle \oplus \cdots \oplus \langle z_s \rangle$ where ann $z_i = (d_i)$ and $d_1 | d_2 | \cdots | d_s$. If we normalize d_i to be non-negative then the order of z_i is d_i if $d_i > 0$ and the order of z_i is infinite if $d_i = 0$. The torsion subgroup ($=$ submodule) of M is the subset of M of elements of finite order. In the foregoing decomposition this coincides with $\langle z_1 \rangle + \cdots + \langle z_r \rangle$ where $d_1 > 0, \ldots, d_r > 0$ but $d_{r+1} = 0$. Since $|\langle z_i \rangle| = d_i$ for $i \leq r$ it is clear that tor M is a finite group of order $\prod_1^r d_i$. The second structure theorem (Theorem 3.10) implies that any finitely generated abelian group is a direct sum of a finite group and a free group. The finite component in any such decomposition is uniquely determined as the torsion subgroup. The free component may not be unique, but its rank is an invariant.

The result on the decomposition of a torsion module as a direct sum of primary cyclic modules specializes in the present case to: any finite abelian group is a direct sum of cyclic groups of prime power orders. The prime powers occurring in such a decomposition counted with their multiplicities are uniquely determined. These are called the *invariants* of the finite abelian group. Clearly, two finite abelian groups are isomorphic if and only if they have the same invariants.

For the sake of reference we summarize the main results on finitely generated abelian groups in the following

THEOREM 3.13. *Any finitely generated abelian group is a direct sum of a finite group, its torsion subgroup, and a free group. The rank of the free component is an invariant. Any finite abelian group is a direct sum of cyclic groups of prime power orders. These orders, together with their multiplicities, are uniquely determined, and constitute a complete set of invariants in the sense that two finite abelian groups are isomorphic if and only if they have the same set of these invariants.*

We apply our results next to the study of a single linear transformation T in a finite dimensional vector space V over a field. Let $(u_1, u_2, \ldots, u_n)$ be a base for V over F and write

$$(31) \qquad Tu_i = \sum_1^n a_{ij}u_j, \qquad i = 1, 2, \ldots, n.$$

Then A is the matrix of T relative to the given base. We recall that if $(v_1, v_2, \ldots, v_n)$ is a second base for V over F and $v_i = \sum s_{ij}u_j$ where $S = (s_{ij})$ is an invertible matrix, then the matrix of T relative to $(v_1, v_2, \ldots, v_n)$ is SAS^{-1}. Matrices related in this way are said to be *similar*. As before (section 3.2), we can make V an $F[\lambda]$-module by defining the action of any polynomial $g(\lambda) = b_0 + b_1\lambda + \cdots + b_m\lambda^m$ on any vector $x \in V$ as

$$g(\lambda)x = b_0x + b_1(Tx) + \cdots + b_m(T^mx).$$

Clearly this action of $F[\lambda]$ is the extension of the action of F such that $\lambda x = Tx$.

We note first that V is a torsion $F[\lambda]$-module. For, let $x \in V$ and consider the sequence of vectors $x, \lambda x, \lambda^2 x, \ldots$. Since V is n-dimensional over F we have an integer $m \leq n$ such that $\lambda^m x$ is a linear combination of $x, \lambda x, \ldots, \lambda^{m-1}x$, say $\lambda^m x = b_0 x + b_1 \lambda x + \cdots + b_{m-1}(\lambda^{m-1}x)$, $b_i \in F$. Then $g(\lambda) = \lambda^m - b_{m-1}\lambda^{m-1} - \cdots - b_0$ is a non-zero polynomial such that $g(\lambda)x = 0$. Thus ann x contains $g(\lambda) \neq 0$ and ann $x \neq 0$.

The base $(u_1, u_2, \ldots, u_n)$ for V over F is evidently a set of generators of V as $F[\lambda]$-module (though, generally, not a base) and we have the homomorphism η of the free module $F[\lambda]^{(n)}$ with base $(e_1, e_2, \ldots, e_n)$ onto V sending $e_i \to u_i$, $1 \leq i \leq n$. Our method of analyzing V as $F[\lambda]$-module calls for a set of generators for $K = \ker \eta$. Such a set is given in the following

LEMMA. *The elements $f_i \equiv \lambda e_i - \sum_{j=1}^n a_{ij}e_j$, $1 \leq i \leq n$, form a base for K.*

Proof.   Since $Tu_i = \sum a_{ij}u_j$ it is clear that $f_i \in K$. We have $\lambda e_i = f_i + \sum a_{ij}e_j$ and these relations permit us to write any element $\sum g_i(\lambda)e_i$ in the form $\sum h_i(\lambda)f_i + \sum b_i e_i$ where the $b_i \in F$. If this element is in K then $\sum b_i e_i \in K$ and so $\sum b_i u_i = 0$. Since the u_i constitute a base for V over F every $b_i = 0$ and our element of K has the form $\sum h_i(\lambda)f_i$. This shows that the f_i generate K. Suppose next that we have a relation $\sum h_i(\lambda)f_i = 0$. Then

$$\sum_{i=1}^{n} h_i(\lambda)\lambda e_i = \sum_{i,j=1}^{n} h_i(\lambda)a_{ij}e_j$$

and, since the e_i form a base for $F[\lambda]^{(n)}$,

$$h_i(\lambda)\lambda = \sum_j h_j(\lambda)a_{ji}.$$

If any $h_i(\lambda) \neq 0$ let $h_r(\lambda)$ be one of maximal degree. Then clearly the relation $h_r(\lambda)\lambda = \sum_j h_j(\lambda)a_{jr}$ is impossible. This proves that every $h_i(\lambda) = 0$ and so the f_i form a base for K. $\square$

The matrix relating the base (f_i) of K to the base (e_i) of $F[\lambda]^{(n)}$ is

(32)
$$\lambda 1 - A = \begin{pmatrix} \lambda - a_{11} & -a_{12} & \cdots & -a_{1n} \\ -a_{21} & \lambda - a_{22} & \cdots & -a_{2n} \\ \cdots\cdots\cdots\cdots\cdots\cdots\cdots\cdots\cdots\cdots \\ -a_{n1} & -a_{n2} & \cdots & \lambda - a_{nn} \end{pmatrix}$$

Hence this is the matrix whose normal form gives the invariant factors of V as $F[\lambda]$-module, and consequently gives the decomposition of this module as direct sum of cyclic ones. The determinant $\det(\lambda 1 - A)$ is called the *characteristic polynomial* of A. It has the form

(33) $$f(\lambda) = \det(\lambda 1 - A) = \lambda^n - a_1\lambda^{n-1} + \cdots + (-1)^n a_n.$$

Here $a_1 = \sum a_{ii}$ is called the *trace*, tr A, of the matrix A, and $a_n = \det A$. In general, a_i is the sum of the i-rowed principal (= diagonal) minors of the matrix A. Since $f(\lambda) \neq 0$ and $f(\lambda)$ is the product of the invariant factors of $\lambda 1 - A$ it is clear that none of these is 0 (which follows also from the fact that V is a torsion module over $F[\lambda]$). Thus a normal form for $\lambda 1 - A$ has the form

(34) $$\text{diag}\{1, \ldots, 1, d_1(\lambda), \ldots, d_s(\lambda)\}$$

where the $d_i(\lambda)$ are monic of positive degree and $d_i(\lambda)\,|\,d_j(\lambda)$ if $i \leq j$. Our results given in section 3.8 show that if P and Q are invertible matrices in $M_n(F[\lambda])$ such that

(35) $$P(\lambda 1 - A)Q = \text{diag}\{1, \ldots, 1, d_1(\lambda), \ldots, d_s(\lambda)\}$$

and if we write $Q^{-1} = (q_{ij}^*)$ and put $v_i = \sum q_{ij}^* u_j$, $z_i = v_{n-s+i}$, then we have

(36) $$V = F[\lambda]z_1 \oplus F[\lambda]z_2 \oplus \cdots \oplus F[\lambda]z_s$$

where ann $z_i = (d_i(\lambda))$.

We shall use (36) for obtaining a certain canonical matrix for the linear transformation T. Suppose first $s = 1$, that is, $V = F[\lambda]z$ is cyclic as $F[\lambda]$-module. Then ann $z = (f(\lambda))$ where $f(\lambda)$ is the characteristic polynomial of A as in (33). Since $f(\lambda)$ is the non-zero polynomial of least degree such that $f(\lambda)z = 0$, z, $\lambda z, \ldots, \lambda^{n-1}z$ are linearly independent. Hence $(z, \lambda z, \ldots, \lambda^{n-1}z)$ is a base for V over F. We have

(37)
$$
\begin{aligned}
Tz &= \lambda z \\
T(\lambda z) &= \lambda(\lambda z) = \lambda^2 z \\
&\ \vdots \qquad \vdots \\
T(\lambda^{n-2}z) &= \lambda^{n-1}z \\
T(\lambda^{n-1}z) &= \lambda^n z = a_1(\lambda^{n-1}z) - a_2(\lambda^{n-2}z) + \cdots + (-1)^{n-1}a_n z.
\end{aligned}
$$

Hence the matrix of T relative to the base $(z, \lambda z, \ldots, \lambda^{n-1}z)$ is

$$
\begin{pmatrix}
0 & 1 & 0 & \cdots & \cdot & 0 \\
0 & 0 & 1 & \cdots & \cdot & 0 \\
\cdots\cdots\cdots\cdots\cdots\cdots\cdots\cdots\cdots\cdots\cdots \\
\cdot & & \cdot & \cdots & 0 & 1 \\
(-1)^{n-1}a_n & \cdot & \cdot & \cdots & -a_2 & a_1
\end{pmatrix}
$$

In general, if $d(\lambda)$ is a monic polynomial, and we write $d(\lambda) = \lambda^m - b_{m-1}\lambda^{m-1} - \cdots - b_0$ then the matrix

(38) $$
B = \begin{pmatrix}
0 & 1 & 0 & \cdots & \cdot & 0 \\
0 & 0 & 1 & \cdots & \cdot & 0 \\
\cdots\cdots\cdots\cdots\cdots\cdots\cdots\cdots\cdots\cdots \\
\cdot & \cdot & \cdot & \cdots & 0 & 1 \\
b_0 & b_1 & \cdot & \cdots & \cdot & b_{m-1}
\end{pmatrix}
$$

is called the *companion matrix* of the given polynomial $d(\lambda)$. Using this terminology we can say that the matrix of T relative to the base $(z, \lambda z, \ldots, \lambda^{n-1}z)$ (in the cyclic case) is the companion matrix of the characteristic polynomial $f(\lambda)$.

We now consider the general case in which we have the decomposition (36). Then we obtain a base for V over F by stringing together F-bases for the cyclic submodules $F[\lambda]z_i$. If deg $d_i(\lambda) = n_i$ then $(z_i, \lambda z_i, \ldots, \lambda^{n_i-1}z_i)$ is a base for $F[\lambda]z_i$ and if

(39) $$d_i(\lambda) = \lambda^{n_i} - b_{i,n_i-1}\lambda^{n_i-1} - \cdots - b_{i,0}$$

then $T(\lambda^{n_i-1}z_i) = b_{i0}z_i + b_{i1}(\lambda z_i) + \cdots + b_{i,n_i-1}(\lambda^{n_i-1}z_i)$. It is clear that the matrix of T relative to the base

$$(z_1, \lambda z_1, \ldots, \lambda^{n_1-1}z_1; z_2, \lambda z_2, \ldots, \lambda^{n_2-1}z_2; \ldots; \ldots \lambda^{n_s-1}z_s)$$

has the form

(40)
$$B = \begin{pmatrix} B_1 & & & 0 \\ & B_2 & & \\ & & \ddots & \\ 0 & & & B_s \end{pmatrix}$$

where B_i is the companion matrix of $d_i(\lambda)$. The matrix B is called the *rational canonical form* for the linear transformation T. Clearly the rational canonical form can be written down as soon as we know the invariant factors of $\lambda 1 - A$, and these can be calculated by performing a series of elementary transformations on the rows and columns of $\lambda 1 - A$.

EXAMPLE

Let T be the linear transformation in $V = \mathbb{Q}^{(3)}$ such that

$$\begin{aligned} Tu_1 &= -u_1 - 2u_2 + 6u_3 \\ Tu_2 &= -u_1 \qquad\quad + 3u_3 \\ Tu_3 &= -u_1 - u_2 + 4u_3. \end{aligned}$$

Here the matrix A is

$$\begin{pmatrix} -1 & -2 & 6 \\ -1 & 0 & 3 \\ -1 & -1 & 4 \end{pmatrix}, \quad \text{and} \quad \lambda 1 - A = \begin{pmatrix} \lambda+1 & 2 & -6 \\ 1 & \lambda & -3 \\ 1 & 1 & \lambda-4 \end{pmatrix}.$$

We have

$$\begin{pmatrix} 0 & 1 & 0 \\ 0 & -1 & 1 \\ 1 & 2-\lambda & -3 \end{pmatrix} (\lambda 1 - A) \begin{pmatrix} 1 & 3 & \lambda-3 \\ 0 & 0 & -1 \\ 0 & 1 & -1 \end{pmatrix} = \begin{pmatrix} 1 & & \\ & \lambda-1 & \\ & & (\lambda-1)^2 \end{pmatrix}$$

and the two matrices flanking $\lambda 1 - A$ have determinants 1 and so are invertible in $M_3(\mathbb{Q}[\lambda])$. Hence the invariant factors of V as $\mathbb{Q}[\lambda]$-module are $\lambda - 1$ and $(\lambda - 1)^2 = \lambda^2 - 2\lambda + 1$, and the rational canonical form is

$$\begin{pmatrix} \boxed{1} & 0 & 0 \\ 0 & 0 & 1 \\ 0 & -1 & 2 \end{pmatrix}.$$

Our method also yields a matrix in $M_3(\mathbb{Q})$ which transforms A into its rational canonical form. Thus, in the above notation we have

$$Q = \begin{pmatrix} 1 & 3 & \lambda - 3 \\ 0 & 0 & -1 \\ 0 & 1 & -1 \end{pmatrix}, \qquad Q^{-1} = \begin{pmatrix} 1 & \lambda & -3 \\ 0 & -1 & 1 \\ 0 & -1 & 0 \end{pmatrix}$$

and

$$v_1 = u_1 + \lambda u_2 - 3u_3 = 0$$
$$v_2 = -u_2 + u_3$$
$$v_3 = -u_2.$$

Then a base (z_1, z_2, z_3) which gives the rational canonical form is

$$z_1 = v_2 = -u_2 + u_3$$
$$z_2 = v_3 = -u_2$$
$$z_3 = \lambda v_3 = u_1 - 3u_3.$$

The matrix relating this to the initial base is

$$\begin{pmatrix} 0 & -1 & 1 \\ 0 & -1 & 0 \\ 1 & 0 & -3 \end{pmatrix}.$$

We can check that

$$\begin{pmatrix} 0 & -1 & 1 \\ 0 & -1 & 0 \\ 1 & 0 & -3 \end{pmatrix} \begin{pmatrix} -1 & -2 & 6 \\ -1 & 0 & 3 \\ -1 & -1 & 4 \end{pmatrix} \begin{pmatrix} 0 & -1 & 1 \\ 0 & -1 & 0 \\ 1 & 0 & -3 \end{pmatrix}^{-1} = \begin{pmatrix} 1 & 0 & 0 \\ 0 & 0 & 1 \\ 0 & -1 & 2 \end{pmatrix}.$$

There is a second canonical form ($=$matrix) for a linear transformation T, the so-called Jordan form, which can be defined if the invariant factors can be factored as products of linear factors $\lambda - r$ in $F[\lambda]$. This will always be the case if F is the field of complex numbers $\mathbb{C}$ (see Chapter 5, p. 309). Under the hypothesis we have made, the elementary divisors of V as $F[\lambda]$-module have the form $(\lambda - r)^e, r \in F$. Corresponding to each of these we have a cyclic direct summand $F[\lambda]w$ with ann $w = ((\lambda - r)^e)$. The F-space $F[\lambda]w$ has the base

(41) $$w, (\lambda - r)w, (\lambda - r)^2 w, \ldots, (\lambda - r)^{e-1}w$$

and we have

$$Tw = \lambda w = rw + (\lambda - r)w$$
$$T(\lambda - r)w = \lambda(\lambda - r)w = r(\lambda - r)w + (\lambda - r)^2 w$$
$$\cdots\cdots\cdots\cdots\cdots\cdots\cdots\cdots$$
$$T(\lambda - r)^{e-2}w = r(\lambda - r)^{e-2}w + (\lambda - r)^{e-1}w$$
$$T(\lambda - r)^{e-1}w = r(\lambda - r)^{e-1}w.$$

Hence the matrix of the restriction of T to $F[\lambda]w$ relative to the base $(w, (\lambda - r)w, \ldots, (\lambda - r)^{e-1}w)$ is

(42)
$$
\begin{pmatrix}
r & 1 & & & & \\
 & r & 1 & & 0 & \\
 & & \ddots & & & \\
 & & & & r & 1 \\
 & 0 & & & & r
\end{pmatrix}
$$

If $V = F[\lambda]w_1 \oplus F[\lambda]w_2 \oplus \cdots \oplus F[\lambda]w_t$ with ann $w_i = ((\lambda - r_i)^{e_i})$ then we can string together bases of the types just indicated for the sequence of cyclic spaces to obtain a base for V over F such that the matrix of T relative to this base is the *Jordan canonical form*

(43)
$$
\begin{pmatrix}
C_1 & & & \\
 & C_2 & & 0 \\
 & & \ddots & \\
 0 & & & C_t
\end{pmatrix}
$$

where

$$
C_1 =
\begin{pmatrix}
r_i & 1 & & & & \\
 & r_i & 1 & & 0 & \\
 & & \ddots & & & \\
 & & & & r_i & 1 \\
 & 0 & & & & r_i
\end{pmatrix}
\qquad (e_i \text{ rows and columns}).
$$

EXAMPLE

In the foregoing example (p. 198) the invariant factors were $(\lambda - 1), (\lambda - 1)^2$. These are also the elementary divisors and the Jordan canonical form is

$$
\begin{pmatrix}
\boxed{1} & 0 & 0 \\
0 & \boxed{\begin{matrix} 1 & 1 \\ 0 & 1 \end{matrix}}
\end{pmatrix}.
$$

Our results can be stated also in terms of matrices rather than linear transformations. Given a matrix $A \in M_n(F)$ we can use this to define the linear transformation T in $V = F^{(n)}$ such that $Tu_i = \sum a_{ij}u_j$, $(u_1, u_2, \ldots, u_n)$ a base for V over F. The various matrices similar to A are the matrices of T relative to the various ordered bases of V over F. We call the rational canonical form of T (or the Jordan canonical form, when this is defined) the *rational canonical form* (the *Jordan canonical form*) of the given matrix A. An immediate consequence of our results is that the matrices A and B are similar if and only if A and B have the same rational canonical forms (or Jordan canonical forms, when defined).

The classical results on characteristic and minimum polynomials of matrices are also consequences of our results. We shall now derive these. Let A, T, and the u_i be as indicated and let the normal form of $\lambda 1 - A$ be $P(\lambda 1 - A)Q = \mathrm{diag}\{1, \ldots, 1, d_1(\lambda), \ldots, d_s(\lambda)\}$, P and Q invertible in $M_n(F[\lambda])$, $d_i(\lambda)$ monic of positive degree. Then P and Q have determinants which are non-zero elements of F and the characteristic polynomial of A is

(44)
$$f(\lambda) = \det(\lambda 1 - A) = d_1(\lambda)d_2(\lambda) \cdots d_s(\lambda).$$

We also have $d_i(\lambda)|d_j(\lambda)$ if $i \leq j$ and $V = F[\lambda]z_1 \oplus F[\lambda]z_2 \oplus \cdots \oplus F[\lambda]z_s$ where ann $z_i = (d_i(\lambda))$. Put $m(\lambda) = d_s(\lambda)$. Since $d_i(\lambda)|m(\lambda)$ we have $m(\lambda)z_i = 0$, and since any $x \in V$ has the form $x = \sum g_i(\lambda)z_i$ we have $m(\lambda)x = 0$. Thus $m(T) = 0$ or, equivalently, $m(A) = 0$ for the matrix A. Since $g(T) = 0$, or $g(A) = 0$ implies $g(\lambda)z_s = 0$, $g(A) = 0$ implies that $m(\lambda)|g(\lambda)$. Thus $m(\lambda)$ is the monic polynomial of least degree such that $m(A) = 0$. It is clear from (44) that $f(A) = 0$. And since every $d_i(\lambda)|m(\lambda)$ it is clear that $f(\lambda)$ and $m(\lambda)$ have the same irreducible factors, differing only in the multiplicities of these factors. Finally, if we recall the formulas for the invariant factors given in Theorem 3.9 (p. 184) we see that $m(\lambda) = d_s(\lambda) = f(\lambda)/\Delta_{n-1}(\lambda)$ where $\Delta_{n-1}(\lambda)$ is the monic g.c.d. of the $(n-1)$-rowed minors of $\lambda 1 - A$. These results can be stated as the following theorem, which is a composite of results due to Hamilton, Cayley, and Frobenius.

THEOREM 3.14   Let $A \in M_n(F)$, F a field, and let $f(\lambda) = \det(\lambda 1 - A)$ be the characteristic polynomial of A. Then $f(A) = 0$. Also let $\Delta_{n-1}(\lambda)$ be the monic g.c.d. of the $(n-1)$-rowed minors of $\lambda 1 - A$ and put $m(\lambda) = f(\lambda)/\Delta_{n-1}(\lambda)$. Then $m(A) = 0$ and $m(\lambda)$ is a factor of every polynomial $g(\lambda)$ such that $g(A) = 0$. Moreover, $m(\lambda)$ and $f(\lambda)$ have the same prime factors in $F[\lambda]$.

EXERCISES

1. Determine the number of non-isomorphic abelian groups of order 360.

2. Let $\mathbb{Z}^{(n)}$ be the free $\mathbb{Z}$-module with base $(e_1, \ldots, e_n)$, K the submodule generated by the elements $f_i = \sum_1^n a_{ij}e_j$ where $a_{ij} \in \mathbb{Z}$ and $d = \det(a_{ij}) \neq 0$. Show that $|\mathbb{Z}^{(n)}/K| = |d|$.

3. Let $a + b\sqrt{-1}$ be a non-zero element of the ring of Gaussian integers $\mathbb{Z}[\sqrt{-1}]$. Show that $|\mathbb{Z}[\sqrt{-1}]/(a + b\sqrt{-1})| = a^2 + b^2$.

4. Verify that the characteristic polynomial of

$$A = \begin{pmatrix} 1 & 0 & 0 & 0 \\ 0 & 1 & 0 & 0 \\ -2 & -2 & 0 & 1 \\ -2 & 0 & -1 & -2 \end{pmatrix}$$

is a product of linear factors in $\mathbb{Q}[\lambda]$. Determine the rational and Jordan canonical forms for A in $M_4(\mathbb{Q})$. Also find matrices which show that A is similar to these canonical forms.

5. Prove that if F is a field, the matrices $A, B \in M_n(F)$ are similar if and only if the matrices $\lambda 1 - A$, $\lambda 1 - B$ are equivalent in $M_n(F[\lambda])$.

6. Prove that any matrix A is similar to its transpose ^tA.

7. Show that the $F[\lambda]$-module determined by a linear transformation T is cyclic if and only if the characteristic polynomial $f(\lambda)$ is the minimum polynomial of T.

8. Prove that any nilpotent matrix in $M_n(F)$ is similar to a matrix of the form

$$\begin{pmatrix} N_1 & & & 0 \\ & N_2 & & \\ & & \ddots & \\ 0 & & & N_s \end{pmatrix}$$

where N_i has the form

$$\begin{pmatrix} 0 & 1 & & & \\ & 0 & 1 & & 0 \\ & & \cdots \cdots & & \\ 0 & & & 0 & 1 \\ & & & & 0 \end{pmatrix}$$

9. Show that a matrix $A \in M_n(\mathbb{C})$ is similar to a diagonal matrix, $\mathrm{diag}\{r_1, r_2, \ldots, r_n\}$, $r_i \in \mathbb{C}$, if and only if the minimum polynomial $m(\lambda)$ has distinct roots.

10. Show that if $A^2 = A$ then A is similar to a matrix $\mathrm{diag}\{1, \ldots, 1, 0, \ldots, 0\}$.

11. (Weyr.) Show that the matrices $A, B \in M_n(\mathbb{C})$ are similar if and only if for every $a \in \mathbb{C}$ and $k = 1, 2, 3, \ldots$ rank $(a1 - A)^k = $ rank $(a1 - B)^k$.

12. Show that the following matrices in $M_p(\mathbb{Z}/(p))$, p a prime, are similar:

$$\begin{pmatrix} 0 & 1 & 0 & \cdot & 0 \\ 0 & 0 & 1 & \cdot & 0 \\ & \cdots \cdots & & & \\ & \cdot & \cdot & 0 & 1 \\ 1 & 0 & \cdot & \cdot & 0 \end{pmatrix}, \quad \begin{pmatrix} 1 & 1 & 0 & \cdot & 0 \\ 0 & 1 & 1 & 0 & \cdot \\ & \cdots \cdots & & & \\ \cdot & \cdot & \cdot & 1 & 1 \\ \cdot & \cdot & \cdot & 0 & 1 \end{pmatrix}.$$

13. Let P be the companion matrix of a monic irreducible polynomial $p(\lambda)$ of degree m and let $N = e_{1m}$. Show that the minimum polynomial of the $em \times em$ matrix.

$$B = \begin{pmatrix} P & N & 0 & \cdot & \cdot & 0 \\ 0 & P & N & \cdot & \cdot & 0 \\ \multicolumn{6}{c}{\cdots\cdots\cdots\cdots\cdots} \\ \cdot & \cdot & 0 & \cdot & P & N \\ \cdot & \cdot & \cdot & \cdot & 0 & P \end{pmatrix}$$

is $p(\lambda)^e$. Hence show that if A is a matrix such that the elementary divisors of $\lambda 1 - A$ are $p_1(\lambda)^{e_1}, p_2(\lambda)^{e_2}, \dots, p_t(\lambda)^{e_t}$, where the $p_i(\lambda)$ are irreducible, then A is similar to

$$\begin{pmatrix} B_1 & & & \\ & B_2 & & 0 \\ & 0 & & \ddots \\ & & & B_t \end{pmatrix}$$

where B_i has the form of B with P the companion matrix of $p_i(\lambda)$ and number of blocks equal e_i.

14. Show that any matrix in $M_n(\mathbb{R})$ is similar to a matrix consisting of diagonal blocks which have one of the following forms:

$$\begin{pmatrix} r & 1 & 0 & \cdot & \cdot \\ 0 & r & 1 & 0 & \cdot \\ \multicolumn{5}{c}{\cdots\cdots\cdots\cdots} \\ 0 & \cdot & 0 & r & 1 \\ 0 & \cdot & \cdot & 0 & r \end{pmatrix}, \quad \begin{pmatrix} 0 & 1 & 1 & 0 & & & & & \\ -b & a & 0 & 1 & & & 0 & & \\ & & 0 & 1 & 1 & 0 & & & \\ & & -b & a & 0 & 1 & & & \\ & & & & \ddots & & \ddots & & \\ & & & & & & 0 & 1 & 1 & 0 \\ 0 & & & & & & -b & a & 0 & 1 \\ & & & & & & & & 0 & 1 \\ & & & & & & & & -b & a \end{pmatrix}$$

where $a^2 < 4b$.

15. Let R be a commutative ring, $R^{(n)}$ the free R-module with base $(e_1, e_2, \dots, e_n)$ and let η be the R-endomorphism of $R^{(n)}$ such that $\eta e_i = \sum a_{ij} e_j$ where $A = (a_{ij}) \in M_n(R)$. Make $R^{(n)}$ an $R[\lambda]$-module, as in the field case, so that ax, $a \in R$, is defined as in $R^{(n)}$ and $\lambda x = \eta x$. Then one has the relations

$$(\lambda - a_{11})e_1 - a_{12}e_2 - \cdots - a_{1n}e_n = 0$$
$$-a_{21}e_1 + (\lambda - a_{22})e_2 - \cdots - a_{2n}e_n = 0$$
$$\cdots\cdots\cdots\cdots\cdots\cdots\cdots\cdots\cdots\cdots\cdots$$
$$-a_{n1}e_1 - a_{n2}e_2 - \cdots + (\lambda - a_{nn})e_n = 0$$

Let A_{ij} be the cofactor of the (i, j)-entry in $\lambda 1 - A$. Multiply the foregoing relations by $A_{1i}, A_{2i}, \dots, A_{ni}$ respectively and add. Show that this gives the relations $f(\lambda)e_i = 0$ where $f(\lambda) = \det(\lambda 1 - A)$. Then $f(\eta) = 0$ and, by the isomorphism of $\operatorname{End}_R R^{(n)}$ with $M_n(R)$ (p. 174), we obtain the Hamilton-Cayley theorem for matrices with entries in $R \colon f(A) = 0$.

3.11   THE RING OF ENDOMORPHISMS OF A FINITELY GENERATED MODULE OVER A P.I.D.

An interesting problem is that of determining the $n \times n$ matrices B with entries in a field F which commute with a given matrix $A \in M_n(F)$. This translates to the geometric problem of determining the linear transformations U in an n-dimensional vector space V over F which commute with a given linear transformation T of V over F. Then U is an endomorphism of the additive group of V such that $U(ax) = a(Ux)$, $a \in F$, and $U(Tx) = T(Ux)$. Regarding V as an $F[\lambda]$-module, as before, the last condition becomes $U(\lambda x) = \lambda(Ux)$, which implies that $U(\lambda^k x) = \lambda^k(Ux)$. Then we have $U(f(\lambda)x) = f(\lambda)(Ux)$ for any polynomial $f(\lambda) \in F[\lambda]$ and so U is an endomorphism of V regarded as an $F[\lambda]$-module. Conversely, this condition is sufficient to insure that U is a linear transformation in V over F which commutes with T, since it includes the facts that U is a group endomorphism, that $U(ax) = a(Ux)$, $a \in F$, and $U(Tx) = U(\lambda x) = \lambda(Ux) = T(Ux)$.

More generally, we now consider the problem of explicitly determining the ring D' of endomorphisms (that is, $\mathrm{Hom}(M, M)$) of a finitely generated module M over a p.i.d. D. We begin with a decomposition $M = Dz_1 \oplus Dz_2 \oplus \cdots \oplus Dz_s$ where ann $z_1 \supset$ ann $z_2 \supset \cdots \supset$ ann z_s and ann $z_i = (d_i) \neq 0$ for $i \leq r$ but ann $z_i = 0$ if $i > r$. Let $\eta \in D'$ and suppose $\eta z_i \ (= \eta(z_i)) = w_i \in M$, $1 \leq i \leq s$. Then if $x \in M$, $x = \sum_1^s a_i z_i$, $a_i \in D$, and hence

$$\eta x = \eta\left(\sum a_i z_i\right) = \sum \eta(a_i z_i) = \sum a_i(\eta z_i) = \sum a_i w_i.$$

This shows (as we know already) that η is determined by its effect on the generators z_i of M. Moreover, $d_i w_i = d_i(\eta z_i) = \eta(d_i z_i) = 0$, which shows that ann $w_i \supset$ ann z_i so if ann $w_i = (g_i)$, then g_i is arbitrary if $i > r$, and $g_i | d_i$ if $i \leq r$.

Conversely, suppose that for each i we pick an element $w_i \in M$ such that ann $w_i \supset$ ann z_i. Suppose $x \in M$ and $x = \sum a_i z_i = \sum b_i z_i$ are two representations of x. Then we have $a_i - b_i \in$ ann z_i. Hence $a_i - b_i \in$ ann w_i and consequently $\sum a_i w_i = \sum b_i w_i$. This shows that $\eta : \sum a_i z_i \to \sum a_i w_i$ is a map of M into M. Direct verification shows also that $\eta \in D'$.

Our result is the following. We have a bijection $\eta \to (w_1, w_2, \ldots, w_s)$ of the ring $D' = \mathrm{Hom}(M, M)$ onto the set of s-tuples of elements of M satisfying ann $w_i \supset$ ann z_i. We now write $w_i = \sum b_{ij} z_j$, $b_{ij} \in D$, and we associate with the ordered set $(w_1, w_2, \ldots, w_s)$ the matrix

$$
(45) \qquad B = \begin{pmatrix} b_{11} & b_{12} & \cdots & b_{1s} \\ b_{21} & b_{22} & \cdots & b_{2s} \\ \cdots & \cdots & \cdots & \cdots \\ b_{s1} & b_{s2} & \cdots & b_{ss} \end{pmatrix}
$$

in the ring $M_s(D)$ of $s \times s$ matrices with entries in D. This matrix may not be uniquely determined since any b_{ij} may be replaced by b'_{ij} such that $b'_{ij} \equiv b_{ij}$ (mod d_j) if $j \leq r$. This is the only alteration which can be made without changing the w_i. The condition that ann $w_i \supset$ ann z_i is equivalent to

$$(46) \qquad\qquad d_i b_{ij} \equiv 0 \pmod{d_j}.$$

This, of course, means that there exist $c_{ij} \in D$ such that $d_i b_{ij} = c_{ij} d_j$. Hence (46) is equivalent to the following condition on the matrix B of (45): there exists a $C \in M_s(D)$ such that

$$(47) \qquad\qquad \mathrm{diag}\{d_1, d_2, \ldots, d_s\} B = C \, \mathrm{diag}\{d_1, d_2, \ldots, d_s\}.$$

The set R of matrices B satisfying (47) is a subring of $M_s(D)$. Any $B \in R$ determines an $\eta \in D'$ such that $\eta z_i = \sum b_{ij} z_j$. It is easy to verify (as in the special case of a free module treated in section 3.4) that the map ${}^t B \to \eta$ is an epimorphism of R onto D'. It is clear that $\eta = 0$ if and only if $b_{ij} \equiv 0 \pmod{d_j}$ for $B = (b_{ij})$. Hence the kernel K of our homomorphism is the set of matrices ${}^t B$ such that,

$$(48) \qquad\qquad B = Q \, \mathrm{diag}\{d_1, d_2, \ldots, d_s\}$$

Where $Q = M_s(D)$. We remark that matrices of this form automatically satisfy (47). This implies

THEOREM 3.15. *Let $M = Dz_1 \oplus Dz_2 \oplus \cdots \oplus Dz_s$ where the order ideals* ann $z_i = (d_i)$ *satisfy* ann $z_1 \supset$ ann $z_2 \supset \cdots \supset$ ann z_s. *Then the ring D' of endomorphisms of the D-module M is anti-isomorphic to R/K where R is the ring of matrices $B \in M_s(D)$ for which there exists a $C \in M_s(D)$ such that* $\mathrm{diag}\{d_1, \ldots, d_s\}$ $B = C \, \mathrm{diag}\{d_1, \ldots, d_s\}$ *and K is the ideal of matrices of the form $Q \, \mathrm{diag}\{d_1, \ldots, d_s\}$, $Q \in M_s(D)$.*

If M is a free module, all the $d_i = 0$. Then $R = M_s(D)$ and $K = 0$. In this case we have the result of section 3.4. If $s = 1$, so that M is cyclic, the condition for $B = (b)$ is trivially satisfied by the commutativity of D. Then $D' \cong D/(d)$ where $d = d_1$.

A more explicit determination of the ring of matrices R can be made if we make use of the conditions on the d_i that $d_i | d_j$ if $i \leq j \leq r$, and $d_i = 0$ if $i > r$. The conditions (46) then imply:

1. b_{ij} is arbitrary if $i \geq j$ since in this case $d_i \equiv 0 \pmod{d_j}$;
2. $b_{ij} = 0$ if $i \leq r$ and $j > r$ since in this case $d_i \neq 0$ and $d_j = 0$;
3. b_{ij} is arbitrary if $i, j > r$ since $d_i = d_j = 0$ in this case;
4. $b_{ij} \equiv 0 \pmod{d_i^{-1} d_j}$ if $i < j \leq r$.

Changing the notation slightly we see that B has the form

(49)
$$\begin{pmatrix} b_{11} & b_{12}d_1^{-1}d_2 & \cdots & b_{1r}d_1^{-1}d_r & 0 & \cdots & 0 \\ b_{21} & b_{22} & \cdots & b_{2r}d_2^{-1}d_r & 0 & \cdots & 0 \\ \cdots & \cdots & \cdots & \cdots & \cdots & \cdots & \cdots \\ b_{r1} & b_{r2} & \cdots & b_{rr} & 0 & \cdots & 0 \\ b_{r+1,1} & b_{r+1,2} & \cdots & b_{r+1,r} & b_{r+1,r+1} & \cdots & b_{r+1,s} \\ \cdots & \cdots & \cdots & \cdots & \cdots & \cdots & \cdots \\ b_{s,1} & b_{s,2} & \cdots & b_{s,r} & b_{s,r+1} & \cdots & b_{ss} \end{pmatrix}$$

Here the upper right-hand corner consists of 0's, all the indicated b_{ij} are arbitrary, and the (i, j)-entry for $i < j \le r$ is $b_{ij}d_i^{-1}d_j$. The conditions that the matrix is in K are that the $b_{ij} = 0$ if $j > r$, that b_{ij} is divisible by d_j if $i \ge j$ and $j \le r$, and that b_{ij} is divisible by d_i if $i < j \le r$. If the module is a torsion module, $r = s$ and (49) reduces to the block of matrix in the upper left-hand corner.

We now specialize $M = V$, where V is the $F[\lambda]$-module determined by a linear transformation T in a finite dimensional vector space V over F. This is a torsion module. Any b_{ij}, $i \ge j$, can be replaced by b'_{ij} in the same coset mod d_j. Hence we may assume deg $b_{ij} < n_j = \deg d_j$ if $i \ge j$. Similarly, we may assume deg $b_{ij} < n_i$ if $i < j$. Matrices $B \in R$ satisfying these conditions will be called *normalized*. It is clear that the map $B \to \eta$ restricted to normalized matrices of R is a bijection into D'. There is a natural way of regarding D' and R as vector spaces over F. For R we obtain a module structure over F simply by multiplying all the entries of $B \in R$ by $a \in F$. For D' we define $a\eta$, $a \in F$, $\eta \in D'$, by $(a\eta)x = a(\eta x) = \eta(ax)$ (cf. exercise 5, p. 175). Using these vector space structures it is immediate that the set S of normalized matrices contained in R is a subspace and $B \to \eta$ is an F-linear isomorphism of S into D'. We are interested in calculating the dimensionality of D' over F, in matrix terms, the dimensionality over F of the vector space of matrices which commute with a given matrix. The isomorphism just established gives us a way of doing this, namely, we may calculate dim S. Let S_{ij}, $1 \le i, j \le s$, denote the subspaces of S of normalized matrices having 0 entries in all places except the (i, j)-position. Then dim $S_{ij} = n_j$ if $i \ge j$, and dim $S_{ij} = n_i$ if $i < j$. Since S is the direct sum of the subspaces S_{ij} we have

$$\dim S = \sum_{j=1}^{s} (s - j + 1)n_j + \sum_{i=1}^{s-1} (s - i)n_i$$

$$= \sum_{j=1}^{s} (2s - 2j + 1)n_j.$$

We can state this result in terms of matrices in the following way:

THEOREM 3.16. (Frobenius.) *Let* $A \in M_n(F)$, *F a field, and let* $d_1(\lambda)$, $d_2(\lambda)$, $\ldots$, $d_s(\lambda)$ *be the invariant factors* $\neq 1$ *of* $\lambda 1 - A$. *Let* $n_i = \deg d_i(\lambda)$. *Then the dimensionality of the vector space over F of matrices commutative with A is given by the formula*

(50)
$$N = \sum_{j=1}^{s} (2s - 2j + 1)n_j.$$

Of course, this can also be stated in terms of linear transformations. In this form it gives the following

COROLLARY. *A linear transformation T is cyclic (that is the corresponding $F[\lambda]$-module is cyclic) if and only if the only linear transformations commuting with T are polynomials in T.*

Proof. T is cyclic if and only if $s = 1$. We also know that $d_s(\lambda)$ is the minimum polynomial $m(\lambda)$ of T and hence n_s is the dimensionality over F of the ring $F[T]$ of polynomials in T with coefficients in F (see exercise 1, p. 133). If $s = 1$ then (50) gives $N = n_1 = \dim F[T]$. Hence the space of linear transformations commuting with T, which, of course, contains $F[T]$, coincides with $F[T]$. If $s > 1$, (50) implies that $N > \sum_1^s n_j > n_s$. Hence there exist linear transformations commuting with T which are not polynomials in T. $\quad\square$

EXAMPLE

Let $F = \mathbb{Q}$ and

$$A = \begin{pmatrix} 1 & 0 & 0 \\ 0 & 0 & 1 \\ 0 & -1 & 2 \end{pmatrix}.$$

If T is the corresponding linear transformation and the vector space is $\mathbb{Q}[\lambda]$-module via T then $V = \mathbb{Q}[\lambda]f_1 \oplus \mathbb{Q}[\lambda]f_2$. The invariant factors are $\lambda - 1$ and $(\lambda - 1)^2$. The normalized matrices of R have the form

(51)
$$\begin{pmatrix} b_{11} & b_{12}(\lambda - 1) \\ b_{21} & b_{22} + b'_{22}\lambda \end{pmatrix}, \qquad b_{ij} \in \mathbb{Q}.$$

Since $\lambda f_1 = f_1$, $\lambda^2 f_2 = (2\lambda - 1)f_2 = -f_2 + 2(\lambda f_2)$, the linear transformation U corresponding to (51) satisfies

$$Uf_1 = b_{11}f_1 - b_{12}f_2 + b_{12}(\lambda f_2)$$
$$Uf_2 = b_{21}f_1 + b_{22}f_2 + b'_{22}(\lambda f_2)$$
$$U(\lambda f_2) = b_{21}f_1 - b'_{22}f_2 + (b_{22} + 2b'_{22})(\lambda f_2).$$

Accordingly, the general form of a matrix which commutes with A is

$$\begin{pmatrix} b_{11} & -b_{12} & b_{12} \\ b_{21} & b_{22} & b'_{22} \\ b_{21} & -b'_{22} & b_{22} + 2b'_{22} \end{pmatrix}.$$

We return to the general case of a finitely generated D-module M for a p.i.d. D, where we have $M = Dz_1 \oplus Dz_2 \oplus \cdots \oplus Dz_s$ as before. Since D is commutative it is clear that the ring of endomorphism D' of M includes all the maps $x \to ax$, $a \in D$. It is clear also that these are contained in the center of D'. We shall now prove

THEOREM 3.17. *The center of* $D' = \mathrm{Hom}(M, M)$ *is the set of maps* $x \to ax$, $a \in D$.

Proof. Our determination of D' shows that for any i, $1 \le i \le s$, there exist an endomorphism ε_{is} such that $\varepsilon_{is} z_s = z_i$, $\varepsilon_{is} z_j = 0$ if $j \ne s$. Now let γ be in the center of D'. Then $\gamma z_s = \gamma \varepsilon_{ss} z_s = \varepsilon_{ss} \gamma z_s = \varepsilon_{ss}(\sum a_i z_i)(a_i \in D) = \sum a_i \varepsilon_{ss} z_i = a_s z_s$. Also $\gamma z_i = \gamma \varepsilon_{is} z_s = \varepsilon_{is} \gamma z_s = \varepsilon_{is}(\sum a_j z_j) = \sum a_j \varepsilon_{is} z_j = a_s z_i$. It follows that γ is the map $x \to a_s x$. $\square$

Specializing Theorem 3.17 to the case of the module determined by a linear tranformation, we obtain

COROLLARY 1. *If U is a linear transformation in a finite dimensional vector space which commutes with every linear tranformation commuting with a given linear transformation T, then U is a polynomial in T.*

An immediate consequence of this corollary obtained by taking $T = 1$ is

COROLLARY 2. *The center of the ring of linear transformations of a finite dimensional vector space over a field F is the set of scalar multiplications $x \to ax$, $a \in F$.*

EXERCISES

1. Let G be a finite abelian group which is a direct sum of cyclic groups of orders $n_1, n_2, \ldots, n_s$ where $n_i | n_j$ if $i \le j$. Show that the number of endomorphisms of G is

$$N = \prod_{j=1}^{s} n_j^{2s - 2j + 1}.$$

2. Determine the matrices in $M_5(\mathbb{Q})$ commuting with

$$\begin{pmatrix} 0 & 0 & 0 & 0 & 0 \\ 1 & 0 & 0 & 0 & 0 \\ 0 & 0 & 0 & 0 & 0 \\ 0 & 0 & 1 & 0 & 0 \\ 0 & 0 & 0 & 1 & 0 \end{pmatrix}.$$

3. Determine the matrices in $M_4(\mathbb{Q})$ commuting with

$$\begin{pmatrix} 1 & 0 & 0 & 0 \\ 0 & 0 & 1 & 0 \\ 0 & 0 & 0 & 1 \\ 0 & 1 & -3 & 3 \end{pmatrix}.$$

4. Prove that a linear transformation T in a finite dimensional vector space over a field is cyclic if and only if the ring of linear transformations commuting with T is a commutative ring.

5. Prove the following extension of Theorem 3.17. The only endomorphisms of M which commute with every idempotent element of D' are the mappings $x \to ax$, $a \in D$.

4

Galois Theory of Equations

The main objective of this chapter is the treatment of two classical problems: solvability of polynomial equations by radicals and constructions with straight-edge and compass. We shall first indicate briefly their history.

In elementary algebra one derives the formula

$$x = \frac{-b \pm \sqrt{b^2 - 4ac}}{2a}$$

for solving the quadratic equation $ax^2 + bx + c = 0$. In essence this was known to the Babylonians. During the period of the Italian Renaissance a considerable effort was directed toward generalizing this to equations of higher degree and this culminated in one of the great achievements of Renaissance mathematics: formulas for the roots of cubic and quartic equations. The first was due to Scipione del Ferro, who was a professor at the University of Bologna from 1496 to 1526. The exact date of his discovery is unknown, however we do know that some time prior to 1541, Niccolo Tartaglia, perhaps aware of the existence of del Ferro's solution, was able to discover it for himself. Tartaglia's solution was published by Geronimo Cardano in *Ars Magna* (1545) and is generally known as "Cardan's formulas" for the solution of cubic equations.[1] It is easy (for us) to see that the solution of cubic equations $x^3 + ax^2 + bx + c = 0$ can be reduced to the "reduced" case of equations of the form $x^3 + px + q = 0$ (by replacing x by $x - \frac{1}{3}a$). Let x_1, x_2, x_3 denote the roots of the reduced equation and put $\delta = -4p^3 - 27q^2$, $\zeta = -\frac{1}{2} - \frac{1}{2}\sqrt{-3}$, $y_1 = x_1 + \zeta^2 x_2 + \zeta x_3$, $y_2 = x_1 + \zeta x_2 + \zeta^2 x_3$. Then Cardan's formulas are

$$y_1 = \sqrt[3]{-\tfrac{27}{2}q + \tfrac{3}{2}\sqrt{-3\delta}}$$
$$y_2 = \sqrt[3]{-\tfrac{27}{2}q - \tfrac{3}{2}\sqrt{-3\delta}}$$

for suitable determinations of the cube roots (see pp. 264–266). The form of the reduced equation implies that $x_1 + x_2 + x_3 = 0$. Hence the determination of its roots x_i is reduced to solving the three linear equations $x_1 + x_2 + x_3 = 0$, $x_1 + \zeta^2 x_2 + \zeta x_3 = y_1$, $x_1 + \zeta x_2 + \zeta^2 x_3 = y_2$.

A general method for solving quartic equations, which was also published by Cardano in *Ars Magna*, is attributed to Cardano's assistant, Ludovico Ferarri. We shall indicate this method later, and note here only that, as in the case of cubics, the solutions are given in terms of root extractions and rational operations performed on the coefficients of the given equation.

From the middle of the sixteenth century to the beginning of the nineteenth century a number of attempts were made by some of the greatest mathematicians of the period (e.g., Euler and Lagrange) to obtain similar results for quintic equations. Lagrange did considerably more than the other would-be solvers of quintic equations: namely, he gave an incisive analysis of the existing solutions of cubics and quartics and showed that the reason these could be solved by radicals was that one could reduce their solution to that of "resolvent" equations of lower degrees. On the other hand, he found that the application of the same method to a quintic led to a resolvent of degree six. This might have suggested strongly that equations of higher degree than the fourth could not generally be solved by radicals. Nevertheless, it was a startling discovery when this was indeed found to be the case. This was established independently by A. Ruffini (published in 1813) and by N. H. Abel (published in 1827). Their result (usually attributed to Abel) states that the "general" equation of nth degree, that is, the equation $x^n + t_1 x^{n-1} + \cdots + t_n = 0$ with indeterminate coefficients t_i is not solvable by radicals. The proofs of Ruffini and of Abel are somewhat obscure and perhaps not complete in all details. For us they are interesting only as history since they were soon superseded by the crowning achievement of this line of research: Galois' discoveries in the theory of equations. Galois

---

[1] For a more detailed discussion of the history of the theory of algebraic equations we refer the reader to C. A. Boyer, *A History of Mathematics*, New York, Wiley, 1968, or to E. T. Bell, *The Development of Mathematics*, New York, McGraw-Hill, 1940.

obtained his results while he was still in his teens: he was killed in a duel in 1832 just before he was twenty-one. Galois' work not only provided a proof of the Ruffini-Abel theorem but it gave a criterion for solvability by radicals of any equation $x^n + a_1 x^{n-1} + \cdots = 0$ (not just the "general" one). Moreover, the main result of Galois' discoveries, which showed that there is a $1-1$ correspondence between the set of subfields of a certain type of field extension and the subgroups of a finite group—the Galois group—has become a central result in all of algebra, whose importance has transcended by far that of the original problem which led to it.[2] Galois' theory has been considerably simplified and refined—mainly by the introduction of more abstract ideas—during the century which followed its publication in 1846, some fifteen years after his death.

The second main problem on which we shall focus our attention had its origin in Greek mathematics. The Greeks were unable to decide whether or not certain geometric constructions were possible using only a straight-edge (unmarked ruler) and compass. The most notable of these were: (1) trisection of any angle; (2) duplication of the cube, that is, construction of the side of a cube whose volume is twice that of the volume of a given cube; (3) construction of a regular heptagon (= regular polygon of seven sides); (4) squaring the circle, that is, construction of a square whose area is that of a given circle. Any problem on straight-edge-compass construction can be formulated as an algebraic problem on fields. Once this is done it is easy to see that the first three of the foregoing problems have negative answers. This can be seen by applying the basic dimensionality formula for fields (Theorem 4.2). The impossibility of squaring the circle follows from the fact, first established by F. Lindemann in 1882, that π is a transcendental number, that is, is not algebraic over $\mathbb{Q}$. The general problem of determining the integers n such that the regular n-gon can be constructed (with straight-edge and compass) was solved by Gauss in his *Disquisitiones Arithmeticae* (1801). A consequence of his results is that the constructions are possible if $n = 17$, 257, or 65537. Gauss' first recorded discovery in mathematics was a method for constructing a regular polygon of 17 sides. This had eluded mathematicians from the time of the Greeks until Gauss— a period of about two thousand years. Gauss' results were obtained by elementary but somewhat lengthy calculations involving the roots of unity. As we shall see, Galois' theory makes it possible to get these rather quickly without calculations.

---

[2] One of the greatest mathematicians of this century, Hermann Weyl, has given the following evaluation of Galois' contribution in his book *Symmetry*, Princeton University Press, 1952, p. 138. "Galois' ideas, which for several decades remained a book with seven seals but later exerted a more and more profound influence upon the whole development of mathematics, are contained in a farewell letter written to a friend on the eve of his death, which he met in a silly duel at the age of twenty-one. This letter, if judged by the novelty and profundity of ideas it contains, is perhaps the most substantial piece of writing in the whole literature of mankind."

Besides the results indicated, we shall be interested in this chapter in some by-products of the Galois theory, one of these being the study of finite fields, which was also initiated by Galois. We shall introduce also some other basic field concepts: norms, traces, primitive elements, and normal bases.

4.1 PRELIMINARY RESULTS, SOME OLD, SOME NEW

We have defined the prime ring of a ring R as the smallest subring of R and we saw that this is the set $\mathbb{Z}1$ of integral multiples of 1 (section 2.7). Moreover, either $\mathbb{Z}1 \cong \mathbb{Z}$ or $\mathbb{Z}1 \cong \mathbb{Z}/(k)$, $(k \neq 0)$. In the first case, R has characteristic 0 and in the second it has characteristic k. If R is a domain, $k = p$ a prime. Now let $R = F$ a field. Then we define the *prime field* of F to be the smallest subfield of F. If F has characteristic $p \neq 0$, the prime subring is a subfield since it is isomorphic to $\mathbb{Z}/(p)$. Hence in this case the prime subring and prime field of F coincide. If F has characteristic 0, we have the monomorphism $m \to m1$ of $\mathbb{Z}$ into the prime field of F, and this can be extended to a monomorphism of $\mathbb{Q}$ into the prime field. It follows that the prime field is isomorphic to $\mathbb{Q}$. In this sense we can say that any field contains either the ring $\mathbb{Z}/(p)$ for some prime p ($=$ the characteristic of the field) or else it contains the field $\mathbb{Q}$ of rational numbers.

Let E be an extension field of the field F (E is a field containing F as subfield). If S is a subset of E, we recall that $F[S]$ denotes the subring of E generated by F and S or, as we shall now say, *the subring of E/F generated by S*. We shall now use the notation $F(S)$ for the *subfield of E/F generated by S* meaning, of course, the subfield of E generated by F and S. As in the ring case, it is immediate that if T is a second subset of E then $F(S)(T) = F(S \cup T)$ (section 2.10). If u is an element of E then we write $F(u)$ for $F(\{u\})$ and, more generally, for a finite set $\{u_1, u_2, \ldots, u_n\}$ we put $F(u_1, u_2, \ldots, u_n) = F(\{u_1, u_2, \ldots, u_n\})$. What does $F(u)$ look like? First, we recall that if x is an indeterminate, then we have the homomorphism $g(x) \to g(u)$ of the polynomial ring $F[x]$ into E, which is the identity of F and sends $x \to u$ (Theorem 2.10, p. 122). If the kernel is 0, then $F[u] \cong F[x]$. Otherwise, we have a monic polynomial $f(x)$ of positive degree such that the kernel is the principal ideal $(f(x))$, and then $F[u] \cong F[x]/(f(x))$. The polynomial $f(x)$ is prime (since, otherwise $F[x]/(f(x))$ is not a domain). Then $F[x]/(f(x))$ is a field (Theorem 2.16, p. 131). Hence, it is clear that in this case, $F(u) = F[u]$. In the other case: $F[x] \cong F[u]$, the homomorphism $g(x) \to g(u)$ is a monomorphism and this has a unique extension to a monomorphism of the field of fractions $F(x)$ of $F[x]$. Then $F(u) \supsetneq F[u]$, $F(u) \cong F(x)$ and $F(u)$ consists of the set of elements $g(u)h(u)^{-1}$ where $g(x), h(x) \in F[x]$ and $h(x) \neq 0$. In this case also, u is transcendental, whereas if $F(u) \cong F[x]/(f(x))$,

$f(x)$ of positive degree, then u is algebraic and, if $f(x)$ is monic, then this is the minimum polynomial of the element u. In any case, if $E = F(u)$, then we say that E is a *simple (field) extension* of F and we call u a *primitive element* ($=$ field generator of E/F).

In studying an extension field E relative to a subfield F it is useful to consider E as vector space (or module) over F. Here the abelian group structure of E is that given by the addition composition and the module composition ay, $a \in F$, $y \in E$, is the product in E. The extensions we shall encounter most frequently in this chapter are finite dimensional extensions over the base field F. We denote the dimensionality as $[E:F]$. We shall show first that an element $u \in E$ is algebraic over F if and only if $[F(u):F] < \infty$ and in this case $[F(u):F]$ is the degree of the minimum polynomial of u over F. We shall call this number the *degree of u over F*. Let u be algebraic, $f(x) = x^n + a_1 x^{n-1} + \cdots + a_n \in F[x]$ the minimum polynomial of u over F. We have $F(u) = F[u]$ and if $g(x) \in F[x]$, $g(x) = f(x)q(x) + r(x)$ where $\deg r(x) < \deg f(x) = n$. Then $g(u) = 0q(u) + r(u)$, which shows that any element of $F[u]$ has the form

$$r(u) = b_0 + b_1 u + \cdots + b_{n-1} u^{n-1}, \qquad b_i \in F,$$

and since $f(x)$ is the polynomial of least degree such that $f(u) = 0$, the only relation of the form $b_0 + b_1 u + \cdots + b_{n-1} u^{n-1} = 0$ which can hold for $b_i \in F$ is the one with all $b_i = 0$. Thus

$$(1, u, \ldots, u^{n-1})$$

is a base for $F(u)/F$. Hence this extension is n dimensional where $n = \deg f(x)$. On the other hand, if u is transcendental the elements $1, u, u^2, \ldots$ of $F(u)$ are linearly independent over F, which implies that $F(u)$ is not finite dimensional over F.

We state a part of our results as

THEOREM 4.1.   *Let u be an element of an extension field E of a field F. Then u is algebraic over F if and only if $F(u)$ is finite dimensional over F. In this case $F(u)$ coincides with the subring $F[u]$ generated by F and u.*

We now suppose we have a two-storied extension of F, that is, we have $F \subset E \subset K$ where K is a field and E and F are subfields. Then we can regard K as vector space over E and over F, and E as vector space over F. We denote these spaces as K/E, K/F, and E/F respectively. We then have the following important relation on dimensionalities.

THEOREM 4.2.   *If $K \supset E \supset F$ are fields then $[K:F]$ is finite if and only if $[K:E]$ and $[E:F]$ are finite. In this case we have the dimensionality relation*

(1) $$[K:F] = [K:E][E:F].$$

Proof. If $[K:F] < \infty$ then $[E:F]$ is finite since E is a subspace of K/F. If $(u_1, u_2, \ldots, u_n)$ is a base for K/F then clearly every element of K is a linear combination of the u_i with coefficients in F and a fortiori with coefficients in E. Hence, by a standard result of linear algebra, we can extract a base for K/E out of the set $\{u_1, u_2, \ldots, u_n\}$. Thus $[K:E] < \infty$. Conversely, suppose $[K:E]$ and $[E:F]$ are finite and $(v_1, \ldots, v_m)$ is a base for K/E, $(w_1, \ldots, w_r)$ a base for E/F. If z is any element of K we have $z = \sum_1^m a_i v_i$ for suitable $a_i \in E$, and $a_i = \sum_{j=1}^r b_{ij} w_j$ for suitable $b_{ij} \in F$. Then $z = \sum_{i,j} b_{ij} w_j v_i$, so every element of K is an F-linear combination of the mr elements $w_j v_i$. Now suppose $\sum_{i,j} b_{ij} w_j v_i = 0$ for $b_{ij} \in F$. Then $\sum a_i v_i = 0$ for $a_i = \sum_j b_{ij} w_j$. Since the v_i form a base for K/E this implies that every $a_i = 0$. Since the w_j form a base for E/F, $a_i = \sum b_{ij} w_j = 0$ implies every $b_{ij} = 0$. Hence we have proved that the mr elements $w_j v_i$ are F-independent, so they constitute a base for K/F. Thus $[K:F] = mr = [K:E][E:F] < \infty$. $\square$

An immediate consequence of the foregoing result is that if $[K:F] < \infty$, then the dimensionality of any subfield E/F is a divisor of the dimensionality of K/F. In particular, if $[K:F]$ is a prime, then the only subfields of K/F are K and F.

EXERCISES

1. Let $E = \mathbb{Q}(u)$ where $u^3 - u^2 + u + 2 = 0$. Express $(u^2 + u + 1)(u^2 - u)$ and $(u - 1)^{-1}$ in the form $au^2 + bu + c$ where $a, b, c \in \mathbb{Q}$.

2. Determine $[\mathbb{Q}(\sqrt{2}, \sqrt{3}):\mathbb{Q}]$.

3. Let p be a prime and let $v \in \mathbb{C}$ satisfy $v \neq 1$, $v^p = 1$ (e.g., $v = \cos 2\pi/p + i \sin 2\pi/p$). Show that $[\mathbb{Q}(v):\mathbb{Q}] = p - 1$. (*Hint:* Use exercise 3, p. 154.)

4. Let $w = \cos \pi/6 + i \sin \pi/6$ (in $\mathbb{C}$). Note that $w^{12} = 1$ but $w^r \neq 1$ if $1 \leq r < 12$ (so w is a generator of the cyclic group of 12th roots of 1). Show that $[\mathbb{Q}(w):\mathbb{Q}] = 4$ and determine the minimum polynomial of w over $\mathbb{Q}$.

5. Let $E = F(u)$ where u is algebraic of odd degree ($=$degree of the minimum polynomial of u). Show that $E = F(u^2)$.

6. Let E_i, $i = 1, 2$, be a subfield of K/F such that $[E_i:F]$ is finite. Show that if E is the subfield of K generated by E_1 and E_2 then $[E:F] \leq [E_1:F][E_2:F]$.

7. Let E be an extension field of F which is *algebraic* over F in the sense that every element of E is algebraic over F. Show that any subring of E/F is a subfield. Hence prove that any subring of a finite dimensional extension field E/F is a subfield.

8. Let $E = F(u)$, u transcendental, and let $K \neq F$ be a subfield of E/F. Show that u is algebraic over K.

9. Let E be an extension field of the field F such that (i) $[E:F] < \infty$, (ii) for any two subfields E_1 and E_2 containing F, either $E_1 \supset E_2$ or $E_2 \supset E_1$. Show that E has a primitive element over F.

4.2 CONSTRUCTION WITH STRAIGHT-EDGE AND COMPASS

The problem of Euclidean construction, that is, construction with straight-edge and compass, can be formulated in the following way. Given a finite set of points $S = \{P_1, P_2, \ldots, P_n\}$ in a plane ω, define a subset S_m, $m = 1, 2, \ldots$, of ω inductively by $S_1 = S$, and S_{r+1} is the union of S_r and (1) the set of points of intersections of pairs of lines connecting distinct points of S_r, (2) the set of points of intersections of the lines specified in (1) with all circles having centers in S_r and radii equal to segments having end points in S_r, (3) the set of points of intersections of pairs of circles defined in (2). Let $C(P_1, P_2, \ldots, P_n) = \bigcup_1^\infty S_i$. Then we shall say that a point P of ω *can be constructed* (*by straight-edge and compass*) *from* $P_1, P_2, \ldots, P_n$ if $P \in C(P_1, P_2, \ldots, P_n)$. Otherwise P cannot be constructed from the P_i.

How does this correspond to constructibility as defined in Euclidean geometry? The given elements in a construction in Euclidean geometry are points, lines, circles, and angles—a finite number of each. Now a line is determined by two of its points, a circle by its center and a point on the circle, and an angle by its vertex and two points on the two sides of the angle equidistant from the vertex. Hence, making these replacements, we may assume that we are given a finite set $S_1 = \{P_1, P_2, \ldots, P_n\}$ in the plane ω. The points of the successive sets $S_2, S_3, \ldots$, which we defined, can certainly be obtained from S_1 by straight-edge–compass construction à la Euclid. We remark also that in Euclidean constructions one sometimes encounters an instruction to use an "arbitrary" point or length restricted only by a condition that the point is contained in a certain region or that the length satisfies a certain inequality. Thus one is instructed to choose points in designated (non-vacuous) open subsets of the plane. We shall see in a moment that if the given set S_1 has at least two distinct points, then the set $C(P_1, P_2, \ldots, P_n)$ we defined is dense in the plane. Hence any instruction involving the choice of a point in a non-vacuous open subset of the plane can be fulfilled by choosing some point in $C(P_1, P_2, \ldots, P_n)$. Consequently, our definition of constructible points—which has the advantage of being precise—is equivalent to what seems to have been intended in Euclidean geometry.

As an example, we consider the problem of trisecting an angle of $60°$. Here we are given the points $P_1 = (0, 0)$ (the vertex), $P_2 = (1, 0)$ and $P_3 = (\cos 60°, \sin 60°) = (\frac{1}{2}, \frac{1}{2}\sqrt{3})$. Is the point $P = (\cos 20°, \sin 20°)$ contained in $C(P_1, P_2, P_3)$? An angle of $60°$ can be trisected using only a straight-edge and compass if and only if this question has an affirmative answer.

We shall now formulate our definition algebraically. We assume $n \geq 2$, since, otherwise, $C(P_1, P_2, \ldots, P_n) = \{P_1\}$. We choose a Cartesian coordinate system so that $P_1 = (0, 0)$, the origin, and $P_2 = (1, 0)$. We associate with the point $P = (x, y)$ the complex number $x + iy$. In this way the plane is identified with the field $\mathbb{C}$ of complex numbers. The given set $\{P_1, P_2, \ldots, P_n\}$ is identified with a set of complex numbers $\{z_1, z_2, \ldots, z_n\}$ such that $z_1 = 0$, $z_2 = 1$. What is the set $C(z_1, z_2, \ldots, z_n)$ of complex numbers corresponding to the set of points $C(P_1, P_2, \ldots, P_n)$? It is natural to call this set *the set of complex numbers which are constructible (by straight-edge and compass) from $z_1, z_2, \ldots, z_n$.* We shall now obtain the following characterization: $C(z_1, z_2, \ldots, z_n)$ is the smallest sub-field of the complex field containing the z_i and closed under square roots and conjugation—that is, containing every z such that z^2 is in the set and containing $\bar{z} = x - iy$ if $z = x + iy$, x, y real, is in the set. By "smallest" we mean, as usual, that $C(z_1, z_2, \ldots, z_n)$ has the indicated closure properties and is contained in every subset of $\mathbb{C}$ having these closure properties.

Suppose z and $z' \in C(z_1, z_2, \ldots, z_n)$. Then $z + z'$ can be constructed by the usual parallelogram method of forming the sum of two vectors:

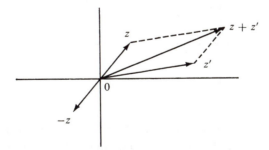

Thus $z + z'$ is obtained as (the obvious) one of the two points of intersection of the circle with center at z and radius $|z'|$ (the length of $0z'$) with the circle centered at z' with radius $|z|$. Also it is clear that $-z \in C(z_1, z_2, \ldots, z_n)$. Hence $C(z_1, z_2, \ldots, z_n)$ is a subgroup of the additive group of $\mathbb{C}$. To see that $C(z_1, z_2, \ldots, z_n)$ is closed under multiplication, inverses, and square roots we use the polar form of $z: z = re^{i\theta}$, where the absolute value $r = (x^2 + y^2)^{1/2}$ if

$z = x + iy$, and θ, the amplitude, is the angle from the x-axis to the line $0z$. If $z' = r'e^{i\theta'}$ then $zz' = rr'e^{i(\theta + \theta')}$ has absolute value rr' equal to the product of the absolute values of z and z', and its amplitude is the sum of the two given amplitudes. It is easy to see that we can construct the ray having amplitude $\theta + \theta'$, and the following figure indicates a construction of rr'.

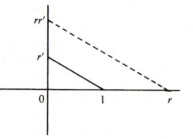

Here the broken line is parallel to $1r'$ and can be constructed by ruler and compass in the same way that the parallels in the first figure were constructed. A reversal of the foregoing construction in which r and r' are placed on the y-axis gives the point r/r' on the x-axis. It follows that $z(z')^{-1}$ can be constructed (if $z' \neq 0$). We see easily (as is well known) that any angle can be bisected with straight-edge and compass. The following diagram indicates how $\sqrt{r}$ can be constructed.

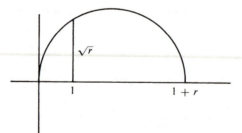

This implies that $z^{1/2} \in C(z_1, z_2, \ldots, z_n)$ if $z \in C(z_1, \ldots, z_n)$. It is clear also that $\bar{z} \in C(z_1, z_2, \ldots, z_n)$ since this point can be obtained by dropping a perpendicular from z to the x-axis (line P_1P_2) and locating $\bar{z}$ as the mirror image of z in the x-axis. This completes the proof that $C(z_1, z_2, \ldots, z_n)$ is a subfield of $\mathbb{C}$ closed under square roots and conjugation.

Next let C' be any subfield of $\mathbb{C}$ containing the z_i, $1 \leq i \leq n$, and closed under square roots and conjugation. If we take into account the inductive definition of $C(z_1, z_2, \ldots, z_n)$ as $\bigcup S_i$ we see that in order to prove that $C' \supset C(z_1, \ldots, z_n)$ it suffices to show that the intersection of any two lines determined by points of

C', or of such a line with a circle having center a point of C' and radius a segment joining two points of C', or of two such circles, all belong to C'. We note first that the fact that C' is closed under conjugation and contains $i = \sqrt{-1}$ implies that if $z = x + iy \in C'$, x, y real, then x, $y \in C'$. It follows from this that the equation of any line through distinct points in C' has the form $ax + by + c = 0$ where a, b, c are real numbers in C' and the equation of a circle with center a point of C' and radius equal to the length of a segment with end points in C' is of the form $x^2 + y^2 + dx + ey + f = 0$ where d, e, f are real numbers in C'. Now, the coordinates of the point of intersection of non-parallel lines $ax + by + c = 0$ and $a'x + b'y + c' = 0$ can be obtained by Cramer's rule as quotients of certain determinants obtained from a, b, c, a', b', c'. Hence the point of intersection of two lines whose coefficients are real numbers in C' has coordinates that are real numbers in C'. The abscissas of the points of intersection of $y = mx + b$ and $x^2 + y^2 + dx + ey + f = 0$ are obtained by solving $x^2 + (mx + b)^2 + dx + e(mx + b) + f = 0$. Using the quadratic formula we see that the solutions are real and in C' if m, b, d, e, and f are real in C' and the line and circle intersect. We handle similarly the case of a line with equation $x = a$ and a circle $x^2 + y^2 + dx + ey + f = 0$. Finally, we note that the points of intersection of the two circles $x^2 + y^2 + dx + ey + f = 0$ and $x^2 + y^2 + d'x + e'y + f' = 0$ are the same as the points of intersection of $x^2 + y^2 + dx + ey + f = 0$ with the line $(d - d')x + (e - e')y + f - f' = 0$. It follows that the points of intersection of lines and circles having real coefficients in C' have coordinates (p, q) expressible rationally or with square roots in terms of the coefficients. Hence $p + qi \in C'$. This completes the proof of our assertion that $C(z_1, z_2, \ldots, z_n)$ is the smallest subfield of $\mathbb{C}$ containing the z_i and closed under conjugation and square roots.

It should be noted that $C(z_1, z_2, \ldots, z_n)$ contains all complex numbers of the form $p + iq$ where p and q are rational, and this subset is dense in $\mathbb{C}$ in the sense that any circular region contains a point of the set. We can now deduce from the characterization of $C(z_1, \ldots, z_n)$ the following

Criterion A.   Let $z_1, z_2, \ldots, z_n \in \mathbb{C}$ and put $F = \mathbb{Q}(z_1, \ldots, z_n, \bar{z}_1, \ldots, \bar{z}_n)$. Then a complex number z is constructible from $z_1, z_2, \ldots, z_n$ if and only if z is contained in a subfield of $\mathbb{C}$ of the form $F(u_1, u_2, \ldots, u_r)$ where $u_1{}^2 \in F$ and every $u_i{}^2 \in F(u_1, \ldots, u_{i-1})$.

A field of the form $F(u_1, u_2, \ldots, u_r)$ where $u_1{}^2 \in F$, $u_i{}^2 \in F(u_1, \ldots, u_{i-1})$ will be called a *square root tower* over F.

Proof of the criterion.   Since $C(z_1, \ldots, z_n)$ is closed under square roots and conjugation it is clear that $C(z_1, \ldots, z_n)$ contains F and every square root tower over F. Hence $C(z_1, \ldots, z_n) \supset C'$ where C' is the set of complex numbers satisfying the stated condition. Let $z, z' \in C'$. Then z is contained in a square root tower $F(u_1, \ldots, u_r)$ and z' is contained in a square root tower $F(u'_1, \ldots, u'_s)$. Then $z + z'$, zz', and z^{-1} if $z \neq 0$, are contained in the square root tower $F(u_1, \ldots, u_r, u'_1, \ldots, u'_s)$. Thus C' is a subfield of $\mathbb{C}$. Clearly C' is closed under square roots, and since $\bar{F} = F$, it is clear that $\overline{F(u_1, \ldots, u_r)} = F(\bar{u}_1, \ldots, \bar{u}_r)$, which implies that C' is closed under conjugation. Hence

$$C' \supset C(z_1, z_2, \ldots, z_n).$$

Thus $C' = C(z_1, z_2, \ldots, z_n)$, which establishes the criterion. $\square$

For the present applications the following easy consequence of the foregoing criterion will be adequate.

COROLLARY. *Let* $F = \mathbb{Q}(z_1, \ldots, z_n, \bar{z}_1, \ldots, \bar{z}_n)$. *Then any complex number* z *which is constructible from* $z_1, \ldots, z_n$ *is algebraic of degree a power of two over* F.

Proof. If L is an extension field of the form $K(u)$ where $u^2 = a \in K$ then it is immediate that either $L = K$ or $[L:K] = 2$. Hence, by iterated application of the dimensionality formula for fields (Theorem 4.2), we see that every square root tower over F has dimensionality a power of two over F. It follows (also by Theorem 4.2) that if z is contained in such an extension then $[F(z):F] = 2^s$ for some $s \geq 0$. $\square$

In many problems on constructibility we are given just two points or, equivalently, a segment. By choosing an appropriate coordinate system, we may take these to be 0 and 1. Then $F = \mathbb{Q}$. In this case we shall call $C \equiv C(z_1, z_2)$ the field of (*Euclidean*) *constructible complex numbers*. The corollary shows that such numbers are algebraic over $\mathbb{Q}$ of degree a power of two.

We shall now use the foregoing corollary to dispose of three of the four classical construction problems stated above. The fourth, on the problem of squaring the circle, will be treated in section 4.12 where we shall prove that π is transcendental.

Trisection of angles. Not every angle can be trisected with straight-edge and compass. In particular, $60°$ cannot be trisected. We have seen that the construction of an angle of $20°$ from one of $60°$ requires the constructibility of the point $P = (\cos 20°, \sin 20°)$ from $P_1 = (0, 0)$, $P_2 = (1, 0)$ and $P_3 = (\cos 60°, \sin 60°) =$

$(\frac{1}{2}, \frac{1}{2}\sqrt{3})$. Then the point $Q = (\cos 20°, 0)$, the foot of the perpendicular from P to $P_1 P_2$, would be constructible. It is easier to apply the criterion and corollary to this. In the present case we have to consider the complex numbers $z_1 = 0$, $z_2 = 1$, $z_3 = \frac{1}{2} + \frac{1}{2}\sqrt{3}i$ and the field $F = \mathbb{Q}(z_1, z_2, z_3, \bar{z}_1, \bar{z}_2, \bar{z}_3) = \mathbb{Q}(\sqrt{-3})$. Applying the corollary we see that success in the trisection of $60°$ requires that $\cos 20°$ has degree a power of two over F, and hence over $\mathbb{Q}$ since $[\mathbb{Q}(\sqrt{-3}):\mathbb{Q}] = 2$. Now we have the trigonometric identity $\cos 3\theta = 4\cos^3\theta - 3\cos\theta$ which gives $4a^3 - 3a = \frac{1}{2}$ for $a = \cos 20°$. Thus the required number a is a root of $4x^3 - 3x - \frac{1}{2} = 0$ and so the minimum polynomial of a over $\mathbb{Q}$ is a factor of this. Hence, if $4x^3 - 3x - \frac{1}{2}$ is irreducible in $\mathbb{Q}[x]$, then this will be the minimum polynomial of a. Then the degree of a will be three and therefore not a power of two. It will follow that $60°$ cannot be trisected. Now, the given polynomial is irreducible if and only if $4(\frac{1}{2}x)^3 - 3(\frac{1}{2}x) - \frac{1}{2} = \frac{1}{2}x^3 - \frac{3}{2}x - \frac{1}{2}$ is irreducible. Multiplying by 2 we get $x^3 - 3x - 1$. Any rational root of this is integral and so must be a divisor of 1. Since 1 and -1 are not roots we see that $x^3 - 3x - 1$ is irreducible.

Duplication of the cube. Here we have to show that $\sqrt[3]{2}$ is not a constructible (complex) number. This follows from $[\mathbb{Q}(\sqrt[3]{2}):\mathbb{Q}] = 3$, since $x^3 - 2$ is irreducible in $\mathbb{Q}[x]$.

Construction of regular p-gons, p a prime. This requires the constructibility of the complex number $z = \cos 2\pi/p + i \sin 2\pi/p$. We have $z^p = 1$ and, since $x^p - 1 = (x - 1)(x^{p-1} + x^{p-2} + \cdots + 1)$, we have $z^{p-1} + z^{p-2} + \cdots + 1 = 0$. Since $x^{p-1} + x^{p-2} + \cdots + 1$ is irreducible in $\mathbb{Q}[x]$ (exercise 3, p. 154) we see that $[\mathbb{Q}(z):\mathbb{Q}] = p - 1$. Hence a necessary condition for constructibility of the regular p-gon by straight-edge and compass is that $p - 1 = 2^s$ for some non-negative integer s. Thus the regular p-gon can be constructed only for primes p of the form $2^s + 1$. Since 6 is not a power of 2 it follows that the regular heptagon cannot be constructed. We now observe that a necessary condition that $2^s + 1$ be a prime is that $s = 2^t$ for some non-negative integer t. For, suppose s is divisible by an odd number u, $s = uv$. Then $2^s + 1 = 2^{uv} + 1 = (2^v + 1) \times (2^{(u-1)v} - 2^{(u-2)v} + \cdots + 1)$ by the identity $x^u + 1 = (x + 1)(x^{u-1} - x^{u-2} + x^{u-3} - \cdots + 1)$ for any odd positive integer u. Then $2^s + 1 = 2^{uv} + 1$ is not a prime. Thus we have the improved necessary condition for constructibility of the regular p-gon, p a prime: p must be of the form $2^{2^t} + 1$. Primes of this form are called *Fermat primes* after Pierre Fermat, who conjectured (wrongly) that any integer of the form $2^{2^t} + 1$ is a prime.[3] The known Fermat primes

---

[3] It was shown by Euler that $2^{32} + 1 = 641 \cdot 6700417$. A simple derivation of this factorization is given in G. H. Hardy and E. M. Wright, *The Theory of Numbers*, New York, Oxford, 1938, p. 14.

are: $p = 3, 5, 17, 257, 65537$, obtained by taking $t = 0, 1, 2, 3, 4$. Based on em-
pirical evidence it has been conjectured that the number of Fermat primes is
finite and it is conceivable that the foregoing list is the complete set.

In section 4.11 we shall give a necessary and sufficient condition for the constructibility of the regular n-gon for any integral n. This will imply the converse
of the foregoing result, namely, that the regular p-gon can be constructed if p is
a Fermat prime. We shall conclude this section by computing $z = \cos 2\pi/17 +
i \sin 2\pi/17$ by a sequence of rational operations and extractions of square roots.
This will show that z is a constructible complex number and hence that the
regular 17-gon can be constructed.

Put $\theta = 2\pi/17$ and let $z = \cos \theta + i \sin \theta$. Then $z^k = \cos k\theta + i \sin k\theta$ and
these are distinct if $1 \le k \le 17$. Also $(z^k)^{17} = (z^{17})^k = 1$, so the z^k furnish 17
distinct 17th roots of unity. Since the equation $x^{17} - 1 = 0$ has at most 17
roots (Theorem 2.17, p. 132), these must be the z^k. Moreover, these constitute a
cyclic subgroup of the multiplicative group $\mathbb{C}^*$ of $\mathbb{C}$. The minimum polynomial
of z over $\mathbb{Q}$ is the irreducible polynomial $f(x) = \sum_0^{16} x^i$ which has the roots
$z, \ldots, z^{16}$ in $\mathbb{C}$. Then $f(x) = \prod_{k=1}^{16} (x - z^k)$ and we have the relation

$$(2) \qquad \sum_1^{16} z^k = -1.$$

Since $z\bar{z} = 1$ gives $\bar{z} = z^{-1}$ we have $(z^k)^{-1} = \bar{z}^k = \cos k\theta - i \sin k\theta$ and

$$(3) \qquad z^k + z^{-k} = 2 \cos k\theta.$$

We recall that the multiplicative group of $\mathbb{Z}/(17)$ is cyclic (Theorem 2.18,
p. 132) and, in fact, $\bar{3} = 3 + (17)$ is a generator, since

$$(4) \qquad 3^2 \equiv 9, \ 3^4 \equiv 13, \ 3^8 \equiv 16 \ (\text{mod } 17)$$

so the order of $\bar{3}$ in $\mathbb{Z}/(17)$ is not $1, 2, 2^2$ or 2^3. Since this order is a divisor of 2^4
it must be 2^4. Now put

$$(5) \qquad \begin{aligned} x_1 &= z + z^{3^2} + z^{3^4} + z^{3^6} + z^{3^8} + z^{3^{10}} + z^{3^{12}} + z^{3^{14}} \\ x_2 &= z^3 + z^{3^3} + z^{3^5} + z^{3^7} + z^{3^9} + z^{3^{11}} + z^{3^{13}} + z^{3^{15}}. \end{aligned}$$

Since $3^0 + 3^8 \equiv 0 \ (\text{mod } 17)$, we have $3^2 + 3^{10} \equiv 0 \ (\text{mod } 17)$, $3^4 + 3^{12} \equiv 0$
$(\text{mod } 17)$ and $3^6 + 3^{14} \equiv 0 \ (\text{mod } 17)$. Hence $x_1 = (z + z^{3^8}) + (z^{3^2} + z^{3^{10}}) +
(z^{3^4} + z^{3^{12}}) + (z^{3^6} + z^{3^{14}}) = (z + z^{-1}) + (z^8 + z^{-8}) + (z^4 + z^{-4}) + (z^2 + z^{-2})$.
Thus

$$(6) \qquad x_1 = 2(\cos \theta + \cos 8\theta + \cos 4\theta + \cos 2\theta)$$

and, similarly,

(7) $$x_2 = 2(\cos 3\theta + \cos 7\theta + \cos 5\theta + \cos 6\theta).$$

We have $x_1 + x_2 = \sum_1^{16} z^k = -1$ and direct multiplication, using the trigonometric identity $2\cos\alpha\cos\beta = \cos(\alpha + \beta) + \cos(\alpha - \beta)$, shows that $x_1 x_2 = 4(x_1 + x_2) = -4$. Hence x_1 and x_2 are roots of $(x - x_1)(x - x_2) = x^2 + x - 4 = 0$. The roots of this equation are $\frac{1}{2}(-1 \pm \sqrt{17})$. Since $\theta = 2\pi/17$, $\cos 3\theta > 0$, $\cos 7\theta < 0$, $\cos 5\theta < 0$ and $\cos 6\theta < 0$. Also $|\cos 6\theta| = \cos(\pi - 6\theta) = \cos 5\pi/17 > \cos 6\pi/17 = \cos 3\theta$. Hence $x_2 < 0$ and so

(8) $$x_1 = \tfrac{1}{2}(-1 + \sqrt{17}), \qquad x_2 = \tfrac{1}{2}(-1 - \sqrt{17}).$$

Next, put

$$y_1 = z + z^{-1} + z^4 + z^{-4} = 2(\cos\theta + \cos 4\theta)$$

$$y_2 = z^8 + z^{-8} + z^2 + z^{-2} = 2(\cos 8\theta + \cos 2\theta)$$

$$y_3 = z^3 + z^{-3} + z^5 + z^{-5} = 2(\cos 3\theta + \cos 5\theta)$$

$$y_4 = z^7 + z^{-7} + z^6 + z^{-6} = 2(\cos 7\theta + \cos 6\theta).$$

Then, directly using the cosine identity noted above, we have

$$y_1 y_2 = 4(\cos\theta + \cos 4\theta)(\cos 8\theta + \cos 2\theta) = 2\sum_1^8 \cos k\theta = -1.$$

Similarly, $y_3 y_4 = -1$, and since $y_1 + y_2 = x_1$ and $y_3 + y_4 = x_2$, y_1 and y_2 are roots of $x^2 - x_1 x - 1 = 0$, and y_3 and y_4 are roots of $x^2 - x_2 x - 1 = 0$. Since $\cos\theta > \cos 2\theta$ and $\cos 4\theta > 0$ but $\cos 8\theta < 0$ we have $y_1 > y_2$. Similarly, $y_3 > y_4$. Hence

(9)
$$y_1 = \tfrac{1}{2}(x_1 + \sqrt{x_1{}^2 + 4}), \qquad y_2 = \tfrac{1}{2}(x_1 - \sqrt{x_1{}^2 + 4}),$$
$$y_3 = \tfrac{1}{2}(x_2 + \sqrt{x_2{}^2 + 4}), \qquad y_4 = \tfrac{1}{2}(x_2 - \sqrt{x_2{}^2 + 4}).$$

Now put

$$z_1 = z + z^{-1} = 2\cos\theta, \qquad z_2 = z^4 + z^{-4} = 2\cos 4\theta.$$

Then $z_1 > z_2$, $z_1 + z_2 = y_1$ and $z_1 z_2 = 4\cos\theta\cos 4\theta = 2(\cos 5\theta + \cos 3\theta) = y_3$. Hence

(10) $$2\cos\theta = \tfrac{1}{2}(y_1 + \sqrt{y_1{}^2 - 4y_3}).$$

Then, using (8) and (9), one can obtain an explicit (somewhat horrendous) formula for $\cos \theta$. Then one obtains $\sin \theta = \sqrt{1 - \cos^2 \theta}$ and $z = \cos \theta + i \sin \theta$. It is clear from this that z is a constructible complex number and consequently the regular 17-gon can be constructed with straight-edge and compass.[4] We have refrained from giving any reasons for the steps in our computations. These will become clear later after we have developed the Galois theory.

EXERCISES

1. Show that the regular pentagon can be constructed with straight-edge and compass.

2. Show that arc cos 11/16 can be trisected with straight-edge and compass.

3. Show that the regular 9-gon cannot be constructed.

4.3   SPLITTING FIELD OF A POLYNOMIAL

The mathematicians of the nineteenth century dealt almost exclusively with the field $\mathbb{C}$ of complex numbers and its subfields. The important fact about $\mathbb{C}$ from the algebraic standpoint is that it is an adequate field for solving algebraic equations in one unknown, that is, it is algebraically closed in the sense that every polynomial equation $x^n + a_1 x^{n-1} + \cdots + a_n = 0$, $a_i \in \mathbb{C}$, has a root in $\mathbb{C}$. The central role played by $\mathbb{C}$ in nineteenth-century algebra can be gleaned from the fact that during this period the result that $\mathbb{C}$ is algebraically closed was called "*the* fundamental theorem of algebra." We still retain this terminology but only out of respect for the past, since the theorem no longer plays a central role in algebra. For one thing, we are interested also in fields of characteristic $p \neq 0$ (for example, because of their usefulness in number theory) and these can not be imbedded in $\mathbb{C}$. Our starting point will be an arbitrary base field F. Given a polynomial $f(x) \in F[x]$ we would like to have at hand an extension field E of F which in some sense contains all the roots of the equation $f(x) = 0$. We recall that $f(r) = 0$ if and only if $f(x)$ is divisible by $x - r$ (p. 130) and we shall say that $f(x)$ (assumed to be monic) *splits* in the extension field E if $f(x) = \prod_1^n (x - r_i)$, that is, is a product of linear (=first degree) factors in $E[x]$. Then

---

[4] For an actual construction, see H. S. M. Coxeter, *Introduction to Geometry*, New York, Wiley, 1961, p. 27.

if r is a root of $f(x)$ in E we have $0 = f(r) = \prod_1^n (r - r_i)$, which implies that r is one of the r_i. We also have the same factorization $f(x) = \prod (x - r_i)$ in the polynomial ring $R[x]$ where R is the field $F(r_1, \ldots, r_n)$. It is clear that in dealing with the single polynomial $f(x)$ it would be a good idea to shift our attention from E to $F(r_1, \ldots, r_n)$, which is tailored to the study of f. We now formulate the following important

DEFINITION 4.1 *Let F be a field, $f(x)$ a monic polynomial in $F[x]$. Then an extension field E/F is called a* splitting field over F of $f(x)$ *if*

(i)
$$f(x) = (x - r_1)(x - r_2) \cdots (x - r_n)$$

in $E[x]$ and

(ii)
$$E = F(r_1, r_2, \ldots, r_n),$$

that is, E is generated by the roots of $f(x)$.

Our first task will be to prove the existence of a splitting field for any polynomial $f(x)$ of positive degree. The proof of this result can be obtained by extending a method used first by A. Cauchy to construct $\mathbb{C}$ from $\mathbb{R}$ (adjunction of $\sqrt{-1}$) and later used by L. Kronecker to construct a single root of an irreducible polynomial. We now give the proof of

THEOREM 4.3. *Any monic polynomial $f(x) \in F[x]$ of positive degree has a splitting field E/F.*

Proof. Let $f(x) = f_1(x)f_2(x) \cdots f_k(x)$ be the factorization of $f(x)$ into monic irreducible factors. Evidently $k \leq n = \deg f(x)$. We use induction on $n - k$. If $n - k = 0$, all the $f_i(x)$ are linear, which means that F itself is a splitting field. Hence assume $n - k > 0$ so that some $f_i(x)$, say $f_1(x)$, is of degree > 1. Put $K = F[x]/(f_1(x))$. Then, since $f_1(x)$ is irreducible, K is a field. K is also an extension field of F (using the identification of $a \in F$ with $a + (f_1(x))$ and $K = F(r)$ where $r = x + (f_1(x))$ is a root of $f_1(x) = 0$). It is now best to forget about the mechanics of all of this and just to keep in mind that we have somehow produced an extension field K/F which is generated over F by a single root r of the irreducible polynomial $f_1(x)$. Since $K \supset F$ and $f(x)$ and the $f_i(x) \in F[x] \subset K[x]$, we obtain the factorization of $f(x)$ into monic irreducible factors in $K[x]$ by factoring every $f_i(x)$ into monic irreducible factors. Also we have $f_1(x) = (x - r)g(x)$ in $K[x]$ since $f_1(r) = 0$. Hence, if l is the number of irreducible factors in the factorization of $f(x)$ in $K[x]$, then $l > k$ so $n - l < n - k$. Hence the induction hypothesis can be applied to $f(x)$ and K to conclude that we have an extension field $E = K(r_1, r_2, \ldots, r_n)$ such that $f(x) = \prod_1^n (x - r_i)$ in $E[x]$. Since $f_1(r) = 0$ and $f_1(x) | f(x)$, we have $f(r) = 0$; hence $r = r_i$ for some i. Then

$$E = K(r_1, r_2, \ldots, r_n) = F(r)(r_1, r_2, \ldots, r_n)$$
$$= F(r, r_1, \ldots, r_n) = F(r_1, r_2, \ldots, r_n)$$

is a splitting field over F of $f(x)$. $\square$

EXAMPLES

1. Let $f(x) = x^2 + ax + b$. If $f(x)$ is reducible in $F[x]$ (F arbitrary) then F is a splitting field. Otherwise, put $E = F[x]/(f(x)) = F(r_1)$ where $r_1 = x + (f(x))$. Then E is a splitting field since $f(r_1) = 0$ so $f(x) = (x - r_1)(x - r_2)$ in $E[x]$. Thus $E = F(r_1) = F(r_1, r_2)$. Since $f(x)$ is the minimum polynomial of r_1 over F, $[E:F] = 2$.

2. Let the base field F be $\mathbb{Z}/(2)$, the field of two elements, and let $f(x) = x^3 + x + 1$. Since $1 + 1 + 1 \not\equiv 0 \pmod 2$ and $0 + 0 + 1 \not\equiv 0 \pmod 2$, $f(x)$ has no roots in F; hence $f(x)$ is irreducible in $F[x]$. Put $r_1 = x + (f(x))$ in $F[x]/(f(x))$ so $F(r_1)$ is a field and $x^3 + x + 1 = (x + r_1)(x^2 + ax + b)$ in $F(r_1)[x]$. (Note that we can write $+$ for $-$ since the characteristic is two.) Comparison of coefficients shows that $a = r_1, b = 1 + r_1^2$. The elements of $F(r_1)$ can be listed as $c + dr_1 + er_1^2$, $c, d, e \in F$. There are eight of these: $0, 1, r_1, 1 + r_1, r_1^2, 1 + r_1^2, r_1 + r_1^2$, and $1 + r_1 + r_1^2$. Substituting these in $x^2 + r_1 x + 1 + r_1^2$ we reach $(r_1^2)^2 + r_1(r_1^2) + 1 + r_1^2 = r_1^4 + r_1^3 + 1 + r_1^2 = 0$ since $r_1^3 = r_1 + 1$ and $r_1^4 = r_1^2 + r_1$. Hence $x^2 + ax + b$ factors into linear factors in $F(r_1)[x]$ and $E = F(r_1)$ is a splitting field of $x^3 + x + 1$ over F.

3. Let $F = \mathbb{Q}$, $f(x) = (x^2 - 2)(x^2 - 3)$. Since the rational roots of $x^2 - 2$ and $x^2 - 3$ must be integral (exercise 1, p. 154), it follows that $x^2 - 2$ and $x^2 - 3$ are irreducible in $\mathbb{Q}[x]$. Form $K = \mathbb{Q}(r_1), r_1 = x + (x^2 - 2)$ in $\mathbb{Q}[x]/(x^2 - 2)$. The elements of K have the form $a + br_1$, $a, b \in \mathbb{Q}$. We claim that $x^2 - 3$ is irreducible in $K[x]$. Otherwise, we have rational numbers a, b such that $(a + br_1)^2 = 3$. Then $(a^2 + 2b^2) + 2abr_1 = 3$ so that $ab = 0$ and $a^2 + 2b^2 = 3$. If $b = 0$ we obtain $a^2 = 3$ which is impossible since $\sqrt{3}$ is not rational, and if $a = 0, b^2 = 3/2$. Then $(2b)^2 = 6$ and since $\sqrt{6}$ is not rational we again obtain an impossibility. Thus $x^2 - 3$ is irreducible in $K[x]$. Now form $E = K[x]/(x^2 - 3)$. Then this is a splitting field over $\mathbb{Q}$ of $(x^2 - 2)(x^2 - 3)$ and $[E:\mathbb{Q}] = [E:K][K:\mathbb{Q}] = 2 \cdot 2 = 4$.

4. Let $F = \mathbb{Q}$, $f(x) = x^p - 1$, p a prime. We have $x^p - 1 = (x - 1)(x^{p-1} + x^{p-2} + \cdots + 1)$ and we know that $x^{p-1} + x^{p-2} + \cdots + 1$ is irreducible in $\mathbb{Q}[x]$. Let $E = \mathbb{Q}(z)$ where $z = x + (x^{p-1} + x^{p-2} + \cdots + 1)$ in $\mathbb{Q}[x]/(x^{p-1} + x^{p-2} + \cdots + 1)$. We have $z^p = 1$ and since $x^{p-1} + \cdots + 1$ is the minimum polynomial of z over $\mathbb{Q}$ the elements $1, z, \ldots, z^{p-1}$ are distinct. Also $(z^k)^p = (z^p)^k = 1$ so every z^k is a root of $x^p - 1$. It follows that $x^p - 1 = \prod_{k=1}^p (x - z^k)$ in $E[x]$. Thus E is a splitting field over $\mathbb{Q}$ of $x^p - 1$, and $[E:\mathbb{Q}] = p - 1$.

Before proceeding to our next main result—the uniqueness up to isomorphism of splitting fields—we note that splitting fields are finite dimensional over the base field. Let E/F be a splitting field over F of $f(x)$. Then $E = F(r_1, r_2, \ldots, r_n)$ where $f(r_i) = 0$, $1 \leq i \leq n$. Then r_i is algebraic over F, hence also over $F(r_1, r_2, \ldots, r_{i-1})$. Then $[F(r_1, \ldots, r_i):F(r_1, \ldots, r_{i-1})] < \infty$ since this is the degree of r_i over $F(r_1, \ldots, r_{i-1})$. Hence, by iterative use of the dimensionality formula for fields (Theorem 4.2), we obtain

$$[E:F] = [F(r_1, \ldots, r_n):F] = \prod_{i=1}^{n} [F(r_1, \ldots, r_i):F(r_1, \ldots, r_{i-1})] < \infty$$

where it is understood that the first term in this product is $[F(r_1):F]$.

We shall now prove that any two splitting fields over F of a polynomial $f(x) \in F[x]$ are isomorphic, and we shall also obtain some important information on the number of isomorphisms between splitting fields of $f(x)$. In order to carry through an inductive argument it is necessary to generalize the considerations slightly as follows. We consider two isomorphic fields F and $\bar{F}$ and an isomorphism $\eta : a \to \bar{a}$ of F onto $\bar{F}$. We know that this can be extended to a unique isomorphism $g(x) \to \bar{g}(x)$ of $F[x]$ onto $\bar{F}[x]$. Let $f(x) \in F[x]$ be monic of positive degree and let E be a splitting field over F of $f(x)$, $\bar{E}$ a splitting field over $\bar{F}$ of $\bar{f}(x)$. Then we have the following important

THEOREM 4.4. *Let $\eta : a \to \bar{a}$ be an isomorphism of a field F onto a field $\bar{F}$, $f(x) \in F[x]$ be monic of positive degree, $\bar{f}(x)$ the corresponding polynomial in $\bar{F}[x]$ (under the isomorphism which extends η and maps $x \to x$), and let E and $\bar{E}$ be splitting fields of $f(x)$ and $\bar{f}(x)$ over F and $\bar{F}$ respectively. Then η can be extended to an isomorphism of E onto $\bar{E}$. Moreover, the number of such extensions is $\leq [E:F]$ and it is precisely $[E:F]$ if $\bar{f}(x)$ has distinct roots in $\bar{E}$.*

Before proceeding to the proof we separate off the following lemma which will serve as the induction step of the proof.

LEMMA. *Let η be an isomorphism of a field F onto a field $\bar{F}$ and let E and $\bar{E}$ be extension fields of F and $\bar{F}$ respectively. Suppose $r \in E$ is algebraic over F with minimum polynomial $g(x)$. Then η can be extended to a monomorphism ζ of $F(r)$ into $\bar{E}$ if and only if $\bar{g}(x)$ has a root in $\bar{E}$, in which case the number of such extensions is the same as the number of distinct roots of $\bar{g}(x)$ in $\bar{E}$.*

Proof. If an extension ζ exists, then we can apply it to the relation $g(r) = 0$ to obtain $\bar{g}(\zeta(r)) = 0$. Thus $\zeta(r)$ is a root of $\bar{g}(x) = 0$ in $\bar{E}$. Conversely, let $\bar{r}$ be such a root. We have the homomorphism $h(x) \rightarrow \bar{h}(\bar{r})$ of $F[x]$ into $\bar{E}$ (Theorem 2.10, p. 122). The kernel contains the ideal $(g(x))$ so we have the induced homomorphism $h(x) + (g(x)) \rightarrow \bar{h}(\bar{r})$ of $F[x]/(g(x))$ into $\bar{E}$. Similarly, we have the homomorphism $h(x) + (g(x)) \rightarrow h(r)$ of $F[x]/(g(x))$ onto $F(r)$. Since $g(x)$ is irreducible, $F[x]/(g(x))$ is a field and so both homomorphisms are monomorphisms and the second one is an isomorphism. If we take the inverse of this isomorphism and follow it with the monomorphism of $F[x]/(g(x))$ into $\bar{E}$ we obtain the monomorphism $h(r) \rightarrow \bar{h}(\bar{r})$ of $F(r) = F[r]$ into $\bar{E}$. Since $F(r)$ is generated by F and r it is clear that this is the only monomorphism of $F(r)$ into $\bar{E}$ extending η and sending $r \rightarrow \bar{r}$. It is now clear that the number of monomorphism extensions is the same as the number of distinct choices of $\bar{r}$, hence, the number of distinct roots of $\bar{g}(x)$ in $\bar{E}$.   □

Proof of theorem 4.4.   We prove the result by induction on $[E:F]$. If $[E:F] = 1$, $E = F$ and $f(x) = \prod (x - r_i)$ in $F[x]$. Applying the isomorphism $h(x) \rightarrow \bar{h}(x)$ of $E[x]$ we obtain $\bar{f}(x) = \prod (x - \bar{r}_i)$ in $\bar{F}[x]$. Thus the $\bar{r}_i$ are the roots of $\bar{f}(x)$ in $\bar{E}$, and, since $\bar{E}$ is generated over $\bar{F}$ by these roots, $\bar{E} = \bar{F}$ and there is just $1 = [E:F]$ extension. Now assume $[E:F] > 1$. Then $f(x)$ is not a product of linear factors in $F[x]$. Let $g(x)$ be a monic irreducible factor of $f(x)$ of degree > 1. Then $\bar{g}(x) | \bar{f}(x)$ in $\bar{F}[x]$. We may also assume $g(x) = \prod_1^m (x - r_i)$, $f(x) = \prod_1^n (x - r_j)$, $\bar{g}(x) = \prod_1^m (x - s_i)$, $\bar{f}(x) = \prod_1^n (x - s_j)$ in $E[x]$ and $\bar{E}[x]$. Put $K = F(r_1)$. Since $g(x)$ is irreducible it is the minimum polynomial of r_1 over F and $[K:F] = m = \deg g(x)$. By the lemma, there exist k monomorphisms $\zeta_1, \ldots, \zeta_k$ of K into $\bar{E}$ which are extensions of η where k is the number of different s_i, $1 \leq i \leq m$. Thus $k = m$ if the s_i, $1 \leq i \leq m$, are distinct. Now it is clear from the definition of a splitting field that E is a splitting field over K of $f(x) \in K[x]$ and $\bar{E}$ is a splitting field over $\zeta_i(K)$ of $\bar{f}(x)$. Since $[E:K] = [E:F]/[K:F] = [E:F]/m < [E:F]$ induction on dimensionality implies that every ζ_i can be extended to an isomorphism of E onto $\bar{E}$, and that the number of such extensions is $\leq [E:K]$ and is $[E:K]$ if $\bar{f}(x)$ has distinct roots in $\bar{E}$. Any of these isomorphisms is an extension of the given isomorphism η of F onto $\bar{F}$. Hence we obtain in this way at least one extension of η to an isomorphism of E onto $\bar{E}$. Moreover, since the extensions of η which are extensions of distinct ζ_i are distinct, we obtain in this way $\leq m[E:K] = [E:F]$ extensions of η and exactly $[E:F]$ such extensions if $\bar{f}(x)$ has distinct roots. Our proof will therefore be complete if we can convince ourselves that our method has accounted for every extension of the isomorphism of F to $\bar{F}$ to one of E to $\bar{E}$. But this is clear, since if ζ is such an extension, the restriction of ζ to K is a monomorphism of K into $\bar{E}$ and so this restriction coincides with one of the ζ_i, $1 \leq i \leq k$.   □

If we specialize the first part of this theorem to the case $F = \bar{F}$ and η the identity mapping on F, we conclude that if E and $\bar{E}$ are two splitting fields over F of $f(x)$, then there exists an isomorphism of E onto $\bar{E}$ which is the identity on F. We refer to such an isomorphism (similarly, monomorphism) as an isomorphism *over F* of E onto $\bar{E}$. The second part of the result applied to $F = \bar{F}$ gives the important information that there are at most $[E:F]$ automorphisms of E/F (E over F) and there are exactly this number if $f(x)$ has n distinct roots.

EXERCISES

1. Show that the dimensionality of a splitting field E/F of $f(x)$ of degree n is at most $n!$.

2. Construct a splitting field over $\mathbb{Q}$ of $x^5 - 2$. Find its dimensionality over $\mathbb{Q}$.

3. Determine a splitting field over $\mathbb{Z}/(p)$ of $x^{p^e} - 1$, $e \in \mathbb{N}$.

4. Let E/F be a splitting field over F of $f(x)$ and let K be a subfield of E/F. Show that any monomorphism of K/F into E/F can be extended to an automorphism of E.

5. Let E be an extension field of F such that $[E:F] = n < \infty$. Let K be any extension field of F. Use the method of the proof of Theorem 4.4 to show that the number of monomorphisms of E/F into K/F does not exceed n.

4.4 MULTIPLE ROOTS

Let $f(x)$ be a monic polynomial of positive degree in $F[x]$ and let E/F be a splitting field. We write the factorization of $f(x)$ in $E[x]$ as

$$(11) \qquad f(x) = (x - r_1)^{k_1}(x - r_2)^{k_2} \cdots (x - r_s)^{k_s},$$

$r_i \in E$, $r_i \neq r_j$ if $i \neq j$, and we say that r_i is a root of *multiplicity* k_i of the equation $f(x) = 0$. If $k_i = 1$, then r_i is called a *simple root*; otherwise r_i is a *multiple root*. If we have a second splitting field $\bar{E}/F$ of $f(x)$, then $f(x) = \prod_1^s (x - \bar{r}_j)^{k_j}$ in $\bar{E}[x]$ where $a \to \bar{a}$ is an isomorphism of E/F onto $\bar{E}/F$. It is clear from this that the multiplicities k_i are independent of the choice of the splitting field. In particular, the fact that $f(x)$ has only simple roots is independent of the choice of E. The last result (Theorem 4.4) shows that there is a distinct advantage in working with polynomials having only simple roots, since in this case we have the exact formula that the number of automorphisms of E/F is $[E:F]$.

We shall show in this section that if F is of characteristic 0 or if F is a finite field, then there is no loss in generality in assuming that all the roots are simple. We observe first that if we factor $f(x) = p_1(x)^{l_1}p_2(x)^{l_2} \cdots p_t(x)^{l_t}$ in $F[x]$ where the $p_i(x)$ are distinct primes, then E/F is a splitting field for $f(x)$ if and only if E/F is a splitting field for $f_0(x) = p_1(x)p_2(x) \cdots p_t(x)$. This is clear from the definition. Hence we may assume at the outset that $f(x)$ is a product of distinct prime polynomials in $F[x]$. We remark also that if $p(x)$ and $q(x)$ are distinct monic prime polynomials in $F[x]$, then $(p(x), q(x)) = 1$ in $F[x]$; hence there exist $a(x), b(x) \in F[x]$ such that $a(x)p(x) + b(x)q(x) = 1$. This precludes that $p(x) = 0$ and $q(x) = 0$ have a common root in E. It follows that if $f(x)$ is a product of distinct primes, then all the roots of $f(x)$ are simple if and only if this is the case for the prime factors of $f(x)$.

We shall now develop a criterion for multiple roots which can be tested in $F[x]$ and thus does not require recourse to a splitting field. This will be based on formal differentiation of polynomials, which we shall now define. We adjoin an indeterminate h to $F[x]$ to obtain the polynomial ring $F[x, h]$ in the two indeterminates x, h. Since $F[x, h] = F[x][h]$ and h is transcendental over $F[x]$, any element of $F[x, h]$ can be written in one and only one way as $f_0(x) + f_1(x)h + \cdots + f_n(x)h^n$, $f_i(x) \in F[x]$. In particular, if $f(x) \in F[x]$ we have $f(x+h) = f_0(x) + f_1(x)h + \cdots + f_n(x)h^n$. Putting $h = 0$ in this (that is, applying the homomorphism of $F[x, h]$ into $F[x]$ such that $a \rightarrow a$ for $a \in F$, $x \rightarrow x$, $h \rightarrow 0$) we obtain $f(x) = f_0(x)$, and so $f(x + h) - f(x)$ is divisible by h. Dividing h out we obtain $f_1(x) + f_2(x)h + \cdots + f_n(x)h^{n-1}$, and putting $h = 0$ in this polynomial we obtain $f_1(x)$, which we define to be the *derivative* $f'(x)$ (or $f(x)'$) of $f(x)$.[5] Clearly $f'(x)$ satisfies the congruence

$$(12) \qquad\qquad f(x + h) \equiv f(x) + f'(x)h \quad (\text{mod } h^2).$$

Moreover, if $g(x) \in F[x]$ satisfies $f(x + h) \equiv f(x) + g(x)h \ (\text{mod } h^2)$ then $f'(x)h \equiv g(x)h \ (\text{mod } h^2)$, and so $f'(x) \equiv g(x) \ (\text{mod } h)$, which gives $g(x) = f'(x)$. Thus $f'(x)$ is characterized by the congruence (12). This characterization permits us to establish quickly the basic properties of the map $f \rightarrow f'$ which we shall call the *standard derivation* in $F[x]$. These are:

(i) F-linearity: $(f + g)' = f' + g'$, $(af)' = af'$ for $a \in F$.

(ii) The product rule:

$$(13) \qquad\qquad\qquad (fg)' = f'g + fg'.$$

(iii) $x' = 1$.

---

[5] The reader should observe the similarity of this process to that defining the derivative of a function of a real variable where h is a variable and the limit is taken of $[f(x + h) - f(x)]/h$ as $h \rightarrow 0$.

Property (i) is immediate from (12). To establish the product rule we multiply (12) by the corresponding congruence for $g(x + h)$. This gives

$$(fg)(x + h) = f(x + h)g(x + h) \equiv [f(x) + f'(x)h][g(x) + g'(x)h] \ (\text{mod } h^2)$$
$$\equiv f(x)g(x) + [f'(x)g(x)$$
$$+ f(x)g'(x)]h \ (\text{mod } h^2).$$

Hence (13) follows from the characteristic property (12) applied to fg. Since $x + h \equiv x + 1h \ (\text{mod } h^2)$, we have $x' = 1$ which is (iii). This and the product rule imply that $(x^k)' = kx^{k-1}$ if $k = 1, 2, 3, \ldots$. Also $1^2 = 1$ gives $1' \ 1 + 1 \ 1' = 1'$ so that $2(1') = 1'$ and $1' = 0$. It now follows from the linearity that if $f(x) = a_0 + a_1 x + \cdots + a_n x^n$, then

$$f'(x) = a_1 + 2a_2 x \cdots + na_n x^{n-1}$$

as in the calculus of functions of a real variable.

We can now prove

THEOREM 4.5. *Let $f(x)$ be a monic polynomial of positive degree in $F[x]$. Then all the roots of f in any splitting field E/F are simple if and only if $(f, f') = 1$.*

Proof. Let $d(x) = (f(x), f'(x))$ in $F[x]$. Suppose $f(x)$ has a multiple root in $E[x]$, so $f(x) = (x - r)^k g(x)$ with $k > 1$. Taking derivatives in $E[x]$ we obtain $f'(x) = (x - r)^k g' + k(x - r)^{k-1} g$ which is divisible by $x - r$ since $k - 1 \geq 1$. Thus $x - r$ is a common factor of $f(x)$ and $f'(x)$ in $E[x]$. It follows that $d(x) \neq 1$. Next suppose all the roots of f are simple. Then we have $f(x) = \prod_1^n (x - r_i)$, $r_i \neq r_j$ if $i \neq j$. The extension of the product rule to more than two factors now gives

$$f'(x) = \sum_1^n (x - r_1) \cdots (x - r_{j-1})(x - r_{j+1}) \cdots (x - r_n).$$

It is clear from this formula that $(x - r_i) \nmid f'(x)$; hence $(f(x), f'(x)) = 1$. $\square$

If $f(x)$ is irreducible in $F[x]$, then $(f, f') \neq 1$ implies that $f | f'$. Since $\deg f' < \deg f$ this forces $f' = 0$. If the characteristic is 0, the formula for the derivative shows that $f' = 0$ if and only if $f \in F$. Hence $f' \neq 0$ if $f(x)$ is irreducible and F is of characteristic 0. If the characteristic is $p \neq 0$ and $f(x) = a_0 + a_1 x + \cdots + a_n x^n$, then $f'(x) = \sum_{i=1}^n i a_i x^{i-1}$ and $f'(x) = 0$ if and only if $ia_i = 0$, $1 \leq i \leq n$. This holds if and only if $a_i = 0$ for every i not divisible by p; hence, if and only if $f(x) = b_0 + b_1 x^p + b_2 x^{2p} + \cdots + b_m x^{mp} = g(x^p)$ where $g(x) = b_0 + b_1 x + \cdots + b_m x^m$.

We shall now construct an example of an irreducible polynomial in characteristic p which has multiple roots. Let F be any field of characteristic p. Then we have $1^p = 1$, and the commutativity of multiplication gives $(ab)^p = a^p b^p$. By the binomial theorem.

$$(a + b)^p = a^p + b^p + \sum_{1}^{p-1} \binom{p}{i} a^i b^{p-i}$$

and since the binomial coefficient $\binom{p}{i} = p!/i!\,(p-i)!$ is an integer, and in the rational form which we have displayed, p occurs in the numerator but not in the denominator, $\binom{p}{i}$ is divisible by p. Hence $\sum_{1}^{p-1}\binom{p}{i} a^i b^{p-i} = 0$ and so $(a+b)^p = a^p + b^p$. Thus we have

(14) $(a + b)^p = a^p + b^p,\ (ab)^p = a^p b^p,\ 1^p = 1$

in F. This shows that the map $a \to a^p$ is an endomorphism of the ring F. Since F is a field this is a monomorphism and the image F^p, the set of pth powers, is a subfield of F.

We now prove the following

LEMMA. *If F has characteristic p and $a \in F$, then $x^p - a$ is either irreducible in $F[x]$ or it is a pth power in $F[x]$.*

Proof. Suppose $x^p - a = g(x)h(x)$ in $F[x]$ where $g(x)$ is a monic polynomial of degree k, $1 \le k \le p - 1$. Let E be a splitting field over F of $x^p - a$ and let $b \in E$ be a root of this polynomial. Then we have $b^p = a$ so $x^p - a = x^p - b^p = (x - b)^p = g(x)h(x)$. Hence $g(x) = (x - b)^k$ and $b^k \in F$. Since $b^p \in F$ also and there exist integers u and v such that $uk + vp = 1$, $b = (b^k)^u (b^p)^v \in F$. Thus we have $x^p - a = (x - b)^p$ in $F[x]$. $\square$

We can now construct our example of an irreducible polynomial which has multiple roots. As base field F we take the field $(\mathbb{Z}/(p))(t)$ of rational expressions in an indeterminate t over the prime field $\mathbb{Z}/(p)$ of p elements, that is, the field of fractions of the polynomial ring $(\mathbb{Z}/(p))[t]$. We claim that t is not a pth power in this field. Suppose $t = (f(t)/g(t))^p$ where $f(t) = a_0 + a_1 t + \cdots + a_n t^n$ and $g(t) = b_0 + b_1 t + \cdots + b_m t^m$. Then $f(t)^p = a_0^p + a_1^p t^p + \cdots + a_n^p t^{np}$, $g(t)^p = b_0^p + b_1^p t^p + \cdots + b_m^p t^{mp}$ so we have a relation

$$(b_0^p + b_1^p t^p + \cdots + b_m^p t^{mp})t = a_0^p + a_1^p t^p + \cdots + a_n^p t^{np}.$$

The linear independence of the powers $1, t, t^2, \ldots$ over $\mathbb{Z}/(p)$ then implies that every $b_i = 0$ contradicting the (tacit) assumption that $g(t) \neq 0$. The foregoing lemma now shows that the polynomial $f(x) = x^p - t \in F[x]$ is irreducible. On the other hand, we see that it is a pth power in $E[x]$, E a splitting field. (We can also see that it has multiple roots by using the derivative criterion and $(x^p - t)' = px^{p-1} = 0$.)

We shall now call a polynomial contained in $F[x]$ *separable* if its irreducible factors have distinct roots. The result we have proved is that if F is of characteristic 0, then every polynomial with coefficients in F is separable and if the characteristic is p there exist inseparable polynomials, at least for certain F. We now look at this question more closely in the characteristic $p \neq 0$ case. We shall call a field (of any characteristic) *perfect* if every polynomial in $F[x]$ is separable. Then we have seen that every field of characteristic 0 is perfect. For characteristic $p \neq 0$ we have the following criterion.

THEOREM 4.6. *A field F of characteristic $p \neq 0$ is perfect if and only if $F = F^p$, the subfield of pth powers of the elements of F.*

Proof. If $F^p \subsetneq F$, let $a \in F$, $\notin F^p$. Then $x^p - a$ is irreducible, by the lemma. Since $(x^p - a)' = 0$, this is an inseparable irreducible polynomial. Hence F is not perfect. Now suppose that $f(x)$ is an inseparable irreducible polynomial $\in F[x]$. Then $(f(x), f'(x)) \neq 1$ and we have seen that this implies that $f(x) = a_0 + a_p x^p + a_{2p} x^{2p} + \cdots$. One of these a_i is not a pth power. For, if every $a_i = b_i^p$ then $f(x) = a_0 + a_p x^p + a_{2p} x^{2p} + \cdots = b_0^p + b_p^p x^p + b_{2p}^p x^{2p} + \cdots = (b_0 + b_p x + b_{2p} x^2 + \cdots)^p$ contrary to the irreducibility of $f(x)$. Hence $F \neq F^p$. $\square$

COROLLARY. *Every finite field is perfect.*

Proof. The characteristic of a finite field is a prime p. The monomorphism $a \to a^p$ of F is an isomorphism since F is finite. Hence $F = F^p$ is perfect by Theorem 4.6. $\square$

EXERCISES

1. Let F be a field of characteristic 0, $f(x) \in F[x]$ be monic of positive degree. Show that if $d(x) = (f(x), f'(x))$ then $g(x) = f(x)d(x)^{-1}$ has the same roots as $f(x)$ and that these are all simple roots of $g(x)$.

2. Let $f(x)$ be irreducible in $F[x]$, F of characteristic p. Show that $f(x)$ can be written as $g(x^{p^e})$ where $g(x)$ is irreducible and separable. Use this to show that every root of $f(x)$ has the same multiplicity p^e (in a splitting field).

3. Let F be of characteristic p. A polynomial $f(x) \in F[x]$ is called a *p-polynomial* if it has the form $x^{p^m} + a_1 x^{p^{m-1}} + \cdots + a_m x$. Show that a monic polynomial of positive degree is a p-polynomial if and only if its roots form a finite subgroup of the additive group of the splitting field and every root has the same multiplicity p^e.

4. Let F be imperfect of characteristic p. Show that $x^{p^e} - a$ is irreducible if $a \notin F^p$ and $e = 0, 1, 2, \ldots$.

5. Let F be of characteristic p and let $a \in F$. Show that $f(x) = x^p - x - a$ has no multiple roots and that $f(x)$ is irreducible in $F[x]$ if and only if $a \neq c^p - c$ for any $c \in F$.

4.5 THE GALOIS GROUP.
THE FUNDAMENTAL GALOIS PAIRING

We shall now derive the central results of Galois' theory. These establish a 1–1 correspondence between the set of subfields of E/F, where E is a splitting field of a separable polynomial in $F[x]$, with the set of subgroups of the group of automorphisms of E/F. The properties of this correspondence serve as the basis of Galois' criterion for solvability of an equation by radicals and for constructibility by straight-edge and compass. Moreover, as we noted in the introduction to this chapter, these results play a fundamental role in many other considerations in algebra and number theory.

First, some definitions and notations. Let E be an extension field of a field F and let G be the set of automorphisms of E/F: that is, the set of automorphisms η of E such that $\eta(a) = a$ for every $a \in F$. G is a group of transformations of E: $1 \in G$ and if $\eta, \zeta \in G$, then $\eta\zeta$ and $\eta^{-1} \in G$. We shall call G *the Galois group of E over F* and denote it as Gal E/F when we wish to indicate the fields E and F.

EXAMPLES

1. $E = F(u)$ where $u^2 = a \in F$ and a is not a square in F. We assume also that the characteristic, char $F \neq 2$. Since a is not a square in F, $x^2 - a$ is irreducible in $F[x]$. Hence this is the minimum polynomial of u over F. Then $[E:F] = 2$ and $(1, u)$ is a base for E/F. Clearly the two maps $c + du \to c + du$, $c, d \in F$, and $c + du \to c - du$ are automorphisms of E/F. These are the only ones. For, if $\eta \in$ Gal E/F, $u^2 = a$ implies $(\eta(u))^2 = a$ and since the roots in E of $x^2 - a = 0$ are u and $-u$, either $\eta(u) = u$ or $\eta(u) = -u$. Then η is either the identity map or the map $c + du \to c - du$. Thus Gal E/F is a cyclic group of order two.

2. $E = \mathbb{Q}(\sqrt{2},\sqrt{3})$. One sees easily that Gal E/F has order 4 and consists of the auto-morphisms $\eta_1 = 1$, η_2, η_3, η_4 such that $\eta_2(\sqrt{2}) = -\sqrt{2}$, $\eta_2(\sqrt{3}) = \sqrt{3}$; $\eta_3(\sqrt{2}) = \sqrt{2}$, $\eta_3(\sqrt{3}) = -\sqrt{3}$; $\eta_4(\sqrt{2}) = -\sqrt{2}$, $\eta_4(\sqrt{3}) = -\sqrt{3}$.

3. Let F be imperfect of characteristic p and let $a \in F$, $\notin F^p$. Then $x^p - a$ is irreducible (Lemma, p. 232). Adjunction of a root u of $x^p = a$ gives an extension $E = F(u)$ such that $[E:F] = p$. Moreover, since $x^p - a = (x - u)^p$ in $E[x]$, then E is a splitting field over F of the inseparable polynomial $x^p - a$. If $\eta \in$ Gal E/F then $\eta(u)^p = a$ so $\eta(u) = u$. It follows that $\eta = 1$ and Gal $E/F = 1$.

4. Let F be a field and let $E = F(t)$ where t is transcendental over F. As shall be indicated in exercise 11 below, $u \in E$ is a generator of E/F if and only if it has the form

(15)
$$u = \frac{at + b}{ct + d}, \qquad ad - bc \neq 0.$$

Since an automorphism of E/F sends generators into generators, it follows that Gal E/F is the set of maps

$$f(t)/g(t) \to f(u)/g(u)$$

where u is as in (15). We can see from this that Gal E/F is isomorphic to the factor group $GL_2(F)/F^*$ where $GL_2(F)$ is the group of invertible 2×2 matrices with entries in F, and F^* is the set of matrices diag$\{a, a\}$, $a \neq 0$.

Now let G be any group of automorphisms of a field E (that is, a subgroup of the automorphism group of E). Let

$$\text{Inv } G = \{a \in E \mid \eta(a) = a, \eta \in G\}$$

in other words, Inv G is the set of elements of E which are not moved by any $\eta \in G$. From the properties

$$\eta(a + b) = \eta(a) + \eta(b), \qquad \eta(ab) = \eta(a)\eta(b), \qquad \eta(1) = 1,$$

$$\eta(a^{-1}) = \eta(a)^{-1}, \qquad a \neq 0$$

of an automorphism of a field, it is clear that Inv G is a subfield of E. We call this the subfield of G-invariants or the G-fixed subfield of E.

If E is a given field then the definitions of Inv G for G a group of automor-phisms in E, and of Gal E/F for F a subfield, provide two maps

$$G \to \text{Inv } G$$

$$F \to \text{Gal } E/F.$$

The first is from the set of groups of automorphisms of E into the set of sub-fields of E, the second from the set of subfields of E to the set of groups of automorphisms. We shall now list the basic properties of these maps:

(i) $G_1 \supset G_2 \Rightarrow \text{Inv } G_1 \subset \text{Inv } G_2$

(ii) $F_1 \supset F_2 \Rightarrow \text{Gal } E/F_1 \subset \text{Gal } E/F_2$

(iii) $\text{Inv } (\text{Gal } E/F) \supset F$

(iv) $\text{Gal } (E/\text{Inv } G) \supset G$.

These are immediate consequences of the definitions and we leave it to the reader to carry out the verifications.

We shall now apply these ideas to splitting fields. Using the present terminology, the remarks following Theorem 4.4. can be restated as follows. If E is a splitting field over F of a polynomial $f(x)$, then Gal E/F is finite and we have the inequality $|\text{Gal } E/F| \leq [E:F]$. Moreover, $|\text{Gal } E/F| = [E:F]$ if $f(x)$ has distinct roots. In section 4.4. we saw that we can replace $f(x)$ by a polynomial $f_1(x)$ which is the product of the distinct prime factors of $f(x)$, and if $f(x)$ is separable then $f_1(x)$ has distinct roots. We therefore have the following important preliminary result.

LEMMA 1.   *Let E/F be a splitting field of a separable polynomial contained in $F[x]$. Then*

(16) $$|\text{Gal } E/F| = [E:F].$$

Our next attack will be from the group side. We begin with an arbitrary field E and any finite group of automorphisms G acting in E. Then we have the following

LEMMA 2.   (Artin.) *Let G be a finite group of automorphisms of a field E and let $F = \text{Inv } G$. Then*

(17) $$[E:F] \leq |G|.$$

Proof.   Let $n = |G|$. Then (17) will follow if we can show that any $m > n$ elements of E are linearly dependent over F. We shall base the proof of this on the well-known result of linear algebra that any system of n homogeneous linear equations in $m > n$ unknowns, with coefficients in a field E, has a non-trivial solution in E. This theorem is often used to prove the invariance of dimensionality of a finite dimensional vector space, so it is very likely familiar to the reader. For the sake of completeness we shall append a proof of the theorem on linear equations after this proof. Let $G = \{\eta_1 = 1, \eta_2, \ldots, \eta_n\}$ and let $u_1, u_2, \ldots, u_m$ be $m > n$ elements of E. Then the theorem on linear equations assures us that we have a non-trivial solution $(a_1, \ldots, a_m)$ of the system of

4.5 The Galois Group. The Fundamental Galois Pairing

237

n equations

$$(18) \qquad \sum_{j=1}^{m} \eta_i(u_j)x_j = 0, \qquad 1 \leq i \leq n$$

in the m unknowns $x_1, \ldots, x_m$. By non-triviality we mean that $(a_1, \ldots, a_m) \neq (0, \ldots, 0)$. Among such solutions we choose one $(b_1, \ldots, b_m)$ with the least number of non-zero b's. By reordering the unknowns we may assume $b_1 \neq 0$ and observing that $b_1^{-1}(b_1, \ldots, b_m)$ is also a solution, we may assume $b_1 = 1$. At this point we claim that every b_j is in $F = \text{Inv } G$, which will prove the F-dependence of the u_i, since the first equation in (18) is $\sum u_j x_j = 0$ $(\eta_1 = 1)$. Suppose some $b_j \notin F$. Without loss of generality we may assume this is b_2 and, by the definition of F, we have an $\eta_k \in G$ such that $\eta_k(b_2) \neq b_2$. Now we apply η_k to the system of equations $\sum_{j=1}^{m} \eta_i(u_j)b_j = 0, 1 \leq i \leq n$. This will give us $\sum_{j=1}^{m} (\eta_k \eta_i)(u_j)\eta_k(b_j) = 0, 1 \leq i \leq n$, and since $(\eta_k \eta_1, \ldots, \eta_k \eta_n)$ is a permutation of $(\eta_1, \ldots, \eta_n)$, we have the equations $\sum_{j=1}^{m} \eta_i(u_j)\eta_k(b_j) = 0, 1 \leq i \leq n$. Thus $(1, \eta_k(b_2), \ldots, \eta_k(b_m))$ is also a solution of (18). Subtracting this from the solution $(1, b_2, \ldots, b_m)$ we obtain the solution $(0, b_2 - \eta_k(b_2), \ldots, b_m - \eta_k(b_m))$ which is non-trivial since $b_2 - \eta_k(b_2) \neq 0$. Clearly this has fewer non-zero entries than $(b_1, b_2, \ldots, b_m)$, contrary to our choice of $(b_1, b_2, \ldots, b_m)$. This completes the proof modulo the

LEMMA ON LINEAR EQUATIONS. *Let*

$$a_{11}x_1 + a_{12}x_2 + \cdots + a_{1m}x_m = 0$$

$$a_{21}x_1 + a_{22}x_2 + \cdots + a_{2m}x_m = 0$$

$$\cdots\cdots\cdots\cdots\cdots\cdots\cdots\cdots\cdots\cdots$$

$$a_{n1}x_1 + a_{n2}x_2 + \cdots + a_{nm}x_m = 0$$

be a system of $n < m$ linear homogeneous equations with coefficients a_{ij} in a field E. Then there exists a solution $(a_1, a_2, \ldots, a_m) \neq (0, 0, \ldots, 0)$ with $a_i \in E$.

Proof. The result is trivial if every $a_{ij} = 0$ so we may assume some $a_{ij} \neq 0$. Since we can reorder the equations and the variables there is no loss in generality in assuming that $a_{nm} \neq 0$. Subtract the last equation multipled by $a_{im}a_{nm}^{-1}$ from the ith, $1 \leq i \leq n - 1$. This gives an equivalent system of equations

$$a'_{11}x_1 + a'_{12}x_2 + \cdots + a'_{1m-1}x_{m-1} = 0$$

$$\cdots\cdots\cdots\cdots\cdots\cdots\cdots\cdots\cdots$$

$$a'_{n-1,1}x_1 \qquad + \cdots + a'_{n-1,m-1}x_{m-1} = 0$$

$$a_{n1}x_1 \qquad + \cdots + a_{n,m-1}x_{m-1} + a_{nm}x_m = 0.$$

Assuming the result for $n - 1$, we have a non-trivial solution $(a_1, \ldots, a_{m-1})$ of the first $n - 1$ equations. Then if we put

$$a_m = -a_{nm}^{-1}(a_{n1}a_1 + \cdots + a_{n,m-1}a_{m-1})$$

we obtain the non-trivial solution $(a_1, a_2, \ldots, a_m)$ of the second system, hence of the first. Since the case $n = 1$ is trivial this proves the result by induction on n. $\square$

It is convenient at this point to introduce two field concepts which are related to concepts for polynomials which we have introduced previously. We recall that an extension field E/F is said to be *algebraic* over F if every element of E is algebraic over F; this will certainly be the case if E is finite dimensional over F, since $F(t)$ is infinite dimensional when t is transcendental. We shall now call E/F *separable* (*algebraic*) if the minimum polynomial of every element of E is separable. The extension field E/F is called *normal* (*algebraic*) if every irreducible polynomial in $F[x]$ which has a root in E is a product of linear factors in $E[x]$. This is equivalent to saying that E contains a splitting field for the minimum polynomial of every element of E. Normality plus separability mean that every irreducible polynomial of $F[x]$ which has a root in E is a product of distinct linear factors in $E[x]$. Also, by the results of the last section, if E is algebraic over F, then E is necessarily separable over F if the characteristic is 0 or if the characteristic is $p \neq 0$ and $F^p = F$.

We are now ready to derive our main results, the first of which gives two abstract characterizations of splitting fields of separable polynomials and some important additional information. We state this as

THEOREM 4.7 *Let E be an extension field of a field F. Then the following conditions on E/F are equivalent:*
  (1) *E is a splitting field over F of a separable polynomial $f(x)$.*
  (2) *$F = \operatorname{Inv} G$ for some finite group of automorphisms of E.*
  (3) *E is finite dimensional normal and separable over F.*

  Moreover, if E and F are as in (1) and $G = \operatorname{Gal} E/F$ then $F = \operatorname{Inv} G$ and if G and F are as in (2), then $G = \operatorname{Gal} E/F$.

Proof. (1) $\Rightarrow$ (2). Let $G = \operatorname{Gal} E/F$ and $F' = \operatorname{Inv} G$. Then F' is a subfield of E containing F. Also it is clear that E is a splitting field over F' of $f(x)$ as well as over F and $G = \operatorname{Gal} E/F'$. Hence, by Lemma 1, $|G| = [E:F]$ and $|G| = [E:F']$. Since $E \supset F' \supset F$ we have $[E:F] = [E:F'][F':F]$. Hence $[F':F] = 1$, and so

$F' = F$. We have proved also that $F = \text{Inv } G$ for $G = \text{Gal } E/F$, which is the first of the two supplementary statements.

(2) $\Rightarrow$ (3). By Artin's lemma, $[E:F] \leq |G|$, and so E is finite dimensional over F. Let $f(x)$ be an irreducible polynomial in $F[x]$ having a root r in E. Let $\{r_1 = r, r_2, \ldots, r_m\}$ be the orbit of r under the action of G. Thus this is the set of distinct elements of the form $\eta(r)$, $\eta \in G$. Hence if $\zeta \in G$, then the set $(\zeta(r_1), \zeta(r_2), \ldots, \zeta(r_m))$ is a permutation of $(r_1, r_2, \ldots, r_m)$. We have $f(r) = 0$ which implies that $f(r_i) = 0$. Then $f(x)$ is divisible by $x - r_i$, and since the r_i, $1 \leq i \leq m$, are distinct, $f(x)$ is divisible by $g(x) \equiv \prod_1^m (x - r_i)$. We now apply to $g(x)$ the automorphism of $E[x]$, which sends $x \to x$ and $a \to \zeta(a)$ for $a \in E$. This gives $\zeta g(x) = \prod_1^m (x - \zeta(r_i)) = \prod_1^m (x - r_i) = g(x)$. Since this holds for every $\zeta \in G$ we see that the coefficients of $g(x)$ are G-invariant. Hence $g(x) \in F[x]$. Since we assumed $f(x)$ irreducible in $F[x]$ we see that $f(x) = g(x) = \prod (x - r_i)$, a product of distinct linear factors in $E[x]$. Thus E is separable and normal over F and (3) holds.

(3) $\Rightarrow$ (1). Since we are given that $[E:F] < \infty$ we can write $E = F(r_1, r_2, \ldots, r_k)$ and each r_i is algebraic over F. Let $f_i(x)$ be the minimum polynomial of r_i over F. Then the hypothesis implies that $f_i(x)$ is a product of distinct linear factors in $E[x]$. It follows that $f(x) = \prod_1^k f_i(x)$ is separable and $E = F(r_1, r_2, \ldots, r_k)$ is a splitting field over F of $f(x)$. Hence we have (1).

It remains to prove the second supplementary statement. We have seen that under the hypothesis of (2) we have $[E:F] \leq |G|$, and that since (3) holds, we have $|\text{Gal } E/F| = [E:F]$. Since $G \subset \text{Gal } E/F$ and $|G| \geq [E:F] = |\text{Gal } E/F|$, evidently $G = \text{Gal } E/F$. $\square$

We are now ready to establish Galois' fundamental group-field pairing. We state this as the

FUNDAMENTAL THEOREM OF GALOIS THEORY. *Let E be an extension field of a field F satisfying any one (hence all) of the equivalent conditions of Theorem 4.7. Let G be the Galois group of E over F. Let $\Gamma = \{H\}$, the set of subgroups of G, and Σ, the set of intermediate fields between E and F (the subfields of E/F). The maps $H \to \text{Inv } H$, $K \to \text{Gal } E/K$, $H \in \Gamma$, $K \in \Sigma$, are inverses and so are bijections of Γ onto Σ and of Σ onto Γ. Moreover, we have the following properties of the pairing:*

(α) $H_1 \supset H_2 \Leftrightarrow \text{Inv } H_1 \subset \text{Inv } H_2$.

(β) $|H| = [E:\text{Inv } H]$, $[G:H] = [\text{Inv } H:F]$.

(γ) *H is normal in G $\Leftrightarrow$ $\text{Inv } H$ is normal over F. In this case*
$$\text{Gal Inv } H/F \cong G/H.$$

Proof. Let H be a subgroup of $G = $ Gal E/F. Since $F = $ Inv G, $F \subset$ Inv H and $K = $ Inv H is thus a subfield of E containing F. Applying the second supplementary result of Theorem 4.7 to H in place of G we see that Gal $E/$Inv $H = H$. In the same way we see that $|H| = |$Gal $E/$Inv $H| = [E:$Inv $H]$. Now let K be any subfield of E/F and let $H = $ Gal E/K. Then $H \subset G = $ Gal E/F so H is a subgroup of G. It is clear also that E is a splitting field over K of a separable polynomial since it is a splitting field over F of a separable polynomial. Hence the first supplementary result of Theorem 4.7 applied to the pair E and K shows that $K = $ Inv $H = $ Inv(Gal E/K). We have now shown that the specified maps between Γ and Σ are inverses. Also we know that if $H_1 \supset H_2$ then Inv $H_1 \subset$ Inv H_2. Moreover, if Inv $H_1 \subset$ Inv H_2, then we have also that $H_1 = $ Gal $E/$ Inv $H_1 \supset$ Gal $E/$Inv $H_2 = H_2$. Hence (α) holds. The first part of (β) was noted before. Since $|G| = [E:F] = [E:$Inv $H][$Inv $H:F] = |H|[$Inv $H:F]$ and $|G| = |H|[G:H]$, evidently $[$Inv $H:F] = [G:H]$. This proves (β). If $H \in \Gamma$ and $K = $ Inv H is the corresponding subfield, then the subfield K' corresponding to the conjugate subgroup $\eta H \eta^{-1}$ is $\eta(K)$. This is clear since the condition $\zeta(k) = k$ is equivalent to $(\eta \zeta \eta^{-1})(\eta(k)) = \eta(k)$. It now follows that H is normal in G if and only if $\eta(K) = K$ for every $\eta \in G$ ($K = $ Inv H). Suppose this holds. Then every $\eta \in G$ maps K onto itself and so its restriction $\bar\eta = \eta|K$ is an automorphism of K/F. Thus we have the restriction homomorphism $\eta \to \bar\eta$ of $G = $ Gal E/F into Gal K/F. The image $\bar G$ is a group of automorphisms in K and clearly Inv $\bar G = F$. Hence $\bar G = $ Gal K/F. The kernel of the homomorphism $\eta \to \bar\eta$ is the set of $\eta \in G$ such that $\eta|K = 1_K$. By the pairing, this is Gal $E/K = H$. Hence the kernel is H and $\bar G = $ Gal $K/F \cong G/H$. Since $F = $ Inv $\bar G$, K is normal over F by Theorem 4.7. Conversely, suppose K is normal over F. Let $a \in K$ and let $f(x)$ be the minimum polynomial of a over F. Then $f(x) = (x - a_1)(x - a_2) \cdots (x - a_m)$ in $K[x]$ where $a = a_1$. If $\eta \in G$ then $f(\eta(a)) = 0$ which implies that $\eta(a) = a_i$ for some i. Thus $\eta(a) \in K$. We have therefore shown that $\eta(K) \subset K$. As before, this implies that $\eta H \eta^{-1} \subset H$ if H is the subgroup corresponding to K in the Galois pairing. Then H is a normal subgroup of G. This completes the proof of (γ). $\square$

As our first example of this theorem we shall consider the field of the 17th roots of unity. This will clear up the mystery in the calculations for $\cos 2\pi/17$ which we gave on pp. 222–224 and reveal the reason for their success.

EXAMPLE

Let $F = \mathbb{Q}$ and let E be the field of the 17th roots of unity over F, that is, a splitting field over $\mathbb{Q}$ of $x^{17} - 1$. Since $(x^{17} - 1)' = 17x^{16}$ is relatively prime to $x^{17} - 1$, $x^{17} - 1$ has distinct roots. These constitute a cyclic subgroup U of the multiplicative group of

E. Let z be a generator of this group. Then $U = \{z, z^2, \ldots, z^{17} = 1\}$ and $E = \mathbb{Q}(z)$. The minimum polynomial of z over $\mathbb{Q}$ is $x^{16} + x^{15} + \cdots + 1$ since this polynomial is irreducible. Hence $[E:\mathbb{Q}] = 16$. Consequently, if $G = \mathrm{Gal}\, E/\mathbb{Q}$ then $|G| = 16$. If $\eta \in G$, $\eta(U) \subset U$ and $\eta | U$ is thus an automorphism of the cyclic group $U = \langle z \rangle$. We have the homomorphism $\eta \to \bar{\eta} = \eta | U$ of G into Aut U. This is a monomorphism since if $\bar{\eta} = 1$ then $\eta(z) = z$, which implies that $\eta = 1$ since $E = \mathbb{Q}(z)$. We know that the group of automorphisms of a cyclic group of order n is isomorphic to the multiplicative group of $\mathbb{Z}/(n)$ and that this has order $\varphi(n)$. In particular, Aut U has order 16 and is isomorphic to the multiplicative group of the field $\mathbb{Z}/(17)$. Moreover, this is a cyclic group with $\bar{3} = 3 + (17)$ as generator. Comparison of orders shows that $\mathrm{Gal}\, E/\mathbb{Q}$ is isomorphic to the multiplicative group of $\mathbb{Z}/(17)$. Hence, the automorphism η of $E/\mathbb{Q}$ such that $z \to z^3$ is a generator of $G = \mathrm{Gal}\, E/\mathbb{Q}$. Thus $G = \{\eta, \eta^2, \ldots, \eta^{16} = 1\}$. We have the following list of subgroups of G:

$$G = G_1 = \langle \eta \rangle \supset G_2 = \langle \eta^2 \rangle \supset G_3 = \langle \eta^4 \rangle \supset G_4 = \langle \eta^8 \rangle \supset G_5 = \{1\}$$

where $\langle \eta^i \rangle$ denotes the subgroup generated by η^i. The respective orders are 16, 8, 4, 2, and 1. Corresponding to these subgroups of $\mathrm{Gal}\, E/\mathbb{Q}$, the Galois pairing gives an increasing sequence of subfields

(19) $$F = F_1 \subset F_2 \subset F_3 \subset F_4 \subset F_5 = E$$

where F_i corresponds to G_i in the Galois pairing. What is F_i? We use the notations which we introduced at the end of section 4.2. As we noted there, $(z, z^2, \ldots, z^{16})$ is a base for $E/\mathbb{Q}$ since $x^{16} + x^{15} + \cdots + 1$ is the minimum polynomial of z over $\mathbb{Q}$. We have $\eta(z) = z^3$, $\eta^i(z) = z^{3^i}$. Putting

$$x_1 = \sum_{i=1}^{8} \eta^{2i}(z)$$

we have $\eta^2(x_1) = x_1$, $\eta(x_1) \neq x_1$. Hence $x_1 \in F_2$, $\notin F_1$. Since $[G:G_2] = [G_1:G_2] = 2$ we have $[F_2:F_1] = 2$. It follows that $F_2 = F_1(x_1)$. Similarly, if we put

$$y_1 = \sum_{i=1}^{4} \eta^{4i}(z), \qquad z_1 = \sum_{i=1}^{2} \eta^{8i}(z)$$

then $F_3 = F_2(y_1)$ and $F_4 = F_3(z_1)$. Thus the chain of subfields (19) is

(19′) $$F \subset F(x_1) \subset F(x_1, y_1) \subset F(x_1, y_1, z_1) \subset E$$

and the calculations we gave before amounted to determination of the minimum polynomials of x_1, y_1, z_1, and z over the fields F_1, F_2, etc., and the calculation of these elements as roots of quadratic equations. We remark also that since the Galois group G is abelian all of its subgroups are normal; hence every subfield of E/F is normal over F.

As a second illustration of the Galois correspondence we shall obtain the theory of symmetric rational expressions, which is similar and related to the results on symmetric polynomials that were obtained in section 2.13.

We begin with a field F and consider the field of fractions $F(x_1, \ldots, x_n)$ of the polynomial ring $F[x_1, \ldots, x_n]$ over F in indeterminates x_i. We recall that if π is any permutation of $\{1, 2, \ldots, n\}$, then we have a unique automorphism $\zeta(\pi)$ of $F[x_1, \ldots, x_n]$ fixing the elements of F and sending $x_i \to x_{\pi(i)}$, $1 \le i \le n$, (Theorem 2.12, p. 125). Moreover, $\zeta(\pi)$ can be extended in one and only one way to an automorphism of the field $F(x_1, \ldots, x_n)$. For the sake of simplicity we denote the extension of $\zeta(\pi)$ to the field $F(x_1, \ldots, x_n)$ by $\zeta(\pi)$ again. For any two permutations π_1 and π_2 we have $\zeta(\pi_1 \pi_2) = \zeta(\pi_1)\zeta(\pi_2)$ in $F[x_1, \ldots, x_n]$, hence also in $F(x_1, \ldots, x_n)$. Thus the set of automorphisms $\{\zeta(\pi)\}$ is a group of automorphisms G in $F(x_1, \ldots, x_n)$ isomorphic to the symmetric group S_n. The fixed elements under the action of G are called *symmetric rational expressions*, and Inv G is the *field of symmetric rational expressions*. We proceed to determine this field by using the Galois correspondence. For this purpose we consider the polynomial ring $E[x]$ where $E = F(x_1, \ldots, x_n)$ and we introduce the polynomial

(20)
$$g(x) = (x - x_1)(x - x_2) \cdots (x - x_n)$$

which we can write as

(21)
$$g(x) = x^n - p_1 x^{n-1} + p_2 x^{n-2} - \cdots + (-1)^n p_n$$

where

(22)
$$p_1 = \sum_1^n x_i, \qquad p_2 = \sum_{i<j} x_i x_j, \qquad \ldots, \qquad p_n = x_1 x_2 \cdots x_n.$$

The automorphism $\zeta(\pi)$ can be extended to an automorphism $\zeta'(\pi)$ of $E[x]$ fixing x. This maps $g(x)$ into $(x - x_{\pi(1)})(x - x_{\pi(2)}) \cdots (x - x_{\pi(n)})$. Since π is a permutation of the indices it is clear that this coincides with $g(x)$. Thus $\zeta'(\pi)(g(x)) = g(x)$ for every π and so $\zeta(\pi)p_i = p_i$ for every π and $i = 1, 2, \ldots, n$. Hence the $p_i \in$ Inv G, and the subfield over F they generate, $F(p_1, p_2, \ldots, p_n) \subset$ Inv G. On the other hand, it is clear from $E = F(x_1, \ldots, x_n) = F(p_1, \ldots, p_n, x_1, \ldots, x_n)$ and (20) that E is a splitting field over $F(p_1, \ldots, p_n)$ of $g(x)$, and $g(x)$ has distinct roots. Let $\zeta \in$ Gal $E/F(p_1, \ldots, p_n)$. Then applying ζ to $g(x_i) = 0$ we obtain $g(\zeta(x_i)) = 0$, and so ζ permutes the x_i. Hence ζ coincides with one of the $\zeta(\pi)$. It follows that Gal $E/F(p_1, \ldots, p_n) = G$ and we have the Galois pairing between the set of subgroups of G and the set of subfields of E containing $F(p_1, \ldots, p_n)$. In particular, this pairs G and $F(p_1, \ldots, p_n)$ and since the subfield corresponding to a subgroup H of G is Inv H, we have Inv $G = F(p_1, \ldots, p_n)$. This proves the field analogue of the first part of Theorem 2.20 on symmetric polynomials: any symmetric rational expression in the indeterminates x_i can be expressed rationally in terms of the elementary symmetric polynomials $p_1, p_2, \ldots, p_n$. This can also be derived easily from Theorem 2.20.

EXERCISES

1. Show that in the subgroup-intermediate subfield correspondence given in the fundamental theorem of Galois theory, the subfield corresponding to the intersection of two subgroups H_1 and H_2 is the subfield generated by the corresponding intermediate fields (Inv H_1 and Inv H_2), and the intersection of two intermediate fields K_1 and K_2 corresponds to the subgroup generated by the corresponding subgroups Gal E/K_1 and Gal E/K_2.

2. Suppose E, F, G are as in the fundamental theorem and E is generated by two intermediate extensions K and L such that $K \cap L = F$ and L/F is normal. Let $N = $ Gal E/L, $H = $ Gal E/K. Show that N is normal in G, $H \cap N = 1$ and $G = HN$, so G is the semi-direct product of N and H (exercise 9, p. 79). Show also that if K/F and L/F are normal then $G = H \times N$.

3. Let $E = \mathbb{Q}(r)$ where $r^3 + r^2 - 2r - 1 = 0$. Verify that $r' = r^2 - 2$ is also a root of $x^3 + x^2 - 2x - 1 = 0$. Determine Gal $E/\mathbb{Q}$. Show that E is normal over $\mathbb{Q}$.

4. Let E be a splitting field over $\mathbb{Q}$ of $x^5 - 2$. Determine the Galois group of $E/\mathbb{Q}$. Show that this is isomorphic to the holomorph of a cyclic group of order 5 (p. 63). Determine the subgroups of Gal $E/\mathbb{Q}$ and the corresponding subfields in the Galois pairing.

5. Let F be a field of characteristic p, a an element of F not of the form $b^p - b, b \in F$. Determine the Galois group over F of a splitting field of $x^p - x - a$.

6. Let $E = \mathbb{C}(t)$ where t is transcendental over $\mathbb{C}$ and let $w \in \mathbb{C}$ satisfy $w^3 = 1, w \neq 1$. Let σ be the automorphism of $E/\mathbb{C}$ such that $\sigma(t) = wt$, and τ the automorphism of $E/\mathbb{C}$ such that $\tau(t) = t^{-1}$. Show that

$$\sigma^3 = 1 = \tau^2, \qquad \tau\sigma = \sigma^{-1}\tau.$$

Show that the group of automorphisms G generated by σ and τ has order 6 and the subfield $F = $ Inv $G = \mathbb{C}(u)$ where $u = t^3 + t^{-3}$.

7. Let $E = (\mathbb{Z}/(p))(t)$ where t is transcendental over $\mathbb{Z}/(p)$. Let G be the group of automorphisms generated by the automorphism of E such that $t \to t + 1$. Determine $F = $ Inv G and $[E{:}F]$.

8. Same as exercise 7 with G replaced by the group of automorphisms such that $t \to at + b, a, b \in \mathbb{Z}/(p), a \neq 0$.

9. Show that $E = \mathbb{Q}(\sqrt{2}, \sqrt{3}, u)$ where $u^2 = (9 - 5\sqrt{3})(2 - \sqrt{2})$ is normal. Determine Gal $E/\mathbb{Q}$.

10. Use the method of proof of Artin's lemma (p. 236) to prove the following result on differential equations. Let $y_1, y_2, \ldots, y_{n+1}$ be real analytic functions which satisfy a linear differential equation $y^{(n)} + a_1 y^{(n-1)} + \cdots + a_n y = 0$ with constant coefficients a_i ($a_i \in \mathbb{R}$). Then the y_i are linearly dependent over $\mathbb{R}$.

11. Let $E = F(t)$ where t is transcendental over F and write any non-zero element of E as $u = f(t)/g(t)$ where $(f(t), g(t)) = 1$. Call the maximum of the degrees of f and g the *degree* of u. Show that if x and y are indeterminates then $f(x) - yg(x)$ is irreducible in $F[x, y]$ and hence is irreducible in $F(y)[x]$. Show that t is algebraic

over $F(u)$ with minimum polynomial the monic polynomial which is a multiple in $F(u)$ of $f(x) - ug(x)$. Hence conclude that $[F(t):F(u)] = 1$, and $F(u) = F(t)$ if and only if deg $u = 1$. Note that this implies that

$$u = (at + b)/(ct + d)$$

where $ad - bc \neq 0$. Hence conclude that Gal E/F is the set of maps $h(t) \to h(u)$ where u is of the form indicated.

12. Let $E = F(x_1, \ldots, x_n)$ where the x_i are indeterminates, and let $\zeta(\pi)$ for a permutation π of $\{1, 2, \ldots, n\}$ be as defined in the text. Write an element of E in the form $f(x_1, \ldots, x_n)/g(x_1, \ldots, x_n)$ where f and g have no common factors. Show that if this element is symmetric then both its numerator and denominator are symmetric. Use this to deduce from Theorem 2.20 that $f/g \in F(p_1, p_2, \ldots, p_n)$, p_i as above.

13. Let F be of characteristic $\neq 2$ and also let H be the subgroup of $G = $ Gal $E/F(p_1, \ldots, p_n)$ corresponding to the alternating group, that is, the set of $\zeta(\pi)$, $\pi \in A_n$. Show that Inv $H = F(p_1, \ldots, p_n, \Delta)$ where $\Delta = \prod_{i<j} (x_i - x_j)$.

4.6   SOME RESULTS ON FINITE GROUPS

We shall now digress briefly to develop some results on finite groups which are needed for the theory of equations. These mainly concern a class of groups, called solvable, which, as we shall see in the next section, correspond to equations which are solvable by radicals.

Let G be a group. We introduce a standard notation $G \rhd H$ or $H \lhd G$ to indicate that H is a normal subgroup of G. A sequence of subgroups

(23) $$G = G_1 \rhd G_2 \rhd \cdots \rhd G_s \rhd G_{s+1} = 1$$

is called a *normal series* for the group G. Here the notation indicates that G_{i+1} is normal in G_i (but not necessarily normal in G). For example, we have $S_3 \rhd A_3 \rhd 1$ and $S_4 \rhd A_4 \rhd V \rhd W \rhd 1$ where

$$V = \{1, (12)(34), (13)(24), (14)(23)\}$$

$$W = \{1, (12)(34)\}.$$

We leave it to the reader to check that V is normal in A_4. Since V is abelian it is clear that W is normal in V. With the normal series (23) we can associate the *sequence of factors*

(24) $$G_1/G_2, G_2/G_3, \ldots, G_s/G_{s+1} \cong G_s.$$

Now we shall call a group G *solvable* if it has a normal series whose sequence of factors are all abelian. The normal sequences we have displayed for S_3 and S_4 show that these groups are solvable. In fact, S_3/A_3 is cyclic of order 2, A_3 is cyclic of order 3, S_4/A_4 is cyclic of order 2, A_4/V is cyclic of order 3, and V/W and W are cyclic of order 2. Of course, any abelian group is solvable since $G \rhd 1$ is a normal series with abelian factor $G = G/1$. Another important class of solvable groups is given in

THEOREM 4.8.   *Any finite group of prime power order is solvable.*

Proof.   Let G be a *p-group*, that is, a group of order p^n, $n \geq 1$, p a prime. We have seen that G has a non-trivial center C. If $G \neq C$, put $C = C_1$ and consider G/C_1. This is a p-group and so it has non-trival center. The center of G/C_1 has the form C_2/C_1 where C_2 is normal in G. If $G \neq C_2$ let C_3 be the subgroup of G such that C_3/C_2 is the center of G/C_2. Continuing in this way we obtain the sequence of normal subgroups $1 \subsetneqq C_1 \subsetneqq C_2 \subsetneqq C_3 \subsetneqq \cdots$. Since G is finite we eventually reach $G = C_{s+1}$. Then $G = C_{s+1} \rhd C_s \rhd \cdots \rhd C_1 \rhd 1$ is a normal series with abelian factors C_{i+1}/C_i. $\square$

We shall now derive some of the basic properties of solvable groups, and we shall begin by giving a test for solvability in terms of a particular series of normal subgroups, the derived series. If $g, h \in G$ we define the *commutator of g and h* as

$$(g, h) = g^{-1}h^{-1}gh.$$

Then $gh = hg(g, h)$, so (g, h) measures the departure from commutativity of the elements g and h. We define the *derived* (or *commutator*) *subgroup* G' to be the subgroup of G generated by all the commutators (g, h), $g, h \in G$. Since $(g, h)^{-1} = (g^{-1}h^{-1}gh)^{-1} = h^{-1}g^{-1}hg = (h, g)$ it is clear that G' coincides with the set of products of the form

(25) $(g_1, h_1)(g_2, h_2) \cdots (g_k, h_k)$, $g_i, h_i \in G.$

Let η be a homomorphism of G into a second group $\bar{G}$. Then $\eta(g, h) = \eta(g^{-1}h^{-1}gh) = \eta(g)^{-1}\eta(h)^{-1}\eta(g)\eta(h) = (\eta(g), \eta(h))$. Hence $\eta(G') \subset \bar{G}'$. Moreover, if η is surjective then this formula shows that every commutator $(\bar{g}, \bar{h})$, $\bar{g}, \bar{h} \in \bar{G}$, is in $\eta(G')$. Hence in this case $\eta(G') = \bar{G}'$. In particular, these remarks apply to any endomorphism η of G. Now suppose $K \lhd G$. Then any inner automorphism $I_a : x \to axa^{-1}$ of G induces an endomorphism of K. Hence we have

$I_a(K') \subset K'$ for every $a \in G$, which means that K' is normal in G. Symbolically this can be stated as

(26) $$K \lhd G \Rightarrow K' \lhd G.$$

Since $G \lhd G$ we see that $G' \lhd G$.

We now define the *second derived group* $G'' = (G')'$ and iterate this to define $G^{(k)} = (G^{(k-1)})'$, $k \geq 1$. By induction, using (26), we see that $G^{(k)} \lhd G$. A weaker statement is that $G \rhd G' \rhd G'' \rhd \cdots$. We shall now show that G is solvable if and only if there exists a $k \geq 1$ such that

(27) $$G^{(k)} = 1.$$

This will follow rather quickly from the following characterization of the derived group.

LEMMA. *G/G' is abelian and G' is contained in every normal subgroup K such that G/K is abelian.*

Proof. It is clear from the definition of G' that G is abelian if and only if $G' = 1$. If $g, h \in G$ and $K \lhd G$, then $(gK, hK) = (gK)^{-1}(hK)^{-1}gKhK = (g, h)K$. Hence $(gK, hK) = 1 (= K) \Leftrightarrow (g, h) \in K$. Thus G/K is abelian if and only if K contains every commutator (g, h), which is the case if and only if $K \supset G'$. Both conclusions follows from this. $\square$

We can now prove

THEOREM 4.9. *A group G is solvable if and only if $G^{(k)} = 1$ for some $k \geq 1$.*

Proof. If the condition holds we have the normal series $G \rhd G' \rhd G'' \rhd \cdots \rhd G^{(k)} = 1$ and every $G^{(i)}/G^{(i+1)}$ is abelian by the lemma. Then G is solvable. Conversely, suppose G is solvable, so we have a normal series $G = G_1 \rhd G_2 \rhd \cdots \rhd G_s \rhd G_{s+1} = 1$ with abelian factors. By the lemma, $G_{i+1} \supset G_i'$, $i = 1, 2, \ldots$, since G_i/G_{i+1} is abelian. In particular, $G_2 \supset G_1' = G'$ and assuming $G_k \supset G^{(k)}$, we have $G_{k+1} \supset G_k' \supset (G^{(k)})' = G^{(k+1)}$. Hence $G_i \supset G^{(i)}$ for all i, and since $G_{s+1} = 1$, we have $G^{(s+1)} = 1$. $\square$

An easy consequence of the foregoing criterion for solvability is the following

THEOREM 4.10. *Any subgroup and any homomorphic image of a solvable group is solvable. If $K \lhd G$ and K and G/K are solvable then G is solvable.*

Proof. If H is a subgroup, it is clear that $H \subset G$ implies $H^{(i)} \subset G^{(i)}$. Hence $G^{(k)} = 1$ implies $H^{(k)} = 1$, and so G solvable implies H solvable. Let η be a surjective homomorphism of G into H. Then $\eta(G') = (\eta(G))'$, so η restricted to G' is a surjective homomorphism of G' onto $(\eta(G))'$. Then $\eta(G'') = \eta((G')') = (\eta(G'))' = (\eta(G))''$. By induction we have $\eta(G^{(i)}) = (\eta(G))^{(i)}$. Hence $G^{(k)} = 1$ implies $(\eta(G))^{(k)} = 1$. This proves that if G is solvable so is any homomorphic image $\eta(G)$. Now assume $K \lhd G$ and G/K is solvable. Let ν be the natural homomorphism of G into G/K. Since this is surjective we have $\nu(G^{(i)}) = (G/K)^{(i)}$. Hence $\nu(G^{(k)}) = 1$ for a suitable $k \geq 1$. This shows that $G^{(k)} \subset K$. If K is also solvable, then we have an $l \geq 1$ such that $K^{(l)} = 1$. Then $G^{(k+l)} \subset K^{(l)} = 1$, and so G is solvable. $\square$

A group G is called *simple* if G and 1 are the only normal subgroups of G. Since every subgroup of an abelian group is normal, an abelian group is simple if an only if it has no subgroups other than itself and 1. Clearly this means that the group is a cyclic group of prime order, a class of simple groups that is generally regarded as trivial. The simplest class of non-abelian simple groups is given by

THEOREM 4.11. *A_n is simple if $n \geq 5$.*[6]

Proof. We shall prove simplicity of A_n by showing that if $A_n \rhd K \neq 1$ then $K = A_n$. It suffices to show that K contains a 3-cycle, say, (123). Then if (ijk) is any 3-cycle we take γ to be a permutation of the form

$$\gamma = \begin{pmatrix} 1 & 2 & 3 & 4 & 5 & \cdots \\ i & j & k & l & m & \cdots \end{pmatrix}$$

of $1, 2, \ldots, n$, which we may assume is even, since if our first choice of γ is odd then we can replace it by the even permutation $(lm)\gamma$. Now assuming γ even, we see that $(ijk) = \gamma(123)\gamma^{-1} \in K$. Thus K contains every 3-cycle and since A_n is generated by the 3-cycles (exercise 2, p. 51), $K = A_n$. Now let α be an element of K such that $\alpha \neq 1$ and α has a maximum number of fixed points among the elements $\neq 1$ in K. By a fixed point of α we mean an i, $1 \leq i \leq n$, such that $\alpha(i) = i$. We claim that α is a 3-cycle. Otherwise, if we write α as a product of

---

[6] This result is due to Galois.

disjoint cycles omitting those of length 1, then in this representation α has either the form

(28) $\alpha = (123 \cdots) \cdots$

or

(29) $\alpha = (12)(34) \cdots$

a product of disjoint transpositions. In the first case α moves two other numbers, say 4 and 5, since α is not one of the odd permutations $(123 \, k)$. Now let $\beta = (345)$ and form $\alpha_1 = \beta \alpha \beta^{-1}$. If α is as in (28) then $\alpha_1 = (124 \cdots) \cdots$, and if α is as in (29) then $\alpha_1 = (12)(45) \cdots$. In either case $\alpha_1 \neq \alpha$ and $\alpha_2 = \alpha_1 \alpha^{-1} \neq 1$. Now any number > 5 is fixed by β so if it is fixed by α, then it is also fixed by $\alpha_2 = \beta \alpha \beta^{-1} \alpha^{-1}$. Moreover, if α is as in (28) then $\alpha_2(2) = 2$ and since in this case α moves 1, 2, 3, 4, and 5, it is clear that α_2 has more fixed points than α contrary to the choice of α. If α is as in (29) we have $\alpha_2(1) = 1$ and $\alpha_2(2) = 2$ so again α_2 has more fixed points than α. This contradiction proves that α is a 3-cycle and $K = A_n$. $\square$

COROLLARY. S_n is not solvable if $n \geq 5$.

Proof.   If it were, then the subgroup A_n would be solvable. Then $A_n' \subset A_n$ and since A_n is simple and $A_n' \lhd A_n$ we have $A_n' = 1$. Then A_n would be abelian. This is certainly not the case (even for $n \geq 4$) since (123) and (234) do not commute. $\square$

We shall now give another criterion for solvability of a finite group. This will be in terms of the concept of a *composition series*, which is an important notion in the theory of finite groups. We define a *composition series* for a group G to be a normal series $G = G_1 \rhd G_2 \rhd \cdots \rhd G_{s+1} = 1$ such that each G_{i+1} is maximal normal in G_i, that is, there exists no normal subgroup H of G_i such that $G_i \gneqq H \gneqq G_{i+1}$. We recall that if $G \rhd K$ then we have the bijective mapping $H \to H/K$ of the set of subgroups of G containing K onto the set of subgroups of G/K. In this, normal subgroups are paired. It follows that K is maximal normal in G if and only if G/K is simple $\neq 1$. Hence a normal series $G = G_1 \rhd G_2 \rhd \cdots \rhd G_{s+1} = 1$ is a composition series if and only if every factor G_i/G_{i+1} is simple $(\neq 1)$. These factors are called the *composition factors* determined by the series.

If G is a finite group, then it is clear that $G = G_1$ contains a maximal normal subgroup G_2, and that G_2 contains a maximal normal subgroup G_3. Continuing in this way we see that any finite group has a composition series. The composition factors are determined up to isomorphism in the following strong sense.

THE JORDAN-HÖLDER THEOREM.   *Let G be a finite group and let* $G = G_1 \rhd G_2 \rhd \cdots \rhd G_{s+1} = 1$, *and* $G = H_1 \rhd H_2 \rhd \cdots \rhd H_{t+1} = 1$ *be two composition series for G. Then* $s = t$ *and there exists a permutation* $i \to i'$ *of* $1, 2, \ldots, s$ *such that* $G_i/G_{i+1} \cong H_{i'}/H_{i'+1}$, $1 \le i \le s$.

Proof.   We shall prove the theorem by induction on $|G|$ and we distinguish two cases: (I) $G_2 = H_2$ and (II) $G_2 \ne H_2$. In I we observe that $G_2 \rhd \cdots \rhd G_{s+1} = 1$ and $H_2 \rhd \cdots \rhd H_{t+1} = 1$ are composition series for the same group $G_2 = H_2$, whose order is less than $|G|$. Hence we may assume that $s - 1 = t - 1$ and we have a permutation $i \to i'$ of $2, \ldots, s$ such that $G_i/G_{i+1} \cong H_{i'}/H_{i'+1}$, $2 \le i \le s$. Since $G_1/G_2 = H_1/H_2$ the result is clear in this case. In II, since $G_2 \lhd G$ and $H_2 \lhd G$, $G_2 H_2 \lhd G$ (exercise 5, p. 57). Since $G_2 H_2$ contains G_2 and H_2 and $G_2 \ne H_2$ we have $G_2 H_2 = G$ by the maximality of G_2 as normal subgroup of G. By the second isomorphism theorem for groups (p. 65), we have $G/G_2 = G_2 H_2/G_2 \cong H_2/(G_2 \cap H_2)$, and, similarly, $G/H_2 \cong G_2/(G_2 \cap H_2)$. Thus we see that $K_3 \equiv G_2 \cap H_2$ is maximal normal in H_2 and in G_2 and we have the isomorphisms

(30) $$G_1/G_2 \cong H_2/K_3, \qquad H_1/H_2 \cong G_2/K_3.$$

Now let $K_3 \rhd K_4 \rhd \cdots \rhd K_{u+1}$ be a composition series for K_3. Then since K_3 is maximal normal in G_2 and H_2 we have the four composition series

(i) $G = G_1 \rhd G_2 \rhd G_3 \rhd \cdots \rhd G_{s+1} = 1$

(ii) $G = G_1 \rhd G_2 \rhd K_3 \rhd \cdots \rhd K_{u+1} = 1$

(iii) $G = H_1 \rhd H_2 \rhd K_3 \rhd \cdots \rhd K_{u+1} = 1$

(iv) $G = H_1 \rhd H_2 \rhd H_3 \rhd \cdots \rhd H_{t+1} = 1$.

By case I we see that $s = u$ and we can permute $i \to i'$, $1 \le i \le s$, to obtain $G_i/G_{i+1} \cong K_{i'}/K_{i'+1}$ where we take $K_1 = G_1$, $K_2 = G_2$. A similar result holds for the last two composition series. The result also holds for (ii) and (iii) since the first two composition factors for these are respectively

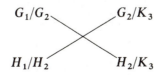

and the indicated pairing pairs isomorphic factors since we have (30). The rest of the factors are K_i/K_{i+1} and these are the same in (ii) and (iii). Putting together the information we have obtained it is clear that we have the theorem also in case II. □

We shall complete our discussion by proving the following useful criterion for solvability of a finite group.

THEOREM 4.12. *A finite group is solvable if and only if every composition factor G_i/G_{i+1} of a composition series $G = G_1 \triangleright G_2 \triangleright \cdots \triangleright G_{s+1} = 1$ is cyclic of prime order.*

Proof. Suppose that G is solvable. Then every composition factor G_i/G_{i+1} is solvable. Being simple, it is abelian, and hence it is cyclic of prime order. Thus every G_i/G_{i+1} has this property. Conversely, assume that we have a composition series $G = G_1 \triangleright G_2 \triangleright \cdots \triangleright G_{s+1} = 1$ with G_i/G_{i+1} cyclic of prime order. Then every G_i/G_{i+1} is abelian and G is solvable by definition. □

EXERCISES

1. Show that an abelian group has a composition series if and only if it is finite.

2. Let G be cyclic of order n $(< \infty)$ and let $G = G_1 \triangleright G_2 \triangleright \cdots \triangleright G_{s+1} = 1$ be a composition series. Put $|G_i| = n_i$. Show that $p_i = n_i/n_{i+1}$ is a prime, and conversely, if $n = n_1, n_2, \ldots, n_{s+1} = 1$ is a sequence of integers such that n_i/n_{i+1} is a prime then we have a composition series for which $|G_i| = n_i$. Use this result to deduce the fundamental theorem of arithmetic for $\mathbb{Z}$.

3. If g and h are elements of a group we write g^h for $h^{-1}gh$. Then $g^{hk} = (g^h)^k$ and by definition of $(g, h) = g^{-1}h^{-1}gh$ we have $g^h = g(g, h)$. Verify that

    (α) $(g, hk) = (g, k)(g, h)^k$

    (β) $(gh, k) = (g, k)^h(h, k)$

    (γ) $(g^h, (h, k))(h^k, (k, g))(k^g, (g, h)) = 1.$

4. If $H \triangleleft G$ and $K \triangleleft G$ define (H, K) to be the subgroup generated by the commutators (h, k), $h \in H$, $k \in K$. Show that $(H, K) = (K, H) \triangleleft G$.

5. Show that if $H \triangleleft G$, $K \triangleleft G$, and $L \triangleleft G$ then

    $$(H, (K, L)) \subset (K, (L, H))(L, (H, K)).$$

6. Define G^i by $G^1 = G$, $G^i = (G^{i-1}, G)$. The sequence of normal subgroups $G^1 \supset G^2 \supset G^3 \supset \cdots$ is called the *lower central series* for G. G is called *nilpotent* if there

exists an integer k such that $G^k = 1$. Show that if G is nilpotent, then it is solvable. Give an example to show that the converse does not hold.

7. Show that every element h of a nilpotent group has the *Engel property*: there

$$\overset{\overset{\textstyle k}{\diagup\diagdown}}{}$$

exists an integer k such that $((\cdots (g, h), h) \cdots h) = 1$ for every $g \in G$.

8. Prove that if H is a proper subgroup of a nilpotent group G, then the normalizer $N(H) \supsetneq H$. (See p. 81 for the definition of $N(H)$.)

9. Show that if G is a finite nilpotent group, then every Sylow subgroup is normal in G. Show that for every prime divisor p of $|G|$ there exists only one Sylow p-subset subgroup.

10. If G is a group define the *upper central series* $1 \subset C_1 \subset C_2 \subset \cdots$ by $C_1 = C(G)$, the center of G, and C_i, the normal subgroup such that C_i/C_{i-1} is the center of G/C_{i-1}. Show that a finite group G is nilpotent if and only if the upper central series ends in a finite number of steps with G ($G = C_k$ for some k). Note that this result and the proof of Theorem 4.8 imply that any p-group is nilpotent.

11. Prove that a finite group is nilpotent if and only if it is a direct product of p-groups.

4.7 GALOIS' CRITERION FOR SOLVABILITY BY RADICALS

Let F be a field and $f(x)$ a polynomial with coefficients in F. It is essential that we have a precise formulation of the statement that the equation $f(x) = 0$ is solvable by radicals over F. We give this in the following

DEFINITION 4.2. *Let $f(x) \in F[x]$ be monic of positive degree. Then the equation $f(x) = 0$ is said to be* solvable by radicals *over F if there exists an extension field K/F which possesses a tower of subfields.*

$$(31) \qquad\qquad F = F_1 \subset F_2 \subset \cdots \subset F_{r+1} = K$$

where each $F_{i+1} = F_i(d_i)$ and $d_i^{n_i} = a_i \in F_i$, and K contains a splitting field over F of $f(x)$. A tower of subfields such as (31) will be called a root tower *over F for K.*

Since each field F_{i+1} is obtained by adjoining a root $\sqrt[n_i]{a_i}$ of an equation $x^{n_i} = a_i$ and all the roots of $f(x)$ are contained in K, this means that every root of $f(x)$ can be obtained by starting with elements of the base field and performing a finite sequence of rational operations and solving equations of the form $x^n = a$.

Now assume $f(x)$ has distinct roots in a splitting field E/F. Then we define the *Galois group of the polynomial $f(x)$*, or *of the equation $f(x) = 0$*, to be the Galois group of the splitting field E/F. Since any two splitting fields are isomorphic over F it is clear that this is essentially independent of the choice of the splitting field. Let $f(x) = \prod_1^n (x - r_i)$ in $E[x]$, so $E = F(r_1, \ldots, r_n)$, and $R = \{r_1, \ldots, r_n\}$ is the set of (distinct) roots of $f(x)$ in E. As we shall now show, one can identify $G = \operatorname{Gal} E/F$ with a permutation group of the set of roots R.[7]

Let $\eta \in G$. Then it is clear that η maps R into itself and hence η induces a permutation of the set R. Hence we have the homomorphism $\eta \to \eta | R$ (the restriction of η to R) of G into the symmetric group S_n of permutations of $R = \{r_1, \ldots, r_n\}$. Since the r_i generate E/F it is clear that $\eta \to \eta | R$ is a monomorphism of G into S_n, and that its image, which we shall denote as G_f, is a subgroup of S_n isomorphic to G. Often we shall not distinguish between G and G_f. For example, if $G_f = S_n$ then we shall say that the Galois group of the equation $f(x) = 0$ is the symmetric group S_n.

We now have within our reach the crowning achievement of Galois' theory: the following criterion for solvability of an equation by radicals.

An equation $f(x) = 0$ is solvable by radicals over a field F of characteristic 0 if and only if its Galois group is solvable.

Besides the fundamental group-field correspondence we shall need some information of a more special type on fields of roots of unity and on cyclic fields, that is, splitting fields whose Galois groups are cyclic. We shall now call a splitting field of $x^n - 1$ over a given base field F a *cyclotomic field of order n over F*. We prove first

LEMMA 1. *The Galois group of the cyclotomic field of order n over F of characteristic 0 is abelian.*

Proof. Since $(x^n - 1)' = nx^{n-1}$ and $x^n - 1$ are relatively prime, $x^n - 1$ has n distinct roots $z_1, z_2, \ldots, z_n$. These constitute a subgroup U of the multiplicative group of the cyclotomic field. We know that U is cyclic. The map $\eta \to \eta | U$ of the Galois group G is a monomorphism of G into $\operatorname{Aut} U$, the group of automorphisms of the cyclic group U of order n. Thus G is isomorphic to a subgroup of $\operatorname{Aut} U$ and we know that the latter is isomorphic to the abelian group of units of the ring $\mathbb{Z}/(n)$. Hence G is abelian. $\square$

Remark. Of course, G need not be isomorphic to the group of units of $\mathbb{Z}/(n)$. For instance, F itself may contain the nth roots of unity, in which case F coincides with the cyclotomic field and $G = 1$.

---

[7] This was Galois' point of view: that is, he defined his group as a certain permutation group of the roots of $f(x) = 0$. The realization that this group could be identified with the group of automorphisms of the splitting field is due to Dedekind, and, as we have seen, this serves as the basis of the modern Galois theory.

We shall now call an extension field E/F *Galois over F* if it satisfies the equivalent conditions given in Theorem 4.7 (e.g., E is finite dimensional, normal, and separable over F). If E/F is Galois we have the fundamental Galois pairing of the subgroups of Gal E/F and the subfields of E/F. E/F is called *abelian* (*cyclic*) if it is Galois over F and $G = $ Gal E/F is abelian (cyclic). Lemma 1 states that any cyclotomic extension of characteristic 0 is abelian. We derive next two results on cyclic extensions under the hypothesis of the existence of certain roots of unity in the base field.

LEMMA 2. *If F contains n distinct nth roots of unity, then the Galois group of* $x^n - a$ *over F is cyclic of order a divisor of n.*

Proof. Let U be the set of nth roots of unity contained in F and let E be the splitting field over F of $x^n - a$. If r is one of the roots of $x^n = a$ in E then this equation has the n roots zr, $z \in U$, so $E = F(r)$. If η, $\zeta \in G = $ Gal E/F, then $\eta(r) = zr$, $\zeta(r) = z'r$ where z, $z' \in U$. Then $\zeta\eta(r) = zz'r$. Thus $\eta \to z$ is a monomorphism of G into the cyclic group U, and G is isomorphic to a subgroup of U. $\square$

We prove next a partial converse to Lemma 2:

LEMMA 3. *Let p be a prime and assume F contains p distinct pth roots of unity. Let E/F be cyclic and p dimensional. Then* $E = F(d)$ *where* $d^p \in F$.

Proof. Let $c \in E$, $\notin F$. Then $E = F(c)$. Let $U = \{z_1, z_2, \ldots, z_p\}$ be the pth roots of unity and let η be a generating automorphism of the Galois group G of E/F. Put $c_i = \eta^{i-1}(c)$, $1 \le i \le p$. Then $c_1 = c$ and $\eta(c_i) = c_{i+1}$ if $1 \le i \le p - 1, \eta(c_p) = c_1$. We introduce the *Lagrange resolvent*

$$(32) \qquad (z_i, c) = c_1 + c_2 z_i + c_3 z_i^2 + \cdots + c_p z_i^{p-1}.$$

Then $\eta(z_i, c) = c_2 + c_3 z_i + \cdots + c_i z_i^{p-1} = z_i^{-1}(z_i, c)$. Hence $\eta(z_i, c)^p = (z_i^{-1}(z_i, c))^p = (z_i, c)^p$ which shows that $(z_i, c)^p \in F$. Now we can express $c_1, c_2, \ldots, c_p$ as linear combinations of $(z_1, c), (z_2, c), \ldots, (z_p, c)$. To do this we regard (32) for $1 \le i \le p$ as a system of linear equations in $c_1, \ldots, c_p$. The determinant of the coefficients is a Vandermonde determinant whose value is well known to be $\prod_{i>j} (z_i - z_j)$. We now see that since $E = F(c)$ we also have $E = F(d_1, d_2, \ldots, d_p)$ where $d_i = (z_i, c)$. Then some $d_i \notin F$, so if we put $d = $ this d_i, we have $E = F(d)$ where $d^p \in F$. $\square$

We shall also need a result which describes what happens to the Galois group of an equation when we extend the base field. (In the older literature the formation of such an extension is called "adjunction of accessory irrationalities".) The result is the following

LEMMA 4. *Let $f(x) \in F[x]$ and let K be an extension field of F. Then the Galois group of $f(x)$ over K is isomorphic to a subgroup of the Galois group of $f(x)$ over F.*

Proof. Let L be a splitting field over K of $f(x)$. Since $K \supset F$, L contains a splitting field E of $f(x)$ over F. In fact, if $f(x) = \prod_1^n (x - r_i)$ in $L[x]$ then $L = K(r_1, \ldots, r_n)$ and $E = F(r_1, \ldots, r_n)$. If $\eta \in \text{Gal } L/K$, η maps $R = \{r_1, \ldots, r_n\}$ into itself and hence it maps E into itself. Since η is determined by its action on R, the restriction homomorphism $\eta \to \eta|E$ is a monomorphism of Gal L/K into Gal E/F. Thus Gal L/K is isomorphic to a subgroup of Gal E/F. $\square$

Let E be a finite dimensional extension field of F. Then E is generated over F by a finite set $\{a_1, \ldots, a_n\}$ of algebraic elements a_i. Let $f_i(x)$ be the minimum polynomial over F of a_i and put $f(x) = \prod f_i(x)$. Then we can construct a splitting field K over E of $f(x)$. Since $K \supset F(a_1, \ldots, a_n)$ and $f_i(a_i) = 0$ it is clear that K is also a splitting field over F of $f(x)$. Now it can be shown that the splitting field of a polynomial is always normal. We shall indicate this in an exercise below. For our present purposes it is enough to consider the case in which $f(x)$ is separable (which includes the characteristic 0 case). In this case the normality (and separability) of K/F follows from Theorem 4.7. It is clear also that every normal extension of E contains a splitting field over F of $f(x)$; hence it contains a subfield isomorphic to K. It follows from this that to within isomorphism the field K is determined by E/F and is independent of the choice of the set of generators. Accordingly, we shall call K/F the *normal closure* of E/F. Again assuming $f(x)$ is separable, let $G = \text{Gal } K/F$. If $\eta \in G$, the subfield $\eta(E)$ is isomorphic over F to E. The subfields $\eta(E)$ are called the *conjugates* of E/F in K. These generate K. For, if we let K' be the subfield of K generated by the $\eta(E)$, $\eta \in G$, then G maps K' into itself and so it determines a finite group of automorphisms G' of K' whose subfield of fixed elements is F. Then K' is normal over F, by Theorem 4.7, and consequently $K' = K$.

We can now show that in the definition of solvability by radicals there is no loss in generality in the separable case in assuming that the field K given in the definition is normal and separable over F. This is a consequence of the following

LEMMA 5. *Let E/F have a root tower over F, say $F = F_1 \subset F_2 \subset \cdots \subset F_{r+1} = E$ with $F_{i+1} = F_i(d_i)$, $d_i^{n_i} \in F_i$, and assume E is generated over F by a finite set of elements whose minimum polynomials are separable. Then the normal closure K/F of E/F has a root tower over F such that the distinct integers n_i for this tower are the same as those occurring in the given tower.*

Proof. The normal closure K/F is generated by the conjugate fields $\eta(E)$, $\eta \in$ Gal K/F. Applying η to the given tower $F = F_1 \subset F_2 \subset \cdots \subset F_{r+1} = E$ with $F_{i+1} = F_i(d_i)$, $d_i^{n_i} \in F_i$, we obtain a root tower over F for $\eta(E)$. Then $K = F(\eta_1(d_1), \ldots, \eta_1(d_r); \eta_2(d_1), \ldots, \eta_2(d_r); \ldots)$ where Gal $K/F = \{\eta_1, \eta_2, \ldots\}$. Obviously we can display a root tower for K over F satisfying the stated condition. $\square$

We are now ready to establish Galois' criterion for solvability of an equation by radicals. Suppose first that $f(x) = 0$ is solvable by radicals over F of characteristic 0. Then we have an extension field K/F of a splitting field of $f(x)$ which has a root tower over F as in (31). By lemma 5 we may assume K normal over F. Since it is automatically separable, it is Galois over F. If n is the least common multiple of the integers n_i associated with this chain, then we can extend the chain from K to $K(z)$ where z is a primitive nth root of unity. Then if K is the splitting field over F of, say, $g(x)$, $K(z)$ is the splitting field over F of $g(x)(x^n - 1)$, and so $K(z)$ is also Galois over F. Moreover, since $z^n = 1 \in F$ we may rearrange the tower for $K(z)$ so that its second term is $F(z)$. Then we have

$$(33) \qquad F = F_1 \subset F_2 = F_1(z) \subset F_3 = F_2(d_1) \subset \cdots \subset K(z).$$

Let G be the Galois group of E/F, H the Galois group of $K(z)$ over F. We now observe that in the arrangement (33) each F_{i+1} is abelian over F_i. This follows from Lemma 1 for $i = 1$ and from Lemma 2 for $i > 1$, since in this case F_i contains the requisite roots of unity. Now let H_i be the subgroup of $H = $ Gal $K(z)/F$ corresponding to the subfield F_i, that is, $H_i = $ Gal $K(z)/F_i$. Since F_{i+1} is normal over F_i, $H_{i+1} \lhd H_i$. Moreover, H_i/H_{i+1} is isomorphic to the Galois group of F_{i+1}/F_i so this is an abelian group. Hence we have a normal series for H with abelian factors and so H is solvable. Since the splitting field E/F is contained in $K(z)/F$, the Galois group G is isomorphic to a factor group of H. Hence G is solvable.

Conversely, assume that the Galois group G of $f(x) = 0$ over F is solvable. Let $n = |G| = [E:F]$ where E is a splitting field over F of $f(x)$. Let $F_1 = F$, $F_2 = F(z)$ where z is a primitive nth root of unity, and let $K = E(z)$. By Lemma 4, the Galois group of K/F_2 is isomorphic to a subgroup H of G. Hence H is solvable and it has a composition series $H = H_1 \rhd H_2 \rhd \cdots \rhd H_{r+1} = 1$ whose composition factor H_i/H_{i+1} is cyclic of prime order p_i for $1 \le i \le r$. Correspondingly, we have an increasing chain of subfields $F_2 \subset F_3 \subset \cdots \subset F_{r+2} = K$ where $H_i = \operatorname{Gal} K/F_{i+1}$. Hence F_{i+1} is normal over F_i with cyclic Galois group of prime order p_i. Since $p_i \mid n \ (= |G|)$ and F_i contains a primitive nth root of unity, F_i contains p_i p_ith roots of 1, so by Lemma 3, $F_{i+1} = F_i(d_i)$ where $d_i^{p_i} \in F_i$. Hence K contains a root tower over F and since K contains the splitting field, $f(x) = 0$ is solvable by radicals over F.

EXERCISES

1. Let p be a prime unequal to the characteristic of the field F. Show that, if $a \in F$, then $x^p - a$ is either irreducible in $F[x]$ or it has a root in F.

2. Assume that $x^p - a$, $a \in \mathbb{Q}$, is irreducible in $\mathbb{Q}[x]$. Show that the Galois group of $x^p - a$ over $\mathbb{Q}$ is isomorphic to the group of transformations of $\mathbb{Z}/(p)$ of the form $y \to ky + l$ where $k, l \in \mathbb{Z}/(p)$ and $k \ne 0$.

3. Let E/F be the cyclotomic field of the pth roots of unity over the field F of characteristic 0. Show that E can be imbedded in a field K which has a root tower over F such that the integers n_i are primes and $[F_{i+1}:F_i] = n_i$. Call such a root tower *normalized*.

4. Obtain normalized root towers over $\mathbb{Q}$ of the cyclotomic fields of 5th and of 7th roots of unity.

5. Prove that, if $f(x) = 0$ has a solvable Galois group over a field F of characteristic 0, then its splitting field can be imbedded in an extension field which has a normalized root tower over F.

6. Let E be a splitting field over F of $f(x) \in F[x]$. Show that E is normal over F. (*Hint*: Let $g(x)$ be an irreducible polynomial having a root s in E. Form a splitting field over E of $g(x)$, say, $K = E(s_1 = s, \ldots, s_m)$ where $g(x) = \prod (x - s_j)$ in $K[x]$. Since s_1 and s_i, $2 \le i \le m$ have the same minimum polynomial $g(x)$ over F, we have an isomorphism of $F(s_1)$ onto $F(s_i)$ over F sending $s_1 \to s_i$. This can be extended to an isomorphism η of $E(s_1)$ onto $E(s_i)$ since both of these are splitting fields over $F(s_1)$ and $F(s_i)$ respectively of $f(x)$. Then $\eta(E(s_1)) = E(s_i)$ and since $s_1 \in E$, $E(s_1) = E$. Hence $E(s_i) = E$ and E contains every s_i. Thus $g(x) = \prod (x - s_j)$ takes place in $E[x]$.)

4.8   THE GALOIS GROUP AS PERMUTATION
GROUP OF THE ROOTS

We are now going to exploit the idea which was introduced at the beginning of section 4.7: that the Galois group of an equation can be identified with a permutation group of the roots. As before, we consider a monic polynomial of positive degree, $f(x) \in F[x]$ with distinct roots $r_1, \ldots, r_n$ in a splitting field $E = F(r_1, \ldots, r_n)$. The group G_f, which is isomorphic to $G = \mathrm{Gal}\, E/F$, is the subgroup of the group S_n of permutations of $R = \{r_1, \ldots, r_n\}$ induced by G. Identifying G with G_f, the Galois correspondence becomes a correspondence between the subgroups of G_f and the subfields of E/F. We consider first the following question. What is the subfield of E/F which corresponds to the subgroup $G_f \cap A_n$, A_n, the alternating subgroup of S_n? For the sake of simplicity we confine our attention to the case in which char $F \neq 2$ and reserve the consideration of the case char $F = 2$ for an exercise. We have the following

THEOREM 4.13. *Let F be a field of characteristic $\neq 2$, $f(x)$ a monic polynomial of positive degree $\in F[x]$ such that $f(x)$ has distinct roots r_i in a splitting field E/F. Put*

$$(34) \qquad\qquad D = \prod_{1 \le i < j \le n} (r_i - r_j)$$

Then the subfield of E/F corresponding to $G_f \cap A_n$ in the Galois pairing is $F(D)$.

Proof. We look first at the ring $F[x_1, \ldots, x_n]$ of polynomials with coefficients in F in the indeterminates x_i, $1 \le i \le n$. We recall that if π is a permutation of $1, 2, \ldots, n$, then we have a unique automorphism $\zeta(\pi)$ of $F[x_1, \ldots, x_n]$ fixing the elements of F and sending $x_i \to x_{\pi(i)}$, $1 \le i \le n$ (Theorem 2.12, p. 125). If π_1 and $\pi_2 \in S_n$, then $\zeta(\pi_1 \pi_2) = \zeta(\pi_1)\zeta(\pi_2)$. Put

$$(35) \qquad\qquad \Delta = \prod_{1 \le i < j \le n} (x_i - x_j)$$

and consider the effect of the automorphism $\zeta(kl)$ on Δ, where (kl) is a transposition and $k < l$. We claim that $\zeta(kl)(\Delta) = -\Delta$. First $\zeta(kl)(x_k - x_l) = x_l - x_k = -(x_k - x_l)$. Next let $k < l < i$. Then $\zeta(kl)$ interchanges $x_k - x_i$ and $x_l - x_i$, both of which are terms in Δ. Similarly, if $i < k < l$ then $\zeta(kl)$ interchanges $x_i - x_k$ and $x_i - x_l$. Now let $k < i < l$. Then $\zeta(kl)$ maps $x_k - x_i$ into $x_l - x_i = -(x_i - x_l)$ and $x_i - x_l$ into $x_i - x_k = -(x_k - x_i)$. Finally, $\zeta(kl)$ fixes every $x_i - x_j$ with $i, j \neq k, l$. These observations imply that $\zeta(kl)(\Delta) = -\Delta$. Then the multiplicative character of ζ implies that $\zeta(\pi)(\Delta) = \Delta$ or $-\Delta$ according as π is even or odd. Now let $\eta \in \mathrm{Gal}\, E/F$ and let π be the corresponding permutation of the roots. Then if we apply the homomorphism of $F[x_1, \ldots, x_n]$ into E, which is the identity on F and sends $x_i \to r_i$, $1 \le i \le n$, to $\zeta(\pi)\Delta$ we obtain $\eta(D)$, where D is as in (34). Hence $\eta(D) = D$ or $-D$ according as π is even or odd. Consequently, the subgroup of $\mathrm{Gal}\, E/F$ which fixes the elements of the subfield $F(D)$ of E is the subgroup of elements η for which the corresponding permutation π of the roots is even. Hence, identifying $G = \mathrm{Gal}\, E/F$ with G_f, we can say that the subgroup of G_f corresponding to $F(D)$ is $G_f \cap A_n$. Then the subfield of E/F corresponding to $G_f \cap A_n$ is $F(D)$. □

Our proof shows also that for any $\eta \in G$, $\eta(D) = \pm D$, so if we put $d = D^2$, then $\eta(d) = d$ for all $\eta \in G$. Then $d \in F$. Since $F(D)$ is the subfield corresponding to $G_f \cap A_n$ it is clear that $F(D) = F$ if and only if $G_f \subset A_n$. Since the two square roots of d in E are D and $-D$ we see that $G_f \subset A_n$ if and only if d is the square of an element of F. Hence we have the

COROLLARY.   *Let F and $f(x)$ be as in Theorem 4.13. Then the Galois group of $f(x)$ over F is a subgroup of the alternating group if and only if the element*

$$(36) \qquad\qquad d = \prod_{1 \leq i < j \leq n} (r_i - r_j)^2$$

is the square of an element of F.

The element $d \in F$ is called the *discriminant* of $f(x)$. We proceed to give a procedure for calculating d. We write

$$(37) \qquad f(x) = x^n - a_1 x^{n-1} + \cdots + (-1)^n a_n = \prod_{1}^{n} (x - r_i).$$

Then

$$(38) \qquad a_1 = \sum_i r_i, \qquad a_2 = \sum_{i<j} r_i r_j, \qquad \ldots, a_n = r_1 r_2 \cdots r_n.$$

Also, we have the well-known Vandermonde determinant formula

$$(39) \qquad \begin{vmatrix} 1 & 1 & \cdots & 1 \\ r_1 & r_2 & \cdots & r_n \\ \cdots & \cdots & \cdots & \cdots \\ r_1^{n-1} & r_2^{n-1} & \cdots & r_n^{n-1} \end{vmatrix} = \prod_{i>j} (r_i - r_j).$$

If we multiply the displayed matrix on the right by its transpose and take the determinant of the resulting matrix, we obtain

$$(40) \qquad \begin{vmatrix} n & s_1 & s_2 & \cdots & s_{n-1} \\ s_1 & s_2 & s_3 & \cdots & s_n \\ s_2 & s_3 & s_4 & \cdots & s_{n+1} \\ \cdots & \cdots & \cdots & \cdots & \cdots \\ s_{n-1} & s_n & s_{n+1} & \cdots & s_{2n-2} \end{vmatrix} = \prod_{i<j} (r_i - r_j)^2 = d.$$

where $s_i = r_1^i + r_2^i + \cdots + r_n^i$. Since these power sums can be expressed as polynomials in the a_i with integer coefficients (Theorem 2.20, p. 139), (40) can be used to obtain a formula for the discriminant $d = d(f)$ as a polynomial in the coefficients a_i with integer coefficients. It is clear from the definition (36) of d that f has multiple roots if and only if $d = 0$.

We shall now calculate d for the cases $n = 2, 3$.

$n = 2$. We have $f(x) = x^2 - a_1 x + a_2 = (x - r_1)(x - r_2)$ and $a_1 = r_1 + r_2$, $a_2 = r_1 r_2$. Then $s_2 = r_1^2 + r_2^2 = (r_1 + r_2)^2 - 2r_1 r_2 = a_1^2 - 2a_2$. The formula (40) gives

$$(41) \qquad\qquad d = 2s_2 - s_1^2 = a_1^2 - 4a_2$$

which is the familiar formula for the discriminant of the quadratic polynomial $x^2 - a_1 x + a_2$.

$n = 3$. Here $f(x) = x^3 - a_1 x^2 + a_2 x - a_3 = (x - r_1)(x - r_2)(x - r_3)$, so $s_1 = r_1 + r_2 + r_3 = a_1$, $r_1 r_2 + r_1 r_3 + r_2 r_3 = a_2$, and $r_1 r_2 r_3 = a_3$. Then $s_2 = r_1^2 + r_2^2 + r_3^2 = (r_1 + r_2 + r_3)^2 - 2(r_1 r_2 + r_1 r_3 + r_2 r_3) = a_1^2 - 2a_2$. To calculate s_3 and s_4 we use the relations $r_k^3 = a_1 r_k^2 - a_2 r_k + a_3$, $r_k^4 = a_1 r_k^3 - a_2 r_k^2 + a_3 r_k$. Then

$$\begin{aligned}
s_3 &= r_1^3 + r_2^3 + r_3^3 \\
&= a_1(r_1^2 + r_2^2 + r_3^2) - a_2(r_1 + r_2 + r_3) + 3a_3 \\
&= a_1(a_1^2 - 2a_2) - a_2 a_1 + 3a_3 \\
&= a_1^3 - 3a_1 a_2 + 3a_3.
\end{aligned}$$

$$\begin{aligned}
s_4 &= a_1 s_3 - a_2 s_2 + a_3 s_1 \\
&= a_1(a_1^3 - 3a_1 a_2 + 3a_3) - a_2(a_1^2 - 2a_2) + a_3 a_1 \\
&= a_1^4 - 4a_1^2 a_2 + 4a_1 a_3 + 2a_2^2.
\end{aligned}$$

Using (40) and these formulas we obtain

$$(42) \qquad \begin{aligned}
d &= 3s_2 s_4 + 2s_1 s_2 s_3 - s_2^3 - 3s_3^2 - s_1^2 s_4 \\
&= -4a_1^3 a_3 + a_1^2 a_2^2 + 18a_1 a_2 a_3 - 4a_2^3 - 27a_3^2.
\end{aligned}$$

We obtain next a criterion on the Galois group regarded as a permutation group of the roots, that $f(x)$ be irreducible in $F[x]$. This is the following

THEOREM 4.14 *Let $f(x) \in F[x]$ have no multiple roots. Then $f(x)$ is irreducible in $F[x]$ if and only if G_f is a transitive permutation group of the roots r_i.*

Proof. We recall that a group G of transformations of a set M is transitive if given any pair of elements (x, y) of M there is an $\eta \in G$ such that $\eta(x) = y$.

Suppose first that $f(x)$ is irreducible and r_i and r_j are two of its roots. Since $f(x)$ is irreducible and $f(r_i) = 0 = f(r_j)$ there exists an isomorphism of $F(r_i)/F$ into $F(r_j)/F$ sending r_i into r_j (p. 227). Since $E = F(r_1, r_2, \ldots, r_m)$ is a splitting field over $F(r_i)$ and over $F(r_j)$ of $f(x) = \prod (x - r_k)$ this isomorphism can be extended to an automorphism η of E/F. Then $\eta \in \text{Gal } E/F$ and $\eta(r_i) = r_j$, which shows that G is transitive on the set of roots. Conversely, suppose G_f is transitive. Let $f_1(x)$ be an irreducible factor of $f(x)$ of positive degree and let r_i be one of its roots. Then if r_j is any other root we have an $\eta \in G_f$ such that $\eta(r_i) = r_j$. Since $f_1(r_i) = 0, 0 = \eta(f_1(r_i)) = f_1(r_j)$. This shows that every root of $f(x)$ is a root of $f_1(x)$. Hence $f(x) = f_1(x)$ is irreducible. $\square$

The two results we have derived make it easy to calculate the Galois groups of quadratic and cubic equations. Similar ideas apply to quartics. We shall look at the first two cases now and will indicate how the quartics can be handled in the exercises which follow.[8] We assume that the characteristric of F is not two and $f(x)$ has distinct roots. If $f(x) = x^2 - a_1 x + a_2$, then its group is the symmetric group S_2 or the alternating group $A_2 = 1$ according as $d = a_1^2 - 4a_2$ is not or is a square in F. Next let $f(x) = x^3 - a_1 x^2 + a_2 x - a_3$. If $f(x) = (x - r)g(x)$ in $F[x]$ then the Galois group of $f(x)$ is the same as that of the quadratic polynomial $g(x)$. Hence we may assume $f(x)$ irreducible in $F[x]$. Since the only transitive subgroups of S_3 are S_3 and A_3, the Galois group G_f is one of these. The corollary to Theorem 4.13 shows that $G_f = A_3$ if $d = -4 a_1^3 a_3 + a_1^2 a_2^2 + 18 a_1 a_2 a_3 - 4a_2^3 - 27 a_3^2$ is a square in F. Otherwise, $G_f = S_3$.

EXERCISES

1. Suppose the discriminant $d(f) \neq 0$. Show that if $f(x) = f_1(x) f_2(x) \cdots f_r(x)$ where $f_i(x)$ is irreducible of degree n_i in $F[x]$, then the set R of roots of f decomposes into orbits under G_f of cardinality n_i, $1 \leq i \leq r$. Hence show that if the Galois group is cyclic, say, $= \langle \eta \rangle$, then by a suitable ordering of R the permutation of R determined by η has the cycle decomposition $(12 \cdots n_1)(n_1 + 1 \cdots n_1 + n_2)(n_1 + n_2 + 1 \cdots n_1 + n_2 + n_3) \cdots$.

2. Let $F = \mathbb{R}$ and let $f(x)$ be a cubic with discriminant d. Show that $f(x)$ has multiple roots, three distinct real roots, or one real root and two non-real roots according as $d = 0$, $d > 0$, or $d < 0$.

3. Let the characteristic of F be arbitrary (including two). Let $f(x)$ have distinct roots $r_1, r_2, \ldots, r_n$. Put

$$D' = \sum_{\pi \in A_n} r_{\pi(1)}^0 r_{\pi(2)}^1 \cdots r_{\pi(n)}^{n-1}.$$

---

[8] For a discussion of quintics consult E. Dehn, *Algebraic Equations*, New York, Columbia Univ. Press, 1930, p. 195, or H. Weber, *Lehrbuch der Algebra*, Vol. I, 1898, p. 670. Some information on quintics will be indicated in the exercises at the end of Section 4.16.

Show that for any odd permutation σ

$$D' - \sum_{\pi \in A_n} r^0_{\sigma\pi(1)} r^1_{\sigma\pi(2)} \cdots r^{n-1}_{\sigma\pi(n)} = \prod_{i>j} (r_i - r_j)$$

Show that the subfield of invariants of $G_f \cap A_n$ is $F(D')$. Determine a quadratic equation with coefficients in F having D' as a root.

In the remainder of these exercises we assume that the characteristic of the base field is $\neq 2$, $f(x) = x^4 - a_1 x^3 + a_2 x^2 - a_3 x^2 + a_4$ has distinct roots r_1, r_2, r_3, r_4, $E = F(r_1, r_2, r_3, r_4)$, $G = \text{Gal } E/F$, and G_f is the corresponding permutation group of the roots.

4. Show that $V = \{1, (12)(34), (13)(24), (14)(23)\}$ is normal in S_4.

5. Show that the subfield of E/F of invariants under $G_f \cap V$ is $F(t_1, t_2, t_3)$ where $t_1 = r_1 r_2 + r_3 r_4$, $t_2 = r_1 r_3 + r_2 r_4$, and $t_3 = r_1 r_4 + r_2 r_3$.

6. Let $g(x) = (x - t_1)(x - t_2)(x - t_3)$. (This polynomial is called the *resolvent cubic* of the quartic $f(x)$.) Verify that

   (43) $$g(x) = x^3 - b_1 x^2 + b_2 x - b_3$$

   where

   (44) $$b_1 = a_2, \quad b_2 = a_1 a_3 - 4a_4, \quad b_3 = a_1^2 a_4 + a_3^2 - 4a_2 a_4$$

   and that $f(x)$ and $g(x)$ have the same discriminant.

7. Show that the transitive subgroups of S_4 are (i) S_4, (ii) A_4, (iii) V, (iv) $C = \{1, (1234), (13)(24), (1432)\}$ and its conjugates, (v) $D = V \cup \{(12), (34), (1423), (1324)\}$ which is a Sylow 2-group (subgroup of order 8 of S_4) and its conjugates.

8. Show that the Galois group G_g of $g(x) = 0$ is isomorphic to $G_f/(G_f \cap V)$. Assume $f(x)$ irreducible and verify that, if (i) $G_f = S_4$ then G_g is of order 6, (ii) $G_f = A_4$, G_g is of order 3, (iii) $G_f = V$, $G_g = 1$, (iv) $G_f = C$ or one of its conjugates (that is, any cyclic subgroup of order 4 of S_4), then G_g is of order 2, (v) $G_f = D$ or one of its conjugates (any Sylow subgroup of order 8 in S_4), then G_g is of order 2. Note that these results identify G_f if we know G_g unless G_f is either as in (iv) or (v).

9. Prove that if G_g is of order 2, then $G_f \cong D$ or $G_f \cong C$ according as $f(x)$ is or is not irreducible in $F(\sqrt{d})$, d the discriminant of $f(x)$.

10. Determine the Galois group of $x^4 + 3x^3 - 3x - 2 = 0$ over $\mathbb{Q}$.

The next four exercises are designed to show that any solvable transitive subgroup of S_p, p a prime, is equivalent to a subgroup of the group of transformations of $\mathbb{Z}/(p)$ of the form $x \rightarrow ax + b$, $a \neq 0$ including all the translations $x \rightarrow x + b$.

11. Let H be a normal subgroup $\neq 1$ of a transitive subgroup G of S_n of transformations of $\{1, 2, \ldots, n\}$. Show that all H-orbits have the same cardinality. Hence show that if $n = p$ is a prime, then H is transitive.

12. Let p be a prime and let L be the group of all transformations of $\mathbb{Z}/(p)$ of the form $x \rightarrow ax + b$, $a \neq 0$ including all the translations $x \rightarrow x + b$. Show that the translations $x \rightarrow x + b$, $b \neq 0$, are the only transformations in L without fixed points and

hence that these are the only transformations in L whose cycle representations are p-cycles.

13. Let G be a group of transformations in $\mathbb{Z}/(p)$ containing the group H of transla-
     tions as normal subgroup. Show that G is a subgroup of L. (*Hint*: Let $\tau: x \to x + 1$
     and let $\eta \in G$. Then, by exercise 12, $\eta\tau\eta^{-1}$ has the form $x \to x + k$. Hence
     $\eta(x + 1) = \eta\tau(x) = \eta(x) + k$, from which one can conclude that $\eta(x) = kx + b$.)

14. Use induction and exercise 13 to prove that any solvable transitive subgroup of
     S_p, p a prime, is equivalent to a subgroup of L containing the subgroup of transla-
     tions.

15. (Galois.) Let $f(x) \in F[x]$ be irreducible of prime degree over F of characteristic 0,
     E a splitting field over F of $f(x)$. Show that $f(x)$ is solvable by radicals over F if
     and only if $E = F(r_i, r_j)$ for any two roots r_i, r_j of $f(x)$.

16. Let E be a splitting field over F of $f(x) \in F[x]$ and let $G = \mathrm{Gal}\, E/F$. Let $x_1, \ldots, x_n$
     be indeterminates and put $\tilde{E} = E(x_1, \ldots, x_n)$, $\tilde{F} = F(x_1, \ldots, x_n)$. Note that $\tilde{E}$ is a
     splitting field over $\tilde{F}$ of $f(x)$ and show that $\tilde{G} = \mathrm{Gal}\, \tilde{E}/\tilde{F}$ is isomorphic to G under
     the restriction map $\tilde{\eta} \to \eta = \tilde{\eta}|E$, $\tilde{\eta} \in \tilde{G}$. Assume that $\deg f(x) = n$ and $f(x)$ has n
     distinct roots $r_1, r_2, \ldots, r_n$ in E. If π is a permutation of $1, 2, \ldots, n$ put

$$u_\pi = \sum_{i=1}^{n} r_{\pi(i)} x_i = \sum_{i=1}^{n} r_i x_{\pi^{-1}(i)}.$$

Observe that $u_{\pi_1} \neq u_{\pi_2}$ if π_1 and π_2 are distinct permutations and hence that the
orbit of u_π under $\tilde{G}$ contains $|\tilde{G}|$ distinct elements. Hence conclude that u_π is a
primitive element of $\tilde{E}$ over $\tilde{F}$ (that is, $\tilde{E} = \tilde{F}(u_\pi)$) and the minimum polynomial
of u_π over $\tilde{F}$ is $\varphi_\pi(x) = \prod_{\sigma \in G_f} (x - \sum r_{\sigma\pi(i)} x_i)$. Let

$$\varphi(x) = \prod_{\pi \in S_n} (x - u_\pi).$$

Show that $\varphi(x) \in F[x_1, \ldots, x_n, x]$ and its irreducible factors in this ring have the
form $\varphi_\pi(x)$. Hence show that G is isomorphic to the subgroup of S_n of the permutations σ such that the automorphism $\eta(\sigma)$ of $F[x_1, \ldots, x_n, x]$ fixing F and x,
and sending $x_i \to x_{\sigma(i)}$, fixes the irreducible factors of $\varphi(x)$.

4.9 THE GENERAL EQUATION OF THE nth DEGREE

By a general equation we mean one whose coefficients are distinct indeterminates. More precisely, let F be a field and let $t_1, t_2, \ldots, t_n$ be distinct indeterminates. Then the equation

(45) $$f(x) = x^n - t_1 x^{n-1} + t_2 x^{n-2} - \cdots + (-1)^n t_n = 0$$

is called a *general equation of the nth degree over F*. This is said to be solvable
by radicals if it is solvable by radicals over the field $F(t_1, \ldots, t_n)$ (the field of
fractions of the polynomial ring $F[t_1, \ldots, t_n]$). For example, the quadratic
formula $x = \frac{1}{2}t_1 \pm \frac{1}{2}\sqrt{t_1^2 - 4t_2}$ shows that the general equation of second de-

gree is solvable by radicals since the roots are contained in $F(t_1, t_2, d)$ where $d^2 = t_1{}^2 - 4t_2 \in F(t_1, t_2)$. To settle the question of solvability by radicals via Galois' criterion, we need to determine the Galois group G_f of $f(x)$ over $F(t_1, \ldots, t_n)$. Let E be the splitting field of $f(x)$ over $F(t_1, \ldots, t_n)$ and suppose $f(x) = (x - y_1)(x - y_2) \cdots (x - y_n)$ in $E[x]$. Then comparing this factorization with (45) we see that $t_1 = \sum y_i, t_2 = \sum_{i>j} y_i y_j, \ldots, t_n = y_1 y_2 \cdots y_n$. Hence

(46) $E = F(t_1, \ldots, t_n, y_1, \ldots, y_n) = F(y_1, \ldots, y_n).$

We shall now obtain G_f by applying a result we obtained in section 4.5 in our discussion of symmetric rational expressions. As before, we introduce new indeterminates $x_1, \ldots, x_n$ and we form the field $F(x_1, \ldots, x_n)$ and its subfield of symmetric rational expressions. We showed that the latter coincides with $F(p_1, \ldots, p_n)$ where the p_i are the elementary symmetric polynomials in the x_i, and that $F(x_1, \ldots, x_n)$ is a splitting field over $F(p_1, \ldots, p_n)$ of

$$g(x) = \prod_1^n (x - x_i)$$

and, moreover, the Galois group G_g is S_n.

We shall now carry over the result we had on the pair of fields $F(x_1, \ldots, x_n) \supset F(p_1, \ldots, p_n)$ to the pair we are really interested in: $F(y_1, \ldots, y_n) \supset F(t_1, \ldots, t_n)$ where the t_i are the indeterminates. We shall do this by establishing an isomorphism of $F(y_1, \ldots, y_n)$ into $F(x_1, \ldots, x_n)$ which carries $F(t_1, \ldots, t_n)$ into $F(p_1, \ldots, p_n)$. Since the t_i are indeterminates we have a homomorphism σ of $F[t_1, \ldots, t_n] \to F[p_1, \ldots, p_n]$ which is the identity on F and sends $t_i \to p_i$, $1 \leq i \leq n$. We claim that σ is a monomorphism. To see this we note that since the x_i are indeterminates, we have a homomorphism τ of $F[x_1, \ldots, x_n]$ into $F[y_1, \ldots, y_n]$ which is the identity on F and sends $x_i \to y_i$, $1 \leq i \leq n$. We have the following diagram

$$F[x_1, \ldots, x_n] \overset{\tau}{\to} F[y_1, \ldots, y_n]$$
$$\cup \qquad\qquad\qquad \cup$$
$$F[p_1, \ldots, p_n] \overset{\sigma}{\leftarrow} F[t_1, \ldots, t_n].$$

It is clear that $\tau\sigma$ is defined. Moreover,

$$\tau\sigma(t_i) = \tau(p_i) = \tau\left(\sum_{j_1 < \cdots < j_i} x_{j_1} x_{j_2} \cdots x_{j_i} \right) = \sum_{j_1 < \cdots < j_i} y_{j_1} y_{j_2} \cdots y_{j_i} = t_i$$

by the formulas relating the p's and the x's and the t's and the y's. It now follows that if $h(t_1, \ldots, t_n) \in F[t_1, \ldots, t_n]$ then $\tau\sigma(h) = h$. This implies that σ is a monomorphism, since $\sigma(h) = 0$ gives $h = \tau\sigma(h) = 0$. It is clear also that σ is surjective and that σ is an isomorphism of $F[t_1, \ldots, t_n]$ into $F[p_1, \ldots, p_n]$. This has a unique extension to an isomorphism, which we shall denote by σ also, of $F(t_1, \ldots, t_n)$ into $F(p_1, \ldots, p_n)$. Moreover, σ extends to an isomorphism σ' of

$F(t_1, \ldots, t_n)[x]$ into $F(p_1, \ldots, p_n)[x]$ fixing x. This maps the polynomial $f(x) = x^n - t_1 x^{n-1} + \cdots + (-1)^n t_n$ into the polynomial $g(x) = x^n - p_1 x^{n-1} + \cdots + (-1)^n p_n$. Since $F(y_1, \ldots, y_n)$ is a splitting field over $F(t_1, \ldots, t_n)$ of $f(x)$ and $F(x_1, \ldots, x_n)$ is a splitting field over $F(p_1, \ldots, p_n)$ of $g(x)$, σ can be extended to an isomorphism ρ of $F(y_1, \ldots, y_n)$ into $F(x_1, \ldots, x_n)$. The existence of the isomorphism ρ which maps $F(t_1, \ldots, t_n)$ into $F(p_1, \ldots, p_n)$ implies that the Galois groups G_f and G_g are isomorphic. In fact, it is immediate that the map $\eta \to \rho \eta \rho^{-1}$ is an isomorphism of G_f into G_g. Since G_g is the symmetric group it follows that $G_f = S_n$. It is clear also that $f(x)$ is irreducible in $F(t_1, \ldots, t_n)[x]$ and that its roots y_i are distinct. The results we have derived can be stated as

THEOREM 4.15.   *The general equation of the nth degree $f(x) = 0$ (as in (45)) is irreducible in $F(t_1, \ldots, t_n)[x]$ and has distinct roots. The Galois group of $f(x) = 0$ is the symmetric group S_n.*

Since S_n is not solvable if $n > 4$, by the Corollary to Theorem 4.11 this implies the celebrated

THEOREM OF RUFFINI-ABEL.   *The general equation of the nth degree is not solvable by radicals if $n > 4$ (characteristic 0).*

Galois' criterion implies also that general cubics and quartics are solvable by radicals. Moreover, the proof of the criterion suggests a procedure for solving these equations. We shall now carry this out for cubics and we shall arrive in this way at Cardan's formulas. The corresponding result for quartics will be indicated in an exercise.

We now assume only that the characteristic of F is $\neq 2$, $\neq 3$ and we consider the general cubic $x^3 - t_1 x^2 + t_2 x - t_3 = (x - x_1)(x - x_2)(x - x_3)$ where the t_i are indeterminates and the x_i are the roots in the splitting field $F(x_1, x_2, x_3)$. The proof of Galois' criterion shows that it will be handy to have available a primitive cube root of unity. These are the roots w and $w^2 = w^{-1}$ of $x^2 + x + 1 = 0$, and so, by the quadratic formula, we have, say, $w = -\frac{1}{2} + \frac{1}{2}\sqrt{-3}$, $w^{-1} = -\frac{1}{2} - \frac{1}{2}\sqrt{-3}$. We assume that these are contained in F. To simplify the calculations we now replace the roots x_i by $y_i = x_i - \frac{1}{3}(x_1 + x_2 + x_3) = x_i - \frac{1}{3}t_1$. Then the equation is replaced by $y^3 + py + q = 0$, whose roots y_1, y_2, y_3 satisfy the relation $y_1 + y_2 + y_3 = 0$. The group of the y-equation is S_3, which has the composition series $S_3 \rhd A_3 \rhd 1$. The subfield corresponding to A_3 is $K(\sqrt{d})$ where $K = F(p, q)$ and d is the discriminant. By (42), we have $d = -4p^3 - 27q^2$. The splitting field $K(y_1, y_2, y_3)$ that we seek is cyclic, three dimensional over $K(\sqrt{d})$, and so it can be obtained by adjoining a cube root of a Lagrange resolvent defined by the y_i (which are permuted cyclically by A_3) and a cube root of unity. The three resolvents are

$$z_1 = y_1 + wy_2 + w^2 y_3$$
$$z_2 = y_1 + w^2 y_2 + wy_3$$
$$z_3 = y_1 + y_2 + y_3 = 0$$

where $w = -\frac{1}{2} + \frac{1}{2}\sqrt{-3}$. Then

(47)
$$z_1{}^3 = \sum y_i{}^3 + 3w(y_1{}^2 y_2 + y_2{}^2 y_3 + y_3{}^2 y_1)$$
$$+ 3w^2(y_1 y_2{}^2 + y_2 y_3{}^2 + y_3 y_1{}^2) + 6y_1 y_2 y_3$$

and $z_2{}^3$ is obtained from this by interchanging w and w^2. Now

$$\sqrt{d} = (y_1 - y_2)(y_2 - y_3)(y_1 - y_3)$$
$$= y_1{}^2 y_2 + y_2{}^2 y_3 + y_3{}^2 y_1 - (y_1 y_2{}^2 + y_2 y_3{}^2 + y_3 y_1{}^2).$$

Hence if we put $u = y_1{}^2 y_2 + y_2{}^2 y_3 + y_3{}^2 y_1 + y_1 y_2{}^2 + y_2 y_3{}^2 + y_3 y_1{}^2$ and use the relations $w + w^2 = -1$, $w - w^2 = \sqrt{-3}$ we obtain from (47):

(48)
$$z_1{}^3 = \sum y_i{}^3 - \tfrac{3}{2}u + \tfrac{3}{2}\sqrt{-3}\sqrt{d} + 6y_1 y_2 y_3.$$

Now $0 = (y_1 + y_2 + y_3)^3 = \sum y_i{}^3 + 3u + 6y_1 y_2 y_3$ and $0 = (y_1 + y_2 + y_3) \times (y_1 y_2 + y_2 y_3 + y_1 y_3) = u + 3y_1 y_2 y_3$. Also $y_1 y_2 y_3 = -q$. Hence

$$z_1{}^3 = -\tfrac{9}{2}u + \tfrac{3}{2}\sqrt{-3}\sqrt{d} = -\tfrac{27}{2}q + \tfrac{3}{2}\sqrt{-3d}.$$

Since $z_2{}^3$ is obtained from $z_1{}^3$ by interchanging w and w^2 and $w^2 - w = -\sqrt{-3}$ we obtain the formula for $z_2{}^3$ by replacing $\sqrt{-3}$ by $-\sqrt{-3}$ in the foregoing formula. Hence we have the formulas

(49)
$$z_1{}^3 = -\tfrac{27}{2}q + \tfrac{3}{2}\sqrt{-3d}, \qquad z_2{}^3 = -\tfrac{27}{2}q - \tfrac{3}{2}\sqrt{-3d}$$

where the same determination of $\sqrt{-3d}$ is used in both formulas. In extracting the cube roots to obtain z_1 and z_2 we have three determinations for these. However, these must be paired appropriately, since

$$z_1 z_2 = \sum y_i{}^2 - \sum_{i<j} y_i y_j = \left(\sum y_i\right)^2 - 3 \sum_{i<j} y_i y_j = -3p.$$

Thus we have

(50)
$$z_1 = \sqrt[3]{-\tfrac{27}{2}q + \tfrac{3}{2}\sqrt{-3d}}, \qquad d = -4p^3 - 27q^2$$
$$z_2 = \sqrt[3]{-\tfrac{27}{2}q - \tfrac{3}{2}\sqrt{-3d}}$$

where in both formulas the same determination of $\sqrt{-3d}$ is used and the cube roots are determined so that $z_1 z_2 = -3p$. Using (47) and the relation $w + w^2 + 1 = 0$ we obtain

$$y_1 = \tfrac{1}{3}(z_1 + z_2)$$

(51)
$$y_2 = \tfrac{1}{3}(w^2 z_1 + w z_2)$$

$$y_3 = \tfrac{1}{3}(w z_1 + w^2 z_2).$$

The formulas (50) and (51) are *Cardan's formulas* for solving the cubic $x^3 + px + q = 0$ with indeterminate coefficients. They can be applied also to cubics with coefficients in any field of characteristic $\neq 2, 3$.

EXERCISES

1. Solve the following over $\mathbb{Q}$ by Cardan's formulas:
    (a) $x^3 - 2x + 4 = 0$,   (b) $x^3 - 15x + 4 = 0$.

2. Assume the characteristic of $F \neq 2, 3$ and consider the general quartic $x^4 - t_1 x^3 + t_2 x^2 - t_3 x + t_4 = \prod_1^4 (x - x_i)$. Replacing x_i by $y_i = x_i - \tfrac{1}{4}t_1$ gives an equation $f(y) = y^4 + py^2 + qy + r = 0$ whose roots y_i satisfy $\sum y_i = 0$. Show that the resolvent cubic of $f(y) = 0$ is $g(z) = z^3 - pz^2 - 4rz + (4pr - q^2) = 0$ (exercise 6, p. 261). Let z_1, z_2, z_3 be the roots of the resolvent cubic. Show that the Galois group of

$$F(x_1, x_2, x_3, x_4) = F(y_1, y_2, y_3, y_4)$$

over $F(z_1, z_2, z_3)$ is $V = \{1, (12)(34), (13)(24), (14)(23)\}$. Obtain formulas for y_1, y_2, y_3, y_4 in terms of z_1, z_2, z_3 and square roots of elements of $F(z_1, z_2, z_3)$. Note that together with Cardan's solution for the resolvent cubic this gives a solution of the quartic by radicals.

3. Apply the method of exercise 2 to solve $x^4 - 2x^3 - 8x - 3 = 0$ over $\mathbb{Q}$.

4. Use the fact that any finite group G is isomorphic to a subgroup of S_n to prove that given any finite group G there exist fields F and E/F such that

$$\text{Gal } E/F \cong G.$$

5. Let E be an extension of $\mathbb{C}$ such that $E = \mathbb{C}(t, u)$ where t is transcendental over $\mathbb{C}$ and u satisfies the equation $u^2 + t^2 = 1$ over $\mathbb{C}(t)$. Determine the Galois group of $\mathbb{C}(t, u)$ over $\mathbb{C}(t^n, u^n)$ for any $n \in \mathbb{N}$. Show that

$$u_n = \tfrac{1}{2}[(t + iu)^n + (t - iu)^n], \qquad i = \sqrt{-1}$$

is contained in $\mathbb{C}(t^n, u^n)$. Use this to prove that the function cos nx is expressible rationally with complex coefficients in terms of $\cos^n x$ and $\sin^n x$. Does this hold for sin nx?

6. Let F be a subfield of $\mathbb{R}$ and let $f(x) \in F[x]$ be an irreducible cubic with discriminant $d > 0$ and $G_f = A_3$. Show that the roots of $f(x)$ in $\mathbb{C}$ are in $\mathbb{R}$. Let p be a prime and let $K = F(r)$ where r is real and $r^p \in F$. Show that K cannot contain a splitting field over F of $f(x)$.

7. Note that if F and $f(x)$ are as in exercise 6, then the roots of $f(x)$ are real but Cardan's formulas give expression of these roots involving non-real numbers. Prove that this is indeed unavoidable, that is, there exists no subfield K/F of $\mathbb{R}/F$ which has a root tower over F and contains a splitting field of $f(x)$ over F. (This is the so-called *Casus irreducibilis* of real equations.)

8. Show that $H = \{1, (12345), (13524), (14253), (15432), (14)(23), (15)(24), (25)(34), (12)(35), (13)(45)\}$ is a solvable subgroup of A_5. Let $f(x) = x^5 - t_1x^4 + \cdots$ be a general quintic equation with roots $x_1, \ldots, x_5$ over $F(t_1, \ldots, t_5)$. Let d be the discriminant and put $K = F(t_1, \ldots, t_5, \sqrt{d})$. Let

$$\chi_1 = x_1x_2 + x_2x_3 + x_3x_4 + x_4x_5 + x_5x_1$$
$$\chi_1' = x_1x_3 + x_1x_4 + x_2x_4 + x_2x_5 + x_3x_5$$

and $\omega_1 = \chi_1 - \chi_1'$. Show that ω_1 is fixed under H and determine the conjugates of ω_1 under the Galois group of $F(x_1, \ldots, x_5)/K$. Show that

$$[K(\omega_1):K] = 6.$$

9. Show that the discriminant of $f(x) = x^5 + px + q$ is $d = 2^8p^5 + 5^5q^4$. Let $\rho_1, \ldots, \rho_5$ be the roots of $f(x)$ and assume $d \neq 0$. Put

$$\lambda_1 = \rho_1\rho_2 + \rho_2\rho_3 + \rho_3\rho_4 + \rho_4\rho_5 + \rho_5\rho_1$$
$$\lambda_1' = \rho_1\rho_3 + \rho_1\rho_4 + \rho_2\rho_4 + \rho_2\rho_5 + \rho_3\rho_5 = -\lambda_1$$

and $\mu_1 = \lambda_1 - \lambda_1' = 2\lambda_1$. Show that μ_1 is a root of the *resolvent sextic* of $f(x)$:

$$g(x) = (x^3 - 5px^2 + 15p^2x + 5p^3)^2 - dx$$

and that $f(x)$ is solvable by radicals over $F(\sqrt{d}, \mu_1)$ (F of characteristic 0).

4.10 EQUATIONS WITH RATIONAL COEFFICIENTS AND SYMMETRIC GROUP AS GALOIS GROUP

The theorem of Ruffini-Abel states that general equations of degree $n \geq 5$ are not solvable by radicals. Roughly this means that it is impossible for $n \geq 5$ to give a general formula in terms of radicals which on substitution of values from a field F gives the roots of any equation of degree n with coefficients in F. In spite of this result it is conceivable that all equations with coefficients in F are solvable by radicals over F. In some cases this is true. For example, it is trivially so if $F = \mathbb{R}$. We shall now show that if $F = \mathbb{Q}$ and p is any prime then there exist $f(x) \in \mathbb{Q}[x]$ having S_p as Galois group. For $p \geq 5$ these are not solvable by radicals. We prove first the following result on permutation groups.

LEMMA.   *If G is a permutation group on a prime number p of elements such that G contains an element of order p and a transposition, then $G = S_p$.*

Proof.   We recall that the order of a cycle $(12 \cdots m)$ is m and the order of a product of disjoint cycles is the l.c.m. of the orders of these cycles (see p. 48). Hence G contains a p-cycle $\sigma = (i_1 i_2 \cdots i_p)$ where the set $\{i_1, i_2, \ldots, i_p\} = \{1, 2, \ldots, p\}$. By re-ordering the elements $1, 2, \ldots$ suitably, we may assume that G contains the transposition (12). Since a suitable power of σ has the form $(12 \cdots)$, further re-ordering of the elements $1, 2, \ldots, p$, if necessary, permits us to assume that G contains (12) and $\sigma = (123 \cdots p)$. Then G contains $\sigma(12)\sigma^{-1} = (23)$, $\sigma(23)\sigma^{-1} = (34), \ldots, \sigma(p-2, p-1)\sigma^{-1} = (p-1, p)$. We see easily that these transpositions generate S_p. Hence $G = S_p$. $\square$

We shall now prove

THEOREM 4.16.   *Let $f(x)$ be a polynomial of prime degree with rational coefficients which is irreducible in the rational field. Suppose $f(x) = 0$ has exactly two non-real roots in $\mathbb{C}$. Then the group G_f of $f(x) = 0$ over $\mathbb{Q}$ is S_p.*

Proof.   We assume the classical result (which will be proved in section 5.1) that $f(x) = \prod_1^p (x - r_i)$ in $\mathbb{C}[x]$, and so $E = \mathbb{Q}(r_1, \ldots, r_p)$ is a splitting field of $f(x)$ over $\mathbb{Q}$ contained in $\mathbb{C}$. Since $E \supset \mathbb{Q}(r_1)$ and $[\mathbb{Q}(r_1):\mathbb{Q}] = \deg f(x) = p$, $[E:\mathbb{Q}]$ is divisible by p. By Sylow's theorem, G_f contains an element of order p. Now consider the conjugation automorphism $u = a + b\sqrt{-1}$, a, b real, $\rightarrow \bar{u} = a - b\sqrt{-1}$ of $\mathbb{C}$. This maps $f(x)$ into itself; hence it permutes the roots r_i of $f(x)$. Let r_1 and r_2 be the non-real roots of $f(x)$. Then $r_2 = \bar{r}_1$ since $f(\bar{r}_1) = 0$ and $\bar{r}_1 \neq r_1$. Thus the conjugation interchanges r_1 and r_2 and fixes all the other roots. Hence the restriction of this automorphism to E is an element of the Galois group G_f which is a transposition. Thus G_f contains an element of order p and a transposition; hence $G_f = S_p$ by the lemma. $\square$

We shall now show how we can construct polynomials satisfying the conditions of the theorem.[9] Let m be a positive even integer, $n_1 < n_2 < \cdots < n_{k-2}$

---

[9] The construction we shall give is due to R. Brauer. A construction of a polynomial whose Galois group over $\mathbb{Q}$ is any S_n will be given in exercise 5, p. 305.

be $k - 2$ even integers where k is odd and > 3. Consider the polynomial

(52)
$$g(x) = (x^2 + m)(x - n_1)(x - n_2) \cdots (x - n_{k-2}).$$

The real roots of $g(x)$ are $n_1, n_2, \ldots, n_{k-2}$ and the graph of $y = g(x)$ has the form:

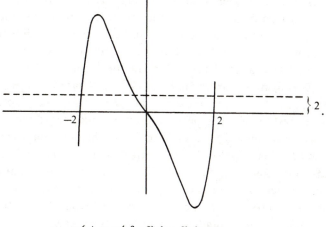

$$g(x) = x(x^2 + 2)(x + 2)(x - 2)$$

This has $(k - 3)/2$ relative maxima and, since $|g(h)| > 2$ for any odd integer h, it is clear that the values of these relative maxima are > 2. This implies that $f(x) = g(x) - 2$ has $(k - 3)/2$ positive relative maxima between n_1 and n_{k-2}. It follows that $f(x)$ has $k - 3$ real roots in the interval (n_1, n_{k-2}). Since $f(n_{k-2}) = -2$ and $f(\infty) = \infty$, there is also a real root $> n_{k-2}$. This gives $k - 2$ real roots for $f(x)$. Let $f(x) = \prod_1^k (x - r_i)$ in $\mathbb{C}[x]$. Since $f(x) = (x^2 + m)(x - n_1) \cdots (x - n_{k-2}) - 2$, equating coefficients of x^{k-1} and x^{k-2} gives the relations:

(53)
$$\sum_1^k r_i = \sum_1^{k-2} n_l, \qquad \sum_{i<j} r_i r_j = \sum_{l<q} n_l n_q + m.$$

Hence

(54)
$$\sum r_i^2 = \left(\sum r_i \right)^2 - 2 \sum_{i<j} r_i r_j = \sum n_l^2 - 2m.$$

If we choose m sufficiently large, (54) shows that $\sum r_i^2 < 0$, which implies that not every r_i is real. If r_1 is a non-real root, then $\bar{r}_1 \neq r_1$ is another one, so that we have at least two non-real roots. Since we saw that we have $k - 2$ real roots we see that $f(x)$ has exactly two non-real roots. We now write $f(x) = x^k + a_1 x^{k-1} + \cdots + a_k$. Clearly the a_i are even integers. Moreover, since the constant term of $g(x)$ is divisible by 4, that of $f(x) = g(x) - 2$ is not divisible by 4. It follows from Eisenstein's criterion applied to the prime 2 (exercise 2, p. 154) that $f(x)$ is irreducible in $\mathbb{Q}[x]$. Thus we see that the condition of Theorem 4.16 can be satisfied for every prime $p = k \geq 5$. Hence we can construct rational equations with Galois groups S_p for any prime $p \geq 5$. Since it is easy to do this also for $p = 2$ and 3, the result holds for every prime p.

The foregoing result suggests an interesting question. Given a finite group G, does there exist an equation with rational coefficients whose Galois group over $\mathbb{Q}$ is isomorphic to G? This turns out to be an extremely difficult problem which, though it was first considered about a hundred years ago, still remains unsolved. The earliest general results on this problem are that the answer is affirmative if $G = S_n$ or A_n for any n. In 1954, I. R. Šafarewič, using deep arithmetic results, proved that the answer is affirmative for every solvable finite group G. Other results of this type have been obtained more recently.

There is a general method for attacking this problem which was initiated by D. Hilbert and further developed by Emmy Noether.[10] The Hilbert-Noether method leads to the following theorem: The answer to the problem for a group G is yes if the answer to the following question on G is affirmative. Suppose G is realized as a subgroup of S_n and let F be the subfield of G-invariants of $\mathbb{Q}(x_1, x_2, \ldots, x_n)$, x_i indeterminates, where S_n acts on $\mathbb{Q}(x_1, x_2, \ldots, x_n)$ by the set of automorphisms which effect all the permutations on the x's. Is F isomorphic to $\mathbb{Q}(x_1, x_2, \ldots, x_n)$? The result we proved in the last section shows that this is true if $G = S_n$, since in this case $F = \mathbb{Q}(p_1, p_2, \ldots, p_n)$ where the p_i are the elementary symmetric polynomials and the p_i are algebraically independent. Very little is known about this question on subfields of $\mathbb{Q}(x_1, x_2, \ldots, x_n)$. For example, the answer is not known for $G = A_n$. Quite recently, it has been shown by R. Swan that the answer is negative for certain cyclic groups (e.g., G cyclic of order 47).[11] Swan's negative result does not give a negative answer to the original question on rational equations with given Galois groups. However, it does show that the Hilbert-Noether method cannot yield affirmative answers in all cases. We shall not discuss any of these results here. They have been mentioned primarily to dispel any notion the reader may have had that the Galois theory, because of its long history, has become a closed subject.[12]

---

[10] This is discussed in N. Tschebotaröw, *Grundzüge der Galoischen Theorie* (translated from Russian into German by H. Schwerdtfeger) Groningen, 1950, p. 399.

[11] R. Swan, "Invariant rational functions and a problem of Steenrod." *Inventiones Mathematicae*, vol. 7 (1969), pp. 148–158.

[12] The fundamental theorem of Galois theory has been generalized in a number of ways and this continues to be a subject of research. The literature on this is too voluminous to indicate here. The interested reader may consult the reviewing journal, *Mathematical Reviews*. See also a survey paper by Swan "Noether's problem in Galois theory" in the volume *Emmy Noether in Bryn Mawr* edited by J. Sally and B. Srinivasan, Springer-Verlag, New York, 1982.

EXERCISE

1. (Masuda.) Let F be a field and n a positive integer and suppose F contains n distinct nth roots of 1. (This implies char $F \nmid n$.) Let $K = F(x_1, x_2, \ldots, x_n)$ where the x_i are indeterminates and let σ be the automorphism of K/F that permutes the x_i cyclically: $\sigma x_i = x_{i+1}, 1 \leq i \leq n-1, \sigma x_n = x_1$. Put $G = \langle \sigma \rangle$, $E = \text{Inv } G$, so K/E is Galois with Gal $K/E = G$ (Theorem 4.7, p. 238). Let ζ be a primitive nth root of 1 and put

$$ y_j = \sum_{i=1}^{n} \zeta^{ij} x_i $$

(cf. Lemma 3, p. 253). Define $c_{jk} = y_j y_k y_{j+k}^{-1}$. Show that $E = F(c_{11}, c_{12}, \ldots, c_{1n})$.

4.11 CONSTRUCTIBLE REGULAR *n*-GONS. CYCLOTOMIC FIELDS OVER ℚ

We return to the problem of Euclidean constructibility which we considered in section 4.2. Our main result there was the criterion that the complex number z is constructible by straight edge and compass from the complex numbers $z_1, z_2, \ldots, z_n$ if and only if z is contained in a subfield K over

$$ F = \mathbb{Q}(z_1, \ldots, z_n, \bar{z}_1, \ldots, \bar{z}_n) $$

that possesses a square root tower over F. We shall now improve this to the following more precise

Criterion B.   The complex number z is constructible with straight-edge and compass from $z_1, z_2, \ldots, z_n$ if and only if z is algebraic over

$$ F = \mathbb{Q}(z_1, \ldots, z_n, \bar{z}_1, \ldots, \bar{z}_n) $$

and the normal closure K/F of $F(z)/F$ has dimension a power of two over F.

Proof.   Suppose that z is constructible from $z_1, \ldots, z_n$. Then z is contained in a subfield L/F of ℂ that has a square root tower over F. By Lemma 5 (p. 255) we may assume that L is Galois over F. Then L contains the normal closure K/F of $F(z)/F$. Since L has a square root tower over F, $[L:F] = 2^s$. Then $[K:F]$, which is a factor of $[L:F]$, has the form 2^t. Conversely, suppose that $[K:F] = 2^t$ for K the normal closure of $F(z)/F$. Then $|G| = 2^t$ for $G = \text{Gal } K/F$ so G is solvable and G has a composition series $G = G_1 \rhd G_2 \rhd \cdots \rhd G_{t+1} = 1$ such that every G_i/G_{i+1} is cyclic of order 2. Correspondingly, we have $F = F_1 \subset F_2 \subset \cdots \subset F_{t+1} = K$ where $[F_{i+1}:F_i] = 2$. Then $F_{i+1} = F_i(u_i)$ where

$u_i^2 - a_i u_i + b_i = 0$, $a_i, b_i \in F_i$. Replacing u_i by $v_i = u_i - \frac{1}{2}a_i$ we obtain $F_{i+1} = F_i(v_i)$ and $v_i^2 \in F_i$. Thus K has a square root tower over F and since $z \in K$, z is Euclidean constructible from $z_1, \ldots, z_n$ by our first criterion. $\square$

We shall now apply this result to determine the n such that the regular n-gon is constructible with ruler and compass. For this purpose we need to determine $[\Lambda^{(n)}:\mathbb{Q}]$ where $\Lambda^{(n)}$ denotes the cyclotomic field of nth roots of unity over $\mathbb{Q}$. We know that the set U of nth roots of unity is a cyclic group of order n under multiplication. Hence the number of primitive nth roots of 1, that is, the number of generators of U, is $\varphi(n)$ (exercise 4, p. 47). If z is one of these, then $\Lambda^{(n)} = \mathbb{Q}(z)$ so $[\Lambda^{(n)}:\mathbb{Q}]$ is the degree of the minimum polynomial of z over $\mathbb{Q}$. Now put

$$(55) \qquad \lambda_n(x) = \prod_{z \text{ primitive}} (x - z).$$

If $\eta \in \operatorname{Gal} \Lambda^{(n)}/\mathbb{Q}$ and z is primitive, then $\eta(z)$ is primitive. Hence $\eta(\lambda_n(x)) = \lambda_n(x)$ and so $\lambda_n(x) \in \mathbb{Q}[x]$. It is clear that $\lambda_n(x) | (x^n - 1)$ and, in fact, since any root of unity has an order $d | n$ we see that

$$(56) \qquad x^n - 1 = \prod_{d|n} \lambda_d(x).$$

We shall now prove

THEOREM 4.17. *The polynomial $\lambda_n(x)$ is irreducible in $\mathbb{Q}[x]$.*

Proof. We observe first that $\lambda_n(x)$ has integer coefficients. This holds for $n = 1$ and assuming it holds for every $\lambda_d(x)$, $d < n$, we have $x^n - 1 = \lambda_n(x)g(x)$ where $g(x) = \prod_{d|n;d<n} \lambda_d(x)$ is a monic polynomial with integer coefficients. The division algorithm gives integral polynomials $q(x)$ and $r(x)$ with $\deg r(x) < \deg g(x)$ such that $x^n - 1 = q(x)g(x) + r(x)$. Since $q(x)$ and $r(x)$ are unique in $\mathbb{Q}[x]$ and $x^n - 1 = \lambda_n(x)g(x)$ in $\mathbb{Q}[x]$, we see that $\lambda_n(x) = q(x) \in \mathbb{Z}[x]$. Now suppose that

$$(57) \qquad \lambda_n(x) = h(x)k(x)$$

where $h(x), k(x) \in \mathbb{Z}[x]$ and $h(x)$ is irreducible in $\mathbb{Z}[x]$, hence, in $\mathbb{Q}[x]$ (p. 153). We may also assume that $h(x)$ and $k(x)$ are monic and so $\deg h(x) \geq 1$. Let p be a prime integer not dividing n and let z be a root of $h(x)$. Since $(p, n) = 1$, z^p is a primitive nth root of 1 and, if z^p is not a root of $h(x)$, z^p is a root of $k(x)$; consequently z is a root of $k(x^p)$. Since $h(x)$ is irreducible and has z as a root also, $(h(x), k(x^p)) \neq 1$ and thus $h(x)|k(x^p)$. It follows (as at the beginning of the proof) that $k(x^p) = h(x)l(x)$, where $l(x)$ is monic with integral coefficients. Since $x^n - 1 = \lambda_n(x)g(x)$, we have $x^n - 1 = h(x)k(x)g(x)$. We now pass to congruences modulo p or, what is the same thing, to equations in $(\mathbb{Z}/(p))[x]$. This gives

$$(58) \qquad\qquad x^n - \bar{1} = \bar{h}(x)\bar{k}(x)\bar{g}(x)$$

where, in general, if $f(x) = a_0 x^m + a_1 x^{m-1} + \cdots + a_m \in \mathbb{Z}[x]$, then $\bar{f}(x) = \bar{a}_0 x^m + \bar{a}_1 x^{m-1} + \cdots + \bar{a}_m$, $\bar{a}_i = a_i + (p)$ in $\mathbb{Z}/(p)$. Similarly, we have $\bar{k}(x^p) = \bar{h}(x)\bar{l}(x)$. Now, using $\bar{a}^p = \bar{a}$ for any $a \in \mathbb{Z}$, we see that

$$
\begin{aligned}
\bar{f}(x)^p &= (\bar{a}_0 x^m + \bar{a}_1 x^{m-1} + \cdots + \bar{a}_m)^p \\
&= \bar{a}_0{}^p x^{pm} + \bar{a}_1{}^p x^{p(m-1)} + \cdots + \bar{a}_m{}^p \\
&= \bar{a}_0 x^{pm} + \bar{a}_1 x^{p(m-1)} + \cdots + \bar{a}_m = \bar{f}(x^p)
\end{aligned}
$$

for any $f(x) \in \mathbb{Z}[x]$. Thus $\bar{k}(x)^p = \bar{k}(x^p) = \bar{h}(x)\bar{l}(x)$ which implies that $(\bar{h}(x), \bar{k}(x)) \neq 1$. Then (58) shows that $x^n - \bar{1}$ has multiple roots in its splitting field over $\mathbb{Z}/(p)$. Since the derivative $(x^n - \bar{1})' = \bar{n}x^{n-1}$ and $\bar{n} \neq 0$, we have $(x^n - \bar{1}, (x^n - \bar{1})') = \bar{1}$, contrary to the derivative criterion for multiple roots. This contradiction shows that z^p is a root of $h(x)$ for every prime $p \nmid n$. A repetition of this shows that z^r is a root of $h(x)$ for every integer r prime to n. Since every primitive nth root of 1 has the form z^r, $(r, n) = 1$, we see that $h(x)$ is divisible by every $x - z'$, z' primitive. Then $h(x) = \lambda_n(x)$ and $\lambda_n(x)$ is irreducible $\mathbb{Q}[x]$. $\qquad\qquad\square$

It is now clear that $\lambda_n(x)$ is the minimum polynomial of any primitive nth root of 1. Hence $\varphi(n) = \deg \lambda_n(x) = [\Lambda^{(n)}:\mathbb{Q}]$. We remark also that the foregoing theorem generalizes a result which we proved earlier that if p is a prime then $\lambda_p(x) = (x^p - 1)/(x - 1) = x^{p-1} + x^{p-2} + \cdots + 1$ is irreducible in $\mathbb{Q}[x]$.

We now write $n = 2^{e_1} p_2^{e_2} \cdots p_s^{e_s}$ where the p_i are distinct odd primes, $e_1 \geq 0$ and $e_i > 0$ if $i > 1$. Then

$$\varphi(n) = \varphi(2^{e_1})\varphi(p_2^{e_2}) \cdots \varphi(p_s^{e_s})$$

$$(59) \qquad = \begin{cases} 2^{e_1 - 1} p_2^{e_2 - 1}(p_2 - 1) \cdots p_s^{e_s - 1}(p_s - 1) & \text{if } e_1 > 0 \\ p_2^{e_2 - 1}(p_2 - 1) \cdots p_s^{e_s - 1}(p_s - 1) & \text{if } e_1 = 0. \end{cases}$$

It is clear from this that $\varphi(n)$ is a power of two if and only if the odd primes which are factors of n are Fermat primes and these have multiplicity 1 in the factorization. The constructibility of a regular n-gon with straight-edge and compass is equivalent to the constructibility of the primitive nth root of unity $z = e^{2\pi i/n}$. Since $\mathbb{Q}(z)$ is Galois and $[\mathbb{Q}(z):\mathbb{Q}] = \varphi(n)$ we obtain from Criterion B the following result, which is due to Gauss.

THEOREM 4.18. *A regular n-gon is constructible with straight-edge and compass if and only if n has the form $n = 2^e p_2 \cdots p_s$ where $e \geq 0$ and the p_i are distinct Fermat primes.*

The formula (56) provides us with an algorithm for calculating the polynomial $\lambda_n(x)$, which we shall now call a *cyclotomic polynomial*. To begin with we have

$$\lambda_1(x) = x - 1$$

and assuming we already know the $\lambda_d(x)$ for proper divisors d of n then (56) gives us $\lambda_n(x)$. For example, we have

$$\lambda_2(x) = (x^2 - 1)/\lambda_1(x) = x + 1$$
$$\lambda_3(x) = (x^3 - 1)/\lambda_1(x) = x^2 + x + 1$$
$$\lambda_4(x) = (x^4 - 1)/\lambda_1(x)\lambda_2(x) = x^2 + 1$$
$$\lambda_6(x) = (x^6 - 1)/\lambda_1(x)\lambda_2(x)\lambda_3(x) = x^2 - x + 1$$
$$\lambda_{12}(x) = (x^{12} - 1)/\lambda_1(x)\lambda_2(x)\lambda_3(x)\lambda_4(x)\lambda_6(x) = x^4 - x^2 + 1.$$

We shall now round out our results on cyclotomic fields over $\mathbb{Q}$ by determining the structure of the Galois groups of these fields or, equivalently, of the cyclotomic polynomials $\lambda_n(x)$ over $\mathbb{Q}$. We now see that the order of this group is the degree of the irreducible polynomial $\lambda_n(x)$ and this is $\varphi(n)$. It follows from the proof of Lemma 1 (p. 252) that the Galois group of the cyclotomic field of order n over $\mathbb{Q}$ is isomorphic to the multiplicative group U_n of units of the ring $\mathbb{Z}/(n)$. If n is a prime then we know that this is a cyclic group of order $\varphi(p) = p - 1$. We proceed to determine the structure of U_n for composite n.

As indicated in exercises 9 and 10 on p. 110, it is easy to see that if $n = p_1^{e_1} p_2^{e_2} \cdots p_s^{e_s}$, p_i distinct primes, then U_n is isomorphic to the direct product of the groups $U_{p_i^{e_i}}$. Hence it suffices to determine the structure of any U_{p^e}, p a prime. We treat first the case of any odd prime power in

THEOREM 4.19. *If p is an odd prime, the multiplicative group U_{p^e} of units of $\mathbb{Z}/(p^e)$ is cyclic.*

Proof. Since the order of $G = U_{p^e}$ is $p^{e-1}(p - 1)$, G is a direct product of its subgroup H of order p^{e-1} consisting of the elements which satisfy $x^{p^{e-1}} = 1$ and the subgroup K of order $p - 1$ of the elements satisfying $x^{p-1} = 1$.[13] It

---

[13] Here and in the remainder of this section we require the structure theorem on finite abelian groups which we derived in section 3.10, p. 195.

suffices to show that both H and K are cyclic since the direct product of cyclic groups having relatively prime orders is cyclic. Since U_p is cyclic we can choose an integer a such that $a + (p), a^2 + (p), \ldots, a^{p-1} + (p)$ are distinct in $\mathbb{Z}/(p)$. Put $b = a^{p^{e-1}}$. Since $(a, p) = 1$, $(b, p^e) = 1$, and $b + (p^e)$ and $a + (p^e) \in G$. Also $b^{p-1} = (a^{p^{e-1}})^{p-1} = a^{\varphi(p^e)} \equiv 1 \pmod{p^e}$ and so $b + (p^e) \in K$. Since $b = a^{p^{e-1}} \equiv a \pmod{p}$, $b + (p), b^2 + (p), \ldots, b^{p-1} + (p)$ are distinct. Hence also $b + (p^e)$, $b^2 + (p^e), \ldots, b^{p-1} + (p^e)$ are distinct. This implies that the order of $b + (p^e)$ is $p - 1$. Since $|K| = p - 1$, it follows that K is cyclic with generator $b + (p^e)$. It remains to prove that H is cyclic, and we now assume $e \geq 2$, since otherwise $H = 1$ and the result is clear. Assuming $e \geq 2$, we see that H is a direct product of $k > 1$ cyclic groups of orders p^{e_i}, $e_i \geq 1$. Then the number of solutions in H of $x^p = 1$ is p^k. Hence it is enough to show that the number of integers n, $0 < n < p^e$, satisfying $n^p \equiv 1 \pmod{p^e}$ does not exceed p. If n satisfies these conditions, then, since $n^p \equiv n \pmod{p}$, we have $n \equiv 1 \pmod{p}$. Then if $n \neq 1$, we may write $n = 1 + yp^f + zp^{f+1}$ where $1 \leq f \leq e - 1$, $0 < y < p$, and z is a non-negative integer. Then

$$n^p = 1 + \binom{p}{1}(y + zp)p^f + \binom{p}{2}(y + zp)^2 p^{2f} + \cdots + (y + zp)^p p^{pf}$$

$$\equiv 1 + yp^{f+1} \pmod{p^{f+2}}.$$

If $n^p \equiv 1 \pmod{p^e}$ and $f < e - 1$, this gives $yp^{f+1} \equiv 0 \pmod{p^{f+2}}$, and $y \equiv 0 \pmod{p}$ contrary to $0 < y < p$. Hence we see that, if $1 < n < p^e$ satisfies $n^p \equiv 1 \pmod{p^e}$, then $n = 1 + yp^{e-1}$, $0 < y < p$. This gives at most p solutions, including 1, and completes the proof of the theorem. $\square$

We consider next the case of the prime 2 in the following

THEOREM 4.20. *U_2 and U_4 are cyclic and, if $e \geq 3$, then U_{2^e} is a direct product of a cyclic group of order 2 and one of order 2^{e-2}.*

Proof. The order of $G = U_{2^e}$ is $\varphi(2^e) = 2^{e-1}$. If $e = 1$, $|G| = 1$, and if $e = 2$, $|G| = 2$, so in these cases G is cyclic. Now suppose $e \geq 3$. We show first that we have four distinct solutions of $x^2 = 1$ in G. This will imply that G is a direct product of at least two distinct cyclic groups $\neq 1$. Put $a_1 = 1$, $a_2 = -1$, $a_3 = 1 + 2^{e-1}$, $a_4 = -1 + 2^{e-1}$, and $x_i = a_i + (2^e)$. Then the x_i are distinct elements of G satisfying $x_i^2 = 1$, which is what we wanted. Moreover, since G is a direct product of at least two cyclic groups $\neq 1$ and $|G| = 2^{e-1}$, we see that, if $x \in G$, then $x^{2^{e-2}} = 1$ or, what is the same thing, if a is an odd integer, then $a^{2^{e-2}} \equiv 1 \pmod{2^e}$. The proof will be completed by displaying an x such that

$x^{2^{e-3}} \neq 1$. Then we shall have a cyclic subgroup of order 2^{e-2} and this can happen only if G is a direct product of a cyclic group of order 2^{e-2} and one of order 2. We now take $x = 5 + (2^e)$. We note first that, if $e = 3$, then $5^{2^{e-3}} \equiv 5 \not\equiv 1 \pmod{8 = 2^e}$ but $5^{2^{e-3}} \equiv 1 \pmod{2^{e-1} = 4}$. Now let $f \geq 3$ and let $k(f)$ be the largest integer such that $5^{2^{f-3}} \equiv 1 \pmod{2^k}$. Then we have $k(3) = 2$. Since for any $f \geq 3$ we have $5^{2^{f-3}} = 1 + y2^{k(f)}$ where y is odd, this gives

$$5^{2^{(f+1)-3}} = (5^{2^{f-3}})^2 = 1 + y2^{k(f)+1} + y^2 2^{2k(f)}$$

which shows that $k(f + 1) \geq k(f)$, so $k(f) \geq 2$ if $f \geq 3$. Then the displayed relation shows that $5^{2^{(f+1-3)}} = 1 + z2^{k(f)+1}$ where $z = y + 2^{k(f)-1} y^2$ is odd. Hence $k(f + 1) = k(f) + 1$. This and $k(3) = 2$ imply that $k(f) = f - 1$ for all $f > 3$. Thus $5^{2^{e-3}} \not\equiv 1 \pmod{2^e}$ if $e \geq 3$, which is what we needed. This completes the proof. $\square$

The last two theorems give a description of the Galois group of the cyclotomic field of p^eth roots of unity over the rationals. The result is the following

THEOREM 4.21.   *The Galois group G of the cyclotomic field of p^eth roots of unity over $\mathbb{Q}$ = Galois group of $\lambda_{p^e}(x) = 0$ is cyclic unless $p = 2$ and $e \geq 3$, in which case G is a direct product of a cyclic group of order 2 and one of order 2^{e-2}.*

EXERCISES

1. Use the Möbius inversion formula (exercise 18, p. 151) to prove that

$$\lambda_n(x) = \prod_{d|n} (x^d - 1)^{\mu(n/d)}.$$

2. Let $f(x)$ have distinct roots $r_1, r_2, \ldots, r_n$. Show that the discriminant

$$d = \prod_{1 \leq i < j \leq n} (r_i - r_j)^2 = (-1)^{n(n-1)/2} \prod_1^n f'(r_i)$$

f', the derivative of f. Let $\lambda_p(x) = x^{p-1} + x^{p-2} + \cdots + 1$, p a prime. Differentiate $x^p - 1 = (x - 1)\lambda_p(x)$ to obtain

$$px^{p-1} = \lambda_p(x) + (x - 1)\lambda'_p(x).$$

Use this and the foregoing formula to show that the discriminant of the cyclotomic polynomial $\lambda_p(x)$ is

$$d = (-1)^{p(p-1)/2} p^{p-2}.$$

3. Let Λ_p be the field of the pth roots of unity over $\mathbb{Q}$ where p is an odd prime. Show that Λ_p has a unique quadratic subfield $E/\mathbb{Q}$ and E is real (subfield of $\mathbb{R}$) or not real according as p has the form $4n + 1$ or $4n + 3$.

4. Use exercise 2, page 243 and Theorem 4.21 to show that for any finite cyclic group G there exists a subfield of some cyclotomic field over $\mathbb{Q}$ having G as Galois group.

Remark A classical theorem of Dirichlet states that any arithmetic progression $a + kd$ where a and d are relatively prime positive integers and $k = 0, 1, 2, \ldots$ contains an infinite number of primes. The special case of this in which $a = 1$ and $d = p$, p prime, and the fundamental structure theorem on finite abelian groups (Theorem 3.13) can be used to prove that any finite abelian group is a homomorphic image of the group of units U_n of some $\mathbb{Z}/(n)$ where n is square-free. This result and Theorem 4.21 can be used to prove the existence of a Galois field extension $E/\mathbb{Q}$ with prescribed finite abelian group as Galois group. Dirichlet's theorem requires function theory for its proof. However, the special case of progressions of the form $1 + kn$ has an elementary proof (see, T. Nagell, *Introduction to Number Theory*, Wiley, New York 1951, p. 118.)

4.12 TRANSCENDENCE OF e AND π. THE LINDEMANN-WEIERSTRASS THEOREM

In this section we shall prove that π is transcendental, that is, not algebraic over $\mathbb{Q}$. This will imply that π and $\sqrt{\pi}$ are not constructible numbers and hence that it is impossible to construct with straight-edge and compass a length equal to the circumference of a circle of given radius, or a length equal to the side of a square whose area is that of a given circle. With a little more effort we can prove a considerably more general result than the transcendence of π, namely,

THE LINDEMANN-WEIERSTRASS THEOREM. *If $u_1, u_2, \ldots, u_n$ are algebraic numbers (that is, complex numbers which are algebraic over $\mathbb{Q}$) which are linearly independent over $\mathbb{Q}$, then the complex exponentials $e^{u_1}, e^{u_2}, \ldots, e^{u_n}$ are algebraically independent over the field of algebraic numbers.*

If $u \in \mathbb{C}$ we can define $e^u = 1 + u + u^2/2! + u^3/3! + \cdots$, which is (absolutely) convergent for every u. We also have the functional equation $e^u e^v = e^{u+v}$, which follows easily from the power series definition of e^u. It should be noted also that the set of algebraic numbers constitute a subfield of $\mathbb{C}$. (We shall prove a more general result in a moment.) Taking $n = 1$ in the foregoing theorem we see that if u is a non-zero algebraic number, then e^u is transcendental. In particular e is transcendental and, since $e^{\pi i} = \cos \pi + i \sin \pi = -1$, πi is transcendental. Since i is algebraic this implies the transcendence of π. Similarly, if u is a real algebraic number $\neq 0$, 1, then the relation $e^{\log u} = u$ implies that $\log u$ is transcendental.

We shall now show that the Lindemann-Weierstrass theorem is equivalent to another theorem, which is sometimes also called the Lindemann-Weierstrass theorem (or the Generalized Lindemann theorem). We state this as

THEOREM 4.22. *If $u_1, u_2, \ldots, u_n$ are distinct algebraic numbers, then the complex exponentials are linearly independent over the field of algebraic numbers.*

Suppose this holds and let $u_1, u_2, \ldots, u_n$ be algebraic numbers which are linearly independent over $\mathbb{Q}$. Let $(k_1, k_2, \ldots, k_n)$ and $(l_1, l_2, \ldots, l_n)$ be distinct sequences of non-negative integers. Then $\prod (e^{u_i})^{k_i} = e^{\sum k_i u_i}$ and $\prod (e^{u_i})^{l_i} = e^{\sum l_i u_i}$, and the exponents $\sum k_i u_i$ and $\sum l_i u_i$ are distinct. It now follows from Theorem 4.22 that if we have r distinct sequences $(k_{1i}, k_{2i}, \ldots, k_{ni})$ of non-negative integers, then the r complex numbers $(e^{u_1})^{k_{1i}}(e^{u_2})^{k_{2i}} \cdots (e^{u_n})^{k_{ni}}$ are linearly independent over algebraic numbers. Clearly this means that the exponentials $e^{u_1}, e^{u_2}, \ldots, e^{u_n}$ are algebraically independent over algebraic numbers. Thus Theorem 4.22 implies the Lindemann-Weierstrass theorem. On the other hand, suppose the Lindemann-Weierstrass theorem holds, and let $u_1, u_2, \ldots, u_n$ be distinct algebraic numbers. The subgroup of the additive group of $\mathbb{C}$ generated by the u_i is a free $\mathbb{Z}$-module (since it has no torsion), so there exist r, $1 \leq r \leq n$, complex numbers $v_1, v_2, \ldots, v_r$ which are linearly independent over $\mathbb{Q}$ such that $u_i = \sum_{j=1}^{r} a_{ij} v_j$, $a_{ij} \in \mathbb{Z}$, $1 \leq i \leq n$. Then $e^{u_i} = \prod_{j=1}^{r} (e^{v_j})^{a_{ij}}$ and since the u_i are distinct, the vectors $(a_{i1}, \ldots, a_{ir})$ are distinct. Hence if we had a nontrivial linear relation with algebraic number coefficients connecting the e^{u_i}, then on multiplying this by a suitable power $(e^{v_1} e^{v_2} \cdots e^{v_r})^a$ with positive integral a we would obtain a non-trivial algebraic relation with algebraic number coefficients connecting the exponentials $e^{v_1}, \ldots, e^{v_r}$. Since the v_i are linearly independent over $\mathbb{Q}$, this would contradict the Lindemann-Weierstrass theorem. Thus we have the equivalence of the two theorems.

We shall prove Theorem 4.22. For this purpose we shall require some results on algebraic elements of fields and on integral algebraic complex numbers. We can obtain most of these simultaneously for the two cases by considering the following general situation. We suppose E to be a field and R to be a subring of E. Then we shall call an element $u \in E$ *integral over R* or *R-integral* if there exists a monic polynomial $f(x) \in R[x]$ such that $f(u) = 0$. If $R = F$ is a subfield this reduces to the concept that u is algebraic over F. If $E = \mathbb{C}$ then the elements of $\mathbb{C}$ which are algebraic over $\mathbb{Q}$ are called *algebraic numbers*. On the other hand, the complex numbers which are $\mathbb{Z}$-integral are called *algebraic integers*. Obviously these are necessarily algebraic numbers.

If $u \in E$ is R-integral, then we have a relation of the form

$$(60) \qquad u^n = a_0 + a_1 u + \cdots + a_{n-1} u^{n-1}.$$

Let $M = R1 + Ru + \cdots + Ru^{n-1}$, the R-submodule of E generated by $1, u, \ldots, u^{n-1}$ (regarding E as an R module in which the addition and 0 are as in the ring E and the module action by $r \in R$ is the multiplication as defined in E). We have $uM \subset Ru + Ru^2 + \cdots + Ru^n$, and since (60) shows that $u^n \in M$ we see that $uM \subset M$. This gives the "only if" part of the following criterion.

LEMMA 1. *The element $u \in E$ is R-integral if and only if there exists a finitely generated R-submodule of E containing 1 and satisfying $uM \subset M$.*

Proof. To prove the sufficiency of the condition let $M = Ru_1 + Ru_2 + \cdots + Ru_n$ satisfy, (a) $1 \in M$, (b) $uM \subset M$. Then we have $uu_i = \sum_{j=1}^n a_{ij} u_j$, $1 \leq i \leq n$, where the $a_{ij} \in R$. Hence the system of linear homogeneous equations

$$(a_{11} - u)x_1 + a_{12}x_2 + \cdots + a_{1n}x_n = 0$$

$$(61) \qquad a_{21}x_1 + (a_{22} - u)x_2 + \cdots + a_{2n}x_n = 0$$

$$\cdots\cdots\cdots\cdots\cdots\cdots\cdots\cdots\cdots\cdots\cdots$$

$$a_{n1}x_1 + a_{n2}x_2 + \cdots + (a_{nn} - u)x_n = 0$$

has the solution $(x_1, \ldots, x_n) = (u_1, \ldots, u_n)$. Since $1 \in M$ some $u_i \neq 0$, and so the solution is not the trivial one $(0, \ldots, 0)$. Hence, by a standard result of linear algebra we have $\det(A - u1) = 0$ where $A = (a_{ij})$. Thus u is a root of the characteristic polynomial $\det(x1 - A)$ of the matrix A. Clearly this polynomial is monic and has coefficients in R, so u is R-integral. $\square$

If M and N are R-submodules of E we let MN denote the submodule generated by all the products uv, $u \in M$, $v \in N$. It is clear that this is the set of elements of E of the form $\sum u_i v_i$, $u_i \in M$, $v_i \in N$. If M and N are finitely generated, then so is MN. In fact, it is immediate that if $M = \sum_1^m Ru_i$ and $N = \sum_1^n Rv_j$, then $MN = \sum Ru_i v_j$. Also, if $w \in E$ satisfies $wM \subset M$ then

$$w(MN) = (wM)N \subset MN.$$

By induction, if $M_1, M_2, \ldots, M_r$ are finitely generated, then the submodule $M_1 M_2 \cdots M_r$ generated by all products $x_1 \cdots x_r$, $x_i \in M_i$, is finitely generated, and if $wM_i \subset M_i$ for one of the M_i, then $wM_1M_2 \cdots M_r \subset M_1M_2 \cdots M_r$. We shall use these remarks and Lemma 1 to prove

THEOREM 4.23.   *If E is a field and R is a subring of E, the set A of R-integral elements of E is a subring of E containing R. Moreover, any element of E which is A-integral is R-integral (and so is contained in A).*

Proof.   Let u and $v \in A$ so that there exist finitely generated R-submodules M and N of E containing 1 such that $uM \subset M$ and $vN \subset N$. Then $(u \pm v)MN \subset u(MN) + v(MN) \subset MN$. Also $(uv)MN \subset MN$. Since $1 \in MN$ the conditions of Lemma 1 are satisfied for $u \pm v$, 1 and uv. Hence these elements are R-integral and A is thus a subring of E. It is clear also that $A \supset R$. Now let u be A-integral. Then we have an $M = Au_1 + \cdots + Au_n$ containing 1 and satisfying $uM \subset M$. We may as well assume $u_1 = 1$. Since $uM \subset M$ there exist $a_{ij} \in A$ such that $uu_i = \sum a_{ij}u_j$. Now there exists a finitely generated R-submodule N_{ij} such that $a_{ij}N_{ij} \subset N_{ij}$ and $1 \in N_{ij}$. Multiplying together the N_{ij} we obtain a finitely generated module $N = Rv_1 + \cdots + Rv_m$ with $v_1 = 1$ satisfying $a_{ij}N \subset N$ for every a_{ij}. Let $P = \sum_{i,j} Ru_iv_j$. Then $1 = u_1v_1 \in P$ and $u(u_iv_k) = \sum a_{ij}u_jv_k = \sum u_ja_{ij}v_k$. Since $a_{ij}v_k \in N$ this is an R-linear combination of the elements u_jv_k. It follows that $uP \subset P$, and so u is R-integral, by Lemma 1. $\square$

In the case in which $R = F$ is a subfield this result states that the elements of E which are algebraic over F constitute a subring. Moreover, in this case, if u is algebraic, then $F(u) = F[u]$, and u^{-1} is therefore algebraic for $u \neq 0$. Hence the set of elements of E which are algebraic over F constitute a subfield A of E and every element of E which is algebraic over A is contained in A.

We now specialize $E = \mathbb{C}$ and $R = \mathbb{Q}$ or $\mathbb{Z}$. Then the $\mathbb{Q}$-integers are the algebraic numbers and the $\mathbb{Z}$-integers are algebraic integers. We have the following criterion for a complex number to be an algebraic integer:

LEMMA 2.   *A complex number u is an algebraic integer if and only if u is an algebraic number and its minimum polynomial $\in \mathbb{Z}[x]$.*

Proof.   The condition is, of course, sufficient. Now assume u is an algebraic integer. Then we have a monic polynomial $f(x)$ in $\mathbb{Z}[x]$ such that $f(u) = 0$. If $\mu(x)$ is a minimum polynomial of u then $\mu(x)|f(x)$. Since $\mathbb{Z}$ is factorial it follows easily that $\mu(x) \in \mathbb{Z}[x]$ (Corollary to Theorem 2.25). $\square$

We can now prove the following important result:

THEOREM 4.24. *A rational number is an algebraic integer if and only if it is an integer. If u is any algebraic number, then there exists a $b \in \mathbb{Z}$ such that bu is an algebraic integer.*

Proof. If $a \in \mathbb{Z}$ it is $\mathbb{Z}$-integral. On the other hand, if $a \in \mathbb{Q}$ its minimum polynomial over $\mathbb{Q}$ is $x - a$, so if a is an algebraic integer then $a \in \mathbb{Z}$. Now let $u \in \mathbb{C}$ be algebraic over $\mathbb{Q}$ and let $f(x) = x^n + \alpha_1 x^{n-1} + \cdots + \alpha_n \in \mathbb{Q}[x]$ be a polynomial such that $f(u) = 0$. If $b \in \mathbb{Z}$, $b \neq 0$, then bu is a root of $b^n f(b^{-1}x) = 0$ and $b^n f(b^{-1}x) = b^n(b^{-n}x^n + b^{-(n-1)}\alpha_1 x^{n-1} + \cdots + \alpha_n) = x^n + b\alpha_1 x^{n-1} + \cdots + b^n\alpha_n$. If we choose b to be the product of the denominators of the rational numbers α_i we obtain a monic polynomial in $\mathbb{Z}[x]$ having bu as a root. Then bu is an algebraic integer. $\square$

We shall need to use the so-called fundamental theorem of algebra, which states that any polynomial in $\mathbb{C}[x]$ of positive degree has a root in $\mathbb{C}$. This result, which will be proved in section 5.1, implies that every monic polynomial of positive degree with coefficients in $\mathbb{C}$ factors as a product $\prod (x - r_i)$ in $\mathbb{C}[x]$. In other words, $\mathbb{C}$ contains a splitting field for every monic polynomial $\neq 1$ in $\mathbb{C}[x]$. It follows that if S is a finite set of algebraic numbers we can imbed $\mathbb{Q}(S)$ in a Galois extension $K/\mathbb{Q} \subset \mathbb{C}$.

We are now ready to begin the

Proof of Theorem 4.22. We assume, contrary to the assertion, that we have distinct algebraic numbers $u_1, u_2, \ldots, u_n$ and algebraic numbers $v_1, v_2, \ldots, v_n$ not all 0 such that

(62) $$v_1 e^{u_1} + v_2 e^{u_2} + \cdots + v_n e^{u_n} = 0.$$

We shall show that this implies that we have a relation of the same sort with rational v_i and then that we have one of the form

(63) $$v_0 + v_1 \sum_{j=1}^{r} e^{\eta_j(u_1)} + \cdots + v_n \sum_{j=1}^{r} e^{\eta_j(u_n)} = 0$$

where the v_i are integers, $v_0 \neq 0$, and the η_j are the elements of the Galois group of a Galois extension field $K/\mathbb{Q}$ containing the u_i and contained in $\mathbb{C}$. Then, by an analytic argument, we shall show that (63) is impossible.

In order to make clearer the formal arguments which give the passage from (62) to (63), we introduce the group algebra of the additive group of algebraic numbers over the field of algebraic numbers. This is a special case of the group algebras which were defined in exercise 8, p. 127. In order to distinguish between the field of algebraic numbers A and its additive group, we now denote the latter as A' and its elements as u', where $u \to u'$ is an isomorphism of $(A, +, 0)$ onto A'. We write the composition in A' as multiplication. Then $a \to a'$ is $1\text{--}1$ and $a'b' = (a + b)'$, $0'$ is the unit of A' and $(-a)'$ is the inverse of a'. The group algebra $A[A']$ we are interested in, is the set of sums $\sum v_i u_i'$, $v_i \in A$, $u_i' \in A'$, where addition is the obvious one, and multiplication is given by the distributive law, and $(v_1 u_1')(v_2 u_2') = v_1 v_2 (u_1 + u_2)'$. Moreover, if $u_1, \ldots, u_n$ are distinct elements of A, then the elements $u_1', u_2', \ldots, u_n'$ are linearly independent over A: that is, $\sum v_i u_i' = 0$ for $v_i \in A$ implies that every $v_i = 0$. Now, in $\mathbb{C}$ we have $e^{u_1} e^{u_2} = e^{u_1 + u_2}$. Hence, by the "universal" property of group algebras given in exercise 8, p. 127, we have a homomorphism ε of $A[A']$ into $\mathbb{C}$ sending $\sum v_i u_i'$ into $\sum v_i e^{u_i}$. Theorem 4.22 can now be restated as: ε is a monomorphism.

The group algebra $A[A']$ is commutative. We shall now show that it is a domain. To see this we introduce an ordering in $\mathbb{C}$ which is compatible with addition, the so-called lexicographic ordering of $\mathbb{C}$. If $x = a + bi$ and $y = c + di$ where a, b, c, d are real, then we say that $x > y$ if $a > c$ or if $a = c$ and $b > d$. This ordering satisfies the trichotomy law: for any pair (x, y) either $x > y$, $x = y$, or $y > x$. Moreover, if $x > y$ and $z > t$ then $x + z > y + t$. Now let $\sum_1^n v_i u_i'$, $\sum_1^m z_j t_j'$ be two non-zero elements of $A[A']$. Then we may assume that $v_1 \neq 0$, $z_1 \neq 0$, and $u_1 > u_2 > \cdots > u_n$, $t_1 > t_2 > \cdots > t_m$. Then $(\sum v_i u_i')(\sum z_j t_j') = v_1 z_1 (u_1 + t_1)' +$ a sum of terms of the form wq' where $q < u_1 + t_1$. Clearly this is not zero, so $A[A']$ is a domain.

Suppose $\sum v_i u_i' \in \ker \varepsilon$. We can imbed the u_i and v_i in a Galois subfield $K/\mathbb{Q}$ of $\mathbb{C}$. Then the subset of elements of the form $\sum x_i y_i'$ with $x_i, y_i \in K$ is a subring $K[K']$ of $A[A']$, and if $\eta \in G = \mathrm{Gal}\, K/\mathbb{Q}$, then η defines two automorphisms in $K[K']$. The first of these, which we shall denote as $\sigma(\eta)$, is $\sum x_i y_i' \to \sum \eta(x_i)y_i'$, and the second is $\tau(\eta) : \sum x_i y_i' \to \sum x_i(\eta(y_i))'$. The fact that these are automorphisms is clear. Now suppose $\sum v_i u_i' \neq 0$. Then if $G = \{\eta_1, \eta_2, \ldots, \eta_m\}$ every $\sigma(\eta_j)(\sum v_i u_i') \neq 0$ and hence

$$U = \prod_{j=1}^m \sigma(\eta_j)\left(\sum v_i u_i'\right) = \prod_{j=1}^m \left(\sum \eta_j(v_i)u_i'\right) \neq 0.$$

Since U contains the factor $\sum v_i u_i'$, $U \in \ker \varepsilon$. It is clear from the commutativity of $A[A']$ that $\sigma(\eta)U = U$ for every $\eta = \eta_k \in G$. Hence if we write U as $\sum z_i t_i'$ with distinct t_i', then $\eta U = U$, that is, $\sum \eta(z_i)t_i' = \sum z_i t_i'$, implies $\eta(z_i) = z_i$ for every z_i and every $\eta \in G$. Then $z_i \in \mathbb{Q} = \mathrm{Inv}\, G$. We have therefore shown that if we have a non-zero element in $\ker \varepsilon$ then we have one of the form $\sum v_i u_i'$ with rational v_i. Now apply $\tau(\eta_j)$ to this element and form

$$V = \prod_{j=1}^m \tau(\eta_j)\left(\sum v_i u_i'\right) = \prod_{j=1}^m \left(\sum v_i(\eta_j(u_i))'\right).$$

Then this is a non-zero element of ker ε satisfying $\tau(\eta)V = V$ for all $\eta \in G$. We can write $V = \sum z_i t_i'$ where the $z_i \in \mathbb{Q}$ and we have $\sum z_i(\eta(t_i)') = \sum z_i t_i'$, $\eta \in G$. We now average these various expressions for V to obtain

$$V = \frac{1}{m}\left(\sum_{j=1}^{m}\sum_{i} z_i(\eta_j(t_i))'\right)$$

$$= \frac{1}{m}\sum_{i} z_i\left(\sum_{j=1}^{m} (\eta_j(t_i))'\right)$$

$$= \sum_{i}\left(\frac{z_i}{m}\right)T'(t_i)$$

where, in general, for $t \in K$, we define

$$T'(t) = \sum_{j=1}^{m} (\eta_j(t))'.$$

We have now shown that ker $\varepsilon \neq 0$ implies that we have a non-zero element in ker ε of the form $\sum v_i T'(t_i)$ where the $v_i \in \mathbb{Q}$. Also, by combining terms we may assume that $T'(t_i) \neq T'(t_j)$ for $i \neq j$, which implies that $t_j \neq \eta(t_i)$ for every $\eta \in G$.

Let $s, t \in K$ and consider

$$T'(s)T'(t) = \left(\sum_{j=1}^{m} (\eta_j(s))'\right)\left(\sum_{k=1}^{m} (\eta_k(t))'\right)$$

$$= \sum_{j,k} (\eta_j(s))'(\eta_k(t))'$$

$$= \sum_{j,k} (\eta_j(s) + \eta_k(t))'$$

$$= \sum_{j,k} (\eta_j(s + \eta_j^{-1}\eta_k(t)))'$$

$$= \sum_{k}\left(\sum_{j} \eta_j(s + \eta_k(t))\right)'$$

$$= \sum_{k} T'(s + \eta_k(t)).$$

This relation shows that if $\sum v_i T'(t_i) \in$ ker ε with $v_i \in \mathbb{Q}$, $v_1 \neq 0$, and $t_j \neq \eta(t_i)$ for every $i \neq j$ and $\eta \in G$, then multiplication by $T'(-t_1)$ gives an element in ker ε of the form

(64) $$v_0 + v_1 T'(u_1) + \cdots + v_n T'(u_n)$$

with $v_i \in \mathbb{Q}$, $v_0 \neq 0$, $u_i \neq 0$. Multiplication by a suitable integer allows us to assume the v_j are integers. The fact that (64) $\in \ker \varepsilon$ implies that we have the relation (63).

So far the argument has been purely algebraic. We now come to the analytic part of the proof, which will consist of establishing a contradiction to a relation of the form (63) where the v_i are integers, $v_0 \neq 0$, and the u_i are algebraic numbers $\neq 0$. We assume that all the u_i and hence all $\eta_j(u_i)$ are roots of a polynomial $f(x) = \sum_0^t a_k x^k \in \mathbb{Z}[x]$, $a_0 \neq 0$. Let p be a prime and introduce

$$h(x) = x^{p-1}f(x)^p = \sum_{p-1}^{s} b_j x^j$$

where $s = tp + p - 1$. Then the $b_j \in \mathbb{Z}$, $b_{p-1} = a_0{}^p$ and for $p - 1 \leq j \leq s$ we have

$$
\begin{aligned}
(65) \qquad j!b_j e^x = {}& \left[j!b_j + \frac{j!}{1!} b_j x + \cdots + \frac{j!}{(j-p)!} b_j x^{j-p} \right] \\
&+ \left[\frac{j!}{(j-p+1)!} b_j x^{j-p+1} + \cdots + jb_j x^{j-1} \right] \\
&+ b_j x^j \left[1 + \frac{x}{j+1} + \frac{x^2}{(j+1)(j+2)} + \cdots \right].
\end{aligned}
$$

It is understood here that the first bracket is 0 if $j = p - 1$. Moreover, if $j \geq p$, then $\dfrac{j!}{(j-p)!} = p!\dbinom{j}{p}$ so this is $p!$ times an integer. A fortiori, $\dfrac{j!}{k!}$, $0 \leq k < j - p$, is $p!$ times an integer and hence the first bracket in (65) is $p!$ times an integral polynomial. Now put

$$(66) \qquad N_p = \frac{1}{(p-1)!} \sum_{p-1}^{s} j!b_j.$$

Then summing (65) for $j = p - 1, \ldots, s$ we obtain

$$
\begin{aligned}
(67) \qquad (p-1)!N_p e^x = {}& \sum_{p-1}^{s} j!b_j e^x \\
= {}& p!g_p(x) + \sum_{p-1}^{s} \left[\frac{j!}{(j-p+1)!} b_j x^{j-p+1} + \cdots + jb_j x^{j-1} \right] \\
&+ \sum_{p-1}^{s} b_j x^j \left[1 + \frac{x}{j+1} + \frac{x^2}{(j+1)(j+2)} + \cdots \right]
\end{aligned}
$$

where $g_p(x) \in \mathbb{Z}[x]$. We observe next that

$$h'(x) = \sum_{p-1}^{s} jb_j x^{j-1},$$

$$h''(x) = \sum_{p-1}^{s} \frac{j!}{(j-2)!} b_j x^{j-2},$$

$$\cdots\cdots\cdots\cdots\cdots\cdots\cdots\cdots\cdots\cdots\cdots\cdots$$

$$h^{(p-1)}(x) = \sum_{p-1}^{s} \frac{j!}{(j-p+1)!} b_j x^{j-p+1}$$

and these are all divisible by $f(x)$ since $h(x) = x^{p-1}f(x)^p$. Hence the first summation in (67), which is $h'(x) + h''(x) + \cdots + h^{(p-1)}(x)$, is divisible by $f(x)$ and this becomes 0 when we put $x = u_i$. Next we need to estimate $|R(u_i)|$ where $R(x)$ is the second summation in (67). We now assume that the prime p is chosen so that $p > 2|u_i|$ for all u_i. Then since $j + 1 \geq p$ also, we have

$$\left| 1 + \frac{u_i}{j+1} + \frac{u_i^2}{(j+1)(j+2)} + \cdots \right| < 2$$

so

$$|R(u_i)| < 2 \sum_{p-1}^{s} |b_j| |u_i|^j \leq 2|u_i|^{p-1} \left(\sum_{0}^{t} |a_k| |u_i| k \right)^p < 2M^p$$

if M is the largest of the $2n$ numbers $\sum_0^t |a_k| |u_i|^k$ and $\sum_0^t |a_k| |u_i|^{k+1}$ for $i = 1, 2, \ldots, n$. Hence if $p > 2|u_i|$ then we have

(68) $$\left| N_p e^{u_i} - p g_p(u_i) \right| < \frac{2M^p}{(p-1)!}.$$

Moreover, if in addition $p > |a_0|$, then N_p, which is given by (66), is not divisible by p since $N_p \equiv b_{p-1} = a_0^p \equiv a_0$ (mod p.) We therefore have the following

LEMMA 3. *Let u_i, $1 \leq i \leq n$, be non-zero algebraic numbers, $f(x) = \sum_0^t a_k x^k \in \mathbb{Z}[x]$, $a_0 \neq 0$, be a polynomial such that $f(u_i) = 0$ for all i. Let M be the maximum of the $2n$ numbers $\sum_{k=0}^{t} |a_k| |u_i|^k$ and $\sum_{k=0}^{t} |a_k| |u_i|^{k+1}$ and let p be a prime > $\max(|a_0|, 2|u_1|, \ldots, 2|u_n|)$. Then there exists an integer N_p not divisible by p and a polynomial $g_p(x) \in \mathbb{Z}[x]$ of degree < tp such that the inequalities (68) hold.*[14]

We now return to the relation (63) where the v_i are integers, $v_0 \neq 0$, and the u_i are non-zero algebraic numbers which are roots of the polynomial $f(x) \in \mathbb{Z}[x]$

---

[14] I am indebted to E. Schenkman for several communications on the subject of the proof of Theorem 4.22. In particular, I am indebted to him for the form of this lemma.

as in Lemma 3. The numbers $\eta_j(u_i)$, $\eta_j \in G = \text{Gal } K/\mathbb{Q}$, are also roots of $f(x)$. Hence, by Lemma 3, for all sufficiently large primes p there exists an integer N_p not divisible by p and an integral polynomial $g_p(x)$ of degree $< pt$, $t = \deg f$, such that $\left| N_p e^{\eta_j(u_i)} - pg_p(\eta_j(u_i)) \right| < 2M^p/(p-1)!$ for all $i = 1, \ldots, n$, $j = 1, \ldots, r(=|G|)$. Now let k be a positive integer such that ku_i^l is an algebraic integer for every u_i and every $l \le t$. The existence of such a k is assured by Theorem 4.24. Then $k^p g_p(u_i)$ is an algebraic integer and hence every $k^p g_p(\eta_j(u_i))$ is an algebraic integer. Also $\sum_{j=1}^r k^p g_p(\eta_j(u_i))$ is an algebraic integer, but since it is fixed by G, it is a rational number. Hence this is an integer, by Theorem 4.23.

Now we have

$$N_p k^p v_0 + N_p k^p v_1 \sum_1^r e^{\eta_j(u_1)} + \cdots + N_p k^p v_n \sum_1^r e^{\eta_j(u_n)} = 0$$

so

$$\left| N_p k^p v_0 + pk^p v_1 \sum_1^r g_p(\eta_j(u_1)) + \cdots + pk^p v_n \sum_1^r g_p(\eta_j(u_n)) \right|$$

$$= \left| (N_p k^p v_0 - N_p k^p v_0) + k^p v_1 \left(\sum_1^r pg_p(\eta_j(u_1)) - \sum_1^r N_p e^{\eta_j(u_1)} \right) \right.$$

$$\left. + \cdots + k^p v_n \left(\sum_1^r pg_p(\eta_j(u_n)) - \sum_1^r N_p e^{\eta_j(u_n)} \right) \right|$$

$$\le k^p \sum_{j=1, i=1}^{r,n} |v_i| \left| pg_p(\eta_j(u_i)) - N_p e^{\eta_j(u_i)} \right| \le \frac{2k^p rnM^p L}{(p-1)!}$$

where M is as before and L is a positive upper bound for the $|v_i|$, $1 \le i \le n$. The numbers $pk^p v_i \sum_{j=1}^r g_p(\eta_j(u_i))$ are integers divisible by p whereas N_p is not. Moreover, if p is sufficiently large then $p \nmid k$ and $p \nmid v_0$ so $p \nmid k^p v_0$. Hence the left-hand side of the inequality

$$\left| N_p k^p v_0 + \sum_{i,j} pk^p v_i g_p(\eta_j(u_i)) \right| \le 2k^p rnM^p L/(p-1)!$$

is a non-zero integer. On the other hand, the right-hand side is positive and < 1 for p sufficiently large. This contradiction shows that (63) is impossible and concludes the proof of Theorem 4.22 and hence of the Lindemann-Weierstrass theorem. □

EXERCISES

1. Show that $\sin u$ is transcendental for all algebraic $u \neq 0$. (*Hint*: Use $\sin u = (1/2i)(e^{iu} - e^{-iu})$ and the transcendence of e^{iu}.)

2. Show that $\csc u$, $\cos u$, $\sec u$, $\tan u$, $\cot u$ are transcendental for any algebraic $u \neq 0$.

3. Let m be an integer without square factors and let $F = \mathbb{Q}(\sqrt{m})$, the subfield of $\mathbb{C}$ generated by $\sqrt{m}$. Show that F is the set of complex numbers of the form $a + b\sqrt{m}$ where $a, b \in \mathbb{Q}$. Let I be the subset of F of integral algebraic numbers. Show that I is a subring of $\mathbb{C}$ and I is the set of elements $a + b\sqrt{m}$ where a and b are rational numbers such that

$$2a \in \mathbb{Z} \qquad \text{and} \qquad a^2 - mb^2 \in \mathbb{Z}.$$

4. Use the same notations as in exercise 3. Show that if $m \equiv 2$ or $3 \pmod 4$ then I is the set of numbers of the form $a + b\sqrt{m}$ where $a, b \in \mathbb{Z}$.

5. Use the same notations as in exercises 3 and 4. Show that if $m \equiv 1 \pmod 4$ then I is the set of numbers of the form $a + b\sqrt{m}$ where a and b are either both integers or both halves of odd integers. Equivalently, show that I is the set of numbers of the form $a + b(1 + \sqrt{m})/2$ where $a, b \in \mathbb{Z}$.

4.13   FINITE FIELDS

We shall now apply the results of Galois theory to derive the main facts about finite fields. We observe first that if F is a finite field then $|F| = p^n$ for some prime p. To begin with we know that the prime field of F can be identified with a field $\mathbb{Z}/(p)$ of residues modulo p for some prime p. We may now regard F as a vector space over $\mathbb{Z}/(p)$ in the usual way. Clearly $[F:\mathbb{Z}/(p)]$ is finite and if $[F:\mathbb{Z}/(p)] = n$, then we have a base $(u_1, u_2, \ldots, u_n)$ for $F/(\mathbb{Z}/(p))$, and every element of F can be written in one and only one way as a linear combination $a_1 u_1 + a_2 u_2 + \cdots + a_n u_n$, $a_i \in \mathbb{Z}/(p)$. Evidently, this implies that $|F| = p^n$. The same method shows that if $E \supset F$, $[E:F] = n$, and $|F| = q < \infty$ then $|E| = q^n$.

The basic facts on finite fields can now be derived very quickly. We have first

THEOREM 4.25.   *The number of elements of a finite field is a power of a prime. Moreover, for any prime power $q = p^m$ there exists one and, in the sense of isomorphism, only one field F with $|F| = q$.*

Proof.   We have already proved the first statement. To prove the second we take the field $P = \mathbb{Z}/(p)$ with p elements and we let F be a splitting field over P of $x^q - x$. We claim that $|F| = q$ and F coincides with the set R of roots of

$x^q - x$ in F. We observe first that since $(x^q - x)' = -1$, $x^q - x$ has q distinct roots in F. Next we shall show that $R = \{u_1, u_2, \ldots, u_q\}$ is a subfield of F. For, using the nice binomial theorem $(a + b)^p = a^p + b^p$ for characteristic p, we see that for any i and j, $(u_i \pm u_j)^q = u_i^q \pm u_j^q = u_i \pm u_j$. Hence $u_i \pm u_j \in R$. Also $1 \in R$ and $(u_iu_j)^q = u_i^qu_j^q = u_iu_j \in R$, and if $u_i \neq 0$, then $(u_i^{-1})^q = (u_i^q)^{-1} = u_i^{-1}$. These results show that R is a subfield of F. Then R contains the prime field P and $R = P(R) = F$.

Next let F and F' be two fields such that $|F| = q = |F'|$. Clearly this implies that both F and F' are extensions of $P = \mathbb{Z}/(p)$. Let F^* be the set of non-zero elements of F, so that F^* is a group under multiplication and $|F^*| = q - 1$. Hence if $u \neq 0$ in F then $u^{q-1} = 1$ and $u^q = u$. Since the last relation holds also for $u = 0$, we see that every element of F is a root of $x^q - x$. Since this equation has no more than q distinct roots in any field it is clear that F is a splitting field over P of $x^q - x$. The same is true of F'. Hence the isomorphism theorem for splitting fields (Theorem 4.4, p. 227) implies that F and F' are isomorphic. $\square$

We shall now consider the relative theory of finite fields, that is, we want to study a finite field relative to a subfield. Let $|F| = q(=p^m)$ and let E be an extension field of F with $[E:F] = n$. Then, as we saw before, $E = q^n$. We have seen also that $a \to a^p$ is an automorphism of E (section 4.4). Hence $\eta : a \to a^q$ is an automorphism of E. Moreover, since $|F| = q$, $b^q = b$ for $b \in F$. Hence $\eta \in$ Gal E/F. We now have

THEOREM 4.26. *Let F be a finite field with $q = p^m$ elements, E an extension field of F such that $[E:F] = n$. Then E is cyclic over F with Galois group $\langle \eta \rangle$ where $\eta : a \to a^q$.*

Proof. We show first that the order $o(\eta) = n$. For, $|E| = q^n$ so $a^{q^n} = a$ for all $a \in E$. Thus $\eta^n = 1$ and if $\eta^{n'} = 1$ for $0 < n' < n$ then $a^{q^{n'}} = a$. This would contradict the fact that the polynomial $x^{q^{n'}} - x$ has no more than $q^{n'}$ roots in E. Hence $o(\eta) = n$ and $|\langle \eta \rangle| = n$. Let $F' = $ Inv $\langle \eta \rangle$. By the Fundamental Theorem of Galois Theory, we know that $[E:F'] = n$ and Gal $E/F' = \langle \eta \rangle$. On the other hand, since $\eta \in$ Gal E/F, $F \subset F' = $ Inv $\langle \eta \rangle$. Since $n = [E:F] = [E:F'][F':F] = n[F':F]$ we have $F' = F$ and so E is Galois over F with Gal $E/F = \langle \eta \rangle$. $\square$

Suppose K is a subfield of E/F. Then $m = [K:F] | n = [E:F]$. On the other hand, let m be any divisor of n. Then the cyclic group Gal E/F has one and only one subgroup of order n/m. Hence, by the Fundamental Theorem of

Galois Theory, we have one and only one subfield K of E/F such that $[K:F] = m$. Also we have $|K| = q^m$. It is now clear that we have the following

COROLLARY 1.   *Let F and E be as in Theorem 4.26. Then if K is a subfield of E/F, $|K| = q^m$ where $m|n$. Conversely, if $m|n$ then E/F has one and only one subfield K with $|K| = q^m$.*

We can apply this result to obtain a formula due to Gauss for the number of monic irreducible polynomials of degree n with coefficients in F. This is given in

COROLLARY 2.   *Let F be a finite field with $|F| = q$ and let $N(n, q)$ denote the number of monic irreducible polynomials in $F[x]$ of degree n. Then*

$$(69) \qquad N(n, q) = \frac{1}{n} \sum_{d|n} \mu\left(\frac{n}{d}\right) q^d$$

where μ is the Möbius function (defined in exercise 17, p. 151).

Proof.   The proof will follow by showing that

$$(70) \qquad x^{q^n} - x = \prod g(x)$$

where the product is taken over all monic irreducible polynomials of degrees dividing n. Since $x^{q^n} - x$ has no multiple roots and hence no multiple factors in $F[x]$, it suffices to show that a monic irreducible polynomial $g(x)$ is a factor of $x^{q^n} - x$ if and only if its degree m is a divisor of n. Let $g(x)$ be a monic irreducible factor of $x^{q^n} - x$ in $F[x]$, deg $g(x) = m$, and let E be an extension field of F with $[E:F] = n$. Then E is a splitting field over F of $x^{q^n} - x$ and hence E contains a root r of $g(x)$. Then $g(x)$ is the minimum polynomial of r over F. Hence $F(r)$ is a subfield of E/F such that $[F(r):F] = m$. Then $m|n$. Conversely, let $g(x)$ be a monic irreducible polynomial in $F[x]$ of degree $m|n$. Then $K' = F[x]/(g(x))$ is an extension field of F with $|K'| = q^m$. Since $m|n$, K' is isomorphic to a subfield K of E/F. Then E contains an element r whose minimum polynomial over F is $g(x)$. Since $r^{q^m} = r$, $g(x)|(x^{q^m} - x)$. This establishes the factorization (70) where $g(x)$ runs through the set of monic irreducible polynomials in $F[x]$ of degree $m|n$. Comparing the degrees of the two sides of (70) we obtain

$$(71) \qquad q^n = \sum_{d|n} N(d, q)d.$$

Applying the Möbius inversion formula (p. 151) we obtain Gauss' formula (69).   □

We have shown in Theorem 2.18, p. 132, that any finite subgroup of the multiplicative group of a field is cyclic. In particular, for the finite field E, the multiplicative group E^* is cyclic. If F is a subfield and z is a generator of E^* then evidently $E = F(z)$. Hence we have

THEOREM 4.27. *If E and F are as in Theorem 4.26, then E has a primitive element over F.*

4.14 SPECIAL BASES FOR FINITE DIMENSIONAL EXTENSION FIELDS

Let E be a finite dimensional extension field of the field F. We consider first the question of the existence of a primitive element for E. There is a very pretty characterization of extensions which have primitive elements, which is due to Steinitz, namely,

THEOREM 4.28. *Let E be a finite dimensional extension field over F. Then E/F has a primitive element if and only if there are only a finite number of intermediate fields between F and E.*

Proof. Suppose first that $E = F(u)$ and let K be a subfield containing F. Let $f(x)$ be the minimum polynomial of u over F, $g(x)$, its minimum polynomial over K. Then $g(x)|f(x)$. Let K' be the subfield of E/F generated by the coefficients of $g(x)$. Then $K' \subset K$, and clearly the minimum polynomial of u over K' is also $g(x)$. Since $E = K(u) = K'(u)$, $[E:K] = \deg g(x) = [E:K']$. This implies $K = K'$. We have therefore shown that the intermediate subfields between F and E are just the subfields over F generated by the coefficients of monic factors of $f(x)$ in $E[x]$. Since there are only a finite number of these we see that there are only a finite number of intermediate fields. Conversely, assume that E/F has only a finite number of subfields. If F is finite we saw in the last section that E has a primitive element over F. Hence we may assume F infinite. The existence of a primitive element will follow by induction on the number of generators in a finite generating set if we can prove that, if u and v are any two elements of E, then $F(u, v)$ has a primitive element. We now consider the subfields $F(u + av)$ where $a \in F$. There are only a finite number of these whereas there are an infinite number of a. Hence there exist $a \neq b$ such that $F(u + av) = F(u + bv)$. Then $v = (a - b)^{-1}(u + av - u - bv) \in F(u + av)$ and $u = u + av - av \in F(u + av)$. Hence $z = u + av$ is a primitive element of $F(u, v)$. $\square$

If E is finite dimensional separable over F, then the normal closure K of E/F is Galois over F (see p. 254). Moreover, there are only a finite number of subfields of K/F since these correspond to the subgroups of Gal K/F. À fortiori, E/F has only a finite number of subfields. Steinitz' criterion therefore implies the

COROLLARY.   *Any finite dimensional separable extension field contains a primitive element.*

We now suppose that K is any Galois extension field of F with Galois group $G = \{\eta_i \,|\, 1 \le i \le n\}$, $n = [K:F]$. If $z \in K$ and $\{z_1, z_2, \ldots, z_m\}$ is the orbit Gz of z under the action of G, then the minimum polynomial for z over F is $\prod_1^m (x - z_i)$. Hence z is a primitive element if and only if the orbit of z contains n elements or, equivalently, the elements $\eta_1(z), \eta_2(z), \ldots, \eta_n(z)$ are distinct. A stronger condition than this is that the *conjugates* $\eta_i(z)$ are linearly independent over F. If this is the case, then $(\eta_1(z), \eta_2(z), \ldots, \eta_n(z))$ is a base for K/F. Such a base is called a *normal base* for K/F. In the remainder of this section we shall prove the existence of such a base for any Galois extension field. The proof will be based on some important independence properties for automorphisms of fields.

We begin with a classical result on linear independence of characters. Let H be a monoid, F a field. Then we define a *character* χ of H in F (or *F-character of H*) to be a homomorphism of H into the multiplicative group F^* of non-zero elements of F. Thus χ is a map of H into F^* such that $\chi(1) = 1$, $\chi(h_1 h_2) = \chi(h_1)\chi(h_2)$ for all $h_i \in H$. An important property of characters which we shall now prove is the

DEDEKIND INDEPENDENCE THEOREM.   *Distinct characters of a monoid into a field are linearly independent, that is, if $\chi_1, \ldots, \chi_n$ are distinct characters of H into a field F, then the only elements $a_1, \ldots, a_n$ in F such that*

$$(72) \qquad\qquad a_1\chi_1(h) + a_2\chi_2(h) + \cdots + a_n\chi_n(h) = 0$$

for all $h \in H$ are $a_1 = a_2 = \cdots = a_n = 0$.

Proof.   We shall prove the result by induction on n. If $n = 1$ the result is clear since $a\chi(h) = 0$ for $a \ne 0$ gives $\chi(h) = 0$, which is impossible since $\chi(1) = 1$. Now let $n > 1$ and assume the theorem for $n - 1$ characters. Suppose we have (72) for all $h \in H$ where the $a_i \in F$. Since we are assuming the result for $n - 1$ characters, we may assume that every $a_i \ne 0$. Since $\chi_1 \ne \chi_2$ there exists an $a \in H$ such that $\chi_1(a) \ne \chi_2(a)$. Now replace h by ah in (72). This gives the relation

$$a_1\chi_1(a)\chi_1(h) + a_2\chi_2(a)\chi_2(h) + \cdots + a_n\chi_n(a)\chi_n(h) = 0$$

since the χ_i are characters. On the other hand, if we multiply (72) by $\chi_1(a)$ we obtain

$$a_1\chi_1(a)\chi_1(h) + a_2\chi_1(a)\chi_2(h) + \cdots + a_n\chi_1(a)\chi_n(h) = 0.$$

Subtracting these two relations we obtain $a_2'\chi_2(h) + \ldots + a_n'\chi_n(h) = 0$ where $a_i' = a_i(\chi_i(a) - \chi_1(a))$, $2 \leq i \leq n$. Since $a_2' = a_2(\chi_1(a) - \chi_2(a)) \neq 0$ this contradicts the validity of the theorem for $n - 1$, and completes the proof. $\square$

One of the main applications of this theorem will be to monomorphisms of one field into another one. For these we have the important

COROLLARY. *Let F_1 and F_2 be fields and let $\eta_1, \eta_2, \ldots, \eta_n$ be distinct monomorphisms of F_1 into F_2. Then these are linearly independent over F_2.*

Proof. Clearly the restrictions of the η_i to the non-zero elements of F_1 are characters of the multiplicative group $H = F_1^*$ into F_2. Hence the result follows from Dedekind's theorem. $\square$

We suppose next that E/F is finite dimensional separable with $[E:F] = n$, and let K/F be the normal closure of E/F. Then we have the following determinant test for a base for E/F.

THEOREM 4.29. *Let E/F be finite dimensional separable, K/F its normal closure. Then the number of monomorphisms of E/F into K/F is $n = [E:F]$, and if these are $\eta_1 = 1, \eta_2, \ldots, \eta_n$, then a sequence of n elements $(u_1, u_2, \ldots, u_n)$, $u_i \in E$ is a base for E/F if and only if*

(73)
$$\begin{vmatrix} u_1 & u_2 & \cdots & u_n \\ \eta_2(u_1) & \eta_2(u_2) & \cdots & \eta_2(u_n) \\ \cdots\cdots\cdots\cdots\cdots\cdots\cdots\cdots\cdots \\ \eta_n(u_1) & \eta_n(u_2) & \cdots & \eta_n(u_n) \end{vmatrix} \neq 0.$$

Proof. Let $G = \text{Gal } K/F$ and let H be the subgroup of G fixing E. Then $n = [E:F] = [G:H]$ and we can write $G = \zeta_1 H \cup \zeta_2 H \cup \cdots \cup \zeta_n H$ where the $\zeta_i H$ are distinct cosets and $\zeta_1 = 1$. Let $\eta_i = \zeta_i | E$. Then η_i is a monomorphism of E/F into K/F, and $\eta_i \neq \eta_j$ if $i \neq j$. For, if $\eta_i = \eta_j$, then $\zeta_i^{-1}(\zeta_j(u)) = u$ for all $u \in E$. Hence $\zeta_i^{-1}\zeta_j \in H$ and $\zeta_i H = \zeta_j H$ contrary to hypothesis. Now let η be any monomorphism of E/F into K/F. Since K is a splitting field over F of a polynomial $f(x) \in F[x]$ it is a splitting field over E and over $\eta(E)$ of $f(x)$. Hence, by

Theorem 4.4, the isomorphism η of E onto $\eta(E)$ can be extended to an automorphism ζ of K/F. Then $\zeta \in \text{Gal } K/F$ and so $\zeta = \zeta_i \lambda$ for some $\lambda \in H$. But then $\eta = \zeta|E = \zeta_i|E = \eta_i$. Hence $\eta_1 = 1, \eta_2, \ldots, \eta_n$ is the list of monomorphisms of E/F into K/F.

Now suppose the elements $u_1, u_2, \ldots, u_n$ of E are linearly dependent over F so we have $a_i \in F$ not all 0 such that $\sum a_i u_i = 0$. Applying η_j we get

$$\sum_{i=1}^{n} a_i \eta_j(u_i) = 0, \qquad 1 \le j \le n.$$

This means that the system of homogeneous linear equations

$$\eta_1(u_1)x_1 + \eta_1(u_2)x_2 + \cdots + \eta_1(u_n)x_n = 0$$
$$\eta_2(u_1)x_1 + \eta_2(u_2)x_2 + \cdots + \eta_2(u_n)x_n = 0$$
$$\cdots \cdots \cdots \cdots \cdots \cdots \cdots \cdots \cdots \cdots \cdots \cdots$$
$$\eta_n(u_1)x_1 + \eta_n(u_2)x_2 + \cdots + \eta_n(u_n)x_n = 0$$

has the non-zero solution $(x_1, \ldots, x_n) = (a_1, \ldots, a_n)$. Then $\det(\eta_j(u_i)) = 0$. Conversely, suppose $\det(\eta_j(u_i)) = 0$. Then, turning things around we see that we have a non-zero solution $(a_1, \ldots, a_n)$, $a_j \in K$, of the system of equations $\sum_{j=1}^{n} \eta_j(u_i)x_j = 0$, $1 \le i \le n$. Thus $\sum_j a_j \eta_j(u_i) = 0$. Then $(u_1, \ldots, u_n)$ is not a base. Otherwise, any u can be written as $\sum c_i u_i$ with the $c_i \in F$ and then $\sum_j a_j \eta_j(u) = \sum_{i,j} a_j c_i \eta_j(u_i) = \sum_i c_i (\sum_j a_j \eta_j(u_i)) = 0$. This contradicts the linear independence of the monomorphisms $\eta_1, \ldots, \eta_n$ of E into K (Corollary to the Dedekind theorem). $\square$

We shall now prove the existence of a normal base for any Galois extension K/F. We distinguish two cases in the proof: (I) K/F is cyclic. (II) F is infinite. If F is finite, K is finite and, as we saw in section 4.13, K is cyclic over F. Hence our two cases cover all possibilities.

We assume first that K/F is cyclic, that is, $G = \text{Gal } K/F = \langle \eta \rangle$ where the order of $\eta = n = [K:F]$. Then $G = \{1, \eta, \ldots, \eta^{n-1}\}$. We observe that η is a linear transformation in K over F, since $\eta(u + v) = \eta(u) + \eta(v)$, and if $a \in F$, then $\eta(au) = a\eta(u)$. We can therefore apply to η the theory of a single linear transformation in a finite dimensional vector space (see section 3.10, pp. 195–197). We recall that the $F[x]$-module $(F[\lambda]$ in the notation of Chapter III) determined by a linear transformation is a direct sum of cyclic ones whose annihilators are $(d_i(x))$ where $d_1(x), \ldots, d_s(x)$ are the invariant factors. The last one is the minimum polynomial, and the product of all the invariant factors is the characteristic polynomial of the transformation. Hence the $F[x]$-module determined by a linear transformation is cyclic if and only if its characteristic polynomial and minimum polynomial coincide or, equivalently, the degree of the minimum polynomial is the dimension of the underlying space. We claim that this is the case for the linear transformation η of K over F. First, we have $\eta^n = 1$, and so η satisfies $x^n - 1 = 0$. On the other hand, if $f(x) = x^m + a_1 x^{m-1} + \cdots + a_m$ with the $a_i \in F$ and $m < n$, then $f(\eta) \neq 0$, since the automorphisms $1, \eta, \ldots, \eta^m$ are distinct and hence are linearly independent over K and so also over F. Now the fact that K is cyclic as $F[x]$-module (via the linear transformation η) means that we have a base for K/F of the form $(u, \eta(u), \eta^2(u), \ldots, \eta^{n-1}(u))$. This is a normal base for K/F.

We now assume F is infinite. Following Artin, we shall prove the existence of a normal base in this case by using a notion of algebraic independence of monomorphisms of fields. If E and K are fields and $\eta_1, \eta_2, \ldots, \eta_n$ are monomorphisms of E into K, then these maps are called *algebraically independent* over K if the only polynomial $f(x_1, \ldots, x_n) \in K[x_1, \ldots, x_n]$, x_i indeterminates, such that $f(\eta_1(u), \eta_2(u), \ldots, \eta_n(u)) = 0$ for all $u \in E$ is $f = 0$. We have the following

THEOREM 4.30. *Let F be an infinite field, E a finite dimensional separable extension of F, K a normal closure of E/F. Let $\eta_1, \ldots, \eta_n$ be the $n = [E:F]$ distinct monomorphisms of E/F into K/F (cf. Theorem 4.29). Then the η_i are algebraically independent over K.*

Proof. Let $f(x_1, \ldots, x_n) \in K[x_1, \ldots, x_n]$ satisfy $f(\eta_1(u), \eta_2(u), \ldots, \eta_n(u)) = 0$ for all $u \in E$. Let $(u_1, u_2, \ldots, u_n)$ be a base for E/F. Then for any $a_i \in F$ we have $0 = f(\eta_1(\sum a_i u_i), \ldots, \eta_n(\sum a_i u_i)) = f(\sum a_i \eta_1(u_i), \ldots, \sum a_i \eta_n(u_i))$. Hence if we put $g(x_1, x_2, \ldots, x_n) = f(\sum \eta_1(u_i)x_i, \ldots, \sum \eta_n(u_i)x_i)$, then $g(a_1, a_2, \ldots, a_n) = 0$ for all choices of $a_i \in F$. Let $(v_1, v_2, \ldots, v_m)$ be a base for K/F. Then we can write $g(x_1, \ldots, x_n) = \sum g_j(x_1, \ldots, x_n)v_j$ where $g_j(x_1, \ldots, x_n) \in F[x_1, \ldots, x_n]$. The condition $g(a_1, \ldots, a_n) = 0$ gives $g_j(a_1, \ldots, a_n) = 0$ for all j. Since this holds for all $a_i \in F$ we conclude from Theorem 2.19 (p. 136) that every $g_j(x_1, \ldots, x_n) = 0$ and $g(x_1, \ldots, x_n) = 0$. By Theorem 4.29, $\det(\eta_i(u_j)) \neq 0$ so the matrix $(\eta_i(u_j))$ has an inverse $(v_{ij}) \in M_n(K)$. Since $g(x_1, \ldots, x_n) = f(\sum \eta_1(u_i)x_i, \ldots, \sum \eta_n(u_i)x_i)$, $g(\sum_{j,k} v_{1j}\eta_j(u_k)x_k, \ldots, \sum_{j,k} v_{nj}\eta_j(u_k)x_k) = f(x_1, \ldots, x_n)$. Since $g(x_1, \ldots, x_n) = 0$ this implies that $f(x_1, \ldots, x_n) = 0$, which proves the algebraic independence of the η_i over K. $\square$

We can now complete the proof of

THE NORMAL BASE THEOREM. *Any (finite dimensional) Galois extension field K/F has a normal base.*

Proof. Since we have proved the result for $G = \text{Gal } K/F$ cyclic, hence for F finite, we may assume F infinite. Then, by the result just proved, the automorphisms $\eta_1, \eta_2, \ldots, \eta_n$ of K/F are algebraically independent over K. We have also seen that if $u \in K$, then $(\eta_1(u), \ldots, \eta_n(u))$ is a base if and only if

$$\det\left((\eta_i\eta_j)(u)\right) \neq 0.$$

Write $\eta_i\eta_j = \eta_{i(j)}$. Then $j \to i(j)$ is a permutation of $1, 2, \ldots, n$. Now consider the polynomial ring $K[x_1, x_2, \ldots, x_n]$ and the matrix X whose (i, j)-entry is $x_{i(j)}$. We claim that $\det X \neq 0$. To see this we specialize $x_1 = 1$, $x_i = 0$ if $i > 1$. Since $j \to i(j)$ is a permutation of $1, 2, \ldots, n$, and for different i these permutations are different, x_1 appears once and only once in each row and column of X. Hence the value of the determinant after putting $x_1 = 1$, $x_i = 0$ for $i > 1$ is ± 1. Thus $d(x_1, \ldots, x_n) \equiv \det X \neq 0$. Hence by the algebraic independence of the η's over K, there exists a $u \in K$ such that $\det\left((\eta_i\eta_j)(u)\right) \neq 0$. Then $(\eta_1(u), \ldots, \eta_n(u))$ is a normal base. □

EXERCISES

1. Find a primitive element of $\mathbb{Q}(\sqrt{2}, \sqrt{3})$ over $\mathbb{Q}$.

2. Find a primitive element for a splitting field over $\mathbb{Q}$ of $x^5 - 2$.

3. Let x and y be indeterminates and let $E = (\mathbb{Z}/(p))(x, y)$, $F = (\mathbb{Z}/(p))(x^p, y^p)$. Show that $[E:F] = p^2$ and E does not have a primitive element over F. Display an infinite number of distinct subfields of E/F.

4. Determine a normal base for the field in exercise 1.

5. Prove that if $E = F(u, v)$ where u and v are algebraic over F and u is separable over F, then E has a primitive element over F.

6. Let E and K be extension fields of the same field F and let $\text{Hom}_F(E, K)$ denote the set of linear maps of the vector space E/F into K/F. Note that $\text{Hom}_F(E, K)$ becomes a vector space over K if we define addition of linear maps as usual and define kl for $k \in K$, $l \in \text{Hom}_F(E, K)$, by $(kl)(x) = k(l(x))$. $x \in E$. Suppose $[E:F] < \infty$ and let $(\xi_1, \ldots, \xi_n)$ be a base for E/F. Let l_i be the element of $\text{Hom}_F(E, K)$ such that $l_i(\xi_j) = 0$ if $j \neq i$ and $l_i(\xi_i) = 1$. Show that $(l_1, \ldots, l_n)$ is a base for $\text{Hom}_F(E, K)$ regarded as a vector space over K. This shows that the dimensionality of $\text{Hom}_F(E, K)$ over K is $[E:F]$.

7. With notations and hypotheses as in exercises 6, use exercise 6 and the Dedekind Independence Theorem to show that there are at most $[E:F]$ monomorphisms of E/F into K/F.

4.15   TRACES AND NORMS

Let E/F be Galois, $G = \text{Gal } E/F = \{\eta_1 = 1, \eta_2, \ldots, \eta_n\}$. If $u \in E$ we define

$$(74) \qquad T_{E/F}(u) = \sum_1^n \eta_i(u), \qquad N_{E/F}(u) = \prod_1^n \eta_i(u)$$

and call these respectively the *trace* and *norm of u in E/F*. Evidently, these are fixed under the Galois group; hence they are contained in F. Thus we have the trace and norm maps

$$T_{E/F} : u \to T_{E/F}(u)$$

$$N_{E/F} : u \to N_{E/F}(u)$$

of E into F. If $u, v \in E$ and $a \in F$, then

$$T_{E/F}(u + v) = \sum \eta_i(u + v) = \sum \eta_i(u) + \sum \eta_i(v) = T_{E/F}(u) + T_{E/F}(v)$$

$$T_{E/F}(au) = \sum \eta_i(au) = a \sum \eta_i(u) = aT_{E/F}(u)$$

$$N_{E/F}(uv) = \prod \eta_i(uv) = \prod \eta_i(u) \prod \eta_i(v) = N_{E/F}(u)N_{E/F}(v)$$

$$N_{E/F}(au) = \prod \eta_i(au) = a^n \prod \eta_i(u) = a^n N_{E/F}(u).$$

The first two of these show that $T = T_{E/F}$ is F-linear, that is, T is a linear function on the vector space E over F. The properties we noted for $N = N_{E/F}$ are that N is a multiplicative map and that it is homogeneous of degree n. Evidently we have $T(0) = 0$, $T(-u) = -T(u)$, $N(0) = 0$. $N(1) = 1$, $N(u) \neq 0$ if $u \neq 0$ and $N(u^{-1}) = N(u)^{-1}$. It is clear that the restriction of N to the multiplicative group E^* of non-zero elements of E defines a homomorphism of E^* into F^*.

As an example let us consider a quadratic extension field $E = \mathbb{Q}(\sqrt{m})$ where m is an integer without square factors. Then any $u \in E$ has the form $a + b\sqrt{m}$, $a, b \in \mathbb{Q}$, and the Galois group consists of the identity map and $a + b\sqrt{m} \to a - b\sqrt{m}$. Hence

$$T(a + b\sqrt{m}) = 2a$$

$$N(a + b\sqrt{m}) = a^2 - b^2 m.$$

Perhaps the most familiar example of traces and norms is obtained by taking $\mathbb{C}$ as quadratic extension of $\mathbb{R}$ by $\sqrt{-1}$. Here, if $u = a + b\sqrt{-1}$, $a, b \in \mathbb{R}$, then $T(u) = 2a$, which is 2 times the real part of u, and $N(u) = a^2 + b^2 = |u|^2$.

Since T and N are homomorphisms, it is natural to seek information on the images $T(E)$ and $N(E^*)$ and the kernels of the maps T and N (as homomorphism of E^* into F^*). The first of these is easy to determine, namely, we have $T(E) \subset F$, and since $T(E)$ is a subspace of the one dimensional space F/F either $T(E) = 0$ or $T(E) = F$. Moreover, $T(E) = 0$ can be ruled out since this amounts to saying that we have $\sum_i \eta_i(u) = 0$ for every $u \in E$, which is contrary to the linear independence of the automorphisms η_i. Information on $N(E^*)$ is usually not easy to obtain. For instance, if $E = \mathbb{Q}(\sqrt{m})$, then the general problem is: for what rational numbers c does the equation $x^2 - my^2 = c$ have a rational solution $(x, y) = (a, b)$? This is a non-trivial arithmetic problem. In the case of $m = -1$, the arithmetic of the ring of Gaussian integers provides an answer which will be indicated in exercise 4 at the end of this section.

There are two general theorems on the kernels of the trace and norm maps which we shall now derive. The first one is universally known as "Hilbert's Satz 90," since it was published by Hilbert in 1897 in his classical report on algebraic number theory, in which it appeared as the ninetieth theorem.[15] The result is our

THEOREM 4.31.   *Let E be a cyclic extension field of the field F, η a generator of the (cyclic) Galois group of E/F. Then $N_{E/F}(u) = 1$ for $u \in E$ if and only if there exists a $v \in E$ such that $u = v(\eta(v))^{-1}$ (sometimes written as $v^{1-\eta}$).*

In one direction the result is trivial: if $u = v(\eta(v))^{-1}$ then

$$N(u) = N(v)N(\eta(v)^{-1}) = N(v)N(v)^{-1} = 1.$$

To prove the converse we shall prove a more general result on Galois extensions which is due to A. Speiser (for matrices), namely,

THEOREM 4.32.   *Let E be a finite dimensional Galois extension field of F, G the Galois group. Let $\eta \to u_\eta$ be a map of G into the multiplicative group E^* satisfying the equations*

(75) $$u_{\zeta\eta} = \zeta(u_\eta)u_\zeta$$

for every $\eta, \zeta \in G$. Then there exists a non-zero $v \in E$ such that

(76) $$u_\eta = v(\eta(v))^{-1}.$$

Proof.   Since the $u_\eta \neq 0$ and the automorphisms $\eta \in G$ are linearly independent

---

[15] D. Hilbert, *Theorie der algebraischen Zahlkörper*, 1897, p. 149.

over E, there exists an element $w \in E$ such that

$$v = \sum_{\eta \in G} u_\eta \eta(w) \neq 0.$$

Then for $\zeta \in G$ we have

$$\zeta(v) = \sum_\eta \zeta(u_\eta)(\zeta\eta)(w) = \sum_\eta u_{\zeta\eta} u_\zeta^{-1}(\zeta\eta)(w)$$

$$= \left(\sum_\eta u_{\zeta\eta}(\zeta\eta)(w) \right) u_\zeta^{-1}$$

$$= \left(\sum_\eta u_\eta \eta(w) \right) u_\zeta^{-1}$$

$$= v u_\zeta^{-1}.$$

Hence $u_\zeta = v\zeta(v)^{-1}$ as required. $\quad\square$

To complete the proof of Hilbert's Satz 90 we now assume G cyclic with generator η, and we suppose $u \in E$ satisfies $N(u) = 1$. Define

(77) $u_{\eta^i} = u\eta(u)\eta^2(u) \cdots \eta^{i-1}(u), \qquad 1 \le i \le n.$

Then for $i + j \le n$, $u_{\eta^i}\eta^i(u_{\eta^j}) = u\eta(u) \cdots \eta^{j-1}(u)\eta^j(u) \cdots \eta^{i+j-1}(u) = u_{\eta^{i+j}}$. The same relation holds for $i + j > n$ since $u_1 = u_{\eta^n} = N(u) = 1$. Thus the equations (75) are satisfied for $G = \langle \eta \rangle$. Hence there exists a v such that $u = u_\eta = v(\eta(v))^{-1}$.

The two results we have proved have additive analogues. The first is the additive form of Theorem 4.32.

THEOREM 4.33.   *Let E, F, G be as in Theorem 4.32 and let $\eta \to d_\eta$ be a map of G into E satisfying*

(78) $d_{\zeta\eta} = d_\zeta + \zeta(d_\eta)$

for every $\eta, \zeta \in G$. Then there exists a $c \in E$ such that

(79) $d_\eta = c - \eta(c), \eta \in G.$

Proof.   We have seen that there exists a $u \in E$ such that $T(u) \neq 0$. Put

$$c = T(u)^{-1} \sum_\eta d_\eta \eta(u).$$

Then

$$c - \zeta(c) = T(u)^{-1} \sum_{\eta} [d_{\eta}\eta(u) - \zeta(d_{\eta})\zeta\eta(u)]$$

$$= T(u)^{-1} \sum_{\eta} [d_{\eta}\eta(u) + d_{\zeta}\zeta\eta(u) - d_{\zeta\eta}\zeta\eta(u)]$$

$$= T(u)^{-1} d_{\zeta} \sum_{\eta} \zeta\eta(u)$$

$$= d_{\zeta} T(u)^{-1} \sum_{\eta} \eta(u)$$

$$= d_{\zeta} T(u)^{-1} T(u)$$

$$= d_{\zeta}$$

as wanted. $\square$

Now let $G = \langle \eta \rangle$ and assume d is an element of E such that $T(d) = 0$. Put

(80) $$d_{\eta^i} = d + \eta(d) + \cdots + \eta^{i-1}(d), \qquad 1 \le i \le n.$$

Then, as in the case of norms, one sees that (79) holds. Hence, as a consequence of Theorem 4.33 one has the following additive analogue of Hilbert's Satz 90.

THEOREM 4.34. *Let E/F be cyclic with Galois group $G = \langle \eta \rangle$. Let d be an element of E of trace 0. Then there exists a c in E such that $d = c - \eta(c)$.*

We shall now apply Hilbert's Satz 90 and its additive analogue to obtain results on the structure of cyclic extensions. First, we shall give an improved proof and extension of an earlier result (Lemma 3 of section 4.7, p. 253):

THEOREM 4.35. *Let F contain n distinct nth roots of 1 and let E/F be an n-dimensional cyclic extension of F. Then $E = F(u)$ where $u^n \in F$.*

Proof. Let z be a primitive nth root of 1. We have $N_{E/F}(z) = z^n = 1$. Hence there exists a $u \in E$ such that $z = u(\eta(u))^{-1}$ where η is a generator of the Galois group. Then we have $\eta(u) = z^{-1}u$ and $\eta(u^n) = \eta(u)^n = (z^{-1}u)^n = u^n$. Accordingly $u^n \in F$. Also $\eta(u) = z^{-1}u$ gives $\eta^i(u) = z^{-i}u$ and shows that there are n-distinct elements in the orbit of u under Gal E/F. Hence the minimum polynomial of u over F has degree n and $E = F(u)$. $\square$

We obtain next the structure of a p-dimensional cyclic extension of characteristic p.

THEOREM 4.36.   *Let F be a field of characteristic $p \neq 0$ and let E/F be a p-dimensional cyclic extension of F. Then $E = F(c)$ where $c^p - c \in F$.*

Proof.   We have $T_{E/F}(1) = 1 + 1 + \cdots + 1$ (p terms) $= 0$. Hence, by Theorem 4.34, we have an element $c \in E$ such that $\eta(c) = c + 1$. Then $\eta^i(c) = c + i$ and the orbit of c under Gal E/F contains p elements. Hence $E = F(c)$. Also $\eta(c^p - c) = (\eta(c))^p - \eta(c) = (c + 1)^p - (c + 1) = c^p - c$. Hence $c^p - c \in F$.   □

EXERCISES

1.   Show that if E is a finite field and F is a subfield, so that E is a cyclic extension of F, then the norm homomorphism $N_{E/F}$ of E^* is surjective on F^*.

2.   (Albert.) Let E be a cyclic extension of dimension n over F and let η be a generator of Gal E/F. Let $r \mid n$, $n = rm$ and suppose c is a non-zero element of F such that $c^r = N_{E/F}(u)$ for some $u \in E$. Show that there exists a v in the (unique) subfield K of E/F of dimensionality m such that $c = N_{K/F}(v)$.

3.   Let E, F, n, u be as in Theorem 4.35. Show that an element $v \in E$ satisfies an equation of the form $x^n = a$ if and only if v has the form bu^k where $b \in F$ and $1 \leq k \leq n$.

4.   Show that a rational number $a \neq 0$ is a norm of an element in $\mathbb{Q}(\sqrt{-1})$ if and only if the odd primes occurring with odd multiplicities in the numerator or denominator of a written in reduced form (b/c, $(b, c) = 1$) are of the form $4n + 1$ (cf. exercise 10, p. 150.)

5.   Assume F has p distinct pth roots of 1, p a prime, and E/F is cyclic of dimension p^f. Let z be a primitive pth root of 1. Show that if E/F can be imbedded in a cyclic field K/F of dimension p^{f+1}, then $z = N_{E/F}(u)$ for some $u \in E$.

6.   Show that if $E = \mathbb{Q}(\sqrt{m})$ where $m \in \mathbb{Z}$ and $m < 0$, then E cannot be imbedded in a cyclic quartic field over $\mathbb{Q}$.

7.   (Albert.) Let the notations be as in exercise 5 and let η be a generator of Gal E/F. Suppose E contains an element u such that $N_{E/F}(u) = z$. Show that there exists a $v \in E$ such that $\eta(v)v^{-1} = u^p$. Show that v is not a pth power in E and that if $K = E(w)$ where $w^p = v$, then K/F is cyclic of dimensionality p^{f+1}.

8.   Note that exercises 5 and 7 imply the following theorem: If F contains p distinct pth roots of 1 (p prime) and E/F is cyclic of dimension $p^f > 1$, then E can be

imbedded in a cyclic extension of dimension p^{f+1} over F if and only if a primitive pth root of 1, z, is a norm in E/F. Use this to prove that if the characteristic of F is $\neq 2$ and $E = F(\sqrt{c}) \neq F$, then E/F can be imbedded in a cyclic quartic extension of F if and only if c is a sum of two squares of elements of F.

9. (Uchida.) Let F_0 be a field of characteristic p, $F = F_0(s, t)$ where s and t are indeterminates. Show that the Galois group of $x^p - sx - t$ over F is isomorphic to the group of maps $x \to ax + b$, $a, b \in \mathbb{Z}/(p)$, $a \neq 0$.

10. (Uchida.) With notations as in exercise 9, show that the Galois group of $x^{p+1} - sx - t$ over F is isomorphic to the linear fractional group of maps $x \to \dfrac{ax + b}{cx + d}$, $a, b, c, d \in \mathbb{Z}/(p)$, $ad - bc \neq 0$.

11. Let $W = F(\zeta)$ where ζ is a primitive pth root of 1, p a prime. Show that W/F is cyclic and $[W:F] = s$ is a divisor of $p - 1$. Show that Gal $W/F = \langle \tau \rangle$ where $\tau(\zeta) = \zeta^t$ and s is the order of $t + (p)$ in the multiplicative group of $\mathbb{Z}/(p)$.

12. Let the notations be as in exercise 11. Let a be an element of W such that a is not a pth power in W but $(\tau a)a^{-t} = b^p$, $b \in W$. Let $K = W[x]/(x^p - a)$. Show that K/F is cyclic with $[K:F] = ps$. Show that K/F contains a unique cyclic subfield E/F with $[E:F] = p$.

13. Let s, t, p be as in exercise 11. Note that $(s, p) = 1 = (t, p)$ so there exist integers s', t' such that $ss' \equiv 1 \pmod{p}$ and $tt' \equiv 1 \pmod{p}$. Put $t_k = s't'^k = t_{k-1}t'$, $0 \leq k \leq s$. Show that

$$\sum_1^s t^k t_k \equiv 1 \pmod{p}$$

$$t_s \equiv s' \pmod{p}.$$

14. Let the notations be as in exercises 11 and 13. Let $a \in W$, $a \neq 0$, and put

$$M(a) = \prod (\tau^k a)^{t_k}.$$

Show that $\tau(M(a))M(a)^{-t}$ is a pth power in W and hence if $M(a)$ is not a pth power then this can be used as the element a of exercise 12 to construct a cyclic extension of dimension p over F.

Note: Exercises 12–14 give results due to Albert, who showed also that every cyclic extension E/F with $[E:F] = p$ can be obtained in this way.

4.16   MOD *p* REDUCTION

It is generally a difficult problem to determine the Galois group of an equation with rational coefficients. If the degree does not exceed 4, the results given on pp. 256–261 are effective. For any degree n we have obtained some information that is relevant for the problem. For example, we have seen that the Galois group G_f of f of degree n is a subgroup of A_n if and only if the discriminant $d(f)$ is a square (p. 257). We showed also that G_f is transitive (on the set of roots) if and only if f is irreducible. We shall now obtain an important result called *reduction mod p* which can be applied for various primes p to obtain information on G_f which together with the results we have indicated are often effective for determining G_f.

It is easily seen that nothing is lost in confining our attention to monic $f \in \mathbb{Z}[x]$ (exercise 1, below). Let p be a prime. Then we have the canonical homomorphism of $\mathbb{Z}[x]$ onto $(\mathbb{Z}/(p))[x]$ obtained by reducing the coefficients modulo p. If $f(x) \in \mathbb{Z}[x]$ we write $f_p(x)$ for the corresponding polynomial in $(\mathbb{Z}/(p))[x]$. We remark that if $f(x)$ is monic of degree n then $f_p(x)$ is monic of degree n. We have seen (pp. 258–259) that the discriminant $d(f)$ is a polynomial in the coefficients a_i with integer coefficients. Hence $d(f) \in \mathbb{Z}$ and $d(f_p) = d(f)_p$ is obtained by reducing $d(f)$ mod p. Thus if $d(f_p) \neq 0$ then $d(f) \neq 0$ and both f and f_p have distinct roots. In this case we shall prove

THEOREM 4.37. (Dedekind.) *Let $f(x) \in \mathbb{Z}[x]$ be monic of degree n, p a prime such that $f_p(x)$ has distinct roots (equivalently $d(f_p) \neq 0$) and let $f_p(x)$ factor in $(\mathbb{Z}/(p))[x]$ as a product of irreducible factors of degree $n_1, n_2, \ldots, n_r$ ($\sum n_i = n$). Then the Galois group G_f contains a permutation (of the roots of f) whose cycle decomposition relative to a suitable ordering of the roots is*

$$(12 \cdots n_1)(n_1 + 1 \cdots n_1 + n_2)(n_1 + n_2 + 1 \cdots n_1 + n_2 + n_3) \cdots.$$

Before giving the proof of this theorem we shall illustrate how it can be used—by applying it to determine G_f for $f(x) = x^6 + 22x^5 + 21x^4 + 12x^3 - 37x^2 - 29x - 15$. We first use reduction mod 2 to obtain $f_2(x) = x^6 + x^4 + x^2 + x + 1$. Checking divisibility by the irreducible polynomials mod 2 of degrees ≤ 3 we see that $f_2(x)$ is irreducible. Hence G_f contains a 6-cycle so G_f is transitive. Next we have $f_3(x) = x(x^5 + x^4 - x + 1)$ and we can show that $x^5 + x^4 - x + 1$ is irreducible mod 3. Hence G_f contains a 5-cycle. Mod 5 we have $f_5(x) = x(x - 1)(x + 1)(x + 2)(x^2 + 2)$ and $x^2 + 2$ is irreducible mod 5. Hence G_f contains a 2-cycle. Now it can be shown (exercise 3 below) that a transitive subgroup of S_n that contains an $(n - 1)$-cycle and a transposition coincides with S_n. Hence $G_f = S_6$.

The proof we shall give of Theorem 4.37 is due to John Tate and is based on the following

THEOREM 4.38. *Let $f(x) \in \mathbb{Z}[x]$ be monic of degree n, E a splitting field of $f(x)$ over $\mathbb{Q}$, p a prime such that $f_p(x)$ has distinct roots in its splitting field E_p*

over $\mathbb{Z}/(p)$. Let D be the subring of E generated by the roots of $f(x)$. Then

  (a) *There exist homomorphisms ψ of D into E_p.*

  (b) *Any such homomorphism gives a bijection of the set R of roots of $f(x)$ in E onto the set R_p of roots of $f_p(x)$ in E_p.*

  (c) *If ψ and ψ' are two such homomorphisms then $\psi' = \psi\sigma$ where $\sigma \in \mathrm{Gal}\ E/\mathbb{Q}$.*

Proof. We have $E = \mathbb{Q}(r_1, \ldots, r_n)$ and $f(x) = \prod_1^n (x - r_i)$ in $E[x]$. The r_i are distinct since $d(f_p) \neq 0$ and hence $d(f) \neq 0$. We have $D = \mathbb{Z}[r_1, \ldots, r_n]$. Put $D' = \sum_{0 \le e_i \le n-1} \mathbb{Z} r_1^{e_1} \cdots r_n^{e_n}$, the set of $\mathbb{Z}$-linear combinations of the elements $r_1^{e_1} \cdots r_n^{e_n}$, $0 \le e_i \le n - 1$. Since $f(r_i) = 0$, r_i^n is a $\mathbb{Z}$-linear combination of 1, $r_i, \ldots, r_i^{n-1}$. Hence $r_i D' \subset D'$ and by iteration, $r_1^{f_1} \cdots r_n^{f_n} D' \subset D'$ for any positive integral f_i. Then D' is a subring of D containing the r_i and hence $D' = D$. This shows that D is a finitely generated $\mathbb{Z}$-module. Since E has characteristic 0, the torsion submodule of D is 0. Hence D is a free $\mathbb{Z}$-module with base, say, $(u_1, \ldots, u_N)$: $D = \mathbb{Z} u_1, \oplus \cdots \oplus \mathbb{Z} u_N$ (p. 190). We claim that the u_i constitute a base for $E/\mathbb{Q}$ and hence $N = [E:\mathbb{Q}]$. The linear independence over $\mathbb{Q}$ of the u_i is clear since a non-trivial $\mathbb{Q}$-linear relation among the u_i gives rise, on multiplying by a non-zero integer, to a non-trivial $\mathbb{Z}$-linear relation. Now consider $\mathbb{Q}D = \sum \mathbb{Q} u_i$. This is a subring of E containing $\mathbb{Q}$. Hence, by exercise 7, p. 216, $\mathbb{Q}D$ is a subfield of E. Since it contains the r_i, $\mathbb{Q}D = E$. Hence the u_i span the vector space $E/\mathbb{Q}$ and the u_i form a base for $E/\mathbb{Q}$.

Consider $pD = \sum_1^N \mathbb{Z}(pu_i)$. Evidently pD is an ideal in D and $|D/pD| = p^N$. Since D/pD is finite, it contains a maximal (proper) ideal. This has the form M/pD where M is a maximal ideal of D containing pD. Then D/M is a field which is a homomorphic image of D/pD (since $(D/pD)/(M/pD) \cong D/M$). Since D/pD has characteristic p so has the field D/M, so its prime field is $\mathbb{Z}/(p)$ and $|D/M| = p^m$ where $m \le N$. The canonical homomorphism ν of D onto D/M maps $\mathbb{Z}$ onto the prime field $\mathbb{Z}/(p)$, and, since $D = \mathbb{Z}[r_1, \ldots, r_n]$ and $f(x) = \prod_1^n (x - r_i)$ in $D[x]$, we obtain $D/M = (\mathbb{Z}/(p))[\bar{r}_1, \ldots, \bar{r}_n]$ where $\bar{r}_i = \nu(r_i) = r_i + M$. Also $\bar{f}(x) = \prod_1^n (x - \bar{r}_i)$. Since $f(x) \in \mathbb{Z}[x]$, the coefficients of $\bar{f}(x)$ are in $\mathbb{Z}/(p)$ and $\bar{f}(x) = f_p(x)$. Thus D/M is a splitting field over $\mathbb{Z}/(p)$ of $f_p(x)$. Since E_p was chosen initially to be such a splitting field, we have an isomorphism of D/M onto E_p. If we take the composite of ν with this isomorphism we obtain a homomorphism ψ of D onto E_p. This proves (a).

(b). Let ψ be a homomorphism of D into E_p. Then $\psi|\mathbb{Z}$ is a homomorphism of $\mathbb{Z}$ onto the prime field of E_p and since it maps 1 into the unit of E_p it is the canonical homomorphism of $\mathbb{Z}$ onto $\mathbb{Z}/(p)$. Then $f_p(x) = \psi(f(x))$ (ψ applied to the coefficients) $= \prod (x - \psi(r_i))$. Thus the $\psi(r_i)$ are the roots of $f_p(x)$ in E_p and $\psi|R$ is a bijection of R onto R_p.

(c). We fix a homomorphism ψ of D into E_p. Let $\sigma \in G = \mathrm{Gal}\ E/\mathbb{Q}$. Then σ permutes the r_i and hence σ maps D into itself. Then $\sigma|D$ is a homomorphism of D into D (actually an automorphism) and $\psi\sigma$ (applied to D) is a homomorphism of D into E_p. Distinct σ, $\sigma' \in G$ give distinct permutations of the roots r_i and since $\psi|R$ is bijective onto R_p, $\psi\sigma$ and $\psi\sigma'$ are distinct. In this way we obtain $N = [E:\mathbb{Q}]$ distinct homomorphisms $\psi_j = \psi\sigma_j$ where $G = \{\sigma_1, \ldots, \sigma_N\}$. We claim that there are no more such homomorphisms. For, let ψ_{N+1} be one distinct from the ψ_j, $1 \leq j \leq N$. By the Dedekind independence theorem, applied to the multiplicative monoid H of the domain D and the field $F = E_p$, the ψ's, now including ψ_{N+1}, are linearly independent over E_p. On the other hand, consider the system of equations

$$\sum_{i=1}^{N+1} x_i\psi_i(u_j) = 0, \qquad 1 \leq j \leq N.$$

Since there are more x_i than equations, this system of linear homogeneous equations with coefficients $\psi_i(u_j) \in E_p$ has a non-trivial solution $(a_1, \ldots, a_{N+1})$, $a_i \in E_p$. Now let $y \in D$. Then $y = \sum_1^N n_j u_j$, $n_j \in \mathbb{Z}$. Then $\psi_i(y) = \sum \bar{n}_j\psi_i(u_j)$, $\bar{n}_j = n_j + (p)$ and $\sum a_i\psi_i(y) = \sum \bar{n}_j a_i\psi_i(u_j) = 0$. This contradicts the independence of the ψ_i and completes the proof of (c). $\square$

We can use this result to give the

Proof of Theorem 4.37.   Since E_p is a field with p^m elements, the map $\pi : a \to a^p$ is an automorphism of E_p. Hence if ψ is any homomorphism of D into E_p then so is $\pi\psi$. Accordingly, we have a unique $\sigma(\psi) \in G$ such that $\pi\psi = \psi\sigma(\psi)$. The automorphism $\sigma = \sigma(\psi)$ is called the *p-Frobenius automorphism* of $E/\mathbb{Q}$ corresponding to ψ. If we restrict ψ and σ to R and use the fact that ψ is bijective of R onto R_p we obtain the relation $\sigma = \psi^{-1}\pi\psi$. This implies that the orbits of R_p relative to $\langle\pi\rangle$ are mapped by ψ^{-1} into the orbits of R relative to $\langle\sigma\rangle$. Now the orbits of R_p relative to $\langle\pi\rangle$ are the sets of roots of the irreducible factors of $f_p(x)$ in $(\mathbb{Z}/(p))[x]$. If these have degrees $n_1, \ldots, n_r$, then the cardinality of the orbits of R relative to $\langle\sigma\rangle$ are $n_1, \ldots, n_r$ and hence σ, as a permutation of R, has the cycle decomposition $(12 \cdots n_1)(n_1 + 1 \cdots n_1 + n_2) \cdots$ for a suitable ordering of the roots. $\square$

EXERCISES

1.   Let $f(x) \in \mathbb{Q}[x]$ be monic and write $f(x) = x^n - a_1x^{n-1} + a_2x^{n-2} - \cdots + (-1)^n a_n$, $a_i = b_i d^{-1}$, $b_i, d \in \mathbb{Z}$. Show that $d^n f(d^{-1}x) \in \mathbb{Z}[x]$ is monic and has the same splitting field over $\mathbb{Q}$ as $f(x)$.

2. Let $f(x) \in \mathbb{Z}[x]$ be monic and assume $f(x)$ has distinct roots. Show that Theorem 4.37 is applicable to all but a finite number of primes.

3. Show that if G is a transitive subgroup of S_n containing an $(n-1)$-cycle and a transposition, then $G = S_n$.

4. Determine G_f for $f(x) = x^6 - 12x^4 + 15x^3 - 6x^2 + 15x + 12$ (over $\mathbb{Q}$).

5. (Tate.) Show that for any prime p and any positive integer n there exists an irreducible monic polynomial of degree n in $(\mathbb{Z}/(p))[x]$. (Use (69) or its proof.) For given n let $g(x)$ be irreducible monic in $(\mathbb{Z}/(2))[x]$ of degree n, $h(x)$ irreducible monic in $(\mathbb{Z}/(3))[x]$ of degree $n-1$, $k(x)$ irreducible monic quadratic in $\mathbb{Z}/(p)$ where p is a prime $> n - 2$. Use the Chinese remainder theorem (exercise 10, p. 110) to show that there exists a monic $f(x) \in \mathbb{Z}[x]$ such that $f_2(x) = g(x)$, $f_3(x) = xh(x)$, $f_p(x) = x(x+1) \cdots (x+n-3)k(x)$. Show that $G_f = S_n$.

6. Show that any transitive subgroup of A_5 is isomorphic to one of the following three groups: (a) the cyclic group Z_5, (b) the dihedral group D_5, (c) A_5.

7. (Jensen and Yui.) Let $f(x) = x^5 - 5x + 12$. Then $f(x)$ is irreducible in $\mathbb{Q}[x]$ and $d(f) = (2^6 5^3)^2$. If $r_1, \ldots, r_5$ are the roots of f, let $P(x) = \prod_{1 \le i < j \le 5} (x - (r_i + r_j))$. Then $P(x)$ is a product of two different monic irreducible polynomials in $\mathbb{Q}[x]$:

$$P(x) = (x^5 - 5x^3 - 10x^2 + 30x - 36)(x^5 + 5x^3 + 10x^2 + 10x + 4).$$

Use this information, exercise 6, and f_3 to show that $G_f \cong D_5$.

8. (Jensen and Yui.) Let $f(x) = x^5 + 20x + 16$. Then $d(f) = (2^8 5^3)^2$ and if $P(x)$ is defined by $f(x)$ as in exercise 7, then

$$P(x) = x^{10} - 60x^6 - 176x^5 - 1600x^2 + 1280x - 256$$

is irreducible in $\mathbb{Q}[x]$. Use this information to show that $G_f \cong A_5$.

Real Polynomial Equations
and Inequalities

The principal objective of this chapter is the theory of polynomial equations and inequalities in several unknowns in the field $\mathbb{R}$ of real numbers. The basic properties of $\mathbb{R}$ that serve as take-off point for the development of analysis are contained in the statement that $\mathbb{R}$ is a complete ordered field: that is, we have a relation $>$ in $\mathbb{R}$ satisfying the axioms of an ordered field (given in section 5.1), and the completeness axiom that every subset of $\mathbb{R}$ which has an upper bound has a least upper bound. Since we shall be concerned only with polynomial functions, it is not surprising that the full force of these properties is not required here. We shall see that it will suffice for our purposes to assume that we have an ordered field R such that: (1) positive elements of R have square roots in R and (2) every equation of odd degree in one unknown with coefficients in R has a root in R. An ordered field satisfying these conditions will be called real closed. It is clear that $\mathbb{R}$ has these properties. Moreover, the subset of elements of $\mathbb{R}$ which are algebraic over $\mathbb{Q}$ is also an instance of a real closed field and this ordered field lacks the classical completeness property.

We shall show that if R is real closed then $R(\sqrt{-1})$ is algebraically closed. Taking $R = \mathbb{R}$ we obtain $\mathbb{C} = \mathbb{R}(\sqrt{-1})$, so this will prove as a corollary the "fundamental theorem of algebra" that the field of complex numbers is algebraically closed.

Our main concern will be the development of algorithms to decide whether or not a given system of polynomial equations, inequations ($\neq$), and inequalities ($>$) with coefficients in R has a solution in R. The first definitive result of this sort is a classical theorem which was proved by J. C. F. Sturm in 1836. The most general result in this direction is a far-reaching extension of Sturm's theorem which was proved by A. Tarski around 1930. We shall give an alternative proof of this theorem which is due to A. Seidenberg. Before passing from Sturm's theorem to Tarski's we shall consider the theory of elimination of variables in systems of equations and inequations with coefficients in any field.

5.1   ORDERED FIELDS. REAL CLOSED FIELDS

We shall give a definition of an ordered field in terms of its set P of positive elements. This is the following

DEFINITION 5.1   *An* ordered field (F, P) *is a field F together with a subset P (the set of* positive *elements) of F such that: (1) $0 \notin P$, (2) if $a \in F$ then either $a \in P$, $a = 0$, or $-a \in P$, (3) if $a, b \in P$ then $a + b$ and $ab \in P$. A field F is called* orderable *if it is possible to specify a subset P in F having the foregoing properties.*

Since any field contains more than one element, it is clear that if (F, P) is an ordered field, then P is not vacuous. If N denotes the subset $\{-a \,|\, a \in P\}$, then (2) states that $F = P \cup \{0\} \cup N$. Moreover, it is clear from (1) that $P \cap \{0\} = \varnothing$ and $N \cap \{0\} = \varnothing$. Also $P \cap N = \varnothing$ since, if $a \in P \cap N$, then $-a \in P \cap N$ and hence $0 = a + (-a) \in P$ contrary to (1). Hence the decomposition $F = P \cup \{0\} \cup N$ is one into disjoint subsets. It is clear that N is closed under addition since $-a + (-b) = -(a + b) \in N$ if $a, b \in P$. On the other hand, $ab = (-a)(-b) \in P$ if $a, b \in N$.

We can introduce an order relation $a > b$ in (F, P) by defining $a > b$ to mean that $a - b \in P$. Then if a, b are any two elements of F, we have the trichotomy: one and only one of the three alternatives $a > b$, $a = b$, $b > a$ holds. If $a > b$ then $a + c > b + c$ for any c and $ap > bp$ for any positive p. Conversely, we could start with a relation $>$ in a field satisfying the trichotomy law, transitivity, and the two properties that $a > b$ implies $a + c > b + c$ and $ap > bp$ if $p > 0$. Then we put $P = \{p \,|\, p > 0\}$ and it is clear that (F, P) is an ordered field as defined above and that the associated relation $>$ defined in (F, P) is the given one.

As usual, it is convenient to write $a < b$ for $b > a$. The elementary properties of inequalities in the field $\mathbb{R}$ of real numbers are readily established. We list some of these: $a > 0$ implies $a^{-1} > 0$ and $a > b > 0$ implies $b^{-1} > a^{-1} > 0$. If $a > b$, then $-a < -b$, and if $a > b$ and $c > d$, then $a + c > b + d$. As usual, we define $|a| = a$ if $a \geq 0$ and $|a| = -a$ if $a < 0$ and we prove that $|a + b| \leq |a| + |b|$ and $|ab| = |a| \, |b|$.

If F' is a subfield of (F, P) then (F', P') is an ordered field for $P' = F' \cap P$. We call this the *induced ordering* in F'. If (F, P) and (F', P') are any two ordered fields, then an isomorphism η of F onto F' is called an *order isomorphism* if $\eta(P) \subset P'$. Then also $\eta(0) = 0$ and $\eta(N) \subset N'$, so $\eta(P) = P'$.

In any ordered field (F, P), $a \neq 0$ implies $a^2 > 0$. Hence, if $a_1, a_2, \ldots, a_r \neq 0$, then $\sum a_i^2 \neq 0$. In particular, $1 + 1 + \cdots + 1 = 1^2 + \cdots + 1^2 \neq 0$ which shows that any ordered field must be of characteristic 0. Also, we can not have $-1 = \sum a_i^2$ in F since this would give $1^2 + \sum a_i^2 = 0$. In particular, $-1 \neq a^2$, $a \in F$, so F does not contain a square root of -1. It is clear from this that the field $\mathbb{C}$ of complex numbers is not orderable.

In the field $\mathbb{R}$ of real numbers it is easy to establish, using the completeness axiom, that we have the following two properties:

(i) Any positive element has a square root in $\mathbb{R}$.
(ii) Any polynomial equation $f(x) = 0$ where $f(x) \in \mathbb{R}[x]$ and is of odd degree has a root in $\mathbb{R}$.

Both of these are consequences of the intermediate value theorem that if f is a continuous function and $f(a)f(b) < 0$ for $a < b$, then there exists a number c, $a < c < b$, such that $f(c) = 0$. We shall now call an ordered field (R, P) *real closed*[1] if it has the properties (i) and (ii) (with R replacing $\mathbb{R}$). We have the following

THEOREM 5.1. *A real closed field has a unique ordering endowing it with the structure of an ordered field. Any automorphism of such a field is an order isomorphism. If R is real closed, then its subfield of elements which are algebraic over $\mathbb{Q}$ ($\subset R$) is real closed.*

Proof. Let (R, P) be real closed and let (R, P') be any ordered field structure on R. If $a \in P$ then $a = b^2$, $b \neq 0$. Hence $a \in P'$. Thus $P \subset P'$ and this implies that $P = P'$. The second statement follows in the same way. Now let R be real closed and let R_0 be the subfield of elements which are algebraic over $\mathbb{Q}$ (cf. section 4.12, p. 280). If $a \in R_0$ and $a > 0$, then we have a $b \in R$ such that $b^2 =$

[1] This notion is equivalent to another one which is central in the theory of formally real fields which is due to Artin-Schreier. An account of this is given in Chapter 11 of Volume 2 of this book.

a. Thus *b* is algebraic over R_0 and so $b \in R_0$. Hence condition (i) holds in R_0. In the same way we see that (ii) holds, so R_0 is real closed. $\square$

In particular, we see that the field of real algebraic numbers, that is, the subfield of $\mathbb{R}$ of numbers which are algebraic over $\mathbb{Q}$ is real closed. Of course, this subfield is not complete. Hence it is clear that the axioms we are using are weaker than the completeness axiom.

We prove next the analogue for real closed fields of the "fundamental theorem of algebra."

THEOREM 5.2   *If R is real closed then* $R(\sqrt{-1})$ *is algebraically closed.*

Proof.   The proof we shall give is due to Artin and is patterned rather closely after one of Gauss' proofs of the classical result. We note first that $\sqrt{-1} \notin R$ and we have the automorphism $r = a + b\sqrt{-1} \to \bar{r} \equiv a - b\sqrt{-1}$, $a, b \in R$, in $C \equiv R(\sqrt{-1})$. If $f(x) \in C[x]$ then $f(x)\bar{f}(x) \in R[x]$ and if this has a root in C, then f has a root in C. Hence to prove that C is algebraically closed it suffices to show that every monic polynomial with coefficients in R has a root in C. This holds by (ii) if the polynomial has odd degree. We show next that every element of C has a square root in this field. This follows from (i) for the elements $a \geq 0$ of R, and if $a \in R$ and $a < 0$ and b satisfies $b^2 = -a$, then $(\sqrt{-1}b)^2 = a$. Now let $r = a + b\sqrt{-1}$, $a, b \in R$, $b \neq 0$. Put $\sqrt{-1} = i$ and let $x, y \in R$. Then $(x + iy)^2 = r$ is equivalent to:

(1) $$x^2 - y^2 = a, \qquad 2xy = b.$$

Since $b \neq 0$ we may (by multiplying by a suitable element of R—which has a square root in C) assume that $b = 2$, so the second equation becomes $xy = 1$. This holds if $y = x^{-1}$. Then the first equation becomes $x^2 - x^{-2} = a$ or $z - z^{-1} = a$ for $z = x^2$. Then we have $z^2 - az - 1 = 0$ which has the solution $(a + \sqrt{a^2 + 4})/2$ in R since $a^2 + 4 > 0$.[2] Also $a + \sqrt{a^2 + 4} > 0$ since $a + \sqrt{a^2 + 4} \leq 0$ leads to $4 \leq 0$. Hence there exists an $x \neq 0$ in R such that $x^2 = \frac{1}{2}(a + \sqrt{a^2 + 4})$. Then $x^4 - ax^2 = 1$ and $x^2 - x^{-2} = a$. Hence x and $y = x^{-1}$ satisfy (1) with $b = 2$. We have therefore proved that every element of C has a square root in this field. Consequently, there exists no extension field E/C with $[E:C] = 2$. We proceed to use this fact to prove that every monic polynomial with coefficients in R has a root in C. Let $f(x)$ be such a polynomial. Let E be a splitting field over R of $f(x)(x^2 + 1)$ which we assume contains C. Since the

---

[2] We use the convention, which is standard for $\mathbb{R}$, that $\sqrt{}$ denotes the positive square root.

310 5 Real Polynomial Equations and Inequalities

characteristic is 0, E is Galois over R. Let $G = \text{Gal } E/R$ and $|G| = 2^e m$ where m is odd. By Sylow's theorem G has a subgroup H with $|H| = 2^e$. If D is the corresponding subfield of E/R, then $[E:D] = 2^e$ and $[D:R] = m$. Since R has no proper odd dimensional extension field we must have $m = 1$, and so $D = R$ and $[E:R] = 2^e$ and its Galois group is $G = H$, a group of order 2^e. Such a group is solvable. If $e > 1$, it follows easily from the Galois theory (cf. section 4.11, p. 271) that E contains a subfield F containing C such that $[F:C] = 2$. This contradicts what we proved before. Hence $e = 1$ and so $E = C$. Thus C contains a root of $f(x)$ and C is algebraically closed. $\square$

It is immediate from the foregoing theorem that the monic irreducible polynomials in $R[x]$ are either of first or second degree. It is also clear from the formula for solving a quadratic equation that $x^2 + ax + b$ is irreducible in $R[x]$ if and only if $a^2 < 4b$.

The algebraic closure of $R(\sqrt{-1})$ permits us to establish for polynomial functions on R a number of basic properties of continuous and differentiable functions of a real variable. One of these which we shall need is the intermediate value theorem for polynomials.

THEOREM 5.3 *Let R be a real closed field, $f(x) \in R[x]$. Suppose a, b are elements of R such that $f(a)f(b) < 0$. Then there exists a c between a and b such that $f(c) = 0$.*

Proof. We may assume $f(x)$ is monic. Then $f(x)$ factors in $R[x]$ as

$$f(x) = (x - r_1) \cdots (x - r_m)g_1(x) \cdots g_s(x)$$

where

$$g_i(x) = x^2 + c_i x + d_i \quad \text{and} \quad c_i^2 < 4d_i.$$

Then

$$g_i(x) = \left(x + \frac{c_i}{2}\right)^2 + \tfrac{1}{4}(4d_i - c_i^2)$$

$$= \left(x + \frac{c_i}{2}\right)^2 + e_i^2, \qquad e_i = \tfrac{1}{2}\sqrt{4d_i - c_i^2}.$$

Then $g_i(u) > 0$ for all $u \in R$. If a and b are $< r_i$, $1 \le i \le m$, then $f(a)f(b) = \prod_{i,j} (a - r_i)(b - r_i)g_j(a)g_j(b) > 0$. Similarly, if $a, b > r_i$ for all i, then $f(a)f(b) >$

0. Since we are assuming that $f(a)f(b) < 0$ it follows that one of the r_i is caught between a and b. Since $f(r_i) = 0$ the result is clear.   □

EXERCISES

1. (Veblen.) Let F be a field satisfying the following two axioms: (i) -1 is not a square in F, (ii) the sum of any two non-squares of F is a non-square. Show that F can be ordered to become an ordered field in one and only one way.

2. Show that $\mathbb{Q}(\sqrt{2})$ has exactly two orderings making it an ordered field.

3. Let F be an ordered field and x an indeterminate over F. Show that $F(x)$ is ordered if one defines

$$(a_0x^n + a_1x^{n-1} + \cdots + a_n)/(b_0x^m + b_1x^{m-1} + \cdots + b_m) > 0$$

    if and only if $a_0b_0 > 0$.

4. Let F be an ordered field, $f(x) = x^n + a_1x^{n-1} + \cdots + a_n$ a polynomial with coefficients in F. Put

$$M = \max(1, |a_1| + |a_2| + \cdots + |a_n|).$$

    Show that if $|u| > M$ then $|f(u)| > 0$. Hence show that every root of $f(x)$ in F is contained in the interval $-M \leq x \leq M$.

In the next three exercises R is a real closed field.

5. Prove Rolle's theorem for polynomials $f(x)$: If $f(a) = 0 = f(b)$ and $a < b$, then there exists a c, $a < c < b$, such that $f'(c) = 0$.

6. Prove the mean value theorem for polynomials: If $a < b$ then there exists a c, $a < c < b$, such that $f(b) - f(a) = (b - a)f'(c)$.

7. Prove that $f(x)$ has a maximum on every closed finite interval, $a \leq x \leq b$.

5.2 STURM'S THEOREM

In this section we shall derive a classical result, Sturm's theorem, which gives a method of determining the exact number of roots in a real closed field of a polynomial equation $f(x) = 0$. In deriving this we shall follow rather closely Weber's exposition in *Lehrbuch der Algebra* (1898), Vol. 1, pp. 301–313.

Let R be a real closed field and let $f(x)$ be a polynomial of positive degree with coefficients in R. Following Weber, we shall say that a sequence of polynomials

(2) $$f_0(x) = f(x), f_1(x), \ldots, f_s(x)$$

is a *Sturm sequence* of polynomials for $f(x)$ for the closed interval $[a, b]$ (that is, $a \leq x \leq b$) if the $f_i(x) \in R[x]$ and satisfy the following conditions:

(i) $f_s(x)$ has no roots in $[a, b]$.

(ii) $f_0(a)f_0(b) \neq 0$.

(iii) If $c \in [a, b]$ is a root of $f_j(x)$, $0 < j < s$, then $f_{j-1}(c)f_{j+1}(c) < 0$.

(iv) If $f(c) = 0$ for $c \in [a, b]$, then there exist open intervals (c_1, c) (that is, $c_1 < x < x$) and (c, c_2) such that $f_0(u)f_1(u) < 0$ for any u in the first of these and $f_0(u)f_1(u) > 0$ for any u in the second.

We shall establish the existence of such sequences for any polynomial with distinct roots, but first we shall see how such a sequence can be used to determine the number of roots of $f(x)$ in the open interval (a, b). We consider the number of variations in sign of the sequences

$$f_0(a), f_1(a), \ldots, f_s(a)$$

$$f_0(b), f_1(b), \ldots, f_s(b)$$

of elements of R. If $c = \{c_1, c_2, \ldots, c_m\}$ is a finite sequence of non-zero elements of R, then we define the *number of variations in sign of* c to be the number of i, $1 \leq i \leq m - 1$, such that $c_i c_{i+1} < 0$. If $c = \{c_1, c_2, \ldots, c_m\}$ is an arbitrary sequence of elements of R, then we define the number of variations in sign of c to be the number of variations in the sign of the subsequence c' obtained by dropping the 0's in c. For example

$$\{1, 0, 0, 2, -1, 0, 3, 4, -2\}$$

has three variations in sign.

We can now state

THEOREM 5.4. *Let $f(x)$ be a polynomial of positive degree with coefficients in a real closed field R and let $f_0(x) = f(x), f_1(x), \ldots, f_s(x)$ be a Sturm sequence for $f(x)$ for the interval $[a, b]$. Then the number of distinct roots of $f(x)$ in (a, b) is $V_a - V_b$ where, in general, V_c denotes the number of variations in sign of the sequence $\{f_0(c), f_1(c), \ldots, f_s(c)\}$.*

Proof. The interval $[a, b]$ is decomposed into subintervals by the roots of the polynomials $f_j(x)$ of the given Sturm sequence. Thus we have a sequence

$a = a_0 < a_1 < \cdots < a_m = b$ such that none of the $f_j(x)$ has a root in (a_i, a_{i+1}). First, let $c \in (a_0, a_1)$ so no f_j has a root in (a_0, c). Then, by the intermediate value theorem (Theorem 5.3), $f_j(a_0)f_j(c) \geq 0$ for $0 \leq j \leq s$. Hence if none of the $f_j(a_0) = 0$, then $f_j(a_0)f_j(c) > 0$ which implies that $V_{a_0} = V_c$. Now suppose $f_k(a_0) = 0$ for some k. Since $f_0(a) \neq 0$, $f_s(a) \neq 0$ by the properties of Sturm sequences, we have $0 < k < s$. Then $f_{k-1}(a_0)f_{k+1}(a_0) < 0$ by property (iii). Since $f_{k-1}(x)$ and $f_{k+1}(x)$ have no roots in (a_0, c) we have $f_{k-1}(a_0)f_{k-1}(c) > 0$ and $f_{k+1}(a_0)f_{k+1}(c) > 0$. It follows that $f_{k-1}(c)f_{k+1}(c) < 0$. Thus $f_{k-1}(a_0)$, 0, $f_{k+1}(a_0)$ and $f_{k-1}(c)$, $f_k(c)$, $f_{k+1}(c)$ each contribute one variation of sign to V_{a_0} and V_c respectively. Taking into account all the k we see that $V_{a_0} = V_c$. A similar argument shows that if $d \in (a_{m-1}, a_m)$, then $V_d = V_{a_m}$. Now let $c \in (a_{i-1}, a_i)$, $d \in (a_i, a_{i+1})$ where $1 < i < m - 1$. Then the same argument shows that $V_c = V_d$ provided that $f(a_i) \neq 0$. Now suppose $f(a_i) = 0$. Then, by (iv), we have $f_0(c)f_1(c) < 0$ and $f_0(d)f_1(d) > 0$. Then the sequence $f_0(c), f_1(c)$ has one variation in sign whereas the sequence $f_0(d), f_1(d)$ has none. The argument used before shows that $f_{j-1}(c), f_j(c), f_{j+1}(c)$ and $f_{j-1}(d), f_j(d), f_{j+1}(d)$ have the same number of variations in sign if $j > 1$. Hence $V_c - V_d = 1$ if $f(a_i) = 0$. Now choose $a_i' \in (a_{i-1}, a_i)$. Then

$$V_a - V_b = (V_a - V_{a_i}) + \sum_{1}^{m-1} (V_{a_i} - V_{a_{i+1}}) + (V_{a_m} - V_b)$$

and our determination of each parenthesis on the right-hand side shows that these are either 0 or 1 and that the number of occurrences of 1 coincides with the number of a_i, $1 \leq i \leq m$, which are roots of $f(x)$. Thus $V_a - V_b$ is the number of roots of $f(x)$ in (a, b). $\square$

Now let $f(x)$ be any polynomial in $R[x]$ of positive degree. We define the *standard sequence* for $f(x)$ by

(3)
$$f_0(x) = f(x), \quad f_1(x) = f'(x) \quad \text{(formal derivative of } f(x))$$
$$f_0(x) = q_1(x)f_1(x) - f_2(x), \quad \deg f_2 < \deg f_1$$
$$\vdots \qquad\qquad \vdots \qquad\qquad \vdots$$
$$f_{i-1}(x) = q_i(x)f_i(x) - f_{i+1}(x), \quad \deg f_{i+1} < \deg f_i$$
$$\vdots \qquad\qquad \vdots \qquad\qquad \vdots$$
$$f_{s-1}(x) = q_s(x)f_s(x) \quad \text{(that is, } f_{s+1}(x) = 0).$$

Thus the $f_i(x)$ are obtained by modifying the Euclid algorithm for finding the g.c.d. of $f(x)$ and $f'(x)$ in such a way that the last polynomial obtained at each stage is the negative of the remainder in the division process. For example, if

$f(x) = x^3 + x + 1$, $f_0(x) = f(x)$, $f_1(x) = 3x^2 + 1$, $f_0(x) = (\frac{1}{3}x)f_1(x) - (-\frac{2}{3}x - 1)$
so $f_2(x) = -\frac{2}{3}x - 1$ and $f_1(x) = (-\frac{9}{2}x + \frac{27}{4})f_2(x) - (-\frac{31}{4})$ so $f_3(x) = -\frac{31}{4}$. Then
the standard sequence for $f(x)$ is

$$x^3 + x + 1, \ 3x^2 + 1, \ -\tfrac{2}{3}x - 1, \ -\tfrac{31}{4}.$$

In the general case it is clear from (3) that $f_s(x)$ is a factor of every $f_i(x)$ and
this is a g.c.d. of $f(x)$ and $f'(x)$. Now put $g_i(x) = f_i(x)f_s(x)^{-1}$ and consider the
sequence

(4) $g_0(x), g_1(x), \dots, g_s(x).$

We proceed to show that this is a Sturm sequence for $g_0(x)$ for any interval
$[a, b]$ such that $g_0(a) \neq 0$, $g_0(b) \neq 0$. Clearly condition (ii) holds, and (i) holds
since $g_s(x) = 1$. Dividing the polynomials in (3) by $f_s(x)$ gives the relation

(5) $g_{j-1}(x) = q_j(x)g_j(x) - g_{j+1}(x).$

Now suppose $g_j(c) = 0$. Then (5) shows that $g_{j-1}(c)g_{j+1}(c) \leq 0$ and $g_{j-1}(c) = 0$
if and only if $g_{j+1}(c) = 0$. In the latter case we obtain $0 = g_{j-1}(c) = g_j(c) =
g_{j+1}(c) = \cdots$ contrary to $g_s = 1$. Hence we see that $g_{j-1}(c)g_{j+1}(c) < 0$, which
establishes (iii). Next suppose $g_0(c) = 0$ for c in $[a, b]$. Then we have $f(x) =
(x - c)^e h(x)$, $e > 0$, $h(c) \neq 0$ and $f'(x) = (x - c)^e h'(x) + e(x - c)^{e-1}h(x)$. Also
$f_s(x) = (x - c)^{e-1}k(x)$ where $k(c) \neq 0$. Hence $h(x) = k(x)l(x)$ where $l(c) \neq 0$ and
$h'(x) = k(x)m(x)$ These relations give

(6)
$$g_0(x) = (x - c)l(x), \qquad l(c) \neq 0$$
$$g_1(x) = (x - c)m(x) + el(x)$$

so $g_1(c) = el(c) \neq 0$. Now choose an interval $[c_1, c_2]$ containing c in its interior
such that $g_1(x)l(x) \neq 0$ in $[c_1, c_2]$. Then, by the intermediate value theorem and
$g_1(c) = el(c) \neq 0$, $g_1(x)l(x) > 0$ in $[c_1, c_2]$. Hence $g_0(x)g_1(x) = (x - c)g_1(x)l(x)$
has the same sign as $x - c$ in $[c_1, c_2]$ so $g_0(x)g_1(x) < 0$ for $c_1 \leq x < c$ and
$g_0(x)g_1(x) > 0$ for $c < x \leq c_2$. This shows that (iv) holds and so (4) is a Sturm
sequence for $g_0(x)$.

If $f(x)$ has no multiple roots, then the g.c.d. of $f(x)$ and $f'(x)$ is 1. Then the sequence $\{f_0(x), f_1(x), \dots, f_s(x)\}$ differs from $\{g_0(x), g_1(x), \dots, g_s(x)\}$ by a non-zero
multiple in R. Hence the sequence of $f_i(x)$ is a Sturm sequence for $f(x) = f_0(x)$.
If $f(x)$ has multiple roots, then the standard sequence (4) will not be a Sturm
sequence for an interval containing a multiple root of $f(x)$. Nevertheless, we can
still use the standard sequence to determine the number of distinct roots of
$f(x)$ in (a, b). This is the content of

STURM'S THEOREM.   *Let $f(x)$ be a polynomial of positive degree with coeffi-cients in a real closed field R and let $\{f_0(x) = f(x), f_1(x) = f'(x), \ldots, f_s(x)\}$ be the standard sequence (3) for $f(x)$. Assume $[a, b]$ is an interval such that $f(a) \neq 0$, $f(b) \neq 0$. Then the number of distinct roots of $f(x)$ in (a, b) is $V_a - V_b$ where V_c denotes the number of variations in sign of $\{f_0(c), f_1(c), \ldots, f_s(c)\}$.*

Proof.   Let $g_i(x) = f_i(x)f_s(x)^{-1}$ as above. Then apart from multiplicities, the polynomials $f(x)$ and $g_0(x)$ have the same roots in $[a, b]$ (exercise 1, p. 233). Since $\{g_i(x)\}$ is a Sturm sequence for $g_0(x)$, the number of these roots is $V_a(g) - V_b(g)$ where $V_c(g)$ is the number of variations in sign in $\{g_i(c)\}$. Since

$$f_i(c) = g_i(c)f_s(c) \text{ and } f_s(a) \neq 0, f_s(b) \neq 0$$

it is clear that $V_a(g) = V_a$, $V_b(g) = V_b$. Hence $V_a - V_b$ is the number of distinct roots of $f(x)$ in (a, b). $\square$

We have indicated (exercise 4, p. 311) that the roots of $x^n + a_1 x^{n-1} + \cdots + a_n$ in R are in the interval $[-M, M]$ where $M = \max(1, |a_1| + \cdots + |a_n|)$. If we put $\mu = 1 + |a_1| + \cdots + |a_n|$, then the roots of $f(x)$ in R are in $(-\mu, \mu)$. Hence if $f_0(x) = f(x), f_1(x), \ldots, f_s(x)$ is the standard sequence for $f(x)$, then the number of roots of $f(x)$ in R is $V_{-\mu} - V_\mu$ where, as usual, V_c is the number of variations of sign of $\{f_0(c), f_1(c), \ldots, f_s(c)\}$. This gives a constructive way of determining the number of roots of $f(x)$ in R. Sometimes it is preferable to use instead of μ a bound η which is a polynomial in the a_j. Such a bound can be obtained by observing $1 + a_i^2 > |a_i|$, so we can take

(7) $$\eta = 1 + \sum(1 + a_i^2) = (n + 1) + \sum a_i^2.$$

Then the roots in R lie in $(-\eta, \eta)$.

EXERCISES

1. Apply Sturm's theorem to show that $x^3 - 7x - 7$ has two real roots in $(-2, -1)$.

2. Apply the theorem to determine the number of real roots of $x^4 + 12x^2 + 5x - 9$.

3. Let $f(x) = x^3 + px + q, p \neq 0$. Show that

$$f_0 = f, \qquad f_1 = 3x^2 + p, \qquad f_2 = -2px - 3q, \qquad f_3 = -4p^3 - 27q^2$$

is a Sturm sequence for $f(x)$ for any $[a, b]$ with $f(a)f(b) \neq 0$. Note that $f_3 = d$, the discriminant of $f(x)$ (p. 259). Use Sturm's theorem to prove that f has a single real root or three distinct real roots according as $d < 0$ or $d > 0$.

4. Let $f(x) = x^4 + qx^2 + rx + s$, $L = 8qs - 2q^3 - 9r^2$, d the discriminant of $f(x)$. Prove that

   if $d < 0$ then the number of real roots of f is two;
   if $d > 0$, $q < 0$, $L > 0$ then f has four distinct real roots;
   if $d > 0$ and either $q \geq 0$ or $L \leq 0$, then f has no real roots.

5. Define the sequence of *Legendre polynomials* $P_0, P_1, \ldots, P_n, \ldots$ by the recursion formula

   $$nP_n(x) - (2n - 1)xP_{n-1}(x) + (n - 1)P_{n-2}(x) = 0$$

   where $P_0(x) = 1$, $P_1(x) = x$. Show that $\{P_m, P_{m-1}, \ldots, P_0\}$ is a Sturm sequence for P_m for the interval $[-1, 1]$. Show that P_m has m distinct real roots in $(-1, 1)$.

6. Let $f(x) \in R[x]$ where R is a real closed field and assume $\deg f(x) = n$. Let W_c denote the number of variations of sign in the sequence $\{f(c), f'(c), \ldots, f^{(n)}(c)\}$. Prove *Budan's theorem*: if $a < b$ and $f(a)f(b) \neq 0$, then $W_a - W_b$ exceeds the number of roots of $f(x)$ in (a, b), counting the multiplicities, by a non-negative even integer.

7. Deduce from exercise 6 *Descartes' rule of signs*. Let $f(x) = a_0 x^n + a_1 x^{n-1} + \cdots + a_l x^{n-l}$, $a_0 a_l \neq 0$, $a_i \in R$. Let P be the number of variations of sign in $\{a_0, a_1, \ldots, a_l\}$. Show that P exceeds the number of positive roots of $f(x)$, counting multiplicities, by a non-negative even integer.

5.3 FORMALIZED EUCLIDEAN ALGORITHM AND STURM'S THEOREM

In the last part of this chapter we shall develop a method for testing the solvability in a real closed field R of any finite system of polynomial equations, inequations ($F \neq 0$), and inequalities ($F > 0$) in several unknowns. The main result (Tarski's theorem) will be that, given such a system, we can determine in a finite number of steps a finite number of systems of polynomial equations, inequations, and inequalities in the coefficients of the given system, such that the given system will have a solution in R if and only if every equation, inequation, and inequality of *one* of the derived systems is satisfied by the coefficients. As an illustration of the type of result we shall obtain, we consider the case of a "reduced" quartic equation $x^4 + qx^2 + rx + s = 0$, q, r, s in R. Here it can be shown that this has a root in R if and only if one of the following alternatives involving the discriminant

$$d = 4\left(4s + \frac{q^2}{3}\right)^3 - 27\left(\frac{8}{3}qs - r^2 - \frac{2}{27}q^3\right)^2$$

and the expression

$$L = 8qs - 2q^3 - 9r^2$$

is satisfied:

    I. $d < 0$
    II. $d > 0, q < 0, L > 0$
    III. $d = 0, r \neq 0$
    IV. $d = 0, r = 0, q \leq 0$.

This follows quite easily from exercise, 4, p. 316 and the fact that $d = 0$ if and only if the equation has multiple roots.

We shall show in this section that we can obtain a similar version of Sturm's theorem for any equation whose coefficients are parameters that take on values in a real closed field. This will be based on a parameterized version of the Euclidean algorithm for determining the g.c.d. of polynomials, which we shall now derive. We begin with a coefficient ring of the form $A = K[t_1, \ldots, t_r]$ where the t_i are indeterminates and K is either $\mathbb{Z}$ or one of the fields $\mathbb{Z}/(p)$, p a prime. Let $F(t_1, \ldots, t_r; x)$, $G(t_1, \ldots, t_r; x) \in A[x]$, so

$$F(t_i; x) = u_n x^n + u_{n-1} x^{n-1} + \cdots + u_0$$

$$G(t_i; x) = v_m x^m + v_{m-1} x^{m-1} + \cdots + v_0$$

where the $u_i, v_j \in A$. We assume $G(t_i; x) \neq 0$ and we take a "section" $G_k(t_i; x) = v_k x^k + v_{k-1} x^{k-1} + \cdots + v_0$ with $v_k \neq 0$ obtained by dropping the terms $v_j x^j$ with $j > k$. Thus the x-degree of G_k, $\deg_x G_k = k$ and k takes on some of the values between 0 and m. The division algorithm can be carried out to write

(8) $$v_k(t_i)^{e_k} F(t_i; x) = Q_k(t_i; x) G_k(t_i; x) - R_k(t_i; x)$$

where $\deg R_k < k$ and e_k is a non-negative integer which is the larger of 0 and $n - k + 1$ (p. 129). Note that we have displayed $-R_k$, the negative of the usual remainder. This is preferable for the application to Sturm's theorem and is as good as the usual remainder in other applications. For Sturm's theorem it is also necessary to have e_k even. Hence we shall now fix e_k to be 0 if $n \leq k$ and otherwise to be the smallest even integer $\geq n - k + 1$. With this definite choice of e_k we can obtain Q_k and R_k satisfying (8). Moreover, degree considerations show that these are unique.

Now let R be any field extension of K, so that R is a field whose prime ring is the ring K ($= \mathbb{Z}$ or $\mathbb{Z}/(p)$). Let $(c_1, \ldots, c_r) \in R^{(r)} = R \times \cdots \times R$ (r times) and let F, G and the other notations be as in the last paragraph. Either $v_j(c_1, \ldots, c_r) = 0$ for all $j = 0, \ldots, m$, in which case $G(c_i; x) = 0$, or there exists a k such that $v_k(c_1, \ldots, c_r) \neq 0$ but $v_j(c_1, \ldots, c_r) = 0$ for $j > k$. Then $G(c_i; x) = G_k(c_i; x)$ is a polynomial of degree k in x with coefficients in R. By (8), we have

$$v_k(c_i)^{e_k}F(c_i; x) = Q_k(c_i; x)G(c_i; x) - R_k(c_i; x)$$

and since $v_k(c_i)^{e_k} \neq 0$ and deg $R_k(c_i; x) <$ deg $G(c_i; x)$ it is clear that $Q_k(c_i; x)$ and $-R_k(c_i; x)$ differ by a non-zero multiplier $(v_k(c_i)^{-e_k})$ in R from the quotient and remainder obtained by dividing $F(c_i; x)$ by $G(c_i; x) \neq 0$. We note also that the multiplier is positive if R is real closed.

We now introduce the following sets of equations and inequations defined by polynomials in $A = K[t_1, \ldots, t_r]$:

(9)
$$\Gamma_{-\infty} = \{v_0 = 0, v_1 = 0, \ldots, v_m = 0\}$$
$$\Gamma_k = \{v_j = 0, j > k, v_k \neq 0\} \quad \text{if} \quad v_k(t_1, \ldots, t_r) \neq 0.$$

The set $\gamma = \{\Gamma_{-\infty}, \Gamma_k\}$ for the k satisfying $0 \leq k \leq m$ and $v_k(t_1, \ldots, t_r) \neq 0$ is a *cover* of A in the sense that if R is any extension field of K and $\Gamma_j(R)$ is defined to be the subset of $R^{(r)}$ of $(c_1, \ldots, c_r)$ satisfying all the conditions in Γ_j, then $R^{(r)} = \bigcup \Gamma_j(R)$. In the present instance $\Gamma_j(R)$ is the set of $(c_1, \ldots, c_r)$ such that $\deg_x G(c_i; x) = j$ ($= -\infty$ if and only if $G(c_i; x) = 0$). In general, the terms Γ_j of a cover are finite sets of equations and inequations whose left hand members are in A. If we have a number of inequations $l_1 \neq 0, \ldots, l_h \neq 0$ we can replace them by a single one $l \neq 0$ where $l = l_1 l_2 \cdots l_h$ since R has no zero divisors $\neq 0$. For the sake of uniformity we append the trivial equation $0 = 0$ (inequation $1 \neq 0$) if Γ_j contains no equation (inequation). Hence we may assume that Γ_j consists of a finite non-vacuous set of equations and a single inequation. We observe also that for real closed R a set of equations $d_1 = 0, \ldots, d_h = 0$ is equivalent to a single equation $d = 0$ where $d = \sum d_i^2$. Hence if we are dealing exclusively with real closed extension fields of K (necessarily $= \mathbb{Z}$), then we may assume Γ_j consists of a single equation and a single inequation.

We can now summarize our results in the following way. Given the polynomials F and $G \in A[x]$ with $G = v_m x^m + v_{m-1}x^{m-1} + \cdots + v_0$, $v_j \in A$, let γ be the cover defined by the v_j as in (9). Then for each $k \neq -\infty$ appearing in (9), we have polynomials $Q_k, R_k \in A[x]$ such that if R is any extension field of K and $(c_1, \ldots, c_r) \in \Gamma_k(R)$, then $G(c_i; x) \neq 0$ and $Q_k(c_i; x)$ and $-R_k(c_i; x)$ differ by a non-zero multiplier in R from the quotient and remainder obtained by dividing $F(c_i; x)$ by $G(c_i; x)$ in $R[x]$. The multiplier is positive if R is real closed.

Let Γ be a finite set of equations and a single inequation determined by elements of A and let $\delta = \{\Delta_1, \Delta_2, \ldots, \Delta_s\}$ be a cover of A. Let $\Gamma^{(j)}$, $1 \leq j \leq s$, be a set of equations and a single inequation such that the set of equations is the union of the sets of equations for Γ and for Δ_j and the inequation is the product of the inequation of Γ and the inequation of Δ_j. Then it is clear that

$\Gamma^{(j)}(R) = \Gamma(R) \cap \Delta_j(R)$ for any R. Since $\bigcup_1^s \Delta_j(R) = R^{(r)}$ it follows that $\Gamma(R) = \bigcup_1^s \Gamma^{(j)}(R)$. Hence if $\gamma = \{\Gamma_1 = \Gamma, \Gamma_2, \dots, \Gamma_q\}$ is a cover for A, then so is $\gamma' = \{\Gamma^{(1)}, \dots, \Gamma^{(s)}, \Gamma_2, \dots, \Gamma_q\}$. The covers obtained in this way and by finite iteration of this process will be called *refinements* of γ.

We recall that if $f(x)$ and $g(x) \neq 0 \in R[x]$ where R is a field, then the Euclidean algorithm for determining a g.c.d. of $f(x)$ and $g(x)$ in $R[x]$ consists of constructing by successive divisions the sequence of polynomials

$$(10) \qquad f_0 = f, \qquad f_1 = g, \qquad f_2, \dots, f_s, \qquad f_{s+1} = 0$$

such that $\deg f_{i+1} < \deg f_i$ for $i \geq 1$ and there exist q_i such that $f_{i-1} = q_i f_i - f_{i+1}$ (cf. exercise 11, p. 150, and equation (3)). It follows that $f_s \neq 0$ and f_s is a g.c.d. of f and g in $R[x]$. We shall call (10) the *Euclidean sequence for the pair* (f, g) if $g \neq 0$. It is convenient also to extend this to the pair $(f, 0)$ by saying that $(f, 0, 0)$ is the Euclidean sequence for $(f, 0)$.

We shall now prove the

LEMMA. *Let F and $G \neq 0 \in A[x]$. Then we can construct in a finite number of steps a cover $\delta = \{\Delta_1, \Delta_2, \dots, \Delta_h\}$ which is a refinement of the cover (9) defined by the coefficients of G, and sequences of polynomials $F_{j0} = F$, $F_{j1}, \dots, F_{js_j} \in A[x]$, $1 \leq j \leq h$, such that for any field extension R of K and any $(c_1, \dots, c_r) \in \Delta_j(R)$, the terms of the sequence*

$$F_{j0}(c_i; x), F_{j1}(c_i; x), \dots, F_{js_j}(c_i; x), F_{j,s_j+1}(c_i; x) = 0$$

differ by non-zero multipliers in R from those of the Euclidean sequence for $(F(c_i; x), G(c_i; x))$. Moreover, the multipliers are positive if R is real closed.

Proof. As above, we determine the polynomials $Q_k, R_k \in A[x]$ for the $k \neq -\infty$ appearing in (9), by the division algorithm applied to F and $G_k = v_k x^k + \cdots + v_0$. If $R_k = 0$ the sequence of polynomials $F_{k0} = F$, $F_{k1} = G_k$, $F_{k2} = 0$ satisfy the stated condition. We now assume $R_k \neq 0$ ($k \neq -\infty$). Then the sum of the degrees of G_k and R_k is less than the sum of the degrees of F and G. Using induction on the sum of the degrees we may assume the result for the pair of polynomials $(G_k, -R_k)$. Thus we have a cover $\delta_k = \{\Delta_{k1}, \Delta_{k2}, \dots, \Delta_{kh_k}\}$ and sequences of polynomials $\{F_{kl0} = G_k, F_{kl1}, \dots, F_{kls_{kl}}\}$, $1 \leq l \leq h_k$, satisfying the conditions of the theorem for $(G_k, -R_k)$. We now refine the cover γ by replacing each set Γ_k, $k \neq -\infty$ by the sets $\Gamma_k^{(1)}, \Gamma_k^{(2)}, \dots, \Gamma_k^{(h_k)}$ where $\Gamma_k^{(l)}$ has as equations the equations of Γ_k and of Δ_{kl} and has the inequation which is the product of the inequation of Γ_k and that of Δ_{kl}. Let δ be the cover obtained by making these replacements for every Γ_k. Then we associate with the set $\Gamma_k^{(l)}$ the sequence

$\{F = F_{kl0}, F_{kl1}, \ldots, F_{kls_{kl}}\}$. Moreover, for the term $\Gamma_{-\infty}$ we take the sequence of polynomials $\{F, 0, 0\}$. It is easily seen that δ together with these sequences satisfies the conditions. $\square$

As an illustration of this result we consider the reduced cubic $F(p, q; x) = x^3 + px + q$, p and q indeterminates, and its derivative $G(p, q; x) = F'(p, q; x) = 3x^2 + p$. We take $K = \mathbb{Z}$. We have $v_2 = 3, v_1 = 0, v_0 = p$. Hence $\Gamma_{-\infty} = \{3 = 0, 0 = 0, p = 0; 1 \neq 0\}$; that is, $\Gamma_{-\infty}$ consists of the equations $3 = 0, 0 = 0, p = 0$ and the trivial inequation $1 \neq 0$. Also $\Gamma_2 = \{0 = 0; 3 \neq 0\}$, $\Gamma_0 = \{3 = 0; p \neq 0\}$. Evidently $\Gamma_{-\infty}(R) = \varnothing$ and $\Gamma_0(R) = \varnothing$ for any R, so we may take $\gamma = \{\Gamma = \Gamma_2\}$. The division algorithm we specified for F and G yields the remainder $-R = -(6px + 9q)$ so we have to repeat the process with $G = 3x^2 + p$ and $-R = -(6px + 9q)$. We leave it to the reader to verify that the result obtained is that we have the cover $\delta = \{\Delta_1, \Delta_2, \Delta_3, \Delta_4\}$ where

$$\Delta_1 = \{0 = 0; \ p(27q^2 + 4p^3) \neq 0\}$$

$$\Delta_2 = \{27q^2 + 4p^3 = 0; \ p \neq 0\}$$

$$\Delta_3 = \{p = 0; \ q \neq 0\}$$

$$\Delta_4 = \{p = 0, q = 0; \ 1 \neq 0\}$$

and the corresponding sequences of F's are

I. $x^3 + px + q, 3x^2 + p, -(6px + 9q), -9(27q^2 + 4p^3), 0$
II. $x^3 + px + q, 3x^2 + p, -(6px + 9q), 0$
III. $x^3 + px + q, 3x^2 + p, -9q, 0$
IV. $x^3 + px + q, 3x^2 + p, 0.$

We shall now give the parameterized version of Sturm's theorem. This is

THEOREM 5.5.   *Let* $F(t_i; x) = u_n x^n + u_{n-1} x^{n-1} + \cdots + u_0 \in A[x]$ *where* $u_j = u_j(t_1, \ldots, t_r) \in A = \mathbb{Z}[t_1, \ldots, t_r]$, t_i *and* x *indeterminates. Then we can determine in a finite number of steps a finite collection* $\{\Gamma_1, \Gamma_2, \ldots, \Gamma_s\}$ *where each* Γ_k *is a finite set of polynomial relations of the form* $C = 0, C > 0, C \neq 0$ *where* $C \in A$, *such that for any real closed field* R, *the statement that* $F(c_i; x) = 0$ *for* $c_i \in R, 1 \leq i \leq r$, *has a root in* R *is equivalent to the validity for* $t_i = c_i$ *of every relation in one of the* Γ_k.

Proof.   We put $G(t_i; x) = F'(t_i; x) = nu_n x^{n-1} + (n-1)u_{n-1}x^{n-2} + \cdots + u_1$. We may assume $G \neq 0$ since otherwise $F(t_i; x) = u_0(t_1, \ldots, t_r)$ and the result is trivial. Then we can apply the lemma to obtain (by a finite process) the cover

$\delta = \{\Delta_1, \Delta_2, \ldots, \Delta_h\}$ and corresponding sequences of polynomials $F_{j0} = F$, $F_{j1}, \ldots, F_{js_j} \in A[x]$, $1 \leq j \leq h$. Then if R is a real closed field and $(c_1, \ldots, c_r) \in \Delta_j(R)$, $F_{j0}(c_i; x)$, $F_{j1}(c_i; x)$, $\ldots$, $F_{js_j}(c_i; x)$ differ by positive multipliers from the terms of the standard Sturm sequence for $F(c_i; x)$. To simplify the notation we now write Δ for any one of the Δ_j and $F_0 = F, F_1, \ldots, F_s$ for the corresponding sequence of polynomials. Since δ is a refinement of the cover γ associated with G, for all $(c_1, \ldots, c_r) \in \Delta(R)$ either $u_n(c_1, \ldots, c_r) = \cdots = u_1(c_1, \ldots, c_r) = 0$ or there exists an m, $1 \leq m \leq n$, such that $u_m(c_1, \ldots, c_r) \neq 0$ and $u_j(c_1, \ldots, c_r) = 0$ for all $j > m$. In the first case, $F(c_i; x) = 0$ has a root in R if and only if $u_0(c_1, \ldots, c_r) = 0$. In the second case we know that the roots of $F(c_i; x) = 0$ in R all lie in the interval $(-\eta, \eta)$ where

$$(11) \qquad \eta = (m + 1) + \sum_0^{m-1} u_j(c_1, \ldots, c_r)^2 u_m(c_1, \ldots, c_r)^{-2}.$$

Since the terms of the sequence $F_0(c_i; x), F_1(c_i; x), \ldots, F_s(c_i; x)$ are positive multiples of those of the standard sequence for $F(c_i; x)$, it follows from Sturm's theorem that $F(c_i; x) = 0$ has a root in R if and only if the number of variations in sign of the sequence $F_0(c_i; \eta), F_1(c_i; \eta), \ldots, F_s(c_i; \eta)$ exceeds that of $F_0(c_i; -\eta), F_1(c_i; -\eta), \ldots, F_s(c_i; -\eta)$. To express this as polynomial conditions on the c_i we associate with each $F_k(t_1, \ldots, t_r; x)$ the pair of polynomials

$$g_k(t_1, \ldots, t_r) = u_m^{2n_k} F_k\left(t_i; m + 1 + \sum_0^{m-1} u_j^2 u_m^{-2}\right)$$

$$h_k(t_1, \ldots, t_r) = u_m^{2n_k} F_k\left(t_i; -(m + 1) - \sum_0^{m-1} u_j^2 u_m^{-2}\right)$$

where $u_j = u_j(t_1, \ldots, t_r)$ and n_k is the degree in x of F_k. Thus $g_k, h_k \in A$ and $g_k(c_1, \ldots, c_r)$ and $h_k(c_1, \ldots, c_r)$ differ from $F_k(c_i; \eta)$ and $F_k(c_i; -\eta)$ respectively by positive multipliers. Hence for the elements $(c_1, \ldots, c_r) \in \Delta(R)$, $F(c_i; x) = 0$ has a root in R if and only if the number of variations in sign in the sequence $g_0(c_i), g_1(c_i), \ldots, g_s(c_i)$ exceeds that of the sequence $h_0(c_i), h_1(c_i), \ldots, h_s(c_i)$. We now consider all possible systems of relations of the form

$$(12) \qquad \begin{array}{ccc} g_0 \gtrless 0, & g_k \gtreqless 0, & g_s \gtrless 0 \\[6pt] h_0 \gtrless 0, & h_k \gtreqless 0, & h_s \gtrless 0, \quad 1 \leq k \leq s - 1. \end{array}$$

We pair off all such relations on the g's with those on the h's so that the number of variations of sign (in the obvious sense) of the sequence of g's exceeds that of the sequence of h's. Then it is clear that for $(c_1, \ldots, c_r) \in \Delta(R)$, $F(c_i; x) = 0$ will be solvable in R if and only if the c_i satisfy one of these paired systems of equations and inequalities. If we append to each of these the relations Δ we obtain one of the systems Γ we require. Doing this for all the pairs and all the Δ's we obtain a finite set of Γ's which satisfy the requirements of the theorem. $\square$

We remark that we can apply the same method to obtain a similar result for the existence of a root in a given interval $(-c, c)$ where $c > 0$. Also we may replace $A = \mathbb{Z}[t_1, \ldots, t_r]$ by any ring $F[t_1, \ldots, t_r]$ where F is a subfield of some real closed field.

5.4   ELIMINATION PROCEDURES. RESULTANTS

Before proceeding to the extension of Theorem 5.5 to systems of equations, inequalities, and inequalities in several unknowns it seems appropriate to consider the simpler problem of developing a test for the solvability of a system of polynomial equations and inequations in some extension field of a given field. The basic theorem we wish to prove is:

THEOREM 5.6.   *Let* $K = \mathbb{Z}$ *or* $\mathbb{Z}/(p)$, p *a prime, and let* $A = K[t_1, \ldots, t_r]$, $B = A[x_1, \ldots, x_n]$ *where the t's and x's are indeterminates. Let*

$$\Gamma = \{F_1, \ldots, F_m; G\} \subset B.$$

Then we can determine in a finite number of steps a finite collection $\{\Gamma_1, \Gamma_2, \ldots, \Gamma_s\}$ *where*

$$\Gamma_j = \{f_{j1}, \ldots, f_{jm_j}; g_j\} \subset A$$

such that for any extension field F of K and any $(c_1, \ldots, c_r) \in F^{(r)}$ *the system of equations and inequation*

(13)
$$\Gamma(c_1, \ldots, c_r) : F_1(c_1, \ldots, c_r; x_1, \ldots, x_n) = \cdots$$
$$= F_m(c_1, \ldots, c_r; x_1, \ldots, x_n) = 0; \quad G(c_1, \ldots, c_r; x_1, \ldots, x_n) \neq 0$$

is solvable for the x's in some extension field E/F if and only if the c_i satisfy one of the systems

(14)
$$\Gamma_j(c_1, \ldots, c_r) : f_{j1}(c_1, \ldots, c_r) = \cdots = f_{jm_j}(c_1, \ldots, c_r) = 0;$$
$$g_j(c_1, \ldots, c_r) \neq 0,$$

$1 \leq j \leq s$. *Moreover, when one of these systems is satisfied then a solution exists for (13) in some algebraic extension field E/F.*

Before proceeding to the proof we prove a lemma which is due to Tarski.

LEMMA. *Let $f(x)$, $g(x)$ be non-zero polynomials contained in $F[x]$, F a field, and let $h = \deg f(x)$. If $f(x) \mid g(x)^h$ then there exists no a in any extension field E/F satisfying $f(a) = 0$, $g(a) \neq 0$. On the other hand, if $f(x) \nmid g(x)^h$ then there exists such an a in some algebraic extension field E/F.*

Proof. The first statement is clear. Now assume $f(x) \nmid g(x)^h$. We claim that there exists an irreducible factor $p(x)$ of $f(x)$ which is not a factor of $g(x)$. Otherwise, if $p_1(x), \ldots, p_m(x)$ are the disinct irreducible factors of $f(x)$ then $p_1(x) \cdots p_m(x) \mid g(x)$ and $(p_1(x) \cdots p_m(x))^h \mid g(x)^h$. Since $f(x) \mid (p_1(x) \cdots p_m(x))^h$ we have the contradiction that $f(x) \mid g(x)^h$. Now let $p(x)$ be an irreducible factor of $f(x)$ which is not a factor of $g(x)$ and let $E = F[x]/(p(x))$. This is an algebraic extension of F containing $a = x + (p(x))$ satisfying $f(a) = 0$, $g(a) \neq 0$. $\square$

We now proceed to the

Proof of Theorem 5.6. We consider first the case in which $n = 1$ and we use induction on the sum of the degrees in x of the F_i and G (where we define $\deg 0 = 0$). If this sum is 0 there is nothing to prove. We proceed to give a series of reductions if one of the F_i or G has positive degree in x.

Case I. $\deg F_i > 0$ for $i = 1, 2$. Let $b_0 x^h$ and $d_0 x^k$ be the leading terms of F_1 and F_2 respectively, that is, $F_1 = b_0 x^h + b_1 x^{h-1} + \cdots$, $b_0 \neq 0$, $F_2 = d_0 x^k + d_1 x^{k-1} + \cdots$, $d_0 \neq 0$. Assume $h \geq k$. We define

$$\Gamma' = \{d_0, F_1, F_2' = F_2 - d_0 x^k, F_3, \ldots, F_m; G\}$$

$$\Gamma'' = \{F_1'' = d_0 F_1 - b_0 x^{h-k} F_2, F_2, \ldots, F_m; d_0 G\}.$$

Suppose $(c_1, \ldots, c_r) \in F^{(r)}$ satisfies $d_0(c_1, \ldots, c_r) = 0$. Then $\Gamma(c_1, \ldots, c_r)$ is solvable in an extension field E/F if and only if $\Gamma'(c_1, \ldots, c_r)$ is solvable in E. On the other hand, suppose $d_0(c_1, \ldots, c_r) \neq 0$. Then $\Gamma(c_1, \ldots, c_r)$ is solvable in E if and only if $\Gamma''(c_1, \ldots, c_r)$ is solvable in E. Since the sum of the x-degrees of the polynomials in Γ' and in Γ'' is less than that of Γ, the theorem for $n = 1$ with the condition in Case I follows by the degree induction.

Case II. $\deg F_1 > 0$, $\deg G > 0$, $\deg F_i = 0$ if $i > 1$. Let $b_0 x^h$ and $d_0 x^k$ be the leading terms of F_1 and G respectively. By long division we can obtain polynomials Q and $R \in B$ such that $b_0^{h^2} G^h = Q F_1 + R$ where $R = r_0 x^{h-1} + r_1 x^{h-2} + \cdots + r_{h-1} \in A$. Now define

$$\Gamma' = \{G_0, F_1' = F_1 - b_0 x^h, F_2, \ldots, F_m; G\}$$

$$\Gamma_i = \{F_2, \ldots, F_m; b_0 r_i\} \subset A, \quad 0 \leq i \leq h - 1.$$

Let $(c_1, \ldots, c_r) \in F^{(r)}$ satisfy $b_0(c_1, \ldots, c_r) = 0$. Then $\Gamma(c_1, \ldots, c_r)$ has a solution in an extension field E/F if and only if $\Gamma'(c_1, \ldots, c_r)$ has a solution in E. Next suppose $b_0(c_1, \ldots, c_r) \neq 0$. Then $F_1(c_1, \ldots, c_r; x) \nmid G(c_1, \ldots, c_r; x)$ if and only if $R(c_1, \ldots, c_r; x) \neq 0$, hence, if and only if $r_i(c_1, \ldots, c_r) \neq 0$ for some $i = 0, 1, \ldots, h-1$. By the lemma, $F_1(c_1, \ldots, c_r; x) = 0$, $G(c_1, \ldots, c_r; x) \neq 0$ is solvable in an extension field E if and only if $R(c_1, \ldots, c_r; x) \neq 0$ and in this case a solution exists in an algebraic extension field E/F. It follows that if $b_0(c_1, \ldots, c_r) \neq 0$ then $\Gamma(c_1, \ldots, c_r)$ is solvable in an extension field if and only if one of the systems of equations and inequation $\Gamma_i(c_1, \ldots, c_r)$ is satisfied. Moreover, in this case $\Gamma(c_1, \ldots, c_r)$ is solvable in an algebraic extension field E/F. Since the sum of the x-degrees of the polynomials in Γ' or in any Γ_i is less than that of Γ, the result for $n = 1$ follows in case II by induction on degree.

Case III. $\deg F_1 > 0$, $\deg F_i = \deg G = 0$ if $i > 1$. Let the leading term of F_1 be $b_0 x^h$. Define

$$\Gamma' = \{b_0, F_1' = F_1 - b_0 x^h, F_2, \ldots, F_m; G\}$$

$$\Gamma'' = \{F_2, \ldots, F_m; Gb_0\}.$$

If $(c_1, \ldots, c_r)$ satisfies $b_0(c_1, \ldots, c_r) = 0$ then $\Gamma(c_1, \ldots, c_r)$ has a solution in an extension field E/F if and only if $\Gamma'(c_1, \ldots, c_r)$ has a solution in E/F. Now let $b_0(c_1, \ldots, c_r) \neq 0$. Then $F_1(c_1, \ldots, c_r; x)$ has a solution in an algebraic extension E/F. It follows that $\Gamma(c_1, \ldots, c_r)$ has a solution in the algebraic extension E/F if $\Gamma''(c_1, \ldots, c_r)$ holds for $(c_1, \ldots, c_r)$. Since the x-degree of F_1' is less than that of F_1, Case III follows by the degree induction.

Case IV. $\deg F_i = 0$, $1 \leq i \leq m$, $\deg G > 0$. Let $d_0 x^k$ be the leading coefficient of G. Define

$$\Gamma' = \{d_0, F_1, \ldots, F_m; G' = G - d_0 x^k\}$$

$$\Gamma'' = \{F_1, \ldots, F_m; d_0\}.$$

If $d_0(c_1, \ldots, c_r) = 0$ then $\Gamma(c_1, \ldots, c_r)$ has a solution in an extension field E/F if and only if $\Gamma'(c_1, \ldots, c_r)$ has a solution in E/F. If $d_0(c_1, \ldots, c_r) \neq 0$ then $G(c_1, \ldots, c_r; x) \neq 0$ has a solution in any field of cardinality exceeding k. Hence $\Gamma(c_1, \ldots, c_r)$ is solvable in an algebraic extension E/F if $\Gamma''(c_1, \ldots, c_r)$ is satisfied. Since the x-degree of G' is less than that of G, the result follows in this case.

The cases listed take care of all possibilities in which the sum of the x-degrees of the F_i and G is positive. Hence the theorem holds if $n = 1$. We now prove the theorem by induction on n and we assume $n > 1$. We treat $x_1, \ldots, x_{n-1}$ as additional t's and apply the result just proved to obtain in a finite number of steps sets Λ_k, $1 \leq k \leq u$, where $\Lambda_k = \{F_{k1}, \ldots, F_{kh_k}; G_k\}$ and the F_{kj} and $G_k \in K[t_1, \ldots, t_r, x_1, \ldots, x_{n-1}]$ such that the following two properties hold:

(1) If F is an extension field of K and $(c_1, \ldots, c_{r+n-1}) \in F^{(r+n-1)}$ satisfies one of the sets $\Lambda_k(c_1, \ldots, c_{r+n-1})$ (in the sense of (14)) then $\Gamma(c_1, \ldots, c_{r+n-1})$ is solvable for x_n in some algebraic extension field E/F, (2) If $\Gamma(c_1, \ldots, c_{r+n-1})$ is solvable for x_n in any extension field E/F then $\Lambda_k(c_1, \ldots, c_{r+n-1})$ is satisfied for one of the k. Next we use induction on the number of x's to obtain for each Λ_k a set $\{\Gamma_{kl_k} | 1 \le l_k \le u_k\}$ where Γ_{kl_k} is a finite set of polynomials contained in A such that $\{\Gamma_{kl_k}\}$ satisfies the statement of the theorem for the given set of polynomials Λ_k (in $n-1$ x's). We now claim that the set $\{\Gamma_{kl_k} | 1 \le k \le u, 1 \le l_k \le u_k\}$ satisfies the conditions of the theorem for the given set of polynomials Γ. First, suppose $(c_1, \ldots, c_r)$ satisfies $\Gamma_{kl_k}(c_1, \ldots, c_r)$ for some k, l_k. Then we have an algebraic extension field E/F such that $\Lambda_k(c_1, \ldots, c_r)$ is solvable for $x_1, \ldots, x_{n-1}$ in E. Denote a solution by $(c_{r+1}, \ldots, c_{r+n-1})$. Then applying statement (1) we see that $\Gamma(c_1, \ldots, c_r)$ has a solution in an algebraic extension E'/F and this is an algebraic extension of F. Conversely, suppose $\Gamma(c_1, \ldots, c_r)$ is solvable in some extension field E/F. Denote such a solution as $(c_{r+1}, \ldots, c_{r+n-1}, c)$. Then $(c_1, \ldots, c_{r+n-1}) \in E^{(r+n-1)}$ and $\Gamma(c_1, \ldots, c_{n+r-1})$ is solvable for x_n in E. Hence there is a k such that $\Lambda_k(c_1, \ldots, c_{n+r-1})$ is satisfied. This in turn implies that there is an l_k such that $\Gamma_{kl}(c_1, \ldots, c_r)$ is satisfied. This completes the proof. $\square$

There is a second, more classical, method of elimination of unknowns which is based on resultants. We now give the main result of this method and we shall indicate extensions of it in the exercises below. We wish to obtain a criterion for the existence of a common factor of positive degree of two polynomials. We consider the polynomials $f(x) = a_n x^n + a_{n-1} x^{n-1} + \cdots + a_0$, $g(x) = x_m x^m + b_{m-1} x^{m-1} + \cdots + b_0$ in $F[x]$ where F is a field. We assume $m > 0$, $n > 0$, but we shall allow $a_n = 0$ or $b_m = 0$. The result we wish to establish is the following

THEOREM 5.7. Let $f(x) = a_n x^n + a_{n-1} x^{n-1} + \cdots + a_0$, $g(x) = b_m x^m + b_{m-1} x^{m-1} + \cdots + b_0$ where $m, n > 0$ and put

$$
(15) \quad \text{Res}(f, g) = \left|
\begin{array}{cccccccccccc}
a_n & a_{n-1} & \cdot & & \cdot & & \cdot & a_0 & 0 & \cdot & \cdots & \cdot \\
0 & a_n & a_{n-1} & \cdot & & \cdot & & \cdot & a_0 & 0 & \cdots & \cdot \\
\multicolumn{12}{c}{\dotfill} \\
\cdot & & \cdot & 0 & a_n & a_{n-1} & \cdot & \cdot & & \cdots & & a_0 \\
b_m & b_{m-1} & \cdot & & \cdot & b_0 & 0 & \cdot & & \cdots & & \cdot \\
0 & b_m & b_{m-1} & \cdot & & \cdot & b_0 & 0 & \cdot & \cdots & & \cdot \\
\multicolumn{12}{c}{\dotfill} \\
\cdot & & \cdot & 0 & b_m & b_{m-1} & \cdot & \cdot & & \cdot & \cdots & b_0
\end{array}
\right|
\begin{array}{l}
\left.\rule{0pt}{2.5em}\right\} m \text{ rows} \\
\left.\rule{0pt}{2.5em}\right\} n \text{ rows}
\end{array}
$$

Then $\text{Res}(f, g) = 0$ if and only if either $a_n = 0 = b_m$ or $f(x)$ and $g(x)$ have a common factor of positive degree in $F[x]$.

Proof. If $a_n = 0 = b_m$, then the first column of the determinant is 0, so Res $(f, g) = 0$. Next assume that $f(x)$ and $g(x)$ have a common factor $h(x)$ of positive degree and either $a_n \neq 0$ or $b_m \neq 0$. Then $f(x) = f_1(x)h(x)$, $g(x) = g_1(x)h(x)$, and either $f_1(x) \neq 0$ or $g_1(x) \neq 0$, according as $a_n \neq 0$ or $b_m \neq 0$. By symmetry, we may assume $a_n \neq 0$, $f_1(x) \neq 0$. If deg $h(x) = r$, then deg $f_1(x) = n - r$. If $g(x) = 0$, we have $g_1(x) = 0$; otherwise, the relation $f(x)g_1(x) = g(x)f_1(x)$ gives deg $g_1(x) \leq m - r$. In either case we may write $f_1(x) = -c_{n-1}x^{n-1} - c_{n-2}x^{n-2} - \cdots - c_0$, $g_1(x) = d_{m-1}x^{m-1} + d_{m-2}x^{m-2} + \cdots + d_0$ where some $c_i \neq 0$, and we have the relation

(16)
$$(a_n x^n + \cdots + a_0)(d_{m-1}x^{m-1} + \cdots + d_0)$$
$$+ (b_m x^m + \cdots + b_0)(c_{n-1}x^{n-1} + \cdots + c_0) = 0.$$

If we equate the coefficients of x^{m+n-1}, x^{m+n-2}, ..., 1 in (16) we obtain the following equations:

$$a_n d_{m-1} + b_m c_{n-1} = 0$$

$$a_n d_{m-2} + a_{n-1}d_{m-1} + b_m c_{n-2} + b_{m-1}c_{n-1} = 0$$

(17)
$$\cdots\cdots\cdots\cdots\cdots\cdots\cdots\cdots\cdots\cdots\cdots\cdots$$

$$a_0 d_0 + b_0 c_0 = 0.$$

Considering this as a system of linear equations in the c's and d's taken in the order $d_{m-1}, d_{m-2}, \ldots, d_0, c_{n-1}, \ldots, c_0$, we see that the determinant of the coefficients of the c's and d's appearing in (17) is 0, since not all the c's and d's are 0. This determinant $=$ Res (f, g). Hence Res $(f, g) = 0$. Conversely, assume Res $(f, g) = 0$. Then we can retrace the steps through (17) and (16) and conclude that there exist $f_1(x)$, $g_1(x)$ such that $f(x) g_1(x) = g(x)f_1(x)$ where deg $f_1 \leq n - 1$, deg $g_1 \leq m - 1$, and either $f_1 \neq 0$ or $g_1 \neq 0$. Assume $f_1 \neq 0$. If $g_1 = 0$, then $g = 0$ and $b_m = 0$, and either $f(x)$ is a non-zero common factor of f and g or $a_n = 0$. If $g_1 \neq 0$ and $g = 0$ the same argument applies to show that either $a_n = 0 = b_m$ or f and g have a common factor of positive degree. Now assume $g_1 \neq 0$ and $g \neq 0$. Then the relations $f(x)g_1(x) = g(x)f_1(x)$, $f_1 \neq 0, g_1 \neq 0, g \neq 0$, imply $f \neq 0$. Either $a_n = 0 = b_m$, or we may assume $a_n \neq 0$ which implies that deg $f(x) = n$. Since deg $f_1(x) \leq n - 1$, the equation $f(x)g_1(x) = g(x)f_1(x)$ and the factorization of f, f_1, g, g_1 into irreducible factors imply that $f(x)$ and $g(x)$ have a common factor of positive degree. $\square$

We shall call Res (f, g) the *resultant* of f and g (relative to x). If the highest coefficient of f or of g is $\neq 0$, then the vanishing of Res (f, g) is a polynomial equation with integer coefficients in the a_i and b_j, which is equivalent to the statement that f and g have a common factor of positive degree.

EXERCISES

1. Show that if $f(x) = x^n + a_{n-1}x^{n-1} + \cdots + a_0$ and $f'(x) = nx^{n-1} + (n-1)a_{n-1}x^{n-2} + \cdots + a_1$, then Res $(f, f') = (-1)^{n(n-1)/2}d$ where d is the discriminant of $f(x)$ (see section 4.8, p. 258).

2. Use the theorem on resultants to obtain a proof of Theorem 5.6 for the case of two equations $F_1(t_1, \ldots, t_r; x)$ and $F_2(t_1, \ldots, t_r; x)$ and $G = 1$.

3. Let $f_1(x), \ldots, f_m(x) \in F[x]$ and write $f_i(x) = a_{n_i,i}x^{n_i} + a_{n_i-1,i}x^{n_i-1} + \cdots + a_{0i}$. Let n be an integer $\geq$ every n_i and let

$$g_1(x) = x^{n-n_1}f_1(x), \ldots, g_m(x) = x^{n-n_m}f_m(x)$$
$$g_{m+1}(x) = (x-1)^{n-n_1}f_1(x), \ldots, g_{2m}(x) = (x-1)^{n-n_m}f_m(x).$$

Show that the $f_i(x)$, $1 \leq i \leq m$, have a common factor of positive degree if and only if the $g_j(x)$, $1 \leq j \leq 2m$, have such a factor. Adjoin $4m$ indeterminates $u_1, \ldots, u_{2m}, v_1, \ldots, v_{2m}$ and let $E = F(u_1, \ldots, u_{2m}, v_1, \ldots, v_{2m})$. Let $u(x) = \sum_1^{2m} u_j g_j(x)$, $v(x) = \sum_1^{2m} v_j g_j(x)$ and form Res $(u(x), v(x)) \in F[u_1, \ldots, u_{2m}, v_1, \ldots, v_{2m}]$. Prove that Res $(u(x), v(x)) = 0$ if and only if either all $a_{n_i} = 0$, $1 \leq i \leq m$, or $f_1(x), \ldots, f_m(x)$ have a common factor of positive degree.

4. Show that the system of equations and inequations of the form (13) is solvable if and only if the following system involving $x_1, \ldots, x_{n+1}$ is solvable:

$$F_1(c_1, \ldots, c_r; x_1, \ldots, x_n) = \cdots = F_m(c_1, \ldots, c_r; x_1, \ldots, x_n)$$
$$= x_{n+1}G(c_1, \ldots, c_r; x_1, \ldots, x_n) - 1 = 0.$$

(Note that this procedure gets rid of inequations.)

5.5 DECISION METHOD FOR AN ALGEBRAIC CURVE

In this section we shall give a method, due to A. Seidenberg, for deciding whether or not a given equation $f(x, y) = 0$, $f(x, y) \in R[x, y]$, has a solution in $R^{(2)}$. In other words, does the algebraic curve $f(x, y) = 0$ have real points? (E.g., $x^2 + y^2 = 0$ has, but $x^2 + y^2 = -1$ does not.) The underlying idea of Seidenberg's method is based on the following simple observation: If $f(x, y) = 0$ has a real point, then it has a real point (a, b) nearest the origin. Then it can be shown that (a, b) is also a solution of $g(x, y) = x(\partial f/\partial y) - y(\partial f/\partial x) = 0$ and this implies that a is a root of the polynomial $h(x)$, which is the resultant with respect to y of $f(x, y)$ and $g(x, y)$. Hence the existence of a solution in $R^{(2)}$ of $f(x, y) = 0$ implies the existence of a root in R of $h(x) = 0$, a fact that can be decided by Sturm's method. We shall see also that the argument can be reversed provided we replace the origin by a suitable point and x, y by another pair of generators x', y'—in other words, if we make a suitable affine transformation of coordinates in the vector space $R^{(2)}$. In this way we shall obtain an algorithm for testing the solvability in $R^{(2)}$ of $f(x, y) = 0$.

The first two steps we have indicated are readily attained if $R = \mathbb{R}$ the field of real numbers. In this case, if $f(x, y)$ has a solution (x_0, y_0), then we consider the set of points (x, y) in $\mathbb{R}^{(2)}$ such that $f(x, y) = 0$ and $x^2 + y^2 \leq x_0^2 + y_0^2$. This is a closed and bounded subset of $\mathbb{R}^{(2)}$; hence it is compact and consequently it contains a point (a, b) nearest the origin. By calculus, either (a, b) is a point at which $(\partial f/\partial x)_{(a,b)} = 0 = (\partial f/\partial y)_{(a,b)}$ or the line joining the origin to (a, b) is normal to the given curve:

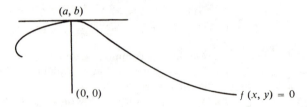

In this case (a, b) is a multiple of the normal vector $((\partial f/\partial x)_{(a,b)}, (\partial f/\partial y)_{(a,b)})$ to the curve. In any case we have $a(\partial f/\partial y)_{(a,b)} - b(\partial f/\partial x)_{(a,b)} = 0$, so (a, b) is a root of $g(x, y) = x(\partial f/\partial y) - y(\partial f/\partial x) = 0$.

We now proceed to establish these results, in two lemmas, for any real closed field.

LEMMA 1. *Let $f(x, y) \in R[x, y]$, x, y indeterminates, R a real closed field. Then if $f(x, y) = 0$ has a solution in R, it has a solution (a, b) nearest the origin.*

Proof. We consider the intersection in the space $R^{(2)}$ of the locus C of $f(x, y) = 0$ with circles $x^2 + y^2 = c^2$, $c \geq 0$. Our hypothesis implies that for some c we have a non-vacuous intersection, and we have to show that the set S of $c \geq 0$ such that C meets $x^2 + y^2 = c^2$, has a minimum. We now consider the polynomials $f(x, y)$ and $x^2 + y^2 - t^2$ in $R[x, y, t]$ where x, y, t are indeterminates, and we form their resultant with respect to y (that is, regarding these as polynomials in y). This resultant $g(t, x) \in R[t, x]$. We claim that the set S defined before is the same as the set of $c \geq 0$ such that $g(c, x)$ has a root in the interval $[-c, c]$. First, if $c \in S$ and (a, b) is a point of intersection of the circle $x^2 + y^2 = c^2$ and the curve C, then $f(a, y)$ and $y^2 + a^2 - c^2$ have the common factor $y - b$. Hence $g(c, a) = 0$ so $g(c, x)$ has the root $a \in R$. Moreover, $-c \leq a \leq c$. Conversely, assume that for $c \geq 0$, $g(c, x)$ has a root a in R satisfying $-c \leq a \leq c$. Since the leading coefficient of y in $y^2 + a^2 - c^2$ is 1, it follows from Theorem 5.7 that $y^2 + a^2 - c^2$ and $f(a, y)$ have a common factor in $R[y]$. Since the factors of $y^2 + a^2 - c^2$ are $y \pm b$ where $b = (c^2 - a^2)^{1/2}$, it follows that (a, b) or $(a, -b)$ is a point of intersection of C and $x^2 + y^2 = c^2$. Hence $c \in S$. Let S' be the subset of S of c such that $g(c, \pm c) \neq 0$. Thus S' is the set of $c \in R$ such that $g(c, x)$ has a root in the open interval $(-c, c)$. By the remarks following Theorem 5.5, we see that S' is the union of a finite number of sets defined by finite systems of polynomial equations $p(c) = 0$, inequations $q(c) \neq 0$, and inequalities $r(c) > 0$, where $p(t), q(t), r(t) \in R[t]$. If we examine the loci in R of $p(t) = 0$, $q(t) \neq 0$ and $r(t) > 0$, we see that the set of points c satisfying the system of conditions $c \geq 0$, $p(c) = 0$, $q(c) \neq 0$, $r(c) > 0$ is a union (possibly vacuous) of a finite number of intervals which may be open, closed, half open, single points, or extend to $+\infty$. It follows that S' is a subset of R of this type. Since the set of $c \geq 0$ such that $g(c, \pm c) = 0$ is either finite or all $c \geq 0$ it is clear that S has the same structure as S'. The result will now follow by showing that the complement of S in the set of non-negative elements of R is the union of open intervals; for this will imply that S is the union of a finite number of closed intervals and hence has a minimal element. Thus let $d \geq 0$, $d \notin S$. Then $g(d, x) = 0$, $-d \leq x \leq d$, has no solution in R. Write $g(t, x)$ as a polynomial in x and $t - d$: $g(x) = g_0(x) + g_1(x)(t - d) + \cdots + g_m(x)(t - d)^m$ where $g_i(x) \in R[x]$. Then $g_0(u) \neq 0$ if $-d \leq u \leq d$ and hence there exists a $d' > d$ such that $g_0(u) \neq 0$ if $-d' \leq u \leq d'$. Then there exist $b > 0$, $B > 0$ such that $|g_0(u)| \geq b$, $|g_i(u)| \leq B$ if $i \geq 1$ for every u in $[-d', d']$ (exercise 7, p. 311). Then if $|c - d| < \frac{1}{2}$, $|c - d| < b/4B$ and $u \in [-d', d']$, we have

$$|g(c, u)| \geq |g_0(u)| - |g_1(u)(c - d) + \cdots + g_m(u)(c - d)^m|$$

$$\geq b - 2B|c - d| > b - \frac{b}{2} = \frac{b}{2}.$$

It follows that every c such that $c \geq 0$, $|c - d| \leq \frac{1}{2}$, $|c - d| < b/4B$, $c \leq d'$ is in the complement of S in the set of non-negative numbers. Thus we see that if d is any point in this complement, then there exists an open interval containing d that is contained in the complement. This completes the proof of the lemma. $\square$

As in the classical case of the field of real numbers, a point (a, b) on $C : f(x, y) = 0$ is called a *simple point* if

$$\left(\left(\frac{\partial f}{\partial x} \right)_{(a,b)}, \left(\frac{\partial f}{\partial y} \right)_{(a,b)} \right) \neq (0, 0).$$

Then the normal vector to C at (a, b) is $((\partial f/\partial x)_{(a,b)}, (\partial f/\partial y)_{(a,b)})$ and the *tangent line* to the curve at (a, b) has the equation

$$\left(\frac{\partial f}{\partial x}\right)_{(a,b)} (x - a) + \left(\frac{\partial f}{\partial y}\right)_{(a,b)} (y - b) = 0.$$

Now let (a, b) be a point on C nearest the origin. We wish to show that $b(\partial f/\partial x)_{(a,b)} - a(\partial f/\partial y)_{(a,b)} = 0$. This is clear if $(a, b) = (0, 0)$ or if (a, b) is not a simple point. Otherwise, the equation states that the vector joining $(0, 0)$ to (a, b) and the normal vector to C at (a, b) are linearly dependent; equivalently, C and the circle with center at the origin and radius $(a^2 + b^2)^{1/2}$ have the same tangent line at the point (a, b). If this were not the case, then the tangent to C at (a, b) would contain interior points of the circle while C itself does not (since (a, b) is the point on C nearest to $(0, 0)$). We shall show that this situation is impossible and this will prove that $b(\partial f/\partial x)_{(a,b)} - a(\partial f/\partial y)_{(a,b)} = 0$. Thus our result will follow from

LEMMA 2. *Let p be a point of intersection (in $R^{(2)}$) of a circle and a curve $C\!:\!f(x, y) = 0$. Assume that p is a simple point of C and the tangent at p to C contains points interior to the circle. Then C itself has points interior to the circle.*

Proof. By a suitable choice of axis we may take $p = (0, 0)$ and the tangent to C at p to be the x-axis. Then $f(0, 0) = 0$ and $(\partial f/\partial x)_{(0,0)} = 0$, and we may assume that $(\partial f/\partial y)_{(0,0)} = 1$. The center of the circle is not on the y-axis, so we may denote it as (a, b) with $a \neq 0$. We have

$$f(x, y) = f(0, 0) + \left(\frac{\partial f}{\partial x}\right)_{(0,0)} x + \left(\frac{\partial f}{\partial y}\right)_{(0,0)} y$$
$$+ \frac{1}{2!}\left[\left(\frac{\partial^2 f}{\partial x^2}\right)_{(0,0)} x^2 + 2\left(\frac{\partial^2 f}{\partial x\, \partial y}\right)_{(0,0)} xy + \left(\frac{\partial^2 f}{\partial y^2}\right)_{(0,0)} y^2\right] + \ldots,$$

taking into account the conditions on $f(x, y)$ we can write $f(x, y) = y(1 + h(x, y)) + g(x)$ where $h(0, 0) = 0$ and $g(x)$ is a polynomial in x divisible by x^2. Since $h(0, 0) = 0$ we may choose a $\delta > 0$ such that $|h(x, y)| \leq \frac{1}{2}$ if $|x| \leq \delta$ and $|y| \leq \delta$. Then $\frac{1}{2} \leq 1 + h(x, y) \leq \frac{3}{2}$ and $\delta(1 + h(x, \delta))$ lies between $\frac{1}{2}\delta$ and $\frac{3}{2}\delta$, and $-\delta(1 + h(x, -\delta))$ is between $-\frac{1}{2}\delta$ and $-\frac{3}{2}\delta$ for all x satisfying $|x| \leq \delta$. Since $g(0) = 0$ there exists a δ', $0 < \delta' < \delta$ such that $f(x, \delta) = \delta(1 + h(x, \delta)) + g(x) > 0$ and $f(x, -\delta) < 0$ if $|x| \leq \delta'$. Then for every x_0, $|x_0| \leq \delta'$ there exists a y_0 in $[-\delta, \delta]$ such that $f(x_0, y_0) = 0$. Then $y_0 = -g(x_0)(1 + h(x_0, y_0))^{-1}$ and

$$(a - x_0)^2 + (b - y_0)^2$$

$$= (a - x_0)^2 + \left(b + \frac{g(x_0)}{1 + h(x_0, y_0)}\right)^2$$

$$= a^2 + b^2 - 2ax_0 + x_0{}^2 + \frac{2bg(x_0)}{1 + h(x_0, y_0)} + \frac{g(x_0)^2}{(1 + h(x_0, y_0))^2}.$$

Since $g(x_0)$ is divisible by $x_0{}^2$, it is clear that if we take x_0 sufficiently small and of the same sign as a (so that $ax_0 > 0$), then $(a - x_0)^2 + (b - y_0)^2 < a^2 + b^2$. Then (x_0, y_0) is a point on C interior to the given circle. $\quad\square$

We have now shown that if $C:f(x, y) = 0$ contains a point in $R^{(2)}$, then the curve C and the curve $D: y(\partial f/\partial x) - x(\partial f/\partial x) = 0$ have a common point in $R^{(2)}$. If we replace the origin by the point (c, d), then we see also that if C has a point in $R^{(2)}$ then C and $D:(y - d)(\partial f/\partial x) - (x - c)(\partial f/\partial y) = 0$ have a common point in $R^{(2)}$.

We shall now apply this to obtain Seidenberg's method for deciding the solvability in $R^{(2)}$ of $f(x, y) = 0$. First, we determine by the Euclidean algorithm a g.c.d. $d(x)$ of the coefficients of the powers of y in $f(x, y)$ and write $f(x, y) = d(x)f_1(x, y)$ where $f_1(x, y)$ is not divisible by a polynomial of positive degree in x alone. Evidently, $f(x, y) = 0$ is solvable if and only if either $d(x) = 0$ or $f_1(x, y)$ is solvable. This reduces the consideration to polynomials in $R[x, y] = (R[x])[y]$ that are primitive as polynomials in y (over $R[x]$) in the sense that they are not divisible by polynomials of positive degree in x alone. We obtain next a reduction to polynomials without multiple factors. For this purpose we calculate, by the Euclidean algorithm, a g.c.d. in $R(x)[y]$ of $f(x, y)$ and $(\partial/\partial y)f(x, y)$ where $R(x)$ is the field of fractions of $R[x]$. We can write the g.c.d. as $u(x)v(x)^{-1}d(x, y)$ where $d(x, y) \in R[x, y]$ and is y-primitive. Then it follows from Gauss' lemma (p. 152) that $d(x, y)$ is a factor of $f(x, y)$ in $R[x, y]$. Moreover, $g(x, y) = f(x, y)d(x, y)^{-1}$ has the same irreducible factors as $f(x, y)$ and has no multiple factors (exercise 1, p. 233). We may now assume that $f(x, y)$ is y-primitive and has no multiple factors. The latter condition implies that $f(x, y)$ and $(\partial/\partial y)f(x, y)$ have no common factors in $R(x)[y]$ of positive y-degree.

Let t be an additional indeterminate and form the resultant $h(t, x)$ of $f(x, y)$ and $g(t, x, y) = y(\partial f/\partial x) - (x - t)(\partial f/\partial y)$ regarded as polynomial in y. It is clear from the definition (15) in Theorem 5.7 that this is in $R[t, x]$. We claim that $h(t, x) \neq 0$. Otherwise, $h(c, x) = 0$ for all $c \in R$ and hence $f(x, y)$ and $g(c, x, y) = y(\partial f/\partial x) - (x - c)(\partial f/\partial y)$ have a common factor in $R(x)[y]$ of positive y-degree. This follows from the theorem on resultants (p. 325) since we may assume that the coefficient of the highest power of y is a non-zero element of $R[x]$. The fact that $f(x, y)$ and $g(c, x, y)$ have a non-trivial common factor in $R(x)[y]$ implies that they have a non-trivial factor in $R[x, y]$. Since up to unit multipliers $f(x, y)$ has only a finite number of irreducible factors in $R[x, y]$ we see that there exist $c_1 \neq c_2$ such that $g(c_1, x, y)$, $g(c_2, x, y)$, and $f(x, y)$ have a common factor $d(x, y)$ in $R[x, y]$ of positive degree. Then $f(x, y)$ and $(\partial f/\partial y) = (c_1 - c_2)^{-1}[g(c_1, x, y) - g(c_2, x, y)]$ have a non-trivial common factor. This contradicts our hypothesis. Hence $h(t, x) \neq 0$.

We now choose a $c \in R$ so that $h(x) = h(c, x) \neq 0$ and we write $g(x, y) = g(c, x, y) = y(\partial f/\partial x) - (x - c)(\partial f/\partial y)$. Since $h(x)$ is the resultant of $f(x, y)$ and $g(x, y)$ these two polynomials have no common factor of positive degree in y, and since $f(x, y)$ is primitive they have no common factor of positive degree in x alone. Hence $f(x, y)$ and $g(x, y)$ have no common factors other than units in $R[x, y]$. It now follows also that if $k(y)$ denotes the resultant in $R(y)[x]$ of $f(x, y)$ and $g(x, y)$ then $k(y)$ is a non-zero polynomial in y.

We have seen that if $f(x, y) = 0$ has a solution in $R^{(2)}$, then $f(x, y) = 0$ and $g(x, y) = 0$ have a common solution $(a, b) \in R^{(2)}$. Then $f(a, y)$ and $g(a, y)$ have the common factor $y - b$ and this implies that the resultant $h(a)$ of $f(a, y)$ and $g(a, y)$ is 0. Thus we see that if $f(x, y) = 0$ has a solution in $R^{(2)}$ then $h(x) = 0$ has a root in R. What about the converse? We shall now show that this is the case, provided that we choose the generators x and y of $R[x, y]$ suitably. We remark that in place of x and y we can use any $x' = a_{11}x + a_{12}y$, $y' = a_{21}x + a_{22}y$ where the $a_{ij} \in R$ and $\det(a_{ij}) \neq 0$.

To achieve our objective of reducing the problem of deciding the solvability of $f(x, y) = 0$ in $R^{(2)}$ to that of $h(x) = 0$ in R we now work in the algebraically closed field $A = R(\sqrt{-1})$. Let V be the intersection in $A^{(2)}$ of $f(x, y) = 0$ and $g(x, y) = 0$. If $(a, b) \in V$ then, as above, $h(a) = 0$ for the resultant relative to y of $f(x, y)$ and $g(x, y)$. Similarly, $k(b) = 0$ for the resultant $k(y)$ of $f(x, y)$ and $g(x, y)$ relative to x. Since $h(x) \neq 0$ and $k(y) \neq 0$ the equations $h(x) = 0$ and $k(y) = 0$ have only a finite number of roots in A. Hence V is a finite set. We have seen that if $C: f(x, y) = 0$ has a point in $R^{(2)}$, then V has such a point and $h(x)$ has a root in R. Conversely, suppose $h(x)$ has a root a in R. If a is not a root of the polynomial $l(x)$, which is the coefficient of the highest power of y in $f(x, y)$, then $h(a) = 0$ implies the existence of a $b \in A$ such that $(a, b) \in V$. If $b \in R$ then the point (a, b) is on V and hence on C. Otherwise, $(a, \bar{b}) \in V$ where $\bar{b}$ is the conjugate of b under the automorphism $\neq 1$ of A/R. Since $\bar{b} \neq b$ we have two points on V, (a, b) and $(a, \bar{b})$, with the same abscissa. Thus we see that if no (a, b) on V satisfies $l(a) = 0$ and no two distinct points of V have the same abscissa then the solvability of $h(x) = 0$ in R implies that of $f(x, y) = 0$ in $R^{(2)}$.

We shall now arrange, by a suitable choice of coordinates, that these two conditions are fulfilled. Let m be a non-zero element of R and put $x' = m^{-1}x - y$, $y' = y$, so $x = m(x' + y')$ and $y = y'$; hence $R[x, y] = R[x', y']$ and $f(x, y) = f(m(x' + y'), y')$. Let $f_n(x, y)$ be the homogeneous part of highest degree n (>0) in x and y in the polynomial $f(x, y)$. Then the coefficient of y'^n in $f(m(x' + y), y')$

is $f_n(m, 1)$. Since $f_n(x, 1) \neq 0$ we can avoid the roots of $f_n(x, 1) = 0$ and choose $m \in R$ so that $f_n(m, 1) \neq 0$. Since the total degree of $f(x, y)$ is n and $f_n(x, y)$ is the homogeneous part of degree n it follows that the coefficient of the highest power of y', that is, of y'^n in $f(m(x' + y'), y')$ is the constant $f_n(m, 1) \neq 0$. This takes care of the first condition. To take care of the second we calculate, via the Euclidean alogorithm, a g.c.d. $d(x)$ of $h(x)$ and its derivative $h'(x)$. Dividing $h(x)$ by $d(x)$ we obtain a polynomial $h_1(x)$ having simple roots $r_1, r_2, \ldots, r_u$, the same as those of $h(x)$. Similarly, we calculate a polynomial $k_1(y)$ having simple roots s_1, $s_2, \ldots, s_v$, the same as those of $k(y)$. Dividing out by the leading coefficients we may assume h_1 and k_1 are monic. Now form the polynomial

$$
(18) \qquad \prod_{\substack{i \neq i' \\ j \neq j'}} [(y_j - y_{j'})x - (x_i - x_{i'})]
$$

where the x's and y's are indeterminates and $i, i' = 1, \ldots, u; j, j' = 1, \ldots, v$. We shall now show by a two-fold application of the theorem on symmetric polynomials (Theorem 2.20, p. 139) that we can express the foregoing polynomial as a polynomial in x with coefficients which are polynomials in the elementary symmetric polynomials of the x_i and the y_j with coefficients in $\mathbb{Z}$. First, we consider the following polynomial in indeterminates x_i and t with integer coefficients:

$$
(19) \qquad
\begin{aligned}
l(x_1, \ldots, x_n; t) &= \prod_{i \neq i'} [t - (x_i - x_{i'})] \\
&= t^m - l_1(x_1, \ldots, x_n)t^{m-1} + l_2(x_1, \ldots, x_n)t^{m-2} - \cdots
\end{aligned}
$$

where $m = n(n - 1)$ and the $l_j \in \mathbb{Z}[x_1, \ldots, x_n]$. Clearly $l(x_1, \ldots, x_n; t)$ is invariant under arbitrary permutations of the x_i. Hence the l_i are symmetric polynomials in the x_i with the integer coefficients. Consequently, we can write $l_j(x_1, \ldots, x_n)$ (uniquely) as a polynomial in the elementary symmetric polynomials $p_1 = \sum x_i$, $p_2 = \sum_{i<j} x_i x_j, \ldots, p_n = x_1 \ldots x_n$. Thus $l_j(x_1, \ldots, x_n) = m_j(p_1, \ldots, p_n) \in \mathbb{Z}[p_1, \ldots, p_n]$ and $l(x_1, \ldots, x_n; t) = t^m - m_1(p_1, \ldots, p_n)t^{m-1} + m_2(p_1, \ldots, p_n)t^{m-2} - \cdots$. Next we consider the polynomial (18). Clearly, we can write this as $\prod_{j \neq j'} l(x_1, \ldots, x_u; (y_j - y_{j'})x)$. Using the expression for $l(x_1, \ldots, x_n; t)$ we obtain

$$
\begin{aligned}
\prod_{\substack{i \neq i' \\ j \neq j'}} & [(y_j - y_{j'})x - (x_i - x_{i'})] \\
&= \prod_{j \neq j'} [(y_j - y_{j'})^m x^m - m_1(p_1, \ldots, p_u)(y_j - y_{j'})^{m-1}x^{m-1} \\
&\qquad + m_2(p_1, \ldots, p_u)(y_j - y_{j'})^{m-2}x^{m-2} - \cdots] \\
&= z_0 x^{m^2} - z_1 x^{m^2-1} + z_2 x^{m^2-2} - \cdots
\end{aligned}
$$

where $z_k \in \mathbb{Z}[p_1, \ldots, p_u][y_1, \ldots, y_v]$. Since this polynomial is unchanged under permutation of the y's, the z_k are symmetric in the y's. Hence

$$z_k \in \mathbb{Z}[p_1, \ldots, p_u][q_1, \ldots, q_v] = \mathbb{Z}[p_1, \ldots, p_u, q_1, \ldots, q_v]$$

where $q_1 = \sum y_i, q_2 = \sum_{i<j} y_i y_j, \ldots$ are the elementary symmetric polynomials in the y's. This shows that (18) can be written as a polynomial in x and the p_i and q_j with integer coefficients. Moreover, all of this can be done constructively since the method given in section 2.13 of proving the fundamental theorem on symmetric polynomials was constructive.

If we now replace the p_i and q_i appearing in the formula for (18) by the corresponding coefficients of $h_1(x)$ and $k_1(y)$ respectively we obtain a polynomial $p(x) \in R[x]$ whose roots are the elements

$$(r_i - r_{i'})(s_j - s_{j'})^{-1} \in A$$

where $i \neq i', j \neq j'$ and the ranges of these are as before. We now choose m to avoid also the roots of $p(x)$ (as well as of $f_n(x, 1)$). Consider the set of points (r_i, s_j) in $A^{(2)}$. This contains V, and no two distinct points in this set have the same abscissa in the (x', y')-coordinate system since (x, y) is the point $(m^{-1}x - y, x)$ in the (x', y')-system and $m^{-1}r_i - s_j \neq m^{-1}r_{i'} - s_{j'}$ if $(i, j) \neq (i', j')$.

It now follows that if we replace $f(x, y)$ by $f(m(x + y), y)$ and $g(x, y)$ by $g(m(x + y), y)$ the conditions are fulfilled which insure that (the new) $f(x, y) = 0$ is solvable in $R^{(2)}$ if and only if the resultant $h(x)$ relative to y of $f(x, y)$ and $g(x, y)$ has a root in R. Since this can be decided by Sturm's theorem we have achieved our goal of giving a recipe for deciding the solvability of the original equation.

In the next section we shall use an inductive procedure for polynomials in a number of parameters and variables. The inductive step will require a small extension of the decision procedure we have just described, namely, we shall need to consider an equation $f(x, y) = 0$ restricted by an inequation $g(x) \neq 0$. As before, we may assume $f(x, y)$ is primitive as a polynomial in y. Also to avoid trivialities we assume $\deg_x f(x, y) > 0$ and $\deg g(x) > 0$. Let $t(y)$ be the resultant with respect to x of $f(x, y)$ and $g(x)$. Then $t(y) \neq 0$ since $f(x, y)$ is y-primitive. Choose c in R so that $t(c) \neq 0$ and replace $f(x, y)$ by $f_1(x, y) = f(x, y + c)$. Clearly $f(x, y) = 0$, $g(x) \neq 0$ is solvable if and only if $f_1(x, y) = 0$, $g(x) \neq 0$ is solvable. The resultant relative to x of $f_1(x, y)$ and $g(x)$ is $t(y + c)$ which is $\neq 0$ for $y = 0$. Hence $g(x)$ and $f_1(x, 0)$ are relatively prime in $R[x]$. Now put $f_2(x, y) = f_1(x, g(x)y)$. Then we claim that $f_1(x, y) = 0$, $g(x) \neq 0$ is solvable in $R^{(2)}$ if and only if $f_2(x, y) = 0$ is solvable in $R^{(2)}$. Suppose (a, b) satisfies the first system. Then $f_2(a, g(a)^{-1}b) = f_1(a, b) = 0$. On the other hand, if $f_2(a, c) = f_1(a, g(a)c) = 0$ then $g(a) \neq 0$ since otherwise $g(a) = 0$ and $f_1(a, 0) = 0$ contrary to the fact that $g(x)$ and $f_1(x, 0)$ are relatively prime. Thus $(a, b = g(a)c)$ satisfies $f_1(x, y) = 0$, $g(x) \neq 0$.

5.6 TARSKI'S THEOREM

We now consider a finite system φ of equations, inequations, and inequali-
ties of the form $F(t_1, \ldots, t_r; x_1, \ldots, x_n) = 0$, $G(t_1, \ldots, t_r; x_1, \ldots, x_n) \neq 0$,
$H(t_1, \ldots, t_r; x_1, \ldots, x_n) > 0$ where the F, G, and H are polynomials with integer coefficients. We wish to show that we can replace φ by a finite set of systems ψ_j of the same type involving no x's, such that if R is any real closed field,
then φ has a solution for the x's in R for the values $t_i = c_i \in R$ if and only if
the c_i satisfy all the conditions of one of the systems ψ_j. We shall prove this by
eliminating all but one of the x's one by one, using the method of the last section. Then we can apply the parameterized version of Sturm's theorem. To
begin with, however, we reverse the direction we wish to take and replace the
system φ by a single equation $F = 0$ at the expense of introducing additional
x's. We observe first that we can replace a finite set of equations $F_i = 0$ by a
single one, $\sum F_i^2 = 0$, and a finite set inequations $G_i \neq 0$ by a single one,
$\prod G_i \neq 0$. An inequation $G \neq 0$ is equivalent to $G^2 > 0$ and the solvability of
$H > 0$ is equivalent to that of $H^2 z^2 - 1 = 0$ where z is a new indeterminate.
Using these reductions we may assume that φ consists of a single equation
$F(t_1, \ldots, t_r; x_1, \ldots, x_n) = 0$. For the inductive step of the proof we need to
carry along an inequation as well as an equation. This appears in the following

THEOREM 5.8.   *Let* $F(t_i; x, y) \in \mathbb{Z}[t_1, \ldots, t_r; x, y]$, $G(t_i; x) \in \mathbb{Z}[t_1, \ldots, t_r; x]$,
t_i, x, y *indeterminates. Then we can determine in a finite number of steps a finite
set of pairs of polynomials* $(F_j(t_i; x), G_j(t_i))$, $F_j \in \mathbb{Z}[t_i; x]$, $G_j \in \mathbb{Z}[t_i]$, $1 \leq j \leq h$,
such that if R is any real closed field, then the point $(c_1, \ldots, c_r) \in R^{(r)}$ *has the
property that*

$$(20) \qquad\qquad F(c_i; x, y) = 0, \qquad G(c_i; x) \neq 0$$

*is solvable for x and y in R if and only if one of the systems of equations and
inequations*

$$(21) \qquad\qquad F_j(c_i; x,) = 0, \qquad G_j(c_i) \neq 0, \qquad 1 \leq j \leq h$$

is solvable in R.

The proof will consist of a finite sequence of constructions of covers and
polynomials corresponding to the steps in Seidenberg's decision method. Since

we are interested exclusively in real closed fields we may assume that the members Δ of a cover δ of $A = \mathbb{Z}[t_1, \ldots, t_r]$ consist of a single equation $d = 0$ and a single inequation $l \neq 0$ where $l, d \in A$. If $\gamma = \{\Gamma_k\}$ is a cover we can use it to define a refinement δ' of δ in which the term Δ is replaced by $\Delta^{(1)}, \Delta^{(2)}, \ldots$ where $\Delta^{(k)}$ has as equation the sum of the squares of the equations of Δ and of Γ_k and the inequation which is the product of that of Δ and of Γ_k (see p. 319). Then, for any real closed field R, $\Delta^{(k)}(R) = \Delta(R) \cap \Gamma_k(R)$ and $\Delta(R) = \bigcup \Delta^{(k)}(R)$. The individual steps of the proof will be of the following type: we are given a cover δ and for $\Delta \in \delta$ a pair of polynomials $(F(t_i; x, y), G(t_i; x))$ in $A[x, y]$ and $A[x]$ respectively. Then we construct a cover δ', as indicated, and for each $\Delta^{(k)}$, a finite set of pairs of polynomials $(F_{kj}(t_i; x, y), G_{kj}(t_i; x))$ such that for $(c_1, \ldots, c_r) \in \Delta^{(k)}(R)$, $F(c_i; x, y) = 0$, $G(c_i; x) \neq 0$ is solvable in R if and only if one of the pairs $F_{jk}(c_i; x, y) = 0$, $G_{jk}(c_i; x) \neq 0$ is solvable in R. This permits us to replace the triple (Δ, F, G) by the various triples $(\Delta^{(k)}, F_{jk}, G_{jk})$. After a finite number of steps of this type we eventually obtain a cover $\omega = \{\Omega_j\}$ and a finite set of pairs of polynomials (F_{jk}, G_{jk}) such that $F_{jk} \in A[x]$, $G_{jk} \in A$, and if $(c_1, \ldots, c_r) \in \Omega_j(R)$, then the initially given system $F(c_i; x, y) = 0$, $G(c_i; x) \neq 0$ is solvable in R if and only if one of the systems $F_{jk}(c_i; x) = 0$, $G_{jk}(c_i) \neq 0$ is solvable. Then we put $F_{jk}^* = F_{jk}^2 + d_j^2$, $G_{jk}^* = G_{jk}l_j$. It is easily seen that the set of pairs (F_{jk}^*, G_{jk}^*) satisfy the conditions for the pairs (F_j, G_j) stated in the theorem.

We observe next that given a finite set of polynomials $\{F, G, \ldots, H\} \subset A[x]$ we can construct a cover $\delta = \{\Delta\}$ of A and corresponding polynomials which are *appropriate for the* g.c.d. *of* $\{F, G, \ldots, H\}$ in the following sense: (i) For each $\Delta \in \delta$ we have a polynomial $D(t_i; x) \in A[x]$ such that for any real closed field R and any $(c_1, \ldots, c_r) \in \Delta(R)$, $D(c_i; x)$ is a g.c.d. in $R[x]$ of $F(c_i; x)$, $G(c_i; x)$, $\ldots$, $H(c_i; x)$. (ii) For any Δ, either $D(c_i; x) = 0$ for all $(c_1, \ldots, c_r) \in \Delta(R)$, or $D(c_i; x) \neq 0$ for all such $(c_1, \ldots, c_r)$. In the latter case we have polynomials $F_1, G_1, \ldots, H_1 \in A[x]$ such that for $(c_1, \ldots, c_r) \in \Delta(R)$, $F_1(c_i; x)$, $G_1(c_i; x), \ldots$, $H_1(c_i; x)$ differ by a nonzero multiplier in R from $F(c_i; x)D(c_i; x)^{-1}$, $G(c_i; x)D(c_i; x)^{-1}, \ldots$, $H(c_i; x)D(c_i; x)^{-1}$ respectively. To obtain (i) we note that the result follows by induction on the number of polynomials if any of the given polynomials is 0. Also the result is clear if there is just one polynomial and it follows from the lemma on p. 319 if there are just two non-zero polynomials. Now assume the number of non-zero polynomials exceeds two. Using induction, we may assume that we have constructed a cover γ of A and corresponding polynomials E appropriate for the g.c.d. of all but the polynomial H in the given set. Next for each of the sets $\{E, H\}$ we can construct a cover and polynomials appropriate for the g.c.d. of $\{E, H\}$. Then we can obtain (i) by refinement as in the proof of the lemma on p. 319. Moreover, we may assume, by refining a cover satisfying (i) that for any Δ in the refined cover either

$D(c_i; x) = 0$ for all $(c_1, \ldots, c_r) \in \Delta(R)$, or we have $D(t_i; x) = v_k(t_1, \ldots, t_r)x^k + v_{k-1}(t_1, \ldots, t_r)x^{k-1} + \cdots + v_0(t_1, \ldots, t_r)$ and $v_k(c_1, \ldots, c_r) \neq 0$ for all $(c_1, \ldots, c_r) \in \Delta(R)$. By the division algorithm, we can obtain a non-negative integer e and polynomials $F_1, G_1, \ldots, H_1; S, T, \ldots, U$ in $A[x]$ such that $v_k^e F = F_1 D - S$, $v_k^e G = G_1 D - T, \ldots, v_k^e H = H_1 D - U$ and $\deg_x S, \deg_x T, \ldots, \deg_x U$ are all $< k$. Since $D(c_i; x)|F(c_i; x), D(c_i; x)|G(c_i; x), \ldots, S(c_i; x) = T(c_i; x) = \cdots = 0$. Hence $F_1, G_1, \ldots, H_1$, satisfy the condition given in (ii).

We shall require also an extension of this result to the case of two polynomials in two indeterminates x and y (in addition to the t_i). Suppose we are given two polynomials $F(t_i; x, y)$, $G(t_i; x, y)$ in $A[x, y]$, $A = \mathbb{Z}[t_1, \ldots, t_r]$. Then we can construct a cover $\delta = \{\Delta\}$ and polynomials which are *appropriate for the* g.c.d. *of F and G in* $R(x)[y]$ in the sense that: (1) For any $\Delta \in \delta$ we have a polynomial $D(t_i; x, y)$ such that if $(c_1, \ldots, c_r) \in \Delta(R)$, then $D(c_i; x, y)$ is a g.c.d. in $R(x)[y]$ of $F(c_i; x, y)$ and $G(c_i; x, y)$. (2) For any Δ, either $D(c_i; x, y) = 0$ for all $(c_1, \ldots, c_r) \in \Delta(R)$ or $D(c_i; x, y) \neq 0$ for all such $(c_1, \ldots, c_r)$, in which, case, we have polynomials $F_1, G_1 \in A[x, y]$ such that $F_1(c_i; x, y)$ and $G_1(c_i; x, y)$ differ by a nonzero multiplier in $R[x]$ from $F(c_i; x, y)D(c_i; x, y)^{-1}$ and $G(c_i; x, y)D(c_i; x, y)^{-1}$ respectively (that is, we have $l(x) \neq 0$ such that $l(x)F(c_i; x, y) = F_1(c_i; x, y)D(c_i; x, y)$ and similarly for G). We observe that the condition on a polynomial $v(t_i; x) \in A[x]$ that $v(c_i; x) = 0$ is equivalent to the vanishing for $t_i = c_i$ of the sum of the squares of the coefficients and $v(c_i; x) \neq 0$ is equivalent to the non-vanishing for $t_i = c_i$ of the sum of the squares of the coefficients. This remark enables us to carry over the results on the division algorithm in the lemma on p. 319 to the case of two indeterminates. The foregoing argument can then be used to obtain the stated result for $F, G \in A[x, y]$. We leave it to the reader to fill in the details.

There is another formal device we shall need, which corresponds to choosing an $\eta \in R$ such that $g(\eta) \neq 0$ for a given $g(x) \neq 0$. Suppose we are given a polynomial $G(t_i; x) = v_m x^m + v_{m-1}x^{m-1} + \cdots + v_0$ where $v_j = v_j(t_1, \ldots, t_r) \in A$ and assume $G \neq 0$. Form the cover (9) and take one of the $k \neq -\infty$. Then $G(c_i; x) = v_k(c_1, \ldots, c_r)x^k + v_{k-1}(c_1, \ldots, c_r)x^{k-1} + \cdots + v_0(c_1, \ldots, c_r)$ and $v_k(c_1, \ldots, c_r) \neq 0$ if $(c_1, \ldots, c_r) \in \Gamma_k(R)$. Let $L_k(t_1, \ldots, t_r)$ be the rational expression $(k + 1) + \sum_0^{k-1} v_j(t_1, \ldots, t_r)^2 v_k(t_1, \ldots, t_r)^{-2}$. Then $L_k(c_1, \ldots, c_r)$ is defined for the $(c_1, \ldots, c_r) \in \Gamma_k(R)$ and $G(L_k(c_1, \ldots, c_r)) \neq 0$, by (7).

We shall need an analogous result also for two indeterminates x and y. Suppose $G(t_j; x, y) \neq 0$. Then we can construct a cover $\gamma = \{\Gamma\}$ and elements $L(t_i) \in \mathbb{Z}(t_1, \ldots, t_r)$ (that is, rational expressions in the t_i with integer coefficients) such that for any $\Gamma \in \gamma$ either $G(c_i; x, y) = 0$ for all $(c_i, \ldots, c_r) \in \Gamma(R)$, or, for one of the $L(t_i)$, $L(c_i)$ is defined and $G(c_i; L(c_i), y) \neq 0$ for every $(c_1, \ldots, c_r) \in \Gamma(R)$. The proof is an immediate extension of the foregoing argument.

We shall now give the

Proof of Theorem 5.8.   We first obtain a reduction from the case of the pair of relations $F(t_i; x, y) = 0$, $G(t_i; x) \neq 0$ to a single one $K(t_i; x, y) = 0$. (This corresponds to the last part of the argument of the preceding section.) We construct a cover δ_1 and polynomials appropriate for the g.c.d. of the coefficients of the powers of y in $F(t_i; x, y)$ and the polynomial $G(t_i; x)$. Let Δ_1 denote any member of the cover δ_1, $D_1(t_i; x)$ the associated polynomial such that for $(c_1, \ldots, c_r) \in \Delta_1(R)$, $D_1(c_i; x)$ is a g.c.d. in $R[x]$ of the coefficients of the powers of y in $F(c_i; x, y)$ and $G(c_i; x)$. If $D_1(c_i; x) = 0$ no solution of $F(c_i; x, y) = 0$, $G(c_i; x) \neq 0$ exists. Hence we may assume $D_1(c_i; x) \neq 0$ for all $(c_1, \ldots, c_r) \in \Delta_1(R)$. Then, by condition (2), we obtain polynomials $F_1(t_i; x, y)$ and $G_1(t_i; x) \in A[x, y]$ and $A[x]$ respectively such that $F(c_i; x, y)$ and $G(c_i; x)$ differ by a non-zero multiplier in R from $D_1(c_i; x)F_1(c_i; x, y)$ and $D_1(c_i; x)G_1(c_i; x)$. Then (ξ, η) satisfies $F(c_i; \xi, \eta) = 0$, $G(c_i; \xi) \neq 0$ if and only if $F_1(c_i; \xi, \eta) = 0$, $G_1(c_i; \xi) \neq 0$. Hence for Δ_1 we have a reduction to the pair F_1, G_1, for which the coefficients of the powers of y in $F_1(c_i; x, y)$ and $G_1(c_i; x)$ are relatively prime. We now refine the cover δ_1 to a cover δ_2 obtained by replacing each Δ_1 by the terms resulting from applying to Δ_1 the cover associated with the coefficients of x in G_1 as in (9). This reduces the consideration to sets Δ_2 of δ_2 and polynomials F_2, G_2 such that for $(c_1, \ldots, c_r) \in \Delta_2(R)$ we have $F_2(c_i; x, y) = F_1(c_i; x, y)$, $G_2(c_i; x) = G_1(c_i; x)$ and $G_2(t_i; x) = v_k x^k + v_{k-1} x^{k-1} + \cdots + v_0$ where $v_j \in A$, $v_k(c_1, \ldots, c_r) \neq 0$. Let $T(t_i; y)$ be the resultant of $F_2(t_i; x, y)$ and $G_2(t_i; x)$ regarded as polynomials in x. Since $v_k(c_1, \ldots, c_r) \neq 0$, $T(c_i; y) = 0$ for $(c_1, \ldots, c_r) \in \Delta_2(R)$ implies that $F_2(c_i; x, y)$ and $G_2(c_i; x)$ have a common factor of positive x-degree in $R(y)[x]$. This can be written as $a(y)b(y)^{-1}h(x, y)$ where $h(x, y) \in R[x, y]$ and is primitive as a polynomial in x with coefficients in $R[y]$. Then $h(x, y) \mid F_2(c_i; x, y)$ and $h(x, y) \mid G_2(c_i; x)$. This implies that $h(x, y) \in R[x]$ and contradicts the fact that the coefficients of the powers of y in $F_2(c_i; x, y)$ and $G_2(c_i; x)$ are relatively prime. Thus we see that $T(c_i; y) \neq 0$ for all $(c_1, \ldots, c_r) \in \Delta_2(R)$. We can now pass to a refinement δ_3 such that for any $\Delta_3 \in \delta_3$ we have a rational expression $L(t_i) = Q(t_i)P(t_i)^{-1}$, P, $Q \in A$, such that for $(c_1, \ldots, c_r) \in \Delta_3(R)$, $P(c_i) \neq 0$ and $T(c_i; L(c_i)) \neq 0$. We now replace the corresponding F_2 by F_3 where $F_3(t_i; x, y) = P(t_i)^f F_2(t_i; x, y + L(t_i))$ where $f = \deg_y F_2(t_i; x, y)$. We write G_3 for G_2. Then the resultant of $F_3(t_i; x, y)$ and $G_3(t_i; x)$ regarded as polynomials in x has the form $P(t_i)^g T(t_i; y + L(t_i))$ and this is $\neq 0$ for $t_i = c_i$, $y = 0$ if $(c_1, \ldots, c_r) \in \Delta_3(R)$. It follows, as in the proof in section 5.5. that $F_3(c_i; x, y) = 0$, $G_3(c_i; x) \neq 0$ is solvable in R if and only if $F_4(c_i; x, y) = 0$ is solvable where $F_4(t_i; x, y) = F_3(t_i; x, G_3(t_i; x)y)$. This reduces the consideration to a single equation with no inequations for the various terms of the cover δ_3.

We may as well make a fresh start and suppose we are given an equation $F(t_i; x, y) = 0$ only (since the result we shall obtain in this case can be applied to the various F_4 and Δ_3 above). We first construct a cover δ_1 and polynomials appropriate for the g.c.d. of the coefficients of y in $F(t_i; x, y)$. Then for $\Delta_1 \in \delta_1$ we have polynomials $D_1(t_i; x), F_1(t_i; x, y) \in A[x]$ and $A[x, y]$ such that for $(c_1, \ldots, c_r) \in \Delta_1(R), D_1(c_i; x)$ is a g.c.d. of the coefficients of y in $F(c_i; x, y)$ and $F(c_i; x, y)$ and $D_1(c_i; x)F_1(c_i; x, y)$ differ by a nonzero multiplier in R. Clearly $F(c_i; x, y) = 0$ is solvable in R if and only if $D_1(c_i; x) = 0$ or $F_1(c_i; x, y) = 0$ is solvable. The first is the kind of condition we are after so we keep it as one of our alternatives. Hence we need to pursue only the second alternative. Here $F_1(c_i; x, y)$ is primitive as a polynomial in y with coefficients in $R[x]$. Next, for each F_1 we obtain a cover appropriate to the g.c.d. of $F_1(t_i; x, y)$ and $(\partial/\partial y)F_1(c_i; x, y)$. We apply these covers to obtain a cover δ_2 such that for any $\Delta_2 \in \delta_2$ which comes from Δ_1 we have polynomials $D_2(t_i; x, y), F_2(t_i; x, y) \in A[x, y]$ such that for $(c_1, \ldots, c_r) \in \Delta_2(R)$ we have a nonzero polynomial $l(x) \in R[x]$ such that $l(x)F_1(c_i; x, y) = D_2(c_i; x, y)F_2(c_i; x, y)$ and $D_2(c_i; x, y)$ is a g.c.d. in $R(x)[y]$ of $F_1(c_i; x, y)$ and $(\partial/\partial y)F_1(c_i; x, y)$. Then $F_1(c_i; x, y)$ and $F_2(c_i; x, y)$ have the same irreducible factors of positive y-degree in $R[x, y]$ and no such factor occurs with multiplicity greater than one in $F_2(c_i; x, y)$. Next we apply the first step to F_2 to obtain a refinement δ_3 of δ_2 such that for any $\Delta_3 \in \delta_3$ which comes from Δ_2 we have a polynomial $F_3(t_i; x, y) \in A[x, y]$ such that for $(c_1, \ldots, c_r) \in \Delta_3(R), F_3(c_i; x, y)$ is primitive as a polynomial in y over $R[x]$ and has the same irreducible factors of positive y-degree as $F_2(c_i; x, y)$ and none of these has multiplicity exceeding one. Then $F_1(c_i; x, y) = 0$ is solvable in R if and only if this is true of $F_3(c_i; x, y) = 0$ (provided $(c_1, \ldots, c_r) \in \Delta_3(R)$). Also, $F_3(c_i; x, y)$ and $(\partial/\partial y)F_3(c_i; x, y)$ have no common factor of positive degree. Put $G_3(t_i, t; x, y) = y(\partial F_3/\partial x) - (x - t)(\partial F_3/\partial y)$ where t is a new indeterminate and let $H(t_i, t; x)$ be the resultant of $G_3(t_i, t; x, y)$ and $F_3(t_i; x, y)$ regarded as polynomials in y. Then it can be argued, as in the decision method itself, that $H(c_i, t; x) \neq 0$. Hence, resorting to another refinement δ_4 and a set $\Delta_4 \in \delta_4$ we obtain a rational expression $L(t_i) = Q(t_i)P(t_i)^{-1}, P, Q \in A$, such that if $(c_1, \ldots, c_r) \in \Delta_4(R)$ then $P(c_i) \neq 0$ and $H(c_i, L(c_i); x) \neq 0$. Then if we replace G_3 by $G_4(t_i; x, y) = P(t_i)G_3(t_i, L(t_i); x, y) \in A[x, y]$ and put $F_4 = F_3$, then the resultant $H(t_i; x)$ of F_4 and G_4 regarded as polynomials in y satisfies $H(c_i; x) \neq 0$ for $(c_1, \ldots, c_r) \in \Delta_4(R)$. The remainder of the proof follows in the same way along the lines of the decision method itself. We leave it to the reader to carry this out. $\square$

We can now combine this elimination theorem with the parameterized version of Sturm's theorem (Theorem 5.5) to prove our main result, which is

TARSKI'S THEOREM.   *Let φ be a finite set of polynomial equations, inequations, and inequalities of the form $F(t_1, \ldots, t_r; x_1, \ldots, x_n) = 0$, $G(t_1, \ldots, t_r;$ $x_1, \ldots, x_n) \neq 0$, $H(t_1, \ldots, t_r; x_1, \ldots, x_n) > 0$ where $F, G, H \in \mathbb{Z}[t_1, \ldots, t_r; x_1,$ $\ldots, x_n]$. Then we can determine in a finite number of steps a finite collection of finite sets ψ_j of polynomial equations, inequations, and inequalities of the same type in the parameters t_i alone such that, if R is any real closed field, then the set φ has a solution for the x's in R for $t_i = c_i$, $1 \leq i \leq r$, if and only if the c_i satisfy all the conditions of one of the sets ψ_j.*

Proof.   As above, we can replace the given system φ by one consisting of a single equation in perhaps more than n x's. Hence we may assume the system has the form

$$F(t_1, \ldots, t_r; x_1, \ldots, x_n) = 0, \qquad G(t_1, \ldots, t_r; x_1, \ldots, x_{n-1}) \neq 0$$

where $n \geq 1$. If $n = 1$ the result follows by applying Theorem 5.5 and adding the parameter condition $G(t_1, \ldots, t_r) \neq 0$ to each of the conditions Γ_k given by this theorem. If $n > 1$ we regard $x_1, \ldots, x_{n-2}$ as parameters $t_{r+1}, \ldots, t_{r+n-2}$ and apply Theorem 5.8. This replaces the given system by a finite set of systems of the form

$$F_j(t_1, \ldots, t_r; x_1, \ldots, x_{n-1}) = 0, \qquad G_j(t_1, \ldots, t_r; x_1, \ldots, x_{n-2}) \neq 0.$$

We can now conclude the proof by applying induction on the number of x's. $\square$

Suppose now that we have two real closed subfields R_1 and R_2 with a common subfield F, and we have a system of equations, inequations, and inequalities with coefficients in F which has a solution in R_1. It is clear that we can introduce parameters and interpret our assertion as one that a certain system involving parameters and having integral coefficients has a solution in R_1 for certain values of the parameters—say $t_i = c_i$ in F. Then Tarski's theorem implies that the c_i satisfy one of a certain system of equations, inequations, and inequalities with rational coefficients which can be determined à priori and are independent of R_1. Going backwards we see that the system given initially has a solution in R_2. In particular, we see that if a given system of equations, inequations, and inequalities with rational coefficients has a solution in one real closed field R_1 (e.g., $\mathbb{R}$), then it has a solution in every real closed field.

More generally, Tarski's theorem implies his metamathematical principle that any "elementary" sentence of algebra which is true in one real closed field (e.g., the field of real numbers) is true in every real closed field. We refer the reader to books on mathematical logic for a precise and detailed account of this result.[3] Here we shall be content to give a sketchy indication of the meaning of Tarski's principle and to illustrate it with a non-trivial application.

We first define an *atomic formula* as an expression of the form $f > 0$ or $f = 0$ where f is a polynomial with rational coefficients. Next we define a *formula* as any expression obtained from a finite number of atomic formulas by applying conjunction ("and"), disjunction ("or"), negation ("not"), and the existential quantifier ("there exists an x such that"). (Other logical concepts such as "implies," "for all x," and so on, can be defined in these terms). We define an *elementary sentence* as a formula involving no free variables.

The trick in applying Tarski's principle is to be able to recognize that a given statement is either an elementary sentence or is equivalent to one. As an illustration of this we shall prove the following extension of Lemma 1 of section 5.5.

Let R be a real closed field and let $f_1(x_1, \ldots, x_n), \ldots, f_m(x_1, \ldots, x_n)$ be polynomials with coefficients in R. Assume that there exists in $R^{(n)}$ a simultaneous solution $(a_1, \ldots, a_n)$ of the equations $f_i(x_1, \ldots, x_n) = 0$. Then there exists a solution nearest the origin (that is, such that $\sum a_i^2$ is minimal).

To prove this we first replace the system by the single equation $f = \sum f_i^2$. Next we replace the coefficients by parameters, and so we have a polynomial $F(t_1, \ldots, t_r; x_1, \ldots, x_n)$ with integral coefficients. Then our assertion can be put in the following elementary form: for $t_i = c_i$ in R either $F(c_1, \ldots, c_r; x_1, \ldots, x_n) = 0$ has no solution or it has a solution $x_j = a_j$ such that $\sum a_j^2 \leq \sum b_j^2$ for every solution $x_j = b_j$. Since this is easily proved for the field $\mathbb{R}$ (using the argument preceding Lemma 1, p. 328) it holds for every real closed field R.

It is worth mentioning also that Tarski's theorem has had an important application to partial differential equations.[4] This is a striking example of the interconnectedness of mathematics in that a result which originated in mathematical logic has an important consequence in one of the most applied parts of mathematics. We note also that Tarski's theorem is used in section 11.4 of Volume 2 as an important element of the proof of a theorem on positive definite rational functions that provides the answer to a famous problem of Hilbert's.

EXERCISE

1. Supply the missing details in the proof of Theorem 5.8.

---

[3] Tarski's original account appears in *A decision method for elementary algebra and geometry*, a publication of RAND Corporation, 1948. A proof of the principle, called "the elimination of quantifiers" in the theory of real closed fields, is given in G. Kreisel and J. L. Krivine, *Elements of Mathematical Logic*. London, North-Holland Pub. Co., 2d rev'd printing, 1971, pp. 60–65. Seidenberg's paper is in *Annals of Math.* vol. 60 (1954), pp. 365–374.

[4] See A. Friedman, *Generalized Functions and Partial Differential Equations*, Englewood Cliffs, N.J., Prentice Hall, 1963, Chapter 7.

Metric Vector Spaces and the Classical Groups

Euclidean geometry viewed analytically is the study of an n-dimensional vector space V over $\mathbb{R}$ relative to a certain symmetric bilinear form which serves to define both the length of a vector and the cosine of the angle between two vectors. Taking $V = \mathbb{R}^{(n)}$ we can take the bilinear form to be the standard one

$$x \cdot y = \sum_{1}^{n} x_i y_i$$

for $x = (x_1, \ldots, x_n)$, $y = (y_1, \ldots, y_n)$. Then $x \cdot x = \sum x_i^2 = |x|^2$, the square of the length of x, and if θ is the angle between x and y, then $\cos \theta = (x \cdot y)/|x||y|$. The function $x \cdot y$, the dot product or scalar product of x and y, is bilinear in the sense that

$$(x + x') \cdot y = x \cdot y + x' \cdot y$$
$$x \cdot (y + y') = x \cdot y + x \cdot y'$$
$$ax \cdot y = a(x \cdot y) = x \cdot (ay)$$

for vectors x, x', y, y' and the real number a, and the dot product is symmetric

$(x \cdot y = y \cdot x)$ and positive definite $(x \cdot x > 0,$ if $x \neq 0)$. All of this is well known in analytic geometry of two and three dimensions and the extension to any n is quite easy and natural.

We can generalize this situation in two ways. First, we can drop the hypothesis of finite dimensionality and replace it by one of completeness. This leads to the study of real Hilbert spaces, and, if we replace $\mathbb{R}$ by $\mathbb{C}$ and the given symmetric bilinear form by a positive definite hermitian one, then we obtain complex Hilbert spaces which play an important role in analysis. We shall not follow this path of generalization here. Instead we shall consider extensions of Euclidean geometry obtained by replacing $\mathbb{R}^{(n)}$ by any finite dimensional vector space V over an arbitrary field F and the dot product by any non-degenerate bilinear form $B(x, y)$ (definition in section 6.1) which is either symmetric, $B(x, y) = B(y, x)$, or alternate, $B(x, x) \equiv 0$. We shall call $B(x, y)$ a metric on V. The geometry obtained by taking $B(x, y)$ symmetric is called orthogonal geometry and that associated with an alternate form is called symplectic geometry. These are the only cases in which orthogonality of vectors, defined by $x \perp y$ if $B(x, y) = 0$, is a symmetric relation.

Associated with a metric $B(x, y)$ we have the group of linear transformations η in V such that $B(\eta x, \eta y) = B(x, y)$ for all $x, y \in V$. If B is symmetric this is called an orthogonal group and if B is alternate it is called a symplectic group. These groups along with the general linear group of bijective linear transformations of a finite dimensional vector space are the "classical" groups, in the terminology of Hermann Weyl. We shall see that they are close to being simple. The proof of the precise result along these lines is one of the major goals of this chapter.

In the first part of this chapter we shall lay the foundations of orthogonal and symplectic geometries. The topics we shall consider are the problem of equivalence of forms, real forms and Sylvester's theorem on the inertia of a real symmetric form (or quadratic form), Witt's cancellation theorem, and the Cartan-Dieudonné theorem on the generation of orthogonal groups by symmetries. After these topics we shall concentrate on the structure theory of the classical geometric groups. In the last section we shall indicate briefly the extension of the theory to hermitian forms.

6.1 LINEAR FUNCTIONS AND BILINEAR FORMS

Let V be a finite dimensional vector space over a field F. We recall that a linear function on V is a map of V into the base field F such that

$$(1) \qquad f(x + y) = f(x) + f(y), \qquad f(ax) = af(x)$$

for $x, y \in V, a \in F$. These constitute a vector space V^*, called *the conjugate space* of V, in which addition and the action by any $a \in F$ are defined by

$$(2) \qquad\qquad (f + g)(x) = f(x) + g(x), \qquad (af)(x) = af(x).$$

If $(e_1, e_2, \ldots, e_n)$ is a base for V over F, then we can define a linear function e_i^* by the conditions

$$(3) \qquad\qquad e_i^* (e_i) = 1, \qquad e_i^* (e_j) = 0 \quad \text{if} \quad j \neq i.$$

Then $(e_1^*, e_2^*, \ldots, e_n^*)$ is a base for V^* over F. For, if $x^* \in V^*$ and $x^*(e_i) = a_i \in F$, then $(\sum a_j e_j^*)(e_i) = a_i = x^*(e_i)$; since a linear map is determined by its restriction to a base, we have $x^* = \sum a_i e_i^*$. If $\sum_i a_i e_i^* = 0$, then $a_j = \sum_i a_i e_i^*(e_j) = 0$ and so the e_i^* are linearly independent. The base $(e_1^*, e_2^*, \ldots, e_n^*)$ is called the *dual or complementary base* of $(e_1, e_2, \ldots, e_n)$. Evidently V^* is n-dimensional.

We now define a *bilinear form* B on V to be a map $(x, y) \to B(x, y)$ of $V \times V$ into F such that for any $y \in V$ the map

$$(4) \qquad\qquad y_R : x \to B(x, y)$$

is a linear function on V and for any $x \in V$ the map

$$(5) \qquad\qquad x_L : y \to B(x, y)$$

is a linear function on V. These conditions amount to the following:

$$(6) \qquad \begin{aligned} B(x + x', y) &= B(x, y) + B(x', y), \qquad B(ax, y) = aB(x, y) \\ B(x, y + y') &= B(x, y) + B(x, y'), \qquad B(x, ay) = aB(x, y) \end{aligned}$$

and we can amalgamate them to the single condition

$$(6') \qquad B(a_1 x_1 + a_2 x_2, b_1 y_1 + b_2 y_2) = \sum_{i,j=1}^{2} a_i b_j B(x_i, y_j).$$

By induction, we can extend $(6')$ to

$$(7) \qquad B\left(\sum_1^m a_i x_i, \sum_1^q b_j y_j \right) = \sum_{i=1, j=1}^{m,q} a_i b_j B(x_i, y_j).$$

Formula (7) suggests a general way of constructing bilinear forms. Let $(e_1, e_2, \ldots, e_n)$ be a base for V over F and for each pair of indices (i, j), $1 \leq i,$

$j \leq n$, choose an element $b_{ij} \in F$. If $x = \sum a_i e_i$, $y = \sum b_i e_i$ we define

$$(8) \qquad\qquad B(x, y) = \sum_{i,j=1}^{n} b_{ij} a_i b_j.$$

Direct verification shows that $B:(x, y) \rightarrow B(x, y)$ is a bilinear form on V. Moreover, it is clear from (7) that every bilinear form on V is obtained in this way. The matrix $(B(e_i, e_j))$ determined by a bilinear form and a base $(e_1, e_2, \ldots, e_n)$ is called *the matrix of B relative to the base* $(e_1, e_2, \ldots, e_n)$. The determinant $\det(B(e_i, e_j))$ is called a *discriminant* of B.

Now suppose we change the base (e_i) to another base (f_i) where $f_i = \sum p_{ij} e_j$ and the matrix $p = (p_{ij})$ has the inverse $q = (q_{ij})$. We have

$$B(f_i, f_j) = B\left(\sum_k p_{ik} e_k, \sum_l p_{jl} e_l \right)$$

$$= \sum_{k,l} p_{ik} B(e_k, e_l) p_{jl}.$$

This shows that if $b = (B(e_i, e_j))$, then the matrix of B relative to the base (f_i) is

$$(9) \qquad\qquad pb\,{}^t p$$

where ${}^t p$ is the transpose of p. Taking the determinant of this matrix we obtain $(\det p)^2 \det b$ so the discriminant $\det b$ is changed to $(\det p)^2 \det b$ on changing the base.

We now consider the maps $x \rightarrow x_L$ and $y \rightarrow y_R$ as in (4) and (5) determined by a bilinear form B. Since x_L and $y_R \in V^*$ these map V into V^*. Moreover, they are linear maps. For, if $x_1, x_2 \in V$ then the corresponding linear functions x_{1L} and x_{2L} are $y \rightarrow B(x_1, y)$ and $y \rightarrow B(x_2, y)$ and their sum $x_{1L} + x_{2L}$ is $y \rightarrow B(x_1, y) + B(x_2, y) = B(x_1 + x_2, y)$ which is $(x_1 + x_2)_L$. Thus $x_{1L} + x_{2L} = (x_1 + x_2)_L$. Also $(ax)_L$ is $y \rightarrow B(ax, y) = aB(x, y)$ which is $a(x_L)$. Hence $x \rightarrow x_L$ is linear. Similarly, $y \rightarrow y_R$ is linear.

Let U be a subspace of the vector space V, B a bilinear form on V. We define

$$U^{\perp L} = \{ v \in V \,|\, B(v, u) = 0, u \in U \}$$

$$U^{\perp R} = \{ v \in V \,|\, B(u, v) = 0, u \in U \}.$$

These are subspaces of V. Moreover, it is clear from the definitions that

$$(10) \qquad\qquad U^{\perp L \perp R} \supset U, \qquad U^{\perp R \perp L} \supset U$$

and if $U_1 \supset U_2$ for subspaces U_1 and U_2 then

$$(11) \qquad\qquad U_1^{\perp L} \subset U_2^{\perp L}, \qquad U_1^{\perp R} \subset U_2^{\perp R}.$$

The subspaces $V^{\perp L}$ and $V^{\perp R}$ are called the *left radical* and *right radical* respectively of B.

THEOREM 6.1 *The following three conditions on a bilinear form B are equivalent: (1) $V^{\perp R} = 0$, (2) $V^{\perp L} = 0$, (3) the matrix of B relative to any base is invertible.*

Proof. Let $(e_1, e_2, \ldots, e_n)$ be a base and let $B(e_i, e_j) = b_{ij}$. Then it is clear from the bilinearity that if $B(e_i, z) = 0$ for $1 \le i \le n$, then $B(x, z) = 0$ for all $x \in V$. Hence $z \in V^{\perp R}$ if and only if $B(e_i, z) = 0$ for all i and, similarly, $z \in V^{\perp L}$ if and only if $B(z, e_i) = 0$ for all i. Now write $z = \sum c_j e_j$. Then $B(e_i, z) = \sum c_j B(e_i, e_j) = \sum b_{ij} c_j$. Hence $z \in V^{\perp R}$ if and only if $(c_1, c_2, \ldots, c_n)$ is a solution of the system of homogeneous linear equations

$$(12) \qquad\qquad b_{i1}c_1 + b_{i2}c_2 + \cdots + b_{in}c_n = 0, \qquad 1 \le i \le n.$$

Similarly, we see that $z = \sum c_j e_j \in V^{\perp L}$ if and only if the c's satisfy

$$(13) \qquad\qquad b_{1i}c_1 + b_{2i}c_2 + \cdots + b_{ni}c_n = 0, \qquad 1 \le i \le n.$$

We know from linear algebra that the condition that (12) or (13) have a solution $(c_1, c_2, \ldots, c_n) \ne (0, 0, \ldots, 0)$ is that $\det (b_{ij}) = 0$. Our result follows from this. $\square$

A bilinear form B is called *non-degenerate* if it satisfies the conditions of Theorem 6.1.

The condition on a vector z that $B(x, z) = 0$ for all x, that is, that $z \in V^{\perp R}$ is equivalent to saying that the linear function $z_R = 0$. Thus $V^{\perp R}$ is the kernel of the linear map R of V into V^*. Hence B is non-degenerate if and only if the kernel of R is 0, which is equivalent to: R is injective. Since $\dim V = \dim V^*$ this is the case if and only if R is surjective, that is, every linear function on V has the form $x_R : y \to B(y, x)$ for some x in V. Similarly, B is non-degenerate if and only if every linear function on V has the form $y \to B(x, y)$ for some x in V.

Still assuming that B is non-degenerate we proceed to show that the maps $U \to U^{\perp R}$ and $U \to U^{\perp L}$ are inverses and hence are bijective maps in the set of subspaces of V. To see this we shall prove the following dimensionality relation:

$$(14) \qquad\qquad \dim U^{\perp R} = n - \dim U = \dim U^{\perp L}.$$

We recall first the well-known formula from linear algebra which states that if T is a linear map of a finite dimensional vector space V into a second vector space then dim $V = $ dim $T(V) + $ dim (ker T). (This can be seen by observing that we have the induced bijective map $x + $ ker $T \to Tx$ of $V/$ker T onto $T(V)$.) Now let U be a subspace of V and let $x \in V$. Then x_R is a linear function on V so its restriction to U, $x_R|U$, is a linear function on U. Thus we have the linear map $x \to x_R|U$ of V into the conjugate space U^* of U. The kernel of this map is the set of x such that $B(y, x) = 0$ for all $y \in U$. Hence the kernel of the map of V into U^* is $U^{\perp R}$. Hence we have $n = $ dim $V = $ dim $U^{\perp R} + $ dim W where W is the set of linear functions on U of the form $y \to B(y, x)$ for $x \in V$. Formula (14) will follow if we can show that $W = U^*$, since dim $U^* = $ dim U. Let g be a linear function on U. Then we can extend g to a linear function on V; for we can obtain a base for V of the form $(f_1, f_2, \ldots, f_n)$ where $(f_1, f_2, \ldots, f_r)$ is a base for U and define g' to be the linear function on V which coincides with g on the f_j, $1 \leq j \leq r$, and maps the remaining f_i into any elements we please in F. Now we have seen that g' has the form $y \to B(y, x)$ for some $x \in V$. Hence this holds also for g. This completes the proof of dim $U^{\perp R} = n - $ dim U and in a similar manner we have dim $U^{\perp L} = n - $ dim U. Applying these twice we obtain dim $U^{\perp L \perp R} = n - $ dim $U^{\perp L} = n - (n - $ dim $U) = $ dim U and dim $U^{\perp R \perp L} = $ dim U. On the other hand, we had $U^{\perp L \perp R} \supset U$ and $U^{\perp R \perp L} \supset U$. Hence

(15) $$U^{\perp L \perp R} = U, \qquad U^{\perp R \perp L} = U$$

for any subspace U (assuming B non-degenerate).

An important point in the proof of the foregoing result is the determination of the form of linear functions on U. Since we shall need to refer to this later we state the result as a

LEMMA. *Let B be non-degenerate and let U be a subspace of V. Then any linear function on U has the form $y \to B(x, y)$ (and also the form $y \to B(y, x)$) for some x in V.*

If $B(x, y) = 0$ we say x is *orthogonal* to y and we indicate this by writing $x \perp y$. It is highly desirable that this be a symmetric relation, that is, $x \perp y$ if and only if $y \perp x$. It is quite easy to determine the conditions for this: namely, we have

THEOREM 6.2. *Let $B(x, y)$ be a bilinear form on V. Then the relation of orthogonality defined by B is a symmetric one if and only if either B is symmetric* *in the sense that $B(x, y) = B(y, x)$ for all x and y or B is alternate in the sense that $B(x, x) = 0$ for all x in V.*

Proof. It is clear that orthogonality defined by a symmetric form is a symmetric relation. Also if B is alternate then $B(x, y) + B(y, x) = B(x + y, x + y) - B(x, x) - B(y, y) = 0$ for all x, y and this skew symmetry of B implies that $x \perp y$ if and only if $y \perp x$. Now suppose B has this last property. Let x, y, and z be arbitrary vectors in V and form $w = B(x, y)z - B(x, z)y$. Then $x \perp w$ and the condition $w \perp x$ is equivalent to

(16) $$B(x, y)B(z, x) = B(y, x)B(x, z)$$

for all x, y, z. Putting $x = y$ we obtain

(17) $$B(x, x)(B(x, z) - B(z, x)) = 0$$

for all x, z. We claim that either $B(x, y) = B(y, x)$ for all x, y or $B(x, x) = 0$ for all x. Otherwise, we have a pair of vectors u, v such that $B(u, v) \neq B(v, u)$ and a vector w such that $B(w, w) \neq 0$. Then, by (17), $B(u, u) = B(v, v) = 0$, and $B(w, u) = B(u, w)$ and $B(w, v) = B(v, w)$. Also since $B(u, v) \neq B(v, u)$ it follows from (16) that $B(u, w) = B(w, u) = 0$ and $B(v, w) = B(w, v) = 0$. Then $B(u, w + v) = B(u, v) \neq B(v, u) = B(w + v, u)$. Hence, by (17), $B(w + v, w + v) = 0$. But $B(w + v, w + v) = B(w, w) + B(w, v) + B(v, w) + B(v, v) = B(w, w)$ so we have contradicted $B(w, w) \neq 0$. □

If B and B' are bilinear forms on vector spaces V and V' respectively, we call B and B' *equivalent* if there exists a bijective linear map $x \to x'$ of V onto V' such that $B(x, y) = B'(x', y')$ for all $x, y \in V$. Evidently this is an equivalence relation and it implies equality of dimensionality of V and V'. It is clear that if B is alternate (symmetric) then B' has the same property.

EXERCISES

1. Show that if B is any bilinear form on V, then $(U_1 + U_2)^{\perp L} = U_1^{\perp L} \cap U_2^{\perp L}$ and $(U_1 + U_2)^{\perp R} = U_1^{\perp R} \cap U_2^{\perp R}$ for any two subspaces U_1 and U_2. Show also that if B is non-degenerate, then $(U_1 \cap U_2)^{\perp L} = U_1^{\perp L} + U_2^{\perp L}$ and $(U_1 \cap U_2)^{\perp R} = U_1^{\perp R} + U_2^{\perp R}$.

2. Let B be an arbitrary bilinear form on V and assume U is a subspace such

that the restriction of B to U is non-degenerate. Show that $V = U \oplus U^{\perp L} = U \oplus U^{\perp R}$.

3. Let B be a non-degenerate bilinear form on V. Show that if C is a bilinear form on V, then there exists a unique linear transformation L_C of V into V such that $C(x, y) = B(L_C x, y)$ for all $x, y \in V$. Show that C is non-degenerate if and only if L_C is bijective. Show that there exists a unique bijective linear transformation P of V onto V such that $B(y, x) = B(Px, y)$ for all $x, y \in V$.

4. Show that if B is non-degenerate, then for every linear transformation T of V into itself there exists a unique linear transformation T' of V into V such that $B(Tx, y) = B(x, T'y)$ for all $x, y \in V$. Determine the matrix of T' in terms of the matrices of T and B relative to a base of V. Show that the map $T \to T'$ is an anti-automorphism in the ring of linear transformations and that $(T')' = T$ for all T if B is either symmetric or skew $(B(y, x) = -B(x, y))$.

5. If B_1 and B_2 are bilinear forms on V define $B_1 + B_2$ by $(B_1 + B_2)(x, y) = B_1(x, y) + B_2(x, y)$, and for $a \in F$, define $(aB_1)(x, y) = a(B_1(x, y))$. Show that these are bilinear forms and that the set of bilinear forms on V is a vector space over F relative to these compositions. Prove that this space is n^2 dimensional over F.

6. Let B be a symmetric (alternate) bilinear form on V so $U^{\perp L} = U^{\perp R}$ for any subspace U of V. Let W be a subspace of $V^{\perp}$. Show that $\bar{B}(x + W, y + W) = B(x, y)$ defines a symmetric (alternate) bilinear form on V/W and that this is non-degenerate if and only if $W = V^{\perp}$.

7. Show that if B is a bilinear form, then there exist bases $(u_1, \ldots, u_n)$, $(v_1, \ldots, v_n)$ for V such that $(B(u_i, v_j)) = \operatorname{diag}\{1, \ldots, 1, 0, \ldots, 0\}$.

8. Let B be a bilinear form. Note that if u and v are fixed vectors then the map $x \to B(x, u)v$ is a linear transformation of V into V. Denote this as $u \otimes v$. Find a formula for the trace $\operatorname{tr} u \otimes v$. Show that if B is non-degenerate then every linear transformation has the form $\sum u_i \otimes v_i$.

6.2   ALTERNATE FORMS

The bilinear forms we shall consider in the remainder of this chapter will be either symmetric or alternate. In either case $U^{\perp L} = U^{\perp R}$ for any subspace so we shall denote this subspace as $U^{\perp}$ and call it the *orthogonal complement* of U. $U \cap U^{\perp} = 0$ if and only if the restriction of B to U is non-degenerate. In this case we shall say that U is a *non-degenerate* subspace. If $(e_1, e_2, \ldots, e_n)$ is a base for V, then we obtain the matrix $b = (B(e_i, e_j))$ of B relative to this base. A change of base replaces b by $pb\,{}^tp$, p invertible. We shall now call the matrices $b, c \in M_n(F)$ *cogredient* if there exists a $p \in GL_n(F)$ (the group of units of $M_n(F)$) such that $c = pb\,{}^tp$. This is an equivalence relation, so with B we have associated a cogredience class of matrices, the set of matrices of B relative to the various ordered bases for V. We have defined a discriminant of B to be $\det b$, b a matrix of B, and we have seen that B is non-degenerate if and only if $\det b \neq 0$. We shall now make the notion of discriminant more precise by defining it to be 0 if B is degenerate, and otherwise to be the coset $(\det b)F^{*2}$ of $\det b$ in the group F^*/F^{*2} where F^* is the multiplicative group of non-zero elements of F, and F^{*2} is the subgroup of squares of elements of F^*. We shall refer to $(\det b)F^{*2}$ as *the discriminant of B*. Then the various discriminants $\det b$, b a matrix of the non-degenerate B, are just representatives of the coset $(\det b)F^{*2}$.

We have seen that if B is alternate, then B is skew symmetric: $B(x, y) = -B(y, x)$. Moreover, if the characteristic, char $F \neq 2$, then skew symmetry implies $2B(x, x) = 0$ and $B(x, x) = 0$. Thus the alternate property and skew symmetry are equivalent if char $F \neq 2$. If B is alternate and $(e_1, e_2, \ldots, e_n)$ is a base, then $B(e_i, e_j) = -B(e_j, e_i)$ and $B(e_i, e_i) = 0$. Hence the matrix $b = (B(e_i, e_j))$ is an *alternate matrix* in the sense that b is skew symmetric, that is ${}^t b = -b$, and the diagonal elements are 0. Conversely, if $b = (b_{ij})$ is any alternate matrix the bilinear form B defined by $B(x, y) = \sum b_{ij} a_i b_j$ for $x = \sum a_i e_i$ and $y = \sum b_i e_i$ is alternate since

$$B(x, x) = \sum_{i \neq j} b_{ij} a_i a_j = \sum_{i < j} (b_{ij} + b_{ji}) a_i a_j = 0.$$

As we shall now show, the structure theory of alternate bilinear forms is extremely simple. Let B be such a form. Then we shall prove that there exists a base

(18) $(u_1, v_1, u_2, v_2, \ldots, u_r, v_r, z_1, \ldots, z_{n-2r})$

for V such that the matrix of B relative to this base has the form

(19) $\text{diag } \{S, S, \ldots, S, 0, \ldots, 0\}$

where

(20) $$S = \begin{pmatrix} 0 & 1 \\ -1 & 0 \end{pmatrix}.$$

Here the notation indicates that we have a string of S's followed by a string of 0's down the diagonal, and that other entries are 0. If $B = 0$ ($B(x, y) = 0$ for all x, y) the result is trivial. Otherwise, we may assume we have u and v such that $B(u, v) = b \neq 0$. Then $u_1 = u$ and $v_1 = b^{-1}v$ satisfy $B(u_1, v_1) = 1 = -B(v_1, u_1)$. Since $B(x, ax) = aB(x, x) = 0$ it is clear that u_1, v_1 are linearly independent. Hence these give us a start in constructing the required base (18). Now suppose that we have already found linearly independent vectors

(21) $(u_1, v_1, u_2, v_2, \ldots, u_k, v_k)$

such that $B(u_i, v_i) = 1 = -B(v_i, u_i)$ and $B(x, y) = 0$ for every other choice of x and y in the set $\{u_i, v_i \mid 1 \le i \le k\}$. Let V_k denote the $2k$ dimensional subspace of V spanned by the vectors u_i, v_i. Then we claim that $V = V_k \oplus V_k^\perp$. Since the matrix of the restriction of B to V_k relative to the base (21) is $\mathrm{diag}\{S, S, \ldots, S\}$ and this is invertible, this bilinear form is non-degenerate, so $V_k \cap V_k^\perp = 0$. Now let $x \in V$ and consider the vector

$$y = x - \sum_1^k B(x, v_i)u_i + \sum_1^k B(x, u_i)v_i.$$

We have

$$B(y, u_j) = B(x, u_j) + B(x, u_j)B(v_j, u_j) = 0$$

$$B(y, v_j) = B(x, v_j) - B(x, v_j)B(u_j, v_j) = 0,$$

which implies that $y \in V_k^\perp$. Since $x = y + \sum B(x, v_i)u_i - \sum B(x, u_i)v_i$ we clearly have $V = V_k + V_k^\perp$. Thus $V = V_k \oplus V_k^\perp$. We now consider the restriction of B to $V_k^\perp$. If this is 0 we choose a base $(z_1, \ldots, z_{n-2k})$ for $V_k^\perp$ and we obtain the base (18) with $r = k$ satisfying our conditions. If B restricted to $V_k^\perp$ is not 0, then we can choose a pair of vectors u_{k+1}, v_{k+1} in this space so that $B(u_{k+1}, v_{k+1}) = 1 = -B(v_{k+1}, u_{k+1})$. Then we can replace the given string of vectors $(u_1, v_1, \ldots, u_k, v_k)$ by $(u_1, v_1, \ldots, u_{k+1}, v_{k+1})$. Continuing in this way we obtain our result, which we state as

THEOREM 6.3. *If B is an alternate bilinear form there exists a base* (18) *for V such that the matrix of B relative to this base has the form* $s = \mathrm{diag}\{S, S, \ldots, S, 0, \ldots, 0\}$ *where* $S = \begin{pmatrix} 0 & 1 \\ -1 & 0 \end{pmatrix}$.

If b is an alternate matrix, b determines an alternate bilinear form B whose matrix is b relative to a given base (e_i) for V. If p is the matrix expressing the base (u_j, v_j, z_k) of Theorem 6.3 in terms of the base (e_i), then $pb\,{}^tp = s$ as given in this theorem. Putting $q = p^{-1}$ we have $b = qs\,{}^tq$. It is clear that the matrices b and s have the same rank since the rank is unchanged on multiplying a matrix on either side by an invertible matrix. Also $\det s = 0$ or 1 and so $\det b = (\det q)^2 \det s = 0$, or $\det b = (\det q)^2$. Hence we have

COROLLARY 1. *The rank of an alternate matrix with entries in a field is even and its determinant is a square.*

It is clear also that we have

COROLLARY 2.   *Two alternate $n \times n$ matrices with entries in a field are cogredient if and only if they have the same rank.*

There is an important sharpening of Corollary 1 which we shall now indicate. Let n be even and let $F = \mathbb{Q}(x_{ij})$ the field of rational expressions with rational coefficients in $n(n-1)/2$ indeterminates x_{ij}, $i < j$. Let X be the alternate matrix in $M_n(F)$ whose (i, j)-entry for $i < j$ is x_{ij}. Then the (i, i)-entry of X is 0 and the (i, j)-entry for $i > j$ is $-x_{ji}$. By Corollary 1, $\det X$ is the square of an element of $F = \mathbb{Q}(x_{ij})$. Clearly, F is the field of fractions of its subring $\mathbb{Z}[x_{ij}]$. Hence there exist $f, g \in \mathbb{Z}[x_{ij}]$ such that $\det X = (f/g)^2$. Evidently, we may cancel common factors of f and g, so we may assume that f and g have no common factors in the factorial ring $\mathbb{Z}[x_{ij}]$ (other than the units ± 1). Then the relation $g^2 \det X = f^2$ implies, by the factoriality of $\mathbb{Z}[x_{ij}]$, that g is a unit so $g = \pm 1$. Thus $\det X = f^2$ and, f is determined to within a sign by this relation.

Now let R be any commutative ring and let $a = (a_{ij})$ be an alternate $n \times n$ matrix with entries in R. There is a unique homomorphism of $\mathbb{Z}[x_{ij}]$ into R sending $x_{ij} \to a_{ij}$, $i < j$. Applying this to the relation $\det X = f(x_{12}, x_{23}, \dots)^2$ we obtain $\det a = f(a_{12} a_{23}, \dots)^2$. In particular, if we specialize $R = \mathbb{Z}$ and $s = \mathrm{diag}\{S, S, \dots, S\}, S = \begin{pmatrix} 0 & 1 \\ -1 & 0 \end{pmatrix}$, we obtain $1 = \det s = f(1, \dots)^2$. We now fix the determination of the sign of f so that $f(1, \dots) = 1$ and we denote this determination as Pf X and call it the Pfaffian of X. Substitution of the a_{ij} for the x_{ij} gives Pf a, the *Pfaffian of the alternate matrix a.* This satisfies

(22) $(\mathrm{Pf}\ a)^2 = \det a.$

We have now established the first part of the following

THEOREM 6.4.   *Let n be even and let X be the alternate $n \times n$ matrix whose (i, j)-entry for $i < j$ is the indeterminate x_{ij}. Then there exists a unique polynomial Pf X in the x_{ij} with integer coefficients such that $(\mathrm{Pf}\ X)^2 = \det X$ and $\mathrm{Pf}\ s = 1$ for $s = \mathrm{diag}\{S, S, \dots, S\}, S = \begin{pmatrix} 0 & 1 \\ -1 & 0 \end{pmatrix}$. For any commutative ring R and alternate matrix $a \in M_n(R)$ we have (22). Moreover, if q is arbitrary in $M_n(R)$, then qa^tq is alternate and*

(23) $\mathrm{Pf}\ (qa\ ^tq) = (\det q)\ \mathrm{Pf}\ a.$

Proof. Let a be alternate and q arbitrary. Then the matrix $qa\,{}'q$ satisfies ${}^t(qa\,{}'q) = -qa\,{}'q$ and if $q = (q_{ij})$, $a = (a_{ij})$, then the (i, i)-entry of $qa\,{}'q$ is

$$\sum_{j,k} q_{ij}a_{jk}q_{ik} = \sum_{j<k} q_{ij}a_{jk}q_{ik} + \sum_{j>k} q_{ij}a_{jk}q_{ik}$$

$$= \sum_{j<k} q_{ij}a_{jk}q_{ik} - \sum_{j>k} q_{ik}a_{kj}q_{ij}$$

$$= \sum_{j<k} q_{ij}a_{jk}q_{ik} - \sum_{j<k} q_{ik}a_{jk}q_{ij} = 0.$$

Hence $qa\,{}'q$ is alternate. To prove (23) we work in the field $\mathbb{Q}(x_{ij}, y_{kl})$ wheie x_{ij} are the $n(n-1)/2$ indeterminates we had previously and y_{kl}, $k, l = 1, 2, \ldots, n$ are n^2 new indeterminates. Let X be as before and let $Y = (y_{kl})$. Then $YX\,{}'Y$ is alternate and $(\mathrm{Pf}\,(YX\,{}'Y))^2 = \det YX\,{}'Y = \det Y^2 X = (\det Y^2)(\mathrm{Pf}\,X)^2$. Hence $\mathrm{Pf}\,(YX\,{}'Y) = \pm(\det Y)(\mathrm{Pf}\,X)$. Specializing $Y = 1$ we see that the sign is $+$, so $\mathrm{Pf}\,(YX\,{}'Y) = (\det Y)(\mathrm{Pf}\,X)$. Specialization then gives (23). $\square$

As a consequence of (23) we have the following result which gives a method of evaluating the Pfaffian of an alternate matrix with entries in a field.

COROLLARY. *Let a be an alternate matrix with entries in a field and let q be an invertible matrix such that $qa\,{}'q = \mathrm{diag}\,\{S, \ldots, S, 0, \ldots, 0\}$ as in Theorem 6.3. Then $\mathrm{Pf}\,a = (\det q)^{-1}$ if a is invertible and $\mathrm{Pf}\,a = 0$ otherwise.*

It is easy to calculate $\mathrm{Pf}\,X$ for $n = 2$ and $n = 4$. These are respectively

$$\mathrm{Pf}\,X = x_{12}, \qquad \mathrm{Pf}\,X = x_{12}x_{34} - x_{13}x_{24} + x_{14}x_{23}.$$

Formulas for $\mathrm{Pf}\,X$ for higher values of n will be indicated in the exercises below.

EXERCISES

1. Show that

$$b = \begin{pmatrix} 0 & 2 & -1 & 3 \\ -2 & 0 & 4 & -2 \\ 1 & -4 & 0 & 1 \\ -3 & 2 & -1 & 0 \end{pmatrix} \qquad s = \begin{pmatrix} 0 & 1 & 0 & 0 \\ -1 & 0 & 0 & 0 \\ 0 & 0 & 0 & 1 \\ 0 & 0 & -1 & 0 \end{pmatrix}$$

are cogredient in $M_4(\mathbb{Q})$ and find a matrix p such that $pb\,{}'p = s$.

2. Assume B is an alternate bilinear form and $(u_1, v_1, \ldots, u_k, v_k)$ satisfy $B(u_i, v_i) = 1 = -B(v_i, u_i)$ with all other $B(x, y) = 0$ for x, y in $(u_1, v_1, \ldots, u_k, v_k)$. Using the notation of exercise 8 (p. 349) let $E_k = \sum_1^k (u_i \otimes v_i - v_i \otimes u_i)$. Verify that $E_k^2 = E_k$ and $B(E_k x, y) = B(x, E_k y)$, x, $y \in V$.

3. Let B be a non-degenerate alternate bilinear form on V, T a linear transformation of V into V. Define the *adjoint of* T *relative to* B as the (unique) linear transformation T' such that $B(Tx, y) = B(x, T'y)$ for all x, $y \in V$. Determine the adjoint of $u \otimes v$ relative to B.

4. Show that Pf a is linear in any one of the rows of the alternate matrix a (for fixed values of the entries in the submatrix obtained by deleting the chosen row and corresponding column).

5. Show that if $a = (a_{ij})$ is alternate, and if one defines $\alpha_{ij} = (-1)^{i+j-1}$ Pf A_{ij}, where A_{ij} is the $(n-2) \times (n-2)$ matrix obtained by striking out the ith and jth rows and ith and jth columns of a, then Pf $a = a_{12}\alpha_{12} + a_{13}\alpha_{13} + \cdots + a_{1n}\alpha_{1n}$.

6. Let $q = (q_{ij})$ be an arbitrary $n \times n$ matrix with entries in a commutative ring. Assume n even and define

$$a_{ij} = \begin{vmatrix} q_{1i} & q_{1j} \\ q_{2i} & q_{2j} \end{vmatrix} + \begin{vmatrix} q_{3i} & q_{3j} \\ q_{4i} & q_{4j} \end{vmatrix} + \cdots + \begin{vmatrix} q_{n-1,\iota} & q_{n-1,j} \\ q_{n,i} & q_{n,j} \end{vmatrix}.$$

Show that $a = (a_{ij})$ is alternate and Pf $a = \det q$.

7. Let $s = \text{diag}\{S, S, \ldots, S\}$, $S = \begin{pmatrix} 0 & 1 \\ -1 & 0 \end{pmatrix}$. Call a matrix $a \in M_n(R)$, R a commutative ring, *symplectic symmetric* if $s^{-1}{}^t as = a$. Show that this condition is equivalent to: sa is skew. Show that a is a root of the equation Pf $(s\lambda - sa) = 0$.

6.3   QUADRATIC FORMS AND SYMMETRIC BILINEAR FORMS

Let V be a vector space with base $(e_1, e_2, \ldots, e_n)$ and let $f(x_1, \ldots, x_n)$ be a polynomial in n indeterminates with coefficients in the base field F of V. This determines the *polynomial function* f on V into F,

$$f : x = \sum a_i e_i \to f(a_1, \ldots, a_n) \qquad \text{(cf. section 2.12)}.$$

If $f(x_1, \ldots, x_n)$ is a homogeneous polynomial of degree r (definition on p. 138), then we call the corresponding function a *form of degree* r. In particular, we have linear forms, quadratic forms, cubic forms, and so on, which are forms of degrees 1, 2, 3, etc. Since a homogeneous polynomial of degree 1 has the form $\sum_1^n c_i x_i$, $c_i \in F$, the concept of a linear form coincides with that of a linear function on V. A homogeneous polynomial of degree 2 has the form $\sum_{i \leq j} c_{ij} x_i x_j$ and hence a quadratic form is a map

(24)
$$x = \sum a_i e_i \to \sum_{i \le j} c_{ij} a_i a_j$$

where the c_{ij} are fixed elements of F. We shall now show that these maps have a simple axiomatic characterization which is given in the following alternatives.

DEFINITION 6.1.   *A quadratic form Q is a map $x \to Q(x)$ of a vector space V into its base field F such that*

1. $Q(ax) = a^2 Q(x)$, $a \in F$, $x \in V$
2. $B(x, y) = Q(x + y) - Q(x) - Q(y)$

is bilinear, that is, $(x, y) \to B(x, y)$ is a bilinear form—which is evidently symmetric.

We claim that the two definitions we have given are equivalent. First, suppose Q is defined by (24). Then $ax = \sum aa_i e_i$ and $Q(ax) = \sum_{i \le j} c_{ij}(aa_i)(aa_j) = a^2 \sum_{i \le j} c_{ij} a_i a_j = a^2 Q(x)$. Also if $y = \sum b_i e_i$, then $x + y = \sum (a_i + b_i) e_i$ and

$$
\begin{aligned}
B(x, y) &= Q(x + y) - Q(x) - Q(y) \\
&= \sum_{i \le j} c_{ij}(a_i + b_i)(a_j + b_j) - \sum_{i \le j} c_{ij} a_i a_j - \sum_{i \le j} c_{ij} b_i b_j \\
&= \sum_{i \le j} c_{ij} a_i b_j + \sum_{i \le j} c_{ij} b_i a_j = \sum_{i,j} d_{ij} a_i b_j
\end{aligned}
$$

where $d_{ii} = 2c_{ii}$, $d_{ij} = c_{ij}$ if $i < j$ and $d_{ij} = c_{ji}$ if $i > j$. It is clear that B is bilinear. Hence Q defined by (24) satisfies the axioms 1 and 2. Conversely, suppose Q satisfies 1 and 2. Then if, $a, b \in F$, $x, y \in V$.

$$
\begin{aligned}
Q(ax + by) &= Q(ax) + Q(by) + B(ax, by) \\
&= a^2 Q(x) + b^2 Q(y) + ab B(x, y).
\end{aligned}
$$

By induction, we have

$$Q\left(\sum a_i e_i \right) = \sum_i a_i^2 Q(e_i) + \sum_{i < j} a_i a_j B(e_i, e_j)$$

so $Q(x) = \sum_{i \le j} c_{ij} a_i a_j$ where $c_{ii} = Q(e_i)$ and $c_{ij} = B(e_i, e_j)$ if $i < j$. Hence Q has the form (24).

The bilinear form B associated with Q ($B(x, y) = Q(x + y) - Q(x) - Q(y)$) is symmetric and we have $B(x, x) = 2Q(x)$. If char $F \ne 2$, then Q is determined by B since $Q(x) = \frac{1}{2} B(x, x)$. If char $F = 2$, then $B(x, x) = 0$, so B is an alternate form. If B is a bilinear form then $Q(x) \equiv B(x, x)$ is quadratic form whose associated bilinear form is $B(x, y) + B(y, x)$.

If B is a symmetric bilinear form, then we have defined the radical of B, rad $B = V^\perp$, and B is non-degenerate if and only if rad $B = 0$. If Q is a quadratic form we define the *bilinear radical*, bilrad Q, to be the radical of the associated bilinear form B. On the other hand, we define the *radical* of Q, rad $Q = \{z \mid Q(x + z) = Q(x),\ x \in V\}$. Since $Q(x + z) = Q(x) + Q(z) + B(x, z)$ it is apparent that $z \in$ rad Q if and only if $Q(z) = 0$ and $B(x, z) = 0$ for all $x \in V$. Thus rad $Q \subset$ bilrad Q and rad Q is the subset of bilrad Q of z such that $Q(z) = 0$. It is clear from this or from the initial definition that rad Q is a subspace of V. If char $F \neq 2$, $B(x, z) = 0$ for all x implies $Q(z) = \frac{1}{2}B(z, z) = 0$, so that in this case rad $Q =$ bilrad Q. In general, we define the *defect* of Q to be the dimensionality of the factor space bilrad $Q/$rad Q or, equivalently, dim bilrad $Q -$ dim rad Q. This is 0 if char $F \neq 2$. On the other hand, if char $F = 2$, then $F^2 = \{a^2 \mid a \in F\}$ is a subfield of F and we can define the dimensionality $[F : F^2]$ of F regarded as a vector space over F^2. Suppose this is finite and let $z_1, z_2, \ldots, z_r \in$ bilrad Q. Since $r > [F : F^2]$ we can choose a_i not all 0 such that $\sum_1^r a_i^2 Q(z_i) = 0$. Then $Q(\sum a_i z_i) = \sum a_i^2 Q(z_i) = 0$ and so $\sum a_i z_i \in$ rad Q. Hence we see that any $r > [F : F^2]$ elements of bilrad Q are linearly dependent modulo rad Q. It follows that the defect of Q does not exceed $[F : F^2]$. In particular, if F is perfect, $F = F^2$, and then the defect of Q is either 0 or 1.

The theory of quadratic forms of arbitrary characteristic is interesting. However, as the foregoing indicates, the characteristic two case adds some complications. We shall therefore confine our attention to the case: char $F \neq 2$.[1] In this case our remarks show that the theory is equivalent to that of symmetric bilinear forms. At times the results will be presented as statements on quadratic forms and at times as statements on symmetric bilinear forms. We prove first the following diagonalization theorem.

THEOREM 6.5 *Let B be a symmetric bilinear form on a vector space V over a field F of characteristic $\neq 2$. Then there exists a base $(u_1, \ldots, u_r, z_1, \ldots, z_{n-r})$ such that the matrix of B relative to this base has the form*

(25) $\mathrm{diag}\{b_1, \ldots, b_r, 0, \ldots, 0\}, \qquad b_i \neq 0,\ 1 \leq i \leq r.$

A base $(u_1, u_2, \ldots, u_n)$ such that $B(u_i, u_j) = 0$ for all $i \neq j$ is called an *orthogonal base* for V (relative to B or a given quadratic form Q). Theorem 6.5 asserts the existence of such a base.

---

[1] Two references for the characteristic two case are: C. Chevalley, *The Algebraic Theory of Spinors*, New York, Columbia University Press, 1954 and J. Dieudonné, *La Géométrie des Groupes Classiques*, 2nd ed., Springer, 1963.

Proof. The method of proof we shall give is a constructive one which is due to Lagrange. We observe first that if $B = 0$, then any base $(z_1, z_2, \ldots, z_n)$ satisfies the condition. Hence suppose $B \neq 0$. We claim that this implies that there exists a $u \neq 0$ such that $B(u, u) \neq 0$. Otherwise, for every u, v

$$2B(u, v) = B(u, v) + B(v, u) = B(u + v, u + v) - B(u, u) - B(v, v) = 0$$

contrary to $B \neq 0$. Now choose u_1 so that $B(u_1, u_1) = b_1 \neq 0$. This gives a start for an inductive construction of the required base. Suppose then that we have already determined linearly independent vectors $(u_1, \ldots, u_k)$ such that $B(u_i, u_j) = \delta_{ij} b_i$, $b_i \neq 0$, $\delta_{ii} = 1$, $\delta_{ij} = 0$ if $i \neq j$, and let V_k be the subspace spanned by these u_i. We shall now show that $V = V_k \oplus V_k^\perp$. Since the matrix of the restriction of B to V_k is $\mathrm{diag}\{b_1, \ldots, b_k\}$ it is clear that this bilinear form is non-degenerate, so $V_k \cap V_k^\perp = 0$. Let $x \in V$ and put

$$y = x - \sum_i B(x, u_i) b_i^{-1} u_i.$$

Then $B(y, u_j) = B(x, u_j) - B(x, u_j) b_j^{-1} B(u_j, u_j) = 0$. Hence $y \in V_k^\perp$ and $x = y + \sum B(x, u_i) b_i^{-1} u_i \in V_k + V_k^\perp$. Thus $V = V_k \oplus V_k^\perp$. If the restriction of B to $V_k^\perp$ is 0 we take $r = k$ and let $(z_1, \ldots, z_{n-k})$ be any base for $V_k^\perp$. Otherwise, we choose a $u_{k+1} \in V_k^\perp$ such that $b_{k+1} = B(u_{k+1}, u_{k+1}) \neq 0$. Then $(u_1, \ldots, u_{k+1})$ is a linearly independent set satisfying the same conditions as $(u_1, \ldots, u_k)$. Repeating the process we finally achieve a base of the required type. $\square$

There are several remarks that are worth making about the proof. First, it really is constructive. To indicate a mechanical way of carrying it out we assume we have a set of vectors $\{e_i\}$ which span V. For example, this could be a base for V. If some $B(e_i, e_i) \neq 0$, then we can choose $u_1 = e_i$. Otherwise, assuming $B \neq 0$, we have $B(e_i, e_j) \neq 0$ for some pair $e_i \neq e_j$. Then $B(e_i + e_j, e_i + e_j) = 2B(e_i, e_j) \neq 0$ and we can take $u_1 = e_i + e_j$. Now suppose we have already determined $(u_1, \ldots, u_k)$ and V_k as in the proof. Then for each e_j in the given set of vectors spanning V we put $f_j = e_j - \sum_i B(e_j, u_i) b_i^{-1} u_i$. Then we see that the f_j span $V_k^\perp$ and we can repeat the process we applied to V. Another point which is of considerable theoretical interest is that b_1 can be taken to be any non-zero element of F which is *represented* by B in the sense that there exists a solution u_1 of the equation $B(u_1, u_1) = b_1$. Similarly, b_{k+1} is any non-zero element represented by the restriction of B to $V_k^\perp$. This generality in the choice of the b_i will enable us in some cases to make a number of the b's equal 1. We remark also that if the base is chosen as in the theorem, then the radical is spanned by the elements $z_1, \ldots, z_{n-r}$. For, if $z = \sum_1^r c_i u_i + \sum_1^{n-r} d_j z_j$, then $B(z, z_j) = 0$ for all z_j and $B(u_i, z) = 0$ if and only if $c_i = 0$. Hence $z \in V^\perp$ if and only if $z = \sum d_j z_j$.

We note finally that as a corollary of the proof we have another proof of the fact that if the restriction of B to a subspace U is non-degenerate, then $V = U \oplus U^\perp$ (exercise 2, p. 348), for we can choose an orthogonal base $(u_1, u_2, \ldots, u_k)$ for U. Then the proof shows that we can supplement this to obtain an orthogonal base $(u_1, \ldots, u_k, u_{k+1}, \ldots, u_n)$ for V. It is clear that we shall have $U^\perp = \sum_{k+1}^n Fu_j$ and $V = U \oplus U^\perp$.

From the matrix point of view, Theorem 6.5 provides a diagonal matrix cogredient to any given symmetric matrix. The difficulty is that there is no uniqueness about this. For example, if we replace u_i by $c_i u_i \neq 0$, then b_i is replaced by $c_i^2 b_i$, so the most we could hope for is that the b_i are determined to within squares. However, even this is not the case, since b_1 can be replaced by any non-zero element of the form $\sum b_i c_i^2$ and this may not be a square times any one of the b_i (see exercise 2 at the end of this section). The problem of classifying symmetric matrices relative to cogredience or, equivalently, symmetric bilinear forms relative to equivalence is generally a very difficult one which depends on arithmetic properties of the underlying field. For the case of $\mathbb{Q}$, or more generally F, an algebraic extension of $\mathbb{Q}$, one does have a complete solution due to Minkowski and Hasse. For $\mathbb{Q}$, the Minkowski result is that cogredience holds if and only if it holds in $\mathbb{R}$ and in certain extensions, the p-adic fields $\mathbb{Q}_p$ of $\mathbb{Q}$, defined for every prime number p.[2] The Minkowski-Hasse theorem is quite deep. On the other hand, there are several important types of fields for which the solution of the cogredience problem is easy. These include the following: (1) algebraically closed fields, (2) real closed fields, (3) finite fields. We shall now consider these.

1. *Algebraically closed fields.* In these fields every b_i can be replaced by 1, and so every symmetric matrix is cogredient to one of the form diag$\{1, \ldots, 1, 0, \ldots, 0\}$. The number of 1's is the rank of the diagonal matrix. Since for invertible p, $ps\,^tp$ and s have the same rank this is also the rank of the given matrix s (and also $n -$ dim rad B). Our result evidently implies

THEOREM 6.6. *If F is algebraically closed of characteristic $\neq 2$, then two symmetric matrices in $M_n(F)$ are cogredient if and only if they have the same rank.*

2. *Real closed fields.*[3] Suppose $F = R$ is real closed. Since positive elements have square roots the positive b_i in (25) can be replaced by 1's and the negative ones by -1's. Re-ordering the u_i we may assume that the canonical matrix is diag$\{1, \ldots, 1, -1, \ldots, -1, 0, \ldots, 0\}$. We shall show that the number of $+1$'s and hence the number of -1's is an invariant. Equivalently, we shall have that

---

[2] These fields are discussed in Vol. II.
[3] These have been defined in section 5.1, p. 308. The most important special case is the field $\mathbb{R}$ of real numbers.

the *signature* defined to be $p - q$ where p is the number of $+1$'s and q is the number of -1's is an invariant. This will follow from

THEOREM 6.7 (Sylvester). *Let F be an ordered field and suppose the diagonal matrices*

$$\text{diag}\{b_1, b_2, \ldots, b_r, 0, \ldots, 0\}, \qquad b_i \neq 0$$

$$\text{diag}\{b'_1, b'_2, \ldots, b'_r, 0, \ldots, 0\}, \qquad b'_i \neq 0$$

are cogredient. Then the number of positive b_i is the same as the number of positive b'_i.

Proof. We may assume that these are matrices of a symmetric bilinear form on an n dimensional vector space over the field F and that the first p b_i and p' b'_i are >0, the remaining ones negative. Let $(u_1, \ldots, u_r, z_1, \ldots, z_{n-r}), (u'_1, \ldots, u'_r, z'_1, \ldots, z'_{n-r})$ be the bases which give the matrices of the theorem. Suppose $z \in \sum_1^p Fu_i$, so $z = \sum_1^p a_i u_i$. Then $B(z, z) = \sum a_i^2 b_i > 0$ if $z \neq 0$. Similarly if $z \in \sum_{p'+1}^r Fu'_j + \sum Fz'_k$ then $B(z, z) \leq 0$. Hence the two spaces $\sum_1^p Fu_i$ and $\sum_{p'+1}^r Fu'_j + \sum Fz'_k$ have only the 0 vector in common. This implies that the sum of their dimensionalities does not exceed n (by the well-known formula $\dim(U_1 + U_2) = \dim U_1 + \dim U_2 - \dim(U_1 \cap U_2)$ for subspaces U_i of V). Thus we have $p + (n - p') \leq n$ and so $p \leq p'$. By symmetry, we have $p' \leq p$ and so $p = p'$. $\square$

This result implies

THEOREM 6.8. *Two diagonal matrices in $M_n(R)$, R a real closed field, are cogredient if and only if they have the same rank and same signature ($=$ number of positive elements minus the number of negative elements).*

Before proceeding to the case of symmetric bilinear forms over a finite field we shall give some definitions and remarks which are of general interest. If B is a non-degenerate symmetric bilinear form, then B is called *isotropic* or a *null form* if there exists a vector $u \neq 0$ such that $B(u, u) = 0$. Such a vector is called *isotropic*. A form which is not isotropic is called *anisotropic*. If B is isotropic, then B is *universal* in the sense that $B(v, v) = b$ has a solution for every $b \neq 0$ in F. For, assuming $B(u, u) = 0$ for $u \neq 0$, non-degeneracy of B implies that we have a w such that $B(u, w) = \frac{1}{2}$. Then for $v = au + w$ we have

$$B(v, v) = a + B(w, w).$$

Hence if we take $a = b - B(w, w)$ we obtain $B(v, v) = b$.

We now consider the case of

3. *Finite fields.* We shall show that these forms can be classified by their discriminants. We prove first the

LEMMA. *Any non-degenerate symmetric bilinear form B on a vector space V of ≥ 2 dimensions over a finite field F of characteristic $\neq 2$ is universal.*

Proof. It is enough to prove the result for binary forms, that is, for the case dim $V = 2$, and since we have proved universality in the isotropic case, we may assume B anisotropic. We may assume also that we have a diagonal matrix for the form. Hence we are reduced to proving that if $ab \neq 0$ and $ax^2 + by^2 \neq 0$ for all $(x, y) \neq (0, 0)$, then $ax^2 + by^2 = c$ is solvable for any $c \neq 0$. Dividing by a we may take $a = 1$. Now $x^2 + by^2 \neq 0$ implies $-b$ is not a square and $x^2 + by^2$ is the norm function for the quadratic extension K/F where $K = F(\sqrt{-b})$. If $|F| = q, |K| = q^2$ and the mapping $u \rightarrow u^q$ is an automorphism $\neq 1$ of K/F. Then $N_{K/F}(u) = uu^q = u^{q+1}$. Hence we have to show that for any $c \neq 0$ in F there exists a $u \in K^*$ such that $u^{q+1} = c$. Now K^* is cyclic of order $q^2 - 1$ and $u \rightarrow u^{q+1}$ is a homomorphism of K^* into F^*. The kernel is the subgroup of u satisfying $u^{q+1} = 1$. This has order $q + 1$ since the group is cyclic. Hence the image has order $(q^2 - 1)/(q + 1) = q - 1$, which implies that the homomorphism is surjective. This completes the proof (cf. exercise 1, p. 300, where a more general result is stated). $\square$

We can now prove

THEOREM 6.9. *Any non-degenerate symmetric bilinear form on a vector space V over a finite field (char $\neq 2$) has a matrix of the form $\mathrm{diag}\{1, 1, \ldots, 1, d\}$ Equivalently, any invertible symmetric matrix with entries in a finite field is co-gredient to one of the form $\mathrm{diag}\{1, 1, \ldots, 1, d\}$. Moreover, two invertible symmetric matrices with entries in a finite field are cogredient if and only if they have the same discriminant.*

Proof. The Lagrange diagonalization process and the foregoing lemma show that we can take $b_1 = b_2 = \cdots = b_{n-1} = 1$. This proves the first statement and the equivalent one on matrices. Since cogredient matrices have the same discriminant (the discriminant of the associated bilinear form) the last statement will follow if we can show that the two diagonal matrices $\mathrm{diag}\{1, 1, \ldots, 1, d_i\}$, $i = 1, 2, d_i \neq 0$, are cogredient if they have the same discriminant. This is clear since the discriminant is $d_i F^{*2}$, and $d_1 F^{*2} = d_2 F^{*2}$ implies that $d_2 = a^2 d_1$, $a \in F^*$, which implies the cogredience of the diagonal matrices. $\square$

EXERCISES

1. Find a diagonal matrix d cogredient in $M_3(\mathbb{Q})$ to

$$ s = \begin{pmatrix} -2 & 3 & 5 \\ 3 & 1 & -1 \\ 5 & -1 & 4 \end{pmatrix}. $$

   Also determine a matrix p such that $ps\,'p = d$.

2. Show that

$$ \mathrm{diag}\{1, 1\}, \qquad \mathrm{diag}\{5, 5\} $$

   are cogredient in $M_2(\mathbb{Q})$.

3. Show that the symmetric bilinear form B in V over $\mathbb{R}$ is *positive definite* in the sense that $B(u, u) > 0$ for all $u \neq 0$ if and only if it has 1 as one of its matrices. Use the Lagrange reduction (in this case called the Schmidt orthogonalization process) to prove that if s is a matrix of a positive definite symmetric bilinear form there exists a triangular matrix p with 0's above the main diagonal such that $ps\,'p = 1$ or $s = q\,'q,\ q = p^{-1}$.

4. Same hypotheses as exercise 3. Call a base $(u_1, u_2, \ldots, u_n)$ *Cartesian* if $B(u_i, u_j) = \delta_{ij}$. Show that if $(v_1, v_2, \ldots, v_n)$ is a second such base then the matrix relating the two is orthogonal ($o\,'o = 1$). Use the result of exercise 3 to show that if m is any invertible matrix in $M_n(\mathbb{R})$, m can be written in the form po where p is triangular and o is orthogonal.

5. Prove that the set of polynomial functions on V can be defined as the subring of the ring of maps from V to F generated by the linear functions. Here addition and multiplication of maps from V to F are the usual ones: $(f + g)(x) = f(x) + g(x)$, $(fg)(x) = f(x)g(x)$. This gives an intrinsic definition of polynomial functions.

6. Let Q be a non-degenerate quadratic form on an $n \geq 3$ dimensional vector space over a finite field. Show that Q is isotropic.

6.4 BASIC CONCEPTS OF ORTHOGONAL GEOMETRY

We shall now introduce the basic definitions of orthogonal geometry (the study of a vector space relative to a non-degenerate symmetric bilinear form). We assume char $F \neq 2$, so it is all the same whether we deal with symmetric bilinear forms B or quadratic forms Q. Given a quadratic form Q we have the associated symmetric bilinear form $B(x, y) = Q(x + y) - Q(x) - Q(y)$, and given a symmetric bilinear form B we have the associated quadratic form $Q(x) = \frac{1}{2}B(x, x)$. We shall call Q non-degenerate if B is non-degenerate.

Let (V_i, Q_i), $i = 1, 2$, be a pair consisting of a vector space V_i and a quadratic form Q_i on V_i. Then we define an *isometry* η of (V_1, Q_1) onto (V_2, Q_2) to be a bijective linear map of V_1 onto V_2 such that $Q_2(\eta x) = Q_1(x)$ for all $x \in V_1$. This implies that if B_i is the corresponding symmetric bilinear form of Q_i, then $B_2(\eta x, \eta y) = B_1(x, y)$, $x, y \in V_1$, since

$$B_2(\eta x, ny) = Q_2(\eta x + \eta y) - Q_2(\eta x) - Q_2(\eta y)$$
$$= Q_2(\eta(x + y)) - Q_2(\eta x) - Q_2(\eta y)$$
$$= Q_1(x + y) - Q_1(x) - Q_1(y)$$
$$= B_1(x, y).$$

The converse is immediate also since $Q_i(x) = \frac{1}{2}B_i(x, x)$. If B_1 is nondegenerate, the requirement of injectivity is superfluous: if η is a linear map of V_1 into V_2 satisfying $Q_2(\eta x) = Q_1(x)$, $x \in V_1$, then η has to be injective; for, $B_2(\eta x, \eta y) = B_1(x, y)$ and $\eta x = 0$ imply $B_1(x, y) = 0$ for all y. Then $x = 0$ by the non-degeneracy of B_1.

If there exists an isometry of (V_1, Q_1) onto (V_2, Q_2) then the quadratic forms Q_1 and Q_2 and the associated symmetric bilinear forms B_1 and B_2 are called *equivalent*.

If Q is a non-degenerate quadratic form on V, an isometry of V onto V is called an *orthogonal transformation* of V (or of (V, Q)). It is clear that any linear transformation of V into itself satisfying $Q(\eta x) = Q(x)$, $x \in V$, is orthogonal, for, we have seen that this implies that η is injective, and since we always assume V finite dimensional, η is also surjective. If η is orthogonal, then so is η^{-1}, and if η_1 and η_2 are orthogonal, then so is $\eta_1\eta_2$. Thus the set $O(V, Q)$ of orthogonal transformations is a subgroup of the group of bijective linear transformations of V. This group is called the *orthogonal group* of V relative to Q.

Let $(e_1, e_2, \ldots, e_n)$ be a base for V and let $\eta \in O(V, Q)$. Then we have $B(\eta e_i, \eta e_j) = B(e_i, e_j)$ for all $i, j = 1, 2, \ldots, n$. Conversely, if these conditions hold for a linear transformation η, then for any $x = \sum a_i e_i$ we have

$$Q(\eta x) = Q\left(\sum a_i(\eta e_i)\right) = \sum_i a_i^2 Q(\eta e_i) + \sum_{i<j} a_i a_j B(\eta e_i, \eta e_j)$$
$$= \sum_i a_i^2 Q(e_i) + \sum_{i<j} a_i a_j B(e_i, e_j) = Q(x).$$

Thus a linear transformation η of V into V is orthogonal if and only if

(26) $$B(\eta e_i, \eta e_j) = B(e_i, e_j), \qquad 1 \le i, j \le n.$$

Now let $b = (B(e_i, e_j))$ the matrix of B relative to the base $(e_1, e_2, \ldots, e_n)$ and let $\eta e_i = \sum h_{ij} e_j$. Then the conditions (26) are that

$$B\left(\sum_k h_{ik} e_k, \sum_l h_{jl} e_l\right) = \sum_{k,l} h_{ik} B(e_k, e_l) h_{jl} = B(e_i, e_j)$$

for all i and j. In matrix form these conditions are

(27) $$hb\,{}^t h = b \quad \text{for} \quad h = (h_{ij}).$$

Hence these are necessary and sufficient conditions on the matrix h of η relative to the base $(e_1, e_2, \ldots, e_n)$ for η to be orthogonal. If we take the determinants of the matrices in (27) we obtain $(\det h)^2 \det b = \det b$, and since $\det b \neq 0$, we see that $\det h = \pm 1$ for the matrix of an orthogonal transformation relative to any base. Since the determinant of the matrix of a linear transformation is unchanged on changing the base it is clear that if we have $\det h = 1$ or -1 relative to one base we shall have the same thing for every other base. An orthogonal transformation is called *proper* or a *rotation* if $\det h = 1$; otherwise, the transformation is *improper*. If we choose an orthogonal base for V, then the matrix b of B is diagonal. Then it is clear that any diagonal matrix with diagonal entries 1 or -1 satisfies (27) and hence determines an orthogonal transformation. Moreover, this is proper or improper according as the number of -1's is even or odd. It is clear that the set $O^+ (V, Q)$ of rotations is a normal subgroup of index two in $O(V, Q)$.

With any vector u such that $Q(u) \neq 0$ we can associate an orthogonal transformation S_u defined by

(28) $$S_u : x \to x - \frac{B(x, u)}{Q(u)} u.$$

Since $x \to x$ and $x \to B(x, u)v$ are linear for any u and v, S_u is linear. Moreover

$$Q(S_u x) = Q\left(x - \frac{B(x, u)}{Q(u)} u\right) = Q(x) + \frac{B(x, u)^2}{Q(u)^2} Q(u) - \frac{B(x, u)}{Q(u)} B(x, u) = Q(x).$$

Hence S_u is orthogonal. Now $S_u u = u - (B(u, u)/Q(u))u = u - 2u = -u$ and if $v \perp u$, then $S_u v = v$. Since $B(u, u) \neq 0$ we have the decomposition $V = Fu \oplus Fu^\perp$ and the result we have just indicated gives a complete description of S_u, namely, this linear transformation is the identity map on $Fu^\perp$ and it sends u into $-u$. We shall call S_u the *symmetry determined by* u. If we choose a base for V consisting of u and a base for $Fu^\perp$, then the matrix of S_u relative to this base is $\text{diag}\{-1, 1, \ldots, 1\}$. Evidently this implies that S_u is improper. It is clear also that $S_u^2 = 1$.

If η is any orthogonal transformation, then we have

(29) $$\eta S_u \eta^{-1} = S_{\eta u}.$$

To see this we calculate

$$\eta S_u \eta^{-1} x = \eta \left(\eta^{-1} x - \frac{B(\eta^{-1} x, u)}{Q(u)} u \right)$$

$$= \eta \left(\eta^{-1} x - \frac{B(\eta \eta^{-1} x, \eta u)}{Q(u)} u \right)$$

$$= x - \frac{B(x, \eta u)}{Q(u)} \eta u$$

$$= x - \frac{B(x, \eta u)}{Q(\eta u)} \eta u$$

$$= S_{\eta u} x.$$

In this verification we have made use of the property $B(\eta x, y) = B(\eta^{-1} \eta x, \eta^{-1} y) = B(x, \eta^{-1} y)$ for any orthogonal transformation. We have defined the adjoint T' of a linear transformation T relative to B by the condition $B(Tx, y) = B(x, T'y)$ (exercise 4, p. 349). Thus the condition we have derived is that the adjoint of an orthogonal transformation coincides with its inverse. It is immediate also that this property, that is, $TT' = 1$, implies that T is orthogonal.

A very important observation about adjoints is that if U is a subspace stabilized by a linear transformation T (that is, $T(U) \subset U$), then $U^\perp$ is stabilized by T'. This is clear, since if $v \in U^\perp$, so $B(u, v) = 0$ for all $u \in U$, then $B(u, T'v) = B(Tu, v) = 0$. It follows that if η is orthogonal and stabilizes U, then $\eta U = U$ and hence $\eta^{-1} U = U$. Since $\eta^{-1} = \eta'$ we see that η' stabilizes U; hence $\eta = (\eta')'$ stabilizes $U^\perp$.

We shall say that V is an *orthogonal direct sum* of the subspaces $U_1, U_2, \ldots, U_r$ if $V = U_1 \oplus U_2 \oplus \cdots \oplus U_r$ and $U_i \perp U_j$ for every $i \neq j$. In this case we write

(30) $$V = U_1 \perp U_2 \perp \cdots \perp U_r.$$

Then if $x = \sum x_i$, $x_i \in U_i$, $Q(x) = \sum Q(x_i) + \sum_{i<j} B(x_i, x_j) = \sum Q(x_i)$. Now, if we have a second decomposition $V = U_1' \perp U_2' \perp \cdots \perp U_r'$ and isometries $\eta_i : U_i \to U_i'$, then the linear transformation such that $\eta | U_i = \eta_i$ satisfies $Q(\eta x) = Q(\sum \eta_i x_i) = \sum Q(\eta_i x_i) = \sum Q(x_i) = Q(x)$. Hence η is an orthogonal transformation. It is clear also that the subspaces U_i in (30) are non-degenerate (that is, the restriction of B to U_i is non-degenerate).

A subspace U is *isotropic* if it contains an isotropic vector ($u \neq 0$, $Q(u) = 0$) and U is *totally isotropic* if the restriction of Q to U is 0, or, equivalently, $U \subset U^{\perp}$.

A two dimensional space V which is non-degenerate and isotropic is called a *hyperbolic plane*. The following theorem says about everything one can say about hyperbolic planes:

THEOREM 6.10. (1) *The following conditions on a two dimensional vector space V equipped with a quadratic form Q are equivalent:* (i) V *is a hyperbolic plane,* (ii) V *has a base* (u, v) *which is a* hyperbolic pair *of vectors in the sense that*

$$(31) \qquad B(u, u) = 0 = B(v, v), \qquad B(u, v) = 1 = B(v, u)$$

(iii) *The discriminant of B is* $-1F^{*2}$. (2) *Any two hyperbolic planes are isometric.* (3) *Any hyperbolic plane contains exactly two one dimensional totally isotropic subspaces.* (4) *The rotation group of a hyperbolic plane V is isomorphic to the multiplicative group F^* of the field F and every improper orthogonal transformation of V is a symmetry.*

Proof. (1) If V is a hyperbolic plane, V contains a vector $u \neq 0$ such that $Q(u) = 0$, and since $V^{\perp} = 0$, V contains a vector v such that $B(u, v) \neq 0$. Since $B(u, u) = 0$, v is not a multiple of u. Hence (u, v) is a base. Replacing v by a multiple of v we may assume $B(u, v) = 1$. Moreover, if $a \in F$ then $Q(v + au) = Q(v) + a$, so if we replace v by $v - Q(v)u$ we shall have $Q(v) = 0$ as well as $B(u, v) = 1$. Then we have (31) for the base (u, v). Thus (i) $\Rightarrow$ (ii). Now assume (ii). Clearly the determinant of the matrix defined by (31) is -1. Hence the discriminant of B is $-1F^{*2}$. Now assume that the discriminant of B is $-1F^{*2}$. Then we have a base (u_1, u_2) such that the matrix of B relative to (u_1, u_2) is $\mathrm{diag}\{b_1, b_2\}$ where $b_1 b_2 = -c^2 \neq 0$, $c \in F$. Let $x = cu_1 + b_1 u_2$. Then $Q(x) = \frac{1}{2}c^2 b_1 + \frac{1}{2}b_1{}^2 b_2 = 0$. Hence V is a hyperbolic plane and we have proved the implication (iii) $\Rightarrow$ (i). (2) This is clear since any two hyperbolic planes have bases (u, v) and (u', v') which are hyperbolic pairs. It is evident that the linear map sending $u \to u'$, $v \to v'$ is an isometry. (3) Let (u, v) be a base which is a hyperbolic pair. Then $Q(au + bv) = ab$. Hence $au + bv$ is isotropic if and only if either $a = 0$, $b \neq 0$ or $a \neq 0$, $b = 0$. Then Fu and Fv are the only one dimensional totally isotropic subspaces of V. (4) Let η be an orthogonal transformation of the hyperbolic plane V and let Fu and Fv be the two totally isotropic one dimensional subspaces. Then either $\eta(Fu) = Fu$ and $\eta(Fv) = Fv$ or $\eta(Fu) = Fv$ and $\eta(Fv) = Fu$. In the first case we have $\eta u = au$ and $\eta v = bv$, and $ab B(u, v) = B(\eta u, \eta v) = B(u, v)$ gives $ab = 1$, since $B(u, v) \neq 0$. Hence $\eta u = au$

and $\eta v = a^{-1}v$, and clearly η is a rotation. Now assume the second possibility. Then $\eta u = av$ and $\eta v = bu$, and again we have $b = a^{-1}$. This time η is improper. Also, it is clear that for any $a \neq 0$ the linear maps such that $u \to au, v \to a^{-1}v$ and $u \to av, v \to a^{-1}u$ are respectively rotations or improper orthogonal transformations. Then the map of $a \in F^*$ into the rotation $u \to au, v \to a^{-1}v$ is an isomorphism of F^* with $O^+(V, Q)$. Finally, if η is an improper orthogonal transformation, so that $\eta u = av$ and $\eta v = a^{-1}u$, then $\eta(u + av) = u + av$ and $\eta(u - av) = -(u - av)$. Hence η is the symmetry $S_{u - av}$. $\quad\square$

EXERCISES

1. Show that if η is an orthogonal transformation and $V_1 = \{x \mid \eta x = x\}$, then dim $V = \dim V_1 + \dim (1 - \eta)V$. Show also that $V_1 = ((1 - \eta)V)^\perp$ and hence $V_1^\perp = (1 - \eta)V$.

2. Let η be an orthogonal transformation such that dim $V_1 \geq \dim V - 1$, where V_1 is as in exercise 1. Show that either $\eta = 1$ or η is a symmetry.

3. Let (u, v) be a hyperbolic pair and let $w \in (Fu + Fv)^\perp$ be non-isotropic. Verify that the linear transformation ρ defined by

    $$u \to u$$
    $$v \to v - Q(w)u - w$$
    $$x \to x + B(x, w)u, \qquad x \in (Fu + Fv)^\perp$$

    coincides with $S_w S_{w - Q(w)u}$. (Note that $Q(w - Q(w)u) \neq 0$.)

4. Let V be equipped with a non-degenerate quadratic form Q and let Σ be a set of linear transformations of V into V such that Σ is closed under adjoints: if $T \in \Sigma$ then its adjoint T' relative to B is contained in Σ. Show that if U is a subspace stabilized by Σ ($TU \subset U$ for all $T \in \Sigma$), then $U^\perp$ is stabilized by Σ.

5. Let Q be anisotropic. Show that 0 is the only nilpotent self-adjoint linear transformation in V relative to Q.

6. Call a linear transformation T *unipotent* if $T - 1 = Z$ is nilpotent. Show that if Q is anisotropic then 1 is the only unipotent orthogonal linear transformation in V relative to Q. Verify that the transformation ρ defined in exercise 3 is unipotent. Hence prove that if dim $V \geq 3$ and Q is isotropic, then $O(V, Q)$ contains unipotent orthogonal transformations $\neq 1$.

7. (Malcev.) Let T be a nilpotent self-adjoint linear transformation in V relative to B (non-degenerate). Show that V is an orthogonal direct sum of subspaces V_i where V_i has a base $(z_i, Tz_i, \ldots, T^{n_i - 1}z_i)$ and the matrix of B relative to this

base has the form

$$\begin{pmatrix} 0 & & b \\ & \ddots & \\ & b & \\ b & & 0 \end{pmatrix}, \quad b \neq 0.$$

Hence show that there exist nilpotent self-adjoint linear transformations $\neq 0$ in (V, Q) if and only if Q is isotropic.

8. Let T be a linear transformation in a finite dimensional vector space V. Show that there exists a non-degenerate symmetric bilinear form B on V such that T is self-adjoint relative to B. (*Hint*: It suffices to assume T is "cyclic", that is, $V = F[T]u$ for some vector u. Then we have a base $(u_1, u_2, \ldots, u_n)$ for V over F such that $Tu_i = u_{i+1}$, $1 \leq i \leq n - 1$, $Tu_n = \sum_1^n \alpha_i u_i$. Let B be the bilinear form on V such that $B(u_i, u_j) = \lambda_{i+j-n} \in F$ where $\lambda_1 = 1$ and $\lambda_k = 0$ if $k \leq 0$. Show that B is symmetric and non-degenerate and the λ's can be chosen so that T is self-adjoint relative to B.)

6.5 WITT'S CANCELLATION THEOREM

We shall now prove a basic theorem on quadratic forms on a vector space V over F, char $F \neq 2$, which oddly enough—in spite of its importance and elementary character—was discovered rather late in the development of the theory. Among its important consequences are a reduction of the classification problem for quadratic forms to anisotropic forms, and a definition of a numerical invariant called the Witt index, which generalizes the notion of the signature of a real quadratic form. The result also implies an extension theorem for isometries to orthogonal transformations. After the preparations of the last section we can begin right in with the proof of this theorem, namely,

WITT'S CANCELLATION THEOREM. *Let Q be a non-degenerate quadratic form on a vector space V over a field F of characteristic $\neq 2$, U_1 and U_2 non-degenerate subspaces which are isometric (that is, there exists an isometry between them). Then $U_1^\perp$ and $U_2^\perp$ are isometric.*

(Since $V = U_1 \oplus U_1^\perp = U_2 \oplus U_2^\perp$ this does appear to be a "cancellation" theorem.)

Proof. We denote isometry by $\sim$, so we are given $U_1 \sim U_2$, and the restrictions of Q to U_1 and U_2 are non-degenerate. We wish to show that $U_1^\perp \sim U_2^\perp$. We shall use induction on dim U_i. Suppose first $U_i = Fu_i$ and $Q(u_i) \neq 0$. We may assume that $Q(u_1) = Q(u_2)$. We have $Q(u_1 \pm u_2) = 2Q(u_1) \pm B(u_1, u_2)$. Hence either $Q(u_1 + u_2) \neq 0$ or $Q(u_1 - u_2) \neq 0$. Suppose first that $Q(u_1 + u_2) \neq 0$ and consider the symmetry $S_{u_1 + u_2}$. Since $B(u_1 + u_2, u_1 - u_2) = 2Q(u_1) - 2Q(u_2) = 0$, $(u_1 - u_2) \perp (u_1 + u_2)$ and so $S_{u_1 + u_2}(u_1 - u_2) = u_1 - u_2$. On the other hand, $S_{u_1 + u_2}(u_1 + u_2) = -(u_1 + u_2)$. Then $S_{u_1 + u_2}u_1 = -u_2$ and consequently $S_{u_1 + u_2}(Fu_1)^{\perp} = (Fu_2)^{\perp}$ so $U_1{}^{\perp} \sim U_2{}^{\perp}$. Similarly, if $Q(u_1 - u_2) \neq 0$, then we can use $S_{u_1 + u_2}$ and note that this maps $u_1 + u_2$ into itself and $u_1 - u_2$ into $u_2 - u_1$. Then $S_{u_1 + u_2}u_1 = u_2$ and $S_{u_1 + u_2}(Fu_1)^{\perp} = (Fu_2)^{\perp}$. Now suppose the result holds for subspaces of dimensionality dim $U_i - 1 \geq 1$. We can choose a non-isotropic vector u_1 in U_1 and write $U_1 = Fu_1 \perp W_1$. Then W_1 is non-degenerate. Applying an isometry of U_1 onto U_2 we obtain $U_2 = Fu_2 \perp W_2$ where $Fu_1 \sim Fu_2$ and $W_1 \sim W_2$. Then $V = Fu_1 \perp W_1 \perp U_1{}^{\perp} = Fu_2 \perp W_2 \perp U_2{}^{\perp}$. Applying the result in the one dimensional case to Fu_1 and Fu_2 we conclude that there is an isometry η sending $W_1 \perp U_1{}^{\perp}$ onto $W_2 \perp U_2{}^{\perp}$. Then we have $W_2 \perp U_2{}^{\perp} = \eta(W_1) \perp \eta(U_1{}^{\perp})$ and $W_2 \sim W_1 \sim \eta(W_1)$. Hence the induction hypothesis applied to the subspaces W_2 and $\eta(W_1)$ of $W_2 \perp U_2{}^{\perp}$ implies that $U_2{}^{\perp} \sim \eta(U_1{}^{\perp}) \sim U_1{}^{\perp}$. □

Suppose, as in the theorem, U_1 and U_2 are non-degenerate subspaces and we have an isometry η of U_1 onto U_2. Then the theorem gives an isometry ζ of $U_1{}^{\perp}$ onto $U_2{}^{\perp}$. Since $V = U_1 \perp U_1{}^{\perp} = U_2 \perp U_2{}^{\perp}$ the linear map of V into V which coincides with η on U_1 and with ζ on $U_1{}^{\perp}$ is an orthogonal transformation which is an extension of the given isometry. We shall now show that this result holds for arbitrary subspaces: any isometry between subspaces of V can be extended to an orthogonal transformation of V. We shall base the proof on a canonical imbedding of a degenerate subspace in a non-degenerate one which will effect a reduction of the proof to the non-degenerate case. The imbedding theorem we require is

THEOREM 6.11 *Let V be equipped with a non-degenerate quadratic form and let U be a subspace such that* rad $U = U \cap U^{\perp} \neq 0$. *Write $U = $ rad $U \oplus U'$ where U' is a subspace and let $(z_1, \ldots, z_r)$ be a base for* rad U. *Then we can imbed U in a non-degenerate subspace $U \oplus W$ where W has a base $(w_1, \ldots, w_r)$ such that (z_i, w_i) is a hyperbolic pair for $1 \leq i \leq r$, and $U + W = U' \perp H_1 \perp H_2 \perp \cdots \perp H_r$, $H_i = Fz_i + Fw_i$, a hyperbolic plane.*

Proof. Let f be the linear function on U such that $f(z_1) = 1$, $f(z_i) = 0$ for $i > 1$, and $f(u') = 0$ for $u' \in U'$. By the lemma on p. 347, there exists a $w_1 \in V$ such that $f(u) = B(u, w_1)$, $u \in U$. Thus $B(z_1, w_1) = 1$, $B(z_i, w_1) = 0$ for $i > 1$, $B(u', w_1) = 0$, $u' \in U'$. Replacing w_1 by a suitable $w_1 + az_1$ we may assume

$Q(w_1) = 0$, and thus (z_1, w_1) is a hyperbolic pair (hence linearly independent). We have $V = (Fz_1 + Fw_1) \oplus (Fz_1 + Fw_1)^\perp$ and $U_1 = U' + \sum_{j>1} Fz_j \subset V_1 \equiv (Fz_1 + Fw_1)^\perp$. The radical of U_1 is $\sum_{j>1} Fz_j$. If $r = 1$ we take $W = Fw_1$ and we have $U + W = U' \perp H_1$ where H_1 is the hyperbolic plane $Fz_1 + Fw_1$. If $r > 1$ we replace the pair of spaces V, U by the pair V_1, U_1 and observe that the dimension of the radical of U_1 is $r - 1$. Hence, using induction on the dimensionality of the radical of the subspace, we obtain vectors $w_2, \ldots, w_r$ in V_1 such that $U_1 + \sum_j Fw_j = U' \perp H_2 \perp \cdots \perp H_r$, $H_j = Fz_j + Fw_j$, a hyperbolic plane. Then $W = \sum_1^r Fw_i$ satisfies our requirements. $\square$

Now suppose we have an isometry η of a subspace U_1 onto a subspace U_2. If U_1 is non-degenerate so is U_2 and we have seen that η can be extended to an orthogonal transformation η mapping U_1 into U. Then $\eta^{-1}U$ is a totally iso- $\oplus U_1'$, U_1' a subspace. By Theorem 6.11, there exists a non-degenerate subspace $U_1 + W_1 = U_1' \perp H_1 \perp \cdots \perp H_r$ where $H_i = Fz_i + Fw_i$ and (z_i, w_i) is a hyperbolic pair. We can imbed $U_2 = \eta U_1$ in $U_2 + W_2 = \eta(U_1') + H_1' \perp \cdots \perp H_r'$ where $H_i' = F(\eta z_i) + Fw_i'$ and $(\eta z_i, w_i')$ is a hyperbolic pair. Now it is clear that the linear map of $U_1 + W_1$ onto $U_2 + W_2$ which coincides with η on U_1 and sends $w_i \to w_i'$, $1 \leq i \leq r$, is an isometry of $U_1 + W_1$ onto $U_2 + W_2$ that coincides with η on U_1. Since $U_1 + W_1$ is non-degenerate this can be extended to an orthogonal transformation. Thus η can be extended to an orthogonal transformation. Hence we have proved

WITT'S EXTENSION THEOREM. *If V is equipped with a non-degenerate quadratic form Q, any isometry of a subspace U_1 onto a subspace U_2 can be extended to an orthogonal transformation.*

This applies in particular to subspaces which are totally isotropic $(Q|U_1 = 0)$. If U_1 and U_2 are two such subspaces of the same dimensionality, then the extension theorem implies that there is an orthogonal transformation mapping U_1 onto U_2. If U is a totally isotropic subspace having maximal dimensionality for such subspaces and U_1 is any totally isotropic subspace, then we have an orthogonal transformation η mapping U_1 into U. Then $\eta^{-1}U$ is a totally isotropic subspace containing U_1. It follows that all maximal totally isotropic subspaces have the same dimensionality. The common dimensionality of maximal totally isotropic subspaces is called the *Witt index* of Q; we shall denote this as $\nu(Q)$.

Now let U be a totally isotropic subspace of dimensionality $\nu = \nu(Q)$. Theorem 6.11 shows that we can imbed U in a subspace $U + W$ which is an orthogonal direct sum of ν hyperbolic planes. Thus $2\nu \leq n = \dim V$ and so $\nu(Q) \leq [n/2]$.

We can also write $V = (U + W) \oplus (U + W)^\perp$ and it is clear that the subspace $X \equiv (U + W)^\perp$ is anisotropic, that is, it contains no isotropic vectors. We have the decomposition

(32) $$V = H_1 \perp H_2 \perp \cdots \perp H_v \perp X$$

where H_i is a hyperbolic plane, $1 \leq i \leq v$, and X is anisotropic. Next suppose $V = H_1' \perp H_2' \perp \cdots \perp H_r' \perp Y$ is a decomposition of V as orthogonal direct sum of hyperbolic planes H_i' and an anisotropic subspace Y. If z_i' is a non-zero vector in H_i' such that $Q(z_i) = 0$, then $\sum_1^r F z_i'$ is an r-dimensional totally isotropic subspace. Hence $r \leq v$. Moreover, there is an orthogonal transformation sending $H_1 \perp \cdots \perp H_r$ into $H_1' \perp \cdots \perp H_r'$. This maps $(\sum_1^r H_i)^\perp = H_{r+1} \perp \cdots \perp H_v \perp X$ onto $(\sum_1^r H_i')^\perp = Y$. Since Y is anisotropic we must have $r = v$ and $X \sim Y$. We shall call any anisotropic subspace Y, such that V is an orthogonal direct sum of Y and hyperbolic planes, an *anisotropic kernel* of V. Our result shows that any two of these are isometric. It is clear that this implies that two non-degenerate quadratic forms are equivalent if and only if their anisotropic kernels (obvious meaning) are equivalent. This reduces the problem of classifying quadratic forms into equivalence classes to the case of anisotropic forms.

EXERCISES

1. Call Q of *maximal Witt index* if $v(Q) = [n/2]$. Show that any two non-degenerate quadratic forms of maximal Witt index are equivalent if n is even, and that they are equivalent if n is odd if and only if the associated bilinear forms have the same discriminant.

2. Show that if Q is a non-degenerate quadratic form in a vector space V over $\mathbb{R}$, then $v(Q) = \frac{1}{2}(n - |\text{sig } Q|)$ where sig Q is the signature of Q.

3. (Cayley.) Let η be a linear transformation in V equipped with a non-degenerate quadratic form Q, and let η' denote the adjoint of η relative to the corresponding symmetric bilinear form B. Let η be orthogonal (so $\eta' = \eta^{-1}$) and suppose $\det(\eta + 1) \neq 0$. Define

   $$\sigma = (1 - \eta)(1 + \eta)^{-1} = (1 + \eta)^{-1}(1 - \eta) \qquad \text{(the *Cayley transform* of η).}$$

   Show that σ is skew relative to B in the sense that $\sigma' = -\sigma$ and that $\det(\sigma + 1) \neq 0$. Show that $\eta = (1 - \sigma)(1 + \sigma)^{-1} = (1 + \sigma)^{-1}(1 - \sigma)$.

4. Use exercise 3 to prove that if V is odd dimensional, then every proper orthogonal transformation has a non-zero fixed point ($\eta x = x$), and if V is even dimensional then every improper orthogonal transformation has a non-zero fixed point.

5. Let Q be a quadratic form of maximal Witt index v on an n-dimensional vector space V, $n = 2v$. Let U be a totally isotropic subspace of V, $(u_1, u_2, \ldots, u_v)$ a base for U. By Theorem 6.11, there exists a base $(u_1, \ldots, u_v, w_1, \ldots, w_v)$ for V such that the matrix of B relative to this base is

(33)
$$\begin{pmatrix} 0 & 1_v \\ 1_v & 0 \end{pmatrix},$$

1_v the $v \times v$ unit matrix. Show that $(u_1, \ldots, u_v, w'_1, \ldots, w'_v)$ is a base such that the matrix of B relative to this base is (33) if and only if $w'_i = w_i + v_i$ where $v_i \in U$ and $B(u_i, v_j) = -B(u_j, v_i)$, $1 \le i, j \le n$. Note that this is equivalent to: $v_i = \sum s_{ij} u_j$ where $S = (s_{ij})$ is skew symmetric.

6. Let Q, V, and U be as in exercise 5. Let G_U be the subgroup of $O(V, Q)$ of η which fix every $u \in U$. Show that G_U is isomorphic to the additive group of $n \times n$ skew symmetric matrices with entries in F and hence that G_U is abelian. Show that $G_U \subset O^+(V, Q)$.

7. Let Q, V, U, $(u_1, \ldots, u_v)$, $(w_1, \ldots, w_v)$ be as in exercise 5 and put $W = \sum Fw_i$. Let $H_{U,W}$ be the subgroup of orthogonal transformations such that $\eta(U) = U$ and $\eta(W) = W$. Show that $H_{U,W} \cong GL_v(F)$, the group of $v \times v$ invertible matrices with entries in F.

8. Use the same notations as in exercise 7. Let $\eta \in G_U$ as defined in exercise 6. Determine the subspace V_1 of vectors fixed under η. Show that there exist η such that $V_1 = U$ if and only if v is even, and then n is divisible by four.

6.6 THE THEOREM OF CARTAN-DIEUDONNÉ

E. Cartan has proved for quadratic forms over the reals or complexes that any orthogonal transformation is a product of at most n symmetries, where n is the dimensionality of the underlying vector space. This result was generalized by Dieudonné to quadratic forms over arbitrary base fields. We shall prove first a cheap version of this theorem, namely:

THEOREM 6.12. *Any orthogonal transformation is a product of symmetries.*

Proof. Let η be orthogonal and let u be a vector with $Q(u) \ne 0$. As in the proof of Witt's cancellation theorem, there exists a symmetry S_w, $w = u + \varepsilon\eta u$, such that $\eta'u = -\varepsilon u$ for $\eta' = S_w\eta$ and ε either 1 or -1. Then η' stabilizes $Fu^\perp$, which is a non-degenerate subspace of dimensionality $n - 1$. Using induction on the dimensionality we see that the restriction $\eta'|Fu^\perp = \bar{S}_{w_1}\bar{S}_{w_2}\cdots\bar{S}_{w_k}$ where $\bar{S}_{w_i}$ is the symmetry in $Fu^\perp$ determined by $w_i \in Fu^\perp$. Then $\bar{S}_{w_i} = S_{w_i}|Fu^\perp$, and since $u \perp w_i$, S_{w_i} fixes u. Then $\eta'' \equiv S_{w_k}S_{w_{k-1}}\cdots S_{w_1}\eta'$ is the identity on $Fu^\perp$ since $(S_{w_1}S_{w_2}\cdots S_{w_k})^{-1} = S_{w_k}S_{w_{k-1}}\cdots S_{w_1}$ (by $S_w^2 = 1$). Also $\eta''u = \eta'u = \pm u$. If

$\eta''u = u$, $\eta'' = 1$, and if $\eta''u = -u$, then $\eta'' = S_u$. In either case η' is a product of symmetries, hence $\eta = S_w\eta'$ is such a product.   □

Before giving the proof of the more precise Cartan–Dieudonné theorem we note a result which we have previously stated in an exercise (exercise 1, p. 366). If η is orthogonal the subspace V_1 of fixed points under η is the orthogonal complement of the range of $1 - \eta$. This means that $\eta x = x$ if and only if $x \perp (1 - \eta)y$ for all y. Now $\eta x = x$ if and only if $\eta^{-1}x = x$ and hence if and only if $\eta'x = x$. Since B is non-degenerate this holds if and only if $B((1 - \eta')x, y) = 0$ for all y. Since $B((1 - \eta')x, y) = B(x, (1 - \eta)\,y)$ the result is clear.

We have called a linear tranformation T unipotent if $T - 1 = Z$ is nilpotent. We now note that this condition implies that det T (the determinant of any matrix of T) $= 1$. This can be seen by using the Jordan canonical matrix of T as on p. 199. However, we can also see it in the following way, which is more elementary. We observe first that T has a non-zero fixed point u. (Take any $v \neq 0$ and let u be the last non-zero vector in the sequence $v, Zv, Z^2v, \dots$.) Then Fu is stabilized by T, and we have the induced linear transformation $\bar{T}$ in $\bar{V} = V/Fu$. This is unipotent, so, by induction on the dimensionality, we have, det $\bar{T} = 1$. Now if we compute the matrix of T relative to a base $(u_1, u_2, \dots, u_n)$ where $u_1 = u$ and $(u_2 + Fu, \dots, u_n + Fu)$ is a base for $\bar{V}$ we see that det $T = 1$. In particular this result shows that any unipotent orthogonal transformation is a rotation.

We shall now give the proof of the

THEOREM OF CARTAN–DIEUDONNÉ.   *If* dim $V = n$, *then any orthogonal transformation η of V is a product of $\leq n$ symmetries.*

Proof.   The proof we shall give is due to Artin. We observe first that the result holds if the subspace V_1 of η-fixed points is not totally isotropic. Then V_1 contains a vector u such that $Q(u) \neq 0$ and η stabilizes $Fu^\perp$. Hence using induction on n, we may assume $\eta\,|\,Fu^\perp$ is a product of $\leq n - 1$ symmetries defined by elements of $Fu^\perp$. It follows that η is a product of $\leq n - 1$ symmetries. We observe next that the result holds if there exists a u with $Q(u) \neq 0$ and $Q(u - \eta u) \neq 0$. Then, as in the proof of Theorem 6.12 (or of Witt's cancellation theorem), we have an S_w such that $\eta' = S_w\eta$ fixes u. Then η' is a product of $\leq n - 1$ symmetries and $\eta = S_w\eta'$ is a product of $\leq n$ symmetries. Next we dispose of the two dimensional case. This is clear by what we have just proved if Q is anisotropic. Hence we may assume V is a hyperbolic plane. Then we have seen that we have a hyperbolic base (u, v) for V and either η maps $u \to au$, $v \to a^{-1}v$, $a \in F$ or $u \to av$, $v \to a^{-1}u$ (Theorem 6.10, p. 365). In the first case we may assume $a \neq 1$, since otherwise $\eta = 1$ and the result is trivial. Then if $w = u + v$, $w - \eta w = (1 - a)u + (1 - a^{-1})v$ satisfies $Q(w) \neq 0$, $Q(w - \eta w) \neq 0$, and so the result holds as before. On the other hand, if $\eta u = av$ and $\eta v = a^{-1}u$, then $w = u + av$ is fixed by η and $Q(w) \neq 0$. Hence the result holds by our first observation.

We now know that the result holds in all cases with the possible exception of the following one: dim $V \geq 3$, the subspace V_1 of η-fixed points is totally isotropic, and $Q(u - \eta u) = 0$ for every u satisfying $Q(u) \neq 0$. We now show that dim $V \geq 3$ and the last condition imply that $Q(u - \eta u) = 0$ for *every* u. It suffices to prove this for the $w \neq 0$ with $Q(w) = 0$. Consider $Fw^{\perp}$. This is an $(n - 1)$-dimensional space, and since $n \geq 3$, $n - 1 > [n/2]$, so $Fw^{\perp}$ is not totally isotropic. Hence there is a vector $u \neq 0$ such that $u \perp w$ and $Q(u) \neq 0$. Then also $(w \pm u) \perp w$ and $Q(w \pm u) = Q(u) \neq 0$. Hence if $\zeta = 1 - \eta$, then we have the three equations $Q(\zeta u) = 0$, $Q(\zeta w + \zeta u) = 0$, $Q(\zeta w - \zeta u) = 0$. These equations imply $Q(\zeta w) = Q(w - \eta w) = 0$. Hence this holds for all vectors in V and so we see that $(1 - \eta)V$ is a totally isotropic subspace of V. Moreover, we have seen that $V_1 = ((1 - \eta)V)^{\perp}$ and hence $(1 - \eta)V = V_1^{\perp}$. Since V_1 and $(1 - \eta)V$ are totally isotropic $V_1 \subset V_1^{\perp} = (1 - \eta)V$ and $(1 - \eta)V \subset ((1 - \eta)V)^{\perp} = V_1$. Thus $V_1 = (1 - \eta)V$, so if x is any vector then $(1 - \eta)^2 x = 0$. Then η is unipotent and hence η is a rotation. Also, since $n = $ dim $V_1 + $ dim $V_1^{\perp}$ and $V_1^{\perp} = (1 - \eta)V = V_1$, $n = 2$ dim V_1 and V is even dimensional.

We can now quickly finish the proof for η as in the last paragraph. We simply form $\eta' = S_w \eta$ where S_w is any symmetry. Then η' is improper and hence this transformation is a product of $k \leq n$ symmetries. Since any symmetry is improper, k is odd, and since n is even, $k \leq n - 1$. Then $\eta = S_w \eta'$ is a product of $\leq n$ symmetries. $\square$

The Cartan–Dieudonné theorem offers a quick dividend: it can be used to prove that any rotation in an odd dimensional vector space and any improper orthogonal transformation in an even dimensional space has a non-zero fixed point. These results have been indicated before to be consequences of Cayley's parametrization of orthogonal transformations by skew linear transformations (exercises 3 and 4, p. 370). To prove the results using the Cartan–Dieudonné theorem we observe first that the well-known dimensionality formula of linear algebra, dim $(U_1 \cap U_2) = $ dim $U_1 + $ dim $U_2 - $ dim $(U_1 + U_2)$ for subspaces of a vector space V, can be used to prove by induction on k that the intersection of k hyperplanes $(=(n - 1)$-dimensional subspaces) has dimensionality $\geq n - k$. Now suppose $\eta = S_{u_1} S_{u_2} \cdots S_{u_k}$, S_{u_i} the symmetry determined by the vector u_i. Since S_{u_i} has a hyperplane of fixed points and the intersection of these hyperplanes is a set of fixed points for η, we see that η has a non-zero fixed point if

η is a product of $k \leq n - 1$ symmetries. Since any symmetry is an improper orthogonal transformation, it is clear that a product of k symmetries is proper or improper according as k is even or odd. Hence it follows from the Cartan–Dieudonné theorem that any rotation is a product of an even number $\leq n$ of symmetries and any improper orthogonal transformation is a product of an odd number $\leq n$ of symmetries. Thus any rotation in an odd dimensional space and any improper orthogonal transformation in an even dimensional space is a product of $k \leq n - 1$ symmetries and so has a non-zero fixed point.

Let V_1 be the set of fixed points of the orthogonal transformation η. Then the argument we have used shows that if η is a product of k symmetries then $\dim V_1 \geq n - k$. Since $\dim V_1 = n - r$ where r is the rank of $1 - \eta$ ($= \dim (1 - \eta)V$) we see that η can not be written as a product of fewer than r symmetries if r is the rank of $1 - \eta$. Can it be written as a product of r symmetries? It has been shown by Scherk that the answer is generally "yes". The exact result is that if r is the rank of $1 - \eta$, then η can be written as a product of r symmetries unless $1 - \eta$ is skew relative to the bilinear form B, in which case the minimum number of symmetries required for η is $r + 2$.[4]

Another important consequence of the Cartan–Dieudonné theorem or even of the weaker Theorem 6.12 is

THEOREM 6.13.   *If $\dim V \geq 3$, then the commutator group $(O(V, Q), O(V, Q))$ coincides with $(O^+(V, Q), O^+(V, Q))$.*

Proof.   We observe that $(O(V, Q), O(V, Q))$ is generated by the commutators $S_u^{-1}S_v^{-1}S_uS_v = (S_uS_v)^2$ of symmetries S_u, S_v. For, since the conjugate $\eta S_u \eta^{-1} = S_{\eta(u)}$ it is clear that the subgroup generated by all $(S_uS_v)^2$ is a normal subgroup $O'(V, Q)$ of $O(V, Q)$. The factor group is generated by the cosets $S_uO'(V, Q)$. Since these generators commute, the group $O(V, Q)/O'(V, Q)$ is commutative. Hence $O'(V, Q) \supset (O(V, Q), O(V, Q))$. On the other hand, it is clear from the definition of the commutator subgroup that the reverse inequality holds. Hence $O'(V, Q) = (O(V, Q), O(V, Q))$. Our result will now follow if we can show that any $(S_uS_v)^2$ is a product of commutators of rotations. If $n = \dim V$ is odd, the linear transformation -1 is an improper orthogonal transformation contained in the center of $O(V, Q)$. Then $-S_u = (-1)S_u$ is a rotation and the commutator of S_u and S_v coincides with the commutator of the two rotations $-S_u, -S_v$. We may now assume n even and so $n \geq 4$. In this case we claim that there exists a vector w with $Q(w) \neq 0$ in $U^\perp$, $U = Fu + Fv$. Otherwise $U^\perp$ is totally isotropic, so $U^\perp \subset U^{\perp\perp} = U$. Since $\dim U^\perp = n - \dim U \geq$

[4] P. Scherk, "On the decomposition of orthogonalities into symmetries," *Proceedings of the American Mathematical Society*, vol. 1 (1950), pp. 481–491.

$n - 2 \geq 2$, we have $U = U^{\perp}$ totally isotropic, contrary to the fact that U contains the vectors u and v and $Q(u) \neq 0$, $Q(v) \neq 0$. Now choose $w \in U^{\perp}$ with $Q(w) \neq 0$. Then $S_u S_w S_u^{-1} = S_{s_u w} = S_w$ so $S_u S_w = S_w S_u$. Similarly, S_v and S_w commute. It follows that the commutator of S_u and S_v coincides with the commutator of the two rotations $S_u S_w$ and $S_v S_w$.

6.7 STRUCTURE OF THE GENERAL LINEAR GROUP $GL_n(F)$

In the remainder of this chapter we shall study the structure of the "classical" geometric groups. By these we mean the *general linear group* $GL_n(F)$ defined to be the group of bijective linear transformations of an n-dimensional vector space V over a field F, the orthogonal groups in V defined by non-degenerate quadratic forms, and the symplectic group, which is defined as the group of isometries of a vector space equipped with a metric given by a non-degenerate alternate bilinear form. The groups $GL_n(F)$ and the symplectic groups over F for $F = \mathbb{Z}/(p)$, p a prime, were first studied by Camille Jordan in his *Traité des Substitutions* (1870). The generalization of these results to the case of an arbitrary finite base field and the study of orthogonal and unitary groups over finite fields was considered by Dickson in his book, *Linear Groups with an Exposition of Galois Field Theory*, which appeared in 1900. Slightly later (in 1901) Dickson initiated the study of the classical groups over an arbitrary base field F and determined the structure of $GL_n(F)$, of the symplectic group $Sp_n(F)$, and of certain orthogonal groups.[5] A simple proof of the results on $GL_n(F)$ was given by Iwasawa in 1941.[6] In his paper Iwasawa also sketched a proof of simplicity of the projective symplectic groups for arbitrary base fields. In 1948 Dieudonné in his monograph *Sur les Groupes Classiques* proved the surprising result that the structure of orthogonal groups for arbitrary base fields differs sharply in the two cases: positive Witt index and Witt index 0 (that is, anisotropic forms). In the first case there is a general theorem for $n \geq 5$: the factor group of the commutator group with respect to its center is simple. This is so for any base field. On the other hand, for anisotropic forms the structure depends on the base field and there exist cases in which the commutator group modulo its center is not simple. In the case of a finite field the Witt index is always positive if $n \geq 3$. The classical groups for finite fields provided the first examples other than the alternating groups, of finite non-abelian simple groups.

We shall begin with $GL_n(F)$, the group of bijective linear transformations of an n-dimensional vector space V over F. Using the correspondence between

[5] L. E. Dickson, "The theory of linear groups in an arbitrary field ," *Transactions of the American Mathematical Society*, vol. 2 (1901), pp. 363–394.
[6] K. Iwasawa, "Über die Einfachheit der speziellen projection Gruppen," *Proceedings of the Imperial Academy of Tokyo*, vol. 17 (1941), pp. 57–59.

linear transformations and their matrices relative to a base for V we can identify $GL_n(F)$ with the group of invertible $n \times n$ matrices with entries in F. To begin with we shall adopt the matrix point of view in studying the group $GL_n(F)$. We have the determinant homomorphism $a \to \det a$ of $GL_n(F)$ into the multiplicative group F^* of non-zero elements of F. The kernel of this homomorphism is the *unimodular group* (or *special linear group*) $SL_n(F)$ of matrices of determinant 1. The main result we shall obtain is that except in the cases in which $n = 2$ and F is the field of two or three elements, $SL_n(F)$ modulo its center is a simple group. We determine first a set of generators for $SL_n(F)$.

LEMMA 1. $SL_n(F)$ *is generated by the elementary matrices* $T_{ij}(b) = 1 + be_{ij}$, $i \neq j, b \in F$.

(Here, as usual e_{ij} is the matrix whose sole non-zero entry is a 1 in the (i,j)-position. We have $e_{ij}e_{kl} = \delta_{jk}e_{il}$.)

Proof. We shall prove the result more generally in the case in which the field F is replaced by a Euclidean domain D. It is clear that for any b, $i \neq j$, $T_{ij}(b)$ has determinant 1 and we shall show that any $A \in M_n(D)$ such that $\det A = 1$, is a product of matrices $T_{ij}(b)$. The proof of Theorem 3.8 (p. 182) shows that there exist matrices P and Q which are products of matrices of the form $T_{ij}(b)$ and of matrices $P_{ij} = 1 + e_{ij} + e_{ji} - e_{ii} - e_{jj}$ such that $PAQ = \mathrm{diag}\{d_1, d_2, \ldots, d_n\}$. Since

$$P_{ij} = (1 + e_{ij})(1 - e_{ji})(1 + e_{ij})(1 - 2e_{ii})$$

we can replace the P_{ij} by $T_{ij}(b)$ and by matrices $1 - 2e_{ii}$. Moreover, since $T_{ij}(b)(1 - 2e_{ii}) = (1 - 2e_{ii})T_{ij}(-b)$, $T_{ji}(b)(1 - 2e_{ii}) = (1 - 2e_{ii})T_{ji}(-b)$, and $1 - 2e_{ii}$ commutes with every $T_{jk}(b), j, k \neq i$, we can gather the factors of the form $1 - 2e_{ii}$ on the left-hand side of P and on the right-hand side of Q. Multiplying by the inverses of these factors we modify the diagonal matrix to obtain another one. Thus we may assume that P and Q are products of matrices of the form $T_{ij}(b)$. This reduces the problem to $A = \mathrm{diag}\{d_1, d_2, \ldots, d_n\}$. The condition $\det A = 1$ gives $d_1 d_2 \cdots d_n = 1$, and so every d_i is invertible. Now by a sequence of elementary transformations, which should be obvious, we can pass in succession from

$$\begin{pmatrix} d^{-1} & 0 \\ 0 & d \end{pmatrix} \to \begin{pmatrix} d^{-1} & 0 \\ d^{-1} & d \end{pmatrix} \to \begin{pmatrix} 0 & -d \\ d^{-1} & d \end{pmatrix} \to \begin{pmatrix} 0 & -d \\ d^{-1} & 0 \end{pmatrix}$$

$$\to \begin{pmatrix} 1 & -d \\ d^{-1} & 0 \end{pmatrix} \to \begin{pmatrix} 1 & 0 \\ d^{-1} & 1 \end{pmatrix}.$$

This implies that $\mathrm{diag}\{d^{-1}, d\}$ is a product of elementary matrices of the form $T_{ij}(b)$, $i, j = 1, 2, i \neq j$. Clearly we have the same result for $\mathrm{diag}\{d_1^{-1}, d_1, 1, \ldots, 1\}$. Right multiplication of $A = \mathrm{diag}\{d_1, d_2, \ldots, d_n\}$ by this gives $\mathrm{diag}\{1, d_1 d_2, d_3, \ldots, d_n\}$. Repeating this process we eventually obtain $\mathrm{diag}\{1, 1, \ldots, 1, d_1, \ldots, d_n\} = 1$, which completes the proof. $\square$

We prove next

LEMMA 2. *Except in the cases $n = 2$ and F the field of two or three elements, $SL_n(F)$, $n \geq 2$, is its own commutator group.*

Proof. It suffices to show that the generators $T_{ij}(b)$ are contained in the commutator group. If $n \geq 3$ choose $k \neq i, j$. Then the result follows from the calculation:

$$
\begin{aligned}
(34) \qquad T_{ij}(b) &= T_{ik}(b) T_{kj}(1) T_{ik}(-b) T_{kj}(-1) \\
&= T_{ik}(b) T_{kj}(1) T_{ik}(b)^{-1} T_{kj}(1)^{-1}.
\end{aligned}
$$

If $n = 2$ we have

$$
(35) \qquad \begin{pmatrix} d & 0 \\ 0 & d^{-1} \end{pmatrix} \begin{pmatrix} 1 & c \\ 0 & 1 \end{pmatrix} \begin{pmatrix} d^{-1} & 0 \\ 0 & d \end{pmatrix} \begin{pmatrix} 1 & -c \\ 0 & 1 \end{pmatrix} = \begin{pmatrix} 1 & c(d^2 - 1) \\ 0 & 1 \end{pmatrix}.
$$

If F has more than three elements we can choose $d \neq 0$ so that $d^2 \neq 1$ and then choose $c = (d^2 - 1)^{-1} b$ for any given b. Then (35) shows that $T_{12}(b)$ is in the commutator group of $SL_2(F)$. Similarly, $T_{21}(b)$ is in the commutator group. $\square$

Since $GL_n(F)/SL_n(F) \cong F^*$ is abelian, $SL_n(F)$ contains the commutator group $GL_n(F)'$ of $GL_n(F)$. On the other hand, Lemma 2 implies that $SL_n(F) = SL_n(F)' \subset GL_n(F)'$. Hence $SL_n(F) = GL_n(F)'$. Since every $T_{ij}(1) = 1 + e_{ij}$, $i \neq j$, is in $SL_n(F)$ it is clear that a matrix which commutes with every element of $SL_n(F)$ commutes with every matrix. Hence it has the form $d1$. It follows that the center C of $SL_n(F)$ is $F^*1 \cap SL_n(F)$ and this is the finite set of matrices $d1$ with $d^n = 1$.

We denote the factor group $SL_n(F)/C$ as $PSL_n(F)$, the *projective unimodular group*, and we shall show that this group is simple if $n \geq 2$ except in the two cases $n = 2$, $|F| = 2$ or 3. The proof we shall give of this result is due to Iwasawa and is based on a natural action of $GL_n(F)$ on a certain set $P_{n-1}(F)$, called the $(n-1)$-*dimensional projective space over F*. This is simply the set of one dimensional subspaces Fx, $x \neq 0$, of the n-dimensional vector space V over F. If T is a bijective linear transformation of V we define an action of T on Fx by $T(Fx) = F(Tx)$. In this way $GL_n(F)$ acts on $P_{n-1}(F)$. The kernel of this

action consists of the T such that $F(Tx) = Fx$ for all $x \neq 0$ in V. We claim
that this is the set of $T = a1$, $a \neq 0$. The condition that T is in the kernel is
equivalent to $Tx = a_x x$, where $a_x \in F^*$, for every $x \neq 0$ in V. If $b \neq 0$ in F we
have $T(bx) = a_{bx}(bx)$ and $T(bx) = bTx = ba_x x = a_x(bx)$. Hence $a_{bx} = a_x$ for
$b \neq 0$. Then $T = a1$, $a = a_x$, if dim $V = 1$. Now suppose dim $V \geq 2$ and let
$(e_1, e_2, \ldots, e_n)$ be a base for V over F. We have $Te_i = a_{e_i} e_i$ and if $i \neq j$,
$T(e_i + e_j) = a_{e_i + e_j}(e_i + e_j)$. Since $T(e_i + e_j) = Te_i + Te_j = a_{e_i} e_i + a_{e_j} e_j$, $a_{e_i + e_j} =$
$a_{e_i} = a_{e_j}$. It follows again that $T = a1$, $a \neq 0$. Conversely, any map of this form
acts as the identity on $P_{n-1}(F)$. Thus if we put $PGL_n(F) = GL_n(F)/F^*1$ then we
have a faithful action of this group on $P_{n-1}(F)$ in which the coset $[T] = F^*T$
acts on Fx by $[T](Fx) = F(Tx)$. The group $PGL_n(F)$ is called *the projective
group*. This contains the subgroup $F^*SL_n(F)/F^*1 \cong SL_n(F)/(F^*1 \cap SL_n(F))$
which we have called the projective unimodular group. This also acts faithfully
on $P_{n-1}(F)$.

We shall now remind the reader of some concepts and results on group
actions which were introduced in section 1.12 and which will be useful in this
simplicity proof and in the subsequent ones that will be given in this chapter.
We recall that if a group G acts on a set S, then the action is called transitive
if, given any $x_1, x_2 \in S$, there exists a $g \in G$ such that $gx_1 = x_2$. If k is a positive
integer then the action of G on S is called k-fold transitive if, given any two
ordered k-tuples of distinct elements $(x_1, x_2, \ldots, x_k)$ and $(y_1, y_2, \ldots, y_k)$ of elements of S, there exists a $g \in G$ such that $gx_i = y_i$, $1 \leq i \leq k$. Clearly 1-fold
transitivity is the same thing as transitivity. The action of G is primitive if the
only partitions of S which are stabilized by the induced action of G on the
power set $\mathscr{P}(S)$ are the two trivial ones: (1) S alone, (2) $S = \bigcup \{x\}$, $x \in S$. Imprimitivity is equivalent to the existence of a proper subset A of S with at least
two elements such that for any $g \in G$ either $gA = A$ or $gA \cap A = \varnothing$. We recall
also the criterion: if G acts transitively on S then G acts primitively if and only
if Stab x, for any x in S, is a maximal subgroup of G (Theorem 1.12, p. 77).

We now prove the following

LEMMA 3. (1) *If the action of a group G on a set S is 2-fold transitive, then
it is primitive.* (2) *If G acts primitively on S and $H \lhd G$ is not contained in the
kernel, then H acts transitively on S.* (3) *If a subgroup H acts transitively on S,
then $G = H$ Stab x for any $x \in S$, where Stab x denotes the stabilizer of x in G.*

Proof. (1) Let A be a proper subset of S containing distinct elements, x, y.
Then if the action of G on S is 2-fold transitive, then there is a $g \in G$ such that
$gx = x$ and $gy \notin A$. Then $gA \neq A$ and $gA \cap A$ contains x, so $gA \cap A \neq \varnothing$.
Hence the action of G is primitive. (2) We have the partition of S into the orbits

of H. Since $H \lhd G$, $g(Hx) = H(gx)$ for any $g \in G$, $x \in S$. Hence G stabilizes the partition of S into the orbits of H. Since G acts primitively and H is not contained in the kernel of the action of G we have just one H-orbit. Thus H acts transitively on S. (3) Let H be a transitive subgroup of G, x an element of S, g an element of G. Then there exists an $h \in H$ such that $hx = gx$. Then $h^{-1}g \in$ Stab x and $g \in H$ Stab x. $\quad\square$

The basic simplicity criterion we shall use is

LEMMA 4.   *Let G act on a set S and let K be the kernel of the action. Then G/K is simple if G satisfies the following conditions.*

    1. *G acts primitively on S.*
    2. *$G = G'$, the commutator group of G.*
    3. *There exists an $x \in S$ such that* Stab *x contains a normal abelian subgroup A_x such that G is generated by the conjugates gA_xg^{-1}, $g \in G$.*

Proof.   Let $H \lhd G$, $H \not\subseteq K$. Then H is transitive on S by Lemma 3(2). Let $x \in S$ satisfy condition 3. Then, by Lemma 3(3), $G = H$ Stab x. Consider $G^* = HA_x$. This is normal in G and so it contains every gA_xg^{-1}. Then $G^* = G$ by condition 3. Thus $G = HA_x$, and $G/H \cong A_x/(H \cap A_x)$ is abelian. Hence H contains $G' = G$. Thus $H = G$. This implies that G/K is simple. $\quad\square$

To apply this to $PSL_n(F)$ we shall need Lemma 2 and two other results which we proceed to establish.

LEMMA 5.   *$SL_n(F)$ is doubly transitive on the projective space $P_{n-1}(F)$ if $n \geq 2$.*

Proof.   We have to show that if $Fx_1 \neq Fx_2$ and $Fy_1 \neq Fy_2$ where $x_i \neq 0$, $y_i \neq 0$, then there exists a linear transformation T of determinant 1 such that $Tx_1 = a_1y_1 \neq 0$, $Tx_2 = a_2y_2 \neq 0$, $a_i \in F$. The given conditions imply that x_1, x_2 and y_1, y_2 are linearly independent. We can choose a base $(x_1, x_2, \ldots, x_n)$ and write $y_1 = \sum a_{1j}x_j$, $y_2 = \sum a_{2j}x_j$. If $n > 2$ we can add $n - 2$ rows to the matrix

$$\begin{pmatrix} a_{11} & a_{12} & \cdots & a_{1n} \\ a_{21} & a_{22} & \cdots & a_{2n} \end{pmatrix}$$

to obtain a matrix (a_{ij}) of determinant 1. Let $y_i = \sum a_{ij}x_j$, $1 \leq i \leq n$, and let T be the linear transformation such that $x_i \to y_i$. Then T satisfies the required conditions. If $n = 2$, then $\det(a_{ij}) = a \neq 0$ if $i, j = 1, 2$, so if we take T to be the linear transformation such that $x_1 \to y_1$, $x_2 \to a^{-1}y_2$ the conditions will be satisfied. $\quad\square$

Let $(e_1, e_2, \ldots, e_n)$ be a base for V and consider Stab e_1 in $SL_n(F)$. This is the set of linear transformations whose matrices have the form

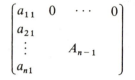

where $a_{11} \det A_{n-1} = 1$. Mapping such a transformation on the matrix A_{n-1} is a homomorphism whose kernel A_{e_1} is the set of linear transformations with matrices

$$\begin{pmatrix} 1 & 0 & \cdots & 0 \\ a_{21} & & & \\ \vdots & & 1_{n-1} & \\ a_{n1} & & & \end{pmatrix}.$$

Multiplication of these matrices shows that A_{e_1} is abelian. Hence this is an abelian normal subgroup of Stab e_1. It is clear that A_{e_1} contains all the linear transformations with matrices $T_{21}(b)$, $b \in F^*$. The formulas

$$(1 - e_{11} - e_{22} + e_{12} - e_{21})T_{21}(b)(1 - e_{11} - e_{22} + e_{12} - e_{21})^{-1} = T_{12}(-b)$$

$$T_{jk}(1)T_{ij}(b)T_{jk}(-1)T_{ij}(-b) = T_{ik}(-b)$$

and the fact that $SL_n(F)$ is generated by the elements $T_{ij}(b)$ implies that $SL_n(F)$ is generated by the conjugates of A_{e_1}. We state our results on Stab e_1 as

LEMMA 6. *Let Stab e_1 be the stabilizer of $e_1 \neq 0$ in $SL_n(F)$. Then Stab e_1 contains an abelian normal subgroup A_{e_1} whose conjugates generate $SL_n(F)$.*

Lemmas 2, 5, and 6 show that the group $SL_n(F)$, except in the cases $n = 2$, $|F| = 2$ or 3, fulfills the requirements for simplicity given in Lemma 4. Hence we have our main result:

THEOREM 6.14. *$PSL_n(F)$ is simple for $n \geq 2$ except in the cases $n = 2$, $|F| = 2$ or 3.*

We now suppose that F is finite and we wish to determine the order of the group $PSL_n(F)$. Let $|F| = q = p^r$ where p is the characteristic of F. We shall first count the number of elements in $GL_n(F)$. Let $(e_1, e_2, \ldots, e_n)$ be a base. Then if

$(f_1, f_2, \ldots, f_n)$ is another base we have one and only one linear transformation sending $e_i \to f_i$, $1 \le i \le n$, and all bijective linear transformations are obtained in this way. Hence $|GL_n(F)|$ is the number of bases for V. Now the first member f_1 of a base can be taken to be any non-zero vector of V. Since the vectors of V can be written in one and only one way in the form $\sum_1^n a_i e_i$ there are q^n of these. Hence we have $q^n - 1$ choices for f_1. Once this choice has been made, then f_2 can be taken to be any vector which is not a multiple of f_1. There are therefore $q^n - q$ choices for f_2. To choose f_3, we have to avoid the q^2 linear combinations $a_1 f_1 + a_2 f_2$ of f_1 and f_2. Hence we have $q^n - q^2$ choices for f_3. Continuing in this way we arrive at

$$N = (q^n - 1)(q^n - q) \cdots (q^n - q^{n-1})$$

as $|GL_n(F)|$. We have the homomorphism $A \to \det A$ of $GL_n(F)$ into F^*. Since this is surjective the image has $|F^*| = q - 1$ elements. The kernel is $SL_n(F)$. Hence

$$|SL_n(F)| = (q^n - 1) \cdots (q^n - q^{n-1})/(q - 1) = (q^n - 1) \cdots (q^n - q^{n-2})q^{n-1}.$$

Finally we want to determine $|PSL_n(F)| = |SL_n(F)|/|C|$. Here $|C|$ is the number of solutions in F^* of $x^n = 1$. Since $|F^*| = q - 1$ we have $x^{q-1} = 1$ for every $x \in F^*$. Hence $|C|$ is the number of solutions of $x^d = 1$ where $d = (n, q - 1)$. Since F^* is a cyclic group the number of these elements is d. Hence we have

$$|PSL_n(F)| = (q^n - 1)(q^n - q) \cdots (q^n - q^{n-2})q^{n-1}/d$$

where $d = (n, q - 1)$.

EXERCISES

1. Determine the structure of $PSL_2(F)$ in the cases $|F| = 2$ and $|F| = 3$.

2. Show that $PSL_n(F)$ is the commutator group of $PGL_n(F)$.

3. Show that $GL_n(F) \rhd SL_n(F) \rhd C \rhd 1$ is a normal series, all of whose factors except $SL_n(F)/C$ are abelian.

4. A linear transformation T is called a *transvection* if there exists a hyperplane U such that $T|U = 1_U$ and for every x, $Tx - x \in U$. Show that the linear transformations corresponding to the matrices $T_{ij}(b)$, $i \ne j$, $b \in F$ are transvections. Show that any transvection τ has the form $x \to x + f(x)u$ where $f(x)$ is a linear function and u is a vector such that $f(u) = 0$. Hence show that there exists a base $(e_1, e_2, \ldots, e_n)$ for V such that the matrix of τ is $T_{12}(1)$.

5. Let $|F| = q = p^r$. Show that the group of upper triangular matrices of the form

form a Sylow p-subgroup of $GL_n(F)$ and of $SL_n(F)$.

6. Determine the normalizer N in $GL_n(F)$ of the subgroup H of diagonal matrices.

Show that $N/H \cong S_n$ the symmetric group on n elements.

6.8 STRUCTURE OF ORTHOGONAL GROUPS

In this section we assume that Q is a non-degenerate quadratic form of positive Witt index on the vector space V. Assume first that dim $V = 2$, so V is a hyperbolic plane. We have seen in Theorem 6.10 (p. 365) and its proof that if (u, v) is a hyperbolic pair in V, then the rotations of V are the linear maps η_a, $a \in F^*$, such that $u \rightarrow au$, $v \rightarrow a^{-1}v$. The map $a \rightarrow \eta_a$ is an isomorphism of the multiplicative group F^* with the rotation group $O^+(V, Q)$. The improper orthogonal transformations are symmetries and have the form τ_b where this is the linear map sending $u \rightarrow bv$, $v \rightarrow b^{-1}u$. Checking for the base (u, v) we see that $\tau_b \eta_a \tau_b^{-1} = \eta_a^{-1}$. Since also $\tau_b^2 = 1$, $O(V, Q) = O^+(V, Q) \cup O^+(V, Q)\tau_b$ is isomorphic to a semi-direct product of a cyclic group of order two and F^* (see exercise 9, p. 79).

From now on we assume dim $V \geq 3$. Let x be an isotropic vector. We proceed to define a certain subgroup H_x of Stab x which will play the role of the subgroup A_x in the simplicity criterion of Lemma 4 of the last section. Let $u \in Fx^\perp$ and consider the map

(36) $\tilde{\rho}_{x,u}: z \rightarrow z + B(z, u)x, \qquad z \in Fx^\perp.$

This is a linear map, and since $x \in Fx^\perp$ it sends $Fx^\perp$ into itself. Moreover, we have

$$Q(z + B(z, u)x) = Q(z) + B(z, B(z, u)x) + B(z, u)^2 Q(x) = Q(z)$$

and for any $u_1, u_2 \in Fx^\perp$,

$$
\begin{aligned}
\tilde{\rho}_{x,u_1 + u_2}z &= z + B(z, u_1 + u_2)x \\
&= z + B(z, u_1)x + B(z, u_2)x \\
&= z + B(z, u_1)x + B(z + B(z, u_1)x, u_2)x \\
&= \tilde{\rho}_{x,u_2}(\tilde{\rho}_{x,u_1}z).
\end{aligned}
$$

Since $\tilde{\rho}_{x,0} = 1$ on $Fx^{\perp}$ it follows that $\tilde{\rho}_{x,u}$ is invertible with $\tilde{\rho}_{x,-u} = \tilde{\rho}_{x,u}^{-1}$. Hence $\tilde{\rho}_{x,u}$ is an isometry of $Fx^{\perp}$ onto itself. By Witt's extension theorem this can be extended to an orthogonal transformation of V. Independently of Witt's theorem we shall now obtain explicitly such an extension and show that it is unique. We choose a vector y such that $B(x, y) = 1$ and $Q(y) = 0$ (as we have done a number of times before). Then $V = Fx^{\perp} \oplus Fy = Fx + Fy + U$ where $U = (Fx + Fy)^{\perp}$. Hence $Fx^{\perp} = U + Fx$. If $u \in Fx^{\perp}$ we can write $u = ax + u'$, $u' \in U$. Since $\tilde{\rho}_{x,ax} = 1$, by (36), we have $\tilde{\rho}_{x,u} = \tilde{\rho}_{x,u'}$, so we may assume $u \in U$. We now define $\rho_{x,u}$ to be the linear transformation on V which coincides with $\tilde{\rho}_{x,u}$ on $Fx^{\perp}$ and maps

$$y \to ax + by + v, \qquad a, b \in F, v \in U$$

where we hope to determine a, b, and v so that $\rho_{x,u} \in O(V, Q)$. Since $V = Fx^{\perp} \oplus Fy$ the conditions for this are:

$$Q(\rho_{x,u}y) = Q(y) = 0$$

$$B(\rho_{x,u}y, \rho_{x,u}z) = B(y, z) \quad \text{if} \quad z \in Fx^{\perp}.$$

The second of these conditions can be replaced by the two conditions

$$B(\rho_{x,u}y, x) = B(y, x) = 1, \qquad B(\rho_{x,u}y, \rho_{x,u}z) = 0, \qquad z \in U.$$

Altogether we obtain the three conditions:

$$ab + Q(v) = 0, \qquad b = 1, \qquad B(z, v) + B(z, u) = 0.$$

These have the unique solution $b = 1$, $a = -Q(v)$, $v = -u$, that is, $b = 1$, $a = -Q(u)$, $v = -u$. Hence we have as definition of $\rho_{x,u}$,

(37)
$$\rho_{x,u}z = z + B(z, u)x, \qquad z \in Fx^{\perp}$$

$$\rho_{x,u}y = -Q(u)x + y - u$$

where y satisfies $B(x, y) = 1$, $Q(y) = 0$ and $u \in (Fx + Fy)^{\perp}$.[7]

The fact that $\rho_{x,u}$ is the only extension of $\tilde{\rho}_{x,u}$ to an orthogonal transformation is easily seen. In the first place, once y is chosen (and we need not change this), then the normalization of u so that $u \in U$ is unique since $Fx^{\perp} = Fx \oplus U$. Then our analysis shows that the form (37) for $\rho_{x,u}$ is unique. The uniqueness of the extension has some important consequences which we shall now note.

---

[7] These transformations seem to have been introduced first by C. L. Siegel in "Über die Zetafunktionen indefiniter quadratische Formen II." *Mathematische Zeitschrift*, vol. 44 (1938), pp. 398–426.

We have seen that if u_1 and $u_2 \in Fx^\perp$, then $\tilde{\rho}_{x, u_1 + u_2} = \tilde{\rho}_{x, u_1} \tilde{\rho}_{x, u_2}$. Now it is clear that $\rho_{x, u_1} \rho_{x, u_2}$ is an orthogonal extension of $\tilde{\rho}_{x, u_1} \tilde{\rho}_{x, u_2} = \tilde{\rho}_{x, u_1 + u_2}$. Since $\rho_{x, u_1 + u_2}$ is also an orthogonal extension of $\tilde{\rho}_{x, u_1 + u_2}$ the uniqueness of the extension gives,

$$\text{(38)} \qquad\qquad \rho_{x, u_1} \rho_{x, u_2} = \rho_{x, u_1 + u_2}.$$

It is clear also that $\rho_{x, u} = 1$ if and only if $\tilde{\rho}_{x, u} = 1$ on $Fx^\perp$. The definition (36) of $\tilde{\rho}_{x, u}$ shows that this holds if and only if $u \in \operatorname{rad} Fx^\perp$. Since this is Fx we have

$$\text{(39)} \qquad\qquad \rho_{x, u} = 1 \Leftrightarrow u \in Fx.$$

Similarly we see that if $\eta \in O(V, Q)$, then

$$\text{(40)} \qquad\qquad \eta \rho_{x, u} \eta^{-1} = \rho_{\eta x, \eta u}.$$

We have seen that there is no loss in generality in choosing $u \in U = (Fx + Fy)^\perp$. With this choice of u we conclude from (39) that $\rho_{x, u} = 1 \Leftrightarrow u = 0$. In view of (38) it is clear that the map $u \to \rho_{x, u}$ is a monomorphism of the additive group of U into $O(V, Q)$. We denote the image $\{\rho_{x, u}\}$ as H_x. Since it is clear from the definition of $\tilde{\rho}_{x, u}$ that $\tilde{\rho}_{ax, u} = \tilde{\rho}_{x, au}$, $a \in F^*$, we have also

$$\text{(41)} \qquad\qquad \rho_{ax, u} = \rho_{x, au}, \qquad a \in F^*.$$

This implies that $H_{ax} = H_x$ for any $a \in F^*$. It is clear that H_x is an abelian subgroup of $O(V, Q)$. Also since $\rho_{x, u} x = x + B(x, u) x = x$ we see that $H_x \subset \operatorname{Stab} x$. Moreover, the formula (40) for $\eta \in \operatorname{Stab} x$ becomes $\eta \rho_{x, u} \eta^{-1} = \rho_{x, \eta u}$ and this shows that $H_x \triangleleft \operatorname{Stab} x$. A part of our results can be stated as

LEMMA 1.   *Let x be an isotropic vector in V and let H_x be the set of linear transformations of V defined by (37), where u ranges over $U = (Fx + Fy)^\perp$, y a vector such that $B(x, y) = 1$, $Q(y) = 0$. Then H_x is a normal abelian subgroup of Stab x (in $O(V, Q)$) and $u \to \rho_{x, u}$ is an isomorphism of the additive group of U with H_x.*

Another point which is worth noticing is that the transformations $\rho_{x, u}$ are unipotent. To see this we put $v_{x, u} = \rho_{x, u} - 1$. Then the definition of $\rho_{x, u}$ gives $v_{x, u} x = 0$, $v_{x, u} z \in Fx$ if $z \in Fx^\perp = Fx + U$ and $v_{x, u} y \in Fx^\perp$. Hence $v_{x, u}^2 z = 0 = v_{x, u}^3 y$. Thus $v_{x, u}^3 = 0$ and $\rho_{x, u} = 1 + v_{x, u}$ is unipotent. We have noted also that a unipotent orthogonal transformation is necessarily a rotation; hence $H_x \subset O^+(V, Q)$.

We shall now introduce the group Ω, which is defined to be the subgroup of $O(V, Q)$ generated by all the subgroups H_x, x isotropic. If η is orthogonal then (40) implies that $\eta H_x \eta^{-1} = H_{\eta x}$. It follows that Ω is a normal subgroup of

$O(V, Q)$. We shall show that Ω coincides with the commutator group of $O(V, Q)$ and we shall see that except for the case $n = 4$, Witt index $v = 2$, Ω modulo its center is a simple group. The proof we shall give is due to Tamagawa[8] and follows Iwasawa's method which is based on the simplicity criterion of section 6.7. In applying this we shall use an action of Ω on a certain quadric cone in the projective space $P_{n-1}(F)$. In the vector space V we define the set C of vectors x such that $Q(x) = 0$. If $(e_1, e_2, \ldots, e_n)$ is a base for V and $x = \sum a_i e_i$, then the condition $Q(x) = 0$ is equivalent to the quadratic equation $\sum b_{ij} a_i a_j = 0$, $b_{ij} = B(e_i, e_j)$, for the coordinates (a_i) of x. Thus C is a quadric cone in V. We let PC be the corresponding quadric cone in $P_{n-1}(F): PC$ is the set of one dimensional subspaces Fx determined by the isotropic vectors $x \in V$. If $\eta \in O(V, Q)$, η permutes the points of C and hence we have an action $Fx \to F(\eta x)$ of $O(V, Q)$ on PC. We proceed to show that the kernel of this action is $\{1, -1\}$. First we prove

LEMMA 2. *The center of $O(V, Q)$ is $\{1, -1\}$.*

Proof. Let γ belong to the center and let S_u be the symmetry determined by the non-isotropic vector u. Since Fu is the set of vectors satisfying $S_u x = -x$, and $S_u(\gamma u) = \gamma(S_u u) = -\gamma u$, $\gamma u \in Fu$. Since γ is orthogonal we have $\gamma u = \pm u$ for every non-isotropic vector u. Let $(u_1, u_2, \ldots, u_n)$ be an orthogonal base for V. Then we have $\gamma u_i = \varepsilon_i u_i$ where $\varepsilon_i = \pm 1$. Let $i \neq j$ and suppose first that $u_i + u_j$ is not isotropic. Then we have $\gamma(u_i + u_j) = \pm (u_i + u_j)$ and also $\gamma(u_i + u_j) = \varepsilon_i u_i + \varepsilon_j u_j$. It follows that $\varepsilon_i = \varepsilon_j$. Now suppose $u_i + u_j$ is isotropic. Then we choose $k \neq i, j$ and consider the vector $u = u_i + u_j + u_k$. This is not isotropic so we have

$$\gamma(u_i + u_j + u_k) = \pm (u_i + u_j + u_k) = \varepsilon_i u_i + \varepsilon_j u_j + \varepsilon_k u_k.$$

Again we obtain $\varepsilon_i = \varepsilon_j$. Thus $\gamma u_i = \varepsilon u_i$ for all i where $\varepsilon = \pm 1$. Since the u_i form a base it follows that $\gamma = \pm 1$. $\square$

Since 1 and -1 produce the identity mapping on PC, it is clear that these maps are contained in the kernel of the action of $O(V, Q)$ on PC. We claim that $\{1, -1\}$ is the kernel of this action. This follows from

LEMMA 3. *If $\eta \in O(V, Q)$ satisfies $\eta x \in Fx$ for every isotropic x, then $\eta = \pm 1$.*

[8] T. Tamagawa, "On the structure of orthogonal groups," *American Journal of Mathematics*, vol. 80 (1958), pp. 191–197. An improved version of this proof appears in some mimeographed notes by Tamagawa.

Proof. Let (u, v) be a hyperbolic pair and let $z \in (Fu + Fv)^{\perp}$. Then $x = z - Q(z)u + v$ is isotropic since $Q(x) = Q(z) - Q(z)B(u, v) = Q(z) - Q(z) = 0$. Thus we have $\eta u = c_u u$, $\eta v = c_v v$, $\eta x = c_x x$ where c_u, c_v, and c_x are non-zero elements of F. Then

$$c_x(z - Q(z)u + v) = \eta x = \eta z - c_u Q(z)u + c_v v.$$

Since $\eta z \in (Fu + Fv)^{\perp}$ it follows that $\eta z = c_x z$ and $c_x = c_u = c_v$. Hence, if $c = c_x$ we have $\eta = c1$. Since η is orthogonal, $c = \pm 1$. $\square$

If we restrict the action of $O(V, Q)$ on PC to Ω we obtain an action of Ω on PC whose kernel is $\Omega \cap \{1, -1\}$. Since $\Omega \subset O^+(V, Q)$ it is clear that if the dimensionality n of V is odd then $-1 \notin \Omega$. In this case Ω acts faithfully on PC. We now study more closely the action of Ω on PC and we prove first

LEMMA 4. *Let* $T_x = C \cap Fx^{\perp}$ *and let* $PT_x = \{Fy \neq 0 | y \in T_x\}$. *Then* H_x *acts transitively on the complement of* PT_x *in* PC.

Proof. What this means is that if y and z are isotropic vectors not orthogonal to x, then there exists a transformation $\rho_{x,u} \in H_x$ such that $\rho_{x,u}y \in Fz$. We may assume that $B(y, x) = 1 = B(z, x)$. We have $V = Fy + Fx^{\perp} = Fy \oplus Fx \oplus U$ where $U = (Fx + Fy)^{\perp}$. Then $z = ay + bx + u$, $u \in U$. Since $B(z, x) = 1$, $a = 1$ and since $Q(z) = 0$ we have $b + Q(u) = 0$. Hence $z = y - Q(u)x + u$. Then $\rho_{x,-u}y = z$ by the definition of $\rho_{x,u}$. $\square$

A pair of points (Fx, Fy) of the set PC will be called *hyperbolic* if $B(x, y) \neq 0$. In this case we may assume that (x, y) is a hyperbolic pair of vectors of V. The preceding lemma shows that if (Fx, Fy) and (Fx, Fz) are hyperbolic, then there exists an element of Ω which fixes Fx and sends Fy into Fz. We now prove

LEMMA 5. Ω *is transitive on* PC *and also on the set of (ordered) hyperbolic pairs of points of* PC.

Proof. Let Fx, Fy be distinct points of PC. We claim that there exists a point Fz in PC such that (Fz, Fx) and (Fz, Fy) are hyperbolic. Suppose first that (Fx, Fy) is hyperbolic so we may assume (x, y) is a hyperbolic pair. Let u be a non-isotropic vector in $U = (Fx + Fy)^{\perp}$ and put $z = x - Q(u)y + u$. Then $Q(z) = -Q(u)B(x, y) + Q(u) = 0$, $B(z, x) = -Q(u) \neq 0$, and $B(z, y) = 1$, so Fz satisfies our requirement. Next assume $B(x, y) = 0$. Since x and y are linearly independent there is a linear function mapping x and y into 1. Hence there is a $z \in V$ such that $B(x, z) = 1 = B(y, z)$. Subtracting a suitable multiple of x from z we can arrange to have $Q(z) = 0$. Then (Fz, Fx) and (Fz, Fy) are hyperbolic, so again we have the required Fz. Having this we can apply Lemma 4 to obtain an $\eta \in \Omega$ such that $\eta(Fx) = Fy$. This gives the transitivity of Ω on PC. Now let (Fx, Fy) and (Fx', Fy') be hyperbolic pairs. Then there exists an $\eta \in \Omega$ such that $\eta(Fx) = Fx'$. Then $(Fx', \eta(Fy))$ is a hyperbolic pair. As we noted above there exists a $\zeta \in \Omega$ such that $\zeta(Fx') = Fx'$ and $\zeta(\eta(Fy)) = Fy'$. Then $\zeta\eta$ maps (Fx, Fy) into (Fx', Fy'), which proves the second statement. $\square$

We can now prove the main result for our purposes on the action of Ω on PC:

LEMMA 6. Ω *acts primitively on PC except when* $\dim V = 4$ *and the Witt index* $v(Q) = 2$.

Proof.   Suppose first that $v(Q) = 1$. Then any pair of distinct points (Fx, Fy) of PC is hyperbolic and so, by Lemma 5, Ω is 2-fold transitive on PC. Then the action of Ω is primitive. We now assume $v \geq 2$, so omitting the case $\dim V = 4$, $v(Q) = 2$, we have $\dim V \geq 5$. Let S be one of the sets of a partition of PC stabilized by Ω and containing more than one point. Primitivity will follow if we can show that $S = PC$. Suppose first that S contains a pair of distinct points Fx, Fy such that $B(x, y) = 0$. Then we can find an isotropic vector z such that $B(x, z) = 1$, $B(y, z) = 0$. We have $V = (Fx + Fz) \oplus U$ where $U = (Fx + Fz)^{\perp}$ is at least three dimensional and is not degenerate. Since $y \in U$ there exists a $w \in U$ such that (y, w) is a hyperbolic pair. We have $\dim U \geq 3$, and U contains isotropic vectors. Hence Lemma 5 can be applied to the space U (relative to the restriction of Q). This implies that there exists an orthogonal transformation η which is a product of $\rho_{u,v}$, u, $v \in U$, such that $\eta(Fy) = Fw$. Since $x \in (Fu + Fv)^{\perp}$, formula (37) shows that $\eta(Fx) = Fx$. Since $Fx \in S$ we have $\eta S = S$, and since $Fy \in S$, $Fw \in S$. Thus Fy and $Fw \in S$, and (Fy, Fw) is hyperbolic. We may therefore assume that S contains two points Fx, Fy with (Fx, Fy) hyperbolic. Let Fz be any point of $PC \neq Fx$. We have seen above that there exists a point Fw such that (Fx, Fw) and (Fz, Fw) are hyperbolic. Applying Lemma 5 to the hyperbolic pairs (Fx, Fy) and (Fx, Fw) we obtain $\eta \in \Omega$ such that $\eta(Fx) = Fx$ and $\eta(Fy) = Fw$. Then $\eta S = S$, and $Fw \in S$ since $Fy \in S$. Next we apply Lemma 5 to (Fx, Fw) and (Fz, Fw) to obtain $\zeta \in \Omega$ such that $\zeta(Fx) = Fz$ and $\zeta(Fw) = Fw$. Since $Fx, Fw \in S$ this implies $\zeta S = S$ and $Fz \in S$. Since Fz was arbitrary in the complement of Fx in PC we see that $S = PC$. $\square$

The final lemma we shall need for the proof of the simplicity theorem is

LEMMA 7. *The group Ω contains the commutator subgroup $(O(V, Q), O(V, Q))$ of $O(V, Q)$.*

Proof. Let (x, y) be a hyperbolic pair and let u be a non-isotropic vector. We claim that there exists a $\rho \in \Omega$ such that $\rho u \in Fx + Fy$. We note first that $u_1 = x + Q(u)y$ satisfies $Q(u_1) = Q(u)$. Hence there exists an $\eta \in O(V, Q)$ such that $\eta u_1 = u$. By Lemma 5, there exists a $\rho \in \Omega$ such that $\rho(F(\eta x)) = Fx$ and $\rho(F(\eta y)) = Fy$. Since $u_1 \in Fx + Fy$, $u = \eta u_1 \in \eta(Fx) + \eta(Fy)$ and hence $\rho u \in \rho F(\eta x) + \rho F(\eta y) = Fx + Fy$ as required. This result implies that if S_u is any symmetry, then there exists a $\rho \in \Omega$ such that $\rho S_u \rho^{-1} = S_{u'}$ where $u' = \rho u \in Fx + Fy$. Let $O_{x,y}$ denote the subgroup of $O(V, Q)$ generated by the symmetries $S_{u'}$, $u' \in Fx + Fy$. Since the restriction of $S_{u'}$ to $U = (Fx + Fy)^\perp$ is the identity, it is clear that $\eta' \to \eta'|Fx + Fy$, $\eta' \in O_{x,y}$ is an isomorphism of $O_{x,y}$ with $O(Fx + Fy, Q)$. This maps the subgroup $O_{x,y}^+$ of $O_{x,y}$ generated by the products of pairs of symmetries determined by $u' \in Fx + Fy$ onto $O^+(Fx + Fy, Q)$. Now let ζ be any rotation in V and write $\zeta = S_{u_1}S_{u_2} \cdots S_{u_{2k}}$, u_i non-isotropic. Then the result we proved shows that there exist $\rho_i \in \Omega$ such that $u_i' = \rho_i u_i \in Fx + Fy$. Then

(42) $$\zeta = S_{u_1}S_{u_2} \cdots S_{u_{2k}} = (\rho_1^{-1}S_{u_1'}\rho_1) \cdots (\rho_{2k}^{-1}S_{u_{2k}'}\rho_{2k}).$$

Since Ω is a normal subgroup of $O(V, Q)$ this relation implies that $\zeta = \rho S_{u_1'} \cdots S_{u_{2k}'}$ where $\rho \in \Omega$. Hence we have

(43) $$O^+(V, Q) \subset \Omega O_{x,y}^+,$$

Since we have seen earlier that $\Omega \subset O^+(V, Q)$ we have

(44) $$O^+(V, Q) = \Omega O_{x,y}^+.$$

Then $O^+(V, Q)/\Omega \cong O_{x,y}^+/(O_{x,y}^+ \cap \Omega)$. Since $O_{x,y}^+ \cong O^+(Fx + Fy, Q)$ and $Fx + Fy$ is a hyperbolic plane, $O_{x,y}^+$ is abelian. Hence $O^+(V, Q)/\Omega$ is abelian and so

(45) $$\Omega \supset (O^+(V, Q), O^+(V, Q)).$$

On the other hand, we have shown earlier (Theorem 6.13, p. 374) that $(O(V, Q), O(V, Q)) = (O^+(V, Q), O^+(V, Q))$. Hence

(46) $$\Omega \supset (O(V, Q), O(V, Q)) \quad \square$$

We are now ready to prove the main structure theorem for orthogonal groups.

THEOREM 6.15 (Dickson–Dieudonné). *Let Q be a non-degenerate quadratic form of positive Witt index v on a vector space V of $n \geq 3$ dimensions. Then the factor group of the commutator group of the orthogonal group $O(V, Q)$ with respect to its center is simple except in the cases $n = 4$, $v = 2$, and $n = 3$, $|F| = 3$.*

Proof. We shall show first that $\Omega = (O(V, Q), O(V, Q))$ and $(\Omega, \Omega) = \Omega$. Since $O(V, Q) \supset \Omega \supset (O(V, Q), O(V, Q))$ the first will follow from the second. To prove $\Omega = (\Omega, \Omega)$ it suffices to show that every $\rho_{x,u} \in (\Omega, \Omega)$ for x isotropic and $u \in Fx^{\perp}$. Choose y so that (x, y) is a hyperbolic pair, and let $O_{x,y}$ be the subgroup of $O(V, Q)$ defined in the proof of Lemma 7. Then $O_{x,y} \cong O(Fx + Fy, Q)$. For any $a \in F^*$ there is an $\eta_a \in O_{x,y}$ such that $\eta_a x = ax$, $\eta_a y = a^{-1}y$, and a $\tau \in O_{x,y}$ such that $\tau x = y$, $\tau y = x$. Then $\tau^{-1} \eta_a^{-1} \tau \eta_a = \eta_a^2 \in (O(V, Q), O(V, Q)) \subset \Omega$. In considering $\rho_{x,u}$ we may assume $u \in U = (Fx + Fy)^{\perp}$ (see p. 383 for this and other results which we shall need on the transformations $\rho_{x,u}$). We have

$$\eta_{a^2} \rho_{x,u} \eta_{a^2}^{-1} \rho_{x,u}^{-1} = \rho_{a^2x, u} \rho_{x, -u} = \rho_{x, a^2 u} \rho_{x, -u} = \rho_{x, (a^2 - 1)u}.$$

If $|F| \geq 4$ we can choose $a \in F^*$ so that $a^2 \neq 1$. Then replacing u by $(a^2 - 1)^{-1}u$ we see that $\rho_{x,u}$ is a commutator of elements of Ω. Hence $\Omega = (\Omega, \Omega) = (O(V, Q), O(V, Q))$ in this case.

Now assume $|F| = 3$. Taking into account the excluded cases, we have $n \geq 4$, and $v = 1$ if $n = 4$. Using the formula $\rho_{x,u_1} \rho_{x,u_2} = \rho_{x, u_1 + u_2}$ and the fact that U has an orthogonal base, it suffices to show that every $\rho_{x,u}$, u non-isotropic in U, is contained in (Ω, Ω). If $n = 4$ and $v = 1$, U is two dimensional anisotropic, and so u can be supplemented to an orthogonal base (u, v) for U. Since $Q(u) = -Q(v)$ implies that U is hyperbolic, we have either $Q(u) = 1 = Q(v)$ or $Q(u) = -1 = Q(v)$. If $n \geq 5$, U is at least three dimensional and is non-degenerate. The orthogonal complement $Fu^{\perp} \cap U$ of Fu in U is at least two dimensional and non-degenerate. Hence the restriction of Q to this space is universal and so again there exists a vector $v \perp u$ with $Q(v) = Q(u)$. Then there exists an orthogonal transformation τ such that $\tau x = x$, $\tau u = -v$, and $\tau v = u$. Then $\tau^2 x = x$, $\tau^2 u = -u$, and $\tau^2 v = -v$; hence

$$\tau^2 \rho_{x,u} \tau^{-2} \rho_{x,u}^{-1} = \rho_{x, -u} \rho_{x, -u} = \rho_{x, -2u} = \rho_{x,u}.$$

We now note that τ^2 is contained in $(O(V, Q), O(V, Q))$. This follows from a general result: if a group G is generated by elements g_i of order two then the square of any element of G is in the commutator group. For, let $\Gamma = \{g_i\}$ be a set of generators such that $g_i^2 = 1$ and hence $g_i^{-1} = g_i$. Then if $g = g_1 g_2 \cdots g_k$,

$$g^2 = g_1 g_2 \cdots g_k g_1 \cdots g_k = g_1 \cdots g_{k-1} g_1 \cdots g_{k-1} (g'^{-1} g_k^{-1} g' g_k)$$

where $g' = g_1 \cdots g_{k-1}$. Then we can conclude by induction on k that $g^2 \in (G, G)$. Since $O(V, Q)$ is generated by symmetries we see that $\tau^2 \in (O(V, Q), O(V, Q)) \subset \Omega$. Hence $\rho_{x,u} \in (\Omega, \Omega)$ so again we have $\Omega = (\Omega, \Omega) = (O(V, Q), O(V, Q))$.

We can now quickly complete the proof of Theorem 6.15 by verifying the conditions of the simplicity criterion (Lemma 4, p. 379) for $\Omega = (O(V, Q), O(V, Q))$. We have seen that this acts primitively on the projective quadric cone PC (Lemma 6) and $\Omega = (\Omega, \Omega)$. Also for any $Fx \in PC$, $\mathrm{Stab}(Fx)$ contains the abelian normal subgroup H_x. Since Ω is transitive on PC (Lemma 5) any two H_x and H_y for isotropic x and y are conjugate in Ω. Hence Ω, which is generated by all the H_x, is generated by the conjugates of one of these subgroups. Thus all the conditions of the simplicity criterion hold, and show that Ω/K is simple for K the kernel of the action. Since $K = \Omega \cap \{1, -1\}$ by Lemma 3, the proof is complete. $\square$

The preceding theorem is the main result on the structure of orthogonal groups. One may wonder whether the hypothesis that Q is of positive Witt index is necessary to insure the simplicity of $P\Omega = \Omega/(\Omega \cap \{1, -1\})$ and whether we need to exclude the cases $n = 4$, $v = 2$, and $n = 3$, $|F| = 3$. That all of these restrictions are needed to insure simplicity will be indicated in the exercises. Another question which is natural to raise is what is the intersection $\Omega \cap \{1, -1\}$, or, equivalently: is $-1 \in \Omega$. If n is odd then -1 is improper, so that in this case $-1 \notin \Omega$. Then Ω is simple. If n is even it can be shown that $-1 \in \Omega$ if and only if the discriminant of the bilinear form associated with Q is $1(=F^{*2})$. This can be established by using Clifford algebras, which constitute an important tool for studying orthogonal groups. This is shown in section 4.8 of Volume II.

EXERCISES

1. Let $F = \mathbb{R}((x))$ the field of formal power series $\sum_{k=r}^{\infty} a_k x^k$, $a_k \in \mathbb{R}$, $r \in \mathbb{Z}$ defined to be the field of fractions of $\mathbb{R}[[x]]$ (see exercise 7, p. 127). Call the *order* of such a power series r if $a_r \neq 0$ and call it *integral* if the order $r \geq 0$. Call the order of 0, ∞. Then

$$\mathrm{ord}\ uv = \mathrm{ord}\ u + \mathrm{ord}\ v$$

$$\mathrm{ord}\ (u + v) \geq \min\ (\mathrm{ord}\ u, \mathrm{ord}\ v)$$

where ord denotes the order and $u, v \in F$. Show that the set I of power series of order ≥ 0 is a subring I of F, and that for any $k > 0$ the set P_k of power series of order $\geq k$ is an ideal in I.

2. Let F be as in exercise 1 and let Q be the quadratic form defined coordinate-wise by $Q = \sum_1^n x_i^2$. Then $O(V, Q)$ is isomorphic to the group of matrices which are orthogonal in the usual sense: $A\,{}^tA = 1$. Show that this implies that A is integral in the sense that $A \in M_n(I)$. For $k > 0$ define G_k to be the set of orthogonal matrices of the form $1 + B$ where $B \in M_n(P_k)$. Verify that G_k is a normal subgroup of the group of orthogonal matrices and that $\bigcap G_k = 1$. Prove that G_k/G_{k+1} is abelian and that $G_k \supsetneqq G_{k+1}$. Use this to prove that $(O(V, Q), O(V, Q))$ modulo its center is not simple.

3. Show that Ω does not act primitively on PC if $n = 4$, $v = 2$, and use this to prove that $P\Omega$ is not simple in this case.

6.9 SYMPLECTIC GEOMETRY. THE SYMPLECTIC GROUP

The study of a finite dimensional vector space V with respect to a non-degenerate alternate bilinear form B is called symplectic geometry. We know that the dimension of such a space is even and we have called the group of bijective linear transformations η of V satisfying $B(\eta x, \eta y) = B(x, y)$, $x, y \in V$, the *symplectic group*. Since any two alternate forms on vector spaces of the same dimensionality are equivalent, it is unnecessary to indicate dependence on B; hence, we denote the symplectic group as $Sp_n(F)$, where F is the underlying field and $n = 2r$ is the dimensionality of V. The study of this group and its associated geometry is similar to and simpler than that of orthogonal groups. For this reason we can give a comparatively brief treatment of the symplectic case.

We develop first the analogue of Witt's extension theorem. We have shown (in section 6.2., p. 351) that V has a base $(u_1, v_1, u_2, v_2, \ldots, u_r, v_r)$ such that

$$(47) \qquad B(u_i, u_j) = 0 = B(v_i, v_j), \qquad B(u_i, v_j) = \delta_{ij} = -B(v_j, u_i).$$

We shall now call such a base a *symplectic base* for V. If (u_i, v_i) is a symplectic base and $\eta \in Sp_n(F)$, then $(\eta u_i, \eta v_i)$ is a symplectic base. Conversely, if (u_i, v_i) and (u_i', v_i') are two symplectic bases, then the linear transformation η such that $u_i \to u_i'$, $v_i \to v_i'$, $i = 1, \ldots, r$, is symplectic. If U is a subspace of V, then we can find a base $(u_1, v_1, \ldots, u_k, v_k, u_{k+1}, \ldots, u_m)$ for U such that $(u_{k+1}, \ldots, u_m)$ is a base for the radical $U \cap U^\perp$ of U and the u_i, v_i, $1 \leq i \leq k$, satisfy (47). As in the orthogonal case (p. 368), we can find vectors $v_{k+1}, \ldots, v_m$ such that the $2m$ vectors u_j, v_j, $1 \leq j \leq m$, satisfy (47). Then the argument used in the orthogonal case carries over to prove that if η is a linear mapping of U into a second subspace U' which is an isometry in the sense that it is bijective and satisfies $B(\eta x, \eta y) = B(x, y)$, $x, y \in U$, then η can be extended to a symplectic transformation of V.

Next we introduce a special set of generators for $Sp_n(F)$. *Let u be a non-zero vector of V and let $c \in F$.* Then we define

(48) $$\tau_{u,c} \colon x \to x + cB(x, u)u.$$

Direct verification shows that $\tau_{u,c}$ satisfies $B(\tau_{u,c}x, \tau_{u,c}y) = B(x, y)$, $x, y \in V$ and

(49) $$\tau_{u,c_1}\tau_{u,c_2} = \tau_{u,c_1 + c_2}.$$

Also $\tau_{u,c} = 1$ if and only if $c = 0$. It follows that $c \to \tau_{u,c}$ is a monomorphism of the additive group of F into $Sp_n(F)$. We call $\tau_{u,c}$ a *symplectic transvection* in the direction u. If $\eta \in Sp_n(F)$ then we have

(50) $$\eta\tau_{u,c}\eta^{-1} = \tau_{\eta u,c}$$

and if $a \in F^*$ then

(51) $$\tau_{au,c} = \tau_{u,a^2c}.$$

It is clear from the definition (48) that $\tau_{u,c}x = x$ if $x \in Fu^{\perp}$, so, in particular, $\tau_{u,c}u = u$. We also have that $\zeta_{u,c} \equiv \tau_{u,c} - 1$ maps any x into Fu; hence $\zeta_{u,c}^2 = 0$. Thus $\tau_{u,c}$ is unipotent and its determinant is 1.

LEMMA 1.   *$Sp_n(F)$ is generated by the symplectic transvections.*

Proof.   We observe first that the lemma will follow if we can show that given two pairs of vectors (u, v), (u', v') which are *hyperbolic* in the sense that $B(u, v) = 1 = B(u', v')$ then there exists a product of symplectic transvections sending $u \to u'$, $v \to v'$. Suppose we have this property and let $\eta \in Sp_n(F)$. Let (u, v) be a hyperbolic pair. Then $(\eta u, \eta v)$ is hyperbolic and so there exists a ζ which is a product of transvections such that $\zeta u = \eta u$, $\zeta v = \eta v$. Then $\eta' = \zeta^{-1}\eta$ fixes u and v, and being symplectic it stabilizes the subspace $U = (Fu + Fv)^{\perp}$. Hence the restriction $\eta'|U$ is a symplectic transformation and since $\dim U = n - 2$, we can use induction to conclude that $\eta'|U$ is a product of transvections in directions given by vectors of U. If we take the product of the transvections in V determined by these same vectors we obtain a symplectic transformation which is the identity on $Fu + Fv$ and coincides with η' on U. Since $V = Fu + Fv + U$ it is clear that this transformation coincides with η'. Hence η' is a product of symplectic transvections and the same thing is true of $\eta = \zeta\eta'$. We have now reduced the proof to showing that if (u, v) and (u', v') are hyperbolic pairs, then there exists a product of symplectic transvections such that $u \to u'$, $v \to v'$. We shall achieve this in two stages:

(52) $$(u, v) \rightarrow (u', v'') \rightarrow (u', v')$$

In the first stage we obtain a product of symplectic transvections sending $u \rightarrow u'$. Suppose first that $B(u, u') \neq 0$. Then $u \neq u'$ and if we put $w = u - u'$, $w \neq 0$ and

$$\tau_{w,c}u = u + cB(u, w)w = u - cB(u, u')(u - u').$$

Hence if we take $c = B(u, u')^{-1}$ we shall have $\tau_{w,c}u = u'$. Next suppose $B(u, u') = 0$. Then we obtain a reduction to the previous case by noting that we can find a vector u'' such that $B(u, u'') \neq 0$ and $B(u', u'') \neq 0$, for there exists a linear function f on V such that $f(u) \neq 0$ and $f(u') \neq 0$. Since B is non-degenerate this can be realized by an element u'' of $V : f(x) = B(x, u'')$ (Lemma, p. 347). Then $B(u, u'') \neq 0$ and $B(u', u'') \neq 0$. Then we can pass by a single symplectic transvection from u to u'' and by another one from u'' to u'. Hence the product of two transvections gets us from u to u'. This accomplishes the first step of (52). To achieve the second we have to show—with a change of notation—that if (u, v) and (u, v') are hyperbolic pairs, then there exists a product of symplectic transvections fixing u and sending v into v'. Again, we begin with the case $B(v, v') \neq 0$ and use a transvection $\tau_{w,c}$ where $w = v - v'$ to move from v to v'. But this fixes u also since we have $B(u, v) = 1 = B(u, v')$, and so $B(u, w) = 0$. Hence we are through if $B(v, v') \neq 0$. Now assume $B(v, v') = 0$. In this case we insert between (u, v) and (u, v') the pair $(u, u + v)$ which is also hyperbolic since $B(u, u + v) = B(u, u) + B(u, v) = 1$. Since $B(v, u + v) = B(v, u) = -1$ and $B(u + v, v') = B(u, v') = 1$, we are in the first situation for the hyperbolic pairs (u, v) and $(u, u + v)$ and the hyperbolic pairs $(u, u + v)$ and (u, v'). Hence we can pass from (u, v) to (u, v') using a product of symplectic transvections. $\square$

An immediate consequence of this lemma and the fact that $\det \tau_{u,c} = 1$ for a symplectic transvection is that $\det \eta = 1$ for every symplectic transformation. We can also use the generation by symplectic transvections to prove

LEMMA 2. *The center of $Sp_n(F)$ consists of the transformations 1 and -1.*

Proof. Let γ be in the center of $Sp_n(F)$. Then γ commutes with every transvection $\tau_{u,c}$. If $c \neq 0$ then the set of fixed points under $\tau_{u,c}$ is $Fu^{\perp}$. Since γ commutes with $\tau_{u,c}$ it maps a $\tau_{u,c}$-fixed point into a $\tau_{u,c}$-fixed point. Hence γ stabilizes $Fu^{\perp}$. Since Fu is the radical of $Fu^{\perp}$ and γ is symplectic, it follows that γ stabilizes every Fu, $u \neq 0$. Then, as we showed on p. 385, this implies that γ is a scalar multiplication. Since for $a \in F$, $a1$ is symplectic if and only if $a = \pm 1$, we see that $\gamma = \pm 1$. $\square$

We shall now study the action of $Sp_n(F)$ on the projective space $P_{n-1}(F)$ of one dimensional subspaces of V. The result we require is

LEMMA 3. *$Sp_n(F)$ acts primitively on the projective space $P_{n-1}(F)$.*

Proof. Let S be a set in a partition of $P_{n-1}(F)$ stabilized by $Sp_n(F)$ such that $|S| > 1$. Suppose first that S contains a pair of points Fx, Fy with $B(x, y) \neq 0$, so that we may assume $B(x, y) = 1$. Let Fz be any point in $P_{n-1}(F)$. If $B(x, z) \neq 0$ we may assume $B(x, z) = 1$ as well as $B(x, y) = 1$. Then by the analogue of Witt's extension theorem there exists an $\eta \in Sp_n(F)$ such that $x \to x$, $y \to z$. Then $\eta S = S$ since $Fx \in S$, and $Fz \in S$ since $Fy \in S$ and $\eta y = z$. Now suppose $B(x, z) = 0$ and $Fz \neq Fx$. Then there exists a $w \in V$ such that $B(x, w) = 1 = B(z, w)$. The result just proved shows that $Fw \in S$. We also have a $\zeta \in Sp_n(F)$ such that $w \to w$, $x \to z$. Then $\zeta S = S$ since $Fw \in S$, and $Fz \in S$ since $Fx \in S$. Thus $S = P_{n-1}(F)$ if S contains a pair of points Fx, Fy with $B(x, y) \neq 0$. Now let Fx, Fy be a pair of distinct points in S such that $B(x, y) = 0$. There exists a $u \in V$ such that $B(x, u) = 1$, $B(y, u) = 0$. Let $U = (Fx + Fu)^{\perp}$ and let G be the subgroup of $Sp_n(F)$ of transformations η which are the identity on $Fx + Fu$. These map the non-degenerate space U into itself and the set of restrictions $\eta \,|\, U$, $\eta \in G$, is the symplectic group on U. Let z be a non-zero vector of U. Since $y \in U$, the analogue of Witt's extension theorem for the symplectic group of U implies that there exists an $\eta \in G$ such that $\eta y = z$. Now $\eta S = S$ since $Fx \in S$, and since $Fy \in S$ we also have $Fz \in S$. This shows that every Fz for $z \neq 0$ in U is contained in S. Since U contains a hyperbolic pair we have a reduction to the case we considered first. $\square$

Before proceeding further in our analysis of symplectic groups we shall dispose of the two dimensional case. Here we have a symplectic base (u, v) and the condition for a linear transformation η to be in the symplectic group boils down to the single condition $B(\eta u, \eta v) = B(u, v) = 1$. If we write $\eta u = au + bv$, $\eta v = cu + dv$, then $B(\eta u, \eta v) = ad - bc$. Hence the condition is that $\begin{pmatrix} a & b \\ c & d \end{pmatrix} \in$ $SL_2(F)$. It follows that $Sp_2(F) = SL_2(F)$ and we have seen that the latter group modulo its center is simple unless $|F| = 2$ or 3. Hence we see that $PSp_2(F) \equiv Sp_2(F)/\{1, -1\}$ is simple unless $|F| = 2$ or 3. We may now assume $n \geq 4$ in the following

LEMMA 4. *$Sp_n(F)$ coincides with its commutator subgroup in all cases except:* $n = 2$, $|F| = 2$ or 3, and $n = 4$, $|F| = 2$.

Proof. We suppose first that $|F| > 3$ and we shall show that any transvection $\tau_{z,c} \neq 1$ is a commutator. Since $|F| > 3$ there exists a d in F such that $d \neq 0$ and $d^2 \neq 1$. Put $b = (1 - d^2)^{-1}c$, $a = -d^2b$. Then $a + b = c$ so $\tau_{z,c} = \tau_{z,a}\tau_{z,b}$. Let η be a symplectic transformation such that $\eta z = dz$. Then

$$\eta \tau_{z,b}^{-1} \eta^{-1} = \eta \tau_{z,-b} \eta^{-1} = \tau_{\eta z,-b} = \tau_{dz,-b} = \tau_{z,-bd^2} = \tau_{z,a}.$$

Hence

$$\tau_{z,c} = \tau_{z,a}\tau_{z,b} = \eta \tau_{z,b}^{-1} \eta^{-1} \tau_{z,b}.$$

We now consider the two cases in which $|F| = 2$ or $|F| = 3$. In both cases it suffices to display a transvection $\neq 1$ which is contained in the commutator group. For, if $\tau_{z,c}$ is such a transvection, then the subgroup $H_z = \{\tau_{z,c} | c \in F\}$, which is cyclic of order two or three, is contained in the commutator group. Hence $H_{\eta z} = \eta H_z \eta^{-1}$ is contained in the commutator group. Thus every H_x is contained, and since $Sp_n(F)$ is generated by the H_x's we shall have $Sp_n(F) = (Sp_n(F), Sp_n(F))$. In both cases, $|F| = 2$ or 3, we begin with a symplectic base $(u_1, v_1, \ldots, u_r, v_r)$ and we introduce a number of linear transformations η whose symplectic character will be clear from the fact that $(\eta u_1, \eta v_1, \ldots, \eta u_r, \eta v_r)$ is again a symplectic base. The motivation for our choices will be explained in the exercises which follow. We now treat the two cases separately.

I. $|F| = 3$, $n \geq 4$. Let σ and η be the linear transformations such that

$$\eta u_1 = u_1 + u_2, \qquad \eta v_1 = v_2, \qquad \eta u_2 = u_1, \qquad \eta v_2 = v_1 - v_2$$

$$\sigma u_1 = u_1 - v_1 + v_2, \qquad \sigma v_1 = v_1, \qquad \sigma u_2 = u_2 + v_1, \qquad \sigma v_2 = v_2$$

$$\eta u_i = u_i, \qquad \eta v_i = v_i, \qquad \sigma u_i = u_i, \qquad \sigma v_i = v_i, \qquad i > 2.$$

These are symplectic and

$$\eta^{-1}u_1 = u_2, \qquad \eta^{-1}v_1 = v_1 + v_2, \qquad \eta^{-1}u_2 = u_1 - u_2, \qquad \eta^{-1}v_2 = v_1$$

$$\sigma^{-1}u_1 = u_1 + v_1 - v_2, \qquad \sigma^{-1}v_1 = v_1, \qquad \sigma^{-1}u_2 = u_2 - v_1, \qquad \sigma^{-1}v_2 = v_2$$

$$\eta^{-1}u_i = u_i, \qquad \eta^{-1}v_i = v_i, \qquad \sigma^{-1}u_i = u_i, \qquad \sigma^{-1}v_i = v_i, \qquad i > 2.$$

We also have

$$\eta\sigma\eta^{-1}\sigma^{-1}u_1 = u_1 + v_1, \qquad \eta\sigma\eta^{-1}\sigma^{-1}u_i = u_i, \qquad i > 1$$

$$\eta\sigma\eta^{-1}\sigma^{-1}v_j = v_j, \qquad 1 \leq j \leq r.$$

Hence $\eta\sigma\eta^{-1}\sigma^{-1} = \tau_{v_1,1}$.

II. $|F| = 2$, $n \geq 6$. In this case we define η and σ by

$$\eta u_1 = u_1 + u_3, \qquad \eta v_1 = v_3, \qquad \eta u_2 = u_1, \qquad \eta v_2 = v_1 + v_3$$

$$\eta u_3 = u_2, \qquad \eta v_3 = v_2, \qquad \eta u_i = u_i, \qquad \eta v_i = v_i, \qquad i > 3$$

$$\sigma u_1 = u_1 + v_2, \qquad \sigma v_1 = v_1, \qquad \sigma u_2 = u_2 + v_1 + v_2 + v_3, \qquad \sigma v_2 = v_2$$

$$\sigma u_3 = u_3 + v_2 + v_3, \qquad \sigma v_3 = v_3, \qquad \sigma u_i = u_i, \qquad \sigma v_i = v_i, \qquad i > 3.$$

These are symplectic and

$$\eta^{-1} u_1 = u_2, \qquad \eta^{-1} v_1 = v_1 + v_2, \qquad \eta^{-1} u_2 = u_3, \qquad \eta^{-1} v_2 = v_3$$

$$\eta^{-1} u_3 = u_1 - u_2, \qquad \eta^{-1} v_3 = v_1, \qquad \eta^{-1} u_i = u_i, \qquad \eta^{-1} v_i = v_i, \qquad i > 3$$

$$\sigma^{-1} = \sigma$$

from which we obtain

$$\eta \sigma \eta^{-1} \sigma^{-1} u_1 = u_1 + v_1, \qquad \eta \sigma \eta^{-1} \sigma^{-1} u_i = u_i, \qquad i > 1$$

$$\eta \sigma \eta^{-1} \sigma^{-1} v_j = v_j, \qquad 1 \leq j \leq r$$

Then $\eta \sigma \eta^{-1} \sigma^{-1} = \tau_{v_1,1}$. This completes the proof of the lemma. $\square$

We have now shown that in all the cases which we are considering $Sp_n(F)$ coincides with its commutator group, that $Sp_n(F)$ acts primitively on $P_{n-1}(F)$ with kernel $\{1, -1\}$, and that the subgroup H_x, $x \neq 0$, is an abelian normal subgroup of Stab (Fx). Moreover, as we saw in the foregoing proof, the conjugates of H_x generate $Sp_n(F)$. Hence, by the simplicity criterion, we have

THEOREM 6.16. $PSp_n(F) = Sp_n(F)/\{1, -1\}$ is simple except in the cases $n = 2$, $|F| = 2$ or 3, and $n = 4$, $|F| = 2$.

EXERCISES

1. Let (u_i, v_i) be a symplectic base for V and let U and U' be the subspaces spanned by the u_i and the v_i respectively. Let K be the subset of $Sp_n(F)$ of η which stabilize U and U'. Show that a linear transformation $\eta \in K$ if and only if its matrix relative to the base

(53) $(u_1, \ldots, u_r, v_1, \ldots, v_r)$

has the form

$$(54) \qquad \qquad \begin{pmatrix} A & 0 \\ 0 & ({}^t A)^{-1} \end{pmatrix}, \qquad A \in GL_r(F).$$

Note that K is a subgroup of $Sp_n(F)$.

2. Let the notations be as in exercise 1. Let L be the subgroup of $Sp_n(F)$ of σ's which fix every $v \in U'$. Show that a linear transformation $\sigma \in L$ if and only if the matrix relative to $(u_1, \ldots, u_r, v_1, \ldots, v_r)$ has the form

$$(55) \qquad \qquad \begin{pmatrix} 1 & S \\ 0 & 1 \end{pmatrix}$$

where ${}^t S = S$. Show that the map $\sigma \to S$ is a monomorphism of L into the additive group of $r \times r$ symmetric matrices. Show that if $S = e_{ii}$, $1 \le i \le r$, then the corresponding σ is a transvection.

3. Let $\sigma \in L$ and $\eta \in K$ (as in exercises 1 and 2). Verify that $\eta \sigma \eta^{-1} \in L$. Verify that if the matrices of η and σ are (54) and (55) respectively, then the matrix of the commutator $\eta \sigma \eta^{-1} \sigma^{-1}$ is

$$(56) \qquad \qquad \begin{pmatrix} 1 & S_1 \\ 0 & 1 \end{pmatrix}$$

where

$$(57) \qquad \qquad S_1 = AS({}^t A) - S.$$

4. Apply exercises 1–3 to the verifications of the statements of the last part of the proof of Lemma 4 on p. 394.

5. Prove that if $A \in M_n(F)$, then the linear transformation of $M_n(F)$ defined by

$$X \to AX({}^t A) - X$$

is invertible if and only if no two characteristic roots of A are inverses. (Note that this transformation stabilizes the space of symmetric and of skew matrices.)

6. Give an example of a symplectic transformation having no fixed points $\neq 0$.

7. Let (u_i, v_i) be a symplectic base arranged as in (53) and let

$$A = \begin{pmatrix} A_{11} & A_{12} \\ A_{21} & A_{22} \end{pmatrix}, \qquad A_{ij} \in M_r(F).$$

Show that A is the matrix of a symplectic transformation if and only if the A_{ij} satisfy

$${}^t A_{11} A_{22} - {}^t A_{21} A_{12} = 1_r = {}^t A_{22} A_{11} - {}^t A_{12} A_{21}$$
$${}^t A_{11} A_{21} - {}^t A_{21} A_{11} = 0_r = {}^t A_{22} A_{12} - {}^t A_{12} A_{22}.$$

8. Prove that the characteristic polynomial of a symplectic transformation has the form $g(x) x^r g(x^{-1})$ where $g(x)$ is a polynomial of degree $n (= 2r)$.

9. Show that every element of $Sp_n(F)$ is a product of at most $2n$ symplectic transvections.

6.10   ORDERS OF ORTHOGONAL AND SYMPLECTIC GROUPS OVER A FINITE FIELD

We shall first count the number of elements in orthogonal groups over a finite field F. This will be based on some formulas for the number of points on quadric surfaces in a vector space over F. We have seen in section 6.3 that there are two equivalence classes of non-degenerate quadratic forms and that these are distinguished by their discriminant. Let d be a non-square in F^*. Then for even $n = 2r$ we can take as representatives the quadratic forms associated with the matrices

(58) $$\mathrm{diag}\{1, -1, 1, -1, \ldots, 1, -1\}$$

(59) $$\mathrm{diag}\{1, -1, \ldots, 1, -1, 1, -d\}$$

For odd $n = 2r + 1$ $(r \geq 0)$ we can take the representatives to be

(60) $$\mathrm{diag}\{1, -1, \ldots, 1, -1, -1\}$$

(61) $$\mathrm{diag}\{1, -1, \ldots, 1, -1, -d\}$$

The discriminants in the four cases (58), (59), (60), and (61) are respectively $(-1)^r, (-1)^r d, (-1)^{r+1}$, and $(-1)^{r+1}d$. Since we are interested in the associated orthogonal groups and since $O(V, Q) = O(V, dQ)$, we may replace the last quadratic form by d times this form. This multiplies the discriminant by d^{2r+1} and so gives us a form equivalent to the one associated with (60). Hence it suffices to consider the three cases (58), (59), and (60). Our enumeration will be based on some formulas for the number of points on quadric surfaces in V over F.

LEMMA.   *Let $|F| = q$. Then the number of solutions of*

(62) $$x_1{}^2 - y_1{}^2 + \cdots + x_r{}^2 - y_r{}^2 = b$$

is

(63) $$N(2r, b) = \begin{cases} q^{2r-1} + q^r - q^{r-1} & \text{if } b = 0 \\ q^{2r-1} - q^{r-1} & \text{if } b \neq 0. \end{cases}$$

The number of solutions of

(64) $$x_1{}^2 - y_1{}^2 + \cdots + x_{r-1}^2 - y_{r-1}^2 + x_r{}^2 - dy_r{}^2 = b$$

is

(65) $$N(2r, d, b) = \begin{cases} q^{2r-1} - q^r + q^{r-1} & \text{if } b = 0 \\ q^{2r-1} + q^{r-1} & \text{if } b \neq 0. \end{cases}$$

The number of solutions of

$$(66) \qquad x_1{}^2 - y_1{}^2 + \cdots + x_r{}^2 - y_r{}^2 - x_{r+1}^2 = b \qquad (r \geq 0)$$

is

$$(67) \qquad N(2r + 1, b) = \begin{cases} q^{2r} & \text{if } b = 0 \\ q^{2r} - q^r & \text{if } -b \text{ is not a square} \\ q^{2r} + q^r & \text{if } -b \text{ is a square} \neq 0. \end{cases}$$

Proof. We consider first the two cases in which $n = 2r = 2$. The equation $x_1{}^2 - y_1{}^2 = b$ is equivalent to $uv = b$ for $u = x_1 - y_1$, $v = x_1 + y_1$, and this has $2q - 1$ solutions if $b = 0$ and $q - 1$ solutions if $b \neq 0$. This accords with (63) for $r = 1$. We now prove (63) by induction on r. We write $b = a + c$ and we have q choices for a. Then (62) is equivalent to the two equations

$$x_1{}^2 - y_1{}^2 + \cdots + x_{r-1}^2 - y_{r-1}^2 = a, \qquad x_r{}^2 - y_r{}^2 = c.$$

If $b = 0$ then the case $a = 0 = c$ contributes $(2q - 1)(q^{2r-3} + q^{r-1} - q^{r-2})$ solutions and each of the cases $a \neq 0$, $c = -a$ contributes $(q - 1)(q^{2r-3} - q^{r-1})$. Altogether we obtain

$$(2q - 1)(q^{2r-3} + q^{r-1} - q^{r-2}) + (q - 1)^2(q^{2r-3} - q^{r-1})$$

solutions. This reduces to $N(2r, 0)$ as given by (63). In a similar manner we obtain the second part of (63) and the two parts of (65). To handle (66) we write this as $x_1{}^2 - y_1{}^2 + \cdots + x_r{}^2 - y_r{}^2 = b + x_{r+1}^2$. If $b = 0$ the choice $x_{r+1} = 0$ gives $q^{2r-1} + q^r - q^{r-1}$ solutions and the $q - 1$ choices of $x_{r+1} \neq 0$ give $(q - 1)(q^{2r-1} - q^{r-1})$ solutions. Altogether we obtain q^{2r} solutions, which is in accord with (67). In a similar way we obtain the other cases in (67). $\square$

We can now establish the formulas for the orders of the orthogonal groups. In these we shall denote the groups associated with the matrices (58) and (60) for a field F of q elements as $O_n(q)$ where n is even in the first case and odd in the second. If d is a non-square in F then the orthogonal group associated with (59) will be denoted as $O_n(q, d)$. The corresponding rotation groups will be denoted as $O_n{}^+(q)$ and $O_n{}^+(q, d)$. Then we have

THEOREM 6.17. *The orders of $O_n(q)$ and $O_n(q, d)$ are given by the formulas:*

$$(68) \qquad |O_n(q)| = 2(q^{2r-1} - q^{r-1})(q^{2r-2} - 1)q^{2r-3} \cdots (q^2 - 1)q, \qquad n = 2r.$$

$$(69) \qquad |O_n(q, d)| = 2(q^{2r-1} + q^{r-1})(q^{2r-2} - 1)q^{2r-3} \cdots (q^2 - 1)q, \qquad n = 2r.$$

$$(70) \qquad |O_n(q)| = 2(q^{n-1} - 1)q^{n-2}(q^{n-3} - 1)q^{n-4} \cdots (q^2 - 1)q, \qquad n = 2r + 1.$$

Proof. For $n = 1$ the orthogonal groups consist of 1 and -1, so the order is 2. We now use induction on n and assume $n \geq 2$. Choose $x \in V$ so that $Q(x) = 1$ and consider the orbit Gx, where G denotes the orthogonal group in question. By Witt's theorem, Gx is the set of vectors y such that $Q(y) = 1$. We now use the formula

$$|G| = |Gx| |\text{Stab } x|$$

((40). p. 76) to obtain $|G|$. Suppose first that we have $G = O_n(q)$, $n = 2r$. Then Stab x is isomorphic to the orthogonal group in $Fx^{\perp}$ relative to the restriction of Q. It is clear that this subgroup is isomorphic to $O_{n-1}(q)$. Also the number of elements y such that $Q(y) = 1$ is the number of solutions of (62) for $b = 1$. By (63), this is $q^{2r-1} - q^{r-1}$. Using induction we may assume formula (70) for $n - 1 = 2r - 1$. Multiplication by $q^{2r-1} - q^{r-1}$ gives (68). Formula (69) is obtained in the same way by multiplying by $q^{2r-1} + q^{r-1}$, which, by (65), is the number of y satisfying $Q(y) = 1$. The remaining case $n = 2r + 1$ is obtained in the same way going down to the case $O_{n-1}(q)$, $n - 1 = 2r$. $\square$

The orders of the corresponding rotation groups are obtained by dropping the 2 in the formulas. It can be shown that $O_n{}^+/(O_n, O_n)$ is a group of order two and that $-1 \in \Omega_n(q) \equiv (O_n(q), O_n(q))$ for even n but $-1 \notin \Omega_n(q, d) \equiv (O_n(q, d), O_n(q, d))$. (These results are established in Vol. II.) Using these one obtains formulas for the orders of the groups $P\Omega_n(q)$ and $P\Omega_n(q, d)$ whose simplicity we proved in section 6.8.

We consider next the symplectic groups $Sp_n(q) \equiv Sp_n(F)$ where F is a field of q elements. Here we allow F to have characteristic 2, in which case q is a power of 2. How many hyperbolic pairs of vectors (x, y) are there in V? The first vector in such a pair can be taken to be any non-zero vector in V. Hence we have $q^n - 1$ choices for this vector. Moreover, if we have a hyperbolic pair (x, y), any other hyperbolic pair beginning with x has the form (x, y') where $y' = y + z$ and $z \in Fx^{\perp}$. Since there are q^{n-1} vectors in the $(n - 1)$-dimensional space $Fx^{\perp}$, we have q^{n-1} choices for z. Thus we have $(q^n - 1)q^{n-1}$ hyperbolic pairs (x, y). By the analogue of Witt's theorem, any two of these can be mapped into each other by a symplectic transformation. Also the stabilizer of (x, y), that is, the set of $\sigma \in Sp_n(q)$ satisfying $(\sigma x, \sigma y) = (x, y)$ is a subgroup isomorphic to $Sp_{n-2}(q)$. Hence as in the proof of the preceding theorem,

$$|Sp_n(q)| = q^{n-1}(q^n - 1)|Sp_{n-2}(q)|.$$

Evidently this implies

THEOREM 6.18. *The order of $Sp_n(q)$ for a field F of q elements is*

$$(71) \qquad |Sp_n(q)| = q^{n-1}(q^n - 1)q^{n-3}(q^{n-2} - 1) \cdots q(q^2 - 1).$$

The orders of the corresponding projective groups $PSp_n(q)$ are again (71) if $q = 2^t$, and are $\frac{1}{2}|Sp_n(q)|$ if q is odd. Here we have to take $n \geq 6$ if $q = 2$ and $n \geq 4$ if $q = 3$. The orders obtained for $q = 2$, $n = 6$ and $q = 3$, $n = 4$ are respectively $2^9 \cdot 3^4 \cdot 5 \cdot 7$ and $2^7 \cdot 3^5 \cdot 5$.

EXERCISES

1. State the lemma as a result on the number of vectors x such that $Q(x) = 0$ or 1 where Q is any non-degenerate quadratic form in an n-dimensional vector space over a field of q elements.

2. Determine the number of hyperbolic pairs in V as in exercise 1.

3. Let p be the characteristic of F, so q is a power of p. Determine a Sylow p-group for the orthogonal groups considered in Theorem 6.17 and $Sp_n(q)$ as in Theorem 6.18.

4. Consider the series of simple groups: (a) alternating groups ($n \geq 5$), (b) projective unimodular groups, (c) the groups $P\Omega_n(q)$, $P\Omega_n(q, d)$, (d) $PSp_n(q)$. Determine all the groups in these series having orders less than one million.

6.11 POSTSCRIPT ON HERMITIAN FORMS AND UNITARY GEOMETRY

Let K be a separable quadratic extension of the field F. Then there exists an automorphism $a \to \bar{a}$ of K over F whose fixed field is F and whose order is two ($\bar{\bar{a}} = a$). A good example to keep in mind in this connection is $F = \mathbb{R}$ and $K = \mathbb{C}$. We shall call an $n \times n$ matrix $h = (h_{ij})$, with entries $h_{ij} \in K$ *hermitian* if $h_{ij} = \bar{h}_{ji}$, or, in matrix notation: ${}^t\bar{h} = h$ where $\bar{h} = (\bar{h}_{ij})$. Now let V be an n-dimensional vector space over K with base $(e_1, e_2, \ldots, e_n)$. If $x = \sum a_i e_i$, $y = \sum b_i e_i$, then we define

$$(72) \qquad H(x, y) = \sum_{i,j} h_{ij} a_i \bar{b}_j.$$

What are the properties of the map $H : (x, y) \to H(x, y)$? First, this is bi-additive:

$$(73) \qquad \begin{aligned} H(x + x', y) &= H(x, y) + H(x', y) \\ H(x, y + y') &= H(x, y) + H(x, y'). \end{aligned}$$

Next, it satisfies

(74) $$H(ax, y) = aH(x, y) \qquad H(x, ay) = \bar{a}H(x, y).$$

The first equation in (73) together with the first part of (74) state that for fixed y, $x \to H(x, y)$ is a linear function on V; the second parts of (73) and (74) state that for fixed x, $y \to H(x, y)$ is *anti-linear* (or *conjugate linear*) in y. We observe also that $H(y, x) = \sum_{i,j} h_{ij}b_i\bar{a}_j = \sum_{i,j} \bar{h}_{ji}b_i\bar{a}_j$ and $\overline{H(x, y)} = \sum \bar{h}_{ij}\bar{a}_i b_j$. Hence, interchanging the summation indices, we have

(75) $$H(y, x) = \overline{H(x, y)}.$$

A mapping H of $V \times V$ into K satisfying the conditions (73)–(75) is called a *hermitian form* on the vector space V/K. The construction we have given of hermitian forms from hermitian matrices catches all such forms. For, if H is any hermitian form and we put $h_{ij} = H(e_i, e_j)$, then $\bar{h}_{ji} = \overline{H(e_j, e_i)} = H(e_i, e_j) = h_{ij}$. Hence $h = (h_{ij})$ is a hermitian matrix. Moreover, it follows from (73) and (74) that $H(\sum a_i e_i, \sum b_j e_j) = \sum H(e_i, e_j)a_i\bar{b}_j = \sum h_{ij}a_i\bar{b}_j$, which shows that H is the hermitian form associated with the hermitian matrix h.

With a bit of care these concepts can be developed also for division rings. It should be noted first that linear algebra can be generalized to vector spaces (or modules) over division rings. The theory of linear dependence and invariance of dimensionality can be extended without using determinants. Now let Δ be a division ring which possesses an *involution*, that is, an anti-automorphsim $a \to \bar{a}$ such that $\bar{\bar{a}} = a$ (see section 2.8, p. 112). A good example to keep in mind here is that of Hamilton's division ring of quaternions $\mathbb{H}$. Let $h = (h_{ij})$ be an $n \times n$ hermitian matrix: $\bar{h}_{ij} = h_{ji}$ and let V be a vector space over Δ with base $(e_1, e_2, \ldots, e_n)$. Then, if $x = \sum a_i e_i$, $y = \sum b_i e_i$, we put

(72′) $$H(x, y) = \sum_{i,j} a_i h_{ij}\bar{b}_j.$$

Then we have (73), and

(74′) $$H(ax, y) = aH(x, y), \qquad H(x, ay) = H(x, y)\bar{a}$$

and (75). These properties define a *hermitian form* on V/Δ. It is clear that we have a bijection between the set of hermitian matrices and the set of hermitian forms.

Clearly the case of a division ring Δ includes the case we considered first: K a separable quadratic extension of F. Since we are not insisting that the map $a \to \bar{a}$ is different from the identity, we allow also the case in which $\Delta = F$ and $\bar{a} \equiv a$. Then the notion of a hermitian form reduces to that of a symmetric bilinear form. A good deal of the theory of symmetric bilinear forms, including the structure theory for the corresponding groups, called unitary groups, carries over to hermitian forms. We shall indicate some of the elementary results and refer the reader to Dieudonné's *La Géometrie des Groupes Classiques* for the group theory. We prove first the existence of an orthogonal base:

THEOREM 6.19. *Let H be a hermitian form on a finite dimensional vector space V over a division ring Δ. Then unless H is alternate and Δ is a field of characteristic two, V has a base $(u_1, \ldots, u_r, z_1, \ldots, z_{n-r})$ such that*

(76)
$$H(u_i, u_j) = \delta_{ij}b_j \quad \text{where} \quad b_i = \bar{b}_i \neq 0$$
$$H(z_k, z_l) = H(u_i, z_k) = 0.$$

Proof. The proof of Theorem 6.3 will carry over if we can show that if $H \neq 0$, then there exists a vector u such that $H(u, u) \neq 0$. Hence suppose $H(u, u) = 0$ for all u. This implies that $H(u, v) + H(v, u) = 0$ for all u, v. Since $H(v, u) = \overline{H(u, v)}$ we have $H(u, v) + \overline{H(u, v)} = 0$ for all u, v. Now assume $H \neq 0$. Then we can choose u and v so that $H(u, v) = 1$. For any $a \in \Delta$ we have $H(au, v) + \overline{H(au, v)} = 0$; hence, $aH(u, v) + \overline{H(u, v)}\bar{a} = 0$, so $a + \bar{a} = 0$. Since $\bar{1} = 1$ this implies that the characteristic is two, and that $\bar{a} = a$ for all $a \in \Delta$. Since the identity map is an anti-automorphism only if Δ is commutative, we see that Δ is a field. Since $\bar{a} = a$, H is bilinear and since $H(u, u) = 0$, we see that H is an alternate form on a vector space over a field of characteristic two. This case was excluded; hence the result follows. $\square$

Two important special cases of this theorem are (1) $\Delta = \mathbb{C}$, $a \to \bar{a}$ as usual, (2) $\Delta = \mathbb{H}$, $a \to \bar{a}$ as usual. In these two cases we can take the $b_i = \pm 1$. In general, if we replace a u_i by $c_i u_i$ then b_i is replaced by $c_i b_i \bar{c}_i$. In both cases $\bar{b}_i = b_i$ implies that b_i is real. Then we can choose c_i so that $c_i \bar{c}_i = |b_i|^{-1}$. Using this choice we replace b_i by ± 1. The proof of Sylvester's theorem carries over to show that the number of $b_i > 0$ is independent of the choice of the orthogonal case. The difference in the number of positive b_i and the number of negative b_i is called the *signature* in both cases.

The next thing we might look at is Witt's theorem, which is valid also for hermitian forms suitably restricted. We refer the reader to Volume II of our *Lectures in Asbtract Algebra*[9] or to Dieudonné's *La Géométrie des Groupes Classiques* for this result.

---

[9] Springer-Verlag, New York, Heidelberg, Berlin. First published in 1953 by D. Van Nostrand Company.

EXERCISES

1. Let K be quadratic extension of F of char. $\neq 2$, $a \to \bar{a}$ the automorphism $\neq 1$ of K, and let $H(x, y)$ be a hermitian form on the n-dimensional vector space V/K. Let $B(x, y) = H(x, y) + H(y, x)$. Show that $B(x, y)$ is a symmetric bilinear form on V/F ($2n$-dimensional) satisfying

    (77) $$B(ax, ay) = N(a)B(x, y), \qquad (N(a) = a\bar{a}).$$

    Suppose $K = F(i)$, $i^2 = b \in F$, and let B be a symmetric bilinear form on V/F satisfying (77). Define

    (78) $$H(x, y) = B(x, y) - b^{-1}iB(x, iy).$$

    Show that H is hermitian. Verify that the two maps $H \to B$ and $B \to H$ defined here are inverses.

2. Let the notations be as in exercise 1. Show that two hermitian forms H_1 and H_2 are equivalent (definition as for bilinear forms) if and only if the associated symmetric bilinear forms B_1 and B_2 defined in exercise 1 are equivalent.

3. Let $\mathbb{H}$ be Hamilton's quaternion algebra with base $(1, i, j, k)$ over $\mathbb{R}$ such that $i^2 = j^2 = k^2 = -1$, $ij = k = -ji$, $jk = i = -kj$, $ki = j = -ik$. Let V be an n-dimensional vector space over $\mathbb{H}$, H a hermitian form on $V/\mathbb{H}$. Let $B(x, y) = H(x, y) + H(y, x)$. Show that B is symmetric on the $4n$-dimensional vector space $V/\mathbb{R}$ and B satisfies (77). Conversely let B be a symmetric bilinear form on $V/\mathbb{R}$ satisfying (77). Define

    (79) $$H(x, y) = B(x, y) - iB(x, iy) - jB(x, jy) - kB(x, ky).$$

    Show that H is hermitian on $V/\mathbb{H}$ and that the maps $B \to H$, $H \to B$ are inverses.

4. Use the same notations as in exercise 3. Show that two hermitian forms H_1 and H_2 on $V/\mathbb{H}$ are equivalent if and only if the corresponding symmetric bilinear forms B_1 and B_2 on $V/\mathbb{R}$ are equivalent.

Algebras Over a Field

The concept of an associative algebra is obtained by combining that of a ring and that of a vector space, together with certain relations connecting these structures. In this chapter we give an introduction to associative algebras as well as to certain classes of non-associative algebras, namely, Lie, Jordan, and alternative algebras. In one way or another these three classes of non-associative algebras are closely related to associative ones. The first two arise in making simple modifications of the product composition in an associative algebra. On the other hand, alternative algebras constitute a mild generalization of associative ones. From the point of view of applications to broad areas of mathematics and physics, the classes we have singled out: associative, alternative, Lie, and Jordan algebras are *the* important classes of algebras.

Associative algebras occur frequently in algebra. A prime example is the algebra of linear transformations of a finite dimensional vector space. This plays the role of "catch-all" algebra, which is analogous to that played by the symmetric group in group theory: any finite dimensional associative algebra is isomorphic to an algebra of linear transformations of a finite dimensional vector space. It is natural to consider homomorphisms of associative algebras into algebras of linear transformations, or, equivalently, into algebras of matrices. Of particular interest are the regular representations. These give rise to the trace and norm maps of an associative algebra into its base field which generalize notions for fields which were introduced in section 4.15.

Alternative algebras originated in the discovery, due independently to J. J. Graves and A. Cayley, of the algebra $\mathbb{O}$ of octonions, an eight dimensional algebra over $\mathbb{R}$ containing Hamilton's quaternion algebra $\mathbb{H}$. This has many of the properties of $\mathbb{H}$; for example, it is a division algebra. However, it does not satisfy the associative law of multiplication. Instead, it satisfies the laws $(xx)y = x(xy)$ and $y(xx) = (yx)x$ which are weaker than associativity. Octonions can be used to coordinatize certain "exceptional" geometries—more exactly, certain non-Desarguesian projective planes.

Lie algebras are named after the great Norwegian mathematician of the late nineteenth century, Sophus Lie. These are obtained from associative algebras by replacing the given associative product by the Lie product, or additive commutator, $[xy] = xy - yx$. Once this is done, one is interested in subalgebras with respect to the composition $[xy]$ and these need not be subalgebras of the given associative algebra. Lie algebras are the fundamental objects of study in Lie's theory of continuous groups. Lie's great achievement was the reduction of the study of local properties of continuous groups to that of associated Lie algebras.

Jordan algebras are of comparatively recent origin. These were introduced in 1931 by a physicist, P. Jordan, with a view of applying them to quantum mechanics. These algebras arise in seeking to formulate simple laws for the Jordan product (or "anti-commutator") $x \cdot y = \frac{1}{2}(xy + yx)$ in an associative algebra over a field of characteristic $\neq 2$. Jordan algebras have applications to analysis, to geometry, and to Lie groups.

For the associative theory we shall consider the regular matrix representations and properties of the trace and norm maps. We shall also introduce the exterior algebra $E(V)$ of a vector space and apply this to give quick and incisive derivations of some of the main properties of determinants. Finally, we shall prove the theorems of Frobenius and of Wedderburn on associative division algebras. For us, alternative algebras will arise in connection with a problem on quadratic forms which was first considered by A. Hurwitz. We shall treat Lie and Jordan algebras very lightly—not much beyond the basic definitions.

A generalization of the concept of an associative algebra over a field to associative algebras over commutative rings is given in Volume II. There we consider also the structure theory of rings and associative algebras.

7.1 DEFINITION AND EXAMPLES
OF ASSOCIATIVE ALGEBRAS

Though we have not yet given a formal definition of the concept of an associative algebra, we have, in fact, already encountered a number of instances of this notion. In the theory of fields we studied a field E relative to a subfield F, and in this connection we considered E as a vector space over F in which the product au, $a \in F$, $u \in E$, is the product as defined in E. We also have the product uv of any two elements of E, which together with the addition in E, 0, and 1 give the ring structure in E. The connection between the two structures of vector space and of ring can be described by noting that the additive groups are the same for the two, and that $a(uv) = (au)v = u(av)$ for $a \in F$, u, $v \in E$.

We have a similar situation in dealing with the ring of polynomials $F[x]$ in an indeterminate x over the field F. In addition to the ring structure we have the vector space structure over the field F in which $af(x)$, $a \in F$, $f(x) \in F[x]$, is the ring product. Again, the addition and 0 are the same for the two structures and $a(f(x)g(x)) = (af(x))g(x) = f(x)(ag(x))$.

Still another example of this kind is obtained in considering the set $M_n(F)$ of $n \times n$ matrices with entries taken from a field F. Here, in addition to the ring structure we also have a vector space over F where, if $M = (m_{ij})$ is the $n \times n$ matrix with m_{ij} as (i, j)-entry and $a \in F$, then $aM = (am_{ij})$. This is identical with the product of M by the matrix $a1 = \text{diag}\{a, a, \ldots, a\}$. Since $a1$ is in the center of the ring, that is, $a1$ commutes with every matrix, it is clear that we have $a(MN) = (a1)(MN) = ((a1)M)N = (aM)N$ and $a(MN) = (a1)(MN) = M(a1)N = M(aN)$. In the first example we considered (the field over field case) the underlying vector space may or may not be finite dimensional. In the second, $F[x]$, it is definitely not finite dimensional since 1, x, $x^2, \ldots$ are linearly independent over F. On the other hand, $M_n(F)$ with the vector space structure we defined is finite dimensional. A particularly useful base for $M_n(F)$ over F is the base consisting of the n^2 *matrix units* $\{e_{ij} | i, j = 1, \ldots, n\}$ where e_{ij} denotes the matrix with a 1 in the (i, j)-position and 0's elsewhere (see section 2.3). We have the multiplication table

$$(1) \qquad\qquad e_{ij}e_{kl} = \delta_{jk}e_{il}$$

and the following expression for the unit 1 in terms of the base:

$$(2) \qquad\qquad 1 = e_{11} + e_{22} + \cdots + e_{nn}.$$

We shall now give the formal definition of an associative algebra.

DEFINITION 1. *An* (associative) *algebra over a field F is a pair consisting of a ring $(A, +, \cdot, 0, 1)$ and a vector space A over F such that the underlying set A and the addition and 0 are the same in the ring and vector space, and*

$$(3) \qquad\qquad a(xy) = (ax)y = x(ay)$$

holds for $a \in F$, x, $y \in A$. If A is finite dimensional over F, then we shall say that the algebra is finite dimensional *(or has a finite base). We shall usually denote the algebra by the letter (e.g., A) used to designate the underlying set.*

It is clear that the foregoing examples are algebras. We give next another important example. Let G be a finite group, say, $G = \{s_1 = 1, s_2, \ldots, s_n\}$, F a field, and let $F[G]$ denote the vector space over F with base G. Thus $F[G]$ consists of the elements $\sum_1^n a_i s_i$, $a_i \in F$, where $\sum a_i s_i = 0$ for distinct s_i if and only if every $a_i = 0$, and addition and multiplication by elements of F are the obvious ones. We define a product in $F[G]$ by

$$\left(\sum_i a_i s_i \right) \left(\sum_j b_j s_j \right) = \sum_{i,j} a_i b_j (s_i s_j)$$

where $s_i s_j$ is as defined in the group G (see exercise 8, p. 127). Using the associative law in G it is trivial to verify that the product defined in $F[G]$ is associative. Also, the distributive laws in F give these laws in $F[G]$ and $1 = s_1$ is the unit for multiplication. Finally, the equation (3) relating multiplication in $F[G]$ and multiplication by elements in F is clear. Hence we have an algebra. This is called the *group algebra over F of the finite group G*.

Now suppose A is any algebra over the field F. Let $a \in F$ and consider the element $a1 \in A$. By (3), we have $(a1)x = a(1x) = ax$ and $x(a1) = a(x1) = ax$. This shows that the vector space product ax coincides with the ring product $(a1)x$ and $a1$ is in the center of A, that is, $a1$ commutes with every $x \in A$. Also, we have the map $a \to a1$ of F into A. Since $1 \to 1$, $(a + b)1 = a1 + b1$ and $(ab)1 = (ab)1^2 = (a1)(b1)$ (by $(ax)(by) = (b(ax))y = ((ba)x)y = (ba)(xy) = (ab)(xy)$), $a \to a1$ is a ring homomorphism. If $A \neq 0$, $1 \neq 0$ in A and then $a \to a1$ is a monomorphism since F is a field. Conversely, suppose we have a ring R and a subring F of the center of R, which is itself a field. Then we can regard R as an algebra over F simply by defining ax for $a \in F$, $x \in R$, as the ring product ax. Then (3) is immediate, and so we have an algebra. All the examples which we gave at the beginning were obtained in this way. These remarks show that an algebra over a field F is essentially the same thing as a pair consisting of a ring together with a distinguished subfield of the center of the ring.[1] The slightly more abstract definition we have given, however, has some advantages—for example, it makes more natural the concept of homomorphism which we shall give in a moment.

By now the reader probably has enough experience to formulate for himself the basic concepts related to that of an algebra.[2] We enumerate these in the style of a shopping list as: (1) subalgebras, (2) ideals, (3) quotient algebras,

---

[1] A given ring R may not have any such subfield. For example $\mathbb{Z}/(6)$ cannot be regarded as an algebra over any field.

[2] We suggest as an exercise that the reader break off the reading at this point and formulate for himself what he regards as the fundamental concepts related to that of an algebra. After that he can check back with the material in the text.

(4) homomorphism, (5) kernel of a homomorphism. We shall now check these items off in succession giving some brief comments on some of them.

1. *Subalgebras*. A subset B of an algebra A is a subalgebra if it is a subring of the ring A and a subspace of the vector space of A. The intersection of subalgebras is a subalgebra. If S is a subset of A one defines the subalgebra $F[S]$ generated by S to be the intersection of all subalgebras of A containing S. It is easily seen that $F[S]$ is the set of F-linear combinations of 1 and the monomials $s_{i_1} s_{i_2} \cdots s_{i_r}$, $s_{i_k} \in S$. These look like $a_0 1 + \sum a_{i_1 \cdots i_r} s_{i_1} \cdots s_{i_r}$, $a_0, a_{i_1 \cdots i_r} \in F$.

2. *Ideals*. A subset I of A is an ideal in the algebra A if I is an ideal in A as ring and a subspace of A as vector space over F.

3. *Quotient algebras*. Let I be an ideal in the algebra A. Then we obtain the quotient ring A/I and the vector space A/I. Together these constitute an algebra which is called the *quotient* (or *difference*) algebra of A with respect to the ideal I.

4. *Homomorphism*. A map of an algebra A into an algebra B (over the same field) is an *algebra homomorphism* if it is both a ring homomorphism and a linear mapping. *Monomorphisms, epimorphisms, endomorphisms,* and *automorphisms* for algebras are special cases of homomorphisms defined in the usual way. If I is an ideal we have the canonical epimorphism $v: a \to a + I$ of A into A/I. It is easy to see that if S is a set of generators for A and η_1 and η_2 are (algebra) homomorphisms of A into B such that $\eta_1(s) = \eta_2(s)$ for all $s \in S$, then $\eta_1 = \eta_2$.

5. *Kernel of a homomorphism*. If η is an algebra homomorphism of A into B, then $K = \eta^{-1}(0)$, the subset of A of elements k such that $\eta(k) = 0$, is an ideal in A called the *kernel* of η. If I is an ideal contained in K we have the induced homomorphism $\bar{\eta}$ of A/I into B such that $\bar{\eta}(a + I) = \eta(a)$, that is, $\eta = \bar{\eta}v$, where v is the canonical homomorphism of A onto A/I.

EXERCISES

1. If S is a subset of an algebra A we let $C_A(S)$ be the subset of A of elements c such that $cs = sc$, $s \in S$. Verify that $C_A(S)$ is a subalgebra. (Note that this proves in particular that the center $C = C_A(A)$ is a subalgebra.)

2. Let $\mathbb{H}$ be the division ring of real quaternions as defined in section 2.4, p. 98. Note that $\mathbb{H}$ is an algebra over $\mathbb{R}$ with base $(1, i, j, k)$ having the multiplication table

$$i^2 = j^2 = k^2 = -1$$
$$ij = -ji = k, \qquad jk = -kj = i, \qquad ki = -ik = j.$$

Note also that $G = \{\pm 1, \pm i, \pm j, \pm k\}$ is a subgroup of the multiplicative group. This subgroup is called a *quaternion group*. Write $1' = -1, i' = -i, j' = -j, k' = -k$ in G and let $\mathbb{R}[G]$ be the group algebra over $\mathbb{R}$ of $G = \{1, 1', i, i', j, j', k, k'\}$. Show that there exists a homomorphism of $\mathbb{R}[G]$ into $\mathbb{H}$ such that $1' \to -1, i \to i, i' \to -i, j \to j, j' \to -j, k \to k, k' \to -k$. Determine the kernel.

3. Let $A = F[a]$, an algebra generated by a single element a. Show that $A \cong F[x]/(f(x))$ where either $f(x) = 0$ or $f(x)$ is a monic polynomial. Note that in the first case $F[a] \cong F[x]$, so $F[a]$ is infinite dimensional. Show that if $f(x)$ is monic of degree n then $(1, a, \ldots, a^{n-1})$ is a base for $F[a]$. In this case a is called *algebraic* and $f(x)$ is its *minimum polynomial*.

4. Let $A = F[a]$ as in exercise 3 with a algebraic with minimum polynomial $f(x)$. Let $f(x) = q_1(x) \cdots q_r(x)$ be the factorization of $f(x)$ into factors $q_i(x) = p_i(x)^{k_i}$ where $p_i(x)$ is monic and prime and $p_i(x) \neq p_j(x)$ if $i \neq j$. Show that if $s_i(x) = \prod_{j \neq i} q_j(x)$, then there exist polynomials $t_i(x)$ such that

$$s_1(x)t_1(x) + s_2(x)t_2(x) + \cdots + s_r(x)t_r(x) = 1.$$

Put $e_i = s_i(a)t_i(a)$. Show that

$$e_i^2 = e_i, \qquad e_i e_j = 0 \qquad \text{if } i \neq j$$

(4)
$$e_1 + e_2 + \cdots + e_r = 1$$

and hence that

(5)
$$F[a] = F[a]e_1 \oplus F[a]e_2 \oplus \cdots \oplus F[a]e_r,$$

in the sense that every element of $F[a]$ can be written in one and only one way in the form $b_1 + b_2 + \cdots + b_r$, $b_i \in F[a]e_i$. Put $a_i = ae_i$. Show that $F[a]e_i$ is an algebra with unit e_i, and that a_i is algebraic in this algebra with $q_i(x)$ as minimum polynomial.

5. Let $A = F[a]$ be as in exercise 4 and let

$$f(x) = (x - b_1)(x - b_2) \cdots (x - b_r)$$

in $F[x]$ where the b_i are distinct. Let $e_i(x)$ be the *Lagrange interpolation polynomial*

(6)
$$e_i(x) = \frac{f(x)}{(x - b_i)f'(b_i)}, \qquad 1 \le i \le r.$$

Show that if $e_i = e_i(a)$ then we have (4) and (5) with $F[a]e_i = Fe_i$, one dimensional.

6. Let $F[G]$ be the group algebra of the finite group $G = \{s_1 = 1, s_2, \ldots, s_n\}$. If $a = \sum \alpha_i s_i$ define $T(a) = \sum \alpha_i$. Show that $a \to T(a)$ is a homomorphism of $F[G]$ into F and determine the kernel.

7. Let $F[G]$ be as in exercise 6. Show that there exists a homomorphism of $F[G]$ into $F[G \times G]$ sending every $s_i \in G$ into $s_i \times s_i$.

8. Let $F[G]$ be as in exercise 6 with G a group of order p^m, p a prime, and F a field of characteristic p. Show that if $K = \ker T$, then every element of K is nilpotent.

9. Show that the matrices $A = \sum_1^{n-1} e_{i,i+1}$ and $B = \sum_1^{n-1} e_{i+1,i}$ generate the algebra $M_n(F)$ of $n \times n$ matrices over F.

10. Show that if $c_1, c_2, \ldots, c_n$ are n distinct elements of F, then $C = \sum c_i e_{ii}$ and $D = \sum_1^{n-1} e_{i,i+1} + e_{n,1}$ generate $M_n(F)$.

11. Prove that if $C \in M_n(F)$ and the characteristic polynomial of C has n distinct roots in a splitting field, then there exists a $D \in M_n(F)$ such that C and D generate $M_n(F)$.

7.2 EXTERIOR ALGEBRAS. APPLICATION TO DETERMINANTS

We shall now define some algebras which have important applications in geometry and which can be used to derive in a transparent fashion the main properties of determinants. These algebras, now called exterior algebras, were introduced in 1844 by H. G. Grassmann.[3] They arise in considering the following problem for vector spaces. Given a finite dimensional vector space V over F, one wants to enlarge this to an algebra A which is generated by V and has the further property that $v^2 = 0$ in A for every $v \in V$. Moreover, one wishes to do this in the most general way possible, that is, no further conditions except the consequences of the ones that have been set down are to be imposed.

We shall now try to carry out this program. To see what is involved we suppose we have an algebra A containing a subspace V such that A is generated as an algebra by V, and $v^2 = 0$ for every $v \in V$. As we saw in section 7.1, the fact that V generates A amounts to saying that every element of A is an F-linear combination of 1 and monomials $v_1 v_2 \cdots v_k$, $k \geq 1$, where the $v_i \in V$. Now let $(u_1, u_2, \ldots, u_n)$ be a base for V over F. Then any v is a linear combination of the u_i; hence any monomial in v's $\in V$ is a linear combination of monomials in the u_i. Hence A is also generated by $(u_1, u_2, \ldots, u_n)$. We now consider the set of monomials $u_{i_1} u_{i_2} \cdots u_{i_r}$ in the elements of the base (u_i). We shall call such a monomial *standard* if $i_1 < i_2 < \cdots < i_r$, and we shall prove that every element of A is a linear combination of 1 and standard monomials in the u_i. Of course, it suffices to prove this for the monomials in the u_i. For these we shall prove a stronger result, namely, any monomial in the u_i which contains (a particular) u_i more than once is 0, and if $i_1 < i_2 < \cdots < i_r$, then

$$(7) \qquad u_{i_1} u_{i_2} \cdots u_{i_{r'}} = (sg\ \sigma) u_{i_1} u_{i_2} \cdots u_{i_r}$$

where σ is the permutation $\begin{pmatrix} 1 & 2 & \cdots & r \\ 1' & 2' & \cdots & r' \end{pmatrix}$ and $sg\ \sigma = 1$ or -1 according as σ is even or odd. We note first that as a consequence of the property $v^2 = 0$,

---

[3] See H. G. Grassmann, *Ausdehnungslehre*. The first edition was published in 1844. A second, expanded and improved edition appeared in 1862.

$v \in V$ we have

$$0 = (u + v)^2 - u^2 - v^2 = uv + vu.$$

Hence we have $uv = -vu$, $u, v \in V$. In particular, we have

(8) $$u_i^2 = 0, \qquad u_i u_j = -u_j u_i, \qquad 1 \le i, j \le n.$$

It is clear from the second of these relations that we may interchange consecutive u_i in a monomial at the expense of a change in sign. A succession of such moves can be used to bring any u_i appearing in a monomial next to any other one. Then, by the first relation, it follows that the monomial is 0 if there is more than one occurrence of a u_i in it. Now consider a product $u_{i_1} u_{i_2} \cdots u_{i_r}$, where $\sigma = \begin{pmatrix} 1 & 2 & \cdots & r \\ 1' & 2' & \cdots & r' \end{pmatrix}$ is a permutation of $1, 2, \ldots, r$. If $i_{j'} > i_{(j+1)'}$ we have $u_{i_1}, u_{i_2}, \cdots u_{i_{r'}} = -u_{i_{1'}} \cdots u_{i_{(j-1)'}} u_{i_{(j+1)'}} u_{i_{j'}} \cdots u_{i_{r'}}$ and the new permutation of $1, 2, \ldots, r$ differs from σ by a transposition. A finite number of moves of the type indicated allows us to pass from $u_{i_1}, u_{i_2}, \cdots u_{i_{r'}}$ to $\pm u_{i_1} \cdots u_{i_r}$. The number of these moves is the number of transpositions in a factorization of σ as a product of transpositions. Hence we have formula (7).

We now see that every element of A is a linear combination of the elements

(9) $$1, u_{i_1} u_{i_2} \cdots u_{i_r} \quad \text{with} \quad i_1 < i_2 < \cdots < i_r.$$

The number of such elements does not exceed the number of subsets $\{i_1, i_2, \ldots, i_r\}$ of the set $N = \{1, 2, \ldots, n\}$ including the vacuous set.[4] Thus we see something which we might not have predicted at the outset: A is finite dimensional and, in fact, $\dim A \le |\mathscr{P}(N)| = 2^n$. We can also derive a formula for the product of any two of the monomials in (9). For this purpose we consider the subsets $S = \{i_1, i_2, \ldots, i_r\}$ of N and we put $u_S = u_{i_1} u_{i_2} \cdots u_{i_r}$ if $i_1 < i_2 < \cdots < i_r$. If $s, t \in N$ we define

(10) $$\varepsilon_{s,t} = \begin{cases} 1 & \text{if } s < t \\ 0 & \text{if } s = t \\ -1 & \text{if } s > t \end{cases}$$

and if S and T are subsets of N we put

(11) $$\varepsilon_{S,T} = \begin{cases} \displaystyle\prod_{s \in S, t \in T} \varepsilon_{s,t} & \text{if } S \ne \varnothing, T \ne \varnothing \\ 1 & \text{if } S \text{ or } T = \varnothing. \end{cases}$$

---

[4] Conceivably some of the elements displayed in (9) could be equal and even if distinct they could be linearly dependent.

It is clear from this definition that if $T_1 \neq \emptyset$, $T_2 \neq \emptyset$, and $T_1 \cap T_2 = \emptyset$, then $\varepsilon_{S,T_1 \cup T_2} = \varepsilon_{S,T_1}\varepsilon_{S,T_2}$ and $\varepsilon_{T_1 \cup T_2,S} = \varepsilon_{T_1,S}\varepsilon_{T_2,S}$. From this one sees easily that

(12) $$u_S u_T = \varepsilon_{S,T} u_{S \cup T}.$$

After this analysis we are ready to construct the exterior algebra $E(V)$ of the vector space V. We consider the set of subsets $\mathscr{P}(N)$ of $N = \{1, 2, \ldots, n\}$, and we let $E(V)$ be the 2^n-dimensional vector space with $\mathscr{P}(N)$ as base. Thus the elements of $E(V)$ have the form $\sum_{S \in \mathscr{P}(N)} a_S S$ where $a_S \in F$. Also we identify $S \in \mathscr{P}(N)$ with the element $1S \in E(V)$. We now define a product in $E(V)$ by defining

(13) $$ST = \varepsilon_{S,T}(S \cup T)$$

$\in E(V)$, and extending this linearly over $E(V)$, by defining

(14) $$\left(\sum a_S S\right)\left(\sum b_T T\right) = \sum \varepsilon_{S,T} a_S b_T (S \cup T).$$

It is clear from (14) that the distributive laws hold for the multiplication and the vector space addition thus defined, and we have for $a \in F$ that $a(XY) = (aX)Y = X(aY)$ if $X = \sum a_S S$, $Y = \sum b_T T$. Since $\varepsilon_{\emptyset,S} = 1 = \varepsilon_{S,\emptyset}$ by (11), $S\emptyset = S = \emptyset S$ by (13) and, by (14), $\emptyset$ is a unit for the multiplication in $E(V)$. From now on we write 1 for $\emptyset$. We wish to verify that if $R, S, T \in \mathscr{P}(N)$, then $(RS)T = R(ST)$. This is clear if any one of R, S, T is 1 so we assume $R \neq \emptyset$, $S \neq \emptyset$, $T \neq \emptyset$. Since $ST = 0$ if $S \cap T \neq \emptyset$ we have $(RS)T = 0 = R(ST)$ by (13) unless $R, S,$ and T are disjoint. In this case we have

$$(RS)T = \varepsilon_{R,S}(R \cup S)T = \varepsilon_{R,S}\varepsilon_{R \cup S,T}(R \cup S \cup T)$$
$$= \varepsilon_{R,S}\varepsilon_{R,T}\varepsilon_{S,T}(R \cup S \cup T)$$
$$R(ST) = \varepsilon_{S,T}R(S \cup T) = \varepsilon_{S,T}\varepsilon_{R,S \cup T}(R \cup S \cup T)$$
$$= \varepsilon_{S,T}\varepsilon_{R,S}\varepsilon_{R,T}(R \cup S \cup T).$$

This implies the associative law in $E(V)$ and proves that $E(V)$ is an algebra.

We shall now identify the base element u_i of V with the base element $\{i\}$ of $E(V)$. This imbeds V in $E(V)$ as the subset of elements $\sum a_i u_i$. Moreover (13) gives $u_i^2 = 0$, $u_i u_j = -u_j u_i$ and if $i_1 < i_2 < \cdots < i_r$, then $u_{i_1} u_{i_2} \cdots u_{i_r} = \{i_1, i_2, \ldots, i_r\}$. If $v = \sum a_i u_i$, then $v^2 = \sum a_i^2 u_i^2 + \sum_{i<j} a_i a_j (u_i u_j + u_j u_i) = 0$. Thus we see that V is a subspace of $E(V)$ which generates $E(V)$ as an algebra and $v^2 = 0$ for every $v \in V$. Also dim $E(V) = 2^n$ since the elements 1 and $\{i_1, \ldots, i_r\} = u_{i_1} u_{i_2} \cdots u_{i_r}$ constitute a base for $E(V)$. We shall call $E(V)$ the *exterior algebra of the vector space V*. The actual mechanics of our construction is not important. What is important is the following property which characterizes the end product (see exercise 1 below).

THEOREM 7.1. *Let L be a linear map of V into an algebra A such that $(Lv)^2 = 0$ for every $v \in V$. Then L can be extended in one and only one way to a homomorphism $\eta(L)$ of the exterior algebra $E(V)$ into A.*

Proof. Put $\bar{v} = Lv$ so we have $\bar{v}^2 = 0$ which implies as before that $\bar{u}\bar{v} = -\bar{v}\bar{u}$, $u, v \in V$. Also, the argument which led to (12) can be repeated verbatim to show that if we define $\bar{u}_S = \bar{u}_{i_1} \cdots \bar{u}_{i_r}$ for $S = \{i_1, i_2, \ldots, i_r\}$, $i_1 < i_2 < \cdots < i_r$, then $\bar{u}_S\bar{u}_T = \varepsilon_{S,T}\bar{u}_{S \cup T}$. We now let $\eta(L)$ be the linear map of $E(V)$ into A whose action on the base $\{S \mid S \in \mathscr{P}(N)\}$ is given by $\eta(L)\varnothing = 1$ and $\eta(L)S = \bar{u}_S$. Then $\eta(L)u_i = \bar{u}_i = Lu_i$, so $\eta(L)v = Lv$ if $v \in V$. Hence $\eta(L)$ is an extension of L. Also $\eta(L)(ST) = (\eta(L)S)(\eta(L)T)$, since

$$\eta(L)(ST) = \eta(L)\varepsilon_{S,T}(S \cup T) = \varepsilon_{S,T}\bar{u}_{S \cup T}$$

and

$$(\eta(L)S)(\eta(L)T) = \bar{u}_S\bar{u}_T = \varepsilon_{S,T}\bar{u}_{S \cup T}.$$

This implies that if $X = \sum a_S S$ and $Y = \sum b_T T$ then we have $\eta(L)(XY) = (\eta(L)X)(\eta(L)Y)$ and since $\eta(L)1 = 1$, we see that $\eta(L)$ is an algebra homomorphism. The uniqueness of $\eta(L)$ is clear. $\square$

COROLLARY 1. *Let U be a subspace of V. Then the subalgebra of $E(V)$ generated by U is isomorphic to $E(U)$.*

Proof. If $(u_1, u_2, \ldots u_n)$ is any base for V then, as we saw for the algebra A at the beginning of our discussion, every element of $E(V)$ is a linear combination of 1 and the standard monomials $u_{i_1} u_{i_2} \cdots u_{i_r}$. Since $\dim E(V) = 2^n$, these 2^n elements are linearly independent. Now if U is a subspace of V we may suppose the base $(u_1, u_2, \ldots, u_n)$ is chosen so that $(u_1, \ldots, u_m)$ is a base for U. This shows that the standard monomials in $u_1, \ldots, u_m$ together with 1 are linearly independent. Since these are contained in the subalgebra B of $E(V)$ generated by U we see that $\dim B \geq 2^m$. On the other hand, since $u^2 = 0$, $u \in U$, holds in $E(V)$, Theorem 7.1 shows that we have a homomorphism of $E(U)$ into $E(V)$ sending the element $u \in U \subset E(U)$ into $u \in U \subset E(V)$. The image of this homomorphism is a subalgebra of $E(V)$ containing U, hence it is B. Since $\dim E(U) = 2^m$ and $\dim B \geq 2^m$ we see that $\dim B = 2^m$. This implies that our homomorphism is an isomorphism. $\square$

COROLLARY 2. *If L is a linear transformation in V and $\eta(L)$ is the endomorphism of $E(V)$ defined by L, then*

$$(15) \qquad \eta(1) = 1, \qquad \eta(L_1 L_2) = \eta(L_1)\eta(L_2)$$

and $\eta(L)$ is an automorphism if L is bijective.

Proof. Since the identity automorphism in $E(V)$ is the identity on V it is clear that $\eta(1) = 1$. Since $\eta(L_1 L_2)$ and $\eta(L_1)\eta(L_2)$ are endomorphisms of $E(V)$ having the same restrictions $L_1 L_2$ to V, we have $\eta(L_1 L_2) = \eta(L_1)\eta(L_2)$. If L is bijective we have the inverse linear transformation L^{-1} of V into itself. Then $LL^{-1} = 1 = L^{-1}L$ gives $\eta(L^{-1})\eta(L) = 1 = \eta(L^{-1})\eta(L)$ so $\eta(L)$ is bijective, hence, an automorphism. □

Before proceeding to the applications there is one more important fact about $E(V)$ which we should note, namely, we have a direct decomposition of $E(V)$ into subspaces

$$(16) \qquad E(V) = F1 \oplus V \oplus V^2 \oplus V^3 \oplus \cdots \oplus V^n$$

where V^r is the space spanned by all the products $v_1 v_2 \cdots v_r$, $v_i \in V$. This is clear since $v_1 v_2 \cdots v_r$ is the linear combination of monomials $u_{i_1} \cdots u_{i_r}$ where $(u_1, u_2, \ldots, u_n)$ is a base and any monomial $u_{i_1} \cdots u_{i_r}$ is either 0 or it is $\pm a$ standard monomial. Hence V^r is the space spanned by the standard monomials of degree r. Since these form a base we have (16). We see also that

$$(17) \qquad \dim V^r = \binom{n}{r}$$

since this is the number of standard monomials $u_{i_1} \cdots u_{i_r}$ of degree r. In particular, we have $\dim V^n = 1$ and $u_1 u_2 \cdots u_n$ is a base of this space.

Now let L be a linear transformation of V into itself and let $\eta(L)$ be the extension of L to an endomorphism of $E(V)$. Then it is clear from the definition of V^r that

$$(18) \qquad \eta(L)V^r \subset V^r.$$

In particular, we have $\eta(L)V^n \subset V^n$ and since $V^n = Fu_1 u_2 \cdots u_n$ we have $\eta(L)(u_1 u_2 \cdots u_n) = \Delta u_1 \cdots u_n$ where $\Delta \in F$. Suppose

$$(19) \qquad Lu_i = \sum l_{ij} u_j, \qquad 1 \le i \le n$$

so $\Lambda = (l_{ij})$ is the matrix of L relative to the base $(u_1, u_2, \ldots, u_n)$. Then

$$\Delta u_1 \cdots u_n = \eta(L)(u_1 \cdots u_n) = \eta(L)u_1 \eta(L)u_2 \cdots \eta(L)u_n$$

$$= \sum_{j_1, \ldots, j_n} l_{1j_1} l_{2j_2} \cdots l_{nj_n} u_{j_1} u_{j_2} \cdots u_{j_n}$$

$$= \sum sg \begin{pmatrix} 1 & 2 & \cdots & n \\ j_1 & j_2 & \cdots & j_n \end{pmatrix} l_{1j_1} \cdots l_{nj_n} u_1 \cdots u_n, \text{ by (7),}$$

$$= (\det \Lambda) u_1 \cdots u_n,$$

by the definition of $\det \Lambda = \sum sg \begin{pmatrix} 1 & 2 & \cdots & n \\ j_1 & j_2 & \cdots & j_n \end{pmatrix} l_{1j_1} \cdots l_{nj_n}$. Thus

(20) $$\eta(L)(u_1 \cdots u_n) = (\det \Lambda) u_1 \cdots u_n.$$

If L_1 and L_2 are linear maps of V into itself and Λ_i is the matrix of L_i relative to $(u_1 u_2, \ldots, u_n)$, then the matrix of $L_1 L_2$ relative to this base is $\Lambda_2 \Lambda_1$. Hence we have $\eta(L_1 L_2)(u_1 \cdots u_n) = \det \Lambda_2 \Lambda_1 u_1 \cdots u_n$. On the other hand

$$\eta(L_1 L_2)(u_1 \cdots u_n) = \eta(L_1)\eta(L_2)(u_1 \cdots u_n) = \eta(L_1)(\det \Lambda_2)(u_1 \cdots u_n)$$

$$= \det \Lambda_1 \det \Lambda_2 u_1 \cdots u_n.$$

This proves the multiplicative property of determinants of matrices in $M_n(F)$:

(21) $$\det \Lambda_1 \Lambda_2 = (\det \Lambda_1)(\det \Lambda_2).$$

We shall use this method next to derive Laplace's formula for expanding a determinant by the minors of a certain set of rows and their corresponding cofactors. We fix a subset $S = \{i_1, i_2, \ldots, i_k\}$, $i_1 < i_2 < \cdots < i_k$ of N, and consider the element $\eta(L)(u_{i_1} \cdots u_{i_k})$. If $T = \{j_1, j_2, \ldots, j_k\}$, $j_1 < j_2 \cdots < j_k$, then we let $\Lambda_{S,T}$ denote the minor obtained from the $i_1, \ldots, i_k$ rows and $j_1, \ldots, j_k$ columns. Then one sees, using (7), that

(22) $$\eta(L)(u_{i_1} \cdots u_{i_k}) = \sum_{|T|=k} \Lambda_{S,T} u_{j_1} \cdots u_{j_k}.$$

Let $S' = \{i_{k+1}, \ldots, i_n\}$, $i_{k+1} < \cdots < i_n$, be the complement of S in N. Then we have

(23) $$\eta(L)(u_{i_{k+1}} \cdots u_{i_n}) = \sum_{|T'|=n-k} \Lambda_{S',T'} u_{j_{k+1}} \cdots u_{j_n}$$

where $T' = \{j_{k+1}, \ldots, j_n\}$ and $j_{k+1} < \cdots < j_n$. Now $ST' = 0$ unless $T' = S'$, in which case, SS' is

(24) $$u_{i_1} \cdots u_{i_k} u_{i_{k+1}} \cdots u_{i_n} = \varepsilon_{S,S'} u_1 \cdots u_n$$

where

(25)
$$\begin{aligned}
\varepsilon_{S,S'} &= (-1)^{\lambda_{S,S'}}, \\
\lambda_{S,S'} &= (i_1 - 1) + (i_2 - 2) + \cdots + (i_k - k).
\end{aligned}$$

The last formula follows by observing that there are $i_1 - 1$ numbers in N less than i_1 and all of these occur in $\{i_{k+1}, \ldots, i_n\}$, there are $i_2 - 1$ numbers in N less than i_2 and all but one of these occur in $\{i_{k+1}, \ldots, i_n\}$, etc. The relation

$$\eta(L)(u_{i_1} \cdots u_{i_k} u_{i_{k+1}} \cdots u_{i_n}) = \eta(L)(u_{i_1} \cdots u_{i_k})\eta(L)(u_{i_{k+1}} \cdots u_{i_n})$$

now gives the formula

$$(-1)^{\sum_1^k (i_q - q)} \det \Lambda = \sum_T \Lambda_{S,T} \Lambda_{S',T'} (-1)^{\sum_1^k (j_q - q)}$$

where $T = \{j_1, j_2, \ldots, j_k\}, j_1 < j_2 < \cdots < j_k$ as before. We now define the co-factor $\Lambda'_{S,T}$ of the minor $\Lambda_{S,T}$ to be $(-1)^{\Sigma(i_q + j_q)}$ times the complementary minor $\Lambda_{S',T'}$ of $\Lambda_{S,T}$. Then the foregoing formula gives Laplace's expansion

(26) $$\det \Lambda = \sum_T \Lambda_{S,T} \Lambda'_{S,T}$$

of the determinant by the minors obtained from the $i_1, i_2, \ldots, i_k$-th rows. In particular, if we take $k = 1$, we obtain the formula for a determinant in terms of the elements of a particular row and the cofactor of these elements.

We now order the subsets S with $|S| = k$ lexicographically. Using this ordering we obtain from (22) the matrix $C_k(\Lambda) = (\Lambda_{S,T})$ of the linear transformation induced in V^k by the endomorphism $\eta(L)$. This is a matrix of $\binom{n}{k}$ rows and columns called the kth *compound* of Λ. Since $\eta(L_1 L_2) = \eta(L_1)\eta(L_2)$ we obtain the multiplicative property

(27) $$C_k(\Lambda_1)C_k(\Lambda_2) = C_k(\Lambda_1 \Lambda_2)$$

which generalizes the multiplicative property of determinants.

It is well known that the results we have obtained on determinants of matrices with entries in a field are valid for matrices with entries in any commutative ring. We shall now show by a method of indeterminates how we can obtain the general case from the case of fields, in fact, form a particular field of the form $\mathbb{Q}(x_1, \ldots, x_m)$, $\{x_k\}$ a sufficiently large set of indeterminates.

For the multiplication theorem for determinants we take $m = 2n^2$ and denote the x_k by x_{ij}, y_{ij}, $1 \leq i, j \leq n$. Then for $X = (x_{ij})$, $Y = (y_{ij})$ we have det $XY =$ (det X)(det Y) in $M_n(\mathbb{Q}(x_{ij}, y_{ij}))$ and hence in $M_n(\mathbb{Z}[x_{ij}, y_{ij}])$. Now suppose R is any commutative ring. Then we have a homomorphism of $\mathbb{Z}[x_{ij}, y_{ij}]$ into R sending the x_{ij} and y_{ij} into any $2n^2$ prescribed elements of R.[5] Now let Λ_1 and Λ_2 be two matrices in $M_n(R)$ and let η be the homomorphism of $\mathbb{Z}[x_{ij}, y_{ij}]$ sending x_{ij}, $1 \leq i, j \leq n$, into the (i,j)-entry of Λ_1 and y_{ij} into the (i,j)-entry of Λ_2. Applying η to the relation det $XY = $ det X det Y we obtain det $\Lambda_1\Lambda_2 = $ (det Λ_1)(det Λ_2) for any two matrices $\Lambda_i \in M_n(R)$. A similar argument applies to the other theorems.

We shall show next that the multiplicative property of determinants can be used to characterize the map $A \to$ det A among the polynomial functions on $M_n(F)$ (see section 2.12, p. 134). For this purpose we require the following

THEOREM 7.2. *Let $F[x_{ij}]$ be the polynomial ring over a field F in the n^2 indeterminates x_{ij}, $1 \leq i,j \leq n$ and let $X = (x_{ij})$ in $M_n(F[x_{ij}])$. Then det X is irreducible in $F[x_{ij}]$.*

Proof. We shall use induction on n, and we recall that $F[x_{ij}]$ is factorial (p. 154). Write det $X = \sum x_{1i}X_{1i}$ where X_{ij} denotes the cofactor of x_{ij} in X. Let D be the subring of polynomials in the $x_{ij} \neq x_{11}$ so $F[x_{ij}] = D[x_{11}]$ and we have det $X = x_{11}X_{11} + Y$, $Y \in D$. We may assume that X_{11} is irreducible as a polynomial in the x_{ij}, $i, j > 1$; hence X_{11} is irreducible in D. Now, det $X = x_{11}X_{11} + Y$ is of degree one in x_{11} so any factorization of det X in $D[x_{11}]$ has the form $(Px_{11} + Q)R$ where $P, Q, R \in D$. This gives $PR = X_{11}$, and since X_{11} is irreducible in D, replacing P, Q, and R by associates, we may assume either $P = 1$ or $P = X_{11}$. If $P = X_{11}$, $R = 1$ and the factorization is trivial. Hence if det X is reducible, then $P = 1$ and hence $R = X_{11}$. Thus reducibility of det X in $F[x_{ij}]$ implies that $X_{11}|$det X. Similarly, it implies that $X_{ii}|$det X and since the cofactors X_{ii}, $1 \leq i \leq n$, are distinct and irreducible we have $X_{11}X_{22} \cdots X_{nn}|$det X. The left hand side has (total) degree $n(n-1)$ and the right hand side has degree n. Hence this is impossible if $n > 2$. If $n = 2$ the relation becomes $x_{11}x_{22}|(x_{11}x_{22} - x_{12}x_{21})$ which is also impossible. $\square$

We can now prove the following result.

THEOREM 7.3. *Let F be an infinite field and let $Q(x_{ij})$ be a homogeneous polynomial of degree q in $F[x_{ij}]$, x_{ij} indeterminates, $1 \leq i, j \leq n$. Assume that for the*

---

[5] We have a homomorphism of $\mathbb{Z}$ into R ($m \to m1$) and this can be extended to $\mathbb{Z}[x_{ij}, y_{ij}]$ sending x_{ij} and y_{ij} into any chosen elements of R.

corresponding polynomial function $A = (a_{ij}) \to Q(a_{ij}) = Q(A)$ *on* $M_n(F)$ *we have* (i) $Q(1) = 1$, (ii) $Q(AB) = Q(A)Q(B)$. *Then* $Q(x_{ij})$ *is a power of det* X.

Proof. If $A = (a_{ij}) \in M_n(F)$ we define the adjoint matrix adj $A = (A_{ij})$ as usual (p. 96). Then we have $A(\text{adj } A) = (\det A)1$ in $M_n(F)$. Since $Q(x_{ij})$ is homogeneous of degree q this gives $Q(A)Q(\text{adj } A) = Q((\det A)1) = (\det A)^q Q(1) = (\det A)^q$. Now let $X = (x_{ij})$, adj $X = (X_{ij})$, so X_{ij} is a polynomial of degree $n - 1$ in the x's. Hence $P(x_{ij}) \equiv Q(x_{ij})Q(X_{ij}) - (\det X)^q \in F[x_{i.}]$. Since F is infinite and $P(a_{ij}) = 0$ for all choices of the $a_{ij} \in F$ it follows from Theorem 2.19 (p. 136) that $P(x_{ij}) = 0$. Thus $Q(x_{ij})Q(X_{ij}) = (\det X)^q$ in $F[x_{ij}]$; hence $Q(x_{ij})|(\det X)^q$. Since det X is irreducible and det $1 = 1 = Q(1)$, we have $Q(x_{ij}) = (\det X)^m$ for some m, $1 \le m \le q$. □

Clearly this gives a characterization of det X as the polynomial of least degree having the properties stated in the theorem. We can use this to derive some further results on determinants.

COROLLARY 1. *If* $A \in M_n(R)$, R *a commutative ring, then* det $A =$ det $\,^t A$ ($^t A$ *the transpose of* A).

Proof. The method of indeterminates used above shows that it is enough to prove this for $R = \mathbb{Q}(x_{ij})$. Hence we may assume $R = F$, an infinite field. Now consider the polynomial $Q = \det \,^t X \in F[x_{ij}]$. We have $Q(1) = \det 1 = 1$ and $Q(AB) = \det \,^t(AB) = \det \,^t B^t A = \det \,^t B \det \,^t A = Q(B)Q(A) = Q(A)Q(B)$. Hence, by Theorem 7.3, $Q(x_{ij})$ is a power of det X. Since the degrees are the same we have $Q(x_{ij}) = \det X$, and so the theorem holds for F, and hence for any R. □

This result enables us to obtain a Laplace's expansion by columns from the result we proved on Laplace's expansion by rows. Theorem 7.3 and degree consideration yield also the following result whose proof is left to the reader.

COROLLARY 2. *If* $C_r(A)$ *is the* rth *compound of the matrix* A, *then* det $C_r(A) = (\det A)^{\binom{n-1}{r-1}}$.

The exterior algebra $E(V)$, more particularly its subspace V^r, can be used to coordinatize the set $\Gamma_r(V)$ of r-dimensional subspaces of the vector space V. We have seen in Corollary 1 to Theorem 7.1 that if U is a subspace of V, then $E(U)$ can be identified with the subalgebra of $E(V)$ generated by U. If $(v_1, \ldots, v_r)$

is a base for U, then this subalgebra is generated by the v_i. We have the decomposition $E(U) = F \oplus U \oplus U^2 \oplus \cdots \oplus U^r$ and $U^r = Fv_1 \cdots v_r = E(U) \cap V^r$. The element $v_1 \cdots v_r$ is determined up to a non-zero multiplier in F by the subspace U, that is, another choice of base $(v'_1, \ldots, v'_r)$ for U gives $v'_1 \cdots v'_r = \rho v_1 \cdots v_r$ where $\rho \neq 0$ in F.

We shall call an element of $E(V)$ *decomposable* if it has the form $v_1 v_2 \cdots v_r$ where the v_i are linearly independent elements of V. Our result shows that an r-dimensional subspace U of V determines a one dimensional subspace Fu where $u = v_1 v_2 \cdots v_r$ is a decomposable element defined by the base $(v_1, v_2, \ldots, v_r)$ for U. Distinct U gives rise in this way to distinct subspaces Fu. For, if U' is a second subspace $\neq U$, we can choose a base $(v_1, \ldots, v_n)$ for V such that $(v_1, \ldots, v_r)$ is a base for U, $(v_1, \ldots, v_s)$, $0 \leq s < r$, is a base for $U \cap U'$, and $(v_1, \ldots, v_s, v_{r+1}, \ldots, v_{2r-s})$ is a base for U'. Then the consideration of the base for V^r determined by the base $(v_1, \ldots, v_n)$ for V shows that $v_1 \cdots v_r$ and $v_1 \cdots v_s v_{r+1} \cdots v_{2r-s}$ are linearly independent.

We now see that we have a 1–1 correspondence between the set $\Gamma_r(V)$ of r-dimensional subspaces of V and the set of one dimensional subspaces Fu, where u is a decomposable element of $E(V)$ contained in V^r. If $U \in \Gamma_r(V)$ has base $(v_1, \ldots, v_r)$ then the corresponding Fu is obtained from $u = v_1 \cdots v_r$, and if $u = v_1 \cdots v_r$ is decomposable then the corresponding subspace U is $\sum Fv_i$.

We now choose a base $(u_1, u_2, \ldots, u_n)$ for V. Then V^r has the base consisting of the $\binom{n}{r}$ products $u_{i_1} u_{i_2} \cdots u_{i_r}$ where $i_1 < i_2 < \cdots < i_r$. We order this base lexicographically. Then if $v_1 \cdots v_r$ is a decomposable element, $v_1 \cdots v_r = \sum \lambda_{i_1 \ldots i_r} u_{i_1} \cdots u_{i_r}$. The coordinates $(\lambda_{i_1 \ldots i_r})$ are determined up to a non-zero multiplier in F by the subspace $U = \sum_1^r Fv_i$. These are called the *Plücker coordinates* of the subspace U (relative to the base $(u_1, \ldots, u_n)$ for V).

EXERCISES

1. Show that Theorem 7.1 gives a characterization of $E(V)$ by proving that if $E'(V)$ is a second algebra having the stated property then there is an isomorphism of $E(V)$ into $E'(V)$ which is the identity map on V.

2. Prove the following addendum to Corollary 2 to Theorem 7.1: $\eta(L)$ is an automorphism only if L is bijective.

3. Let x_{ij}, $i \leq j$, be indeterminates over F and let $X = (x_{ij})$ where $x_{ji} = x_{ij}$ (so X is a "generic" symmetric matrix). Show that $\det X$ is irreducible in $F[x_{ij}]$.

4. Let x_{ij}, $i < j = 1, 2, \ldots, n$, be indeterminates over a field F and let $X = (x_{ij})$ where x_{ij} is as indicated if $i < j$, $x_{ii} = 0$ and $x_{ji} = -x_{ij}$ if $i < j$. Show that the Pfaffian Pf X is irreducible in $F[x_{ij}]$ (see section 6.2, p. 352).

5. Use Laplace's expansion to prove that if

$$M = \begin{pmatrix} M_1 & & & 0 \\ & M_2 & \cdot & \\ & & & \cdot \\ * & & & \cdot & M_r \end{pmatrix}$$

    where the diagonal blocks are square matrices, then det $M = \prod \det M_i$.

The following set of exercises (6–10) outlines an alternative proof of Theorem 7.3 in which $Q(x_{ij})$ is not assumed to be homogeneous. This proof has been communicated to me by George Seligman.

6. Let $Q(x_{ij})$ be as in Theorem 7.3 but without the assumption of homogeneity. Restrict the map first to the group D of diagonal matrices diag$\{a_1, a_2, \ldots, a_n\}$ with $\prod a_i \neq 0$. Then the map coincides with that determined by the polynomial $Q(x_{11}, \ldots, x_{nn}, 0, \ldots, 0)$—that is, the polynomial obtained by setting $x_{ij} = 0$, for every $i \neq j$ in $Q(x_{ij})$. Use the theorem on linear independence of distinct characters of a group (see p. 291) to show that $Q(x_{11}, \ldots, x_{nn}, 0, \ldots, 0) = x_{11}{}^{m_1} x_{22}{}^{m_2} \cdots x_{nn}{}^{m_n}$, $m_i \geq 0$.

7. Show that the m_i in exercise 6 are equal by using the relation

$$P \text{ diag}\{a_1, a_2, \ldots, a_n\} P^{-1} = \text{diag}\{a_2, \ldots, a_n, a_1\}$$

    where $P = e_{12} + e_{23} + \cdots + e_{n-1,n} + e_{n,1}$.

8. Let K be an extension field of F and use Q to define a map of $M_n(K)$ into K. Show that $Q(AB) = Q(A)Q(B)$ holds also in $M_n(K)$.

9. Use the foregoing exercises to show that there exists an $m = 0, 1, 2, \ldots$ such that $Q(A) = (\det A)^m$ for every invertible A with distinct characteristic roots.

10. Let $X = (x_{ij})$ and let $\Delta \in F[x_{ij}]$ be the discriminant of the characteristic polynomial det $(\lambda 1 - X)$. Show that the polynomial

$$\Delta \det X((\det X)^m - Q(X))$$

    vanishes for all matrices $X = A$ in $M_n(F)$. Use this to prove that $Q(X) = (\det X)^m$.

11. Let $\mathbb{Z}[x_{ij}]$ be the polynomial ring over $\mathbb{Z}$ in n^2 indeterminates x_{ij} and let $X = (x_{ij}) \in M_n(\mathbb{Z}[x_{ij}])$. Write the characteristic polynomial $f(\lambda) = \det(\lambda 1 - X) = \lambda^n - p_1 \lambda^{n-1} + \cdots + (-1)^n p_n$ where $p_i \in \mathbb{Z}[x_{ij}]$ and let $s_i = \text{tr } X^i$. Use diagonalization of X in a suitable extension field of $\mathbb{Z}[x_{ij}]$ and Newton's identities (p. 140) to show that $n! p_i \in \sum_1^n s_j \mathbb{Z}[s_1, \ldots, s_n]$.

12. Use exercise 11 and the Hamilton-Cayley theorem to show that if R is a commutative ring and $A \in M_n(R)$ satisfies tr $A = \text{tr } A^2 = \cdots = \text{tr } A^n = 0$, then $n! A^n = 0$.

13. If $A_1, \ldots, A_r \in M_n(R)$, R a commutative ring, define the *standard polynomial in the* A_i as

$$[A_1, \ldots, A_r] = \sum_{\pi \in S_r} (sg\ \pi) A_{\pi 1} A_{\pi 2} \cdots A_{\pi r}.$$

Show that $\text{tr}[A_1, \ldots, A_r] = 0$ if r is even.

14. Let $E^+(V) = F1 + V^2 + V^4 + \cdots$ (see (16)). Show that $E^+(V)$ is a commutative subalgebra of $E(V)$.

15. (Amitsur-Levitzki theorem.) Show that if $A_1, \ldots, A_{2n} \in M_n(R)$, R a commutative ring, then $[A_1, \ldots, A_{2n}] = 0$. (*Sketch of a proof by S. Rosset:* Since $[A_1, \ldots, A_{2n}]$ is multilinear in the A_i it suffices to prove the identity $[A_1, \ldots, A_{2n}] = 0$ for all choices of the A_i in the base $\{e_{ij}\}$ of matrix units. Hence it suffices to assume $R = \mathbb{Z}$ or $\mathbb{Q}$. Take $R = \mathbb{Q}$ and let V be the vector space over $\mathbb{Q}$ with base $(u_1, u_2, \ldots, u_{2n})$. Consider the matrix $A = \sum_1^{2n} A_i u_i \in M_n(E(V))$. The relation $[A_1, \ldots, A_{2n}] = 0$ is equivalent to $A^{2n} = 0$ and, since the characteristic is 0, to $n! A^{2n} = 0$. Note that $A^2 \in M_n(E^+(V))$ and $\text{tr}\ A^{2k} = 0$ for $k = 1, 2, \ldots$ follows from exercise 13. Then $n! A^{2n} = n!(A^2)^n = 0$ follows from exercise 12.)

16. Show that $[A_1, \ldots, A_k] = 0$ if $k \geq 2n$ for any choices of the A_i in $M_n(R)$, but that if $k < 2n$ then there exist $A_i \in M_n(R)$ such that $[A_1, \ldots, A_k] \neq 0$. (*Hint for the second part*: Take $A_1 = e_{11}$, $A_2 = e_{12}$, $A_3 = e_{22}$, $A_4 = e_{23}$, etc.)

7.3 REGULAR MATRIX REPRESENTATIONS OF ASSOCIATIVE ALGEBRAS. NORMS AND TRACES

Let V be a vector space over a field F and let $\text{End}_F V = \text{Hom}_F(V, V)$ be the set of linear transformations of V into itself. We have seen in section 3.3 (p. 169) that $\text{End}_F V$ can be endowed with a ring structure in which addition, multiplication, 0, and 1 are defined by: $(L + M)x = Lx + Mx$, $(LM)x = L(Mx)$, $0x = 0$, and $1x = x$ for $L, M \in \text{End}_F V$, $x \in V$. Since F is commutative, the multiplications by "scalars" (elements of F) are linear transformations. Such a map has the form $x \to ax$ where a is an element of F. Clearly $a(x + y) = ax + ay$ and if $b \in F$ then $a(bx) = b(ax)$, so $x \to ax$ is contained in $\text{End}_F V$. Since $a(Lx) = L(ax)$ for every $L \in \text{End}_F V$ it is clear also that the map $a_V : x \to ax$ is contained in the center of $\text{End}_F V$. It is immediate that $a \to a_V$ is a monomorphism of F into $\text{End}_F V$ whose image $F_V = \{a_V | a \in F\}$ is a subring of the center of $\text{End}_F V$.[6] This fact permits us to endow $\text{End}_F V$ with an algebra structure in which the ring structure is the usual one and the vector space structure is given by the usual addition and $aL = a_V L = La_V$ (cf. section 7.1). From now on in dealing with $\text{End}_F V$ we shall treat this as an algebra in this manner, and we shall call $\text{End}_F V$ the (*associative*) *algebra* of *linear transformations* of the vector space V over F.

Now suppose V over F is finite dimensional with base $(u_1, u_2, \ldots, u_n)$. If $L \in \text{End}_F V$ we obtain the matrix $\Lambda = (l_{ij})$ of L relative to the (ordered) base

---

[6] Actually F_V is the center of $\text{End}_F V$. See Corollary 2 to Theorem 3.17, p. 208.

$(u_1, u_2, \ldots, u_n)$ by writing $Lu_i = \sum_{j=1}^n l_{ij} u_j,\ 1 \le i \le n$. A change of base to $(v_1, v_2, \ldots, v_n)$, where $v_i = \sum c_{ij} u_j$ and $\Gamma = (c_{ij})$ is invertible, results in the matrix $\Gamma \Lambda \Gamma^{-1}$ for L relative to $(v_1, \ldots, v_n)$. We recall also the definition of the characteristic polynomial of Λ as

(28) $$f(\lambda) = \det(\lambda 1 - \Lambda)$$

(p. 196). Since

(29) $$\lambda 1 - \Lambda = \begin{pmatrix} \lambda - l_{11} & -l_{12} & \cdots & -l_{1n} \\ -l_{21} & \lambda - l_{22} & \cdots & -l_{2n} \\ \cdots\cdots\cdots\cdots\cdots\cdots\cdots\cdots\cdots \\ -l_{n1} & -l_{n2} & \cdots & \lambda - l_{nn} \end{pmatrix}$$

we see that

(30) $$f(\lambda) = \lambda^n - \left(\sum l_{ii} \right) \lambda^{n-1} + \cdots + (-1)^n \det \Lambda.$$

The element $\sum l_{ii}$ is called the *trace* tr Λ, of the matrix Λ. Evidently, we have

(31) $$\operatorname{tr}(\Lambda_1 + \Lambda_2) = \operatorname{tr} \Lambda_1 + \operatorname{tr} \Lambda_2, \qquad \operatorname{tr} a\Lambda = a \operatorname{tr} \Lambda.$$

that is, $\Lambda \to \operatorname{tr} \Lambda$ is a linear function on $M_n(F)$. Also, we have

(32) $$\det \Lambda_1 \Lambda_2 = \det \Lambda_1 \det \Lambda_2, \qquad \det a\Lambda = a^n \det \Lambda.$$

If M is similar to Λ, that is, $M = \Gamma \Lambda \Gamma^{-1}$, then M and Λ have the same characteristic polynomials. Hence they have the same traces and determinants. It follows that these are determined by the linear transformation L, so that we may define the characteristic polynomial of L, the trace of L, and the determinant of L to be these objects determined by any matrix of L.

We recall also that the map $L \to \Lambda$ of $\operatorname{End}_F V$ into $M_n(F)$ determined by the choice of a base is a ring anti-isomorphism (p. 111). Moreover, since the matrix of aL, $a \in F$, is $a\Lambda$, $L \to \Lambda$ is an algebra anti-isomorphism. It is psychologically advantageous to deal with isomorphisms rather than anti-isomorphisms. In the present situation we can go over to isomorphisms by considering the map $L \to {}^t\Lambda$ in place of $L \to \Lambda$. Since the characteristic polynomials of Λ and ${}^t\Lambda$ are the same we can calculate the characteristic polynomial of L from ${}^t\Lambda$ as well as from Λ. A change of base replaces ${}^t\Lambda$ by ${}^tM = \Delta^{-1} \Lambda \Delta$, $\Delta = {}^t\Gamma$.

Now suppose A is an (associative) algebra over the base field F. We proceed to show—using the same method of proof as that used for Cayley's theorem, and the corresponding result for rings—that A is isomorphic to an algebra of linear transformations.

THEOREM 7.4. *Any (associative) algebra A is isomorphic to a subalgebra of the algebra $\operatorname{End}_F A$ of linear transformations of the vector space A over F.*

Proof. As in the ring case (Theorem 3.2, p. 162), a monomorphism of A into $\text{End}_F A$ is the map $u \to u_L$ where u_L is the left multiplication $x \to ux$ in A: Since the algebra conditions give $u_L(ax) = u(ax) = a(ux) = au_L x$ for $a \in F$, $u_L \in \text{End}_F A$. Moreover, $u \to u_L$ is a ring monomorphism by Theorem 3.2. Since $(au)_L x = aux$ and $a(u_L x) = aux$, $u \to u_L$ is an algebra monomorphism. The image A_L is a subalgebra of $\text{End}_F A$ isomorphic to A. $\square$

A homomorphism of an algebra A over F into an algebra $\text{End}_F V$ of linear transformations of a vector space V over F is called a *representation* of A. The particular representation $u \to u_L$ we used in the foregoing proof is called *the regular* representation of A. If we have a representation of A by linear transformations in a finite dimensional vector space V, then we can combine this with an isomorphism $L \to {}^t\Lambda$ of $\text{End}_F V$ with $M_n(F)$ determined by a base $(u_1, \ldots, u_n)$ as before, to obtain a homomorphism $u \to \rho(u)$ of A into $M_n(F)$. Such a homomorphism is called a *matrix representation* of A. A change of base gives rise to an *equivalent* (or *similar*) representation $u \to \Delta^{-1}\rho(u)\Delta$. The matrix representations of a finite dimensional algebra associated with the regular representation are called *the regular matrix representations* of A.

Let $u \to \rho(u)$ be a regular matrix representation of A (finite dimensional over F). Then we define the *trace* and *norm* function T and N on A by

$$(33) \qquad T(u) = \text{tr } \rho(u), \qquad N(u) = \det \rho(u), \qquad u \in A.$$

Since similar matrices have the same traces and norms it is clear that these functions are unchanged on changing from one regular matrix representation to another. Since ρ is an algebra homomorphism it is clear from the trace and determinant of matrices that we have the following properties of T and N:

$$(34) \qquad T(u + v) = T(u) + T(v), \qquad T(au) = aT(u), \qquad a \in F$$

$$(35) \qquad N(uv) = N(u)N(v), \qquad N(au) = a^n N(u)$$

where n is the dimensionality $[A:F]$. In the next section we shall see that if A is a Galois extension field of F, then the foregoing definitions yield the same functions as those we defined in section 4.15.

We shall now look at some

EXAMPLES

1. Let $\mathbb{H}$ be Hamilton's quaternion algebra over $\mathbb{R}$, with base $(1, i, j, k)$ and the multiplication table

$$i^2 = j^2 = k^2 = -1, \qquad ij = -ji = k, \qquad jk = -kj = i, \qquad ki = -ik = j.$$

We determine the regular matrix representation of $\mathbb{H}$ given by the base $(1, i, j, k)$. Let $u = a_0 + a_1 i + a_2 j + a_3 k$. Then

$$u1 = a_0 + a_1 i + a_2 j + a_3 k$$
$$ui = -a_1 + a_0 i + a_3 j - a_2 k$$
$$uj = -a_2 - a_3 i + a_0 j + a_1 k$$
$$uk = -a_3 + a_2 i - a_1 j + a_0 k.$$

The corresponding matrix representation is obtained by taking the transpose of the matrix of the coefficients of the right-hand side of these equations, that is, it is

(36)
$$u \rightarrow \begin{pmatrix} a_0 & -a_1 & -a_2 & -a_3 \\ a_1 & a_0 & -a_3 & a_2 \\ a_2 & a_3 & a_0 & -a_1 \\ a_3 & -a_2 & a_1 & a_0 \end{pmatrix}.$$

2. Let $A = F[u]$ where u is algebraic with minimum polynomial $f(\lambda)$. Then $A \cong F[\lambda]/(f(\lambda))$. Suppose

$$f(\lambda) = \lambda^n - a_{n-1}\lambda^{n-1} - \cdots - a_0.$$

We have the base $(1, u, \ldots, u^{n-1})$ and

$$u1 = u$$
$$uu = u^2$$
$$\vdots$$
$$uu^{n-2} = u^{n-1}$$
$$uu^{n-1} = u^n = a_0 + a_1 u + \cdots + a_{n-1}u^{n-1}.$$

This implies that for the regular matrix representation determined by the base $(1, u, \ldots, u^{n-1})$ we have

(37)
$$\rho(u) = \begin{pmatrix} 0 & 0 & \cdot & \cdots & \cdot & \cdot & a_0 \\ 1 & 0 & \cdot & \cdots & \cdot & \cdot & a_1 \\ 0 & 1 & 0 & \cdots & \cdot & \cdot & \cdot \\ \cdot & \cdot & \cdot & \cdots & \cdot & \cdot & \cdot \\ 0 & 0 & \cdot & \cdots & 0 & 1 & a_{n-1} \end{pmatrix}.$$

3. As a special case of the last example we take $f(\lambda) = \lambda^n - 1$. Then it is easy to calculate $\rho(x)$ for $x = x_0 + x_1 u + x_2 u^2 + \cdots + x_{n-1}u^{n-1}$. One obtains

(38)
$$\rho(x) = \begin{pmatrix} x_0 & x_{n-1} & x_{n-2} & \cdots & x_1 \\ x_1 & x_0 & x_{n-1} & \cdots & x_2 \\ x_2 & x_1 & x_0 & \cdots & x_3 \\ \cdot & \cdot & \cdot & \cdots & \cdot \\ x_{n-1} & x_{n-2} & x_{n-3} & \cdots & x_0 \end{pmatrix}.$$

We have $N(x) = \det \rho(x)$. A determinant of this form is called a *circulant determinant*. A formula for calculating this is given in exercise 6 below.

EXERCISES

1. Let A be the algebra with base $(e_1, e_2, \ldots, e_n)$ such that $e_i^2 = e_i$, $e_i e_j = 0$ if $i \neq j$. Show that if $x = \sum_1^n x_i e_i$, then $\rho(x)$ determined by the given base is $\mathrm{diag}\{x_1, x_2, \ldots, x_n\}$.

2. Verify that if $u = a_0 + a_1 i + a_2 j + a_3 k$ in $\mathbb{H}$ then $T(u) = 4a_0$ and $N(u) = (a_0^2 + a_1^2 + a_2^2 + a_3^2)^2$.

3. Determine $\rho(x)$ for $x = x_{11} e_{11} + x_{12} e_{12} + x_{21} e_{21} + x_{22} e_{22}$ using the base $(e_{11}, e_{21}\ e_{12}, e_{22})$ for $M_2(F)$.

4. Let $A = M_n(F)$. Prove that if $X = (x_{ij}) \in M_n(F)$ then $T(X) = n\,\mathrm{tr}\,X$ and $N(X) = (\det X)^n$.

5. Let A be a finite dimensional extension field of F. Let $u \in A$ have minimum polynomial $m(\lambda)$ and let $f(\lambda)$ be the characteristic polynomial of $\rho(u)$, ρ a regular matrix representation. Show that $f(\lambda) = m(\lambda)^{[A:F(u)]}$. Suppose $m(\lambda) = \prod(\lambda - u_i)$ in a splitting field. Show that $N(u) = (\prod u_i)^{[A:F(u)]}$.

6. Assume F has n distinct nth roots of $1 = \zeta_1, \zeta_2, \ldots, \zeta_n$. Show that if $\rho(x)$ is as in (38) then $\det \rho(x) = \prod_{i=1}^{n} (x_0 + x_1 \zeta_i + \cdots + x_{n-1} \zeta_i^{n-1})$.

7. Verify that if Λ_1 and $\Lambda_2 \in M_n(F)$ then

(39) $\mathrm{tr}\,\Lambda_1 \Lambda_2 = \mathrm{tr}\,\Lambda_2 \Lambda_1$.

Hence conclude that

(40) $T(uv) = T(vu)$

holds for the trace function on an algebra.

7.4 CHANGE OF BASE FIELD.
TRANSITIVITY OF TRACE AND NORM

Let A be an algebra over the field F and let K be a subfield of F such that $[F:K] < \infty$. Then, as in the case in which A is a field (Theorem 4.2, p. 215), it is easily seen that the dimensionality $[A:K] = [A:F][F:K]$, and if $(u_1, \ldots, u_n)$ is a base for A over F and $(v_1, \ldots, v_r)$ is a base for F over K, then the nr elements $v_j u_i$ constitute a base for A over K. Clearly A can be regarded as an algebra over K as well as over F, and F is an algebra over K. We therefore have norm and trace functions from A to F, regarding A as an algebra over F, and from A to K as well as norms and traces from F to K. We denote these as $N_{A/F}$, $T_{A/F}$, $N_{A/K}$, $T_{A/K}$, $N_{F/K}$, and $T_{F/K}$ respectively. We shall now proceed to establish the following transitivity formulas for these functions. If $u \in A$, then

(41) $T_{A/K}(u) = T_{F/K}(T_{A/F}(u))$

(42) $N_{A/K}(u) = N_{F/K}(N_{A/F}(u))$.

The first of these is easy. All we have to do is look at the matrices relative to suitable bases. As before, let $(u_1, \ldots, u_n)$ be a base for A/F, $(v_1, \ldots, v_r)$ a base for F/K. Then we have the base

(43)
$$(v_1 u_1, \ldots, v_r u_1; v_1 u_2, \ldots, v_r u_2; \ldots; v_1 u_n, \ldots, v_r u_n)$$

for A over K. Let ρ be the regular matrix representation of A determined by the base $(u_1, \ldots, u_n)$ and μ the regular matrix representation of F over K determined by $(v_1, \ldots, v_r)$. Write $\rho(u) = (\rho_{ij}(u))$ for $u \in A$, $\mu(v) = (\mu_{kl}(v))$ for $v \in F$. If we recall the definitions, including the use of the transpose matrix, we see that we have the following relations:

(44)
$$uu_i = \sum_{j=1}^{n} \rho_{ji}(u)u_j, \quad 1 \le i \le n,$$

(45)
$$vv_k = \sum_{l=1}^{r} \mu_{lk}(v)v_l, \quad 1 \le k \le r.$$

Then

$$u(v_k u_i) = v_k(uu_i) = v_k \sum_{j=1}^{n} \rho_{ji}(u)u_j = \sum_{j} v_k \rho_{ji}(u)u_j$$

(46)
$$= \sum_{j} \rho_{ji}(u)v_k u_j = \sum_{j=1}^{n} \sum_{l=1}^{r} \mu_{lk}(\rho_{ji}(u))v_l u_j.$$

Accordingly, the regular matrix representation of A over K determined by the base (43) is

(47)
$$u \to \begin{bmatrix} \mu(\rho_{11}(u)) & \cdots & r(\rho_{1n}(u)) \\ \cdot & \cdots & \cdot \\ \cdot & \cdots & \cdot \\ \cdot & \cdots & \cdot \\ \mu(\rho_{n1}(u)) & \cdots & \mu(\rho_{nn}(u)) \end{bmatrix}.$$

In other words, we obtain a regular matrix representation of A/K by taking one of A/F and replacing the entries, which are elements of F, by the matrices representing them in a regular matrix representation of F/K. It is clear from (46) that $T_{A/K}(u) = \sum_{i=1}^{n} \operatorname{tr}(\mu(\rho_{ii}(u))) = \sum T_{F/K} \rho_{ii}(u) = T_{F/K}(\sum \rho_{ii}(u)) = T_{F/K}(T_{A/F}(u))$. This proves (41).

To prove (42) we shall establish a general transitivity property of determinants. We suppose we have an $nr \times nr$ matrix M with entries in a field K and we assume that if we partition this into $n \times n$ blocks of $r \times r$ matrices A_{ij}, then these $r \times r$ matrices all commute. This is precisely the situation we have for the matrix in (47) in which the $n \times n$ blocks $\mu(\rho_{ij}(u))$ commute, since the $\rho_{ij} \in F/K$

and $v \rightarrow \mu(v)$ is a homomorphism. Since the A_{ij} commute they are contained in a commutative subring B of the ring $M_r(K)$ of $r \times r$ matrices with entries in K. We can regard M as an $n \times n$ matrix with entries in B, that is, as an element of $M_n(B)$, and we can calculate the determinant of M as element of $M_n(B)$:

(48) $$\det_B M = \sum_\sigma (sg\ \sigma) A_{1,\sigma(1)} \cdots A_{n,\sigma(n)}.$$

This is an element of the subring generated by the A_{ij} and so is independent of the choice of the commutative subring B. Moreover, being an element of $M_r(K)$ it has a determinant which is an element of K. The result we wish to prove is:

(49) $$\det (\det_B M) = \det M.[7]$$

We assume first that $\det A_{11} \neq 0$, so A_{11}^{-1} exists in $M_r(K)$. Since A_{11}^{-1} commutes with every A_{ij} we may adjoin it to B obtaining a larger commutative subring of $M_r(K)$. Replacing B by this subring we may assume that $A_{11}^{-1} \in B$. Then we have the following calculation in $M_n(B)$:

$$
\begin{pmatrix} A_{11} & A_{12} & \cdots & A_{1n} \\ A_{21} & A_{22} & \cdots & A_{2n} \\ \cdots\cdots\cdots\cdots\cdots\cdots \\ A_{n1} & A_{n2} & \cdots & A_{nn} \end{pmatrix}
\begin{pmatrix} 1 & -A_{11}^{-1}A_{12} & \cdots & -A_{11}^{-1}A_{1n} \\ 0 & 1 & \cdots & \cdot \\ \cdots\cdots\cdots\cdots\cdots\cdots\cdots\cdots \\ 0 & 0 & \cdots & 1 \end{pmatrix}
$$
$$
= \begin{pmatrix} A'_{11} & 0 & \cdots & 0 \\ A'_{21} & A'_{22} & \cdots & A'_{2n} \\ \cdots\cdots\cdots\cdots\cdots\cdots \\ A'_{n1} & A'_{n2} & \cdots & A'_{nn} \end{pmatrix}
$$

($A'_{11} = A_{11}$). Calling the last matrix M' we have $\det_B M = \det_B M'$ by the multiplicative property of $\det_B$ and the fact that $\det_B$ of the triangular matrix is clearly 1. Also we have $\det M = \det M'$. Hence (49) will follow if we can prove $\det (\det_B M') = \det M'$. We have

$$\det_B M' = A'_{11} \det_B N', \qquad N' = (A'_{ij}), \qquad 2 \leq i, j \leq n,$$

so $\det (\det_B M') = \det A'_{11} \det (\det_B N')$. Also $\det M' = \det A'_{11} \det N'$ (exercise 5, p. 421). Now we can use induction on n to assume that $\det (\det_B N') = \det N'$. This gives the required relation $\det (\det_B M') = \det M'$.

To prove the result when A_{11} is not invertible we extend the base field K to $K(\lambda)$, λ an indeterminate, and we replace the matrix M by the matrix $M(\lambda)$ which is obtained by replacing the entry A_{11} by $A_{11} - \lambda 1$. Let $B(\lambda)$ be a commutative

---

[7] This result seems to be due to M. H. Ingraham, *Bulletin of the American Mathematical Society*, vol. 43, (1937) pp. 579–580.

subring of $M_r(K(\lambda))$ containing the entries of $M(\lambda)$. Since

$$\det (A_{11} - \lambda 1) = (-1)^r \lambda^r + \cdots \neq 0$$

the result just proved shows that det $(\det_{B(\lambda)} M(\lambda)) = \det M(\lambda)$. This is an identity in the polynomial ring $K[\lambda]$ which specializes to (49) by putting $\lambda = 0$.

We now apply this to norms. If M denotes the matrix on the right-hand side of (47), then $N_{A/K}(u) = \det M$ and taking B to be the commutative ring of $r \times r$ matrices $\mu(v)$, $v \in F$, we have det $M = \det (\det_B M)$. Since $v \to \mu(v)$ is a homomorphism, $\det_B M = \mu(\det (\rho_{ij}(u))) = \mu(N_{A/F}(u))$. Also det $\mu(v) = N_{F/K}(v)$ for $v \in F$. Hence det $(\det_B M) = \det \mu(N_{A/F}(u)) = N_{F/K}(N_{A/F}(u))$. Hence we have the transitivity property (42).

We now specialize everything to the case in which $A = E$ is a finite dimensional extension field of the field F. Let $u \in E$ and write the minimum polynomial of u over F as

(50) $m_u(\lambda) = \lambda^m - a_1 \lambda^{m-1} + a_2 \lambda^{m-2} + \cdots + (-1)^m a_m.$

Then in the regular matrix representation of $F(u) = F[u]$ over F using the base $(1, u, \ldots, u^{m-1})$, the matrix representing u is

(51)
$$\begin{pmatrix} 0 & 0 & \cdots & \cdot & (-1)^{m-1}a_m \\ 1 & 0 & \cdots & \cdot & (-1)^{m-2}a_{m-1} \\ 0 & 1 & \cdots & \cdot & (-1)^{m-3}a_{m-2} \\ \cdot & \cdot & \cdots & \cdot & \cdot \\ 0 & 0 & \cdots & 1 & a_1 \end{pmatrix}.$$

The trace and determinant of this matrix are respectively a_1 and a_m. Hence $T_{F(u)/F}(u) = a_1$, $N_{F(u)/F}(u) = a_m$. Also we have $m = [F(u):F]$ and $r \equiv [E:F(u)] = [E:F]/m$. Since $u \in F(u)$ we have $T_{E/F(u)}(u) = ru$ and $N_{E/F(u)}(u) = u^r$. Hence $T_{E/F}(u) = T_{F(u)/F}(T_{E/F(u)}(u)) = T_{F(u)/F}(ru) = ra_1$ and similarly $N_{E/F}(u) = a_m{}^r$. Thus we have

(52) $T_{E/F}(u) = [E:F(u)]a_1$

(53) $N_{E/F}(u) = a_m^{[E:F(u)]}.$

Suppose $m_u(\lambda) = \prod_1^m (\lambda - u_i)$ is a factorization of the minimum polynomial $m_u(\lambda)$ in a splitting field. Then $a_1 = \sum u_i$ and $a_m = \prod u_i$ and we can substitute these in the foregoing formulas.

Finally, suppose E is Galois over F with Galois group

$$G = \{\eta_1 = 1, \eta_2, \ldots, \eta_n\}.$$

Then the factorization $m_u(\lambda) = \prod_1^m (\lambda - u_i)$ takes place in $E[\lambda]$ and the set of roots $\{u_i\}$ is the orbit of $u = u_1$ under G. We see easily that the sequence $\{\eta_1(u),$ $\eta_2(u), \ldots, \eta_n(u)\}$ contains r copies of the orbit of u. Hence $\sum_1^n \eta_i(u) = [E:F(u)]a_1$, $\prod_1^n \eta_i(u) = a_m^{[E:F(u)]}$ and so

$$(54) \qquad\qquad\qquad T_{E/F}(u) = \sum_1^n \eta_i(u)$$

$$(55) \qquad\qquad\qquad N_{E/F}(u) = \prod_1^n \eta_i(u)$$

which were the definitions we gave in section 4.15 (p. 296) for the trace and norm of an element of a Galois extension field E over F.

7.5 NON-ASSOCIATIVE ALGEBRAS.
LIE AND JORDAN ALGEBRAS

One way of trying to create new mathematics from an existing mathematical theory, especially one presented in an axiomatic form, is to generalize the theory by dropping or weakening some of its hypotheses. If we play this axiomatic game with the concept of an associative algebra, we are likely to be led to the concept of a non-associative algebra, which is obtained simply by dropping the associative law of multiplication. If this stage is reached in isolation from other mathematical realities, it is quite certain that one would soon abandon the project, since there is very little of interest that can be said about non-associative algebras in general. What have turned out to be interesting are certain classes of non-associative algebras that have been brought to the attention of algebraists because of real or hoped for applications to other fields.

We shall look first at the two most important examples—Lie and Jordan algebras—and we begin with the former. These were introduced under the name of "infinitesimal groups" by Sophus Lie in connection with his studies of continuous groups, or more precisely, what are nowadays called Lie groups, which are suitably restricted continuous groups. We shall refrain from giving any precise definitions here but will try to suggest only that a continuous group is a composite notion involving a group and a topological space.[8] An example is the group $GL_n(\mathbb{R})$ of $n \times n$ invertible matrices with real-number entries. Here, besides the group structure, we have the added structure of a topological space which comes from the imbedding of $GL_n(\mathbb{R})$ in $M_n(\mathbb{R})$ and the fact that $M_n(\mathbb{R})$ can be regarded as a Euclidean space of n^2 dimensions. The connection between

---

[8] A good introduction to Lie theory can be found in P. M. Cohn, *Lie Groups*, Cambridge Tract in Mathematics no. 46, 1957, or in L. Pontrjagin, *Topological Groups*, Princeton University Press, 1939.

the algebraic and topological structures is that the group multiplication and the map $X \to X^{-1}$ are continuous. Another example of a continuous group is the real orthogonal group $O_n(\mathbb{R})$. Still another example is the Lorentz group, which is fundamental in relativity theory.

The great achievement of Sophus Lie was the reduction of local problems on Lie groups to problems on Lie algebras. With each Lie group there is an associated Lie algebra. For $GL_n(\mathbb{R})$ this is the Lie algebra $M_n(\mathbb{R})^-$ of all $n \times n$ real matrices. The Lie algebra structure on $M_n(\mathbb{R})^-$ is that given by the vector space structure of $M_n(\mathbb{R})$ and the Lie or (additive) commutator composition

$$(56) \qquad\qquad [X, Y] = XY - YX.$$

The Lie algebra of $O_n(\mathbb{R})$ is the set $\mathrm{Sk}_n(\mathbb{R})$ of $n \times n$ skew symmetric matrices. Just as $O_n(\mathbb{R})$ is a subgroup of $GL_n(\mathbb{R})$, $\mathrm{Sk}_n(\mathbb{R})$ is a subalgebra of $M_n(\mathbb{R})^-$, that is, a subspace of the vector space $M_n(\mathbb{R})$ closed under commutation.

The examples we have just given are special cases of a general process for obtaining Lie algebras from associative ones. In general we begin with an associative algebra A over any field F. We then obtain a new structure by replacing the given associative product by the *Lie* or *commutator* product.

$$(57) \qquad\qquad [x, y] = xy - yx.$$

In this way we obtain the Lie algebra A^-. Besides this it is natural to consider also subalgebras of the algebra A^-. An important class of example is obtained if the associative algebra A has an involution j: that is, an anti-automorphism $x \to \bar{x}$ of A such that $j^2 = 1$. This is the case if A is the matrix algebra $M_n(F)$ and $j = t$ the transpose map $X \to {}^tX$. Let $\mathrm{Sk}(A, j)$ denote the set of j-*skew* elements of A, that is, the elements s such that $\bar{s} = -s$. If $s_1\, s_2 \in \mathrm{Sk}(A, j)$ and a_1, $a_2 \in F$, then $s = a_1 s_1 + a_2 s_2 \in \mathrm{Sk}(A, j)$ since $\bar{s} = \overline{(a_1 s_1 + a_2 s_2)} = a_1 \bar{s}_1 + a_2 \bar{s}_2 = a_1(-s_1) + a_2(-s_2) = -(a_1 s_1 + a_2 s_2) = -s$. Also, $\overline{[s_1, s_2]} = \overline{s_1 s_2 - s_2 s_1} = \bar{s}_2 \bar{s}_1 - \bar{s}_1 \bar{s}_2 = (-s_2)(-s_1) - (-s_1)(-s_2) = s_2 s_1 - s_1 s_2 = -(s_1 s_2 - s_2 s_1) = -[s_1, s_2]$. Hence $\mathrm{Sk}(A, j)$ is a subalgebra of A^-.

What are the properties of the Lie product $[x, y]$ in an associative algebra which we can discover easily? First, it is immediate that if x, x_1, x_2, y, y_1, y_2 are elements of an associative algebra A over F and $a \in F$, then

$$(58) \qquad [x_1 + x_2, y] = [x_1, y] + [x_2, y], \qquad [x, y_1 + y_2] = [x, y_1] + [x, y_2],$$
$$a[x, y] = [ax, y] = [x, ay].$$

We omit the verification, which is trivial. We note next that

$$(59) \qquad\qquad [x, x] = 0$$

since $[x, x] = x^2 - x^2$, and we ask: is the product $[xy] (\equiv [x, y])$ associative? In terms of the associative product xy in A we have

$$[[xy]z] = (xy - yx)z - z(xy - yx)$$
$$= xyz - yxz - zxy + zyx$$
$$[x[yz]] = x(yz - zy) - (yz - zy)x$$
$$= xyz - xzy - yzx + zyx.$$

Hence associativity of $[,]$ is equivalent to $yxz + zxy = xzy + yzx$, or to $y(xz - zx) - (xz - zx)y = 0$, or $[y[xz]] = 0$. Thus associativity will hold only if $[y[xz]] = 0$ for all x, y, z in A. The first example we might test, $A = M_2(F)$, will show that this is not the case. For instance, if we take $x = e_{12}$, $y = e_{12}$, $z = e_{21}$, then $[y[xz]] = -2e_{12}$. If we look again at the foregoing calculations we obtain a positive result on the iterated commutators, namely, the calculations show that $[[xy]z] - [x[yz]] = [[xz]y]$. Since $[xx] = 0$ we have $[x + y, x + y] = [xx] + [xy] + [yx] + [yy] = 0$ so $[xy] = -[yx]$. Using this and the last relation we obtain the *Jacobi identity* for $[xy]$:

(60) $$[[xy]z] + [[yz]x] + [[zx]y] = 0.$$

This states that if we take a product $[[xy]z]$, permute the three elements cyclically and add, we obtain 0.

The properties we have just derived will be used in a moment to define abstract Lie algebras. Before doing this we consider the second class of nonassociative algebras which we wish to define in this section: Jordan algebras. Here we begin with any associative algebra A over a field F of characteristic $\neq 2$. We introduce the *Jordan product* (called the *anti-commutator* by physicists):

(61) $$x \cdot y = \tfrac{1}{2}(xy + yx).$$

We replace the associative product xy by the Jordan product $x \cdot y$ to obtain the Jordan algebra A^+. Then we obtain Jordan algebras also as subalgebras of the algebras A^+. For example, if A has an involution j then the set $\mathrm{Sym}(A, j)$ of j-symmetric elements is such a subalgebra. For, it is clear as with $\mathrm{Sk}(A, j)$, that $\mathrm{Sym}(A, j)$ is a subspace and if $h_1, h_2 \in \mathrm{Sym}(A, j)$, then $\overline{(h_1 h_2 + h_2 h_1)} = \overline{h}_2 \overline{h}_1 + \overline{h}_1 \overline{h}_2 = h_2 h_1 + h_1 h_2 \in \mathrm{Sym}(A, j)$.

The Jordan product $x \cdot y$ is commutative:

(62) $$x \cdot y = y \cdot x.$$

Since this holds we have only three distinct products of three elements $(x \cdot y) \cdot z$,

7.5 Non-Associative Algebras. Lie and Jordan Algebras

433

$(y \cdot z) \cdot x$ and $(z \cdot x) \cdot y$. Moreover, direct calculation gives

$$4(x \cdot y) \cdot z = xyz + yxz + zxy + zyx.$$

Hence

(63) $$4((x \cdot y) \cdot z - x \cdot (y \cdot z)) = [y[xz]]$$

and, as in the case of $[,]$, it follows that the Jordan product $x \cdot y$ is not associative. The formula (63) and the Jacobi identity for commutators does give the relation $(x \cdot y) \cdot z - x \cdot (y \cdot z) + (y \cdot z) \cdot x - y \cdot (z \cdot x) + (z \cdot x) \cdot y - z \cdot (x \cdot y) = 0$. However, this is a trivial consequence of the commutative law. We seek identities which do not follow in this way and we note first that $x^{\cdot 2} \equiv x \cdot x = \frac{1}{2}(x^2 + x^2)$. By induction, if we define $x^{\cdot k} = x^{\cdot k-1} \cdot x$, $x^{\cdot 1} = x$, then $x^{\cdot k} = x^k$. This implies that

(64) $$x^{\cdot k} \cdot x^{\cdot l} = x^{\cdot k+l}$$

a property which is called *power associativity*. Next, we compute

$$2x^{\cdot 2} \cdot y = x^2 y + yx^2, \qquad 4x \cdot (x \cdot y) = x^2 y + 2xyx + yx^2.$$

Taking $x = e_{12} + e_{21}$, $y = e_{11}$ in $A = M_2(F)$, we obtain $x^2 y + yx^2 = 2e_{11}$, $xyx = e_{22}$ which implies that $x^{\cdot 2} \cdot y$ and $x \cdot (x \cdot y)$ are linearly independent. Hence we have no relation of the form $ax^{\cdot 2} \cdot y = bx \cdot (x \cdot y)$ for non-zero $a, b \in F$, valid in every A^+. On the other hand, we have

$$4(x^{\cdot 2} \cdot y) \cdot x = x^3 y + xyx^2 + x^2 yx + yx^3$$

$$4x^{\cdot 2} \cdot (y \cdot x) = x^2 yx + x^3 y + yx^3 + xyx^2$$

which shows that we have the *Jordan identity*

(65) $$x^{\cdot 2} \cdot (y \cdot x) = (x^{\cdot 2} \cdot y) \cdot x$$

in every A^+.

We shall now make a fresh start and give formal definitions of the concepts of non-associative ($=$not necessarily associative) algebras, Lie algebras, and Jordan algebras.

DEFINITION 7.2. *We define a* non-associative algebra A *over a field* F *as a vector space equipped with a binary product* $(x, y) \to xy$ *which is* bilinear *in the sense that*

$$(x_1 + x_2)y = x_1 y + x_2 y, \qquad x(y_1 + y_2) = xy_1 + xy_2,$$

$$a(xy) = (ax)y = x(ay), \qquad x, x_i, y, y_i \in A, a \in F.$$

Note that we do not assume the existence of a unit, as we do in associative algebras. The main reason for not doing so is that units cannot exist in the most important special case of non-associative algebras, namely: Lie algebras. Their definition is given in

DEFINITION 7.3. *A Lie algebra is a non-associative algebra whose product, which we shall denote as* $[xy]$ *(or* $[x, y]$*), satisfies the following two laws:*

$$[xx] = 0, \qquad [[xy]z] + [[yz]x] + [[zx]y] = 0.$$

An immediate consequence of the first of these is *anti-commutativity*:

$$[xy] = -[yx].$$

We remark that anti-commutativity implies that $2[xx] = 0$, so if the base field does not have characteristic 2, then anti-commutativity is equivalent to $[xx] = 0$, and may be used in place of this law in the definition of a Lie algebra of characteristic $\neq 2$.

The result we obtained above (equations (58), (59), and (60)) is that if A is any associative algebra, then A defines a Lie algebra A^- with the same underlying vector space and the Lie product $[xy] = xy - yx$. We have seen also that if A is an associative algebra with an involution j, then the set $Sk(A, j)$ of j-skew elements is a subalgebra (in the obvious sense) of the Lie algebra A^-. Of course, in general, this will not be a subalgebra of A. We shall give next another way in which Lie algebras arise, namely, as derivation algebras of algebras. Let A now be any non-associative algebra. Then we define a *derivation* D of A to be a linear map of A into A such that

$$D(xy) = (Dx)y + x(Dy), \qquad x, y \in A.$$

If D_1 and D_2 have this property, then it is clear that $D_1 + D_2$ has, and if $a \in F$ then

$$(aD)(xy) = a(D(xy)) = a[(Dx)y + x(Dy)] = (a(Dx))y + x(a(Dy))$$
$$= (aD(x))y + x(aD(y)).$$

Hence the set Der A of derivations is a subspace of the vector space $\mathrm{End}_F\, A$ of linear transformations of A over F. If D_1 and D_2 are derivations then D_1D_2 is a linear transformation and

$$D_1D_2(xy) = D_1((D_2x)y + x(D_2y))$$
$$= (D_1D_2x)y + (D_1x)(D_2y) + (D_2x)(D_1y) + x(D_1D_2y).$$

This indicates that D_1D_2 may not be a derivation. However, $(D_1x)(D_2y) +$

$(D_2x)(D_1y)$ is symmetric in D_1 and D_2, so if we interchange these and subtract, we obtain 0. Consequently, we have

$$[D_1D_2](xy) = ([D_1D_2]x)y + x([D_1D_2]y)$$

which shows that $[D_1, D_2] \in \text{Der } A$. We have therefore shown that Der A is a subalgebra of the Lie algebra $\text{End}_F A^-$ of linear transformations in the vector space A. This is called the *derivation algebra* of the non-associative algebra A.

DEFINITION 7.4. *A Jordan algebra is a non-associative algebra over a field of characteristic* $\neq 2$ *whose product, denoted as* $x \cdot y$, *satisfies the laws:*

$$(66) \qquad x \cdot y = y \cdot x, \qquad (x^2 \cdot y) \cdot x = x^2 \cdot (y \cdot x) \quad \text{where} \quad x^2 = x \cdot x.$$

What we showed in our preliminary discussion is that any associative algebra A over a field of characteristic $\neq 2$ determines a Jordan algebra A^+ having the same vector space as A and the product $x \cdot y = \frac{1}{2}(xy + yx)$. We saw also that if A has an involution j, then the set $\text{Sym}(A, j)$ of j-symmetric elements is a subalgebra of A^+.

We shall now show that the other property we noted for A^+, power associativity (equation (64)), is a consequence of the definition of a Jordan algebra, that is, this holds in every Jordan algebra. In any non-associative algebra we define the *associator* $[x, y, z]$ of x, y, z in the algebra by

$$(67) \qquad\qquad\qquad [x, y, z] = (xy)z - x(yz).$$

This is additive in every argument, $[x_1 + x_2, y, z] = [x_1, y, z] + [x_2, y, z]$ etc., and satisfies the following rule for scalars: $a[x, y, z] = [ax, y, z] = [x, ay, z] = [x, y, az]$, $a \in F$. These two properties can be expressed by saying that the associator $[x, y, z]$ is a *trilinear* function of its arguments. The last condition defining a Jordan algebra can be written as the associator condition

$$(68) \qquad\qquad\qquad [x^2, y, x] = 0.$$

This condition is a cubic condition on x. From it we shall derive a multilinear identity by a process of linearization or polarization (which goes back to ancient times). There are a number of ways of doing this. The most direct, but perhaps not the shortest, is to calculate

$$
\begin{aligned}
0 = [(x_1 + x_2)^2, y, x_1 + x_2] &= [x_1^2 + 2x_1 \cdot x_2 + x_2^2, y, x_1 + x_2] \\
&= [x_1^2, y, x_1] + [x_1^2, y, x_2] + 2[x_1 \cdot x_2, y, x_1] + 2[x_1 \cdot x_2, y, x_2] \\
&\quad + [x_2^2, y, x_1] + [x_2^2, y, x_2] \\
&= [x_1^2, y, x_2] + 2[x_1 \cdot x_2, y, x_1] + 2[x_1 \cdot x_2, y, x_2] + [x_2^2, y, x_1].
\end{aligned}
$$

Replacing x_2 by $x_2 + x_3$ in this we obtain

$$0 = [(x_1 + x_2 + x_3)^2, y, x_1 + x_2 + x_3] = [x_1{}^2, y, x_2] + [x_1{}^2, y, x_3]$$
$$+ 2[x_1 \cdot x_2, y, x_1] + 2[x_1 \cdot x_3, y, x_1] + 2[x_1 \cdot x_2, y, x_2]$$
$$+ 2[x_1 \cdot x_3, y, x_2] + 2[x_1 \cdot x_2, y, x_3] + 2[x_1 \cdot x_3, y, x_3]$$
$$+ [x_2{}^2, y, x_1] + 2[x_2 \cdot x_3, y, x_1] + [x_3{}^2, y, x_1].$$

Subtracting the preceding relation and the one obtained from it by replacing x_2 by x_3 we obtain the desired multilinear identity:

$$2[x_1 \cdot x_2, y, x_3] + 2[x_2 \cdot x_3, y, x_1] + 2[x_3 \cdot x_1, y, x_2] = 0.$$

Cancelling the 2 (since the characteristic is $\neq 2$) we obtain

(69) $$[x_1 \cdot x_2, y, x_3] + [x_2 \cdot x_3, y, x_1] + [x_3 \cdot x_1, y, x_2] = 0.$$

In any non-associative algebra A we denote the linear map $y \to yx$ by x_R (the right multiplication by x) and the linear map $y \to xy$ by x_L (left multiplication by x). Using these we can formulate the commutative law by $x_L = x_R$ for all x, and, using this, the Jordan identity (68) by

(68') $$(x^2)_L x_L = x_L(x^2)_L \quad \text{or} \quad [(x^2)_L, x_L] = 0, \qquad (x^2 = x \cdot x).$$

The identity (69) is clearly equivalent to

(69') $$[(x_1 \cdot x_2)_L, x_{3L}] + [(x_2 \cdot x_3)_L, x_{1L}] + [(x_3 \cdot x_1)_L, x_{2L}] = 0.$$

Moreover, we can also derive another operator identity equivalent to (69) by writing this out and choosing one of the x_i as the element on which we operate. We have

$$((x_1 \cdot x_2) \cdot y) \cdot x_3 + ((x_2 \cdot x_3) \cdot y) \cdot x_1 + ((x_3 \cdot x_1) \cdot y) \cdot x_2$$
$$= (x_1 \cdot x_2) \cdot (y \cdot x_3) + (x_2 \cdot x_3) \cdot (y \cdot x_1) + (x_3 \cdot x_1) \cdot (y \cdot x_2).$$

Interchanging y and x_2 we obtain

70
$$x_{1L} x_{2L} x_{3L} + x_{3L} x_{2L} x_{1L} + ((x_1 \cdot x_3) \cdot x_2)_L$$
$$= x_{1L}(x_2 \cdot x_3)_L + x_{2L}(x_3 \cdot x_1)_L + x_{3L}(x_1 \cdot x_2)_L.$$

We now define the powers x^k (or $x^{\cdot k}$) by $x^1 = x$, $x^k = x \cdot x^{k-1}$. Then (70) gives the recursion formula

(71) $$x^{k+2}{}_L = (x^k)_L(x^2)_L + 2x_L(x^{k+1})_L - (x_L)^2(x^k)_L - (x^k)_L(x_L)^2.$$

Now x_L and $x^2{}_L$ commute so they generate a commutative algebra X of linear transformations. The recursion formula implies that every $x^k{}_L$, $k \geq 1$, is contained in X. Hence we have

(72)
$$[(x^k)_L, (x^l)_L] = 0, \qquad k, l \geq 1$$

which is equivalent to

(72')
$$[x^k, y, x^l] = 0.$$

This implies power associativity $x^k \cdot x^l = x^{k+l}$. For, this holds for all l and $k = 1$, by definition. Assuming it for all l and a fixed k, we have

$$x^{k+1} \cdot x^l = (x \cdot x^k) \cdot x^l = x \cdot (x^k \cdot x^l) = x \cdot x^{k+l} = x^{k+l+1}.$$

EXERCISES

1. Verify that the following associator identity holds in every non-associative algebra

$$v[x, y, z] + [v, x, y]z = [vx, y, z] - [v, xy, z] + [v, x, yz].$$

2. If A is a non-associative algebra one defines the *nucleus* $N(A)$ to be the subset of elements v which associate with everything, that is, every associator in which one of the arguments is v is 0. Use exercise 1 to show that $N(A)$ is an associative subalgebra of A.

3. The *center* $C(A)$ of a non-associative algebra A is the subset of $N(A)$ of elements c such that $cx = xc$, $x \in A$. Show that this is a commutative associative subalgebra of A.

4. Let D be a derivation of $F[\lambda]/F$, λ an indeterminate and let $D\lambda = f(\lambda)$. Show that D is determined by $f(\lambda)$. Show that the map $D \to f(\lambda)$ is an isomorphism of vector spaces of Der $F[\lambda]$ and $F[\lambda]$. Show that if $D \to f(\lambda)$ and $E \to g(\lambda)$, then $[D, E] \to f(\lambda)g'(\lambda) - f'(\lambda)g(\lambda)$, $f'(\lambda)$, the formal derivative of $f(\lambda)$ (see section 4.4, p. 230).

5. Generalize exercise 4 to $F[\lambda_1, \lambda_2, \ldots, \lambda_r]$, λ_i indeterminates.

6. Let A be an associative algebra over F which is a commutative domain. Show that any derivation D in A has a unique extension to the field of fractions of A.

7. Let A be a finite dimensional separable extension of F. Show that Der $A = 0$.

8. Determine Der A if F is of characteristic $p \neq 0$ and $A = F[\lambda]/(\lambda^p - a)$, $a \in F$.

9. If A is an algebra, the set $M_n(A)$ of $n \times n$ matrices with entries in A is an algebra with the usual vector space structure and usual matrix multiplication. Let D be

a map of A into itself and let $\eta(D)$ be the map

$$x \rightarrow \begin{pmatrix} x & Dx \\ 0 & x \end{pmatrix}$$

of A into $M_2(A)$. Show that $\eta(D)$ is a homomorphism of A into $M_2(A)$ if and only if D is a derivation.

10.   Show that a linear transformation D in a non-associative algebra A is a derivation if and only if either one of the following conditions holds:

$$[D, x_L] = (Dx)_L, \qquad x \in A$$
$$[D, x_R] = (Dx)_R, \qquad x \in A.$$

Show that if A is associative then $x_R - x_L$ is a derivation for every $x \in A$. Show that if A is Lie then $x_L = -x_R$ is a derivation. Show that if A is Jordan then $[x_L, y_L] (= [x_R, y_R])$ is a derivation for any $x, y \in A$.

11.   Let $B(u, v)$ be a symmetric bilinear form on a vector space V over F of characteristic $\neq 2$. Let $A = F1 \oplus V$ the vector space direct sum of V and a one dimensional space with base 1. Define a product $x \cdot y$ for $x = a1 + u$, $y = b1 + v$, $a, b \in F$, $u, v \in V$ by

$$(a1 + u) \cdot (b1 + v) = (ab + B(u, v))1 + (av + bu).$$

Verify that A with this product is a Jordan algebra with 1.

12.   Let $E(V)$ be the exterior algebra over V. Show that $F1 + V$ is a subalgebra of $E(V)^+$. Show that if $B \equiv 0$ in exercise 11, the resulting Jordan algebra is isomorphic to the subalgebra $F1 + V$ of $E(V)^+$.

7.6   HURWITZ'S PROBLEM. COMPOSITION ALGEBRAS

The following problem was considered by A. Hurwitz in 1898. For what values of n do there exist identities of the form

(73)
$$\left(\sum_1^n x_i^2 \right)\left(\sum_1^n y_i^2 \right) = \sum_1^n z_i^2$$

where the z_i have the form

(74)
$$z_i = \sum_{j,k=1}^n a_{ijk} x_j y_k,$$

a_{ijk} complex numbers? At the time Hurwitz posed and solved this problem a number of identities of this type were known and there had been a number of abortive attempts to find others. The known ones were identities for $n = 1, 2, 4,$ and 8. The first one of these is the trivial one: $x_1^2 y_1^2 = (x_1 y_1)^2$. The next two are already non-trivial, namely,

$$(x_1{}^2 + x_2{}^2)(y_1{}^2 + y_2{}^2) = (x_1 y_1 - x_2 y_2)^2 + (x_1 y_2 + x_2 y_1)^2$$
$$(x_1{}^2 + x_2{}^2 + x_3{}^2 + x_4{}^2)(y_1{}^2 + y_2{}^2 + y_3{}^2 + y_4{}^2) = (z_1{}^2 + z_2{}^2 + z_3{}^2 + z_4{}^2),$$

where

$$z_1 = x_1 y_1 - x_2 y_2 - x_3 y_3 - x_4 y_4$$
$$z_2 = x_1 y_2 + x_2 y_1 + x_3 y_4 - x_4 y_3$$
$$z_3 = x_1 y_3 - x_2 y_4 + x_3 y_1 + x_4 y_2$$
$$z_4 = x_1 y_4 + x_2 y_3 - x_3 y_2 + x_4 y_1.$$

These can be verified directly, or better still, they can be deduced from properties of the multiplication of complex numbers and of quaternions (see exercise 3, p. 100). It is rather tedious to write down the corresponding identity for $n = 8$. A somewhat less explicit form of this, from which we could write out the explicit identity if we wished, will be given later. It is not known who first discovered the foregoing identity for $n = 2$. The one for $n = 4$ seems to be due to Euler and, according to L. E. Dickson, the one for $n = 8$ was found by C. F. Degen in 1822. The sum of squares identity for $n = 4$ plays an important role in the proof of a beautiful theorem of Lagrange which states that every positive integer can be expressed as a sum of four squares of integers.[9] Hurwitz's theorem, which we shall prove and generalize in this section, is that identities of the form (73)–(74) exist only if $n = 1, 2, 4$, and 8.

The Hurwitz problem can be viewed either from the formal or the functional point of view. In the first we consider the x's and y's as indeterminates and (73) as a relation in the polynomial ring of these indeterminates over $\mathbb{C}$. From the functional point of view the starting point is the function $(x_1, x_2, \ldots, x_n) \to \sum_1^n x_i{}^2$ whose domain is the n-dimensional vector space of n-tuples $(x_1, x_2, \ldots, x_n)$ over $\mathbb{C}$. Clearly this function is a quadratic form and (73)–(74) is a functional relation. It is not difficult to see that a solution of the problem from either point of view implies the solution from the other one. This is trivial in the direction formal $\Rightarrow$ functional. The direction functional $\Rightarrow$ formal follows in the usual way from Theorem 2.19 (p. 136), since $\mathbb{C}$ is an infinite field. We shall adopt the functional point of view. Accordingly, we consider the n-dimensional vector space $\mathbb{C}^{(n)}$ of n-tuples of complex numbers $x = (x_1, x_2, \ldots, x_n)$ on which we have defined the quadratic form $x \to \sum_1^n x_i{}^2$, which is non-degenerate. If $y = (y_1, \ldots, y_n)$ and $z = (z_1, \ldots, z_n)$, then we have the binary composition $(x, y) \to z$

---

[9] See G. H. Hardy and E. M. Wright, *An Introduction to the Theory of Numbers*, 5th ed. *Oxford University Press*, 1975, p. 302.

where the z_i are given by (74) in terms of the fixed complex numbers a_{ijk}. It is clear from the form of (74) that this composition is bilinear. Hence if we denote z as xy, we obtain a non-associative algebra. Also, denoting $\sum x_i^2$ as $Q(x)$, we have $Q(x)Q(y) = Q(xy)$.

We shall now sharpen and generalize Hurwitz's problem. We suppose we have a finite dimensional vector space A over a field F of characteristic $\neq 2$ equipped with a non-degenerate quadratic form Q.[10] We shall say that Q *permits composition* if it is possible to define a bilinear product xy on A such that

$$(75) \qquad\qquad Q(x)Q(y) = Q(xy)$$

for all $x, y \in A$. We have a non-associative algebra defined by the vector space and the product xy, and we shall now show that by modifying the product we may assume that our algebra has a unit. For this purpose we choose an element v such that $Q(v) \neq 0$ and put $u = Q(v)^{-1}v^2$. Then $Q(u) = 1$ and hence $Q(xu) = Q(x) = Q(ux)$ for all x. Thus the multiplications u_R and u_L are orthogonal transformations of A relative to Q, and so these are invertible and their inverses are also orthogonal. We now define a new product $x * y$ in A by

$$x * y = (u_R^{-1}x)(u_L^{-1}y)$$

and we have $Q(x * y) = Q(u_R^{-1}x)Q(u_L^{-1}y) = Q(x)Q(y)$. Also $u_L^{-1}u^2 = u_L^{-1}(u_L u) = u$ and $u_R^{-1}u^2 = u_R^{-1}(u_R u) = u$, which implies that

$$u^2 * x = (u_R^{-1}u^2)(u_L^{-1}x) = u(u_L^{-1}x) = x$$

$$x * u^2 = (u_R^{-1}x)(u_L^{-1}u^2) = (u_R^{-1}x)u = x.$$

Thus u^2 is a unit relative to the $*$ multiplication. We shall now revert to the original notation xy for $x * y$, and so we assume at the outset that the algebra A has a unit, which we denote as 1. We state

DEFINITION 7.5. *A composition algebra is a pair consisting of a non-associative algebra A with unit 1 and a non-degenerate quadratic form Q on A such that $Q(xy) = Q(x)Q(y)$.*

We can now proclaim our objective in this section: to determine all the composition algebras. At the moment this may appear to be an overly ambitious goal. However, it turns out that we can achieve it in a surprisingly elementary fashion.

---

[10] Neither restriction—finite dimensionality or characteristic $\neq 2$—is essential. This will be indicated in exercises below.

Let (A, Q) be a composition algebra. We observe first that the composition law (75) gives $Q(x)Q(1) = Q(x)$ so we have

(76) $$Q(1) = 1.$$

Next we linearize the composition law in the variable x by replacing x by $x + z$ to obtain

$$Q(x + z)Q(y) = Q((x + z)y) = Q(xy + zy).$$

Since $Q(x + y) - Q(x) - Q(y) = B(x, y)$, the symmetric bilinear form associated with the quadratic form Q, the foregoing gives

$$Q(x)Q(y) + B(x, z)Q(y) + Q(z)Q(y) = Q(xy) + B(xy, zy) + Q(zy)$$
$$= Q(x)Q(y) + B(xy, zy) + Q(z)Q(y).$$

Hence we have

(77) $$B(x, z)Q(y) = B(xy, zy).$$

Similarly, if we linearize with respect to the variable y we obtain

(78) $$Q(x)B(y, w) = B(xy, xw).$$

Next we linearize (77) with respect to y by replacing y by $y + w$ in this equation. This leads to the relation

(79) $$B(x, z)B(y, w) = B(xy, zw) + B(zy, xw).$$

Since the left-hand side of this is unchanged if we interchange x and y and z and w we obtain also

(80) $$B(x, z)B(y, w) = B(yx, wz) + B(yz, wx).$$

We now introduce the map $j = -S_1$ where S_1 is the symmetry in the hyperplane orthogonal to 1,

$$j : x \to \frac{B(x, 1)}{Q(1)} 1 - x = B(x, 1)1 - x,$$

and we abbreviate

$$\bar{x} = j(x), \qquad T(x) = B(x, 1).$$

Then we have

(81) $$Q(\bar{x}) = Q(x), \qquad \bar{\bar{x}} = x$$

and we wish to prove

LEMMA 1. *We have the following properties:*

$$(82) \qquad \bar{x}x = Q(x)1 = x\bar{x}$$

$$(83) \qquad \bar{x}(xy) = (\bar{x}x)y = Q(x)y$$

$$(84) \qquad (yx)\bar{x} = y(x\bar{x}) = Q(x)y$$

$$(85) \qquad \overline{xy} = \bar{y}\bar{x}.$$

Proof. We note first that $B(x, \bar{y}z) = B(x, B(y, 1)z - yz) = B(y, 1)B(x, z) - B(x, yz) = B(yx, z) + B(yz, x) - B(x, yz)$ (by (79)) $= B(yx, z)$. Thus

$$(86) \qquad B(yz, x) = B(z, \bar{y}x)$$

and, similarly,

$$(87) \qquad B(xy, z) = B(x, z\bar{y}).$$

By (78) and (86), we have $Q(x)B(1, y) = B(x, xy) = B(\bar{x}x, y)$. Hence $B(Q(x)1, y) = B(\bar{x}x, y)$ and $B(Q(x)1 - \bar{x}x, y) = 0$ for all y. Since B is non-degenerate, this implies that $\bar{x}x = Q(x)1$. Also, replacing x by $\bar{x}$ we obtain $x\bar{x} = \bar{\bar{x}}\bar{x} = Q(\bar{x})1 = Q(x)1$. Hence (82) is proved. Next we have $B(\bar{x}(xy), z) = B((B(x, 1)1 - x)(xy), z) = B(x, 1) B(xy, z) - B(x(xy), z) = B(x(xy), z) + B(xy, xz) - B(x(xy), z)$ (by (79)) $= B(xy, xz) = Q(x)B(y, z) = B(Q(x)y, z)$. By the non-degeneracy of B, this gives $\bar{x}(xy) = Q(x)y$. Since $\bar{x}x = Q(x)1$, we have (83). Similarly one establishes (84). To prove (85) we compute

$$B(\bar{x}\bar{y}, z) = B(\bar{x}\bar{y}1, z) = B(1, (xy)z)$$

$$
\begin{aligned}
B(\bar{y}\bar{x}, z) &= B((B(y, 1)1 - y)(B(x, 1)1 - x), z) \\
&= B(B(x, 1)B(y, 1)1 - B(y, 1)x - B(x, 1)y + yx, z) \\
&= B(x, 1)B(y, 1)B(z, 1) - B(y, 1)B(x, z) - B(x, 1)B(y, z) \\
&\quad + B(yx, z) \\
&= B(xy, 1)B(z, 1) + B(x, y)B(z, 1) - B(yz, x) \\
&\quad - B(xy, z) - B(xz, y) \\
&= B((xy)z, 1) + B(xy, z) + B(xz, y) + B(yz, x) \\
&\quad - B(yz, x) - B(xy, z) - B(xz, y) \\
&= B((xy)z, 1).
\end{aligned}
$$

Hence $B(\bar{x}\bar{y}, z) = B(\bar{y}\bar{x}, z)$ and so again by non-degeneracy we obtain (85). $\qquad \square$

Since $j: x \to \bar{x}$ is linear, $\bar{\bar{x}} = x$, and (85) holds, j is an involution in A. Also, by definition of $\bar{x}$, we have $x + \bar{x} = T(x)1$. The relations (83) and (84) give the associator relations

(88) $$[\bar{x}, x, y] = 0 = [y, x, \bar{x}].$$

Since $[1, x, y] = 0 = [y, x, 1]$ and $x = T(x)1 - \bar{x}$ these relations imply

(89) $$[x, x, y] = 0 = [y, x, x].$$

Thus A is an alternative algebra in the sense of

DEFINITION 7.6.   *An algebra is* alternative *if the identities* (89) *hold for all x, y in the algebra.*

We have now shown that if (A, Q) is a composition algebra, then A is alternative with involution $j: x \to \bar{x}$ such that $\bar{x}x = Q(x)1$. It turns out that these conditions are also sufficient for a composition algebra. Before we can prove this we shall need to derive a few basic properties of alternative algebras.

We now suppose A is any alternative algebra. We note first that linearization of the alternative laws (89) gives the relations $[x, z, y] + [z, x, y] = 0 = [y, x, z] + [y, z, x]$. These imply that the associator $[x, y, z]$ is an alternating function of its arguments, that is, it is unchanged under even permutations $([x, y, z] = [y, z, x] = [z, x, y])$ and changes sign under odd permutations of the arguments. It follows also that $[x, y, x] = -[y, x, x] = 0$. Hence we have the laws

(90) $$x^2 y = x(xy), \qquad (xy)x = x(yx), \qquad yx^2 = (yx)x.$$

We shall abbreviate $(xy)x = x(yx)$ to xyx. We establish next the following important identity for alternative algebras which is due to R. Moufang:

(91) $$(ux)(yu) = u(xy)u.$$

Our starting point is the relation

(92) $$(ux)y + x(yu) = u(xy) + (xy)u$$

which is equivalent to $[u, x, y] = [x, y, u]$. We replace successively x by ux then y by yu in (92) and add the resulting equations. This gives

$$
\begin{aligned}
(u^2 x)y &+ 2(ux)(yu) + x(yu^2) \\
&= u((ux)y) + ((ux)y)u + u(x(yu)) + (x(yu))u \\
&= u[(ux)y + x(yu)] + [(ux)y + x(yu)]u \\
&= u[u(xy) + (xy)u] + [u(xy) + (xy)u]u \qquad \text{(by (92))} \\
&= u^2(xy) + 2u(xy)u + (xy)u^2.
\end{aligned}
$$

If we subtract from this the relation obtained from (92) by replacing u by u^2 we obtain Moufang's identity.[11]

Now suppose A is an alternative algebra with 1 and involution $j:x \to \bar{x}$ such that $\bar{x}x = Q(x)1$ where $Q(x)$ is a non-degenerate quadratic form. Then we have by linearization

$$(93) \qquad\qquad\qquad \bar{x}y + \bar{y}x = Q(x, y)1.$$

Putting $y = 1$ in this we obtain $x + \bar{x} = T(x)1$ where $T(x) = Q(x, 1)$. Then the alternative laws $[x, x, y] = 0 = [y, x, x]$ and $x + \bar{x} = T(x)1$ yield (88), so we have $\bar{x}(xy) = (\bar{x}x)y = Q(x)y$. We now have

$$
\begin{aligned}
Q(xy)1 &= (\overline{xy})(xy) = (\bar{y}\bar{x})(xy) = [(T(y)1 - y)\bar{x}](xy) \\
&= (T(y)\bar{x} - y\bar{x})(xy) = T(y)\bar{x}(xy) - (y\bar{x})(xy) \\
&= Q(x)T(y)y - y(\bar{x}x)y \qquad \text{(Moufang)} \\
&= Q(x)[T(y)1 - y]y = Q(x)(\bar{y}y) \\
&= Q(x)Q(y)1.
\end{aligned}
$$

Hence $Q(xy) = Q(x)Q(y)$ and (A, Q) is a composition algebra.

We have now achieved the first important step in our analysis, namely,

THEOREM 7.5. *Any composition algebra (A, Q) is alternative and has an involution $j:x \to \bar{x}$ such that $\bar{x}x = Q(x)1$. Conversely, let A be an alternative algebra with unit and involution $j:x \to \bar{x}$ such that $\bar{x}x = Q(x)1$, where $Q(x)$ is a non-degenerate quadratic form. Then (A, Q) is a composition algebra.*

We shall give next a construction of composition algebras. This will constitute an almost trivial generalization of the familiar construction of complex numbers as pairs of real numbers. For the moment we drop the alternative law and we assume only that A is a non-associative algebra with a unit 1 and an involution j such that $\bar{x}x = Q(x)1$ where $Q(x)$ is a non-degenerate quadratic form. Then we have (93) and $x + \bar{x} = T(x)1$, $T(x) = B(x, 1)$. Let c be a non-zero element of the base field F. From A, j, and c we shall now construct an algebra D satisfying the same conditions as A and having dimensionality 2 dim A. Let $D = A^{(2)}$, the vector space of pairs (x, y), $x, y \in A$, with the usual direct sum vector space structure. We introduce a binary product in D by the formula

$$(94) \qquad\qquad\qquad (u, v)(x, y) = (ux + c\bar{y}v, yu + v\bar{x}).$$

---

[11] The proof we have given makes use of the restriction that the characteristic is $\neq 2$. The result is valid without this restriction.

It is immediate that this is bilinear, so along with the vector space structure it defines an algebra on D. It is clear from (94) that $(1, 0)$ is a unit in D so we write $1 = (1, 0)$. We also have $(u, 0)(x, 0) = (ux, 0)$, from which it follows that $u \to (u, 0)$ is a monomorphism of A into D. Thus we may identify A with the sub-algebra of D made up of the elements $(u, 0)$, $u \in A$. We now extend the involution j on A to the linear map

(95) $$j : (x, y) \to (\overline{x, y}) = (\bar{x}, -y).$$

Clearly $j^2 = 1$. Direct verification, which we leave to the reader, shows that j is an involution in D. Moreover, we have

$$(\overline{x, y})(x, y) = (\bar{x}, -y)(x, y) = ((Q(x) - cQ(y))1, 0) = (Q(x) - cQ(y))1.$$

Now $(x, y) \to Q(x) - cQ(y)$ is a quadratic form on D. The corresponding sym-metric bilinear form is $((u, v), (x, y)) \to B(u, x) - cB(v, y)$. One sees easily that this is non-degenerate. Hence D and its involution satisfy the same conditions as A. We shall call D the *c-double* of A and we now prove

LEMMA 2.   (1) *The c-double D is commutative and associative if and only if A is commutative and associative and $j = 1$. (2) D is associative if and only if A is commutative and associative. (3) D is alternative if and only if A is associative.*

Proof.   Write $X = (x, y)$, $U = (u, v)$, $Z = (z, t)$ for x, y, etc. in A. Then

(96) $$[U, X] = ([u, x] + c(\bar{y}v - \bar{v}y), y(u - \bar{u}) + v(\bar{x} - x))$$

(97) $$[U, X, Z] = ([u, x, z]) + c\{\bar{t}(yu) - u(\bar{t}y) + \bar{t}(v\bar{x}) - (\bar{x}\bar{t})v$$
$$+ (\bar{y}v)z - (z\bar{y})v\}, t(ux) - (tx)u + (yu)\bar{z} - (y\bar{z})u$$
$$+ (v\bar{x})\bar{z} - v(\bar{z}\bar{x}) + c\{t(\bar{y}v) - v(\bar{y}t)\}).$$

Since A is a subalgebra of D (under the identification $x \to (x, 0)$), it is clear that if D is commutative or associative, then A is respectively commutative or asso-ciative. Also (96) with $u = 0 = x$, $v = 1$ shows that $[U, X] = 0$ implies $\bar{y} = y$; hence $j = 1$. Conversely, it is clear from (96) that if A is associative and com-mutative, and $j = 1$, then D is commutative. Also if we put $v = x = z = 0, t = 1$ in (97) we obtain the necessary condition $yu = uy$, y, $u \in A$, for associativity of D. Thus D associative implies A associative and commutative. Conversely, (97) shows that if A is associative and commutative, then D is associative. This proves (1) and (2). To prove (3) we note that D is alternative if and only if $[\bar{X}, X, Z] = 0$ for all X, Z: since $X + \bar{X} = T(X)1$ this is equivalent to

$[X, X, Z] = 0$. Applying the involution to this relation we obtain $[\bar{Z}, \bar{X}, \bar{X}] = 0$, since $[X, Y, Z] = (\overline{XY})Z - X(\overline{YZ}) = \bar{Z}(\bar{Y}\bar{X}) - (\bar{Z}\bar{Y})\bar{X} = -[\bar{Z}, \bar{Y}, \bar{X}]$. Hence we have $[Z, X, X] = 0$ for all Z, X. Now assume A is alternative. Then taking $U = \bar{X} = (\bar{x}, -y)$ in (97) and using $[\bar{x}, x, y] = 0$, $(\bar{y}y)z = Q(y)z = (z\bar{y})y$, etc., we obtain

$$[\bar{X}, X, Z] = (c[\bar{x}, \bar{t}, y], -[y, \bar{z}, \bar{x}])$$

which shows that $[\bar{X}, X, Z] = 0$ for all X, Z if and only if A is associative. This proves (3). □

Recalling that the algebras we are considering in Lemma 2 are composition algebras if and only if they are alternative (Theorem 7.5), we can obtain a hierarchy of examples as follows. We begin with $A = F$ which satisfies the conditions trivially. Doubling this gives a commutative associative algebra (by Lemma 2,(1)) which is two dimensional. These composition algebras will be called *quadratic algebras*. Among them are included the quadratic field extensions of F. A double of a quadratic algebra is associative but not commutative since the involution in the quadratic algebra is not the identity mapping. The doubles of quadratic algebras are called *(generalized) quaternion algebras* over F. These are four dimensional over F. Doubling again we obtain eight dimensional algebras which are alternative. These composition algebras are called *octonion algebras* (or *Cayley algebras*). Since the quaternion algebras are not commutative, the octonions are not associative. Hence, as far as composition algebras are concerned, we have reached the end of the road.

We shall now prove that our constructions yield all the composition algebras. To see this we need the following

LEMMA 3. *Let (A, Q) be a composition algebra, C a proper subalgebra containing 1 stabilized by the involution j of A ($\bar{C} \subset C$) such that C is a non-degenerate subspace of A relative to B. Then C can be imbedded in a subalgebra D of A satisfying the same conditions as C and isomorphic to a double of C.*

Proof. Since C is non-degenerate we have $A = C \oplus C^{\perp}$ and we can choose an element $t \in C^{\perp}$ such that $Q(t) = -c \neq 0$. Since $1 \in C$, $T(t) = B(1, t) = 0$, so $\bar{t} = -t$ and hence

(i) $$t^2 = -\bar{t}t = -Q(t)1 = c1.$$

If $x \in C$, $B(x, t) = 0$. Then, by (93), $\bar{x}t + \bar{t}x = 0$, and

(ii) $$tx = \bar{x}t, \qquad x \in C.$$

If $x, y \in C$ then $\bar{y}x \in C$, so $B(x, yt) = B(\bar{y}x, t) = 0$. Hence the subspace $Ct = \{yt \mid y \in C\} \subseteq C^{\perp}$ and, consequently, $D \equiv C + Ct = C \oplus Ct$. In A we have the relation $\bar{x}(xy) = Q(x)y$ which linearizes to

$$(98) \qquad\qquad \bar{x}(zy) + \bar{z}(xy) = B(x, z)y.$$

Taking $x, y \in C$, $z = t$ this gives $\bar{x}(ty) = t(xy)$. Then, by (ii), $\bar{x}(\bar{y}t) = (\bar{y}\bar{x})t$. Thus we have

(iii) $\qquad\qquad\qquad x(yt) = (yx)t, \qquad x, y \in C.$

Applying the involution and (ii) we obtain also

(iv) $\qquad\qquad\qquad\qquad (yt)x = (y\bar{x})t.$

Finally, $(xt)(yt) = (t\bar{x})(yt) = t(\bar{x}y)t$ (by Moufang's identity) $= (\bar{y}x)t^2 = c\bar{y}x$. Hence

(v) $\qquad\qquad\qquad\quad (xt)(yt) = c\bar{y}x, \qquad x, y \in C.$

The formulas (i)–(v) show that if $u, v, x, y \in C$, then

(vi) $\qquad\qquad (u + vt)(x + yt) = (ux + c\bar{y}v) + (yu + v\bar{x})t.$

Hence $D = C + Ct$ is a subalgebra of A containing C. Also $\overline{u + vt} = \bar{u} + \bar{t}\bar{v} = \bar{u} - t\bar{v} = \bar{u} - vt$ so $D^j \subset D$ and $Q(xt) = Q(x)Q(t) = -cQ(x)$. This implies that $x \to xt$ is a bijective linear map of C onto Ct. Hence C and Ct are isomorphic as vector spaces. Moreover, Ct is non-degenerate. Hence D, which is an orthogonal direct sum of C and Ct, is non-degenerate. Comparison of (vi) and (94) shows that $(x, y) \to x + yt$ is an isomorphism of the c-double of C with D. This completes the proof of the lemma. $\square$

We can now prove the main result.

THE GENERALIZED HURWITZ THEOREM. *The following is a complete list of the composition algebras over a field F of characteristic $\neq 2$: (I) $F1$, (II) quadratic algebras, (III) quaternion algebras, (IV) octonion algebras.*

Proof. We have seen that the algebras listed are composition algebras (with Q as defined in the construction). Now let (A, Q) be a composition algebra. If $A = F1$ we have case I. Otherwise, $F1 \subset A$, so (by Lemma 3) A contains a quadratic subalgebra that is non-degenerate and is stable under j. If A coincides with this subalgebra we have II. Otherwise, A contains a quaternion subalgebra stable under j and non-degenerate. If A coincides with this we have III. Otherwise, A contains an octonion subalgebra stable under j and non-isotropic. Then A coincides with this subalgebra since, otherwise, A contains a double of an octonion algebra. Such a double is not alternative. Since A is alternative this is impossible, and so we have case IV. $\square$

We shall now derive in explicit form the bases and multiplication tables which are provided naturally by the doubling process. These can be used to write out the composition laws for the quadratic form.

First, we have F with base $i_0 = 1$ and multiplication $i_0^2 = i_0$. Let A_1 denote the c_1-double of this. The base we choose for A_1 is $i_0 = 1$ and $i_1 = (0, 1)$. Omitting the products involving i_0 the multiplication is described by

(99) $$i_1^2 = c_1 1.$$

Next we form the c_2-double A_2 of A_1 and write $i_2 = (0, 1)$ in this. Then we have the base $(i_0, i_1, i_2, i_3 = i_1 i_2)$ since $A_2 = A_1 \oplus A_1 i_2$. The essential part of the multiplication table for this base of the quaternion algebra A_2 is

(100)
$$i_1^2 = c_1 1, \qquad i_2^2 = c_2 1, \qquad i_3^2 = -c_1 c_2 1$$
$$i_1 i_2 = i_3 = -i_2 i_1$$
$$i_2 i_3 = -c_2 i_1 = -i_3 i_2$$
$$i_3 i_1 = -c_1 i_2 = -i_1 i_3.$$

These all follow from the associative law and $i_1^2 = c_1 1$, $i_2^2 = c_2 1$, $i_1 i_2 = -i_2 i_1$, (i) and (ii) above. Finally, we consider the octonion algebra A_3 which is the c_3-double of A_2 and hence has the base

(101) $\quad i_0 = 1, i_1, i_2, i_3 = i_1 i_2, i_4, i_5 = i_1 i_4, i_6 = i_2 i_4, i_7 = (i_1 i_2) i_4.$

The multiplication for this base which one deduces from (i)–(v) is:

(102)

	i_0	i_1	i_2	i_3	i_4	i_5	i_6	i_7
i_0	i_0	i_i	i_2	i_3	i_4	i_5	i_6	i_7
i_1	i_1	$c_1 i_0$	i_3	$c_1 i_2$	i_5	$c_1 i_4$	$-i_7$	$-c_1 i_6$
i_2	i_2	$-i_3$	$c_2 i_0$	$-c_2 i_1$	i_6	i_7	$c_2 i_4$	$c_2 i_5$
i_3	i_3	$-c_1 i_2$	$c_2 i_1$	$-c_1 c_2 i_0$	i_7	$c_1 i_6$	$-c_2 i_5$	$-c_1 c_2 i_4$
i_4	i_4	$-i_5$	$-i_6$	$-i_7$	$c_3 i_0$	$-c_3 i_1$	$-c_3 i_2$	$-c_3 i_3$
i_5	i_5	$-c_1 i_4$	$-i_7$	$-c_1 i_6$	$c_3 i_1$	$-c_1 c_3 i_0$	$c_3 i_3$	$c_1 c_3 i_2$
i_6	i_6	i_7	$-c_2 i_4$	$c_2 i_5$	$c_3 i_2$	$-c_3 i_3$	$-c_2 c_3 i_0$	$-c_2 c_3 i_1$
i_7	i_7	$c_1 i_6$	$-c_2 i_5$	$c_1 c_2 i_4$	$c_3 i_3$	$-c_1 c_3 i_2$	$c_2 c_3 i_1$	$c_1 c_2 c_3 i_0$

If $x = x_0 i_0 + x_1 i_1$ in A_1, then $\bar{x}x = (x_0^2 - c_1 x_1^2)1$, so

(103) $$Q(x) = x_0^2 - c_1 x_1^2.$$

Since for $y = y_0 i_0 + y_1 i_1$ we have $xy = (x_0 y_0 + c_1 x_1 y_1) i_0 + (x_0 y_1 + x_1 y_0) i_1$, the composition law for Q in this case is

(104) $$(x_0^2 - c_1 x_1^2)(y_0^2 - c_1 y_1^2) = (x_0 y_0 + c_1 x_1 y_1)^2 - c_1 (x_0 y_1 + x_1 y_0)^2.$$

Similarly, taking $x = x_0 i_0 + x_1 i_1 + x_2 i_2 + x_3 i_3$, $y = y_0 i_0 + y_1 i_1 + y_2 i_2 + y_3 i_3$ in A_2 we obtain

(105) $$Q(x) = x_0^2 - c_1 x_1^2 - c_2 x_2^2 + c_1 c_2 x_3^2$$

and the composition law:

(106)
$$\begin{aligned}
(x_0^2 &- c_1 x_1^2 - c_2 x_2^2 + c_1 c_2 x_3^2)(y_0^2 - c_1 y_1^2 - c_2 y_2^2 + c_1 c_2 y_3^2) \\
&= (x_0 y_0 + c_1 x_1 y_1 + c_2 x_2 y_2 - c_1 c_2 x_3 y_3)^2 \\
&\quad - c_1 (x_0 y_1 + x_1 y_0 - c_2 x_2 y_3 + c_2 x_3 y_2)^2 \\
&\quad - c_2 (x_0 y_2 + x_2 y_0 + c_1 x_1 y_3 - c_1 x_3 y_1)^2 \\
&\quad + c_1 c_2 (x_0 y_3 + x_3 y_0 + x_1 y_2 - x_2 y_1)^2.
\end{aligned}$$

Taking the $c_i = -1$ we obtain the identities we listed at the beginning of our discussion. We could also write down the quadratic form Q provided by the octonion algebra and the corresponding composition law. We refrain from doing this because of the length of the formulas.

If the base field $F = \mathbb{R}$ and $c_1 = -1$, then the quadratic algebra A_1 has base (i_0, i_1) with i_0 as unit and $i_1^2 = -1$. Clearly this is the field $\mathbb{C}$ of complex numbers. Taking $c_2 = -1$ we obtain the quaternion algebra with base (i_0, i_1, i_2, i_3) such that $i_0 = 1$, $i_j^2 = -1$ for $1 \leq j \leq 3$, $i_1 i_2 = i_3 = -i_2 i_1$, $i_2 i_3 = i_1 = -i_3 i_2$, $i_3 i_1 = i_2 = -i_1 i_3$. Clearly this is Hamilton's quaternion algebra $\mathbb{H}$. Taking $c_3 = -1$ we obtain the classical octonion algebra $\mathbb{O}$ which was discovered independently by J. J. Graves before 1844 and by A. Cayley in 1845. The definition of this algebra as a double of a quaternion algebra is due to L. E. Dickson.[12] The Cayley-Graves algebra $\mathbb{O}$ is a division algebra in the sense that any $x \neq 0$ in $\mathbb{O}$ has an inverse x^{-1} such that $xx^{-1} = 1 = x^{-1}x$. If $x = \sum_0^7 x_j i_j$ then $\bar{x} = x_0 i_0 - \sum_1^7 x_k i_k$ and $Q(x) = \sum x_i^2 \neq 0$. Then we may take $x^{-1} = Q(x)^{-1}\bar{x}$. More generally, one sees that any composition algebra whose quadratic form is anisotropic is a division algebra.

---

[12] L. E. Dickson, *Linear Algebras*, Cambridge Tract in Mathematics, no. 16, 1914, p. 15; or his *Algebras and Their Arithmetics*, University of Chicago Press, 1923, p. 62.

EXERCISES

1. Show that every x in a composition algebra satisfies the quadratic equation

$$x^2 - T(x)x + Q(x)1 = 0.$$

2. Use Witt's theorem and the doubling construction to prove that, if (A, Q) and (A', Q') are composition algebras such that Q and Q' are equivalent quadratic forms and C and C' are isomorphic non-degenerate subalgebras of A and A', respectively, then any isomorphism of C onto C' can be extended to an isomorphism of A onto A'. Hence prove that composition algebras (A, Q) and (A', Q') are isomorphic if and only if Q and Q' are equivalent.

3. Show that if (A, Q) is a composition algebra which is not a division algebra, then Q has maximal Witt index ($=n/2, n=2, 4,$ or 8). Such a composition algebra is called *split*. Show that any two such algebras of the same dimension are isomorphic.

4. Show that if F is a finite field, then any composition algebra of dimension 4 or 8 is split. Does this hold for $n = 2$?

5. Define a *quadratic algebra* over a field F of any characteristic to be an algebra $F[\lambda]/(\lambda^2 - \lambda + a)$ such that $4a \neq 1$, together with the quadratic form $Q(b + cu) = b^2 + bc + c^2 a$, $u = \lambda + (\lambda^2 - \lambda + a)$. Show that this has an involution such that $u \to \bar{u} = 1 - u$ and that $\bar{x}x = Q(x)$ for $x = b + cu$. Show that for characteristic $\neq 2$ this is isomorphic to the quadratic algebras defined in the text.

6. Define a *quaternion algebra* over a field F of any characteristic as a double of a quadratic algebra as defined in exercise 5, and an octonion algebra as a double of a quaternion algebra. Define composition algebras over F as in the text where it is understood that non-degeneracy means that the only z such that $Q(z) = 0 = Q(x, z)$ for all x is $z = 0$. Prove the following generalized Hurwitz theorem for arbitrary F. The composition algebras over F are: (I) F, (II) quadratic algebras, (III) quaternion algebras, (IV) octonion algebras, (V) for char $F = 2$, a finite dimensional extension field A of F such that for every $x \in A$, $x^2 = Q(x) \in F$.

7. Show that if composition algebras are defined as in Definition 7.5, then the Generalized Hurwitz theorem is still valid if the (implicit) finite dimensionality hypothesis is dropped.

8. Prove the following Moufang identities for alternative algebras of any characteristic:

$$(uvu)x = u(v(ux))$$
$$x(uvu) = ((xu)v)u$$
$$u(xy)u = (ux)(yu).$$

9. Show that the second of the foregoing identities is equivalent to the associator identity:

$$[x, uv, u] = -[x, u, v]u$$

and that this linearizes to

$$[x, uv, w] + [x, wv, u] = -[x, u, v]w - [x, w, v]u.$$

Use these to prove Artin's theorem: the subalgebra generated by any two elements of an alternative algebra is associative. (Note that this implies that alternative algebras are power associative.)

7.7 FROBENIUS' AND WEDDERBURN'S THEOREMS ON ASSOCIATIVE DIVISION ALGEBRAS

In volume II we consider the structure theory of rings and of associative algebras. One of the main results of this theory is the reduction of the study of some quite general classes of rings to division rings. In the case of finite dimensional algebras we have a reduction to division algebras. What can be said about these? The answer to this depends considerably on the underlying field. In this section we shall consider the three simplest cases, those in which F is either algebraically closed, the field $\mathbb{R}$ of real numbers, or a finite field.[13]

Let A be a finite dimensional associative algebra over F which is a division algebra in the sense that every $x \neq 0$ in A has an inverse in A. If $m_x(\lambda)$ is the minimum polynomial of x and $m_x(\lambda) = m_1(\lambda)m_2(\lambda)$ in $F[\lambda]$, then $m_1(x)m_2(x) = 0$ which implies that either $m_1(x) = 0$ or $m_2(x) = 0$. It follows that the minimum polynomial of every $x \in A$ is irreducible. We shall identify F with the subalgebra $F1$ of multiples $a1$, $a \in F$. Then it is clear that $m_x(\lambda)$ is linear if and only if $x = a \in F$.

The determination of the finite dimensional division algebras over an algebraically closed field F is trivial; for, we have the following

THEOREM 7.6. · *If F is algebraically closed, then the only finite dimensional division algebra over F is F itself.*

Proof. Let A be a finite dimensional division algebra over the algebraically closed field F and let $x \in A$. Then the minimum polynomial of x is linear, since it is irreducible and F is algebraically closed. Hence $F[x] = F$, so $x \in F$. Since this holds for all $x \in A$, we have $A = F$. $\square$

We consider next the case $F = \mathbb{R}$ and A a finite dimensional division algebra over $\mathbb{R}$. The monic irreducible polynomials in $\mathbb{R}[\lambda]$ are the linear ones $\lambda - a$ or the quadratic ones $\lambda^2 - 2a\lambda + b$ with $a^2 < b$. If $x \notin \mathbb{R}$, its minimum polynomial has the second of these forms and $x = y + a$, where the minimum polynomial of y is $\lambda^2 + (b - a^2)$. It follows that every element of A has the form

---

[3] The case $F = \mathbb{Q}$ turns out to be surprisingly difficult, requiring deep arithmetic results.

$a + y$ where $a \in \mathbb{R}$ and either $y = 0$ or $y^2 = b \in \mathbb{R}$ with $b < 0$. We shall use this simple remark to prove

FROBENIUS' THEOREM.   *The only finite dimensional division algebras over* $\mathbb{R}$ *are:* (1) $\mathbb{R}$, (2) $\mathbb{C}$, *and* (3) $\mathbb{H}$.

Proof.   The proof we shall give is a somewhat polished version of one which has been given by Dickson.[14] We let A' denote the subset of A consisting of the elements u whose squares are elements ≤ 0 in $\mathbb{R}$. We claim that A' is a subspace of A. Since it is clear that if $u \in A'$ and $a \in \mathbb{R}$, then $au \in A'$, it suffices to show that if u and v are linearly independent elements of A', then $u + v \in A'$. We observe first that we cannot have a relation of the form $u = av + b$, $b \in \mathbb{R}$. For, we have $u^2 = c < 0$, $v^2 = d < 0$ (since $u \neq 0$, $v \neq 0$), so $u = av + b$ gives the relation $c = (av + b)^2 = a^2 d + 2abv + b^2$. Since $v \notin \mathbb{R}$ we have $ab = 0$, and $a = 0$ or $b = 0$. The first alternative implies that $u \in \mathbb{R}$, the second that u is a multiple of v. Since both of these have been ruled out, it follows that we cannot have $u = av + b$. Thus we see that 1, u, and v are linearly independent. Now consider $u + v$ and $u - v$. Both are roots of quadratic equations. Hence we have $p, q, r, s \in \mathbb{R}$ such that

$$(u + v)^2 = p(u + v) + q$$
$$(u - v)^2 = r(u - v) + s.$$

Since $(u \pm v)^2 = u^2 \pm (uv + vu) + v^2$ and $u^2 = c$, $v^2 = d$, these give the relations

$$c + d + (uv + vu) = p(u + v) + q$$
$$c + d - (uv + vu) = r(u - v) + s.$$

Adding, we get $(p + r)u + (p - r)v + (q + s - 2c - 2d) = 0$. Since u, v, 1 are linearly independent, this implies that $p = r = 0$. Then $(u + v)^2 = q \in \mathbb{R}$ and since $u + v \notin \mathbb{R}$, $q < 0$. Thus $u + v \in A'$ and A' is a subspace. We saw above that any element of A has the form $a + y$, $a \in \mathbb{R}$, $y \in A'$. Hence we have $A = \mathbb{R} \oplus A'$.

If $u \in A'$ we now write $u^2 = -Q(u)$ where $Q(u) \in \mathbb{R}$ and $Q(u) \geq 0$. Moreover, $Q(u) = 0$ if and only if $u = 0$. Clearly, $Q(au) = a^2 Q(u)$ if $a \in \mathbb{R}$ and $B(u, v) \equiv Q(u + v) - Q(u) - Q(v) = -(u + v)^2 + u^2 + v^2 = -(uv + vu)$. The formula $B(u, z) = -(uv + vu)$ shows that $B(u, v)$ is a symmetric bilinear form. Hence we see that $Q(u)$ is a quadratic form and $B(u, v)$ is its associated symmetric bilinear form. Moreover, Q is positive definite.

---

[14] L. E. Dickson, *Linear Algebras*, pp. 10–12.

We can now complete the proof. If $A = \mathbb{R}$ we have the first possibility we listed. Suppose $A \supseteq \mathbb{R}$. Then $A' \neq 0$ and we can choose a vector i in A' such that $Q(i) = 1$. Then $i^2 = -1$ and $\mathbb{R}[i] = \mathbb{C} = \mathbb{R} + \mathbb{R}i$. If $A = \mathbb{C}$ we have our second alternative. Now suppose $A \supsetneq \mathbb{C}$. Then $A' \supsetneq \mathbb{R}i$ and we can choose $j \perp \mathbb{R}i$ such that $Q(j) = 1$. Then $j^2 = -1$ and $ij + ji = -Q(i, j) = 0$, so $ij = -ji$. Putting $k = ij$ we obtain $k^2 = -1$, $ik + ki = 0 = kj + jk$. Hence $k \in A'$ and $k \perp i, j$. It follows that $1, i, j, k$ are linearly independent and $\mathbb{R} + \mathbb{R}i + \mathbb{R}j + \mathbb{R}k = \mathbb{H}$. Now $A = \mathbb{H}$. Otherwise, there exists an $l \in A'$ such that $Q(l) = 1$ and $l \perp i, j, k$. Then $li = -il$, $lj = -jl$, $lk = -kl$, $k = ij$. However, the first two of these gives $l(ij) = (li)j = -(il)j = -i(lj) = i(jl) = (ij)l$, so $lk = kl$. This contradiction shows that $A = \mathbb{H}$ and we have the third alternative. $\square$

We now turn to the case of a finite field F. If $|F| = q$ and V is an n-dimensional vector space over F, then $|V| = q^n$. In particular, if A is a finite dimensional division algebra over F, then A is a finite division ring. Conversely, let A be a finite division ring and let F be the center of A. Then F is a finite field and A is a finite dimensional algebra over F. In 1905 J. H. M. Wedderburn discovered the surprising fact that every finite division ring is commutative. Wedderburn's theorem has a striking consequence for projective geometry. For, it is known that Desarguesian projective geometries—that is, projective geometries in which the theorem of Desargues holds— can be coordinatized by division rings, and that these are commutative if and only if the theorem of Pappus holds.[15] It therefore follows from Wedderburn's theorem that the theorem of Pappus is valid in any finite projective geometry in which Desargues' theorem holds. We shall now prove

WEDDERBURN'S THEOREM. *Every finite division ring is commutative.*

Proof.   Let F be the center of the finite division ring A and let $|F| = q$, $[A:F] = n$. Then $|A| = q^n$. We have to show that $n = 1$. Let A^* be the multiplicative group of non-zero elements of A. Then we have the class equation

(107) $$|A^*| = q^n - 1 = \sum_i [A^* : \mathrm{Stab}\, x_i]$$

where the x_i range over a set of representatives of the conjugacy classes of A^* (one element from each class). If $x_i \in F$, $\mathrm{Stab}\, x_i = A^*$, so that we have a contribution of 1 in the above sum coming from such an x_i. The number of $x_i \in F \cap A^*$ is $q - 1$, so altogether we get the contribution $q - 1$ in this way. Next let $x_i \notin F$.

---

[15] See E. Artin, *Geometric Algebra*, New York, Wiley, 1957, p. 73.

Then the subset of elements of A which commute with x_i is a division subring F_i of A containing F. Hence $|F_i| = q^{d_i}$ where $d_i = [F_i : F]$ and $d_i < n$ since $x_i \notin F$, and so $F_i \subset A$. It is clear that Stab $x_i = F_i \cap A^*$. Hence $|\text{Stab } x_i| = q^{d_i} - 1$. We can now rewrite (107) as

$$(108) \qquad\qquad q^n - 1 = (q - 1) + \sum_i \frac{q^n - 1}{q^{d_i} - 1}$$

where every $d_i < n$. We observe also that every $d_i | n$. This follows since A can be regarded in the obvious way as a (left) vector space over F_i. When this is done and F_i is regarded in the usual way as vector space over F, then we have the product formula as for fields (Theorem 4.2, p. 215). Thus the dimensionality n of A over F is divisible by the dimensionality d_i of F_i. Thus far the proof is Wedderburn's. It remains to show that (108) is impossible when the $d_i | n$ and $d_i < n$, unless $n = 1$. The argument we shall give for this is due to E. Witt. We look at the $\lambda^n - 1$, $\lambda^{d_i} - 1$, λ an indeterminate. We recall that if we define the nth cyclotomic polynomial $l_n(\lambda) = \prod (\lambda - z)$, z running over the primitive nth roots of 1 in $\mathbb{C}$, then $\lambda^n - 1 = \prod_{d | n} l_d(\lambda)$ (section 4.11, p. 272). Also we saw that the $l_d(\lambda)$ are monic polynomials with integer coefficients. It is clear from this that if $d_i | n$ and $d_i < n$, then $(\lambda^n - 1)/(\lambda^{d_i} - 1)$ is a polynomial with integer coefficients divisible in $\mathbb{Z}[\lambda]$ by $l_n(\lambda)$. Hence $(q^n - 1)/(q^{d_i} - 1)$ as well as $q^n - 1$ is divisible by the integer $l_n(q)$. Then it follows from (108) that $l_n(q) | q - 1$. Now suppose $n > 1$ and consider the factorization $l_n(\lambda) = \prod (\lambda - z)$, z ranging over the primitive nth roots of unity. Since $n > 1$, no $z = 1$ and the distance from the point q on the real axis to any one of the z's exceeds the distance from q to the point 1. Hence $|q - z| > q - 1$ and therefore $|l_n(q)| = \prod |q - z| > q - 1$ contrary to $l_n(q) | q - 1$. Thus $n = 1$ and $A = F$ is commutative. $\square$

EXERCISE

1. Prove the following extension of Frobenius' theorem to alternative division algebras. The only finite dimensional alternative division algebras over $\mathbb{R}$ are (a) $\mathbb{R}$, (b) $\mathbb{C}$, (c) $\mathbb{H}$, (d) $\mathbb{O}$. (*Hint:* Apply the generalized Hurwitz theorem.)

Lattices and Boolean Algebras

Associated with a set S one has the power set $\mathscr{P}(S)$, the set of its subsets, and the algebra (in the non-technical sense) of $\mathscr{P}(S)$ based on intersection $A \cap B$ and union $A \cup B$. When one attempts to set down the basic properties of the structure $(\mathscr{P}(S), \cap, \cup)$ one is led to the abstract concept of a Boolean algebra. It was George Boole who first realized that this type of algebra could be used to analyze the calculus of propositions in logic and that it played a basic role in probability theory.[1] A more general concept than that of a Boolean algebra is that of a lattice, which was introduced by Dedekind in studying divisibility in commutative rings and the combinatorial properties of ideals with respect to intersection $A \cap B$ and sum $A + B$.[2]

In this chapter we shall give an introduction to lattices and Boolean algebras. Our purpose will be to acquaint the reader with the concepts and elementary results on lattices and Boolean algebras which are applicable to other parts of

[1] George Boole, *The Mathematical Analysis of Logic*, 1847, (Barnes and Noble reprint, 1965) and his *Investigation of the Laws of Thought*, 1854 (Dover reprint, 1953).

[2] Richard Dedekind, "Über Zerlegungen von Zahlen durch ihre grössten gemeinsamen Teiler," in his *Gesamelte Matematische Werke*, vol. 2, 1931, pp. 103–147, and "Über die von drei Moduln erzeugte Dualgruppe," ibid., pp. 236–272.

algebra. Some of these will be needed in Volume II in connection with the study of universal algebra.

Besides the basic definitions, the main topics we shall treat in this chapter are: the Jordan-Hölder theorem for semi-modular lattices, the "fundamental theorem of projective geometry," which determines the isomorphisms between the lattices of subspaces of vector spaces, the equivalence of Boolean algebras and Boolean rings, and the Möbius function of a partially ordered set.

8.1 PARTIALLY ORDERED SETS AND LATTICES

The most general concept we shall consider in this chapter is that of a partially ordered set. We recall that a binary relation on a set S is a subset R of the product set $S \times S$ (Introduction, p. 10). We say that a is in the relation R to b and write aRb if and only if $(a, b) \in R$. We now give

DEFINITION 8.1.   *A partially ordered set is a set S together with a binary relation $a \geq b$ satisfying the following conditions:*

   PO1 $a \geq a$ *(reflexivity)*.
   PO2 *If $a \geq b$ and $b \geq a$, then $a = b$ (anti-symmetry).*
   PO3 *If $a \geq b$ and $b \geq c$, then $a \geq c$ (transitivity).*

If $a \geq b$ and $a \neq b$, then we write $a > b$. Also we write $a \leq b$ as an alternative for $b \geq a$ and $a < b$ for $b > a$. In general we may have neither $a \geq b$ nor $b \geq a$ for a pair of elements $a, b \in S$. If we do have $a \geq b$ or $b \geq a$ for every pair (a, b), then we call S *totally ordered* (or a *chain*).

We have encountered quite a few examples of partially ordered sets: the set $\mathscr{P}(S)$ of subsets of a set S where $A \geq B$ for subsets A and B means $A \supset B$, the set of subrings of a ring, the set of subgroups of a group, the set of ideals of a ring, and so on—all partially ordered by inclusion as defined for subsets. In general, if $S, \geq$ is a partially ordered set, then any subset T of S is partially ordered by the relation $\geq$ of S restricted to T. Other interesting examples of partial orderings arise in discussing divisibility in monoids and rings. For example, in the multiplicative monoid of positive integers we can define $a \geq b$ to mean $a|b$ (a is a divisor of b). Then PO1–PO3 hold. More generally, let S be a commutative monoid satisfying the cancellation law. We say that S is *reduced* if 1 is the only invertible element in S. In this case $a|b$ and $b|a$ imply $a = b$. Then S is partially ordered if we define $a \geq b$ by $a|b$. If S is not reduced we obtain a non-trivial congruence relation in S by defining $a \sim b$ if $a = bu, u$ invertible. The quotient monoid $\bar{S}$ relative to this congruence relation is reduced and can be partially ordered by the divisibility relation.

In a finite partially ordered set the relation $>$ can be expressed in terms of a relation of covering. We say that a_1 is a *cover* of a_2 if $a_1 > a_2$ and there exists no u such that $a_1 > u > a_2$. It is clear that $a > b$ in a finite partially ordered set if and only if there exists a sequence $a = a_1, a_2, \ldots, a_n = b$ such that each a_i is a cover of a_{i+1}. The notion of cover suggests a way of representing a finite partially ordered set S by a diagram. We represent the elements of S by dots. If a_1 is a cover of a_2 then we place a_1 above a_2 and connect the two dots by a straight line. Then $a > b$ if and only if there is a descending broken line connecting a to b. If no line connects a and $b \neq a$, then a and b are not comparable, that is, we have neither $a \geq b$ nor $b \geq a$. Some examples of diagrams of partially ordered sets are

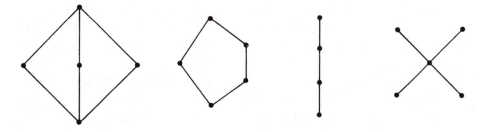

the third one of these representing a totally ordered set.

An element u of a partially ordered set S is an *upper bound* of a subset A of S if $u \geq a$ for every $a \in A$. The element u is a *least upper bound* or sup of A if u is an upper bound of A and $u \leq v$ for every upper bound v of A. It is clear from PO2 that if a sup A exists, then it is unique. In similar fashion one defines lower bounds and greatest lower bounds or infs of a set A. Also if inf A exists, then it is unique. We now introduce the following

DEFINITION 8.2.   *A lattice is a partially ordered set in which any two elements have a least upper bound and a greatest lower bound.*

We denote the least upper bound of a and b by $a \vee b$ ("a cup b" or "a union b") and the greatest lower bound by $a \wedge b$ ("a cap b" or "a meet b"). If a, b, c are elements of a lattice L, then $(a \vee b) \vee c \geq a, b, c$ and if $v \geq a, b, c$, then $v \geq (a \vee b)$, c so $v \geq (a \vee b) \vee c$. Hence $(a \vee b) \vee c$ is a sup of a, b, c. By induction, one shows that any finite set of elements of a lattice have a sup. Similarly, any finite subset has an inf. We denote the sup and inf of $a_1, a_2, \ldots, a_n$ by

$$a_1 \vee a_2 \vee \cdots \vee a_n, \qquad a_1 \wedge a_2 \wedge \cdots \wedge a_n$$

respectively.

Any totally ordered set is a lattice. For, if a and b are two elements of such a set we have either $a \geq b$ or $b \geq a$. In the first case, $a \vee b = a$ and $a \wedge b = b$. If $b \geq a$ then $a \vee b = b$ and $a \wedge b = a$.

A partially ordered set is called a *complete lattice* if every subset $A = \{a_\alpha\}$ has a sup and an inf. We denote these by $\bigvee a_\alpha$ and $\bigwedge a_\alpha$ respectively. If the set $\{a_\alpha\}$ coincides with the underlying set of the lattice L then $0 \equiv \bigwedge a_\alpha$ is the least element of L and $1 \equiv \bigvee a_\alpha$ is the greatest element of L: $0 \leq a$ and $1 \geq a$ for every $a \in L$. The following is a very useful criterion for recognizing that a given partially ordered set is complete lattice.

THEOREM 8.1.   *A partially ordered set with a greatest element 1 such that every non-vacuous subset $\{a_\alpha\}$ has a greatest lower bound is a complete lattice. Dually, a partially ordered set with a least element 0 such that every non-vacuous subset has a least upper bound is a complete lattice.*

Proof.   Assuming the first set of hypotheses we have to show that any $A = \{a_\alpha\}$ has a sup. Since $1 \geq a_\alpha$ the set B of upper bounds of A is non-vacuous. Let $b = \inf B$. Then it is clear that $b = \sup A$. The second statement follows by symmetry. $\square$

EXAMPLES

1. For any set S, $\mathscr{P}(S)$ is a complete lattice. Here $1 = S$ and $0 = \varnothing$.

2. The set of subgroups of a group G ordered by inclusion. Since G is a subgroup and the intersection of any set of subgroups is a subgroup, the set of subgroups is a complete lattice. The proof of Theorem 8.1 shows that the sup of a set of subgroups is the intersection of all subgroups containing the given set $\{H_\alpha\}$. Clearly this is the subgroup generated by all the H_α.

The next four examples are similar to 2. They are complete lattices in which "$\geq$" means inclusion.

3. The set of normal subgroups of a group. The sup of a set of normal subgroups is the subgroup they generate.

4. The set of subspaces of a vector space ordered by inclusion. The inf is the set intersection and the sup is the subspace spanned by the given set of subspaces.

5. The set of ideals of a ring R. Inf is the set intersection, sup is the ideal generated. For two ideals I_1, I_2 this is $I_1 + I_2$, the set of sums $b_1 + b_2$, $b_i \in I_i$.

6. The set of left (right) ideals of a ring.

7. The set of positive integers partially ordered by divisibility: $a \geq b \Leftrightarrow a|b$. Here $a \vee b$

is the greatest common divisor of a and b and $a \wedge b$ is the least common multiple of a and b. This is a lattice but it is not complete.

8. All the diagrams above except the last one represent lattices (necessarily complete since they are finite).

9. The set $\mathbb{Q}$ of rational numbers with $a \geq b$ having the usual significance. This is totally ordered and hence, as we noted above, $\mathbb{Q}$ is a lattice. However, $\mathbb{Q}$ is not complete.

10. Even the subset of $\mathbb{Q}$ of rationals between 0 and 1 is not complete. On the other hand, the real interval $[0, 1]$ (with the usual order) is a complete lattice.

It is useful to sort out the basic properties of the binary compositions $a \wedge b$ and $a \vee b$ in a lattice L. This will lead us to an alternative definition of a lattice in terms of conditions on two binary compositions on a set. We note first that it follows from the definitions that $a \vee b$ and $a \wedge b$ are symmetric in the two arguments. Hence we have the commutative laws $a \vee b = b \vee a$ and $a \wedge b = b \wedge a$. Also we saw that $(a \vee b) \vee c$ is the sup of $a, b,$ and c. Since the sup is a symmetric function of $a, b,$ and c, it follows that $(a \vee b) \vee c = a \vee (b \vee c)$ and similarly, $(a \wedge b) \wedge c = a \wedge (b \wedge c)$. It is clear that every a is idempotent relative to $\vee$ and to $\wedge : a \vee a = a, a \wedge a = a$. Also it is clear that if $a \geq b$, then $a \vee b = a$ and $a \wedge b = b$. Hence, for any a and b we have $(a \vee b) \wedge a = a$ and $(a \wedge b) \vee a = a$.

Conversely, let L be any set in which there are defined two binary compositions $\vee$ and $\wedge$ satisfying the conditions we have noted:

L1 $a \vee b = b \vee a, \qquad a \wedge b = b \wedge a.$
L2 $(a \vee b) \vee c = a \vee (b \vee c), \qquad (a \wedge b) \wedge c = a \wedge (b \wedge c).$
L3 $a \vee a = a, \qquad a \wedge a = a.$
L4 $(a \vee b) \wedge a = a, \qquad (a \wedge b) \vee a = a.$

We shall show that L is a lattice relative to a suitable definition of $\geq$ and that $a \vee b$ and $a \wedge b$ are the sup and inf of a and b in this lattice.

Before proceeding to the proof we remark that we have made precisely the same assumptions on the two compositions $\vee$ and $\wedge$. Hence, we have the important *principle of duality* that states that, if S is a statement which can be deduced from our axioms, then the *dual statement* S' obtained by interchanging $\vee$ and $\wedge$ throughout S can also be deduced.

We note next that, if $a, b \in L$ (satisfying L1–L4), then the conditions $a \vee b = a$ and $a \wedge b = b$ are equivalent. We shall now define a relation $\geq$ in L by specifying that $a \geq b$ means that $a \vee b = a$, hence $a \wedge b = b$. Evidently, in dualizing, a statement $a \geq b$ has to be replaced by $b \geq a$.

We shall now verify that the $\geq$ we have introduced satisfies PO1–PO3. Since $a \vee a = a$ we have $a \geq a$ so PO1 holds. If $a \geq b$ and $b \geq a$, then we have $a \vee b = a$ and $b \vee a = b$. Since $a \vee b = b \vee a$ this gives $a = b$, which proves PO2. Next assume that $a \geq b$ and $b \geq c$. Then $a \vee b = a$ and $b \vee c = b$. Hence

$$a \vee c = (a \vee b) \vee c = a \vee (b \vee c) = a \vee b = a$$

which means that $a \geq c$. Hence PO3 is valid.

Since $(a \vee b) \wedge a = a$, by L4, $a \vee b \geq a$. Similarly, $a \vee b \geq b$. Now let c be an element such that $c \geq a$ and $c \geq b$. Then $a \vee c = c$ and $b \vee c = c$. Hence

$$(a \vee b) \vee c = a \vee (b \vee c) = a \vee c = c$$

so $c \geq a \vee b$. Thus $a \vee b$ is a sup of a and b in L. By duality, $a \wedge b$ is an inf of a and b. This completes the verification that a set L with binary compositions satisfying L1–L4 is a lattice and $a \vee b$ and $a \wedge b$ are the sup and inf in this lattice.

A subset M of a lattice L is called a *sublattice* if it is closed under the compositions $\vee$ and $\wedge$. It is evident that a sublattice is a lattice relative to the induced compositions. On the other hand, a subset of a lattice may be a lattice relative to the partial ordering $\geq$ defined in L without being a sublattice. For example, the lattice of subgroups of a group G is not a sublattice of the set $\mathscr{P}(G)$ since $H_1 \cup H_2$ is generally not a subgroup.

If a is a fixed element of a lattice L, then the subset of elements x such that $x \geq a$ ($x \leq a$) is evidently a sublattice. If $a \leq b$, the subset of elements $x \in L$ such that $a \leq x \leq b$ is a sublattice. We call such a sublattice an *interval* and we denote it as $I[a, b]$.

The definition of a lattice by means of the axioms L1–L4 makes it natural to define a *homomorphism* of a lattice L into a lattice L' to be a map $a \to a'$ such that $(a \vee b)' = a' \vee b'$ and $(a \wedge b)' = a' \wedge b'$. In this case if $a \geq b$ then we have $a \vee b = a$; hence $a' \vee b' = a'$ and $a' \geq b'$. A map between partially ordered sets having this property is called *order preserving*. Thus we have shown that a lattice homomorphism is order preserving. However, the converse need not hold. A bijective homomorphism of lattices is called an *isomorphism*. These can be characterized by order preserving properties, as we see in the following

THEOREM 8.2.   *A bijective map of a lattice L onto a lattice L' is a lattice isomorphism if and only if it and its inverse are order preserving.*

Proof.   We have seen that if $a \to a'$ is a lattice isomorphism, then this map is order preserving. It is clear also that the inverse map is an isomorphism of L' into L so it is order preserving. Conversely, suppose $a \to a'$ is bijective and it and its inverse are order preserving. This means that $a \geq b$ in L if and only if $a' \geq b'$ in L'. Let $d = a \vee b$. Then $d \geq a, b$, so $d' \geq a', b'$. Let $e' \geq a', b'$ and let e be the inverse image of e'. Then $e \geq a, b$. Hence $e \geq d$ and $e' \geq d'$. Thus we have shown that $d' = a' \vee b'$. In a similar fashion we can show that $(a \wedge b)' = a' \wedge b'$. $\square$

EXERCISES

1. Show that the lattice of subgroups of a cyclic group of prime power order is totally ordered.

2. What about the converse of exercise 1?

3. Obtain the diagrams for the following partially ordered sets: (i) $\mathscr{P}(S)$ where $S = \{1, 2, 3\}$, (ii) the lattice of subgroups of a cyclic group of order 6, (iii) the lattice of subgroups of the symmetric group S_3.

4. Let S be the set of real valued continuous functions on $[0, 1]$. Define $f \geq g$ if $f(x) \geq g(x)$ for all x in $[0, 1]$. Show that S is a lattice with this definition of $\geq$. Is this complete?

5. Define the *dual* of a partially ordered set S, $\geq$ as S, $\geq'$ where $a \geq' b$ if and only if $a \leq b$. Describe the relation of the diagram of the dual $S' (= S, \geq')$ of a finite partially ordered set S to the diagram of S.

6. Determine all the lattices of ≤ 5 elements by constructing the diagrams. Which are *self-dual*, that is, isomorphic to their duals?

7. Let L_1 and L_2 be partially ordered sets. Then one defines a partial order on $L_1 \times L_2$ by agreeing that $(a_1, a_2) \geq (b_1, b_2)$ if and only if $a_1 \geq b_1$ and $a_2 \geq b_2$. Show that if L_1 and L_2 are lattices, then $L_1 \times L_2$ is a lattice.

8. Let S be a set and L, $\geq$ a partially ordered set. Consider the set L^S of maps $s \rightarrow f(s)$ of S into L. Define $f \geq g$ for $f, g \in L^S$ by $f(s) \geq g(s)$ for all s. Show that this defines a partial ordering on L^S, and that L^S is a lattice if L is a lattice.

9. Give an example of a pair of lattices L_1 and L_2 for which there exists a bijective order preserving map of L_1 onto L_2 which is not an isomorphism.

8.2 DISTRIBUTIVITY AND MODULARITY

One of the compositions of a lattice may be viewed as the analogue of addition in a ring, and the other can be taken as the analogue of multiplication. Depending on which we use for addition and which for multiplication we can formulate the following two distributive laws:

D $$a \wedge (b \vee c) = (a \wedge b) \vee (a \wedge c)$$

and its dual

$$D' \qquad\qquad a \vee (b \wedge c) = (a \vee b) \wedge (a \vee c).$$

It is a bit surprising that—as we shall now show—these two conditions are equivalent. Suppose D holds. Then

$$
\begin{aligned}
(a \vee b) \wedge (a \vee c) &= ((a \vee b) \wedge a) \vee ((a \vee b) \wedge c) \\
&= a \vee ((a \vee b) \wedge c) \\
&= a \vee ((a \wedge c) \vee (b \wedge c)) \\
&= (a \vee (a \wedge c)) \vee (b \wedge c) \\
&= a \vee (b \wedge c)
\end{aligned}
$$

which is D'. Dually D' implies D. A lattice in which these distributive laws hold is called *distributive*. There are some important examples of this. First, as we showed in the Introduction (p. 4), the lattice $\mathscr{P}(S)$ of subsets of a set S is distributive. Second, we have the following

LEMMA. *Any totally ordered set is a distributive lattice.*

Proof. We wish to establish D for any three elements a, b, c and we distinguish two cases (1) $a \geq b, a \geq c$, (2) $a \leq b$ or $a \leq c$. In (1) we have $a \wedge (b \vee c) = b \vee c$ and $(a \wedge b) \vee (a \wedge c) = b \vee c$. In (2) we have $a \wedge (b \vee c) = a$ and $(a \wedge b) \vee (a \wedge c) = a$. Hence in both cases (D) holds. $\square$

This lemma can be used to show that the set of positive integers ordered by divisibility is a distributive lattice. In this example, $a \vee b = (a, b)$ the g.c.d. of a and b and $a \wedge b = [a, b]$ the l.c.m. of a and b. Also, if we write $a = p_1^{a_1} p_2^{a_2} \cdots p_k^{a_k}$, $b = p_1^{b_1} p_2^{b_2} \cdots p_k^{b_k}$ where the p_i are distinct primes and the a_i and b_i are non-negative integers, then $(a, b) = \prod p_i^{\min(a_i, b_i)}$, $[a, b] = \prod p_i^{\max(a_i, b_i)}$. Hence if $c = p_1^{c_1} p_2^{c_2} \cdots p_k^{c_k}$, c_i non-negative integral, then

$$[a, (b, c)] = \prod p_i^{\max(a_i, \min(b_i, c_i))}$$

and

$$([a, b], [a, c]) = \prod p_i^{\min(\max(a_i, b_i), \max(a_i, c_i))}.$$

Now the set of non-negative integers with the natural order is totally ordered and $\max(a_i, b_i) = a_i \vee b_i$, $\min(a_i, b_i) = a_i \wedge b_i$ in this lattice. Hence, the distributive law D' in this lattice gives the relation

$$\max(a_i, \min(b_i, c_i)) = \min(\max(a_i, b_i), \max(a_i, c_i)).$$

Then we have

$$[a, (b, c)] = ([a, b], [a, c])$$

which is D for the lattice of positive integers ordered by divisibility.

The same reasoning applies to any reduced factorial monoid (cf. section 2.14, p. 140).

Another remark on distributivity which is worth noting is that in any lattice we have $a \wedge (b \vee c) \geq a \wedge b$ and $a \wedge (b \vee c) \geq a \wedge c$. Hence

$$a \wedge (b \vee c) \geq (a \wedge b) \vee (a \wedge c).$$

Thus in order to establish distributivity it suffices to establish the reverse inequality

(1) $$a \wedge (b \vee c) \leq (a \wedge b) \vee (a \wedge c).$$

The most important lattices which occur in algebra (e.g., the lattice of submodules of a module, the lattice of normal subgroups of a group) are not distributive. For instance, let $L(V)$ denote the lattice of subspaces of a vector space V over a field F. Assume dim $V \geq 2$ and let x and y be linearly independent vectors in V. Then $F(x + y) \cap (Fx + Fy) = F(x + y)$ but $F(x + y) \cap Fx = 0$ and $F(x + y) \cap Fy = 0$ so $F(x + y) \cap (Fx + Fy) \neq (F(x + y) \cap Fx) + (F(x + y) \cap Fy)$. As we shall see in a moment, the lattice $L(V)$ satisfies a weakening of the distributive condition, which was first formulated by Dedekind. This is the condition:

M $\qquad\qquad$ If $a \geq b$, then $a \wedge (b \vee c) = b \vee (a \wedge c)$.

Since $b = a \wedge b$ the right hand side can be replaced by $(a \wedge b) \vee (a \wedge c)$. Hence the condition M is equivalent to D in the special case in which $a \geq b$ (or $a \geq c$). Condition M is called *modularity* and a lattice satisfying it is said to be *modular*. The dual condition M′ reads: If $a \leq b$ then $a \vee (b \wedge c) = b \wedge (a \vee c)$. Clearly this is the same thing as M. It follows that, as for distributive lattices, the principle of duality is valid in modular lattices.

The importance of modular lattices in algebra stems from the following

THEOREM 8.3. *The lattice of normal subgroups of a group is modular. The lattice of submodules of a module is modular.*

Proof. The normal subgroup generated by two normal subgroups H_1 and H_2 of a group G is $H_1 H_2 = H_2 H_1$. Hence we have to prove that if H_i, $i = 1, 2, 3$, are normal subgroups such that $H_1 \supset H_2$ then

$$H_1 \cap (H_2 H_3) = H_2(H_1 \cap H_3).$$

The remark above about the distributive law shows that it is enough to prove that

$$H_1 \cap (H_2 H_3) \subset H_2(H_1 \cap H_3).$$

Suppose $a \in H_1 \cap (H_2 H_3)$. Then $a = h_1 = h_2 h_3$, $h_i \in H_i$, and $h_3 = h_2^{-1} h_1 \in H_1$, since $H_1 \supset H_2$. Thus $h_3 \in H_1 \cap H_3$ and $a = h_2 h_3 \in h_2(H_1 \cap H_3)$. This proves the required inclusion. The argument for modules is similar and simpler so we omit it. $\square$

An alternative definition of modularity which is sometimes useful can be extracted from the following

THEOREM 8.4.   *A lattice L is modular if and only if whenever $a \geq b$ and $a \wedge c = b \wedge c$ and $a \vee c = b \vee c$ for some c in L, then $a = b$.*

Proof.   Let L be modular and let a, b, c be elements of L such that $a \geq b$, $a \vee c = b \vee c$, $a \wedge c = b \wedge c$. Then

$$a = a \wedge (a \vee c) = a \wedge (b \vee c) = b \vee (a \wedge c) = b \vee (b \wedge c) = b.$$

Conversely, suppose that L is any lattice satisfying the condition stated in the theorem. Let $a, b, c \in L$ and $a \geq b$. We know that $a \wedge (b \vee c) \geq b \vee (a \wedge c)$. Also

$$(a \wedge (b \vee c)) \wedge c = a \wedge ((b \vee c) \wedge c) = a \wedge c$$

and

$$a \wedge c = (a \wedge c) \wedge c \leq (b \vee (a \wedge c)) \wedge c \leq a \wedge c.$$

Hence

$$(b \vee (a \wedge c)) \wedge c = a \wedge c.$$

Since $b \leq a$ the dual of our first relation is

$$(b \vee (a \wedge c)) \vee c = b \vee c$$

and the dual of the second one is

$$(a \wedge (b \vee c)) \vee c = b \vee c.$$

Thus we have

$$(a \wedge (b \vee c)) \wedge c = (b \vee (a \wedge c)) \wedge c$$

$$(a \wedge (b \vee c)) \vee c = (b \vee (a \wedge c)) \vee c.$$

Hence the assumed property implies that $a \wedge (b \vee c) = b \vee (a \wedge c)$, which is the modular axiom. □

We shall prove next an analogue for modular lattices of the second isomorphism theorem for groups (Theorem 1.9, p. 65), namely,

THEOREM 8.5. *If a and b are elements of a modular lattice, then the map $x \rightarrow x \wedge b$ is an isomorphism of the interval $I[a, a \vee b]$ onto $I[a \wedge b, b]$. The inverse isomorphism is $y \rightarrow y \vee a$.*

Proof. We note first that in any lattice the maps $x \rightarrow x \vee a$ and $x \rightarrow x \wedge a$ are order preserving. For, we have $x \geq y$ if and only if $x \vee y = x$ and if and only if $x \wedge y = y$. Then $x \vee y = x$ implies $(x \vee a) \vee (y \vee a) = (x \vee y) \vee (a \vee a) = (x \vee y) \vee a = x \vee a$. Hence $x \geq y$ implies $x \vee a \geq y \vee a$. Similarly, we have $x \wedge a \geq y \wedge a$. Now if $a \leq x \leq a \vee b$, then $a \wedge b \leq x \wedge b \leq b = (a \vee b) \wedge b$, and if $a \wedge b \leq y \leq b$, then $a = a \vee (a \wedge b) \leq y \vee a \leq a \vee b$. Hence $x \rightarrow x \wedge b$ and $y \rightarrow y \vee a$ map $I[a, a \vee b]$ into $I[a \wedge b, b]$ and $I[a \wedge b, b]$ into $I[a, a \vee b]$ respectively. Since these maps are order preserving the theorem will follow from Theorem 8.2 if we can show that they are inverses. Let $x \in I[a, a \vee b]$. Then, since $x \geq a$, by modularity

$$(x \wedge b) \vee a = x \wedge (a \vee b)$$

and since $x \leq a \vee b$, this gives $(x \wedge b) \vee a = x$. Dually, we have that if $y \in I[a \wedge b, b]$, then $(y \vee a) \wedge b = y$. This proves the two maps are inverses. □

This theorem leads us to introduce a notion of equivalence for intervals which in modular lattices is stronger than isomorphism. First, we define the intervals $I[u, v]$ and $I[w, t]$ to be *transposes* if there exist a and b in the lattice such that one of these coincides with $I[a, a \vee b]$ and the other with $I[a \wedge b, b]$. The intervals $I[u, v]$ and $I[w, t]$ are *projective* if there exists a finite sequence

$$I[u, v] = I[u_1, v_1], I[u_2, v_2], \ldots, I[u_n, v_n] = I[w, t]$$

such that consecutive pairs $I[u_k, v_k], I[u_{k+1}, v_{k+1}]$ are transposes. It is immediate that this is an equivalence relation. Also it is clear from Theorem 8.5 that in a modular lattice projective intervals are isomorphic.

EXERCISES

1. Show that the lattice of subgroups of A_4 is not modular.

2. Let G be a group with two generators x, y such that $x^{p^m} = 1$, $y^{p^r} = 1$, $y^{-1}xy = x^m$ where $m^{p^r} \equiv 1 \pmod{p^m}$, p a prime. Show that if H_1 and H_2 are subgroups of G, then $H_1 H_2 = H_2 H_1$. Hence show that the lattice of subgroups of G is modular.

3. Show that if a lattice is not distributive then it contains a sublattice of five elements whose diagram is either the first or second diagram on p. 457. Show that if a lattice is not modular then it contains a sublattice whose diagram is the second one on p. 457

8.3   THE THEOREM OF JORDAN-HÖLDER-DEDEKIND

A partially ordered set S is said to be of *finite length* if the lengths (number of distinct terms) of its chains (= totally ordered subsets) are bounded. If a and b are elements of a partially ordered set of finite length and $a > b$, then we can find a finite sequence of elements $a = a_1, a_2, \ldots, a_n = b$ such that each a_i is a cover of a_{i+1}. A sequence of elements having this property is called a *connected chain from a to b*. A desirable property is that any two connected chains from a to b ($a > b$) have the same length. We shall now show that this property is assured for a lattice L of finite length if L is *semi-modular* in the sense that if a and b are a pair of elements in L such that $a \vee b$ covers a and b, then a and b cover $a \wedge b$. We have seen that if L is modular, then $I[a \wedge b, a]$ and $I[b, a \vee b]$ are isomorphic. Hence it is clear that modularity implies semi-modularity. The following theorem is the lattice analogue of the Jordan-Hölder theorem for finite groups (p. 249).

THEOREM OF JORDAN-HÖLDER-DEDEKIND.   *Let L be a semi-modular lattice of finite length. Then any two connected chains from a to b, $a > b$, have the same length. Moreover, if L is modular and*

(2) $$a = a_1 > a_2 > \cdots > a_{n+1} = b$$

(3) $$a = a_1' > a_2' > \cdots > a_{m+1}' = b$$

are two connected chains from a to b then the corresponding intervals $I[a_{i+1}, a_i]$ and $I[a_{j+1}', a_j']$ can be paired so that the paired ones are projective.

Proof.   The proof imitates the proof of the group result. We use induction on n where $n + 1$ is the length of one of the connected chains from a to b. If $n = 1$, then a is a cover of b and the result is clear. If $a_2 = a_2'$, then we have two connected chains from a_2 to b and the theorem follows by induction on n. Now suppose $a_2 \neq a'_2$. Then a_1 is a cover of a_2 and of $a'_2 \neq a_2$, which implies that $a_2 \vee a'_2 = a_1$. Then the semi-modularity implies that a_2 and a'_2 are covers of $a''_3 \equiv a_2 \wedge a'_2$. Also $a''_3 \geq b$. If $b = a''_3$ we have the diagram

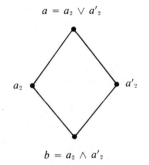

$$a = a_2 \vee a'_2$$

$$a_2 \qquad\qquad a'_2$$

$$b = a_2 \wedge a'_2$$

In this case $m = n = 2$ and, in the modular case, $I[a_2, a_1]$ and $I[b, a'_2]$, and $I[a'_2, a_1]$ and $I[b, a_2]$ are transposes. If $a''_3 > b$, then we can find a connected chain $a''_3, a''_4, \dots, a''_{q+1} = b$. Then the result follows by induction on n applied to $a_2, a_3, \dots, a_{n+1} = b$ and $a_2, a''_3, \dots, a''_{q+1} = b$ as well as to $a'_2, a''_3, \dots, a''_{q+1} = b$ (using $q = n$) and $a'_2, a'_3, \dots, a'_{m+1} = b$. Also in the modular case we have to use the fact that $I[a_2, a_1]$ and $I[a''_3, a'_2$ and, $I[a'_2, a_1]$ and $I[a''_3, a_2]$ are transposes as in the proof of the group result. The remaining details are left to the reader. $\square$

Assume now that L is modular with a least element 0, and that L is of finite length. If we have a connected chain $a_1 = a, a_2, \dots, a_{n+1} = b$ from a to b, then we shall call the number n (uniquely determined by a and b) the *length* of the interval $I[b, a]$. We denote the length of $I[0, a]$ as $d(a)$ and call this the *rank* of a. If $a \geq b$, then it is clear that

$$d(a) = d(b) + \text{length } I[b, a].$$

Hence for any a and b in L we have

$$d(a \vee b) = d(a) + \text{length } I[a, a \vee b]$$

$$d(b) = d(a \wedge b) + \text{length } I[a \wedge b, b].$$

Since $I[a, a \vee b]$ and $I[a \wedge b, b]$ are isomorphic, they have the same lengths. Hence

$$d(a \vee b) - d(a) = d(b) - d(a \wedge b)$$

or

(4)
$$d(a \vee b) = d(a) + d(b) - d(u \wedge b)$$

which is analogous to the dimensionality formula for the subspaces of a finite dimensional vector space.

EXERCISES

1. Verify that the lattice whose diagram is

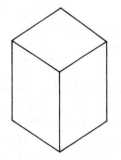

is semi-modular but not modular.

2. Note that the definition of rank requires only that L has a 0 and is of finite length and satisfies the Jordan-Hölder-Dedekind condition (the first conclusion of the J-H-D theorem). Let L be a lattice with 0 of finite length. Then the following conditions are equivalent:

    (i) L is modular.
    (ii) L and its dual are semi-modular.
    (iii) L satisfies the J-H-D condition and the rank condition (4).

8.4   THE LATTICE OF SUBSPACES OF A VECTOR SPACE. FUNDAMENTAL THEOREM OF PROJECTIVE GEOMETRY

We consider first the basic properties of the lattice $L(V)$ of subspaces of a vector space V. Here we shall assume that V is finite dimensional and, since it adds nothing to the difficulty and considerably to the generality, we consider vector spaces over division rings rather than, as has been usual in this book, vector spaces over fields. Thus we assume that V is a vector space (or left module) over a division ring Δ and that V has a base $(e_1, e_2, \ldots, e_n)$ over Δ. We have already seen that the lattice $L(V)$ is modular (Theorem 8.3). If U is a subspace, U has a base $(f_1, f_2, \ldots, f_m)$ with $m \le n$ and the dimensionality $m = n$ if and only if $U = V$. It follows that $L(V)$ is of finite length. Also $L(V)$ has a greatest element $1 = V$ and a least element $0 = 0$ (the subspace consisting of 0 only). Another important property of $L(V)$ is that of existence of complements for subspaces:

given any subspace U there exists a subspace U' such that $V = U + U'$, $U \cap U' = 0$. More briefly, we indicate these two conditions by the single one: $V = U \oplus U'$. To prove the existence of a complement we choose a base $(f_1, f_2, \ldots, f_m)$ and supplement this to a base $(f_1, \ldots, f_m, f_{m+1}, \ldots, f_n)$ for V. Then it is immediate that if $U' = \sum_{j=m+1}^{n} \Delta f_j$, then $V = U \oplus U'$. We remark that U' is not unique if $U \neq 0$, V. For, then $0 < m < n$ and $(f_1, \ldots, f_m, f_m + f_{m+1}, \ldots, f_n)$ is a base, and so $U'' = \Delta(f_m + f_{m+1}) + \cdots + \Delta f_n$ is also a complement of U. It is clear that $U'' \neq U'$.

A lattice L with 0 and 1 is said to be *complemented* if for any $a \in L$ there exists an a' such that $1 = a \vee a'$, $a \wedge a' = 0$. The element a' is called a *complement* of a. Thus we have shown that $L(V)$ is complemented.

We shall now consider the problem of obtaining conditions for the isomorphism of the lattices $L(V_i)$, $i = 1, 2$, where V_i is a vector space over a division ring Δ_i. We shall see that the lattices are isomorphic if and only if the division rings Δ_i are isomorphic and the vector spaces have the same dimensionality. We shall also determine all the lattice isomorphisms between the lattices $L(V_i)$ when these exist. A number of problems, e.g., the problem of determining the automorphisms of the group of bijective linear maps of V over Δ (equivalently of the group of invertible matrices $GL_n(\Delta)$) lead to these lattice problems.

We assume first that $\Delta_1 \cong \Delta_2$ and V_1 over Δ_1 and V_2 over Δ_2 have the same dimensionality n. Let $(e_1, \ldots, e_n)$, $(f_1, \ldots, f_n)$ be bases for V_1 and V_2 respectively and let σ be an isomorphism of Δ_1 onto Δ_2. If $x \in V_1$ we can write x in one and only one way as $x = \sum a_i e_i$, $a_i \in \Delta_1$ and we can define the map

$$\eta : \sum a_i e_i \to \sum \sigma(a_i) f_i.$$

Clearly this is additive:

(5)
$$\eta(x + y) = \eta x + \eta y$$

and for $a \in \Delta_1$ we have $ax = \sum a a_i e_i$ so

$$\eta(ax) = \sum \sigma(a a_i) f_i = \sum \sigma(a) \sigma(a_i) f_i = \sigma(a)(\eta x).$$

Then we have

(6)
$$\eta(ax) = \sigma(a)(\eta x)$$

$a \in \Delta_1$, $x \in V_1$. A map of V_1 over Δ_1 into V_2 over Δ_2 satisfying (5) and (6) is called a *semi-linear map of V_1 into V_2 with associated isomorphism σ*, or a *σ-semi-linear map of V_1 into V_2*. If η is as defined before, $\sum a_i e_i \to \sum \sigma(a_i) f_i$, then we have the inverse map, $\sum b_i f_i \to \sum \sigma^{-1}(b_i) e_i$, which is σ^{-1}-semi-linear.

Now let η be any bijective σ-semi-linear map of V_1 onto V_2. Then we claim that η induces a lattice isomorphism of $L(V_1)$ onto $L(V_2)$. We note first that if

U_1 is a subspace of V_1, then its image $\eta(U_1)$ is a subspace of V_2. For, any pair of vectors of $\eta(U_1)$ have the form ηx and ηy where $x, y \in U_1$. Then $\eta x + \eta y = \eta(x + y) \in \eta(U_1)$ since $x + y \in U_1$. Also, if $b \in \Delta_2$ then $b = \sigma(a)$ for some $a \in \Delta_1$ and $b(\eta x) = \sigma(a)(\eta x) = \eta(ax) \in \eta(U_1)$ since $ax \in U_1$. Hence $\eta(U_1)$ is a subspace of V_2 and we have the map

$$(7) \qquad\qquad [\eta] : U_1 \to \eta(U_1)$$

of the lattice $L(V_1)$ into the lattice $L(V_2)$. Clearly $[\eta]$ is order preserving: if $U_1 \subset W_1$ then $[\eta]U_1 \subset [\eta]W_1$. Now consider η^{-1}. We can check that this is a σ^{-1}-semi-linear mapping of V_2 onto V_1, so it gives rise to the mapping $[\eta^{-1}]$ of $L(V_2)$ into $L(V_1)$. It is evident that if $U_i \in L(V_i)$ then $[\eta^{-1}][\eta]U_1 = U_1$ and $[\eta][\eta^{-1}]U_2 = U_2$. Hence $[\eta]$ is bijective and so, by Theorem 8.2, $[\eta]$ is a lattice isomorphism of $L(V_1)$ onto $L(V_2)$. To summarize: if Δ_1 and Δ_2 are isomorphic, and V_1 and V_2 have the same dimensionality, then $L(V_1)$ and $L(V_2)$ are isomorphic. Moreover, if η is a bijective semi-linear mapping of V_1 over Δ_1 onto V_2 over Δ_2 then $[\eta]$ defined by (7) is a lattice isomorphism of $L(V_1)$ onto $L(V_2)$. We shall now show that the converses of these results hold—at any rate if the dimensionalities are ≥ 3. This fact is an old result which first appeared in a somewhat different form in projective geometry.[3] There it was called the

FUNDAMENTAL THEOREM OF PROJECTIVE GEOMETRY.

Let V_i, $i = 1, 2$, be a vector space over a division ring Δ_i and assume the lattice of subspaces $L(V_1) \cong L(V_2)$ and $\dim V_1 \geq 3$. Then $\Delta_1 \cong \Delta_2$ and $\dim V_1 = \dim V_2$. Moreover, any isomorphism of $L(V_1)$ onto $L(V_2)$ has the form $[\eta]$ as in (7), where η is a bijective semi-linear map of V_1 onto V_2.

Proof (Artin). Let $(e_1, e_2, \ldots, e_n)$ be a base for V_1 and put $V_{1i} = \sum_{j \geq i} \Delta_1 e_j$, $i = 1, 2, \ldots, n$. Then

$$(8) \qquad\qquad V_1 = V_{11} \supset V_{12} \supset \cdots \supset V_{1n} \supset 0$$

is a connected chain in $L(V_1)$ from $1 = V_1$ to 0. If ζ is an isomorphism of $L(V_1)$ onto $L(V_2)$, then ζ maps the connected chain (8) into the connected chain

$$V_2 = V_{21} \supset V_{22} \supset \cdots \supset V_{2n} \supset 0.$$

Then the V_{2i} are subspaces and there are no subspaces properly between V_{2i} and $V_{2,i+1}$. Hence $\dim V_{2i} = \dim V_{2,i+1} + 1$ and $\dim V_2 = n$. Similarly we see that if U is an m-dimensional subspace, then $\dim \zeta(U) = m$. In particular, if U

---

[3] Cf. E. Artin, *Geometric Algebra*, New York, Wiley, 1957, p. 88, or R. Baer, *Linear Algebra and Projective Geometry*, New York, Academic Press, 1952, p. 44.

is one dimensional so is $\zeta(U)$. Hence we have that $\zeta(\Delta_1 e_i) = \Delta_2 e_i' \neq 0$. Since $V_{1i} = \sum_{j \geq i} \Delta_1 e_j$, $V_{2i} = \sum_{j \geq i} \Delta_2 e_j'$, and $V_2 = \sum_1^n \Delta_2 e_j'$, which implies that $(e_1', e_2', \ldots, e_n')$ is a base for V_2. Let $a \neq 0$ in Δ_1. Then $\Delta_1(e_1 + ae_2) \subset \Delta_1 e_1 + \Delta_1 e_2$ and $\Delta_1(e_1 + ae_2) \neq \Delta_1 e_1, \neq \Delta_1 e_2$. Hence $\zeta(\Delta_1(e_1 + ae_2)) = \Delta_2(e_1' + a'e_2')$, $a' \neq 0$, and a' is uniquely determined since $\Delta_2(e_1' + a'e_2') \neq \Delta_2(e_1' + b'e_2')$ if $a' \neq b'$. This defines a map $a \to a'$ which can be extended by $0 \to 0$ to a map of Δ_1 into Δ_2. If we replace e_2', as we may, by $1'e_2'$, we may assume that $1' = 1$. Similarly, we have a map $a \to a''$ of Δ_1 into Δ_2 such that $\Delta_1(e_1 + ae_3) \to \Delta_2(e_1' + a''e_3')$ and $0'' = 0$, $1'' = 1$. We claim that $a' = a''$. To see this we note that the linear independence of e_1, e_2, and e_3 implies that if $a \neq 0$,

$$\Delta_1(e_2 - e_3) = (\Delta_1(e_1 + ae_2) + \Delta_1(e_1 + ae_3)) \cap (\Delta_1 e_2 + \Delta_1 e_3).$$

The image of $\Delta_1(e_2 - e_3)$ is the intersection

$$(\Delta_2(e_1' + a'e_2') + \Delta_2(e_1' + a''e_3')) \cap (\Delta_2 e_2' + \Delta_2 e_3')$$

which contains $a'e_2' - a''e_3'$. Hence

$$\Delta_1(e_2 - e_3) \to \Delta_2(a'e_2' - a''e_3').$$

Since the left-hand side is independent of a and $1' = 1'' = 1$ we have $a'' = a'$. Similarly, we see that

(9) $$\Delta_1(e_1 + ae_i) \to \Delta_2(e_1' + a'e_i'), \qquad 2 \leq i \leq n.$$

We prove next by induction on $r = 2, \ldots, n$ that

(10) $$\Delta_1(e_1 + a_2 e_2 + \cdots + a_r e_r) \to \Delta_2(e_1' + a_2' e_2' + \cdots + a_r' e_r').$$

Assume this for some r and consider $\Delta_1(e_1 + a_2 e_2 + \cdots + a_{r+1} e_{r+1})$. This is the intersection of $\Delta_1(e_1 + a_2 e_2 + \cdots + a_r e_r) + \Delta_1 e_{r+1}$ and $\Delta_1(e_1 + a_{r+1} e_{r+1}) + \Delta_1 e_2 + \cdots + \Delta_1 e_r$, so its image is the intersection of $\Delta_2(e_1' + a_2' e_2' + \cdots + a_r' e_r') + \Delta_2 e_{r+1}'$ with $\Delta_2(e_1' + a_{r+1}' e_{r+1}') + \Delta_2 e_2' + \cdots + \Delta_2 e_r'$ and this is $\Delta_2(e_1' + a_2' e_2' + \cdots + a_{r+1}' e_{r+1}')$. Hence (10) holds for all r. Then $\Delta_1(e_1 + a_2 e_2 + \cdots + a_n e_n) \to \Delta_2(e_1' + a_2' e_2' + \cdots + a_n' e_n')$. The same type of argument based on the observation that the intersection of $\Delta_1(e_1 + a_2 e_2 + \cdots + a_n e_n) + \Delta_1 e_1$ and $\Delta_1 e_2 + \cdots + \Delta_1 e_n$ is $\Delta_1(a_2 e_2 + \cdots + a_n e_n)$ shows that

$$\Delta_1(a_2 e_2 + \cdots + a_n e_n) \to \Delta_2(a_2' e_2' + \cdots + a_n' e_n').$$

We can now prove that $a \to a'$ is an isomorphism. We observe that

$$\Delta_1(e_1 + (a + b)e_2 + e_3) \subset \Delta_1(e_1 + ae_2) + \Delta_1(be_2 + e_3);$$

hence $\Delta_2(e_1' + (a + b)'e_2' + e_3') \subset \Delta_2(e_1' + a'e_2') + \Delta_2(b'e_2' + e_3')$. Now the only vector of the form $e_1' + ce_2' + e_3'$ contained in the right-hand side is $e_1' + (a' + b')e_2' + e_3'$. It follows that $(a + b)' = a' + b'$. Similarly, using the fact that $\Delta_1(e_1 + abe_2 + ae_3) \subset \Delta_1 e_1 + \Delta_1(be_2 + e_3)$, we can conclude that $(ab)' = a'b'$. Hence $a \to a'$ is a homomorphism, and since Δ_1 is a division ring it is a monomorphism. The one dimensional subspaces of V_1 have one of the forms $\Delta_1(e_1 + a_2e_2 + \cdots + a_ne_n)$ or $\Delta_1(a_2e_2 + \cdots + a_ne_n)$ and their respective images are $\Delta_2(e_1' + a_2'e_2' + \cdots + a_n'e_n')$ and $\Delta_2(a_2'e_2' + \cdots + a_n'e_n')$. Since ζ is surjective, for any $b \in \Delta_2$ the subspace $\Delta_2(e_1' + be_2')$ is the image of a one dimensional subspace of V_1, and clearly this subspace is of the first type. Thus we have $\Delta_2(e_1' + a_2'e_2' + \cdots) = \Delta_2(e_1' + be_2')$ which implies that $b = a_2'$ and so $a \to a'$ is an epimorphism. Hence this is an isomorphism σ of Δ_1 onto Δ_2 and we have the bijective semi-linear map $\sum_1^n a_ie_i \to \sum_1^n a_i'e_i'$. It is clear that ζ and $[\eta]$ have the same effect on one dimensional subspaces. Since any subspace is a sum of one dimensional subspaces it is clear that $\zeta = [\eta]$. This completes the proof. $\square$

Remarks. The hypothesis $\dim V_1 \geq 3$ is essential for the validity of the main conclusion of the fundamental theorem. It is easy to sort out what happens if $\dim V_i \leq 2$. In the first place, $L(V)$ has exactly two elements if and only if $\dim V = 1$. Moreover, if $\dim V = 2$, then the intersection of distinct subspaces U_1 and U_2 that are different from 0 and V is 0, and $U_1 + U_2 = V$. Hence, if $\dim V_1 = 2 = \dim V_2$, any bijective map of the set of one dimensional subspaces of V_1 onto the set of one dimensional subspaces of V_2 can be supplemented by $V_1 \to V_2$, $0 \to 0$ to an isomorphism of $L(V_1)$ onto $L(V_2)$. It follows that if $\dim V_i \leq 2$, then $L(V_1) \cong L(V_2)$ if and only if either $\dim V_1 = 1 = \dim V_2$ or if $\dim V_1 = 2 = \dim V_2$ and $|L(V_1)| = |L(V_2)|$. It is easy to see that the last condition holds if and only if $|\Delta_1| = |\Delta_2|$.

We now consider the special case of the fundamental theorem in which $V = V_1 = V_2$. In this case, we are considering lattice automorphisms of $L(V)$. These form a group of transformations of $L(V)$. We also have the group $GS(V)$ of bijective semi-linear transformations of the vector space V; for, it is immediate that if η_1 is a σ-semi-linear map of V_1 into V_2 and η_2 is a τ-semi-linear map of V_2 into V_3, then $\eta_2\eta_1$ is a $\tau\sigma$-semi-linear map of V_1 into V_3. Moreover, if η_1 is bijective, then η_1^{-1} is a σ^{-1}-semi-linear map. Clearly, these results imply that the set $GS(V)$ of bijective semi-linear transformations of an n-dimensional vector space over Δ is a transformation group. If $a \neq 0$ is in Δ then the scalar multiplication $x \to ax$ satisfies $a(bx) = (aba^{-1})ax$, and so this map is a bijective semi-linear transformation corresponding to the inner automorphism $b \to aba^{-1}$ in Δ. Clearly, the map $x \to ax$ induces the identity in the lattice $L(V)$. We have the homomorphism $\eta \to [\eta]$ of $S_n(\Delta)$ into the group of lattice automorphisms of $L(V)$. By the fundamental theorem of projective geometry, this homomorphism is surjective if dim $V \geq 3$. As we have just shown, the kernel contains all the scalar multiplications. On the other hand, the argument used on p. 378 implies that the kernel is the set of scalar multiplications $\neq 0$. Denoting the latter set as Δ_L^* we see that the group of lattice automorphisms of $L(V)$ is isomorphic to $GS(V)/\Delta_L^*$. We state this as a

COROLLARY. *If dim $V \geq 3$ the group of lattice automorphisms of the lattice of subspaces of a vector space V over a division ring Δ is isomorphic to $GS(V)/\Delta_L^*$ where $GS(V)$ is the group of bijective semi-linear transformations of V and Δ_L^* is the set of non-zero scalar multiplications.*

EXERCISES

1. Define an anti-isomorphism of a lattice L onto a lattice L' to be a bijective map $a \to a'$ such that $(a \wedge b)' = a' \vee b'$, $(a \vee b)' = a' \wedge b'$. Note that this is the same as an isomorphism of the dual lattice of L onto L' and hence, by Theorem 8.2, a lattice anti-isomorphism can be characterized as a bijective order inverting map whose inverse is also order inverting ($a \leq b \Leftrightarrow a' \geq b'$). Let V be a finite dimensional vector space over a division ring Δ, V^* the right vector space of linear functions on V. If U is a subspace of V let ann $U = \{f \in V^* | f(y) = 0, \; y \in U\}$. Prove that $U \to$ ann U is a lattice anti-isomorphism of $L(V)$ onto $L(V^*)$.

2. Show that if dim $V_1 \geq 3$ than V_1 and V_2 have anti-isomorphic lattices of subspaces if and only if the underlying division rings are anti-isomorphic and the dimensionalities are the same.

3. Show that for dim $V \geq 3$, if $L(V)$ has an anti-automorphism ζ, then Δ has an anti-automorphism $a \to \bar{a}$, and there exists a map g of $V \times V$ into Δ with the following properties:

   (i) $\qquad\qquad g(x_1 + x_2, y) = g(x_1, y) + g(x_2, y)$

   (ii) $\qquad\qquad g(x, y_1 + y_2) = g(x, y_1) + g(x, y_2)$

   (iii) $\qquad\qquad g(ax, y) = ag(x, y)$

   (iv) $\qquad\qquad g(x, ay) = g(x, y)\bar{a}$

   (v) $\qquad\qquad g(y, x) = \overline{g(x, Qy)}$

     where Q is bijective and σ-semi-linear for σ, the inverse of $a \to \bar{a}$.

   (vi) g is non-degenerate in the sense that $g(z, x) = 0$ for all x if and only if $z = 0$.

   (vii) For every subspace U, $\zeta(U) = \{v \in V | g(v, u) = 0, \; u \in U\}$.

4. Let V be two dimensional over a field F of q elements. Count the number of one dimensional subspaces of V and the order of the group $S_2(F)$. Hence conclude that there exist automorphisms of $L(V)$ which do not come from bijective semi-linear maps of V onto V.

5. Let V be a three-dimensional vector space over $\mathbb{Z}/(p)$, p a prime. Determine the number of lattice automorphisms of $L(V)$.

8.5 BOOLEAN ALGEBRAS

DEFINITION 8.3.   *A Boolean algebra[4] is a lattice with a greatest element* 1 *and least element* 0 *which is distributive and complemented.*

The most important instances of Boolean algebras are the lattices of subsets of any set S. More generally any *field of subsets* of S, that is, a collection of subsets of S which is closed under union and intersection, contains S and $\varnothing$, and the complement of any set in the collection is a Boolean algebra.

The following theorem gives the most important elementary properties of complements in a Boolean algebra.

THEOREM 8.6.   *The complement a' of any element a of a Boolean algebra B is uniquely determined. The map $a \to a'$ is an anti-automorphism of period ≤ 2: $a \to a'$ satisfies*

(11) $$(a \vee b)' = a' \wedge b', \qquad (a \wedge b)' = a' \vee b'$$

(12) $$a'' = a.$$

Proof.   Let $a \in B$ and let a' and a_1 satisfy $a \vee a' = 1$, $a \wedge a_1 = 0$. Then

$$a_1 = a_1 \wedge 1 = a_1 \wedge (a \vee a') = (a_1 \wedge a) \vee (a_1 \wedge a') = a_1 \wedge a'.$$

Hence, if in addition, $a \vee a_1 = 1$ and $a \wedge a' = 0$, then $a' = a' \wedge a_1$, and so $a' = a_1$. This proves the uniqueness of the complement. It is clear that a is the complement of a'. Hence $a'' \equiv (a')' = a$ and $a \to a'$ is of period one or two; hence bijective. Now let $a \leq b$. Then $a \wedge b' \leq b \wedge b' = 0$, so

$$b' = b' \wedge 1 = b' \wedge (a \vee a') = (b' \wedge a) \vee (b' \wedge a') = b' \wedge a'.$$

---

[4] Because of the conflict with the notion of an algebra, a better term for this would be "Boolean lattice." However, since Boolean "algebra" is most commonly used we have chosen this terminology.

Hence $b' \leq a'$. Since $a \to a'$ is its own inverse and is order inverting it follows from Theorem 8.2 (see exercise 1, p. 473) that $a \to a'$ is a lattice anti-isomorphism. □

Historically, Boolean algebras were the first lattices to be studied. They were introduced by Boole to formalize the calculus of propositions. For a long time it was supposed that the type of algebra represented by these systems was of a different character from that involved in number systems and their generalizations (algebras in the technical sense and rings). However, it was discovered rather late in the day by M. H. Stone that this is not the case. In fact, any Boolean algebra, if properly viewed, becomes a special type of ring.

In order to make a ring out of a Boolean algebra B we introduce the new composition

$$a + b = (a \wedge b') \vee (a' \wedge b)$$

which is called the *symmetric difference* of a and b. We have

$$
\begin{aligned}
(a \vee b) \wedge (a \wedge b)' &= (a \vee b) \wedge (a' \vee b') \\
&= ((a \vee b) \wedge a') \vee ((a \vee b) \wedge b') \\
&= ((a \wedge a') \vee (b \wedge a')) \vee ((a \wedge b') \vee (b \wedge b')) \\
&= (b \wedge a') \vee (a \wedge b') \\
&= a + b.
\end{aligned}
$$

(13)

The first formula shows that in the Boolean algebra of subsets of a set, $U + V$ is the set of elements contained in U or in V but not in both:

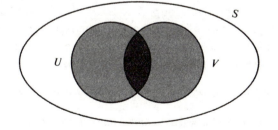

We shall now show that B is a ring with $+$ as just defined, the product $ab = a \wedge b$, and 1 as the unit of B.

Evidently $+$ is commutative. To prove associativity we note first that, by (13),

$$(a + b)' = (a \vee b)' \vee (a \wedge b) = (a \wedge b) \vee (a' \wedge b').$$

Hence

$$(a + b) + c = [((a \wedge b') \vee (a' \wedge b)) \wedge c'] \vee [((a \wedge b) \vee (a' \wedge b')) \wedge c]$$
$$= [(a \wedge b' \wedge c') \vee (a' \wedge b \wedge c')]$$
$$\vee [(a \wedge b \wedge c) \vee (a' \wedge b' \wedge c)]$$
$$= (a \wedge b' \wedge c') \vee (a' \wedge b \wedge c') \vee (a \wedge b \wedge c) \vee (a' \wedge b' \wedge c).$$

This is symmetric in a, b, and c. In particular, $(a + b) + c = (c + b) + a$. Commutativity therefore implies the associative law for $+$. Evidently,

$$a + 0 = (a \wedge 1) \vee (a' \wedge 0) = a$$

and

$$a + a = (a \wedge a') \vee (a' \wedge a) = 0.$$

Hence $(B, +, 0)$ is a commutative group.

We know that $\cdot (= \wedge)$ is associative and commutative. Also $a \cdot 1 = 1 \cdot a = a \wedge 1 = a$ for all a in B. It remains to check one of the distributive laws. Now we have

$$(a + b)c = ((a \wedge b') \vee (a' \wedge b)) \wedge c$$
$$= (a \wedge b' \wedge c) \vee (a' \wedge b \wedge c)$$
$$ac + bc = ((a \wedge c) \wedge (b \wedge c)') \vee ((a \wedge c)' \wedge (b \wedge c))$$
$$= ((a \wedge c) \wedge (b' \vee c')) \vee ((a' \vee c') \wedge (b \wedge c))$$
$$= (a \wedge c \wedge b') \vee (a' \wedge b \wedge c).$$

Comparison shows that $(a + b)c = ac + bc$. Hence $(B, +, \cdot, 0, 1)$ is a ring.

We have noted also that the ring B is commutative and every element is of order ≤ 2 in the additive group. Also every element is idempotent: $a^2 = a \wedge a = a$. These properties of a ring are not independent; for, as we now note, if every element of a ring is idempotent, then the ring is commutative and $2a = 0$ for every a. To prove this we observe that

$$a + b + ab + ba = a^2 + b^2 + ab + ba = (a + b)^2 = a + b.$$

Hence $ab + ba = 0$. Then $2a = 2a^2 = aa + aa = 0$ and so $a = -a$. Then $ab = -ba = ba$. These considerations lead us to introduce the following

DEFINITION 8.4.   *A ring called Boolean if all of its elements are idempotent.*

We have seen that such a ring is of characteristic two. We shall prove next that any Boolean ring B defines a Boolean algebra, and that, in fact, these two concepts are equivalent. Suppose $(B, +, \cdot, 0, 1)$ is a Boolean ring. In order to reverse the process we used to go from a Boolean algebra to a Boolean ring we now define

$$a \lor b = a + b - ab = 1 - (1 - a)(1 - b).$$

The second expression for $a \lor b$ shows that if we introduce the map $\sigma : x \to 1 - x$ in B, then $a \lor b = \sigma^{-1}(\sigma(a)\sigma(b))$, since $\sigma^2 = 1$. It is clear from this and the associative law of multiplication in B that $\lor$ is associative and, of course, this composition is commutative. Also $a \lor a = 2a - a^2 = -a^2 = a$. We now define $a \land b = ab$. Then associativity and commutativity are clear, and $a \land a = a$ since every element of B is idempotent. Also we have $(a \lor b) \land a = (a + b - ab)a = a$ and $(a \land b) \lor a = ab + a - a^2b = a$. Thus the defining conditions L1–L4 on $\lor$ and $\land$ for a lattice hold. It is immediate that the ring 1 and 0 are greatest and least elements of the lattice $(B, \lor, \land)$ and that $1 - a$ is a complement of a, since $a \lor (1 - a) = 1$ and $a \land (1 - a) = 0$. The lattice is distributive since

$$(a \lor b) \land c = (a + b - ab)c = ac + bc - abc = ac + bc - acbc$$
$$= (a \land c) \lor (b \land c).$$

Thus $(B, \lor, \land, 0, 1, ')$ is a Boolean algebra.

It remains to show that the process of passing from a Boolean algebra to a ring and the process of passing from a ring to a Boolean algebra are inverses. Thus suppose we begin with a Boolean algebra $(B, \lor, \land, 0, 1, ')$. Then we obtain the ring $(B, +, \cdot, 0, 1)$ in which $a + b = (a \land b') \lor (a' \land b)$ and $ab = a \land b$. An application of the second process to this ring gives a Boolean alegbra in which $1 = 1$, $0 = 0$, $a' = 1 - a$ and the new $\land$ and $\lor$ which we now denote as $\bar{\lor}$ and $\bar{\land}$ respectively are $a \bar{\lor} b = a + b - ab = 1 - (1 - a)(1 - b) = (a' \land b')' = a \lor b$ and $a \bar{\land} b = ab = a \land b$. Hence $\bar{\lor} = \lor$, $\bar{\land} = \lor$ and so we obtain the original Boolean algebra. On the other hand, suppose we start with a Boolean ring $(B, + \cdot, 0, 1)$ and we obtain the Boolean algebra $(B, \lor, \land, 0, 1,')$ in which $a \lor b = a + b - ab, a \land b = ab, 0 = 0, 1 = 1, a' = 1 - a$. Then applying the process we gave yields a ring in which the new addition $\oplus$ and multiplication $\odot$ are

$$a \oplus b = (a \land (1 - b)) \lor ((1 - a) \land b)$$
$$= a(1 - b) \lor (1 - a)b$$
$$= (a - ab) \lor (b - ab)$$
$$= a - ab + b - ab - (a - ab)(b - ab)$$
$$= a - ab + b - ab - ab + ab + ab - ab$$
$$= a + b.$$

$$a \odot b = a \land b = ab.$$

Also $1 = 1, 0 = 0$ so we obtain the original ring. We have therefore proved the following theorem, which is due to Stone.

THEOREM 8.7.   *The following two types of abstract systems are equivalent: Boolean algebra and Boolean ring.*

There is one more remark worth making. In passing from a Boolean algebra to a Boolean ring we could have used $\vee$ for $\wedge$, $\wedge$ for $\vee$, 1 for 0, and 0 for 1 in the construction. This follows from the principle of duality which is applicable to Boolean algebras. Our process then leads to a ring B' with the same underlying set B and with the addition

$$a +' b = (a \vee b') \wedge (a' \vee b)$$

and multiplication

$$a \cdot' b = a \vee b.$$

Also the new 0 and 1 are $0' = 1$, $1' = 0$. In terms of the ring B we have

$$a +' b = (a + (1 - b) - a + ab)(b + (1 - a) - b + ab)$$
$$= (1 - b + ab)(1 - a + ab)$$
$$= 1 - (a + b)$$

$$a \cdot' b = a + b - ab.$$

We define an *ideal* of a Boolean algebra B to be an ideal of the associated Boolean ring $(B, +, \cdot, 0, 1)$. The conditions for a subset I to be an ideal are (1) if $u, v \in I$, then $u + v \in I$, and (2) if a is arbitrary in B, then $ua \in I$. Since $ua = u \wedge a$ and $ua = a$ if and only if $a \leq u$, the second condition is equivalent to: if $u \in I$, then $b \in I$ for every $b \leq u$. Since $u \vee v = u + v + uv$, $u \vee v \in I$ for every $u, v \in I$. Conversely, let I be a subset of B such that if $u, v \in I$, then $u \vee v \in I$ and if $u \in I$, then every $b \leq u$ is in I. Then $u \wedge v'$ and $v \wedge u' \in I$ (u', v' the complements of u and v). Hence $u + v = (u \wedge v') \vee (v \wedge u') \in I$ and so I is an ideal. Thus a subset I of a Boolean algebra is an ideal if and only if it is closed under $\vee$ and contains every $b \leq u$ for any $u \in I$.

An ideal I is called *proper* if $I \neq B$. It is clear that I is proper if and only if $1 \notin I$. If $u \in B$ then $(u) = \{x \in B \mid x \leq u\}$ is an ideal called the *principal ideal* generated by u. An ideal I is *maximal* if I is proper and there is no proper ideal $\bar{I}$ properly containing $I(\bar{I} \supsetneq I)$. We now observe that an ideal I is maximal if and only if I is proper and for every $a \in B$ either a or $a' \in I$. First, suppose I is maximal and let $a \notin I$. Consider the set $\bar{I}$ of elements of the form $u + b$ where

$u \in I$ and $b \leq a$. This is an ideal properly containing I, so, by the maximality of I, it coincides with B. Thus $1 = b + u$ where $b \leq a$ and $u \in I$. Hence $b' = 1 + b = u \in I$. Since $a' \leq b'$ it is also true that $a' \in I$. Conversely, let I be a proper ideal such that for every $a \in B$ either a or $a' \in I$. Let I be any ideal properly containing I and let $a \in \bar{I}, \notin I$. Then $a' \in I$, and so $a' \in \bar{I}$ and $1 = a + a' \in \bar{I}$. Thus $\bar{I} = B$ and I is maximal.

All of this can be dualized by applying the same considerations to the second ring $B' = (B, +\,', \cdot\,', 0', 1')$ associated with the Boolean algebra B. Accordingly, we define a *filter* (*dual ideal*) of B to be an ideal of B'. The foregoing results can be dualized as follows. First, we note that the dual of our criterion for a subset to be an ideal is that a subset F of a Boolean algebra B is a filter if and only if it is closed under $\wedge$ and containing every $b \geq u$ for any $u \in F$. Since $(a \wedge b)' = a' \vee b'$ and $(a \vee b)' = a' \wedge b'$ it is clear that F is a filter if and only if the set F' of complements a', $a \in F$, is an ideal. Condition (1) is equivalent to the finite intersection property: F is closed under finite intersections. A filter is *proper* in the sense that $F \neq B$ if and only if $0 \notin F$. A maximal ideal of B' is called an *ultra filter* of the Boolean algebra B. A filter F is an ultra filter if and only if (1) $0 \notin F$, (2) for any $a \in B$ either a or $a' \in F$. If $a \in B$ the subset of elements $x \geq a$ is a filter called the *principal filter* generated by a.

We conclude our brief introduction to Boolean algebras by giving a couple of examples of filters.

EXAMPLES

1. Let $\mathbb{R}$ be the real line endowed with its usual topology and let S denote the collection of non-vacuous open subsets of $\mathbb{R}$. This has the finite intersection property. The set $\bar{S}$ of subsets which contain open subsets of $\mathbb{R}$ is a filter.

2. Let S be any set, $B = \mathscr{P}(S)$ the set of subsets of S. Let I be the set of finite subsets of S. This is an ideal in B; hence the set F of complements of the finite subsets is a filter.

EXERCISES

1. Show that if e and f are idempotent elements of a ring which commute, then ef and $e \circ f = e + f - ef$ are idempotents. Prove that the idempotent elements contained in the center form a Boolean algebra relative to $e \vee f = e + f - ef$, $e \wedge f = ef$, $e' = 1 - e$.

2. Prove that if R is a ring such that $pa = 0$ and $a^p = a$ for every $a \in R$ where p is a prime, then R is commutative.

3.   Show that the cardinality of a finite Boolean algebra is a power of two.

4.   (Seligman). Let $e_1, e_2, \ldots, e_n$ be commuting idempotents of a ring R and let $s = \sum_{i=1}^{n} e_i$. Show that $\prod_{j=0}^{n} (s - e_j) = 0$.

8.6   THE MÖBIUS FUNCTION OF
A PARTIALLY ORDERED SET

In this section we shall give an application of partially ordered sets to problems of enumeration.[5] The type of problem we shall consider involves a summation over a partially ordered set whose inversion gives the required enumeration formula. The following problem is an instance of this type of problem.

Problem 1.   We wish to count the number of *derangements* of a finite set S, that is, the number of permutations of S which have no fixed points. Let T be a subset of S. We define

$f(T)$ = the number of permutations of S which fix all the elements of T but fix no element of the complement T' of T in S;

$g(T)$ = the number of permutations of S which fix all the elements of T and perhaps some additional elements as well. Then

(14) $$g(T) = \sum_{U \supseteq T} f(U)$$

where, of course, $U \in \mathscr{P}(S)$. The objective is to "invert" (14), that is, to obtain a formula for $f(U)$ in terms of the $g(T)$. This will give $f(\varnothing)$, which is the number of derangements of the set S, since we have trivially that $g(T)$ is the number of permutations of T' and this is $|T'|!$.

In general, one has a finite partially ordered set S and functions f and g on S with values in a commutative group A, such that

(15) $$g(y) = \sum_{\substack{x \in S \\ x \geq y}} f(x) \qquad \text{for all } y \in S.$$

Again, we wish to express f in terms of g. We shall need the following lemma.

LEMMA (Szpilrajn-Marczewski).   *S can be totally ordered; say, as $x_1, x_2, \ldots, x_n$ so that if $x_i < x_j$ in the original partial ordering then $i < j$.*

---

[5] I am indebted to Neil White for the material in this section.

Proof. Since S is finite it contains a minimal element x_1. We continue this process by selecting inductively x_{i+1} minimal in the complement $\{x_1, x_2, \ldots, x_i\}'$. Then S ordered as $x_1, x_2, \ldots, x_n$ satisfies the desired condition: for, it is clear from the procedure that if $x_i < x_j$, then x_j could not have been chosen before x_i in our ordering. Thus we must have $i < j$. $\square$

We now define for $x, y \in S$

(16)
$$\zeta(x, y) = \begin{cases} 1 & \text{if } x \leq y \\ 0 & \text{otherwise} \end{cases}$$

and regard this as defining a function of two variables from S to the integers $\mathbb{Z}$. Using the total ordering given in the lemma we see that (15) can now be written out as a system of equations:

$$g(x_i) = \sum_{j=1}^{n} \zeta(x_i, x_j) f(x_j),$$

or, in matrix form,

(17)
$$\begin{pmatrix} g(x_1) \\ g(x_2) \\ \vdots \\ g(x_n) \end{pmatrix} = \begin{pmatrix} \zeta(x_1, x_1) & \zeta(x_1, x_2) & \cdots & \zeta(x_1, x_n) \\ \zeta(x_2, x_1) & \zeta(x_2, x_2) & \cdots & \zeta(x_2, x_n) \\ \cdots\cdots\cdots\cdots\cdots\cdots\cdots\cdots \\ \zeta(x_n, x_1) & \zeta(x_n, x_2) & \cdots & \zeta(x_n, x_n) \end{pmatrix} \begin{pmatrix} f(x_1) \\ f(x_2) \\ \vdots \\ f(x_n) \end{pmatrix},$$

We recall that the values $f(x), g(x)$ are in the abelian group A, which can be regarded as a $\mathbb{Z}$-module in the natural way (see p. 166). We have $\zeta(x_i, x_i) = 1$, and if $i > j$ we cannot have $x_i \leq x_j$, so $\zeta(x_i, x_j) = 0$. Hence the matrix $Z = (\zeta_{ij}) = (\zeta(x_i, x_j))$ is upper triangular with 1's along the diagonal, that is, Z has the form $1 - N$ where

(18)
$$N = \begin{pmatrix} 0 & & & \\ & 0 & & * \\ & & \ddots & \\ 0 & & & 0 \end{pmatrix}.$$

Here $N = (v_{ij})$ where $v_{ij} = 0$ if $i \geq j$ and $v_{ij} = -\zeta_{ij}$ if $i < j$. It is immediate by induction that every (i, j)-entry of N^k is 0 for $i \geq j - k + 1$ and hence that $N^n = 0$. Thus $Z = 1 - N$ is invertible with inverse

(19)
$$M = 1 + N + N^2 + \cdots + N^{n-1}.$$

The equation (17) has the abbreviated form $G = ZF$ where G and F are the column vectors $(= n \times 1$ matrices) $(g(x_i)), (f(x_i))$. We can invert this and obtain

$F = MG$, so if we write $M = (\mu_{ij})$ we have

(20) $$f(x_i) = \sum \mu_{ij} g(x_j).$$

The matrix $M = (\mu_{ij})$ defines the *Möbius function* of the partially ordered set S (to $\mathbb{Z}$) by

(21) $$\mu(x_i, x_j) = \mu_{ij}, \qquad \text{for all } x_i, x_j.$$

In terms of this function we can rewrite (20) as

(20′) $$f(y) = \sum_{x \in S} \mu(y, x) g(x).$$

We have the following

THEOREM 8.8.   *For any finite partially ordered set S, there exists a unique function μ from $S \times S$ to $\mathbb{Z}$ such that if A is any commutative group and f and g are functions form S to A such that (15):*

$$g(y) = \sum_{\substack{x \in S \\ x \geq y}} f(x), \qquad \text{for all } y \in S$$

then

(22) $$f(y) = \sum_{x \in S} \mu(y, x) g(x), \qquad \text{for all } y \in S.$$

Proof.   The existence of μ has been show. To prove the uniqueness, there is no loss in generality in assuming that the x_i are ordered as in the Szpilrajn-Marczewski lemma. We specialize $A = (\mathbb{Z}, +)$ and we let δ_k be the function from S to $A = \mathbb{Z}$ such that $\delta_k(x_i) = \delta_{ik}$ ($= 1$ if $i = k$ and $= 0$ otherwise). Let ε_k be the corresponding function from S to $\mathbb{Z}$ defined by (15) or, equivalently, (17). Thus $\varepsilon_k(x_i) = \sum_j \zeta(x_i, x_j) \delta_k(x_j) = \zeta(x_i, x_k)$. By (20), we have $\delta_{ik} = \delta_k(x_i) = \sum_j \mu_{ij} \varepsilon_k(x_j) = \sum_j \mu_{ij} \zeta(x_j, x_k)$. Thus we have the matrix equation $MZ = 1$ where $M = (\mu_{ij})$ and $Z = (\zeta(x_i, x_j))$. Hence M is uniquely determined as Z^{-1} and consequently the function μ is uniquely determined. $\square$

In a similar manner we can handle systems of equations of the form

(23) $$g(y) = \sum_{\substack{x \in S \\ x \leq y}} f(x), \qquad \text{for all } y \in S.$$

If we define $Z = (\zeta_{ij})$ as before (using an ordering as in the lemma) then (23) is equivalent to the matrix equation

$$(g(x_1), g(x_2), \ldots, g(x_n)) = (f(x_1), f(x_2), \ldots, f(x_n))Z.$$

Then

$$(f(x_1), f(x_2), \ldots, f(x_n)) = (g(x_1), g(x_2), \ldots, g(x_n))M$$

where $M = Z^{-1}$. We therefore have the following

COROLLARY 1. *Let f and g be functions from S to an abelian group A satisfying*

$$g(y) = \sum_{\substack{x \in S \\ x \le y}} f(x), \qquad y \in S.$$

Then

(24)
$$f(y) = \sum_{x \in S} \mu(x, y)g(x).$$

The Möbius function can be determined by a recursion formula. For, we have

COROLLARY 2. *The Möbius function is the unique function from $S \times S$ to $\mathbb{Z}$ satisfying $\mu(x, y) = 0$ unless $x \le y$ and the recursion formula*

(25)
$$\sum_{\substack{y \in S \\ x \le y \le z}} \mu(x, y) = \delta(x, z)$$

where the delta function

$$\delta(x, z) = \begin{cases} 1 & \text{if } x = z \\ 0 & \text{if } x \ne z. \end{cases}$$

Alternatively, (25) may be replaced by

(26)
$$\sum_{\substack{y \in S \\ x \le y \le z}} \mu(y, z) = \delta(x, z).$$

Proof. These are equivalent to the matrix equations $MZ = 1$ and $ZM = 1$, 1, the $n \times n$ unit matrix. □

COROLLARY 3. *If the intervals $I[x, z]$ and $I[w, t]$ are isomorphic in S then $\mu(x, z) = \mu(w, t)$.*

Proof. From (25) we have

$$\mu(x, z) = \delta(x, z) - \sum_{\substack{y \in S \\ x \le y < z}} \mu(x, y).$$

The result follows from this by induction on the length of the interval. □

We shall apply this enumeration method in a moment to solve the problem posed on p. 480. We now formulate another such problem as

Problem 2. The map coloring problem. A *map* is a plane divided into a finite number of non-overlapping connected regions called *countries* by a finite number of arcs which intersect only at their endpoints. Two countries are *adjacent* if they have a common boundary which is one of the arcs. A *proper coloring* is an assignment of colors to the countries so that no two adjacent countries are given the same color. Given a map and a number k one might ask in how many ways can the map be properly colored using k colors. A famous problem—first posed by DeMorgan in 1850 and recently solved using 1200 hours of computer time—is *the four color problem*: can every map be colored properly with four colors? In other words, is the number of proper colorings by $k = 4$ colors positive for every map? (The answer is "yes".) We shall now show that for any given map there is a polynomial with integer coefficients in k, called the *chromatic polynomial of the map*, which gives the number of proper colorings of the map using k colors.

We define a *submap* Δ of a map Γ to be the map obtained by erasing some of the boundaries. We define a partial ordering in the set S of submaps of Γ by putting $E \le \Delta$ if E is a submap of Δ. If $\Delta \in S$ we define

$f(\Delta) =$ the number of proper colorings of Δ in k colors;

$g(\Delta) =$ the total number of colorings of Δ in k colors.

If $c(\Delta)$ is the number of countries in Δ then

$$g(\Delta) = k^{c(\Delta)}.$$

Moreover, since any coloring of Δ is a proper coloring of some submap E of Δ we clearly have

$$g(\Delta) = \sum_{\substack{E \in S \\ E \le \Delta}} f(E).$$

Hence, by Corollary 1,

(27) $$f(\Delta) = \sum_{\substack{E \in S \\ E \le \Delta}} \mu(E, \Delta) k^{c(E)}$$

where μ is the Möbius function of S. This is the chromatic polynomial of Δ.

We shall now consider the problem of calculating Möbius functions of some partially ordered sets and we prove first

THEOREM 8.9. *Let* $C = \{0, 1, \ldots, n\}$ *be a chain of length n with the natural order, then the Möbius function* μ *of C is given by*

(28)
$$\mu(i, i) = 1, \qquad \mu(i - 1, i) = -1$$
$$\mu(j, i) = 0, \qquad if \ j \neq i, i - 1.$$

Proof. If $g(i) = \sum_{j \leq i} f(j)$ then $g(i) = \sum_{j=0}^{i} f(j)$. Clearly, we have $f(i) = g(i) - g(i - 1)$. Hence, from Corollary 1, we obtain (28). $\square$

We obtain next a way of reducing the calculation of the Möbius function of a product of two partially ordered sets to the Möbius functions of the two sets. We recall (exercise 7, p. 461) that if S_1 and S_2 are partially ordered sets the product $S_1 \times S_2$ is the set $S_1 \times S_2$ partially ordered by $(x_1, x_2) \leq (y_1, y_2)$ if and only if $x_1 \leq y_1$ and $x_2 \leq y_2$. We have

THEOREM 8.10. *Let* $S = S_1 \times S_2$ *where* S_i *are partially ordered and let* μ, μ_1, *and* μ_2 *be the Möbius functions of S, S_1, and S_2 respectively. Then*

(29)
$$\mu((x_1, x_2), (y_1, y_2)) = \mu_1(x_1, y_1)\mu_2(x_2, y_2)$$

for all $x_1, y_1 \in S_1, x_2, y_2 \in S_2$.

Proof. Let $\delta_1, \delta_2, \delta$ be the delta functions of S_1, S_2, and S. Then

$$\delta((x_1, x_2), (y_1, y_2)) = \delta_1(x_1, y_1)\delta_2(x_2, y_2).$$

Also

$$\sum_{\substack{(y_1, y_2) \in S \\ (x_1, x_2) \leq (y_1, y_2) \leq (z_1, z_2)}} \mu_1(y_1, z_1)\mu_2(y_2, z_2)$$

$$= \sum_{\substack{y_1 \in S_1 \\ x_1 \leq y_1 \leq z_1}} \sum_{\substack{y_2 \in S_2 \\ x_2 \leq y_2 \leq z_2}} \mu_1(y_1, z_1)\mu_2(y_2, z_2)$$

$$= \left(\sum_{\substack{y_1 \in S_1 \\ x_1 \leq y_1 \leq z_1}} \mu_1(y_1, z_1) \right) \left(\sum_{\substack{y_2 \in S_2 \\ x_2 \leq y_2 \leq z_2}} \mu_2(y_2, z_2) \right)$$

$$= \delta_1(x_1, z_1)\delta_2(x_2, z_2) = \delta((x_1, x_2), (z_1, z_2)).$$

Hence $\mu_1(y_1, z_1)\mu_2(y_2, z_2)$ and $\mu((y_1, y_2), (z_1, z_2))$ satisfy the same recursion formula as in Corollary 2. It follows from this corollary that $\mu((y_1, y_2), (z_1, z_2)) = \mu_1(y_1, z_1)\mu_2(y_2, z_2)$. $\square$

We can use this result and Theorem 8.9 to calculate the Möbius function of the Boolean algebra $\mathscr{P}(S)$ of subsets of a finite set $S = \{1, 2, \ldots, n\}$.

COROLLARY. *The Möbius function on the Boolean algebra $\mathscr{P}(S)$, $S = \{1, 2, \ldots, n\}$ is given by the formula*

$$\mu(U, V) = \begin{cases} (-1)^{|V - U|} & \text{if } U \subset V \\ 0 & \text{if } U \not\subset V \end{cases}$$

where $V - U = V \cap U'$ is the set of elements in V not in U.

Proof. We observe that $\mathscr{P}(S)$ is isomorphic to a product of n copies of the chain $C = \{0, 1\}$. In fact, if U is a subset of $S = \{1, 2, \ldots, n\}$, then we associate with U its *characteristic function* χ_U which is the map of S to $\{0, 1\}$ defined by

$$\chi_U(s) = \begin{cases} 1 & \text{if } s \in U \\ 0 & \text{if } s \notin U. \end{cases}$$

We can then represent χ_U by the vector $(\chi_U(1), \chi_U(2), \ldots, \chi_U(n))$. The map

$$U \to (\chi_U(1), \ldots, \chi_U(n))$$

is an isomorphism of $\mathscr{P}(S)$ onto $C_1 \times C_2 \times \cdots \times C_n$ where $C_i = \{0, 1\}$ with the natural order. Then, by Theorem 8.10 (iterated) and Theorem 8.9, we have

$$\mu(U, V) = \prod_1^n \mu_i(\chi_U(i), \chi_V(i)) = \begin{cases} (-1)^{|V - U|} & \text{if } U \subset V \\ 0 & \text{if } U \not\subset V. \end{cases} \quad \square$$

The use of the Möbius function of $\mathscr{P}(S)$ is often referred to as the *method of inclusion-exclusion*. We can now give the

Solution of Problem 1. The number of derangements of $S = \{1, 2, \ldots, n\}$ is

$$f(\varnothing) = \sum_{U \in \mathscr{P}(S)} \mu(\varnothing, U)g(U)$$

$$= \sum_U (-1)^{|U|}|U'|!$$

$$= \sum_{i=0}^n (-1)^i \binom{n}{i}(n - i)!$$

$$= n! \sum_{i=0}^n (-1)^i/i!$$

This is asymptotically equal to $n!/e$. Thus the probability that a randomly selected partition is a derangement is very close to $\dfrac{1}{e}$, essentially independent of n.

We consider next the classical example which started all of this:

Problem 3. Let n be a positive integer and let D_n be the lattice of positive integer divisors of n ordered by divisibility ($a \geq b$ means $a | b$). If $n = p_1{}^{e_1} p_2{}^{e_2} \cdots p_h{}^{e_h}$ where the p_i are distinct primes and the $e_i > 0$, then we obtain an isomorphism of D_n with $C_1 \times C_2 \times \cdots \times C_h$ where C_i is the chain $\{0, 1, \ldots, e_i\}$, by mapping

$$d = p_1{}^{d_1} p_2{}^{d_2} \cdots p_h{}^{d_h} \to (d_1, d_2, \ldots, d_h).$$

If $c = p_1{}^{c_1} p_2{}^{c_2} \cdots p_h{}^{c_h} | d$, so that $c_i \leq d_i$, then

$$\mu(c, d) = \prod_{i=1}^{h} \mu(c_i, d_i)$$

$$= \begin{cases} (-1)^l & \text{if } d/c \text{ is a product of } l \text{ distinct primes} \\ 0 & \text{if } d/c \text{ has a square factor.} \end{cases}$$

We note that $\mu(c, d) = \mu(1, d/c)$, which is the classical Möbius function of number theory written as $\mu(d/c)$. The inversion formula based on this is the one which was discovered by Möbius.

EXERCISES

1. Let $\varphi(n)$ be the Euler φ-function: $\varphi(n)$ is the number of positive integers less than and relatively prime to n. Use the inversion method to derive the formula

$$\varphi(n) = n \prod_{\substack{p, prime \\ p|n}} \left(1 - \frac{1}{p}\right).$$

2. Determine the partially ordered set of submaps of the map Γ:

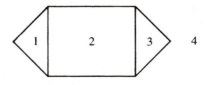

Determine the chromatic polynomial for Γ.

3. Let $L(V)$ be the lattice of subspaces of the n-dimensional vector space V over a
    field of q elements, and let $\begin{bmatrix} n \\ k \end{bmatrix}_q$, *the Gaussian coefficient*, denote the number of
    k-dimensional subspaces of V. Derive the formulas

    (30) $$\begin{bmatrix} n \\ k \end{bmatrix}_q = \frac{(q^n - 1)(q^n - q) \cdots (q^n - q^{k-1})}{(q^k - 1)(q^k - q) \cdots (q^k - q^{k-1})}$$

    (31) $$\begin{bmatrix} n \\ k \end{bmatrix}_q = q^k \begin{bmatrix} n-1 \\ k \end{bmatrix}_q + \begin{bmatrix} n-1 \\ k-1 \end{bmatrix}_q.$$

4. If $X, Y \in L(V)$ as in exercise 3, then $\mu(X, Y)$ for $X \subset Y$ depends only on
    $m = \dim Y - \dim X$ by Corollary 3 to Theorem 8.8. Hence write $\mu(X, Y) = \mu(m)$. Prove that

    (32) $$\mu(m) = (-1)^m q^{m(m-1)/2}.$$

5. Let W be an l-dimensional subspace of $L(V)$ as in exercise 3. Show that the
    number of k-dimensional subspaces U such that $U \cap W = 0$ is given by the
    formula

    (33) $$\sum_{i=0}^{l} (-1)^i q^{i(i-1)/2} \begin{bmatrix} n-i \\ k-i \end{bmatrix}_q \begin{bmatrix} l \\ i \end{bmatrix}_q.$$

6. Find the total number of sets of vectors which generate V (as in exercise 3).

7. Let $G_{n,p}$ denote the abelian group which is the direct product of n cyclic groups
    of prime order p. Let H be a subgroup of $G_{n,p}$ isomorphic to $G_{k,p}$. Find the
    number of injective homomorphisms $\eta: G_{l,p} \to G_{n,p}$ such that $\eta(G_{l,p}) \cap H = 0$.

8. If $\pi(S)$ and $\rho(S)$ are partitions of S, we say that π is a *refinement* of ρ if each
    block (see p. 11) of π is contained in some block of ρ. Let P be the collection
    of the set S, ordered by refinement. Show that P is a lattice and determine μ.

9. Determine the Möbius functions of the following partially ordered sets:

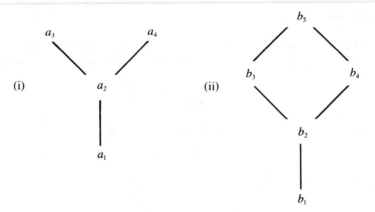

Appendix

SOME TOPICS FOR INDEPENDENT STUDY

1. Euclidean Domains

References: (1) G. H. Hardy and E. M. Wright, *An Introduction to the Theory of Numbers*, 5th ed. *Oxford University Press, New York, 1975*, pp. 212–217. (2) H. Chatland, "On the Euclidean algorithm," *Bulletin of the American Mathematical Society*, vol. 55 (1949), pp. 948–953. (3) T. Motzkin, "The Euclidean algorithm," *Bulletin of the American Mathematical Society*, vol. 55 (1949), pp. 1142–1156. (4) P. Samuel, "About Euclidean rings," *Journal of Algebra*, vol. 19 (1975), 282–301.

The first three references discuss the problem of the determination of the rings I of integral elements of quadratic number fields $Q(\sqrt{m})$ that are Euclidean (see pp. 147–151 and exercises 3–5 on p. 287). The paper by Samuel develops a general theory of Euclidean domains.

2. Non-commutative Principal Ideal Domains

References: (1) N. Jacobson, *Theory of Rings*, American Mathematical Society Surveys, No. 2, Providence, Rhode Island, 1943, Chapter 3. (2) P. M. Cohn, *Free Rings and their Relations*, Academic Press, London and New York, 1971, Chapter 8.

This is an extension of the theory presented in pp. 147–151 to non-commutative domains. Examples of such domains are the ring of polynomials in one indeterminate with coefficients in a division ring and the ring of formal differential polynomials in one indeterminate.

3. The Four Square Theorem and Integral Quaternions

References: (1) G. H. Hardy and E. M. Wright, *An Introduction to the Theory of Numbers*, 5th ed. Oxford University Press, New York, 1975, pp. 300–310. (2) I. N. Herstein, *Topics in Algebra*, 2nd ed., John Wiley and Sons, New York, 1975, pp. 371–377.

The theorem, due to Lagrange, is that every positive integer is a sum of four squares. The ring of integral quaternions is the ring I defined in exercise 5 on p. 100. These form a non-commutative Euclidean, hence principal ideal domain, whose arithmetic can be used to prove the four square theorem. An important step in the proof is that if $N(x)$ is the norm form of a quaternion algebra over a field of characteristic $p \neq 2$ then there exist $x \neq 0$ such that $N(x) = 0$. Herstein proves this by invoking Wedderburn's theorem on the commutativity of finite division rings. A more direct proof of this fact follows from exercise 6, p. 361.

4. Two-Dimensional Crystallographic Groups

References: (1) H. Weyl, *Symmetry*, Princeton University Press, Princeton, New Jersey, 1952, pp. 83–115. (2) H. S. M. Coxeter, *Introduction to Geometry*, John Wiley and Sons, New York, London, and Sidney, 1961, Chapter 4. (3) R. Schwatzenberger, *N-dimensional Crystallography*, Pitman Publishing Program, San Francisco, London, and Melbourne, 1980, pp. 1–10.

These groups are the discontinuous (discrete) groups of Euclidean motions in the plane. Such a motion is given analytically as a map $(x, y) \to (x', y')$ where $x' = ax + by + h$, $y' = cx + dy + k$ where $a, b, c, d, h, k \in \mathbb{R}$ and $\begin{pmatrix} a & b \\ c & d \end{pmatrix}$ is an orthogonal matrix. Discontinuity means that there is a neighborhood of the identity map containing no element $\neq 1$ of the group. Using a natural notion of equivalence (see Weyl's book) the problem can be transformed into that of classifying the groups in which the coefficients $a, b, \ldots$ are integers and the subgroup with $h = k = 0$ is finite. The interesting case is that in which h and k take on all integer values. In the sense of unimodular equivalence there are 17 such groups. These are the possible groups of symmetries of planar ornaments (e.g., wallpaper). Their historical significance can be gleaned from the following quotation from Weyl's book (p. 103):

"Examples for all 17 groups of symmetry are found among the decorative patterns of antiquity, in particular among Egyptian ornaments. One can hardly overestimate the depth of geometric imagination and inventiveness reflected in these patterns.

Their construction is far from being mathematically trivial. The art of ornament contains in implicit form the oldest piece of higher mathematics known to us."

5. Finite Reflection Groups

References: (1) R. Steinberg, *Lectures on Chevalley Groups*, Yale University Lecture Notes, Department of Mathematics, Yale University, New Haven, Connecticut, 1967, Appendix. (2) C. T. Benson and L. C. Grove, *Finite Reflection Groups*, Bogden and Quigley, Tarrytown-on-Hudson, New York and Belmont, California, 1971. (3) R. Carter, *Simple Groups of Lie Type*, Wiley-Interscience, New York, 1972, Chapter 2.

These are the finite groups generated by reflections (that is, symmetries as defined on p. 363) in a Euclidean space. They play a fundamental role under the guise of "Weyl groups" in the theory of simple Lie algebras and simple Lie groups.

6. Mathieu Groups

References: (1) E. Witt, "Die 5-fach transitiven Gruppen von Mathieu" and "Uber Steinerche Systeme," *Abhandlungen aus den Mathematischen Seminar der Hansischen Universität*, vol. 12 (1938), pp. 256–264 and pp. 265–275. (2) J. A. Todd, "On representation of the Mathieu group M_{24} as a collineation group," *Annali di Matematica Pura ed Applicata (IV)*, vol. 71 (1966), pp. 199–238. (3) N. L. Biggs and A. T. White, *Permutation Groups and Combinatorial Structures*, Cambridge University Press, London, 1979, p. 57 and pp. 70–74.

These five groups denoted as M_{11}, M_{12}, M_{22}, M_{23}, and M_{24} are multiply transitive simple groups discovered by E. L. Mathieu in 1861 and 1873. The subscript n indicates the degree of the permutation group (M_n is a subgroup of S_n). These groups were called "sporadic" simple groups by Burnside since they do not belong to any infinite classes of simple groups (e.g. the alternating groups, the simple groups defined by $GL_n(F)$ for finite F, etc.). In the period 1966–1981, 21 additional sporadic simple groups have been found and the complete classification of finite simple groups has been achieved through the efforts of a large number of mathematicians. The list of these groups is: (1) cyclic groups of prime order, (2) the alternating groups A_n for $n \geq 5$, (3) groups of Lie types defined in reference 3 in Section 5, and (4) the 26 sporadic simple groups. A certain lattice, the "Leech lattice," plays an important role in the definition of most of the new sporadic groups. This is related to the Steiner systems that are used in the definitions of the Mathieu groups. The reader may consult a paper by J. Conway in the *Bulletin of the London Mathematical Society*, vol. 1 (1969) for a definition of the Leech lattice and its relation to Steiner systems as well as the definition of the Conway sporadic group.

7. Finite Fields

References: (1) L. E. Dickson, *Linear Groups with an Exposition of Galois Field Theory*, 1900; Dover Publications, 1958, reprint edition, pp. 1–54. (1) A. A. Albert, *Fundamental Concepts of Higher Algebra*, University of Chicago Press, Chicago, 1956, Chapter 5.

These books contain many special properties of finite fields that do not appear in general books on algebra. It should be noted that finite fields have important applications, e.g., to computer science and to cryptography.

8. Hilbert Irreducibility Theorem

References: (1) D. Hilbert, "Über die Irreduzibilitat ganzer rationaler Funktionen mit ganzzahligen Koeffizienten," *Journal für die reine und angewandete Mathematik*, vol. 110 (1892), pp. 104–129; or *Gesammelte Abhandlungen*, vol. 2, Springer-Verlag, Berlin, 1933. (2) S. Lang, *Diophantine Geometry*, Wiley-Interscience, New York, 1962, Chapter 8. (3) C. R. Hadlock, *Field Theory and its Classical Problems*, Carus Mathematical Monographs, Mathematical Association of America, 1978, Chapter 4.

Hilbert's theorem states that if $f(t_1, \ldots, t_n, x) \in D = \mathbb{Q}[t_1, \ldots, t_n, x]$ is irreducible in D then there exist infinitely many choices of $t_i = a_i \in \mathbb{Q}$ such that $f(a_1, \ldots, a_n, x)$ is irreducible in $\mathbb{Q}[x]$. The third reference above has a comparatively simple proof of the theorem. The second reference proves the result for polynomials with coefficients in a field of algebraic numbers over $\mathbb{Q}$. Hilbert used his theorem to prove the existence of infinitely many polynomials with rational coefficients having Galois group S_n or A_n.

9. Galois Groups of Some Classical Polynomials

References: (1) I. Schur, "Gleichungen ohne Affekt," *Sitzungsberichte Preussische Akademie der Wissenschaften–Physicalische–Mathematische Klasse*, 1930, pp. 443–449; or *Gesammelte Abhandlungen* vol. 3, pp. 191–197. (2) I. Schur, "Affektlose gleichungen in der Theorie Laguerreschen und Hermiteschen Polynome," Journal für die reine und angewandte Mathematik, vol. 165 (1931), pp. 52–58; or *Gesammelte Abhandlungen*, vol. 3, pp. 227–233.

These papers determine the Galois groups over $\mathbb{Q}$ of Laguerre, Hermite polynomials, the polynomials $E_n(x) = 1 + x + \dfrac{x^2}{2!} + \cdots + \dfrac{x^n}{n!}$ and related polynomials. In all cases the Galois groups are either S_n or A_n.

10. Plücker Equations

References: (1) W. V. D. Hodge and D. Pedoe, *Methods of Algebraic Geometry*, vol. 1, Cambridge University Press, Cambridge, England, 1947, pp. 286–315. (2) N. Jacobson and D. Saltman, *Finite Dimensional Division Algebras*, a forthcoming book, Chapter 3.

The Plücker equations are algebraic equations on the Plücker coordinates of an element ω of the homogeneous part V^r of the exterior algebra $E(V)$ that are necessary and sufficient conditions that ω is decomposable. They endow the set of decomposable vectors with an algebraic geometric structure. These define a Grassmannian variety corresponding to the set of r dimensional subspaces of the vector space V.

Index

Mathematics

FUNCTIONAL ANALYSIS (Second Corrected Edition), George Bachman and Lawrence Narici. Excellent treatment of subject geared toward students with background in linear algebra, advanced calculus, physics and engineering. Text covers introduction to inner-product spaces, normed, metric spaces, and topological spaces; complete orthonormal sets, the Hahn-Banach Theorem and its consequences, and many other related subjects. 1966 ed. 544pp. 6⅛ x 9¼. 0-486-40251-7

DIFFERENTIAL MANIFOLDS, Antoni A. Kosinski. Introductory text for advanced undergraduates and graduate students presents systematic study of the topological structure of smooth manifolds, starting with elements of theory and concluding with method of surgery. 1993 edition. 288pp. 5⅜ x 8½. 0-486-46244-7

VECTOR AND TENSOR ANALYSIS WITH APPLICATIONS, A. I. Borisenko and I. E. Tarapov. Concise introduction. Worked-out problems, solutions, exercises. 257pp. 5⅝ x 8¼. 0-486-63833-2

AN INTRODUCTION TO ORDINARY DIFFERENTIAL EQUATIONS, Earl A. Coddington. A thorough and systematic first course in elementary differential equations for undergraduates in mathematics and science, with many exercises and problems (with answers). Index. 304pp. 5⅜ x 8½. 0-486-65942-9

FOURIER SERIES AND ORTHOGONAL FUNCTIONS, Harry F. Davis. An incisive text combining theory and practical example to introduce Fourier series, orthogonal functions and applications of the Fourier method to boundary-value problems. 570 exercises. Answers and notes. 416pp. 5⅜ x 8½. 0-486-65973-9

COMPUTABILITY AND UNSOLVABILITY, Martin Davis. Classic graduate-level introduction to theory of computability, usually referred to as theory of recurrent functions. New preface and appendix. 288pp. 5⅜ x 8½. 0-486-61471-9

AN INTRODUCTION TO MATHEMATICAL ANALYSIS, Robert A. Rankin. Dealing chiefly with functions of a single real variable, this text by a distinguished educator introduces limits, continuity, differentiability, integration, convergence of infinite series, double series, and infinite products. 1963 edition. 624pp. 5⅜ x 8½. 0-486-46251-X

METHODS OF NUMERICAL INTEGRATION (SECOND EDITION), Philip J. Davis and Philip Rabinowitz. Requiring only a background in calculus, this text covers approximate integration over finite and infinite intervals, error analysis, approximate integration in two or more dimensions, and automatic integration. 1984 edition. 624pp. 5⅜ x 8½. 0-486-45339-1

INTRODUCTION TO LINEAR ALGEBRA AND DIFFERENTIAL EQUATIONS, John W. Dettman. Excellent text covers complex numbers, determinants, orthonormal bases, Laplace transforms, much more. Exercises with solutions. Undergraduate level. 416pp. 5⅜ x 8½. 0-486-65191-6

RIEMANN'S ZETA FUNCTION, H. M. Edwards. Superb, high-level study of landmark 1859 publication entitled "On the Number of Primes Less Than a Given Magnitude" traces developments in mathematical theory that it inspired. xiv+315pp. 5⅜ x 8½. 0-486-41740-9

CALCULUS OF VARIATIONS WITH APPLICATIONS, George M. Ewing. Applications-oriented introduction to variational theory develops insight and promotes understanding of specialized books, research papers. Suitable for advanced undergraduate/graduate students as primary, supplementary text. 352pp. 5⅜ x 8½.
0-486-64856-7

MATHEMATICIAN'S DELIGHT, W. W. Sawyer. "Recommended with confidence" by *The Times Literary Supplement,* this lively survey was written by a renowned teacher. It starts with arithmetic and algebra, gradually proceeding to trigonometry and calculus. 1943 edition. 240pp. 5⅜ x 8½.
0-486-46240-4

ADVANCED EUCLIDEAN GEOMETRY, Roger A. Johnson. This classic text explores the geometry of the triangle and the circle, concentrating on extensions of Euclidean theory, and examining in detail many relatively recent theorems. 1929 edition. 336pp. 5⅜ x 8½.
0-486-46237-4

COUNTEREXAMPLES IN ANALYSIS, Bernard R. Gelbaum and John M. H. Olmsted. These counterexamples deal mostly with the part of analysis known as "real variables." The first half covers the real number system, and the second half encompasses higher dimensions. 1962 edition. xxiv+198pp. 5⅜ x 8½. 0-486-42875-3

CATASTROPHE THEORY FOR SCIENTISTS AND ENGINEERS, Robert Gilmore. Advanced-level treatment describes mathematics of theory grounded in the work of Poincaré, R. Thom, other mathematicians. Also important applications to problems in mathematics, physics, chemistry and engineering. 1981 edition. References. 28 tables. 397 black-and-white illustrations. xvii + 666pp. 6⅛ x 9¼.
0-486-67539-4

COMPLEX VARIABLES: Second Edition, Robert B. Ash and W. P. Novinger. Suitable for advanced undergraduates and graduate students, this newly revised treatment covers Cauchy theorem and its applications, analytic functions, and the prime number theorem. Numerous problems and solutions. 2004 edition. 224pp. 6½ x 9¼.
0-486-46250-1

NUMERICAL METHODS FOR SCIENTISTS AND ENGINEERS, Richard Hamming. Classic text stresses frequency approach in coverage of algorithms, polynomial approximation, Fourier approximation, exponential approximation, other topics. Revised and enlarged 2nd edition. 721pp. 5⅜ x 8½.
0-486-65241-6

INTRODUCTION TO NUMERICAL ANALYSIS (2nd Edition), F. B. Hildebrand. Classic, fundamental treatment covers computation, approximation, interpolation, numerical differentiation and integration, other topics. 150 new problems. 669pp. 5⅜ x 8½.
0-486-65363-3

MARKOV PROCESSES AND POTENTIAL THEORY, Robert M. Blumental and Ronald K. Getoor. This graduate-level text explores the relationship between Markov processes and potential theory in terms of excessive functions, multiplicative functionals and subprocesses, additive functionals and their potentials, and dual processes. 1968 edition. 320pp. 5⅜ x 8½.
0-486-46263-3

ABSTRACT SETS AND FINITE ORDINALS: An Introduction to the Study of Set Theory, G. B. Keene. This text unites logical and philosophical aspects of set theory in a manner intelligible to mathematicians without training in formal logic and to logicians without a mathematical background. 1961 edition. 112pp. 5⅜ x 8½.
0-486-46249-8

INTRODUCTORY REAL ANALYSIS, A.N. Kolmogorov, S. V. Fomin. Translated by Richard A. Silverman. Self-contained, evenly paced introduction to real and functional analysis. Some 350 problems. 403pp. 5⅜ x 8½.   0-486-61226-0

APPLIED ANALYSIS, Cornelius Lanczos. Classic work on analysis and design of finite processes for approximating solution of analytical problems. Algebraic equations, matrices, harmonic analysis, quadrature methods, much more. 559pp. 5⅜ x 8½.
0-486-65656-X

AN INTRODUCTION TO ALGEBRAIC STRUCTURES, Joseph Landin. Superb self-contained text covers "abstract algebra": sets and numbers, theory of groups, theory of rings, much more. Numerous well-chosen examples, exercises. 247pp. 5⅜ x 8½.   0-486-65940-2

QUALITATIVE THEORY OF DIFFERENTIAL EQUATIONS, V. V. Nemytskii and V.V. Stepanov. Classic graduate-level text by two prominent Soviet mathematicians covers classical differential equations as well as topological dynamics and ergodic theory. Bibliographies. 523pp. 5⅜ x 8½.   0-486-65954-2

THEORY OF MATRICES, Sam Perlis. Outstanding text covering rank, nonsingularity and inverses in connection with the development of canonical matrices under the relation of equivalence, and without the intervention of determinants. Includes exercises. 237pp. 5⅜ x 8½.   0-486-66810-X

INTRODUCTION TO ANALYSIS, Maxwell Rosenlicht. Unusually clear, accessible coverage of set theory, real number system, metric spaces, continuous functions, Riemann integration, multiple integrals, more. Wide range of problems. Undergraduate level. Bibliography. 254pp. 5⅜ x 8½.   0-486-65038-3

MODERN NONLINEAR EQUATIONS, Thomas L. Saaty. Emphasizes practical solution of problems; covers seven types of equations. ". . . a welcome contribution to the existing literature. . . ."–*Math Reviews.* 490pp. 5⅜ x 8½.   0-486-64232-1

MATRICES AND LINEAR ALGEBRA, Hans Schneider and George Phillip Barker. Basic textbook covers theory of matrices and its applications to systems of linear equations and related topics such as determinants, eigenvalues and differential equations. Numerous exercises. 432pp. 5⅜ x 8½.   0-486-66014-1

LINEAR ALGEBRA, Georgi E. Shilov. Determinants, linear spaces, matrix algebras, similar topics. For advanced undergraduates, graduates. Silverman translation. 387pp. 5⅜ x 8½.   0-486-63518-X

MATHEMATICAL METHODS OF GAME AND ECONOMIC THEORY: Revised Edition, Jean-Pierre Aubin. This text begins with optimization theory and convex analysis, followed by topics in game theory and mathematical economics, and concluding with an introduction to nonlinear analysis and control theory. 1982 edition. 656pp. 6⅛ x 9¼.   0-486-46265-X

SET THEORY AND LOGIC, Robert R. Stoll. Lucid introduction to unified theory of mathematical concepts. Set theory and logic seen as tools for conceptual understanding of real number system. 496pp. 5⅝ x 8¼.   0-486-63829-4

Physics

OPTICAL RESONANCE AND TWO-LEVEL ATOMS, L. Allen and J. H. Eberly. Clear, comprehensive introduction to basic principles behind all quantum optical resonance phenomena. 53 illustrations. Preface. Index. 256pp. 5⅜ x 8½.
0-486-65533-4

QUANTUM THEORY, David Bohm. This advanced undergraduate-level text presents the quantum theory in terms of qualitative and imaginative concepts, followed by specific applications worked out in mathematical detail. Preface. Index. 655pp. 5⅜ x 8½.
0-486-65969-0

ATOMIC PHYSICS (8th EDITION), Max Born. Nobel laureate's lucid treatment of kinetic theory of gases, elementary particles, nuclear atom, wave-corpuscles, atomic structure and spectral lines, much more. Over 40 appendices, bibliography. 495pp. 5⅜ x 8½.
0-486-65984-4

A SOPHISTICATE'S PRIMER OF RELATIVITY, P. W. Bridgman. Geared toward readers already acquainted with special relativity, this book transcends the view of theory as a working tool to answer natural questions: What is a frame of reference? What is a "law of nature"? What is the role of the "observer"? Extensive treatment, written in terms accessible to those without a scientific background. 1983 ed. xlviii+172pp. 5⅜ x 8½.
0-486-42549-5

AN INTRODUCTION TO HAMILTONIAN OPTICS, H. A. Buchdahl. Detailed account of the Hamiltonian treatment of aberration theory in geometrical optics. Many classes of optical systems defined in terms of the symmetries they possess. Problems with detailed solutions. 1970 edition. xv + 360pp. 5⅜ x 8½. 0-486-67597-1

PRIMER OF QUANTUM MECHANICS, Marvin Chester. Introductory text examines the classical quantum bead on a track: its state and representations; operator eigenvalues; harmonic oscillator and bound bead in a symmetric force field; and bead in a spherical shell. Other topics include spin, matrices, and the structure of quantum mechanics; the simplest atom; indistinguishable particles; and stationary-state perturbation theory. 1992 ed. xiv+314pp. 6⅛ x 9¼.
0-486-42878-8

LECTURES ON QUANTUM MECHANICS, Paul A. M. Dirac. Four concise, brilliant lectures on mathematical methods in quantum mechanics from Nobel Prize-winning quantum pioneer build on idea of visualizing quantum theory through the use of classical mechanics. 96pp. 5⅜ x 8½.
0-486-41713-1

THIRTY YEARS THAT SHOOK PHYSICS: THE STORY OF QUANTUM THEORY, George Gamow. Lucid, accessible introduction to influential theory of energy and matter. Careful explanations of Dirac's anti-particles, Bohr's model of the atom, much more. 12 plates. Numerous drawings. 240pp. 5⅜ x 8½. 0-486-24895-X

ELECTRONIC STRUCTURE AND THE PROPERTIES OF SOLIDS: THE PHYSICS OF THE CHEMICAL BOND, Walter A. Harrison. Innovative text offers basic understanding of the electronic structure of covalent and ionic solids, simple metals, transition metals and their compounds. Problems. 1980 edition. 582pp. 6⅛ x 9¼.
0-486-66021-4

CATALOG OF DOVER BOOKS

A TREATISE ON ELECTRICITY AND MAGNETISM, James Clerk Maxwell. Important foundation work of modern physics. Brings to final form Maxwell's theory of electromagnetism and rigorously derives his general equations of field theory. 1,084pp. 5⅜ x 8½. Two-vol. set.   Vol. I: 0-486-60636-8   Vol. II: 0-486-60637-6

MATHEMATICS FOR PHYSICISTS, Philippe Dennery and Andre Krzywicki. Superb text provides math needed to understand today's more advanced topics in physics and engineering. Theory of functions of a complex variable, linear vector spaces, much more. Problems. 1967 edition. 400pp. 6½ x 9¼.   0-486-69193-4

INTRODUCTION TO QUANTUM MECHANICS WITH APPLICATIONS TO CHEMISTRY, Linus Pauling & E. Bright Wilson, Jr. Classic undergraduate text by Nobel Prize winner applies quantum mechanics to chemical and physical problems. Numerous tables and figures enhance the text. Chapter bibliographies. Appendices. Index. 468pp. 5⅜ x 8½.   0-486-64871-0

METHODS OF THERMODYNAMICS, Howard Reiss. Outstanding text focuses on physical technique of thermodynamics, typical problem areas of understanding, and significance and use of thermodynamic potential. 1965 edition. 238pp. 5⅜ x 8½.   0-486-69445-3

THE ELECTROMAGNETIC FIELD, Albert Shadowitz. Comprehensive undergraduate text covers basics of electric and magnetic fields, builds up to electromagnetic theory. Also related topics, including relativity. Over 900 problems. 768pp. 5⅝ x 8¼.   0-486-65660-8

GREAT EXPERIMENTS IN PHYSICS: FIRSTHAND ACCOUNTS FROM GALILEO TO EINSTEIN, Morris H. Shamos (ed.). 25 crucial discoveries: Newton's laws of motion, Chadwick's study of the neutron, Hertz on electromagnetic waves, more. Original accounts clearly annotated. 370pp. 5⅜ x 8½.   0-486-25346-5

EINSTEIN'S LEGACY, Julian Schwinger. A Nobel Laureate relates fascinating story of Einstein and development of relativity theory in well-illustrated, nontechnical volume. Subjects include meaning of time, paradoxes of space travel, gravity and its effect on light, non-Euclidean geometry and curving of space-time, impact of radio astronomy and space-age discoveries, and more. 189 b/w illustrations. xiv+250pp. 8⅜ x 9¼.   0-486-41974-6

THE VARIATIONAL PRINCIPLES OF MECHANICS, Cornelius Lanczos. Philosophic, less formalistic approach to analytical mechanics offers model of clear, scholarly exposition at graduate level with coverage of basics, calculus of variations, principle of virtual work, equations of motion, more. 418pp. 5⅜ x 8½.   0-486-65067-7

Paperbound unless otherwise indicated. Available at your book dealer, online at **www.doverpublications.com**, or by writing to Dept. GI, Dover Publications, Inc., 31 East 2nd Street, Mineola, NY 11501. For current price information or for free catalogues (please indicate field of interest), write to Dover Publications or log on to **www.doverpublications.com** and see every Dover book in print. Dover publishes more than 400 books each year on science, elementary and advanced mathematics, biology, music, art, literary history, social sciences, and other areas.